D1614014

Handbook of the Neuroscience of Language

Handbook of the Neuroscience of Language

BRIGITTE STEMMER AND HARRY A. WHITAKER

AMSTERDAM • BOSTON • HEIDELBERG • LONDON • NEW YORK • OXFORD
PARIS • SAN DIEGO • SAN FRANCISCO • SINGAPORE • SYDNEY • TOKYO
Academic Press is an imprint of Elsevier

ELSEVIER

Academic Press is an imprint of Elsevier
84 Theobald's Road, London WC1X 8RR, UK
Radarweg 29, PO Box 211, 1000 AE Amsterdam, The Netherlands
30 Corporate Drive, Suite 400, Burlington, MA 01803, USA
525 B Street, Suite 1900, San Diego, CA 92101-4495, USA

First edition 2008

British Library Cataloguing in Publication Data
A catalogue record for this book is available from the British Library

Library of Congress Cataloging-in-Publication Data
A catalog record for this book is available from the Library of Congress

ISBN: 978-0-08-045352-1

For information on all Academic Press publications
visit our website at books.elsevier.com

Typeset by Charon Tec Ltd (A Macmillan Company), Chennai, India
www.charontec.com

Printed and bound in the United States of America

08 09 10 11   11 10 9 8 7 6 5 4 3 2 1

This book is dedicated to
Harold Goodglass, André Roch Lecours and Klaus Poeck
who inspired and motivated generations
of students and researchers

Contents

4. Architectonic Language Research

Katrin Amunts

5. Microgenesis of Language: Vertical Integration of Linguistic Mechanisms Across the Neuraxis

Don M. Tucker, Gwen Frishkoff and Phan Luu

6. A Brief Introduction to Common Neuroimaging Techniques

Frank A. Rodden and Brigitte Stemmer

PART

II

NEUROIMAGING OF LANGUAGE

7. PET Research of Language

Barry Horwitz and Richard J.S. Wise

PART

III

EXPERIMENTAL NEUROSCIENCE OF LANGUAGE AND COMMUNICATION

22. Neurolinguistic Computational Models

Brian MacWhinney and Ping Li

23. Mirror Neurons and Language

Michael A. Arbib

24. Lateralization of Language across the Life Span

Merrill Hiscock and Marcel Kinsbourne

25. Interhemispheric Interaction in the Lateralized Brain

Joseph B. Hellige

PART

IV

CLINICAL NEUROSCIENCE OF LANGUAGE

A. Language in Special Populations and in Various Disease Processes

26. Acute Aphasias

Claus-W. Wallesch and Claudius Bartels

42. The Role of Electronic Devices in the Rehabilitation of Language Disorders

Brian Petheram and Pam Enderby

PART

V

RESOURCES

43. Resources in the Neuroscience of Language: A Listing

Brigitte Stemmer

Contributors

Katrin Amunts Research Center Juelich, INB-3, Medicine, 52425 Juelich, Germany: Department of Psychiatry and Psychotherapy, RWTH Aachen University, Aachen, Germany

Michael A. Arbib Computer Science Department, Neuroscience Program and USC Brain Project, University of Southern California, Los Angeles, CA, USA

Claudius Bartels Deptartment of Neurology, University of Magdeburg, 39120 Magdeburg, Germany

Anna Basso Department of Neurological Sciences, Milan University, Milan 20122, Italy

Alan Beretta Department of Linguistics, Michigan State University, East Lansing, MI 48824, USA

Stephanie B. Boswell Department of Neurology, Comprehensive Epilepsy Center, Thomas Jefferson University Hospital/Jefferson Medical College, Philadelphia, PA 19107, USA

Hugh W. Buckingham Department of Communication Sciences and Disorders, Linguistics Program, Louisiana State University, LA 70803, USA

Stefano F. Cappa Vita-Salute University and San Raffaele Scientific Institute, DIBIT, Milano 20132, Italy

Sarah S. Christman Department of Communication Sciences and Disorders, The University of Oklahoma Health Sciences Center, Oklahoma City, OK 73190, USA

John F. Connolly Centre de Recherche, Institut Universitaire de Gériatrie de Montréal (CRIUGM), Université de Montréal, Montréal, Québec, Canada; Centre de Recherche en Neuropsychologie et Cognition (CERNEC), Département de Psychologie, Université de Montréal, Montréal, Québec, Canada

Timothy J. Crow SANE POWIC, Warneford Hospital, Oxford OX3 7JX, UK

Gianfranco Denes Department of Linguistics, Università Ca' Foscari, 30100, Venezia, Italy

Joseph T. Devlin Department of Psychology, University College London, London WCIE 6BT, UK

Mayada Elsabbagh Centre for Brain and Cognitive Development, Birkbeck, University of London, London WC1E 7HX, UK

Pam Enderby ScHARR HSR Department, University of Sheffield, Sheffield, UK

Gwen Frishkoff Learning Research and Development Center, University of Pittsburgh, Pittsburgh, PA 15260, USA; NeuroInformatics Center, University of Oregon, Eugene, OR, USA

Guido Gainotti Neuropsychology Service of the Policlinico Gemelli, Catholic University of Rome, 00168 Roma, Italy

Barry Gordon Cognitive Neurology/Neuropsychology Group, Department of Neurology, The Johns Hopkins University School of Medicine, Baltimore, MD, USA; Department of Cognitive Science, The Johns Hopkins University, Baltimore, MD 21231, USA

Murray Grossman Department of Neurology, University of Pennsylvania School of Medicine, Philadelphia, PA 19104-4283, USA

Uri Hasson Department of Neurology, The University of Chicago, Chicago, IL 60637, USA

Joseph B. Hellige Psychology Department, Loyola Marymount University, Los Angeles, CA 90045-2659, USA

Argye E. Hillis Johns Hopkins University School of Medicine, Johns Hopkins Hospital, Baltimore, MD 21287, USA

Merrill Hiscock Department of Psychology and Center for Neuro-Engineering and Cognitive Science, University of Houston, Houston, TX 77204-5022, USA

Barry Horwitz Brain Imaging and Modeling Section, National Institute on Deafness and Other Communications Disorders, National Institutes of Health, Bethesda, MD 20892, USA

Gonia Jarema Department of Linguistics, University of Montreal and Research Center, Institut universitaire de gériatrie de Montréal, Montreal, Quebec, Canada H3W 1W5

Marcel Kinsbourne Department of Psychology, New School for Social Research, New York, NY 10003, USA

Kerry Ledoux Cognitive Neurology/Neuropsychology Group, Department of Neurology, The Johns Hopkins University School of Medicine, Baltimore, MD 21231, USA

Andrew W. Lee Department of Cerebrovascular Neurology, Johns Hopkins University School of Medicine, Baltimore, MD 21287, USA

Ping Li Department of Psychology, University of Richmond, Richmond, VA 23173, USA

Gary Libben Department of Linguistics, University of Alberta, Edmonton, Alberta, Canada T6G 2E5

Phan Luu Department of Psychology, University of Oregon, Eugene, OR, USA; Electrical Geodesics, Inc., Eugene, OR 97403, USA

Claudio Luzzatti Department of Psychology, University of Milano-Bicocca, I-20126 Milano, Italy

Joël Macoir Département de réadaptation, Programme de maîtrise en orthophonie, Université Laval, Québec (QC), G1K 7P4, Canada

Brian MacWhinney Department of Psychology, Carnegie Mellon University, Pittsburgh, PA 15213, USA

Skye McDonald School of Psychology, University of New South Wales, Sydney 2052 NSW, Australia

Kirsten R. Mohn Department of Neurology (CB 7025), School of Medicine, University of North Carolina, Chapel Hill, NC 27599-7025, USA

Stephen E. Nadeau Geriatric Research, Education and Clinical Center, Brain Rehabilitation Research Center, Malcom Randall Veterans Administration Medical Center, and the Department of Neurology, University of Florida College of Medicine, Veterans Administration Medical Center 1601 SW Archer Road Gainesville, FL 32608-1197, USA

Lyndsey Nickels Macquarie Centre for Cognitive Science (MACCS), Macquarie University, Sydney NSW 2109, Australia

Loraine K. Obler Program in Speech, Language and Hearing Sciences, Graduate Center, City University of New York, New York, NY 10016, USA

Michel Paradis Department of Linguistics, McGill University, Montreal, Quebec, Canada H3A 1A7; Cognitive Neuroscience Center, Université du Québec à Montréal, Montreal, Quebec, Canada

Seija Pekkala Department of Speech Sciences, University of Helsinki, FI-00014 Helsinki, Finland

Charles A. Perfetti Learning Research and Development Center, University of Pittsburgh, Pittsburgh, PA 15260, USA

Brian Petheram Faculty of Computing, Engineering, & Mathematical Sciences, University of the West of England, Bristol, UK; Speech & Language Therapy Research Unit, Frenchay Hospital, Bristol BS16 1LE, UK

Frank A. Rodden Section of Experimental MR of the CNS, Department of Neuroradiology, Eberhard-Karls-Universität-Tübingen, Tübingen, Germany

David B. Rosenfield Speech and Language Center, Neurological Institute, The Methodist Hospital/Weill Cornell College of Medicine, Department of Communicative Disorders, University of Houston Shepherd School of Music, Rice University, Houston, TX, USA

Carlo Semenza Department of Psychology, University of Trieste, 34124 Trieste, Italy

Michael Siegal Department of Psychology, University of Sheffield, Western Bank, Sheffield S10 2TP, UK; Department of Psychology, University of Trieste, Trieste, Italy

Steven L. Small Department of Neurology, The University of Chicago, Chicago, IL 60637, USA

Luise Springer Collaborative Research Centre "Media and Cultural Communication", University of Cologne, D-50969 Köln, Germany; School of Speech and Language Therapy, University Hospital, D-52074 Aachen, Germany

Karsten Steinhauer Centre for Research on Language, Mind and Brain (CRLMB) and School of Communication Sciences and Disorders (SCSD), Faculty of Medicine, McGill University, Montréal, Québec, Canada

Brigitte Stemmer Faculty of Arts and Science and Faculty of Medicine, Université de Montréal, Montreal, Quebec, Canada

Luca Surian Department of Cognitive Sciences and Education, University of Trento, 38068 Rovereto (TN), Italy

Joseph I. Tracy Department of Neurology, Comprehensive Epilepsy Center, Thomas Jefferson University Hospital/Jefferson Medical College, Philadelphia, PA 19107, USA

Alexander I. Tröster Department of Neurology (CB 7025), School of Medicine, University of North Carolina, Chapel Hill, NC 27599-7025, USA

Don M. Tucker Electrical Geodesics, Inc., Eugene, OR 97403, USA; Department of Psychology, University of Oregon Eugene, OR, USA

Yves Turgeon Restigouche Health Authority, Campbellton Regional Hospital, Campbellton, NB, Canada E3N 3H3

Michael T. Ullman Brain and Language Laboratory, Departments of Neuroscience, Linguistics, Psychology and Neurology, Georgetown University, Washington, DC 20057-1464, USA

Diana Van Lancker Sidtis New York University Department of Speech Pathology, New York, NY 10003, USA

Claus-W. Wallesch Department of Neurology, University of Magdeburg, Magdeburg D-39120, Germany

Kate E. Watkins Department of Experimental Psychology, University of Oxford, Oxford OX1 3UD, UK

Harry A. Whitaker Department of Psychology, Northern Michigan University Marquette, Marquette, MI 49855, USA

Richard J.S. Wise Faculty of Medicine, Division of Neuroscience and Mental Health and Medical Research Council Clinical Sciences Centre, Imperial College London, Hammersmith Hospital, London W12 ONN, UK

Preface

A decade has passed since the 1998 publication of the Handbook of Neurolinguistics, the first reference book in the field. During these 10 years the field has matured, new theories have been advanced and old ones supported or retracted. Although there is little continuity in the content of cognitive models of brain function from the middle ages with those of the late twentieth and early twenty-first centuries, there is a remarkable continuity in the goal: we are still fascinated with brain geography. The remarkable developments in human brain imaging to date – one of us (H.W.) personally remembers when the EMI company introduced CAT scans of brain-damaged individuals in grand rounds at the Mayo Clinic in early 1972 – and the even more remarkable developments to come, bode well for the future of neurolinguistics. Whereas 10 years ago, neuroimaging was just being explored for neurolinguistic questions, today it constitutes a routine component. Nevertheless, what one should keep in mind, as the present *Handbook of the Neuroscience of Language* clearly demonstrates, is that developments in linguistic and psychological theory are equally important. The image means nothing until and unless it is validly interpreted. Describing language and communication disorders and correlating them with lesion sites was a beginning; studying the neural systems associated with language and communication within the framework of interacting brain systems that mediate affective, cognitive and monitoring systems is the challenge.

While putting together the present handbook, three friends, colleagues and scientists who greatly influenced and advanced the neuroscience of language passed away: Harold Goodglass, André Roch Lecours and Klaus Poeck. We dedicate this handbook to them, beginning with the prologue by Guido Gainotti honoring each of their contributions to the field. The volume is then divided into five sections starting with methods and techniques that introduce the reader to classical (clinical) assessment approaches, methods of mapping the human brain and a theoretical framework for interpreting the multiple levels of neural organization that contribute to language comprehension. The second part provides an overview of the contribution that various imaging techniques (PET, fMRI, ERPs, electrical stimulation of language cortex, TMS) have made to language research. This is followed by part three that discusses experimental approaches to the neuroscience of language and communication, including disorders at different language levels, in reading, writing and number processing; other topics address computational models in neurolinguistics, the role of mirror systems for language, and brain lateralization with respect to language. Part four focuses on language in special populations and in various disease processes, in developmental disorders and on the recovery, treatment and rehabilitation of language and communication. The book ends with a listing of resources in the neuroscience of language and a glossary of items and concepts to help the novice become acquainted with the field.

Our hope is that this handbook will be a standard reference book for many years to come. We have attempted to cover a broad range of topics and be as comprehensive as possible within a single volume. Prominent and creative researchers in the field were invited to contribute to the book; however, as is always the case with any enterprise of this magnitude, not all invited researchers were able to participate. Some dropped out during the process and were replaced by new authors and others had to leave at a stage where replacement was not possible anymore. Despite our efforts, not all the topics one could relate to the neuroscience of language have been covered; as well, individual chapters may not address all possible angles, since many issues require a book unto themselves. The challenge and the limitation was to squeeze a field that keeps on growing into a one volume book. We took several steps to find solutions: The authors were asked to strictly adhere to the space limitations of their chapters, to limit their bibliography to roughly 30–40 references and to refer the reader to review work whenever possible. Not all authors were happy with these limitations and we were most grateful that those unhappy ones in the end did succumb good-naturedly to their fate – although not always without putting up a fight. To adhere to a state-of-the-art account, the authors were also asked to focus on the advancements that were achieved in their field during the last 10 years (for earlier work the reader is referred to the *Handbook of Neurolinguistics*, the precursor to this new handbook). Generally, the field has advanced substantially in the last 10 years and for many authors it was a challenge to cover their topic within these limitations. On the other hand, in a few cases advancements were more limited, and older literature was still relevant and thus cited.

Contributors were encouraged to present their chapters in such manner that the information could be used by a broad audience, not just the specialist in the field. The book thus addresses students, clinicians and scholars in linguistics, neuroscience, medicine, psychiatry, psychology, speech language pathology and so forth – in short, everybody with an interest in the neuroscience of language.

We are grateful to the contributors of the book for their science, creativity, the time and effort they put into their chapters, and last but not least their patience with the idiosyncrasies of the editors. Thanks also go to Bruce Roberts with whom we started the project, and with Nikki Levy and Barbara Makinster of Elsevier who helped us to the finish line.

Life sometimes interferes with one's best intentions and we have to succumb to it. Despite life-turning events, friends and contributors helped us through these times and we would like to thank them for this. Enjoy the read.

Brigitte Stemmer, Montreal
Harry A. Whitaker, Marquette
October 2007

Prologue

GUIDO GAINOTTI

Neuropsychology Service of the Policlinico Gemelli, Catholic University of Rome, Rome, Italy

Several generations of clinicians and neuroscientists have developed in the last century the work of those eminent pioneers of the cognitive neurosciences (such as Paul Broca, Carl Wernicke, Jules Dejerine, Gordon Holmes and John Hughlings Jackson) who had discovered the cortical localization of language and other cognitive functions and proposed theoretical models aiming to explain the meaning of these discoveries. Among the great personalities who developed the investigation of the neuroscience of language and cognition, some authors (such as Jean Babinski, Leonardo Bianchi, Joseph Gerstmann, Russel Brain, Joachim Bodamer and Oliver Zangwill) extended and deepened the exploration of the anatomical correlates of specific cognitive and linguistic disorders, whereas other authors (such as Henry Head, Kurt Goldstein, Eberhard Bay, Henry Hécaen and Alexander Romanoff Luria) offered a comprehensive personal synthesis of this growing body of clinical and experimental data.

The importance and the complexity of the scientific inheritance that these authors left to the generation of Harold Goodglass, Roch Lecours and Klaus Poeck, to whom this handbook is dedicated, pushed these protagonists of the modern neuroscience of language not only to introduce and develop new models, methods and lines of research in the study of the cognitive neurosciences, but also to tackle some basic educational problems. The dimension and the relevance of the fields of knowledge covered by the neuroscience of language and cognition required, indeed, the development of high level schools, capable of expanding everywhere in the world the study of neuropsychology and neurolinguistics, to pass from an elite of scientists and clinicians, mainly working in Europe and North America, to a much greater number of researchers and professionals working all around the world. Harold Goodglass, Roch Lecours and Klaus Poeck were particularly involved in this activity of forming a new generation of behavioral neurologists, and of students of neuropsychology and of neuroscience of language, because (as will be shown by a short illustration of their motivations and of the early stages

of their research activities) each of them had been strongly attracted by these kinds of studies and each belonged to the school of a very influential founder of cognitive neuroscience.

1. THE EARLY STAGES

Harold Goodglass' interest in aphasia went back to an experience during the Second World War, when he was admitted to a military hospital following a war wound and found himself in the same room with other soldiers who presented with language disorders following head injuries. Since then, his interest in this kind of human suffering and for this area of inquiry never weakened. Goodglass began his neuropsychological research in the laboratory of Fred Quadfasel (who had been a resident in the clinic of Kurt Goldstein); with Quadfasel, he published his first landmark paper (Goodglass & Quadfasel, 1954), in which he showed that the left hemisphere is dominant for language not only in right-handers, but also in the majority of left-handers, thus shaking the classical doctrine that language and handedness are controlled by the same hemisphere.

Roch Lecours made his first medical experience as resident in a gynecological ward in Montreal, but was soon captured by the complexity and fascination of brain research. He therefore took a neurological post-doctoral position in Boston, where he worked with Raymond Adams (chief of the Neurological Service at the Massachusetts General Hospital), Hans Lukas Teuber (a psychology professor at MIT) and Ivan Yakovlev (who had previously worked with Pierre Marie in France and had in Boston a unique collection of serially cut brains). A strong friendship rapidly developed between Lecours and Yakovlev (whom Lecours has always considered as his true master) and together they conducted a study on the relations between myelinogenesis and language acquisition. This study gave rise to an important publication (Yakovlev & Lecours, 1967) and definitively introduced Lecours to the world of the neuroscience

Harold Goodglass
(Photo courtesy of Harry A. Whitaker)

André Roch Lecours
(Photo courtesy of Françoise Cot)

Klaus Poeck
(Photo courtesy of Walter Huber)

of language. With this new identity, Lecours went from Boston to "La Salpetrière" in Paris, where classical aphasiology had witnessed its origins and main development. Here he worked in the group of Francois Lhermitte, who, with his extraordinary clinical knowledge, shaped, in association with an outstanding speech pathologist (Blanche Ducarne) Lecours's views about aphasia.

Klaus Poeck's interest in cognitive neurosciences emerged during the early stages of his academic schooling; Poeck took his clinical and research training in Düsseldorf, under the supervision of Eberhard Bay, who was at that time one of the most original leading European aphasiologists, and in Pisa, with the eminent Italian physiologist Giuseppe

Moruzzi (co-discoverer of the functions of the reticular formation with H.W. Magoun). After being an assistant professor of neurology in Freiburg, where he conducted his first studies on body schema (Poeck, 1965), he was appointed professor of neurology at the Aachen medical faculty, where from the early 1970s, his main interest shifted to the pathology of language, convincing him to found the neuropsychology and aphasia unit.

2. THE EDUCATIONAL CHALLENGE

The fact of being rooted in schools of great tradition pushed Goodglass, Lecours and Poeck to make an extraordinary contribution to the educational effort requested by their epoch, devoting an important part of their work to the training of graduate students, postgraduate research and clinical fellows, to the foundation of neuropsychological or aphasiological journals and societies, to the preparation of textbooks of neuropsychology or neurolinguistics and to the construction and validation of important aphasia batteries.

Since I consider this seminal work as fundamental not only for the development of that wonderful domain of research that is cognitive neuroscience, but also for the clinical implications that this kind of knowledge has for diagnosis and rehabilitation, let me quickly mention the specific activities that Goodglass, Lecours and Poeck have devoted to this enterprise.

It is perhaps necessary to begin this survey with Harold Goodglass, because for about 20 years, under his direction, the Aphasia Research Center of the Boston Veterans Administration Hospital (that after his death was named the "Harold Goodglass Aphasia Research Center") became the world's most attractive center for the formation of young researchers. Under the guidance of this superb mentor, many of his pupils and collaborators (e.g., Michael Alexander, Errol Baker, Sheila Blumstein, Joan Borod, Nelson Butters, Laird Cermak, Gianfranco Denes, Rhonda Friedman, Jean Berko-Gleason, Mary Hyde, Edith Kaplan, Theodor Landis, Lise Menn, Gabriele Miceli, Margaret Naeser, Loraine K. Obler, Marlene Oscar-Bermann, Alan Rubens, Carlo Semenza, Lewis Shapiro and Edgar Zurif) have produced landmark papers in various fields and have then gone on to distinguished careers in their respective countries. Furthermore, Goodglass was a founding member of the Academy of Aphasia and of the International Neuropsychological Society (INS). He constructed with Edith Kaplan the "Boston Diagnostic Aphasia Test" (1973) and published his comprehensive book "Understanding Aphasia" (1993), which collects the best empirical contributions and theoretical insights of his outstanding career.

The same educational problem was tackled with different targets, but equally effective results by Roch Lecours, and by Klaus Poeck. Suffice to remind, with respect to Lecours, that the "Research Center of the Côte-de-Neiges Hospital" that he founded in Montreal, has been for many years the reference centre for Canadian and European French-speaking students interested in research on the pathology and rehabilitation of language. One of the main goals pursued by Lecours during the second part of his life has consisted in the development and consolidation of neuropsychology and neurolinguistics in all Latin America, through the foundation of the Latino-American Society of Neuropsychology (SLAN). As an acknowledgment of this passionate work, Lecours received in 2003 a special award from the Mexican Association of Neuropsychology. Furthermore, Lecours wrote, in collaboration with Lhermitte (1979), a textbook on aphasia and developed, in collaboration with Jean-Luc Nespoulous et al. (1992) a protocol for the linguistic exploration of aphasia (MT-86) that is still widely used in French-speaking countries and has been translated into other languages as well. Finally, Lecours was for many years, with his friend and collaborator Harry Whitaker, one of the pillars of *"Brain and Language"*, and he published with Gonia Jarema a special issue of this journal, dealing with functionalism in aphasiology (Jarema & Lecours, 1993).

Klaus Poeck, on the other hand, founded in Aachen the "Neuropsychology and Aphasia Unit," where he trained the most eminent contemporary German neuropsychologists and neurolinguists and many other postgraduate research and clinical fellows, coming from various European countries. Poeck wrote two of the best German textbooks of neurology (Poeck, 1974) and clinical neuropsychology (Poeck, 1982; revisions later published in collaboration with H. Hartje). Finally, his direction and firm determination led to the development and standardization of the "Aachen Aphasia Test (AAT)" (Huber *et al.*, 1984), an aphasia battery that has been subsequently adapted to English, French, Italian, Dutch and several other languages.

3. THE METHODOLOGICAL CHALLENGE

In addition to the educational problem, there was also a methodological change common to all these eminent protagonists of the modern cognitive neuroscience, with respect to their predecessors.

This methodological change consisted of the transition from a time in which a creative individual took into account a critical problem of the extant literature and proposed a personal solution, to a time in which this problem is collectively tackled by an interdisciplinary team, trying to leave intact the autonomy and the originality of the approach typical of the various components of the team (e.g., behavioral neurologist, linguist, neuropsychologist). Leaving aside these common educational and methodological aspects of the activities of Goodglass, Lecours and Poeck, we must acknowledge that their personalities and the contribution that each of them gave to the development of aphasiology, neuropsychology and neurolinguistics were very different.

Goodglass was one of the first authors who used systematically the methods of experimental psychology in the study of brain-damaged patients. More precisely, he was a psycholinguist, who liked elegant, well designed experiments, devised to clarify critical issues in the neuroscience of language, more than very general theories. This rather cautious attitude, considering experimental data as more important than theoretical models, was perhaps, at least in part due to the complementary personality of his friend and collaborator Norman Geschwind, who worked with him in Boston, and was very attracted by comprehensive and well structured general models of the brain-behavior relationships, as is documented by his monumental work on the disconnexion syndromes in animals and man (Geschwind, 1965). It is probably for this tendency to appreciate experimental data more than theoretical models that Goodglass never adhered to the cognitive approach to the study of brain functions, swinging between a moderate interest and a frank skepticism toward this approach.

Among the most beautiful and influential articles published by Goodglass and his eminent collaborators, I would mention his seminal papers on the distinction between fluent and non-fluent forms of aphasia (Goodglass *et al.*, 1964), and on the organization and disorganization of semantic fields (Goodglass & Baker, 1976) in which he investigated, in a very elegant manner all the main components of the semantic networks underlying word meaning. Equally innovative and influential were his papers on disorders of naming (Goodglass *et al.*, 1976; Pease & Goodglass, 1978; Kohn & Goodglass, 1985), on agrammatism (Myerson & Goodglass, 1972; Gleason *et al.*, 1975), on phonological

(Blumstein *et al.*, 1977), semantic (Baker *et al.*, 1981) and syntactic (Goodglass *et al.*, 1979) aspects of auditory comprehension in aphasia. Finally, I would mention his paper on category-specific dissociations in naming (Goodglass *et al.*, 1966) that anticipated the fundamental papers of Warrington and coworkers (Warrington & McCarthy, 1983, 1987; Warrington & Shallice, 1984) opening a very fruitful and still open field of research (Gainotti, 2006) in the area of cognitive neuroscience.

Lecours had a different personality: He was extremely generous and curious, being driven by passion in many of his activities and always searching for new experiences and knowledge. He was therefore constantly exploring new horizons, though being very firmly rooted in the French Canadian culture of his time. As Pierre Marie, who had contributed to the formation of his master Ivan Yakovlev, Lecours can be considered as an "iconoclaste" of currently accepted views in many domains of the neuroscience of language, since there is an original, provocative component in many of his primary research contributions. He was among the first neurologists, after Alajouanine, who clearly understood the importance of linguistics in the study of language pathology. For this reason, after having studied the science of language, he founded with Jean-Luc Nespoulous a strong neurologist–linguist duo, that for more than two decades contributed significantly (in collaboration with Francois Lhermitte, Harry Whitaker, David Caplan, Yves Joanette and other students and clinicians) to different areas of research in neurolinguistics. Among his most innovative contributions, I would mention his studies of phonemic paraphasias, describing their linguistic structure (Lecours & Lhermitte, 1969) and using computer simulation to test the underlying mechanisms (Lecours *et al.*, 1973); his use of a similar approach to make a comparative description of jargonaphasia and schizoaphasia (Lecours & Vanier-Clement, 1976); his investigations on the influence of illiteracy on speech and language disorders after brain damage (Lecours *et al.*, 1987, 1988); his attempts to use paroxismal aphasia to address unsolved issues about language, mind and consciousness (Lecours & Joanette, 1980); and his challenge of the classical views about the anatomo-clinical correlations in aphasia (Lhermitte *et al.*, 1973; Basso *et al.*, 1985).

Poeck was a cultivated, well-balanced personality, considered as a trustworthy friend and a good father by his pupils and collaborators. He had an excellent knowledge not only of the cognitive neurosciences, but also of clinical neurology and (having worked for some time in Pisa with Moruzzi), of neurophysiology. Just as Lecours, he had been one of the first to realize the crucial role that linguistics and cognitive psychology must play in a good aphasiological team. Furthermore, being very interested in rehabilitation, he founded one of the most prominent European schools of neurolinguistic and neuropsychological rehabilitation. His German collaborators (Ria De Bleser, Wolfang Hartjie,

Walter Huber, Bernt Orgass, Luise Springer, Franz-Joseph Stakowiak, Walter Sturm, Dorothea Weniger and Klaus Willmes) and his Italian pupil Claudio Luzzatti are among the most eminent contemporary European neuropsychologists and neurolinguists. His main fields of research concerned classical neuropsychological topics, such as "body schema" (Poeck & Orgass, 1969, 1971), various aspects of apraxia (Poeck, 1983, 1986; Lehmkuhl *et al.*, 1983), slowly progressive aphasia (Poeck & Luzzatti, 1988; Luzzatti & Poeck, 1991) and anatomo-clinical correlations in aphasia (Willmes & Poeck, 1993). He also tried to investigate in a well controlled manner the outcome of aphasia rehabilitation, computing in a first study (Willmes & Poeck, 1984) the rate of spontaneous recovery in various groups of untreated aphasic patients and correcting for these data results obtained in individual subjects (Poeck *et al.*, 1989). Probably, however, his most important contribution to the study of aphasia remains the "AAT" (Huber *et al.*, 1984), which, due to its excellent psychometric foundations, has been extensively used to date in Europe for the diagnosis of aphasic syndromes and for follow-up in language rehabilitation.

4. SHAPING THE NEUROSCIENCE OF LANGUAGE

At the beginning of this introductory chapter, I have mentioned the scientific inheritance that Harold Goodglass, Roch Lecours and Klaus Poeck had received by their great predecessors. I would conclude these short notes by stressing the personal contribution that each of them gave to the development of the contemporary neuroscience of language, thus increasing our common scientific inheritance. The greater contribution of Harold Goodglass has probably consisted in his capacity to identify the critical points both in clinical aphasiology and in controversies between competing theoretical models and to elaborate elegant experimental designs, aiming to clarify these critical issues. Many important classical and other still debated questions have drawn a great benefit from this solid experimental attitude. The main lesson of Roch Lecours, on the other hand, has consisted of a sort of intrepid, passionate attempt to tackle very complex interdisciplinary problems, with the pleasure of exploring new areas of inquiry and of putting again into question commonly accepted general assumptions, but avoiding, at the same time, the risk of less soundly based alternative models. Finally, the greater merit of Klaus Poeck has probably consisted of his capacity to understand the crucial role of linguistics and cognitive neuropsychology in contemporary neuroscience of language, creating a strong interdisciplinary team, where these domains of knowledge integrated with excellent psychometric and methodological competences.

All these considerations, concerning the educational and the scientific activity of Harold Goodglass, Roch Lecours and Klaus Poeck clearly show that the decision of dedicating to them this handbook is certainly not a formal, symbolic act. It is, on the contrary, the sincere acknowledgment of the gratitude that a whole generation of clinicians, researchers and speech pathologists, scattered all around the world, feel for these three outstanding protagonists of the modern neuroscience of language.

Acknowledgments

I am very grateful to Gabriele Miceli and Claudio Luzzatti for their precious comments and contribution to draw faithful portraits of Harold Goodglass and Klaus Poeck.

References

Baker, E., Blumstein, S.E., & Goodglass, H. (1981). Interaction between phonological and semantic factors in auditory comprehension. *Neuropsychologia*, *19*, 1–15.

Basso, A., Lecours, A.R., Moraschini, S., & Vanier, M. (1985). Anatomical correlations of the aphasia as depicted through computer tomography: Exceptions. *Brain and Language*, *26*, 201–229.

Blumstein, S.E., Baker, E., & Goodglass, H. (1977). Phonological factors in auditory comprehension in aphasia. *Neuropsychologia*, *15*, 19–30.

Gainotti, G. (2006). Anatomical, functional and cognitive determinants of semantic memory disorders. *Neuroscience and Biobehavioural Reviews*, *30*, 577–594.

Geschwind, N. (1965). Disconnexion syndromes in animals and man. *Brain*, *88*, 237–294.

Gleason, J.B., Goodglass, H., Green, E., Ackerman, N., & Hyde, M. R. (1975). The retrieval of syntax in Broca's aphasia. *Brain and Language*, *2*, 451–471.

Goodglass, H. (1993). *Understanding aphasia*. San Diego, CA: Academic Press.

Goodglass, H., & Baker, E. (1976). Semantic field, naming, and auditory comprehension in aphasia. *Brain and Language*, *3*, 359–374.

Goodglass, H., & Kaplan, E. (1973). *Boston diagnostic aphasia test*. St Antonio, TX: The Psychological Corporation.

Goodglass, H., & Kaplan, E. (1983). *The assessment of aphasia and related disorders*. Philadelphia, PA: Lippincott Williams and Wilkins.

Goodglass, H., & Quadfasel, F.A. (1954). Language laterality in left-handed aphasics. *Brain*, *77*, 521–548.

Goodglass, H., Quadfasel, F.A., & Timberlake, W.H. (1964). Phrase length and the type and severity of aphasia. *Cortex*, *7*, 133–155.

Goodglass, H., Klein, B., Carey, P., & Jones, K. (1966). Specific semantic word categories in aphasia. *Cortex*, *7*, 74–89.

Goodglass, H., Kaplan, E., Weintraub, S., & Ackerman, N. (1976). The "tip-of-the-tongue" phenomenon in aphasia. *Cortex*, *12*, 145–153.

Goodglass, H., Blumstein, S.E., Gleason, J.B., Green, E., & Statlender, S. (1979). The effect of syntactic encoding on sentence comprehension in aphasia. *Brain and Language*, *7*, 201–209.

Huber, W., Poeck, K., & Willmes, K. (1984). The Aachen aphasia test. *Advances in Neurology*, *42*, 291–303.

Kohn, S.E., & Goodglass, H. (1985). Picture-naming in aphasia. *Brain and Language*, *24*, 266–283.

Jarema, G., Lecours, A.R. (Eds.) (1993). *Functionalism in Aphasiology* (pp. 467–603). Special Issue, *Brain and Language*, *45*.

Lecours, A.R., & Joanette, Y. (1980). Linguistic and other psychological aspects of paroxismal aphasia. *Brain and Language*, *10*, 1–23.

Lecours, A.R., & Lhermitte, F. (1969). Phonemic paraphasias: Linguistic structures and tentative hypotheses. *Cortex*, *5*, 193–228.

Lecours, A.R., & Lhermitte, F. (1979). *L'Aphasie*. Paris: Flammarion.

Lecours, A.R., & Vanier-Clement, M. (1976). Schizoaphasia and jargonaphasia. A comparative description with comments on Chaika's and Fromkin's respective look at "schizophrenic language". *Brain and Language*, *3*, 516–565.

Lecours, A.R., Deloche, G., & Lhermitte, F. (1973). Paraphasies phonémiques: Déscription et simulation sur l'ordinateur. *Information Médicale*, *1*, 311–350.

Lecours, A.R., Mehler, J., Parente, M.A., Caldera, A., Cary, L., Castro, M.J. *et al.* (1987). Illiteracy and brain damage: Aphasia testing in culturally contrasted populations (subjects). *Neuropsychologia*, *25*, 231–245.

Lecours, A.R., Mehler, J., Parente, M.A., Bettrami, M.C., de Tolipan, L.C., Cary, L. *et al.* (1988). Illiteracy and brain damage: A contribution to the study of speech and language disorders in illiterates with unilateral brain damage (initial testing). *Neuropsychologia*, *26*, 575–589.

Lehmkuhl, G., Poeck, K., & Willmes, K. (1983). Ideomotor apraxia and aphasia: An examination of types and manifestations of apraxic symptoms. *Neuropsychologia*, *21*, 199–212.

Lhermitte, F., Lecours, A.R., Ducarne, B., & Escourolle, R. (1973). Unexpected anatomical findings in a case of fluent jargon aphasia. *Cortex*, *9*, 433–446.

Luzzatti, C., & Poeck, K. (1991). An early description of slowly progressive aphasia. *Archives of Neurology*, *48*, 228–229.

Myerson, R., & Goodglass, H. (1972). Transformational grammars of three agrammatic patients. *Language and Speech*, *15*, 40–50.

Nespoulous, J.-L., Lecours, A.R., & Lafond, D. (1992). *Protocole Montréal-Toulouse d'examen linguistique de l'aphasie (MT-86)*. Isbergues: L'Ortho-Edition.

Pease, D.M., & Goodglass, H. (1978). The effects of cuing on picture naming in aphasia. *Cortex*, *14*, 178–189.

Poeck, K. (1965). On orientation with respect to one's own body. *Bibliography in Psychiatry and Neurology*, *127*, 144–167.

Poeck, K. (1974). *Neurologie*. Heidelberg: Springer.

Poeck, K. (Ed.) (1982). *Klinische Neuropsychologie*. Stuttgard: Thieme.

Poeck, K. (1983). Ideational apraxia. *Journal of Neurology*, *230*(1), 1–5.

Poeck, K. (1986). The clinical examination for motor apraxia. *Neuropsychologia*, *24*, 129–134.

Poeck, K., & Luzzatti, C. (1988). Slowly progressive aphasia in three patients. The problem of accompanying neuropsychological deficit. *Brain*, *111*, 151–168.

Poeck, K., & Orgass, B. (1969). An experimental investigation of finger agnosia. *Neurology*, *19*, 801–807.

Poeck, K., & Orgass, B. (1971). The concept of the body schema: A critical review and some experimental results. *Cortex*, *7*, 254–277.

Poeck, K., Huber, W., & Willmes, K. (1989). Outcome of intensive language treatment in aphasia. *Journal of Speech and Hearing Disorders*, *54*, 471–479.

Warrington, E.K., & McCarthy, R. (1983). Category-specific access dysphasia. *Brain*, *106*, 859–878.

Warrington, E.K., & McCarthy, R. (1987). Categories of knowledge: Further fractionations and an attempted integration. *Brain*, *110*, 1465–1473.

Warrington, E.K., & Shallice, T. (1984). Category-specific semantic impairments. *Brain*, *107*, 829–854.

Willmes, K., & Poeck, K. (1984). Ergebnisse einer multizentrischen Untersuchung über eine Spontanprognose von Aphasien vaskularer Ätiologie. *Nervenarzt*, *65*, 62–71.

Willmes, K., & Poeck, K. (1993). To what extent can aphasic syndromes be localized? *Brain*, *116*, 1527–1540.

Yakovlev, P.I., & Lecours, A.R. (1967). The myelogenetic cycles of regional maturation of the brain. In A. Minkowski (Ed.), *Regional development of the brain in early life* (pp. 3–70). Philadelphia, PA: Davis.

METHODS AND TECHNIQUES

1

Classical and Contemporary Assessment of Aphasia and Acquired Disorders of Language

YVES TURGEON[1] and JOËL MACOIR[2]

[1]*Restigouche Health Authority, Campbellton Regional Hospital, Campbellton, NB, Canada*
[2]*Département de réadaptation, Programme de maîtrise en orthophonie, Université Laval, Québec (QC), Canada*

ABSTRACT

The detection and diagnosis of language problems are the starting point for all clinical intervention. The choice of the theoretical approach and selection of assessment tools are thus of utmost importance. Classical and contemporary assessment methods for acquired language impairments are discussed with a comparison of clinical–neuroanatomical and psycholinguistic approaches. Bedside evaluation and screening tools provide some rough information on the patient's global communication profile and may indicate whether more comprehensive assessment is indicated. Comprehensive assessment explores in depth the different aspects of language and depending on the theoretical framework provides information on the type of aphasia or the functional processing components impaired. Specific pathologies such as dementia, traumatic brain injury, and right hemisphere brain damage may require the assessment of specific aspects of language. Finally, the importance of evaluating language in conjunction with other mental functions must be considered.

1.1. INTRODUCTION

The evaluation of language is one of the most important tasks of speech-language pathologists and professionals from a variety of disciplines and backgrounds (neuropsychologists, occupational therapists, physicians, nurses, and so forth). The assessment procedure is often the first contact with clients and also constitutes the starting point of all clinical interventions. Because of the absence of biological markers or simple assessment methods, the early detection or diagnosis of language problems remains dependent on various indirect assessments (i.e., language function must be inferred from

the client's performance on various tasks devised to explore different areas of this function) aimed at identifying specific impairments and eliminating other possible causes.

The main goal of language screening is to determine whether a client has a problem or not; the output of this type of assessment is a "pass" or "fail", based on an established criterion that could lead to a more extensive evaluation or a follow-up assessment. Diagnosis and differential diagnosis assessments are usually performed to label the communication problem and/or differentiate it from other disorders in which similar characteristics are usually reported. Language evaluation provides the clinician with a detailed description of the client's baseline level of functioning in all areas of communication in order to identify affected and preserved components, plan treatment, establish treatment effectiveness, or track progress over time through periodic re-evaluations. The clinician must consider not only the different areas of language, but also important related abilities and components such as cognitive functions, pragmatics, emotions, awareness of deficits, and so forth. The selection of evaluation tools is also influenced by the specific objectives of the assessment. Screening for a language disorder is usually performed with standardized screening measures whereas standardized norm-referenced tests are used for diagnosis and differential diagnosis assessments as well as for clinical treatment purposes (baseline, effectiveness, and progress).

In this chapter, we first outline the nature of acquired language deficits as well as reference models for their assessment. We then briefly report classical methods and tests for the assessment of language impairments in aphasia and other pathological affections. Finally, we address the question of the interface between language and other cognitive functions.

1.2. NATURE OF LANGUAGE DEFICITS

Since higher mental functions depend on specialized cerebral substrates, a disturbance of any of these brain areas may lead to acquired impairments of language and communication. These may involve disorders of articulation, word and sentence comprehension and production, reading, and writing, which are commonly regarded as clinical manifestations of acquired language deficits. The majority of assessment tools are designed to evaluate these problems.

1.2.1. Classification of Aphasic Syndromes and Symptoms

There are various classifications for aphasic syndromes. For example, Goodglass (2001) suggested categorizing language deficits under 10 different types. In each of these aphasia types, particular symptoms can be regarded as signs of comprehension and production problems. For example, deficits at the word, sentence, or discourse levels are common forms of comprehension disorders. In contrast, reduced verbal fluency and word-finding difficulties are common forms of production disorders. Some types of aphasic disturbances, such as the loss of grammar and syntax or semantic deficits, have expressive and receptive aspects, and thus contribute to both comprehension and production disorders.

The classical definition of aphasic syndromes includes various categorization systems. Some experts define aphasic syndromes according to the types of language errors. Others focus on language production and related impairments of spontaneous speech. Despite these diverging views, a few clusters of aphasic symptoms have been proposed over the past few decades. It is conventional to group aphasias into two broadly defined categories: fluent and non-fluent. Fluent aphasias are distinguished by fluent speech and relatively normal articulation but difficulties in auditory comprehension, repetition, and presence of paraphasias. Non-fluent aphasias are characterized by relatively preserved verbal comprehension, but significant articulation and spoken production problems. Table 1.1 presents eight classical aphasia syndromes after Beeson and Rapcsak (2006).

1.2.2. Pure Language Impairments

Aphasia experts discriminate between the aphasias and other disorders that impair communication but do not meet the definition of classical syndromes. A group of these conditions are regarded as "pure" impairments, and include pure alexia, also called letter-by-letter reading, in which the patient reads by spelling the letters out loud. Pure word deafness is a deficit distinct from generalized auditory agnosia in which comprehension and repetition of speech are impaired but reading, writing, and spontaneous speech

TABLE 1.1   Classical Aphasia Syndromes

Syndromes	Language deficits	Key language errors
Fluent aphasias		
Anomic	Normal fluency; good auditory comprehension and repetition	Anomia; may resolve to minimal word-finding difficulties
Conduction	Normal fluency; good auditory comprehension	Phonemic paraphasias; poor repetition
Transcortical sensory aphasia	Normal fluency; preserved repetition; poor comprehension	Verbal paraphasias; anomia
Wernicke	Normal fluency; poor comprehension; poor repetition	Jargon; logorrhea; anomia
Non-fluent aphasias		
Transcortical motor	Reduced fluency; good auditory comprehension; good repetition	Reduced spontaneous speech; better naming than spontaneous speech
Broca	Reduced fluency; relatively good comprehension; poor repetition; agrammatism	Slow, halting speech production; phonetic and phonemic paraphasias; anomia; recurring utterances; articulatory impairment
Mixed transcortical	Reduced fluency; preserved repetition; markedly impaired auditory comprehension	Severely impaired verbal expression; anomia
Global	Severe reduction of fluency, severe comprehension deficit; poor repetition	Slow, halting speech production or mutism; articulatory impairment; severe anomia

After Beeson and Rapcsak (2006)

are preserved. Pure agraphia refers to the inability to program movements necessary to form written words. Finally, agnosia refers to selective impairments of information processing in a single sensory modality (e.g., vision) that is not explained by a primary sensory defect, attention disorder, or language disorder. The most typical example is visual agnosia in which the individual fails to name an object (e.g., a telephone) present in his visual field but can name that same object when allowed to touch it or to hear sounds from that object. Beeson and Rapcsak (2006) view word deafness, pure alexia, and the agnosias as an input problem that may compromise comprehension of spoken and written language. Apraxia of speech, dysarthria, and mutism are considered as production problems. Apraxia of speech is an impairment of motor programming that results in articulatory impairment in the absence of a disturbance of motor control for speech production. Dysarthria is a motor speech disorder in which the speech subsystems (respiration, phonation, resonance, and articulation) are affected. Mutism is the complete inability to produce speech.

A comprehensive language assessment should always be based on typologies of syndromes and symptoms and/or on a theoretical model of language functioning. In the following section, we discuss the two main reference models for language assessment.

1.3. THEORETICAL MODELS FOR THE ASSESSMENT OF LANGUAGE IMPAIRMENT

The choice of a particular assessment method, the selection of evaluation tools as well as the interpretation of results are highly dependent not only on the clinician's own conception of language but also on the reference to an assessment model. The nature and definition of language disorders is related to contemporary approaches to assessment. According to Spreen and Risser (2003), the way a scientist or practitioner designs an assessment procedure is directly influenced by the way she or he conceptualizes language disturbances, which she or he may regard either as disorders of specific language abilities or as a disturbance of the ability to communicate. Furthermore, one's approach to language assessment will depend on how one defines language disturbances, that is, as unitary or heterogeneous concepts. It is not surprising that classical approaches to language assessment have developed around issues pertaining to the measurement of aphasic disturbances.

Classically, assessment procedures have been intended either to identify the cardinal features of a given clinical presentation or to measure the various abilities that are required to communicate. In other words, language assessment tools are either designed to match a given theoretical framework of aphasia or to probe the presence or absence of the psycholinguistic requirements for effective communication. Two contemporary approaches to language assessment are the clinical–neuroanatomical approach that emerges from the clinical observation of brain-injured individuals and the psycholinguistic approach that emerges from the laboratory or experimental setting.

1.3.1. The Clinical–Neuroanatomical Approach to Language Assessment

The clinical–neuroanatomical (or clinical–pathological) approach to aphasia assessment is by far the oldest and thus traditional approach. It is called clinical–pathological because it relies on clinical observation and clinical validation of the neuroanatomical substrata responsible for the clinical manifestations. For example, according to the clinical–neuroanatomical approach, Wernicke's aphasia, which is usually associated with large posterior lesions

TABLE 1.2 Illustration of the Output of Assessments Conducted within Clinical-Neuroanatomical or Psycholinguistic Frameworks

Aphasia type	Clinical-neuroanatomical approach	Psycholinguistic approach*
Wernicke's aphasia	Fluent speech/poor comprehension/poor repetition/severe anomia/impaired reading and written spelling	Semantic deficit
	Lesion of the posterior part of the temporal lobe	Phonological and orthographic input and output deficits Surface dyslexia Surface agraphia
Broca's aphasia	Non-fluent, effortful speech/poor repetition/anomia/preserved comprehension/impaired reading and written spelling	Phonological and orthographic output deficits
	Lesion of Broca's area, frontoparietal operculum, anterior part of the insula	Deep dyslexia
		Deep agraphia

* Characteristics usually but not necessarily associated with the aphasic syndrome

adjacent to the Sylvian fissure, has cardinal features that distinguish it from other fluent and non-fluent aphasic disturbances such as transcortical sensory aphasia or Broca's aphasia (see Table 1.2).

In the clinical–neuroanatomical model, the general assessment process of an aphasic person consists of: (1) gathering case history data (e.g., cerebrovascular accident in the left frontal area); (2) administering a specific test battery (e.g., the Boston Diagnostic Aphasia Examination (BDAE) (Goodglass et al., 2000); (3) comparing the results and description of the behavior (e.g., impaired fluency, impaired articulatory agility, relatively good auditory comprehension, and agrammatism) with the classification of neurogenic acquired language deficits; and (4) specifying the precise aphasic label (Broca's aphasia) that best fits these characteristics.

1.3.2. Psycholinguistic Approach to Language Assessment

The psycholinguistic approach to language assessment derives from information processing theories. This approach, building on linguistics and cognitive psychology, focuses on language processing and provides case examples to explain given language disturbances by analyzing the simple or complex processes that may be disrupted in a given individual's language system. The psycholinguistic approach combines the theoretical viewpoints of linguistics and cognitive psychology to analyze language impairment in terms of processing instead of describing and classifying clinical symptoms. In these models, cognitive functions, including language, are sustained by specialized interconnected processing components represented in functional architectural models. For example, as shown in Figure 1.1, the ability to orally produce a word in picture naming is conceived as a staged process in which the activation flow is initiated in a conceptual–semantic component and ends with the execution of articulation mechanisms.

An assessment process based on cognitive neuropsychological models consists of the identification of the impaired and preserved processing components for each language modality (see Table 1.2). This analysis is performed by the administration of specific tasks or test batteries (e.g., psycholinguistic assessments of language processing in aphasia (PALPA) (Kay *et al.*, 1992) aimed at evaluating each component and path in the model. For example, the evaluation of naming abilities in an aphasic person could be performed by administering tasks exploring the conceptual–semantic (e.g., semantic questionnaire), phonological output lexicon (e.g., picture naming task controlled for frequency, familiarity, and so forth), and phonological output buffer (e.g., repetition of words and non-words controlled for length)

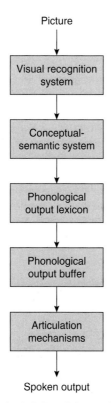

FIGURE 1.1   Schematic depiction of the cognitive neuropsychological model of spoken picture naming.

components. Important information regarding the level of impairments is also derived from error analysis. An anomic error could stem from distinct underlying deficits (e.g., in the activation of conceptual–semantic representations or in retrieving phonological forms of words in the output lexicon), leading to distinct types of errors (e.g., semantic substitutions and phonemic errors). The complete cognitive assessment process should give the clinician an understanding of the client's deficits (i.e., surface manifestations, underlying origins, and affected components) as well as enable him to identify the strengths and weaknesses in the client's communication abilities.

The selection of assessment methods and tools stems directly not just from reference models but also from the purpose of the evaluation. In the following section, we briefly describe bedside and screening tests as well as comprehensive test batteries for aphasia and other language disturbances.

1.4. CLASSICAL TESTS AND APHASIA ASSESSMENT PROCEDURES

Aphasia is the most common communication disorder resulting from brain damage. As mentioned previously, this impairment involves language problems of production and comprehension as well as disturbances in reading and spelling. The reader will find a complete description of the tests mentioned below, and others not reported here, in Haynes and Pindzola (2003), McCauley (2001), Murray and Chapey (2001), Spreen and Risser (2003), and Strauss *et al.* (2006).

1.4.1. Bedside and screening tests

A patient's symptoms typically change rapidly during the first days and weeks following brain damage. Moreover, the patient's health often does not allow an exhaustive aphasia examination to be done, and thus bedside or screening instruments may be useful in order to advise relatives and health care professionals about the global communication profile and the best way to communicate in functional situations. These instruments are also useful in helping clinicians to determine the need to perform a more thorough and extensive language assessment or to prioritize patients on a waiting list. In addition to actual screening tests, clinicians may also administer shortened versions of comprehensive tests of aphasia (i.e., short form of the Token Test, Spellacy & Spreen, 1969). As pointed out by Spreen and Risser (2003), although bedside and screening tests may be used to identify language impairments in moderate and severe aphasics (language is obviously affected, even in simple and natural communication situations), they are inappropriate or of little use in distinguishing the responses of individuals with mild deficits from those with normal language skills.

1.4.2. Comprehensive Examinations and Aphasia Batteries

In contrast to bedside and screening tests, the main purpose of comprehensive examinations of aphasia is to provide an extensive description of language skills through the administration of tests designed to explore the different areas of language.

Depending on the reference assessment model, the output of a comprehensive examination may consist of a particular diagnosis of aphasia with a description of the severity of the deficits in each language area (clinical–neuroanatomical approach), or identifying specific impairments affecting the functional processing components of language skills (psycholinguistic approach). There are several classical comprehensive examinations and aphasia batteries. The most widely used in clinical and research settings in the English language are presented in Box 1.1. All these standardized test batteries comprise different subtests that assess all the dimensions of language in order to diagnose and classify aphasic syndromes according to clinical localization-based classifications (i.e., Broca's, Wernicke's aphasia, and so forth). The PALPA (Kay *et al.*, 1992) is a comprehensive test battery directly derived from the cognitive neuropsychology approach to assessment. This aphasia battery consists of a set of resource materials comprising 60 rigorously controlled tests that enable the user to select tasks that can be used to identify impaired and intact abilities in an individual. The scoring and analysis of errors give the clinician a detailed profile of language abilities, which can be interpreted within current cognitive models of language. Compared to classical aphasia batteries, however, the versatility and flexibility of the PALPA is lessened by the lack of standardization and validity/reliability measures.

1.4.3. Assessment of Specific Aspects of Language

Tests of specific aspects of language are often used to supplement comprehensive batteries but some of them are also administered for screening purposes. These tests, which usually include more items and more levels of difficulty, may provide precise and detailed information about specific language abilities. Their selection also depends on the underlying theoretical assessment model. For example, comprehension may be tested through the administration of specific tests focusing on semantics, syntax, commands, or narrative discourse (see Box 1.1). Other tests are available for measuring verbal expression, spoken and written naming, verbal fluency, reading, writing, gestural abilities, and so forth.

Box 1.1   Most representative assessment instruments according to types of assessment

Bedside and screening tests
- Reitan, R.M. (1991). *Aphasia screening test.* Tucson, AZ: Reitan Neuropsychology Laboratory.
- Whurr, R. (1996). *The aphasia screening test* (2nd edn). San Diego, CA: Singular Publishing Group.

Comprehensive examinations
- Goodglass, H., Kaplan, E., & Barresi, B. (2001). *Boston diagnostic aphasia examination.* Philadelphia, PA: Lippincott, Williams & Wilkins.
- Helm-Estabrooks, N. (1992). *Aphasia diagnostic profiles.* Chicago, IL: Riverside Publishing.
- Kay, J., Lesser, R., & Coltheart, M. (1992). *Psycholinguistic assessments of language processing in aphasia (PALPA).* Hove, England: Lawrence Erlbaum Associates.
- Kertesz, A. (2006). *Western aphasia battery revised.* San Antonio, TX: Harcourt Assessment.

Assessment of specific aspects of language
Auditory and reading comprehension
- Brookshire, R., & Nichols, L.E. (1993). *The discourse comprehension test.* Minneapolis, MN: BRK Publishers.
- De Renzi, E., & Vignolo, L. (1962). The Token Test: A sensitive test to detect receptive disturbances in aphasics. *Brain, 85,* 665–678.
- LaPointe, L.L., & Horner, J. (1998). *Reading comprehension battery for aphasia.* Austin, TX: Pro-Ed.

Semantic processing
- Howard, D., & Patterson, K.E. (1992). *The pyramids and palm trees test.* Oxford: Harcourt Assessment.

Naming
- German, D.J. (2000). *Test of adolescent and adult word finding* (2nd edn). Austin, TX: Pro-Ed.
- Kaplan, E., Goodglass, H., & Weintraub, S. (2001). *Boston naming test* (2nd edn). Philadelphia, PA: Lippincott, Williams & Wilkins.

Syntax
- Bastiaanse, R., Edwards, S., & Rispens, J. (2002). *The verb and sentence test.* Toronto: Harcourt Assessment.

Writing
- Hammill, D.D., & Larson, S.C. (1996). *Test of written language* (3rd edn). Austin, TX: Pro-Ed.

Assessment of functional communication
- Frattali, C.M., Thompson, C.K., Holland, A.L., Wohl, C.B., & Ferketic, M.M. (1995). *Functional assessment of communication skills for adults.* Rockville, MD: American Speech-Language-Hearing Association.
- Holland, A.L., Frattali, C.M., & Fromm, D. (1999). *Communication activities of daily living* (2nd edn). Austin, TX: Pro-Ed.

1.4.4. Assessment of Functional Communication

Although traditional tests provide useful information about linguistic abilities and language impairments in aphasia, performance on these tests does not necessarily predict how a person will communicate in the more naturalistic settings of everyday life. Instead of focusing on the importance and nature of the deficits, the functional communication assessment approach looks at the impact of these deficits on the person's activities and participation in society. Functional communication skills may be assessed with specific structured tests or with rating scales and inventories of communication profiles. Structured tests such as the *Communication activities of daily living* 2 (Holland *et al.*, 1999) and the *Amsterdam–Nijmegen everyday language test* (Blomert *et al.*, 1994) have been devised to explore functional communication skills using role playing in daily life activities (shopping, dealing with a receptionist, and so forth) and have been shown to be useful in tracking progress over time. However, while they are certainly more ecologically valid than comprehensive examinations for specific aspects of language, such structured functional communication tests do not necessarily give a reliable picture of the communication skills of a person in real-life situations. In this respect, rating scales and inventories of communication profiles are closer to functional situations. For example, the *Functional assessment of communication skills for adults* (Frattali *et al.*, 1995) is a rating protocol based on the observations made by a speech-language pathologist or other significant person in the following four domains: social communication (e.g., "refers to familiar people by name"); communication of basic needs (e.g., "expresses need to eat"); reading, writing, and number concepts (e.g., "writes messages"); and daily planning (e.g., "tells time").

1.5. CLASSICAL TESTS FOR THE ASSESSMENT OF LANGUAGE IMPAIRMENT IN SPECIAL POPULATIONS

Referral for language assessment includes individuals of different age groups and aetiologies presenting with various language and communication problems. In adults, referral for a language evaluation may be required for patients with Alzheimer's disease, other forms of dementia, right hemisphere damage (RHD) or traumatic brain injury (TBI), and so forth.

1.5.1. Assessment of Language in Dementia

According to the diagnostic and statistical manual of mental disorders (DSM-IV) (American Psychiatric Association, 1994), a dementia syndrome is characterized by multiple deficits in cognition, including memory impairment, which are the direct consequence of physiological changes. The DSM-IV criteria require that these deficits are of sufficient magnitude to impair social or occupational functioning. Diagnostic classifications for dementia include subtypes based on characteristics such as the presentation of typical symptoms, the progression and course of the disease, psychiatric and behavioral features, as well as presumed causes. The early detection of dementia often depends on various assessment tools, including language and cognitive tests administered to exclude other possible disease processes or to identify specific forms of a given disease. The clinical approach to the diagnosis of dementia usually begins with the recognition of a progressive decline in memory, a decrease in the patient's ability to perform daily living activities, the presence of psychiatric problems, personality changes, or problem behaviors. Except for these screening instruments and severity rating scales, there are few cognitive function assessment tools specially designed for dementia. The classical assessment of language in this population is usually performed with aphasia batteries. Since these were devised to allow clinicians to identify aphasia syndromes, the contribution of language assessment to the differential diagnostic of dementia may be limited.

Many studies have sought to identify the neuropsychological features that distinguish the different forms of dementia. What emerge from these studies are descriptions of cognitive functioning in which some distinctions are useful for the differential diagnosis of dementia. However, there is also significant heterogeneity in the neuropsychological manifestations in the early stages of major forms of dementia (for a review, see Rosenstein, 1998). With respect to language, neurolinguistics studies also contribute to a better characterization of deficits in dementia by specifically identifying the functional localization of impaired and preserved components and subcomponents of the language processing system. For example, patients presenting with Alzheimer's disease may show different patterns of spelling impairment, including written production peripheral deficits, while individuals with semantic dementia, a clinical syndrome that results from a degenerative disease of the temporal lobes (Neary *et al.*, 1998), usually present with surface agraphia (Macoir & Bernier, 2002), a deficit in which patients are better at spelling orthographically regular than irregular words, and tend to produce phonologically plausible spelling errors (e.g., soap → SOPE). Similarly, agrammatism and word-finding problems associated with phonological errors and preservation of semantic processing are prominent characteristics of individuals diagnosed with non-fluent progressive aphasia, another predominantly frontotemporal lobar degeneration. These individuals usually differ from patients diagnosed with Alzheimer's disease, who are more likely to show word-finding problems with semantic errors, semantic deficits, and preservation of phonology and syntax. Therefore, using a cognitive neuropsychological assessment approach, properly controlled evaluation tasks can help to differentiate common disease processes in the elderly population.

1.5.2. Assessment of Language in TBI and RHD

TBI broadly refers to any damage to the brain caused by external forces. In mild TBI, many patients exhibit moderate-to-severe cognitive deficits, including communication problems. For most of them, these deficits disappear quite quickly, however, for a substantial number of patients who present with moderate or severe TBI, the cerebral insult has a significant and permanent effect on cognitive, emotional, and psychosocial functioning as well as on autonomy in activities of daily living. With respect to language, TBI rarely results in aphasia-like syndromes (Ylvisaker *et al.*, 2001). Word-finding difficulties are frequent but often stem from slower information processing rather than from a semantic or lexical access deficit *per se*. In fact, communication problems in TBI are frequently not linguistic in nature but result from the impairment of other cognitive functions. They are usually associated with frontal–limbic damage and are characterized by difficulties at the discourse level (e.g., disorganization or paucity of discourse), word-finding problems, difficulties understanding and expressing abstract concepts and indirect language, and so forth (for a complete description, see McDonald *et al.*, 1999; see Chapter 28, this volume). Different standardized tests were recently developed for the diagnosis of cognitive deficits in TBI, including communication disorders. For example, the *Ross information processing assessment* (Ross-Swain, 1996) includes tests of memory, orientation, problem-solving, abstract reasoning, organization, and auditory processing. (For a description and a critical review of these instruments, see Brookshire, 2003.)

Patients with RHD do not usually present with symptoms of aphasia but show "cognitive-communication" impairments too. For example, they may have difficulty in interpreting figurative language, humor, or irony. They may have problems identifying the overall theme of a message or organizing information in narrative discourse. Pragmatic communication may also be affected in patients with RHD (e.g., initiation of speech, turn-taking, interpretation of speech acts and intents). The assessment of these communication problems can be done with tests directly aimed at identifying specific RHD deficits. For example, the *Right hemisphere language battery* (RHLB-2) (Bryan, 1995) consists of different subtests that assess metaphor comprehension, humor appreciation, discourse production, and so forth (for a review, see Brookshire, 2003).

Although language and communication can be assessed with batteries and tests directly aimed at linguistic abilities, language should also be considered through its interaction with the different cognitive domains, an important consideration which we address in the following section.

1.6. INTERFACE BETWEEN LANGUAGE AND OTHER COGNITIVE FUNCTIONS

In the past, language abilities were considered in isolation from other cognitive functions. The characterization of language processes and language disorders constituted the first goal of clinical studies as well as assessment methods. According to this traditional approach, language is different from memory, attention, executive functions, and so forth, and should therefore be treated separately. The same observation applies to other cognitive functions. Studies conducted in different populations were restricted to specific areas of cognition, for example, studies of language deficits in aphasia, studies of memory deficits in Alzheimer's disease, studies of executive function deficits in TBI, and so forth. This approach is no longer justified: language disorders do not occur in isolation; aphasic disturbances rarely occur in the absence of memory impairment or attention/executive problems, even in milder cases of aphasia and related disorders.

Higher mental functions are closely interrelated and complementary. Capturing the nature of the deficits affecting a particular cognitive function also involves evaluating the integrity or impairment of connected functions. Language plays a central role in human cognition; as such, it is closely related to other higher mental functions as well as to basic functions such as attention and working memory. The occurrence of an altered performance, particular behavior or specific error in a language assessment task may certainly be linguistic in nature but is frequently linked to a primary source external to language.

Three different cognitive domains illustrate the interrelationship between language and cognition. These examples emphasize the importance of performing an integrated language assessment including non-linguistic as well as linguistic tasks.

1.6.1. Working Memory, Executive Functions, and Language

Working memory refers to structures and processes used to temporarily store and manipulate information. This basic function is necessary for a wide range of complex cognitive activities, including language. Its role in language acquisition is generally known and it is also involved in abilities such as the production and comprehension of sentences in adults. For example, Waters and Caplan (1999) showed that patients with working memory storage deficits (e.g., span limited to two or three words) do not have any difficulty interpreting spoken sentences, even when they are syntactically complex such as the passive form (e.g., the dog was washed by the woman). However, patients presenting with a deficit in the central executive component of working memory (e.g., patients with Alzheimer's or Parkinson's disease) show problems in comprehending sentences in which the integration of individual word meanings into a semantic representation is delayed, as in the case of a sentence including several prenominal adjectives (e.g., the old, hairy, nice dog was lying under the table) or with several nouns preceding or following the verb. In this case, the observed impairment is not syntactic *per se* but stems directly from an executive deficit. In fact, the main role of the executive system or central executive in cognition is to control, regulate, and coordinate cognitive processes.

Executive functions are involved in memory, reasoning, and learning. In language, they are involved in sentence and discourse production (e.g., planning how to express an idea), discourse comprehension (e.g., interpreting indirect acts of language), as well as in accessing semantic or lexical representations. For example, recent studies showed that limited executive functioning contributes to semantic categorization deficits in patients with Alzheimer's disease or frontotemporal dementia (Grossman *et al.*, 2003), and leads to deficits in word comprehension and production in patients with aphasia (Jefferies & Lambon Ralph, 2006). For similar reasons, an executive origin was suggested to account for the performance of ANG, a patient with transcortical motor aphasia (Robinson *et al.*, 1998). According to these authors, ANG presents substantial difficulties in the production of words, phrases, and sentences because of an executive impairment of the inhibition mechanisms that normally block the activation of competitors during lexical access. Numerous other language skills require the central executive system to manage attentional resources. As a consequence, the interpretation of results of language tests should be adjusted or modified according to the performance on tests exploring working memory and executive functions (e.g., Stroop Test, Trail Making Test, and so forth). Whether performance on these tests is representative of the executive mechanisms involved in language is, however, still unclear. Nevertheless, it is plausible that a cognitive system devoted to the selection and inhibition of verbal or non-verbal input stimuli is comparable to the executive mechanisms involved in the activation of semantic and/or lexical representations in language production.

1.6.2. Object Recognition, Semantic Processing, and Language

The interface between language and cognitive functions can also be illustrated with an example taken from the object recognition domain. Comprehensive language assessment batteries almost always include picture naming tasks. The quantitative results as well as the qualitative analysis of errors produced on these tasks are informative regarding the lexical access processes in spoken (or written) production. For example, the production of semantic but non-visual errors (e.g., mirror instead of comb on the Boston Naming Test (BNT)) can be attributed to a semantic or lexical deficit. When the error is purely visual (i.e., the response corresponds to a visually similar concept to the target but belongs to a different semantic category: skirt instead of volcano, BNT), the deficit is probably functionally localized in the object recognition system (see Figure 1.1). Particular caution should be exercised when visuo-semantic errors are observed (i.e., the response corresponds to a visually similar concept to the target and belongs to the same semantic category so that the error could be perceptually, semantically, or lexically based: horse instead of camel, BNT). In this particular case, the assessment of language production should be complemented with an evaluation of visuoperceptual processes (e.g., using the Birmingham Object Recognition Battery; Riddoch & Humphreys, 1993).

1.7. CHALLENGES AND FUTURE DIRECTIONS

Language production and comprehension are complex cognitive skills which should not be considered in isolation in assessment procedures. The interrelationship between language and other cognitive functions must be taken into account, particularly with respect to the possible influence of attention, working memory, and executive functions on linguistic abilities. To date, however, very few if any assessment tools have been directly designed to capture the interface between language and other cognitive functions. Further research is thus needed to clarify this interface and ultimately lead to the development of comprehensive tests of "cognitive–linguistic" skills.

The assessment of language and communication is more than just an evaluation of specific skills in terms of the preservation/impairment of processing components and surface structures. The scope of assessment should be broadened to include information about individual and sociocultural factors that could have a major impact on communicative efficacy and competence. In this respect, we have little data and few assessment tools that we can use to appropriately consider these factors as well as the question of the impact of intra-individual differences on performance on language and communication tests.

References

American Psychiatric Association (1994). *Diagnostic and statistical manual of mental disorders* (4th edn). Washington, DC: American Psychiatric Association.

Beeson, P.M., & Rapcsak, S.Z. (2006). The aphasias. In P.J. Snyder & P.D. Nussbaum (Eds.), *Clinical neuropsychology: A pocket handbook for assessment* (2nd edn, pp. 436–459). Washington, DC: American Psychological Association.

Blomert, L., Kean, M.L., Koster, C., & Schokker, J. (1994). Amsterdam–Nijmegen everyday language test: Construction reliability and validity. *Aphasiology, 8*, 381–407.

Brookshire, R.H. (2003). *Introduction to neurogenic communication disorders* (6th edn). St. Louis, MO: Mosby.

Bryan, K. (1995). *The right hemisphere language battery* (2nd edn). London: Whurr.

Frattali, C.M., Thompson, C.K., Holland, A.L., Wohl, C.B., & Ferketic, M.M. (1995). *Functional assessment of communication skills for adults*. Rockville, MD: American Speech-Language-Hearing Association.

Goodglass, H. (2001). *The assessment of aphasia and related disorders* (3rd edn). Baltimore, MD: Lippincott, Williams & Wilkins.

Goodglass, H., Kaplan, E., & Barresi, B. (2000). *Boston diagnostic aphasia examination*. Philadelphia, PA: Lippincott, Williams & Wilkins.

Grossman, M., Smith, E.E., Koenig, P., Glosser, G., Rhee, J., & Dennis, K. (2003). Categorization of object descriptions in Alzheimer's disease and frontotemporal dementia: Limitation in rule-based processing. *Cognitive, Affective, and Behavioral Neuroscience, 3*(2), 120–132.

Haynes, W.O., & Pindzola, R.H. (2003). *Diagnosis and evaluation in speech pathology* (6th edn). Boston, MA: Allyn and Bacon.

Holland, A.L., Frattali, C.M., & Fromm, D. (1999). *Communication activities of daily living* (2nd edn). Austin, TX: Pro-Ed.

Jefferies, E., & Lambon Ralph, M.A. (2006). Semantic impairment in stroke aphasia versus semantic dementia: A case-series comparison. *Brain, 129*, 2132–2147.

Kay, J., Lesser, R., & Coltheart, M. (1992). *Psycholinguistic assessments of language processing in aphasia (PALPA)*. East Sussex: Lawrence Erlbaum Associates.

Macoir, J., & Bernier, J. (2002). Is surface dysgraphia linked to semantic impairment? Evidence from a patient with semantic dementia. *Brain and Cognition, 48*, 452–457.

McCauley, R.J. (2001). *Assessment of language disorders in children*. Mahwah, NJ: Lawrence Erlbaum Associates.

McDonald, S., Togher, L., & Code, C. (1999). *Communication disorders following traumatic brain injury*. New York: Psychology Press.

Murray, L.L., & Chapey, R. (2001). Assessment of language disorders in adults. In R. Chapey (Ed.), *Language intervention strategies in aphasia and related neurogenic communication disorders* (4th edn, pp. 55–126). Philadelphia, PA: Lippincott, Williams & Wilkins.

Neary, D., Snowden, J.S., Gustafson, L., Passant, U., Stuss, D., Black, S., & Freedman, M. *et al.* (1998). Frontotemporal lobar degeneration: A consensus on clinical diagnostic criteria. *Neurology, 51*(6), 1546–1554.

Riddoch, M.J., & Humphreys, G.W. (1993). *Birmingham object recognition battery*. Hove: Lawrence Erlbaum Associates.

Robinson, G., Blair, J., & Cipolotti, L. (1998). Dynamic aphasia: An inability to select between competing verbal responses? *Brain, 121*(1), 77–89.

Rosenstein, L.D. (1998). Differential diagnosis of the major progressive dementias and depression in middle and late adulthood: A summary of the literature of the early 1990s. *Neuropsychology Review, 8*, 109–167.

Ross-Swain, D. (1996). *Ross information processing assessment* (2nd edn). Austin, TX: Pro-Ed.

Spellacy, F., & Spreen, O. (1969). A short form of the Token Test. *Cortex, 5*, 390–397.

Spreen, O., & Risser, A.H. (2003). *Assessment of aphasia*. New York: Oxford University Press.

Strauss, E., Sherman, E.M.S., & Spreen, O. (2006). *A compendium of neuropsychological tests: Administration, norms and commentary* (3rd edn). New York: Oxford University Press.

Waters, G., & Caplan, D. (1999). Verbal working memory and sentence comprehension. *Behavioral and Brain Sciences, 22*, 77–126.

Ylvisaker, M., Szekeres, S.F., & Feeney, T. (2001). Communication disorders associated with traumatic brain injury. In R. Chapey (Ed.), *Language intervention strategies in aphasia and related neurogenic communication disorders* (4th edn, pp. 745–808). Philadelphia, PA: Lippincott, Williams & Wilkins.

Further Readings

Goodglass, H. (2001). *The assessment of aphasia and related disorders* (3rd edn). Baltimore, MD: Lippincott, Williams & Wilkins.
This is the manual for the Boston Diagnostic Aphasia Examination (BDAE), which is a classical language battery to measure the most common aphasic disturbances. The book has been a reference guide to the assessment of aphasia for over three decades. It consists of six chapters, including a discussion on the nature of language deficits as well as an interpretative summary on the major aphasic syndromes.

Spreen, O., & Risser, A.H. (2003). *Assessment of aphasia*. New York: Oxford University Press.
This book was specifically written for the practicing speech clinician and neuropsychologist working with aphasic persons. It could have been named a *Handbook for the assessment of aphasia*. It is broad in scope and provides base knowledge for the graduate student yet comprehensive enough to appeal to even the well-informed and experienced practitioner. The book's main content deals with contemporary methods of assessment: it describes commonly used tests and procedures; it reviews eaczh instruments strengths and weaknesses; and it also offers relevant and valuable technical information on tests and procedures as well (e.g., psychometric properties and normative data).

Whitworth, A., Webster, J., & Howard, D. (2005). *A cognitive neuropsychological approach to assessment and intervention in aphasia: A clinician's guide*. New York: Psychology Press.
In this reference book, Whitworth, Webster, and Howard provide therapists with both a theoretical background and a practical approach for use in the assessment and treatment of people with aphasia. The authors describe the theory and principles of the cognitive neuropsychological approach, together with models for the identification and characterization of impairments. They also offer an extensive and up-to-date review of the literature on studies of aphasia therapy. This book is highly recommended for students and professionals in the field of aphasia therapy.

2

The Hypothesis Testing Approach to the Assessment of Language

LYNDSEY NICKELS

Macquarie Centre for Cognitive Science (MACCS), Macquarie University, Sydney, Australia

ABSTRACT

This chapter discusses the hypothesis testing approach to the assessment of language processing in aphasia. This approach argues that the most effective assessment technique is to prioritize assessment, based initially on hypotheses formed from observation. These hypotheses relate symptoms to the effects of impairment in theoretical models of language processing. Assessment then continues as a cyclical process with continuous revision of the hypotheses, until the hypotheses are sufficiently well supported to provide a basis from which to focus treatment. This hypothesis testing assessment is characteristic of the cognitive neuropsychological/psycholinguistic/neurolinguistic approaches to aphasia.

2.1. INTRODUCTION

Brain damage often leaves individuals, who previously communicated with no difficulty, with marked difficulty in even the most straightforward of communicative situations. For example, Richard was asked whether he needed his glasses for reading and replied "Yeah. Oh, no, hopeless, I'm hopeless, yeah, today, to … today that's what paypoorer, bew, oh, he has hopeless." When asked where he lived, George answered "I live at er …. I live at the er … At the moment now I'm living at the um er, at a what's-er-name home." Eileen, when asked about her difficulties with talking had enormous trouble producing any response, saying "Yes… oh … yes … oh … oh … Sad."

All of these individuals would be diagnosed as having the acquired language disorder known as aphasia. However, clearly they have very different types of problem and respond very differently in conversation. The fascinating puzzle for both clinicians and researchers is why do they respond the way they do? Can we discover why they have such different language symptoms? Can we determine which treatment might best help their language recovery?

Traditionally, the first step in answering such a question would be to use a standard aphasia battery (see Box 2.1). These batteries will give a broad overview of how well the person is doing across a range of language tasks such as picture naming, understanding spoken words, repeating words, reading, writing and so on. However, some clinicians and researchers believe that this is not the most efficient way of learning what is wrong and deciding how best to treat the problem. Instead, they argue for a hypothesis testing approach to assessment. What does this mean?

2.2. WHAT IS THE HYPOTHESIS TESTING APPROACH TO ASSESSMENT?

A hypothesis is simply a suggestion for why a particular phenomena or behavior may have occurred. It can be at a variety of levels of certainty – it may be asserted as a provisional conjecture to guide investigation (a working or initial hypothesis), or accepted as highly probable in the light of established facts.

When trying to work out what the language impairment is for a particular individual, the emphasis is on first generating a "provisional conjecture" to guide the investigation, with the aim of ending up with an explanation that is "highly probable in the light of established facts." This all sounds very grand! However, this is precisely what many

13

Box 2.1 The battery approach to aphasia assessment

Aphasia assessment batteries are groups of tests which sample from a range of language behaviors. They have a core of tests which are to be given in every case, but may also have additional supplementary (optional) tests, which will be used at the clinician's discretion. The choice of battery will be influenced by severity, time post-onset, and often also reflects the training, theoretical perspective and country of origin of the clinician. It is embodied in standardized aphasia assessments such as the Boston Diagnostic Aphasia Examination (BDAE) (Goodglass & Kaplan, 1983), Minnesota Test for Differential Diagnosis of Aphasia (MTDDA) (Schuell, 1965) and Western Aphasia Battery (WAB) (Kertesz, 1982), and screening assessments such as the Aphasia Screening Test (AST) (Whurr, 1996).

An individual with aphasia is assessed using the whole of a particular test battery. This usually results in an overall measure of severity and a profile across the subtests which may be compared to the non-aphasic population and/or the average aphasic. Many of these batteries were not designed primarily to elucidate the underlying nature of the language disorder, but rather to assign a syndrome classification on the basis of the language symptoms. However, Goodglass (1990) argues that syndrome classification is only part of the role of standardized aphasia batteries, they also aim to provide the clinician with an overview of language skills. Indeed, very often this is the more common goal for clinicians performing the assessments.

Goodglass, H. (1990). Cognitive psychology and clinical aphasiology: Commentary. *Aphasiology*, 4, 93–95.
Goodglass, H., & Kaplan, E. (1983). *Boston diagnostic aphasia examination* (2nd edn). Philadelphia, PA: Lea and Febiger.
Kertesz, A. (1982). *Western aphasia battery*. New York: Grune and Stratton.
Schuell, H. (1965). *The Minnesota test for differential diagnosis of aphasia*. Minneapolis, MN: University of Minnesota Press.
Whurr, R. (1996). *The aphasia screening test* (2nd edn). San Diego, CA: Singular Publishing Group.

experienced clinicians do, often without realizing it. What they might attribute to "gut feeling" or "clinical intuition" is in fact nothing of the sort. What these skilled clinicians are doing, often rapidly and unconsciously, is using their observation of the language behavior of the current individual to develop a hypothesis of the language impairment. They do this by comparing the observed behavior with an internal database of similar language behavior from previous cases and recalling what the corresponding impairment was in those cases. This is exactly the process many would advocate should be best practice in the assessment of language disorders (e.g., Murray & Chapey, 2001).

This approach is commonly associated with the cognitive neuropsychological approach to language (see Box 2.2). Of critical importance to this approach is that the hypothesis has to be related to a theory of language processing. Hence, the observed behaviors, or symptoms, are used to generate a hypothesis regarding the underlying cause of the symptoms in terms of a theory of language processing such as that shown in Figure 2.1. The theory depicted specifies the likely stages of processing and types of stored information used to understand and produce spoken and written words. Impairment to any one of these processes or loss of any of the stored information will result in a language disorder. While different levels of breakdown within this theoretical model will result in different language symptoms, the same symptoms can occur from different levels of impairment. Hence, observation of a single symptom (e.g., impaired picture naming, impaired understanding of spoken words) is unlikely to be able to pinpoint the precise nature of the language impairment.

Box 2.2 The cognitive neuropsychological approach to assessment

Cognitive neuropsychology is based on the assumption that the mind's language system is organized in separate modules of processing and that these can be impaired selectively by brain damage. Testing aims to provide information about the integrity of these modules, to find those in which the aphasic person seems to be functioning below normal and those which appear to be functioning normally or near normally.

It comprises systematic assessment of the component processes of a cognitive task (as established in cognitive theories) to establish which of these processes are intact and which impaired. In other words, establishing the "level of breakdown" of a particular skill within a cognitive model (as the result of brain damage).

The basis of this approach is to use the errors patients make on tasks to draw conclusions as to which processing routines are available and which are impaired (modality differences, level of breakdown). However, as many different disorders can underlie the same symptom, it is not sufficient to observe the errors on one task alone. Rather evidence must be accumulated on a number of different tasks tapping different levels of processing.

Why do we need to know exactly what the impairment is? Is it not enough to know that there is impaired understanding of spoken words, or difficulty retrieving words for communication? The proponents of the approach would argue it is not. They argue that treatment will be maximally effective only when the direction of treatment is determined by precise knowledge of the individual's language processing

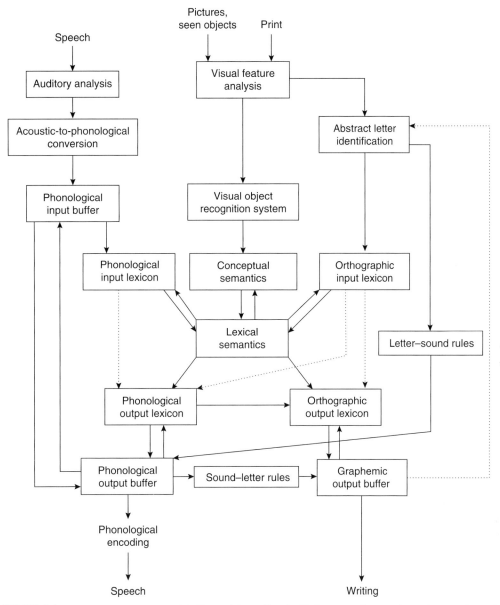

FIGURE 2.1 Language processing model. The model specifies the likely stages of processing and types of stored information used to understand and produce spoken and written words. Dotted lines refer to processing routes where there is less concensus. (© L. Nickels.)

strengths and weaknesses. Hence, as symptoms can arise as a result of various different types of impairment, analysis limited to surface symptoms will not enable construction of effective treatments (but see Kertesz, 1990, for discussion).

2.2.1. How Do We Form Hypotheses?

So far we have mentioned in general terms that we should use "observed behavior" to form hypotheses. It is important to realize that this observed behavior can be of many forms, and the resulting hypotheses of many different levels of specificity. Once again, even those clinicians who mostly choose aphasia batteries use hypotheses they have formed to guide them to the appropriate battery. Even before the clinician has met the person with aphasia, hypotheses are drawn from the comments in the referral (be that from a carer, the individual themselves or a medical professional). For example, Richards' referral might mention very poor understanding and expressive language that was full of jargon. This could lead the clinician to a hypothesis of a severe aphasia and make them less likely to attempt a full language assessment battery. Instead they may choose a short aphasia screening battery. In contrast, Eileen's referral might say that she was producing very few words, but seems to understand conversation, and George's might speak of good understanding of conversation but some problems finding the words he wants to say. These referrals are more likely to lead to a hypothesis of a primarily mild–moderate expressive aphasia and the use of a full language battery (as a screening battery might be insensitive to the milder impairment of these individuals). A clinician using the (overt) hypothesis testing approach will use the same information in the same way to generate hypotheses. However, these hypotheses will be more clearly related to underlying theories, and will lead to different assessment strategies. Thus, an initial hypothesis for Richard (who was reported to have problems in both understanding and producing language) might be that there is an impairment of stored semantic concepts (meanings), which affects every kind of language process. This would lead to initial assessment of semantics (see below for further discussion). For George and Eileen, the initial hypothesis might be of a specific problem with word retrieval, possibly in the context of relatively intact stored semantic concepts (meanings), because of the reportedly generally good understanding of conversation. Hence, the initial assessment might be to confirm that there is no semantic impairment underlying the word retrieval impairment.

In this example, we have demonstrated that even very general comments about behavior can be used as sources of (tentative) hypotheses. But of course these are not the only source of information. Typically, the most reliable source of information on which to form an initial hypothesis is the language behavior observed in the initial meeting with the person with aphasia. For example, the response to the clinician's questions can provide information on comprehension (access to semantics from auditory input), and spoken word retrieval (the stages from semantics to articulation). Later, the results of informal and formal testing provide the evidence for the development of more specific hypotheses.

2.2.2. How Do We Test Hypotheses?

Just as hypotheses can be formed on the basis of different types of evidence, the testing of these hypotheses can also take different forms. The same type of evidence that allows formation of initial hypotheses can be used to accept, reject or revise the hypotheses. For example, on the basis of referral information for Eileen, above, an initial (if tentative) hypothesis was generated that she may have good semantic processing (giving good comprehension in conversation) but impaired retrieval of word forms (giving word finding problems). The initial interview may have initially supported this impression, but as the interview progresses, more detailed questioning leads to doubt about comprehension (e.g., Eileen responds "no" to the question do you have any daughters, when the interviewer is aware she has a daughter; or asked about her son starts talking about their daughter). Hence, a revised hypothesis is of a mild-to-moderate semantic impairment which may also be the cause of Eileen's word retrieval

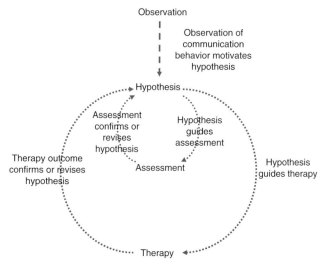

FIGURE 2.2   The cyclical nature of hypothesis testing (adapted from an unpublished figure from Jennifer Cole-Virtue, reproduced with permission). Observation of communicative behavior serves as the initial basis for generation of a hypothesis about the nature of the language impairment. This hypothesis then guides the choice of assessments to test the hypothesis, and the results of the assessment confirm or lead to revision of the hypothesis. The revised hypothesis then guides the choice of the next assessments. This cyclical process of hypothesis generation, testing and revision continues until a point is reached where the hypothesis is sufficiently detailed and confirmed to serve as a basis for treatment planning. However, treatment itself can also serve as a test of the hypothesis, and lead to revision. The hypothesis generation process is therefore continuously cyclical throughout a clinician's relationship with the person with aphasia.

problems. At this stage, the use of a more formal assessment may be indicated to further evaluate the hypothesis (see below for further discussion). This cyclical process of hypothesis generation, testing and revision continues until a point is reached where the hypothesis is sufficiently detailed and certain that it can serve as a basis for treatment planning. However, treatment itself can also serve as a test of the hypothesis, and lead to revision. The hypothesis generation process is therefore continuously cyclical throughout a clinician's relationship with the person with aphasia. This cyclical process is depicted diagrammatically in Figure 2.2.

2.3. HYPOTHESIS TESTING OF LANGUAGE IMPAIRMENT: ASSESSMENT RESOURCES AND CONSIDERATIONS

As discussed above, any interaction with the language impaired person can provide evidence for formation and evaluation of a hypothesis. The critical factor is that the information is related to a theoretical model of language processing. Unless the processing components of the model that are required to perform a particular task are clearly specified, the implications of success or failure cannot be accurately interpreted.

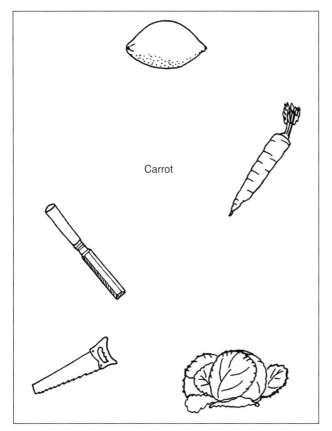

Carrot

FIGURE 2.3 Word–picture matching assessment (item 6, subtest 48 from Kay *et al.*, 1992, reproduced with permission). In this task the person with aphasia has to read the word written in the center of the page and point to the picture that corresponds to this word. There are forty target items and four distractor pictures for each target. For the target word "carrot," depicted here, the distracter pictures are "cabbage" (close semantic), "lemon" (distant semantic), "saw" (visually related) and "chisel" (unrelated). The distracters are related to each other semantically but not to the target item. This control feature has been incorporated to prevent the individual responding on the basis of perceived semantic category.

2.3.1. Interpreting Success and Failure on a Task

It is best to use a concrete example to illustrate this issue. A commonly used task for testing whether the representation of meaning (semantics) is intact is a word-to-picture matching task such as that depicted in Figure 2.3. In this task the person with aphasia has to read the word written in the center of the page and point to the picture that corresponds to this word. If the person responds as accurately (and ideally as swiftly) as non-language impaired control subjects (i.e., "within normal limits"), then it is valid to conclude that the representation of meaning is intact for this individual (or at least sufficiently intact to allow this task to be performed). However, while failure to respond as accurately as non-language impaired control subjects is generally interpreted as indicating a semantic impairment, this conclusion is premature (in the absence of other supporting evidence).

Let us consider the task more carefully and with reference to Figure 2.1. First, the word has to be understood. This involves all those components in Figure 2.1 from "print" to conceptual semantics (visual feature analysis, abstract letter identification, orthographic input lexicon, lexical semantics). If any one of these components or the connections between them is impaired then there will be poor performance on this task. But only if there is impairment to (lexical or conceptual) semantics can we justify a conclusion of a semantic impairment on the grounds of poor performance on this task. Similarly, not only does the word have to be read accurately, but also each picture has to be recognized and the representation of its meaning accessed. This involves all of those processes in Figure 2.1 from "pictures" to conceptual semantics. Once again, poor performance on the task can result from an impairment to any of these levels of representation and processes. However, even if the semantics of both word and picture can be accessed accurately, this is still not enough to guarantee success – there are many other processes involved, including those governing scanning and attending to each picture in turn, comparing the representation of the picture(s) and the word in memory, decision processes and finally generating a pointing response. Only if all of these processes are intact will success be guaranteed. Clearly, failure can result from impairments to many processes other than the semantic impairment that the test is designed to elucidate. How then can we work out whether components are intact or impaired, when even this seemingly simple task draws so many processes into play? The answer lies in converging sources of evidence.

As already noted, failure can be a result of many different impairments, and so failure to succeed on a single task is rarely diagnostic. However, success on a task (defined as performing as well as appropriate non-language impaired controls) is diagnostic – it allows one to conclude that all those processes involved in a task are unimpaired or at least any impairment is not sufficiently great to impair performance on this task. Hence, if someone succeeds on the written word–picture matching task, we can conclude that all the processes from "print" to conceptual semantics and "pictures" to conceptual semantics are intact, or at least any impairment is mild enough not to impact on word–picture matching.

2.3.2. Converging Evidence

If someone shows impaired performance on a task, converging evidence is required to enable conclusions to be drawn regarding level of impairment. To continue with our example, in this case the aim is to evaluate semantic processing, and so the choice might be a task which also uses semantic processing but does not involve other components used in the written word–picture matching task. Such a task is spoken word-to-picture matching. This task has the same format as the written word–picture matching task, how-ever, rather than the target word being written, it is spoken. Hence, while the picture processing components of the task are identical, the processes for understanding the word now involve different components of the language processing model: those from "speech" to conceptual semantics (audi-tory analysis, acoustic-to-phonological conversion, pho-nological input buffer, phonological input lexicon, lexical semantics, conceptual semantics). Critically, the task uses the same items as the written word-to-picture matching task. Hence, differences in performance cannot be attrib-uted to one set of items being "easier" than the other.

Now, suppose that the same individual who was impaired on written word–picture matching could perform this spo-ken word–picture matching task flawlessly. This would lead us to the conclusion that the reason for failure on written word–picture matching was not a semantic problem, nor any problem to do with processing pictures or other task constraints (scanning, attention, memory, decision making): these impairments would have resulted in failure on spoken word–picture matching also. Instead, we are led to the con-clusion that the impairment in written word–picture match-ing is due to an impairment in the processes after visual feature analysis (which are shared with picture processing) and before lexical semantics (which is shared with spoken word comprehension). Further testing could refine the level of impairment further if this was a priority for treatment. In contrast, if the person performed equally poorly on both spoken and written word–picture matching, then it becomes more likely that semantic processing is impaired. However, still more converging evidence from different tasks would be required to ensure that it was not impaired picture processing or other task demands that were the source of impairment.

This example, while seemingly complex, serves to illus-trate key principles of this approach: think carefully about the demands of each task within a theoretical model, inter-pret the implications of results carefully within the same model and use converging evidence from different tasks tapping both the same and different level of processing to support conclusions: it is rarely sufficient to observe the errors on one task alone as many different levels of impair-ment can underlie the same symptom. A worked example is provided in the Appendix to further illustrate this process.

2.3.3. Assessment Materials

It follows from the discussion in the previous section that the assessment materials required by this approach are a series of tasks which are designed to tap different levels of processing within a theoretical model of language process-ing. As this approach was initially used by researchers rather than clinicians, historically, these assessment materials were developed according to need: an assessment was developed

to assess the specific language features of the language impaired person under consideration. Indeed, the majority of the published assessments used were originally developed as research tools (e.g., Howard & Patterson, 1992; Kay et al., 1992). While there are now published resources, it needs to be emphasized that informal and unpublished assessments can be just as valid for testing hypotheses and determining the nature of the language impairment. However, these assessments need to have several key features.

First, they need to be valid: they need to test what you think they are testing. This comes back to the discussion earlier, where we stressed the need to think carefully about what skills and processes the task involves in relation to a theory of language processing (e.g., Figure 2.1).

Second, the assessments need to be reliable and sensitive. To achieve both of these it is important to have relatively large samples of behavior. Consider a confrontation naming test with only 6 items (as was found in some of the assessment batteries in the past). If on this test someone scored 2/6 on one occasion and 4/6 on a second occasion, how would this be interpreted? Is this two item difference in accuracy of two items due to fluctuations from testing session to testing session or does it represent real improvement? Is 2/6 severely impaired performance and 4/6 moderately impaired? I would suggest that it is simply impossible to tell. However, consider the same task with 60 items. The same two item difference (2/60, 4/60) would result in the scores being considered (impres-sionistically and statistically) as identical and most likely to be due to fluctuations in performance, which could be due to many factors unrelated to the language disorder (e.g., level of fatigue, attention, motivation). Similarly the same per-centage difference, from 20/60 to 40/60, clearly represents a marked (and statistically significant) difference between the two testing sessions. Hence, small numbers of items are often unhelpful in determining the degree of impairment and in particular are inadequate for determining change in level of performance (e.g., monitoring the effects of treatment).

Third, it needs to be clear how individuals with unim-paired language would perform on the assessment. For many language tasks (e.g., word repetition, word–picture matching) it can usually be assumed that control participants would be at or close to ceiling (100% accurate). Hence, if a person with aphasia performs poorly on the task, it can safely be concluded that they are impaired (even without control data). However, if someone performs relatively well on a task control data is required to be able to interpret their performance accurately. In addition, many tasks are (perhaps surprisingly) not accurately performed by unimpaired indi-viduals (e.g., rhyme judgments, picture naming). For these tasks it is essential to obtain control data to ensure that the level of accuracy of the person with aphasia can be evalu-ated correctly. A further consideration is the interpretation of the degree of impairment of an individual given a score on a particular test. Some tests, particularly those that controls find "easy," may not be sensitive to anything but the greatest degree of impairment and even a small number of errors may be indicative of a substantial impairment. Cole-Virtue and Nickels (2004), for example, note that 50% of individuals that were within normal limits on the word–picture matching task we have described above, were impaired on a further test of semantic processing (synonym judgments).

Finally, it is important that assessments control for (and manipulate where appropriate) factors that affect performance such as the frequency or length of a word. We have already mentioned that different versions of the same test should use the same items. This is to ensure that performance can be attributed to the different modality of testing (e.g., spoken versus written word–picture matching, word–picture matching versus naming) and could not be due to differences in the difficulty of the stimuli used in each modality. However, in addition, if stimuli are not properly sampled, inappropriate conclusions could be drawn regarding whether some language processes are intact or impaired. For example, some individuals have impairments of semantic processing that are specific to particular semantic categories (often living things). If a test of semantic processing failed to include living things (or included only living things) then the conclusion drawn on the basis of this test would not be accurate. However, even ensuring that both living and non-living things are included in the assessment is not enough. On average, living things tend to be of lower frequency (have word forms that occur less commonly in the language) than non-living things. Hence, some individuals appear to have impaired processing for living things, but in fact when the living and non-living things are matched for frequency there is no longer a difference between the categories. Similarly, some individuals may seem to be worse at naming pictures that have low frequency names. However, low frequency words tend also to be longer (contain more syllables and/or sounds), and when the length of high and low frequency words is equated these individuals no longer show an effect of frequency on accuracy.

These considerations are not intended to dissuade the clinician or researcher from using their own materials and developing their own assessments – far from it. It is meant to reinforce the point that the most important factor is to be acutely aware of what an assessment is testing and what it is valid to conclude on the basis of that assessment.

There are, however, published assessments which are commonly used resources for those taking this approach. Once again, the user needs to beware, even though a test has been published, the same considerations apply. It is important to evaluate assessments using the criteria discussed above.

The most familiar published assessment is what is commonly referred to as PALPA (psycholinguistic assessments of language processing in aphasia) (Kay et al., 1992). This large collection of testing materials can be rather daunting. However, they were developed with the hypothesis testing approach in mind: the idea is not to do every task with the person with aphasia but rather form a hypothesis as to the likely nature of their impairment and test that hypothesis using a selection of these assessments. A more recent assessment resource is the comprehensive aphasia test (CAT) (Swinburn et al., 2004). This includes similar assessments to those from PALPA but limits itself to what may be considered the "core"

Box 2.3   Some selected English language assessment resources

Kay, J., Lesser, R., & Coltheart, M. (1992). *Psycholinguistic assessments of language processing in aphasia (PALPA)*. London: Lawrence Erlbaum.
A collection of tests assessing all aspects of single word processing (auditory comprehension, written comprehension, reading, spelling, repetition, spoken naming, etc.) which are to be selected for use according to the nature of the impairment under investigation. There are also Spanish and Dutch versions of PALPA.

Swinburn, K., Porter, G., & Howard, D. (2004). *Comprehensive aphasia test*. Hove, UK: Psychology Press.
This set of assessments contains a cognitive screen, a language battery and a disability questionnaire. It complements PALPA well, by providing shorter versions of many tests and good normative data.

Howard, D., & Patterson, K.E. (1992). *Pyramids and palm trees*. Bury St. Edmunds, UK: Thames Valley Test Company.
This test determines the degree to which a subject can access meaning from pictures and words. Six different versions of the test are possible and the pattern of results can be used to build up a picture of the subject's ability to access semantic and conceptual information.

Druks, J., & Masterson, J. (2000). *An object and action naming battery*. Hove, UK: Psychology Press.
This is a resource rather than a test, and provides noun and verb materials, and matched lists which can be used for developing test materials.

Bastiaanse, R., Edwards, S., & Rispens, J. (2002). *The verb and sentence test (VAST)*. Bury St. Edmunds, UK: Thames Valley Test Company.
This contains 10 subtests aimed at pinpointing the underlying impairment at the sentence level in aphasia. Subtests include those assessing comprehension of verbs, grammaticality judgment, sentence comprehension, production of verbs as single words and in sentences.

Marshall, J., Black, M., Byng, S., Chiat, S., & Pring, T. (1999). *The sentence processing resource pack*. London: Winslow Press.
This resource kit is comprised of three books: a handbook which introduces the processes involved in sentence production and comprehension, offering suggestions about ways to assess sentence-level disorders and providing guidance for therapy. In addition, there are two assessments: the event perception test and the reversible sentence comprehension test.

assessments for language processing. It also tends to have fewer items per subtest. This somewhat compromises the sensitivity and reliability, but means that it is better suited for an initial test of a hypothesis (particularly with acute clients or those who are unable to tolerate prolonged testing). Further examples of testing resources are provided in Box 2.3.

In sum, the art of the hypothesis testing approach is to use a variety of sources of evidence to make your assessment as efficient as possible. By forming a hypothesis regarding the nature of the breakdown within a theoretical model, you can then guide your investigation of that person with aphasia. Hence, rather than doing a whole assessment battery, only those assessments are performed (perhaps subtests of PALPA) that are needed to confirm or reject your hypothesis.

2.4. SUMMARY AND CONCLUSIONS

It is true that there is no *a priori* reason why an assessment should be hypothesis driven rather than being derived from assessments (Kertesz, 1990). However, the hypothesis testing approach allows assessment to be streamlined in order that assessment determines the underlying impairment as directly as possible, and without testing using nondiscriminatory tasks that do not contribute to this process. This approach is not without its critics. While recent debate has been limited, a series of replies to Byng *et al.* (1990) discussion paper "Aphasia Tests Reconsidered" (David, 1990; Goodglass, 1990; Kay *et al.*, 1990; Kertesz, 1990; Weniger, 1990) and the later forum focused around PALPA (Basso, 1996; Ferguson & Armstrong, 1996; Kay *et al.*, 1996a, b; Marshall, 1996; Wertz, 1996) cover the major points of debate. There is no doubt that standardized aphasia batteries have played an important role in clinical and research aphasiology (Kertesz, 1988). Kertesz (1990) notes that they have contributed, for example, in the study of lesions, behavior correlations, cerebral dominance and recovery patterns. However, the focus here is on assessment in order to target treatment, and in this domain the question is whether performing an assessment battery provides sufficient additional information, over and above informal observation and hypothesis-driven targeted assessment, to warrant the time taken to perform it. In addition, there is the question of whether the overview of language skills provided by the test battery is of value when these test batteries do not provide a theoretical framework within which to interpret the performance. The data from them can, of course, be used to generate a hypothesis regarding language function within other theoretical frameworks.

Hypothesis testing cannot be said to be a short-cut or an easy way out – it is demanding for the clinician (Marshall, 1996). In order to be most effective, hypothesis testing should become an automatic way of thinking for the clinician. Acquiring this way of thinking can be hard and time consuming, but once it is well established it can guide

every assessment and every task used. Moreover, because this approach involves testing components of a theoretical model, it is essential that the investigator understands both the theoretical model and the assumptions of the particular assessments. Without these preconditions, it is not possible for the results of testing to be interpreted correctly.

As discussed above, any task can provide evidence that can help test hypotheses about and determine the level of breakdown within a cognitive model of language processing. These tasks do not have to be published tasks, although the latter do help overcome certain common pitfalls, and may also provide data from a control population.

Hypothesis-driven assessment enables treatment to be directed precisely at the problems that have been identified and the use of those processing abilities that remain (relatively) unimpaired (but for discussion of whether this is necessary see Kertesz, 1990). Similarly, the effects of treatment on the deficits can be precisely monitored with retesting. However, analysis of what is wrong unfortunately does not uniquely determine what to do about it (under any approach to assessment).

In order to effectively assess the individual with language impairments and use those assessments to track change over time, we must be acutely aware of the strengths and limitations of our assessment tools. We have a duty to all those involved in the rehabilitation process to strive to overcome these limitations and critically evaluate the efficacy of our interventions as part of routine clinical practice.

2.5. CHALLENGES AND FUTURE DIRECTIONS

In this chapter, we have focused on impairment level assessment using hypothesis testing, an approach commonly referred to as the cognitive neuropsychological approach. While the examples given here have focused on acquired communication disorders, the same hypothesis testing approach can easily be applied to the assessment of developmental disorders in children and adults.

In addition, hypothesis testing should not be restricted to the impairment level, it can and indeed should be used at all levels of description (language functions, communication activities and participation, quality of life/psychosocial issues). Of course, the specification of a theory in order to guide the formation of hypotheses is critical. Theories are often less well developed in these areas than at the level of impairment. Nevertheless, there is increasing awareness of the fact that functional communication and quality of life batteries are not adequate for the clinician's needs. For example, Worrall *et al.* (2002) note that a single (functional) assessment is unlikely to be appropriate to assess all individuals from all cultures, with all impairments and in all settings. Similarly, Sacchett and Marshall (1992) question the appropriateness and adequacy of a single measure of functional communication. Instead, they argue for assessment that focuses on specific areas motivated by prioritization by clinician, the individual with aphasia and their communicative partners – in other words hypothesis driven and focused assessment.

In sum, the hypothesis testing approach can be applicable to every aspect of an individual and their social context that is, or might be, impacted by the language impairment and can be integrated with the other tools in a clinician's armoury.

APPENDIX

Hypothesis Testing: A Worked Example

Referral: George was a 70-year-old man who reported problems with "memory," specifically forgetting the words he wanted to say.

Hypothesis 1: Individuals often refer to problems retrieving the form of the words they want to say as "memory" problems. Hence, the initial hypothesis is problems retrieving word forms from the phonological lexicon. The theory specified in Figure 2.1 indicates that these word retrieval problems can be due to a problem in semantic processing or a problem in retrieving the word form itself (with semantics being intact). Both of these impairments would predict semantic errors in word retrieval, and possibly circumlocutions (talking around the subject, to avoid the inaccessible word). A semantic impairment would also predict problems with understanding.

Hypothesis Testing 1: Observe word retrieval in conversation: are there clear difficulties? What is the nature of the errors produced? Is there evidence of difficulties understanding conversation?

Results of Hypothesis Testing 1: In conversation, there is clear evidence of word finding difficulties (see e.g. from George at beginning of the text): incomplete utterances where he is unable to retrieve the necessary word and a lot of "talking around" the subject; no phonological errors (errors where wrong sounds are produced). He appears to comprehend conversation well. His reports of other situations that would have necessitated comprehension (e.g., other conversations, meetings, etc.) also imply good comprehension.

Hypothesis 2: Broadly the same as hypothesis 1, but with firmer evidence. The fact that no phonological errors or phonetic errors/distortions are produced makes it unlikely that he has an impairment at any level "below" the phonological output lexicon. The lack of overt semantic errors and good conversational comprehension makes a severe semantic impairment unlikely but a mild–moderate impairment is still a possibility, as is a problem either in the "links" from lexical semantics to phonological form or in the phonological output lexicon itself.

Hypothesis Testing 2: Test picture naming: to confirm the fact that George does indeed have greater difficulty in word finding than unimpaired subjects and to observe error types in further detail.

Results of Hypothesis Testing 2: George names only 52% of a set of 100 pictures correctly (well below the level of control subjects). Errors include circumlocutions and descriptions (e.g., kennel named as "a dog's residence"). Unlike in conversation, in picture naming George produces some erroneous responses which are semantically related to their targets (e.g., saxophone named as clarinet). No phonological or phonetic errors are produced.

Hypothesis 3: Once again, testing has confirmed the previous hypothesis (impairment of semantics or access to phonological output lexicon). However, thus far we cannot distinguish the two possibilities. As different remediation approaches would be taken depending on which is the true level of George's impairment, it is important to continue testing.

Hypothesis Testing 3: Test word comprehension: in order to determine whether George has a semantic impairment. This requires a task involving semantic processing without necessitating word production. George is assessed using the spoken word–picture matching task from PALPA (subtest 47).

Results of Hypothesis Testing 3: George scores 38/40 correct which is within the range of unimpaired controls.

Hypothesis 4: George has unimpaired semantics and his word retrieval impairment is at the level of accessing word forms from the phonological output lexicon. However, as word–picture matching is neither a very sensitive nor a very demanding test of semantic processing, the conclusion of unimpaired semantics remains tentative.

Hypothesis Testing 4: Perform a more stringent test of semantic processing: in order to determine whether George does in fact have a semantic impairment despite scoring within the range of controls on word–picture matching. Synonym judgements task from PALPA. This task involves presentation of two words (in either spoken or written form – we used the spoken version), which have to be judged as to whether they are synonymous/similar in meaning or not, and includes both concrete and abstract stimuli.

Results of Hypothesis Testing 4: George makes only three errors all on the abstract stimuli which is a level of performance easily within the range of unimpaired subjects.

Hypothesis 5: The results confirm hypothesis 4. George's naming impairment is most likely due to an impairment at the level of access/retrieval of the word form from the phonological output lexicon. This level of detail is sufficient for intervention planning which aims to improve access to word forms using techniques suggested from the literature (combining meaning and phonology, see Nickels (2002) for a review).

Note: This cyclical hypothesis testing was restricted to word retrieval as this was George's main concern functionally. It did not address the problems he also reported with reading and writing. However, these can (and were) also investigated using the same approach but intervention for word retrieval could start immediately and further hypothesis testing of other skills proceed in parallel.

Acknowledgment

Lyndsey Nickels was funded by an NHMRC Senior Research Fellowship during the preparation of this manuscript.

References

Basso, A. (1996). PALPA: An appreciation and a few criticisms. *Aphasiology, 10*, 190–193.

Byng, S., Kay, J., Edmundson, A., & Scott, C. (1990). Aphasia tests reconsidered. *Aphasiology, 4*, 67–91.

Cole-Virtue, J.C., & Nickels, L.A. (2004). Why cabbage and not carrot? An investigation of factors affecting performance on spoken word to picture matching. *Aphasiology, 18*, 153–180.

David, R.M. (1990). Aphasia assessment: The acid test. *Aphasiology, 4*, 103–107.

Ferguson, A., & Armstrong, E. (1996). The PALPA: A valid investigation of language? *Aphasiology, 10*, 193–197.

Goodglass, H. (1990). Cognitive psychology and clinical aphasiology: Commentary. *Aphasiology, 4*, 93–95.

Howard, D., & Patterson, K.E. (1992). *Pyramids and palm trees*. Bury St. Edmunds, UK: Thames Valley Test Company.

Kay, J., Byng, S., Edmundson, A., & Scott, C. (1990). Missing the wood and the trees: A reply to David, Kertesz, Goodglass and Weniger. *Aphasiology, 4*, 115–122.

Kay, J., Lesser, R., & Coltheart, M. (1992). *Psycholinguistic assessments of language processing in aphasia*. London: Lawrence Erlbaum.

Kay, J., Lesser, R., & Coltheart, M. (1996a). Psycholinguistic assessments of language processing in aphasia (PALPA): An introduction. *Aphasiology, 10*, 159–179.

Kay, J., Lesser, R., & Coltheart, M. (1996b). PALPA: The proof of the pudding is in the eating. *Aphasiology, 10*, 202–215.

Kertesz, A. (1988). Is there a need for standardized aphasia tests? Why, how, what and when to test aphasics. *Aphasiology, 2*, 313–318.

Kertesz, A. (1990). What should be the core of aphasia tests? (The authors promise but fail to deliver). *Aphasiology, 4*, 97–101.

Marshall, J. (1996). The PALPA: A commentary and consideration of the clinical implications. *Aphasiology, 10*, 197–202.

Murray, L.L., & Chapey, R. (2001). Assessment of language disorders in adults. In R. Chapey (Ed.), *Language intervention strategies in aphasia and related neurogenic communication disorders* (4th edn, pp. 55–126). Philadelphia, PA: Lippincott, Williams, and Wilkins.

Nickels, L.A. (2002). Therapy for naming disorders: Revisiting, revising and reviewing. *Aphasiology, 16*, 935–980.

Sacchett, C., & Marshall, J. (1992). Functional assessment of communication: Implications for the rehabilitation of aphasic people: Reply to Carol Frattali. *Aphasiology, 6*, 95–100.

Swinburn, K., Porter, G., & Howard, D. (2004). *Comprehensive aphasia test*. Hove, UK: Psychology Press.

Weniger, D. (1990). Diagnostic tests as tools of assessment and models of information processing: A gap to bridge. *Aphasiology, 4*, 109–113.

Wertz, R.T. (1996). The PALPA's proof is in the predicting. *Aphasiology, 10*, 180–190.

Worrall, L.E., McCooey, R., Davidson, B., Larkins, B., & Hickson, L. (2002). The validity of functional assessments of communication and the activity/participation components of the ICIDH-2: Do they reflect what really happens in life? *Journal of Communication Disorders, 35*, 107–137.

Further Readings

Whitworth, A., Webster, J., & Howard, D. (2005). *A cognitive neuropsychological approach to assessment and intervention in aphasia: A clinician's guide*. Hove, UK: Psychology Press.
This volume provides a theoretical description of each aspect of language processing and its assessment. Treatment approaches for each type of impairment (as defined within the theory) are reviewed. The book provides an accessible and practical link between theory, research and practice.

Hillis, A.E. (2002). *Handbook of adult language disorders: Integrating cognitive neuropsychology, neurology, and rehabilitation*. New York: Psychology Press.
This book covers each of the basic domains of spoken and written language from three different perspectives: theoretical accounts, neural substrates and diagnosis and treatment.

Rapp, B. (Ed.) (2002). *A handbook of cognitive neuropsychology*. Philadelphia, PA: Psychology Press.
This volume focuses on the contribution that studies of impaired performance have made to our understanding of central issues in the development of theories of cognitive processing.

Nickels, L.A. (2005). Tried, tested and trusted? Language assessment for rehabilitation. In P.W. Halligan & D.T. Wade (Eds.), *The effectiveness of rehabilitation for cognitive deficits* (pp. 169–184). Oxford: Oxford University Press.
This chapter provides a critical review of clinical assessments used to evaluate acquired language impairments. It reviews assessments aimed at examining both language functions ("impairment" based approaches) and language activities ("functional" measures). In particular it discusses the adequacy of these assessments as tools in the rehabilitation process.

3

The Intracarotid Amobarbital Test (Wada Test) and Complementary Procedures to Evaluate Language Before Epilepsy Surgery

ALEXANDER I. TRÖSTER and KIRSTEN R. MOHN

Department of Neurology, School of Medicine, University of North Carolina, Chapel Hill, NC, USA

ABSTRACT

In a substantial percentage of patients with temporal lobe epilepsy (TLE), anticonvulsant medications do not effectively control seizures. Uncontrolled seizures are both psychologically and socially disruptive. Brain surgery affords many patients with intractable TLE the opportunity to eliminate (or reduce) seizure activity, but carries the risk of language and memory dysfunction. To avoid postoperative aphasia, the treatment team first tries to determine whether language will still be supported by the unoperated portions of the brain. This chapter aims to familiarize the reader with the amobarbital test, a method of selective anesthesia to parts of the brain, as a technique to determine the lateralization of language in epilepsy surgery candidates and as a procedure useful in investigating the neural correlates of language. An evaluation of the amobarbital test and the interpretation of its results are discussed. This chapter concludes with a description of alternative, non-invasive procedures to address language lateralization.

epilepsy – most often temporal lobe epilepsy (TLE) – the opportunity to eliminate (or reduce) seizure activity. Although the potential benefits of surgery are appealing (e.g., increased quality of life and reduced cost of treatment), the removal of the tissue involved in seizures within the temporal cortex is not without risk as structures critical to language and memory are embedded within the temporal lobe. Therefore, to avoid postoperative language disruption or aphasia, the treatment team must ascertain whether the unoperated portions of the brain can support language. The most widely used method or "gold standard" for determining which side of the brain is relatively more important for language (language dominance) is the intracarotid amobarbital test (IAT), a method employing selective anesthetization of brain areas. This chapter details the basic elements of the IAT, its methodological pitfalls, and the advent of "less invasive" alternatives to determine hemispheric language dominance.

3.1. INTRODUCTION

The exact frequency of treatment-resistant or medically intractable epilepsy is unknown, but approximately 20–40% of individuals with epilepsy do not respond adequately to treatment with anticonvulsant medications (Aicardi & Shorvon, 1997). As one might imagine, uncontrolled seizures are both psychologically and socially disruptive. Patients often avoid situations in which having a seizure would prove embarrassing or dangerous and many such patients are either underemployed or unemployed. Surgical intervention affords patients with pharmacologically intractable

3.2. HISTORICAL BACKGROUND

IAT was first described by Juhn Wada (1949), and thus, is often referred to as the "Wada test." Wada originally developed this technique to study the spread of epileptiform discharges (abnormal electrical activity) between the left and right halves of the brain in patients undergoing unilateral electroconvulsive therapy (a somewhat controversial treatment for mood disorders wherein a seizure is induced by electrodes placed on the patient's head). Wada observed that when the presumed language-dominant hemisphere was injected with amobarbital, and anesthetized, the result was a disruption of spoken language or expressive aphasia. Clinically, this pharmacological method of determining the language-dominant hemisphere allowed physicians to apply electroconvulsive therapy selectively to the contralateral (opposite, subdominant) hemisphere, thereby reducing the risk of cognitive dysfunction. Based on his observations of psychiatric patients, Wada reasoned that this technique might also be useful in determining hemispheric language dominance (and minimizing cognitive morbidity) in neuro-surgical candidates. Working as a fellow with Rasmussen at the Montreal Neurological Institute, the procedure was first introduced there into the preoperative evaluation of persons with medically refractory epilepsy.

3.3. CURRENT CLINICAL USE OF THE IAT

At present, the IAT is generally used for three purposes in epilepsy surgery centers. The relative importance of these purposes (and the weight given to IAT test results in

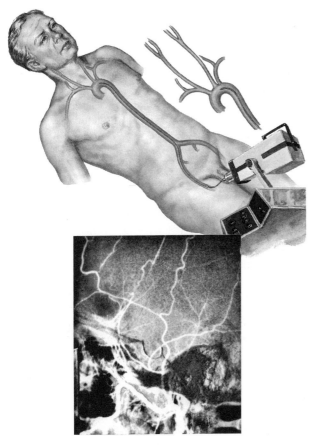

FIGURE 3.1 Femorocerebral angiography is the means by which amo-barbital is delivered to the patient's ICA, allowing anesthetization of one cerebral hemisphere. *Source*: Reprinted from Frank H. Netter. The Netter Collection of Medical Illustrations – Nervous, © 1984, Elsevier Inc. All Rights Reserved. (See Plate 1.)

determinations with respect to whether or not to operate) differs, however, from center to center. These purposes are: First, determination of which brain hemisphere is dominant for language in order to reduce postoperative language impairment. Second, determination of the adequacy of memory performance in the unoperated hemisphere in order to predict which persons are at risk for amnesia (memory loss) after temporal lobe resection. Third, corroboration of the presumed side of seizure onset as determined by other diagnostic methods (e.g., electroencephalography (EEG), video EEG telemetry and ictal semiology, structural and functional radiography). This chapter focuses on the assess-ment of language lateralization. With respect to IAT in the evaluation of memory, there are several recent reviews that are suggested in the further readings section.

As is outlined in the next section, the IAT is an inva-sive procedure and neurological complications during the procedure are possible such as stroke, embolism, and tran-sient global amnesia. This understandably results in con-troversy about its use. Despite the risks involved, the IAT remains an integral component of the comprehensive pre-surgical evaluation of patients with intractable epilepsy. A worldwide survey of 39 epilepsy centers between the years 1995 and 1997 indicated that over 2000 IAT proce-dures were performed with pre-surgical candidates (Lüders, 2001). Moreover, capacity to perform the IAT is one crite-rion required of epilepsy centers under the current National Association of Epilepsy Centers (NAEC, 2001) guidelines.

3.4. RATIONALE UNDERLYING THE IAT PROCEDURE

In the simplest terms, the underlying rationale for employing the IAT procedure is to determine how well one hemisphere of the brain can support cognitive func-tions, such as language, when the other hemisphere is "put to sleep" or anesthetized. The most widely used anesthetic agent is sodium amobarbital, which crosses the blood–brain barrier (a functional obstruction in the brain's circulatory system which usually keeps harmful substances out of brain tissue) easily, allowing for a rapid anesthetic effect. The amobarbital is, however, not injected directly into the patient's head or neck, but rather, in most Wada test pro-cedures, the drug is injected into the internal carotid artery (ICA) via a thin tube (catheter) inserted in the femoral artery (see Figure 3.1 for an illustration of this procedure). Pharmacological inactivation of brain areas occurs in the regions of vascular supply of the ipsilateral (same-sided) anterior and middle cerebral arteries, and the anterior choroidal artery, thereby affecting areas of the frontal and temporal lobes pertinent to speech and language functions; the anterior one-third of the hippocampus is also affected. Thus, if the amobarbital is injected into the left ICA, the left anterior portions of the brain will be anesthetized. When supplementary information is desired about memory function, one can separately anesthetize the occipitoparietal and posterior mesiotemporal regions by similar injection into the posterior cerebral artery (PCA). Potential advantages of selective anesthesia of the PCA supplied areas include the possibility of assessing memory in the absence of significant sedation or aphasia, and the possibility of observing language-related impairments (such as dyslexia) which may not be revealed by ICA injection. Many centers, however, feel that there is possibly a greater risk of stroke or vasospasm with PCA than with ICA injection, and that the additional risk of injury with multiple injections, outweighs the benefits of PCA and other "superselective" injections.

Because injection of amobarbital into an ICA temporarily inactivates only part of one cerebral hemisphere, the IAT allows one to assess independently the cognitive functions served by each hemisphere; that is, it is assumed that disruptions of language and memory during the IAT are a consequence of the temporary "lesioning" of the injected hemisphere, and that IAT mimics the effects that surgery on the injected hemisphere might have. As discussed above, this is particularly important in the surgical treatment of TLE, in which the tissue excised may be adjacent to or embedded within cortical tissue which supports language and memory functions. Moreover, because there is a greater incidence of atypical speech development (i.e., right hemisphere or bilateral representation) in those with known neurologic dysfunction, such as epilepsy, it is important to evaluate, rather than presume language lateralization in individual patients (Box 3.1).

3.5. COMPONENTS OF THE IAT PROCEDURE

IAT procedures vary from center to center, but often share several general features. Before the IAT, a baseline cognitive evaluation may be performed on the day of the IAT or in the preceding days. Such a baseline evaluation provides a basis for comparing cognitive test performances during the IAT relative to "normal" and familiarizes the patient with the test procedures. Before the IAT commences, angiography (radiographic visualization of cerebral blood vessels) is performed to identify any abnormalities or developmental variations in the blood vessels, and to determine whether there is a potential for cross-flow of amobarbital into the other hemisphere (as one might imagine, the presence of anesthetic in both hemispheres would complicate interpretation of test findings). A pre-IAT arteriogram is pictured in Figure 3.2. At centers evaluating memory and language lateralization, stimuli (e.g., words and objects) are often presented before the amobarbital injection so that recall of the items can be tested during the partial anesthesia. Most centers among those carrying out IATs on both cerebral hemispheres first examine the hemisphere for which surgery is planned, while others may inject the left ICA first (given the likelihood that language functions will be lateralized in the left hemisphere).

At the time of injection the supine patient is instructed to hold his or her arms straight up, and to begin counting. Upon injection, the patient's arm opposite the side of the injection becomes limp (if it does not, an inadequate dose of drug was injected, or the tip of the catheter may have slipped down

---

Box 3.1   Why do we need the Wada test: How typical is atypical language lateralization?

Language is lateralized in the left cerebral hemisphere for most humans. There are, however, a small percentage of individuals with "atypical" lateralization, meaning that language is represented either bilaterally or in the right hemisphere. There is disagreement among scholars about what atypical lateralization indicates. Some believe that it a rare variant of "normal," that is, without any prior neurologically relevant incidents (Knecht et al., 2000); others, however, contend that atypical lateralization generally stems from early brain insult or developmental aberration (Miller et al., 2003). Unfortunately, there are too few studies that have examined language lateralization in individuals with normal developmental and neurologic histories to adequately test either hypothesis. That being said, it is also well recognized that individuals with a history of early insult to the left hemisphere are more likely to develop atypical cerebral distribution of language (such as those with left TLE). This is attributed to the human brain's tendency to reorganize in order to accommodate language (albeit sometimes

at the expense of other functions such as spatial abilities). For example, in examining the Wada results from 170 patients with known neurological insults before that age of 15, Miller et al. found that approximately 14% of these patients had atypical speech lateralization (Miller et al., 2005). Interestingly, they found that atypical language lateralization was more frequent in females (19 women versus 5 men). The authors suggest that this could indicate overall greater plasticity in the female brain.

Knecht, S., Deppe, M., Dräger, B., Bobe, L., Lohmann, H., Ringelstein, E.-B., et al. (2000). Language lateralization in healthy right handers. Brain, 123, 74–81.

Miller, J.W., Dodrill, C.B., Born, D.E., & Ojemann, G.A. (2003). Atypical speech is rare in individuals with normal developmental histories. Neurology, 60, 1042–1044.

Miller, J.W., Jaydev, S., Dodrill, C.B., & Ojemann, G.A. (2005). Gender differences in handedness and speech lateralization related to early neurologic insults. Neurology, 65, 1974–1975.

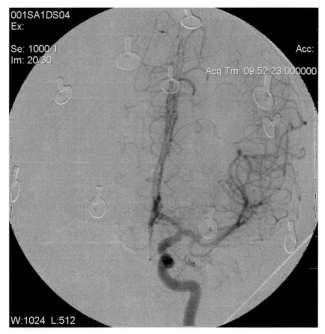

```
001SA1DS04
Ex:
Se: 1000/1                                                      Acc:
Im: 20/30
                                            Acq Tm: 09:52:23.000000
W:1024 L:512
```

FIGURE 3.2   Arteriogram of ICA completed before Wada testing to visualize the patient's vasculature. This procedure helps the team determine
whether there are any abnormalities in the vessels which would increase
the risk of amobarbital flowing into unintended areas of the brain.

resulting in drug delivery to the external carotid). When the
language-dominant hemisphere is affected, a global aphasia typically results and speech is arrested or dysarthric (the
patient will stop counting or slur). Because only one hemisphere is affected, the patient does not become unconscious
(as in general anesthesia), but the extent of alteration in
consciousness or somnolence, varies among patients.

There is significant variability among centers in the extent
to which (and how) speech and language functions are
tested (Benbadis, 2001). These functions are generally evaluated by tasks measuring expressive and receptive (comprehension) abilities including: automatic speech (e.g., months
of the year), visual confrontation naming (e.g., objects,
pictures, or colors), following verbal commands, repetition
of words and phrases, and reading. The most widely used
measure of language is naming to visual presentation (Rausch
et al., 1993), because dysnomia (difficulty naming pictured
or actual objects) is a feature of most aphasic syndromes.

The second ICA is injected 30 min to 24 h after the first
injection to ensure that the amobarbital has cleared from the
first-injected hemisphere. The test procedure is repeated,
typically using test materials that are different (to avoid a
practice effect), but comparable in difficulty to those used
during the first injection.

Some IAT protocols involve the use of functional
imaging (e.g., EEG or single photon emission computed
tomography (SPECT)) to ascertain drug effect and to visualize the distribution of the amobarbital within the injected

hemisphere. Protocols typically involve monitoring the extent
of hemiparesis to determine the duration of the anesthesia
(the effects of the anesthesia generally last between 90 and
300 s) and to ensure that assessment was conducted during
adequate anesthesia (Benbadis, 2001) so that results are valid.

3.6. PROBLEMS AND CONSIDERATIONS IN IAT PROTOCOL ADMINISTRATION AND INTERPRETATION

Although the rationale underlying the IAT seems
straightforward, the performance and interpretation of the
tests are not. Both the determination of speech dominance
and that of memory representation are subjective. Protocols
for the IAT vary from center to center, making meaningful
data comparisons difficult. Additionally, given the invasive-
ness of the procedure, there is a lack of IAT data derived
from subjects without neurological abnormalities that might
alter lateralization and localization of functions (normals).
Consequently, there are no data with which to compare the
results of patients being considered for surgery. While there
has been a recent push toward increasing the uniformity
of the comprehensive pre-surgical evaluation (Haut et al.,
2002), some important between-center differences still exist.
The protocol differences which most affect interpretation of
IAT language test results are discussed below. For a comprehensive review of intersite similarities and differences with
regards to the IAT procedure, we direct the reader to a published survey of surgical centers at the *1992 Palm Desert
International Epilepsy Conference* (Rausch et al., 1993).

3.6.1. Unilateral versus Bilateral IAT

The majority of centers conduct some form of the IAT
before surgically removing the seizure focus, although some
centers perform only unilateral IAT to either confirm left-
hemisphere language presence or to establish the extent of
dysfunction when the hemisphere slated for surgery is anesthetized (Rausch et al., 1993). Many would contend that
candidates who undergo surgery based on results obtained
from a single injection may be at greater risk for postoperative language deficits. Wellmer et al. (2005) argue, however,
that bilateral IAT is often redundant. In their retrospective
review of 107 cases having undergone bilateral IAT, results
from the unilateral IATs performed on the side of intended
surgery would have accurately lateralized language in over
80% of cases. The 20% of patients for whom unilateral IAT
is insufficient to document language lateralization (in this
case roughly 21 individuals) might argue, however, that
avoiding potential loss of language is well worth the possible redundancy of bilateral IAT!

3.6.2. Drugs and Administration Methods

Most frequently, sodium amobarbital is the anesthetic agent of choice. Intermittent shortages in amobarbital, mostly due to manufacturing problems (Buchtel *et al.*, 2002; Grote & Meador, 2005; Jones-Gotman *et al.*, 2005) have lead to delays in scheduling epilepsy surgery. Consequently, and because in some countries amobarbital is not available, several other anesthetic agents have been explored. Methohexital (Brevital) has a similar anesthetic effect, but is shorter-lived and requires reinjection (Buchtel *et al.*, 2002). Patients reportedly demonstrate less sedation with methohexital and less time is required between administrations to each hemisphere (both hemispheres can be evaluated within 2 h). This drug is usually injected along with an anticonvulsant, however, because of its epileptogenic potential. Etomidate has also been successfully used in place of amobarbital. Jones-Gotman *et al.* (2005) described using a constant infusion of this similarly short-lived drug until critical language and memory tests have been completed. Infusion offers two advantages: avoiding changing levels of anesthesia between re-administrations and allowing examiners to determine the length of anesthesia. Unfortunately, renal insufficiency has been a reported side effect of etomidate, causing concerns about its use with critically ill patients (Grote & Meador, 2005). Another short-acting anesthetic used in the course of Wada testing is propofol (also used by some as an anesthetic during epilepsy surgery), although the incidence of complications may be unacceptably high, having occurred in 19 of 58 patients in one series (Mikuni *et al.*, 2005).

Drug parameters – for example, amobarbital dosages and concentration, volume of amobarbital and saline mixture, rate of delivery (steady or incremental), and method of delivery (hand or automated injection) – also vary widely (Rausch *et al.*, 1993) and can lead to discrepant findings. Drug parameters affect the extent and duration of anesthesia. For instance, a faster rate or larger volume of injection will typically perfuse a more extensive vasculature, thus compromising more domains of function whereas smaller volumes or slower rates of injection will lead to a greater concentration of drug in a smaller area, possibly leading to more intense or prolonged drug effects. Further, the desired level of anesthetization is the result of a fine balance; it should not be so sedating or persistent that the patient cannot respond, yet should be sufficient to create a condition modeling as closely as possible the effects of surgery. Finally, drug effect should be long enough to permit presentation of an adequate number of test items, allowing the evaluation team to make valid inferences from test results.

Interpretation of the IAT is predicated on the assumption that brain regions supplied by the anterior and middle cerebral arteries are inactivated during anesthesia. A neuroimaging study by Hart *et al.* (1993) suggests that this assumption is not always warranted. Using SPECT, they found that there is great interindividual variability in the regions actually perfused by amobarbital after ICA injection. Further, there will occasionally be cross-filling of the contralateral hemisphere or the posterior circulation, leaving uncertainty about the neural bases of elicited responses and deficits (Hart *et al.*, 1993). Although some centers estimate likelihood of cross-flow of amobarbital into the contralateral hemisphere or perfusion of other territories by amobarbital during angiography (which traces the perfusion of contrast media) before the IAT procedure, the correlation between contrast medium and amobarbital distribution is limited by differences in methods of injection (e.g., Rausch *et al.*, 1993).

3.6.3. Determination of Adequacy of Anesthesia and Timing of Stimulus Presentation

Among many features of IAT protocols that differ across centers are the determinations of when an adequate drug effect to begin testing is evident, timing of stimulus presentation, types of stimuli and response formats, and criteria used to infer adequacy of language and memory. It is agreed across centers that presentation of stimuli is contingent on sufficient hemispheric anesthesia. In order to achieve a model of how the brain will function if tissue were removed, testing should occur during adequate drug effect. Unfortunately, the means of determining onset and duration of adequate anesthesia differ across centers. One or more of the following might be used to infer acceptable level of anesthesia: contralateral hemiparesis, grip strength, loss of antigravity tone, and marked EEG slowing. Yet, other centers simply present stimuli during a predetermined, standard interval (e.g., within the first 300 s) (Rausch *et al.*, 1993). Even if sufficiency of anesthetic effect were similarly defined across centers, disagreement might remain about the timing of stimulus presentation. When the speech-dominant hemisphere is injected, and speech arrest ensues, some clinicians wait for speech to return before proceeding with testing (*note*: speech arrest may reflect motor speech disruption rather than language deficit and may also interfere with the ability to respond during memory testing), whereas others continue stimulus presentation regardless of speech difficulty.

3.6.4. Criteria to Establish Hemispheric Language Dominance

What exactly constitutes evidence for language representation in a cerebral hemisphere? Benbadis *et al.* (1998) described two popular ways of determining hemispheric dominance (often referred to as the "laterality index") during IAT. One method is based on speech arrest alone and another is based on language task performance. In the calculation relying on speech arrest, duration of muteness after injection of each hemisphere is calculated and then compared across the two hemispheres by an equation: $(L - R)/(L + R)$. Using this method, the laterality index varies from -1 (strong right hemisphere dominance) to $+1$ (strong left hemisphere dominance). Alternatively, laterality indices based on language performance first calculate a percentage correct score on a battery of language tests obtained during anesthetization of each hemisphere (using tests evaluating, e.g., comprehension of commands, naming, phrase repetition, and sentence reading). The following equation is then employed $(P_L - P_R)$, where P represents the percentage correct score after left and right hemisphere injections (testing the right and left hemisphere functions, respectively). Using this approach, scores vary from -100 (strong right hemisphere dominance) to $+100$ (strong left hemisphere dominance).

Although these two methods appear generally comparable, Benbadis *et al.* (1998) warn that this is not the case. In a study of 21 patients, there was no significant correlation between the laterality index as calculated by speech arrest and the laterality index as calculated by performances on measures of language ability. Benbadis *et al.* did, however, find a strong relationship between lateral dominance as determined by language performance (the latter equation) and dominance as visualized by functional magnetic resonance imaging (fMRI). This suggests that speech arrest might not be a sufficient criterion for determining language dominance.

3.7. IAT IN PEDIATRIC POPULATIONS

Research exploring the IAT in pediatric populations is extremely limited and little has been published on this topic (most attention has been focused on less invasive techniques such as fMRI). Discussion of IAT in children and adolescents is frequently couched in terms of adult research findings. The few published empirical studies suggest that the IAT protocol might need to be modified for children and adolescents. Szabo and Wyllie (1993) noted that language dominance was established in all children who had bilateral injections and at least borderline intelligence, but in only about half of the children with mental retardation. Westerveld *et al.* (1994) reported more encouraging data. Using amobarbital doses of 100 or 130 mg, they considered IAT to yield unambiguous data concerning language dominance in children as young as 7 years.

3.8. VALIDATION AND RELIABILITY STUDIES

Despite differences in protocols, a multicenter study of seven epilepsy centers found a high degree of interrater reliability for language lateralization with IAT (Haut *et al.*, 2002) and the validity of the IAT to establish hemispheric language dominance is well accepted (Dodrill, 1993). The interesting question raised is whether it is sufficient to carry out the IAT to avoid postoperative language deficits? Specifically one might ask whether cases undergoing IAT also require intraoperative cortical mapping using electrical stimulation, and whether this leads to better outcomes (fewer postoperative language or cognitive declines) than in cases not undergoing language mapping. Among left-hemisphere language-dominant patients who underwent IAT and then conservative left temporal lobe resection without language mapping, no significant postoperative language decrements were observed

---

Box 3.2   What can the IAT tell us about language representation in bilingual individuals?

Several studies have shown that the interhemispheric organization of both languages in bilingual individuals is complementary, that is, hemispheric dominance for the two languages is similar (Berthier *et al.*, 1990; Gomez-Tortosa *et al.*, 1995). Intrahemispheric organization of the native and second languages, however, is likely different. Berthier (1990) described a case study in which a bilingual patient demonstrated the ability to speak his second language (English) 1 min before he was able to speak his native tongue (Spanish) after injection of amobarbital into the left middle cerebral artery. Based on this observation, the author speculated that the second language might be organized within the central sylvian core, whereas the first language might be represented in more distal perisylvian regions. Electrical stimulation studies (for review, see Ojemann, 1983) offer an opposing view, however, demonstrating that object naming in the second language tends to be more

peripheral from the sylvian fissure and represented in a larger area. Thus, one might conclude, assuming that amobarbital effects dissipate earlier in more distant areas, that Berthier *et al.*'s (1990) findings actually indicate the first language to be more centrally located.

Berthier, M.L., Starkstein, S.E., Lylyk, P., & Leiguarda, R. (1990). Differential recovery of languages in a bilingual patient: A case study using selective amytal test. *Brain and Language, 38,* 449–453.

Gomez-Tortosa, E., Martin, E.M., Gaviria, M., Charbel, F., & Ausman, J.I. (1995). Selective deficit of one language in a bilingual patient following surgery in the left perisylvian area. *Brain and Language, 48,* 320–325.

Ojemann, G.A. (1983). Brain organisation for language from the perspective of electrical stimulation mapping. *Behavioral and Brain Sciences, 6,* 189–230.

Box 3.3   The Wada test and American Sign Language

As in investigations involving bilingual persons, sign language studies in non-deaf individuals also indicate interhemispheric organization of signed and spoken language to be similar, at least in that the same hemisphere is dominant for both forms of communication. The existence of subtle interhemispheric differences in the organization of signed and spoken language, however, has not been settled by amobarbital studies (instead we would direct the reader to numerous studies published using techniques such as cortical stimulation mapping, fMRI, and PET). On the one hand, some studies have found that sign language is characterized by greater bilateral representation than spoken language (e.g., Homan et al., 1982). In contrast, several other studies have found that the interhemispheric organization of signed and spoken language to be complementary and highly similar (e.g., Mateer et al., 1984).

Because a substantial portion of deaf individuals apparently have experienced some cortical reorganization (Wolff et al., 1994), reports of IAT results in deaf individuals are of particular

significance. That is, such studies provide tentative data that speak to the issue of whether findings from normal-hearing individuals concerning the organization of sign language apply to deaf individuals. Complete left-hemisphere dominance has been found in a right-handed individual for American Sign Language, signed English, and finger spelling (Wolff et al., 1994). Thus, evidence of bilateral representation for sign language was not found.

Homan, R.W., Criswell, E., Wada, J.A., & Ross, E.D. (1982). Hemispheric contributions to manual communication (signing and finger-spelling). *Neurology, 32,* 1020–1023.

Mateer, C.A., Rapport, R.L., & Kettrick, C. (1984). Cerebral organization of oral and signed language responses: Case study evidence from amytal and cortical stimulation studies. *Brain and Language, 21,* 123–135.

Wolff, A.B., Sass, K.J., & Keiden, J. (1994). Case report of an intracarotid amobarbital procedure performed for a deaf patient. *Journal of Clinical and Experimental Neuropsychology, 16,* 15–20.

---

by Davies et al. (1995). In a follow-up study of 162 patients who underwent temporal lobectomy without mapping, Hermann et al. (1994) observed a postoperative dysnomia in 7% of left lobectomy patients. Moreover, an association between later age at onset of epilepsy and postoperative dysnomia was observed in the left lobectomy group undergoing language mapping. These findings suggest that mapping after IAT may still be a prudent course of action (even though most patients fare well) given that a subgroup of individuals do experience postoperative dysnomia. Furthermore, intraoperative mapping may be especially important in persons with bilateral language representation, because such patients are more likely than left-hemisphere language-dominant patients to have multiple, non-contiguous language areas in the left hemisphere (Jabbour et al., 2005). Intraoperative right hemisphere language mapping may also be helpful in identifying what are assumed to be accessory language areas in those persons with IAT-demonstrated bilateral language representation (Jabbour et al., 2005).

3.9. SUPPLEMENTARY AND ALTERNATIVE TECHNIQUES FOR ESTABLISHING LANGUAGE LATERALIZATION

Some centers consider surface EEG during IAT helpful in locating and monitoring the slowing of brain activity after amobarbital injection. Some centers also perform the IAT after intracranial electrode implantation so as to permit EEG recording directly from subdural grid and/or strip electrodes,

as well as deep electrodes implanted in the mesial temporal lobes. Electrocortical stimulation mapping of language cortex is also frequently employed before resection, especially when IAT results are ambiguous (Jabbour et al., 2005; Kho et al., 2005).

Because of the IAT's invasive nature and limitations, many clinicians and researchers have explored the use of alternative technologies, such as functional neuroimaging, to provide information about language lateralization in surgical candidates. For example, studies have been published examining the usefulness and accuracy of lateralization determination with near-infrared spectroscopy (NIS), magnetoencephalography (MEG), and positron emission tomography (PET). All of these techniques rely on computing metabolic or blood flow changes correlated with performance of language tasks. Among the studies that compared functional imaging with results of IAT, agreement ranged from 88% to 91% (Hunter et al., 1999; Bowyer et al., 2005).

The most widely explored potential alternative to the IAT is fMRI. fMRI allows visualization of changes in the flow of oxygenated blood within neural tissue. In the brain, changes in blood oxygen level have been shown to be related to neural activity. Consequently, fMRI is often used to infer localization of brain activity by comparing the location and extent of blood oxygenation while persons are engaged in particular language tasks versus activity observed during rest or a comparison task. In general, fMRI indicates hemispheric dominance through the use of laterality indices that compare the ratio of pixels (small areas within an image) activated during language tasks in one hemisphere to those in the other hemisphere. Hemispheric dominance for a function can be

TABLE 3.1   The Wada Test versus fMRI

	Wada	fMRI
Advantages	Mimics temporary lesion Well-established as valid measure of hemispheric language dominance	Non-invasive procedure May establish intrahemispheric language organization Less time restraint, allowing for more extensive batteries
Disadvantages	Invasive procedure Short length of anesthesia precludes extensive batteries Lack of standardization among protocols No normative database with healthy controls	Lack of standardization among protocols Difficulty statistically determining essential versus non-essential language areas Less effective for determining lateralization in individuals with mixed or right dominance

also inferred from comparisons of the activation of specific, homologous regions in each hemisphere (often referred to as "regions of interest" or "ROIs"). It remains unclear which method of inferring language dominance from fMRI yields results more consistent with those of the IAT and electrical stimulation.

fMRI has several advantages over IAT (see Table 3.1). Most importantly, fMRI is non-invasive and therefore does not pose physical risks to patients. This is a particularly attractive feature when dealing with medically compromised patients and pediatric populations. fMRI also allows for repeated test sessions, if necessary, and there are fewer drawbacks to creating a normative database with healthy controls, given less procedural risks. Additionally, most major medical centers have fMRI capability, making it a more appealing alternative than other functional technologies such as PET. Deblaere et al. (2004) reported the ability to determine language lateralization with relatively low field strength (1 Tesla magnetic force), a field strength available in many clinical settings. Finally, fMRI potentially imposes less time constraints (e.g., one is not restricted to 300 s or less of anesthetization), allowing use of a more extensive battery of tests that evaluate more language functions.

In addition to the procedural advantages, fMRI can provide a visual estimate of the cortical areas involved in language for any given individual. Thus, fMRI allows for an intrahemispheric language map that can help determine the functionality of specific areas targeted for surgical resection. This is particularly useful given that usually limited amounts of tissue are removed, rather than the entire temporal lobe or hemisphere. fMRI can also provide information about

the spectrum or continuum of lateralization which is not as well characterized by IAT.

Initial studies comparing IAT and fMRI have been promising, with language lateralization concordance rates exceeding 90% (Binder et al., 1996; Lehéricy et al., 2000; Woermann et al., 2003). Benke et al. (2006) found fMRI to accurately identify language dominance in persons with right TLE; however, concordance rates were poor for individuals with left TLE and left or mixed hemispheric language dominance.

The use of fMRI for determining language lateralization is not without criticism. Most importantly, fMRI cannot mimic a lesion like IAT. Thus, it is difficult if not impossible to ascertain whether unoperated tissue can truly support language. Moreover, functional imaging is based on the assumption that "active areas" are responsible for or involved in concurrent behaviors (such as producing or comprehending language). While many of the associations observed between activities in specific brain regions and particular behaviors in functional imaging are consistent with our knowledge of brain–behavior relationships (often gleaned from lesion studies), functional imaging inferences are based on a correlation from which causation cannot be inferred. In short, when a portion of the brain "lights up" during a specific task on fMRI, we cannot be entirely sure that the association is meaningful. Loring et al. (2002) further explain the challenge of distinguishing between essential and non-essential areas using fMRI in analyzing the location and parameters of "eloquent cortex" in eight adults. The authors noted that they were less likely to find random activations during language tasks if more conservative statistical methods were employed (i.e., raising the threshold for statistical significance). However, using these conservative methods they were also less likely to find activations among areas known to be associated with language functions. Another issue is how to interpret *decreased* activation during language tasks compared to control tasks.

Lack of standardized test batteries to assess language dominance is also a criticism of fMRI studies. Among published studies, the tasks employed are varied and range from semantic decision tasks (i.e., distinguishing between words and non-words) (Binder et al., 1996) to covert word generation (i.e., thinking of as many words from a certain category as possible within 1 min) (Woermann et al., 2003), making comparisons between protocols difficult. Further, lack of standardized test batteries presents a challenge similar to IAT procedures: which task or tasks best demonstrate language dominance? Rutten et al. (2002) have suggested using a combined analysis of several language tasks to improve detection of hemispheric dominance. The authors reported more reliable and robust concordance with IAT findings using combined task analysis compared with analysis of any individual language task (Rutten et al., 2002).

3.10. CHALLENGES AND FUTURE DIRECTIONS

So one might ask: what is the future of the Wada test? Will it be replaced by functional imaging for determination of language lateralization? Certainly, fMRI is a safer technology than IAT, but does it afford less risk of language loss postoperatively? Data suggest a high concordance rate between fMRI and IAT, but mostly in cases of left hemisphere language dominance. Unfortunately, if concordance rates are examined separately for persons with right, left, and mixed hemispheric language representations, fMRI struggles to accurately define language representation in persons with right and mixed dominance. It is possible that functional imaging will surpass the accuracy of IAT in determining language lateralization (it already provides important information about intrahemispheric language mapping); however, it will require the establishment of a standardized protocol. While researchers may achieve this goal, they will be hard pressed to replace the one most impressive feature of the IAT: glimpsing into the patient's future by inflicting a temporary lesion.

References

Aicardi, J., & Shorvon, S.D. (1997). Intractable epilepsy. In J. Engels & T.A. Pedley (Eds.), *Epilepsy: A comprehensive textbook* (pp. 1325–1331). Philadelphia, PA: Lippincott-Raven.

Benbadis, S.R. (2001). Intracarotid amobarbital test to define language lateralization. In H.O. Lüders & Y.G. Comair (Eds.), *Epilepsy surgery,* (pp. 525–529). Philadelphia, PA: Lippincott Williams & Wilkins. 2nd edn.

Benbadis, S.R., Binder, J.R., Swanson, S.J., Fischer, M., Hammeke, T.A., & Morris, G.L. *et al.* (1998). Is speech arrest during Wada testing a valid method for determining hemispheric representation of language? *Brain and Language, 65,* 441–446.

Benke, T., Koylu, B., Visani, P., Kamer, E., Brenneis, C., & Bartha, L. *et al.* (2006). Language lateralization in temporal lobe epilepsy: A comparison between fMRI and the Wada test. *Epilepsia, 47,* 1308–1319.

Binder, J.R., Swanson, S.J., Hammeke, T.A., Morris, G.L., Mueller, W.M., & Fischer, M. et al (1996). Determination of language dominance using functional MRI: A comparison with the Wada test. *Neurology, 46,* 978–984.

Bowyer, S.M., Moran, J.E., Weiland, B.J., Mason, K.M., Greenwald, M.L., & Smith, B.J. *et al.* (2005). Language laterality determined by MEG mapping with MR-FOCUSS. *Epilepsy and Behavior, 6,* 235–241.

Buchtel, H.A., Passaro, E.A., Selwa, L.M., Deveikis, J., & Gomez-Hassan, D. (2002). Sodium methohexital (Brevital) as an anesthetic in the Wada test. *Epilepsia, 43,* 1056–1061.

Davies, K.G., Maxwell, R.E., Beniak, T.E., Destafney, E., & Fiol, M.E. (1995). Language function after temporal lobectomy without stimulation mapping of cortical function. *Epilepsia, 36,* 130–136.

Deblaere, K., Boon, P.A., Vandenmaele, P., Tieleman, A., Vonck, K., & Vingerhoets, G. *et al.* (2004). MRI language dominance assessment in epilepsy patients at 1.0 T: Region of interest analysis and comparison with intracarotid amytal testing. *Neuroradiology, 46,* 413–420.

Dodrill, C.B. (1993). Preoperative criteria for identifying eloquent brain: Intracarotid amytal for language and memory testing. *Neurosurgery Clinics of North America, 4,* 211–216.

Grote, C.L., & Meador, K. (2005). Has amobarbital expired? Considering the future of the Wada. *Neurology, 65,* 1692–1693.

Hart, J., Lewis, P.J., Lesser, R.P., Fisher, R.S., Monsein, L.H., & Schwerdt, P. *et al.* (1993). Anatomic correlates of memory from intracarotid amobarbital injections with technetium Tc 99 mm hexamethylpropylene-amine oxime SPECT. *Archives of Neurology, 50,* 745–750.

Haut, S.R., Berg, A.T., Shinnar, S., Cohen, H.W., Bazil, C.W., & Sperling, M.R. *et al.* (2002). Interrater reliability among epilepsy centers: Multicenter study of epilepsy surgery. *Epilepsia, 43,* 1396–1401.

Hermann, B.P., Wyler, A.R., Somes, G., & Clement, L. (1994). Dysnomia after left anterior temporal lobectomy without functional mapping: Frequency and correlates. *Neurosurgery, 35,* 52–56.

Hunter, K.E., Blaxton, T.A., Bookheimer, S.Y., Figlozzi, C., Gaillard, W.D., & Grandin, C. *et al.* (1999). ^{15}O water positron emission tomography in language localization: A study comparing positron emission tomography visual and computerized region of interest analysis with the Wada test. *Annals of Neurology, 45,* 662–665.

Jabbour, R.A., Hempel, A., Gates, J.R., Zhang, W., & Risse, G. (2005). Right hemisphere mapping in patients with bilateral language. *Epilepsy and Behavior, 6,* 587–592.

Jones-Gotman, M., Sziklas, V., Djordjevic, J., Dubeau, F., Gotman, J., & Angle, M. *et al.* (2005). Etomidate speech and memory test (eSAM): A new drug and improved intracarotid procedure. *Neurology, 65,* 1723–1729.

Kho, K.H., Leijten, F.S.S., Rutten, G., Vermeulen, J., van Rijen, P., & Ramsey, N.F. (2005). Discrepant findings for Wada test and functional magnetic resonance imaging with regard to language function: Use of electrocortical stimulation mapping to confirm results. *Journal of Neurosurgery, 102,* 169–173.

Lehéricy, S., Cohen, L., Bazin, B., Samson, S., Giacomini, E., & Rougetet, L. *et al.* (2000). Functional MR evaluation of temporal and frontal language dominance compared with the Wada test. *Neurology, 54,* 1625–1633.

Loring, D.W., Meador, K.J., Allison, J.D., Pillai, J.J., Lavin, T., & Lee, G.P. *et al.* (2002). Now you see it, now you don't: Statistical and methodological considerations in fMRI. *Epilepsy and Behavior, 3,* 539–547.

Lüders, H.O. (2001). Protocols and outcome statistics from epilepsy surgery centers. In H.O. Lüders & Y.G. Comair (Eds.), *Epilepsy surgery* (pp. 973–977). Philadelphia, PA: Lippincott Williams & Wilkins. 2nd edn,

Mikuni, N., Takayama, M., Satow, T., Yamada, S., Hayashi, N., & Nishida, N. *et al.* (2005). Evaluation of adverse effects in intracarotid propofol injection for Wada test. *Neurology, 65,* 1813–1816.

Rausch, R., Silfevenius, H., Wieser, H., Dodrill, C., Meador, K., & Jones-Gotman, M. (1993). Intraarterial amobarbital procedures. In J. Engel (Ed.), *Surgical treatment of the epilepsies* (pp. 341–357). New York: Raven Press. 2nd edn

Rutten, G.J.M., Ramsey, N.F., van Rijen, P.C., Alpherts, W.C., & van Veelen, C.W.M. (2002). fMRI-determined language lateralization in patients with unilateral or mixed language dominance according to the Wada test. *Neuroimage, 17,* 447–460.

Szabo, C.A. & Wyllie, E. (1993). Intracarotid amobarbital testing for language and memory dominance in children. *Epilepsy Research, 15,* 239–246.

The National Association of Epilepsy Centers (NAEC) (2001). Guidelines for essential services, personnel, and facilities in specialized epilepsy centers in the United States. *Epilepsia, 42,* 804–814.

Wada, J. (1949). A new method for the deterioration of the side of the cerebral speech dominance: A preliminary report on the intracarotid injection of sodium amytal in man. *Igaku Seibutsugaku, 14,* 221–222.

Wellmer, J., Fernandez, G., Linke, D.B., Urbach, H., Elger, C.E., & Kurthen, M. (2005). Unilateral intracarotid amobarbital procedure for language lateralization. *Epilepsia, 46,* 1764–1772.

Westerveld, M., Zawacki, T., Sass, K.J., Spencer, S., Novelly, R.A., & Spencer, D.D. (1994). Intracarotid amytal procedure evaluation of hemispheric speech and memory function in children and adolescents. *Journal of Epilepsy, 7,* 295–302.

Woermann, F.G., Jokeit, H., Luerding, R., Freitag, H., Schulz, R., & Guertler, M. *et al.* (2003). Language lateralization by Wada test and fMRI in 100 patients with epilepsy. *Neurology, 61,* 699–701.

Further Readings

The following two references are suggested for the reader who has an interest in learning more about the use of the Wada test for memory lateralization.

Akanuma, N., Koutroumanidis, M., Adachi, N., Alarcon, G., & Binnie, C.D. (2003). Presurgical assessment of memory-related brain structures: The Wada test and functional neuroimaging. *Seizure, 12,* **346–358.**
Akanuma *et al.* focus on the role of the Wada test in pre-surgical evaluation of memory in epilepsy candidates. They also explore the use of other functional imaging techniques in identifying extratemporal regions important in memory functioning.

Simkins-Bullock, J. (2000). Beyond speech lateralization: A review of the variability, reliability, and validity of the intracarotid amobarbital procedure and its nonlanguage uses in epilepsy surgery candidates. *Neuropsychology Review, 10,* **41–74.**
Simkins-Bullock provides a comprehensive review of the Wada for non-language purposes (such as memory evaluation), addressing the future of the IAT along with her predictions for future research.

CHAPTER

4

Architectonic Language Research

KATRIN AMUNTS

Research Centre Juelich, INB-3, Medicine, Juelich, Germany
Department of Psychiatry and Psychotherapy, RWTH Aachen University Aachen, Germany

ABSTRACT

Broca and Wernicke regions are crucial for language. Yet they cover several cytoarchitectonic areas in cortex, and the involvement of each area in particular components of language processing is still unknown. From an anatomical perspective, the assignment of function to distinct cytoarchitectonic areas on the basis of functional imaging studies is problematic for four different reasons: (1) the relatively low spatial resolution of current techniques; (2) the high intersubject variability in brain anatomy; (3) problems inherent in traditional cortical maps including the subjective nature of the definition of areal borders; and (4) missing 3D information on the extent and localization of the areas. The chapter describes recent progress in cytoarchitectonic mapping, introduces cytoarchitectonic probabilistic maps in stereotaxic space and compares it with brain maps obtained from functional magnetic resonance imaging (fMRI) and with data from quantitative receptoraoutoradiography of classical neurotransmitters. Taken together, these results illustrate how microstructural multimodal mapping of the human brain may contribute to improved understanding of the neural basis of language.

4.1. INTRODUCTION

Developing linguistic and neuropsychological concepts of language together with rapid development of functional imaging techniques have led to increased interest in neural mechanisms underlying language. The segregation of the cerebral cortex into cortical areas with their specific cyto-, myelo-, receptorarchitecture and, finally, connectivity, provides organizational principles that can be correlated with brain function.

Brodmann (1909) was one of the first researchers analyzing the microstructure of the cerebral cortex (Box 4.1). His cortical map is still widely used, via the atlas of Talairach and Tournoux (Talairach & Tournoux, 1988), that serves as the main reference for functional imaging studies. Based on regional differences of the cellular architecture (i.e., cytoarchitecture) analyzed in histological sections of a postmortem brain, Brodmann subdivided the cortex into more than 40 cortical cytoarchitectonic areas, and presented it as a schematic view of a "typical" brain. He assumed that each area subserves a certain brain function, although, with the exception of a few primary areas (e.g., the primary visual cortex), such relationship was not proven at that time. Brodmann did not describe in detail where the borders of the areas are located, and whether and how brains of different individuals vary in this respect.

Borders between cytoarchitectonic areas are functionally relevant since response properties of neurons change at the border between two cytoarchitectonic areas (e.g., Luppino *et al.*, 1991; Nelissen *et al.*, 2005). Correlations between many cortical areas and cognitive systems have been demonstrated. To cite one example, the characteristic anatomy of the primary auditory cortex which receives projections from cochlear hair cells via the medial geniculate body is well defined. Likewise, its role in the generation of conscious hearing of frequencies, and for pattern and spatial analysis is understood. Projections from the medial geniculate body go to the inner granular layer (layer IV) of the primary auditory cortex (Brodmann's area (BA) 41, or Te1, Morosan *et al.*, 2001). As a result, layer IV is broad and contains densely packed granular cells, a cytoarchitectonic feature typical of primary sensory areas (e.g., somatosensory areas 3a, 3b and 1 also receive heavy projections from the thalamus, terminating in a well developed layer IV). At the border of the secondary

Box 4.1   Brodmann's cytoarchitectonic map

The best known cytoarchitectonic map, published by Brodmann nearly a 100 years ago (Brodmann, 1909) is shown in below figure. It relies on extensive studies of cell body-stained (Nissl-stain) histological sections. In describing the cytoarchitecture, Brodmann applied criteria such as the density, arrangement and size of neurons, thickness of cortical layers and the presence of special cells (e.g., giant Betz cells of area 4). To avoid confusion of histological data and unproven evolutionary and functional speculations, he created a "neutral" system of nomenclature by numbering different cytoarchitectonic areas according to their dorso-ventral sequence ("BAs"). Unfortunately, he never described most of the areas in detail.

Brodmann was convinced that the cerebral cortex is composed of numerous cortical areas, each of them characterized by a distinct cytoarchitecture and function. For example, BA 4 was conceptualized as the anatomical equivalent of the primary motor cortex which guided voluntary movements. Brodmann assumed that the Pars opercularis corresponds to Broca's region "in a strict sense" ("Broca'sche Stelle"), the anatomical correlate of the center of speech. Research has shown that Brodmann's map is oversimplified and even wrong for several brain regions, including language-related ones. In addition, his map does not reflect intersubject variability in size and localization of areas. The map is a schematic drawing of the lateral view of a "typical" brain, and does not allow drawing conclusions with respect to the location of areal borders in the depths of the sulci.

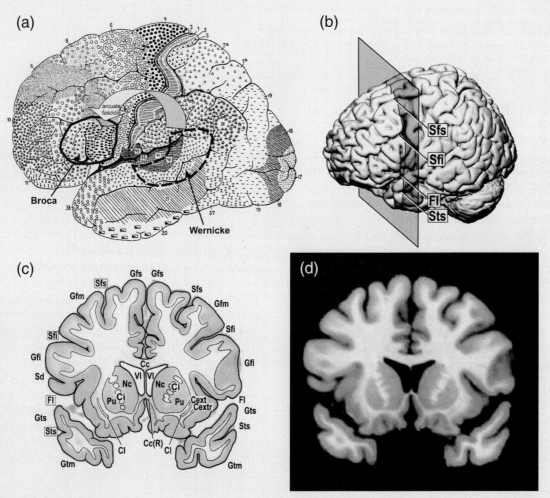

(a) Localization of language-related regions. Schematic view of the localization of language-related areas in Broca (bold line) and Wernicke (dotted bold line) regions, projected to Brodmann's schematic cortical map (1909). The numbers indicate BAs. (b) Surface reconstruction of an in vivo MR data set of a human brain. Note the differences in the sulcal pattern with respect to (a). The sectioning plane of (c) and (d) through the Broca region is marked. (c) and (d) – sections through the MR data set (resolution: 1 mm isotropic) and corresponding drawing of the section with labeled sulci and gyri. Gfi marks the Gyrus frontalis inferior, that is the gyrus at which Broca's region can be found; Sfi – Sulcus frontalis inferior (corresponds to its approximate dorsal border). The structure of the MR data does not enable an identification of the underlying cytoarchitectonic area. The MR data set belongs to the single-subject MNI reference (http://www.bic.mni.mcgill.ca/). This brains serves as a widely used reference both for functional imaging data and postmortem data (see also Figure 4.3).

auditory area, however, the cytoarchitecture changes abruptly: layer IV becomes narrower.

The functional significance and interpretation of other cytoarchitectonic features, for example, the size and shape of neurons, the distribution of neurons within a cortical layer and so forth, is less clear. How such cytoarchitectonic information can be "translated" into brain function remains a project for future research.

The relationship between cortical brain areas involved in language and language processing, however, is less well known compared to many other brain functions (e.g., the motor visual systems). One reason for this difference is that language has to be analyzed in humans, not in experimental animals (although potentially relevant data – e.g., with respect to the mirror system – have been obtained in macaque brains, see Chapter 23). Another reason is that the borders of cytoarchitectonic areas are usually not bound to certain macro-anatomical landmarks such as gyri and sulci. Whereas primary areas show a stable association of (some) borders to sulci or gyri (e.g., the primary auditory is always located on the Heschl gyrus of the temporal lobe (Morosan et al., 2001)), multimodal association cortex, involved in language such as areas 44 and 45 of Broca's region (Amunts et al., 1999), do not. Moreover, the location of cytoarchitectonic borders with respect to sulci and gyri varies, and, in addition, sulci and gyri vary independently from the borders (Ono et al., 1990; Amunts et al., 1999; Tomaiuolo et al., 1999). Thus, the precise localization of the borders of these areas cannot be predicted on the basis of macroscopical brain anatomy as provided, for example, by MR scans using routine protocols. A third reason for the discrepancy may be the complexity of language relative to other cognitive functions, and the lack of agreement among psycho- and neurolinguistics as to how this function should be analyzed, and as a consequence, be related to the underlying anatomy.

How can we bring together microstructural information of language-related areas obtained in postmortem brains and imaging data from living human brains during language tasks? The most straightforward solution requires architectonic and functional/behavioral analyses in one and the same subject. Theoretically, this would be feasible if microstructural parcellation could be performed by ultra-high-resolution anatomical MR imaging in the living human brain. First attempts have already been made for selected regions of the cortex (for a brief overview see Eickhoff et al., 2005). This approach, however, is not practical, in particular for language-related regions, due to technical and methodical restrictions.

As an alternative, data obtained in histological sections of postmortem brains, with excellent spatial, microscopical (in the range of 1 μm) resolution, can be transformed into a brain atlas, for example, an MR data set of an individual reference brain. As we will show later, the combination of high-resolution cytoarchitectonic data obtained in histological sections of postmortem brains, and functional imaging data

of the living human subjects may open new perspectives for analyzing structural–functional relationships in human brain. As a result, conclusions about the involvement of one cytoarchitectonic area or another in a certain function can be drawn on a statistical (probabilistic) basis (Roland & Zilles, 1994; Amunts et al., 2004) (see Section 6 for some examples).

The present chapter reviews the micro- and macro-anatomy of brain areas involved in language. These include Broca's and Wernicke's regions, as well as motor and premotor cortices and subcortical nuclei. It will report a putative anatomical correlate of language dominance – interhemispheric differences in the cytoarchitecture of language-related regions. It will also show that cytoarchitecture, as well as its interhemispheric asymmetry, undergo changes during development. Finally, it will illustrate how cytoarchitectonic probabilistic maps can be applied to analyze the topography of activations obtained in speech and language tasks during functional imaging experiments, designed to test linguistic and neuropsychological hypotheses.

4.2. BROCA'S REGION

Broca's region (the anterior speech region, Broca's area, the motor speech region) occupies the posterior part of the left inferior frontal gyrus (Broca, 1861). Macroscopically, it comprises the opercular and triangular parts of the inferior frontal gyrus. Cytoarchitectonically, BA 44 and 45 are located at the opercular and triangular parts of the inferior frontal gyrus (Figure in Box 4.1). Their localization makes them candidates to be the microstructural correlates of Broca's region.

It should be noted that other cytoarchitectonic areas may also be part of Broca's region: the most ventral part of BA 6 (inferior precentral gyrus), and parts of the cortex hidden in the depths of the Sylvian fissure or the orbital part of the inferior frontal gyrus (e.g., BA 47). However, brain macroscopy lets us neither accept nor exclude them from Broca's region because many cytoarchitectonic areas are located in close neighborhood. The list of BAs which can be found here includes BA 6, 44, 45, 47, 46, 43, 8 and 9. The localization of borders between these areas as well as their number differ between the classical cytoarchitectonic maps of Brodmann (1909), von Economo and Koskinas (1925), the Russian school under Sarkisov (1949), and Riegele (1931). Such differences may arise, among other factors, as a consequence of intersubject variability in brain anatomy.

Given the problems just noted, the observation that the term "Broca's region" (and that of "Wernicke's region") is not consistently used in the literature should come as no surprise. This inconsistency is not just a problem of nomenclature; rather, it is a conceptual one. A large group of papers does not relate this term to a certain cytoarchitectonically defined region, but applies it rather in a historical or clinical context (corresponding to the brain region that is lesioned in Broca's aphasics), or with respect to its participation in language processing. In some studies, Broca's region includes BA 44 and 45. In others, it refers to BA 44, 45 and 47; further papers restrict it to either BA 44 or 45, exclusively. Riegele (1931) distinguished "Broca's region in a strict sense" from an "extended Broca's region," which covers parts of the orbital surface as well. The term "Broca's complex" has been introduced recently in order to characterize a brain region of related but distinct areas of the left prefrontal cortex, at least encompassing BA 47, 45, ventral 6 and 44; these areas subserve more than one function in the language domain and other nonlanguage functions as well (Hagoort, 2006). The common denominator in the language domain has been conceptualized in selection and unification operations (Hagoort, 2006). Whereas such conception is precise in terms of linguistics, it is less well defined with respect to the microstructural correlates. Vice versa, anatomical definitions are often quite imprecise with respect to specific language functions that are processed in the cortical areas. Thus, localizing Broca's region in the context of a functional imaging study analyzing linguistic material, or a lesion study of a Broca aphasic may refer to completely different areas with different cytoarchitecture, connectivity and, ultimately, function.

Finally, we still do not know, for most of the language components, whether they activate an entire cytoarchitectonic area (e.g., "semantics activate area 45"), only a part of it (e.g., "phonological processing is restricted to the dorsal part of area 44") or a group of areas (for example, "areas 44 and 45 of Broca's region as well as area 22 posterior of Wernicke's region are involved in language"). Here, both functional and anatomical analyses are necessary to understand the neural organizational principles of language.

4.2.1. Cytoarchitecture of BA 44 and 45

Let's have a more detailed look into the cytoarchitecture of BA 44 and 45. As parts of the isocortex, both areas show the typical six layers (I–VI; Figure 4.1). They can be distinguished from neighboring cortical areas by the presence of large to giant pyramidal cells in deep layer III. These cells project to other areas of the same and the contralateral hemisphere, which is an important prerequisite to their functioning in complex behavior such as language involving large neural networks and linking to other functional systems (motor, visual, auditory and so forth). Layer IV is a distinguishing feature of areas 44 and 45: while BA 45 is granular (i.e., it has a well defined layer IV), BA 44 is dysgranular (i.e., its layer IV is visible but pyramidal cells from the lower part of layer III and from the upper part of layer V may intermingle with granular cells of layer IV). This blurs the appearance of layer IV as a separate layer containing granular cells exclusively. If the layer IV criterion is taken into account, then the dysgranular

BA 44 seems to have an intermediate position between the more posterior, agranular BA 6, and the more anterior, granular BA 45. Based on the cytoarchitecture, we may speculate on a putative role that BA 44 may have in movement control.

The two cytoarchitectonic areas, BA 44 and 45 are found in the language dominant hemisphere, but also contralaterally, in the non-dominant hemisphere. Although the architecture of these areas is quite similar when looking through the microscope, subtle but significant left–right differences are found when left–right comparisons are performed on a quantitative basis. These are expected, if we assume that a structural–functional relationship for language lateralization exists.

In contrast to classical cytoarchitectonic descriptions, which were mainly based on a purely visual inspection of histological sections and subjective criteria, recent methodological advances have enabled an observer-independent definition of the borders of BA 44 and 45 in histological sections (see Box 4.2). These methods have made border definition reliable and reproducible. The method, though observer-independent, searches for borders along the same lines that Brodmann proposed, namely by searching for abrupt changes in cytoarchitecture. The quantification of these changes employs a multivariate statistical analysis of profiles sampled from neighboring cytoarchitectonic areas at the cortical ribbon, and reflecting changes in the volume fraction of cell bodies from the cortical surface to the white matter border (Schleicher et al., 1999) (Figure 4.1). Blocks of neighboring profiles are compared to each other, and a multivariate distance measure (e.g., the Euclidean distance, or the more complex Mahalanobis distance), is calculated between them. The idea behind this approach is that large cytoarchitectonic differences (e.g., between two neighboring areas) correspond to high values of the multivariate distance measure whereas small cytoarchitectonic differences (e.g., local inhomogeneities within an area) are reflected by low values (Schleicher et al., 1999). The value of the distance measure can be tested for significance, and cytoarchitectonic borders are defined on positions on the cortical ribbon where the distance measure reaches a significant value.

The quantitative cytoarchitectonic analysis via profiles does not only open the possibility to detect borders between cortical areas, but also enables an analysis of similarities and dissimilarities between cortical areas, for example, between homologous areas of the two hemispheres.

4.2.2. Localization of BA 44 and 45 and Their Intersubject Variability

Once the borders of the areas have been defined in serial histological sections as discussed above, these sections can be 3D-reconstructed in order to assess the localization of

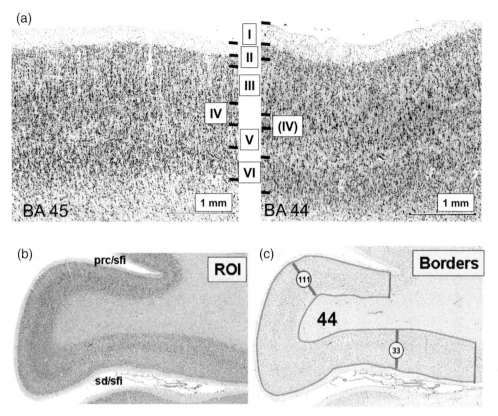

FIGURE 4.1 (a) Cytoarchitecture of BA 44 and 45 of the Broca region. Histological section through a human brain. Cortical layers are numbered. The two areas can be distinguished by their layer IV, which is well developed in BA 45 (granular cortex), and not consistently seen in BA 44 (dysgranular cortex). Both areas show very large pyramidal cells in deep layer III. (b) In order to map the extent of cytoarchitectonic areas, regions of interest (ROI) are scanned under the microscope (b). The inner and outer contours of the cortex are labeled, and "profiles" are extracted perpendicular to the cortical surface. (c) Two of them, at positions 111 and 33 are marked. The positions of these two profiles correspond to significant changes in the cytoarchitecture, as estimated by an observer-independent definition of borders (Amunts *et al.*, 1999; Schleicher *et al.*, 1999). prc/sfi – junction of precentral and inferior frontal sulci, sd/sfi – junction of diagonal and inferior frontal sulci.

borders with respect to the brain surface and its sulci and gyri in more detail as has been done in two previous studies (Amunts *et al.*, 1999; Amunts *et al.*, 2004) (Figure 4.1). The studies showed that BA 44 and 45 were always located at the inferior frontal gyrus, and never reached the free surface of the middle frontal gyrus in a sample of 10 post-mortem brains. However, the dorsal border of BA 44 has been found in the ventral bank of the inferior frontal sulcus in some sections, but also in its dorsal bank. In its posterior part, BA 44 was approximately limited by the ventral part of the inferior precentral sulcus. The precise location relative to the fundi of sulci varied up to 1 cm (Amunts *et al.*, 1999). Moreover, the diagonal sulcus may indicate the border between both areas in some hemispheres; but it was located within BA 44 in other brains, or the sulcus was event absent (see below). That is, although the localization of BA 44 and 45 in the opercular and triangular of the inferior frontal gyrus is consistent across the brains, the localization of their *borders* may vary considerably between the brains

and even between the hemispheres of one and the same brain. The more one moves from the free surface of both parts into the depths of surrounding sulci, the lower the probability for a non-ambiguous definition of the cortical area based on macroscopical landmarks.

In addition to the variability of the borders of the areas with respect to macro-anatomical landmarks such as sulci and gyri, landmarks themselves are also variable. Variability concerns the shape of the sulci, the number of their segments and their presence, and also the relationship between borders of cytoarchitectonic areas and sulci. For example, the inferior frontal sulcus can form one continuing segment or it can be interrupted (e.g., it consists of two, three or even four independent segments). In some hemispheres, it is interconnected with the precentral sulcus, in others it is not (Ono *et al.*, 1990). This also pertains to the precentral sulcus, in particular to its ventral part (Germann *et al.*, 2005). Not only do the sulcal patterns vary between individual brains (Tomaiuolo *et al.*, 1999), but also there is variability

---

Box 4.2   New concepts of architectonic mapping

Whereas the classical cytoarchitectonic maps were mainly based on visual inspection and verbal description of histological sections, later studies laid the ground for a quantitative analysis of the cortex by density profiles, which reach from the pial surface to the white matter quantifying laminar changes of morphometric parameters (Haug, 1979). This approach enabled the definition of the borders of cortical areas in serial histological sections of postmortem brains using a statistical analysis of local changes in architecture (Schleicher *et al.*, 1999).

Areal borders may represent differences not only in cyto- or myeloarchitecture, but also in receptorarchitecture, enzymo- and immunohistology, connectivity, electrophysiology and/or function. For example, the border between the primary visual cortex BA 17 (V1) to BA 18 (V2) is defined by the disappearance of sublayers in lamina IV (cytoarchitecture), the Gennari stripe (myeloarchitecture), a decrease of receptor densities of the muscarinic M2 receptor (receptorarchitecture), differences in cytochrome–oxidase and SMI-32 antibody staining (enzymo- and immunohistology) and callosal fibers originating at the transition zone (connectivity) (Zilles & Clarke, 1997). V1 differs from V2 also in function: V1 is a primary area that performs a first analysis of the visual information and receives fibers from the retina via the lateral geniculate body. V2 is involved in further data processing such as the detection of illusory contours,

face and word processing and so forth. A similar relationship between micro-anatomy and brain function can be assumed for the language regions. In contrast to vision, however, the function of language areas is less well understood.

Thus, rather then relying on one modality exclusively, for example, cytoarchitecture, modern architectonic concepts favour a multimodal approach, combining different types of structural information in order to achieve a functionally relevant parcellation of the cerebral cortex (Zilles *et al.*, 2002).

Haug, H. (1979). The evaluation of cell-densities and nerve-cell size distribution by stereological procedures in a layered tissue (cortex cerebri). *Microscopica Acta*, 82, 147–161.

Schleicher, A., Amunts, K., Geyer, S., Morosan, P., & Zilles, K. (1999). Observer-independent method for microstructural parcellation of cerebral cortex: A quantitative approach to cytoarchitectonics. *Neuroimage*, 9, 165–177

Zilles, K., & Clarke, S. (1997). Architecture, connectivity and transmitter receptors of human extrastriate visual cortex. Comparison with non-human primates. *Cerebral Cortex*, 12, 673–742.

Zilles, K., Palomero-Gallagher, N., Grefkes, C., Scheperjans, F., Boy, C., Amunts, K., & Schleicher, A. (2002). Architectonics of the human cerebral cortex and transmitter receptor fingerprints: Reconciling functional neuroanatomy and neurochemistry. *European Neuropsychopharmacology*, 12, 587–599.

---

in the localization of microstructural borders with respect to the (variable) surrounding sulci and gyri. The intersubject variability makes it difficult or even impossible to relate findings based on MR measurement of the living human brain to cytoarchitectonically defined areas.

However, do we really need to consider cytoarchitectonic parcellations for the understanding of language? Cytoarchitectonic analysis in postmortem brains is only one aspect of brain organization; perhaps other techniques should be used, especially since different modalities may be structured in different ways. For example, the distribution of receptors of classical neurotransmitter systems is a further important aspect of brain organization since neurotransmitters and their receptors are important for signal transduction in the brain.

4.2.3. Receptorarchitecture

BA 44 and 45 differ in receptorarchitecture both from one another, and from neighboring cytoarchitectonic areas. Receptorarchitecture refers to regional differences in the receptor density of neurotransmitter systems such as glutamate, gamma-amino butyric acid (GABA), serotonin or acetylcholine. Receptor density can be analyzed using quantitative receptor autoradiography (Zilles *et al.*, 2002a, b).

In tissue preparations from postmortem brains, we analyzed radioactively labelled ligands that bind specifically to one or more receptor subtypes of neurotransmitters. This approach enables an analysis of the absolute concentration of a receptor binding site in fmol/mg protein (Zilles *et al.*, 2002a). Many types of receptors are expressed in every cortical area, and even within a single neuron or glial cell. However, differences in the expression of a neurotransmitter can be found between the layers of the cortex in a given cortical area, and between cortical areas of different brain regions (see also Figure 4.2). In addition, regional differences can be found in the balance of receptor concentrations of many neurotransmitters. Such differences can be analyzed and visualized using "receptor fingerprints" (Zilles *et al.*, 2002a). Thus, information on receptorarchitecture is another important aspect of human brain anatomy (see also Box 4.2). Differences in a variety of different receptor binding sites enable a parcellation of the human brain according to the molecular organization of the human brain (Figure 4.2).

Interestingly, receptorarchitectonic borders coincide in location with those of cytoarchitecture if the respective receptor binding site is sensitive to the relevant architectonic border. Cyto- and receptorarchitecture, therefore, reflect different, but complementary aspects of cortical microstructure. Finally, analysis of the distribution of selected transmitter receptors in the *living* human brain is also possible, through

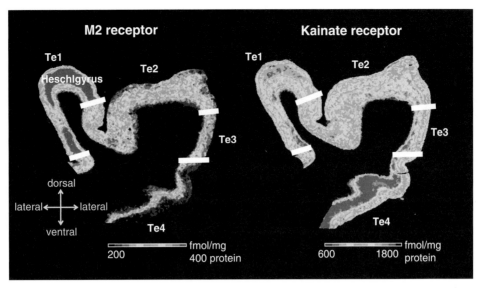

FIGURE 4.2 Receptorarchitecture of the superior temporal gyrus. Distribution of the M2 receptor for acetylcholine and the kainite receptor for glutamate in a coronal section through the superior temporal gyrus and the Heschl gyrus (cryostat section, 20 μm). The color coding indicates receptor densities in fmol/mg protein (Zilles *et al.*, 2002a, b). The white lines indicate the borders between cytoarchitectonic areas Te1–Te4 (Morosan *et al.*, 2005) (see Plate 2).

the analysis of receptor positron emission tomography (rPET). Yet thus far, it has achieved a much lower spatial resolution, and has not proven relevant to language research.

4.3. WERNICKE'S REGION

We know that Broca's and Wernicke's regions are connected, from functional, anatomical, behavioral and lesion studies. Wernicke's region (posterior speech region, Wernicke area, sensory speech center) occupies the posterior part of the superior temporal gyrus, caudally from the Heschl gyrus (transverse gyrus), where the primary auditory cortex, BA 41, is located (Figure in Box 4.1). Macroscopically, this region corresponds to the Planum temporale and surrounding cortical regions. It may also encompass the most inferior parts of the inferior parietal lobule (parts of the angular and supramarginal gyri), and the lateral bulge of the superior temporal gyrus. Based on lesion studies, Wernicke and Lichtheim proposed the first model of language processing, which includes the anterior and posterior centers for speech as well as the arcuate fascicle connecting both centers. More recent studies have suggested that parts of the longitudinal fascicle III and fibres of the extreme capsule are more relevant for language processing (Schmahmann *et al.*, 2007).

Microscopically, the Wernicke's region is not well defined. It is not clear whether BA 42, 39, 40 and 38 are also constituents of this region, and if so, to what extent. Multimodal association cortex is found on the ventral lip of the superior temporal gyrus, which may also belong to this region (reviewed in Calvert, 2001).

Four architectonically distinct areas, Te1–Te4 (Figure 4.2), have been delineated within the human superior temporal gyrus on the basis of cytoarchitecture and the binding patterns of different transmitter receptors for glutamate, serotonin, adrenalin, GABA and dopamine (Morosan *et al.*, 2005). Areas Te1 and Te2 are located on the dorsal surface of the superior temporal gyrus, where they occupy Heschl gyrus and the temporal plane, respectively (Figure 4.2). Area Te3 is confined to the lateral bulge of the superior temporal gyrus, whereas area Te4 is found onto the ventral lip of the gyrus. Area Te1 represents the human primary auditory cortex; it corresponds roughly to BA 41 (Morosan *et al.*, 2001). Areas Te2 and Te3 can be interpreted as higher order auditory areas, whereas area Te4 is an auditory-related area. Te2 roughly corresponds to BA 42 (Morosan *et al.*, 2005); it does not occupy, however, large portions of the lateral surface of the superior temporal gyrus in contrast to BA 42 (Brodmann, 1909). At least two cortical areas, Te3 and Te4 occupy the putative region of BA 22. The border between Te3 and Te4 was regularly found on the ventral or ventro-lateral surface of the superior temporal gyrus; in contrast to Brodmann's map and his area 22, Te3 did never reach the middle temporal gyrus (Morosan *et al.*, 2005).

Areas Te1–Te4 differ between each other in cytoarchitecture (Morosan *et al.*, 2005). Cytoarchitectonic criteria include the thickness of layer IV (a layer well established in Te1, and thinner in the other three areas), the presence and number of relatively large neurons in deep layer III (more frequently found in Te4 than in the other three areas), the distinctiveness of the laminar pattern (most pronounced in Te1) and others. In addition to such qualitative criteria (based on a pure visual inspection of cytoarchitectonic sections), observer-independent mapping of the temporal cortex has made the definitions of cytoarchitectonic borders of the four areas reliable (Schleicher et al., 1999; Morosan et al., 2005).

Recent receptorarchitectonic studies have shown that the human primary auditory cortex, Te1, shows very high concentrations of cholinergic M2 receptors in layers III and IV (Zilles et al., 2002a; Morosan et al., 2005). Concentration decreases from Te1 to higher auditory areas Te2 and Te3 and Te4, where the lowest concentrations are found. Areas Te2, Te3 and Te4 are distinguished from each other based on differences in laminar distribution patterns of the noradrenergic α_1 receptors and serotonergic $5HT_{1A}$ receptors. Interareal comparisons of receptor densities have shown that some transmitter receptors, for example, the cholinergic muscarinic M2 receptors, reach their highest densities in the thalamorecipient layers III–IV of primary auditory area Te1, whereas other transmitter receptors, for example the noradrenergic α_1 receptors, accumulate their highest densities in layers III–VI of multimodal area Te4. Analyses of similarities and dissimilarities in receptor architecture between Te1, Te2, Te3 and Te4 have shown that human auditory and auditory-related areas are organized in a hierarchical system. Given that neurotransmitter receptors play a major role in neurotransmission, these studies may provide a significant link between the anatomical and functional connectivity of the human auditory domain (Morosan et al., 2005).

4.4. OTHER REGIONS INVOLVED IN LANGUAGE

In addition to cortical areas in the vicinity of Broca's and Wernicke's regions, other brain areas participate in language processing as well, for example, the motor cortex. Their role in speech function such as articulation has been demonstrated in electrophysiological studies of Foerster and the Vogts at the beginning of the twentieth century. The involvement of the premotor cortex in language processing has been elucidated by recent functional imaging techniques (e.g., Wilson et al., 2004; Skipper et al., 2005), but also intraoperative mapping in patients during neurosurgery (Ojemann et al., 1989; Duffau et al., 2003). In addition, it has been hypothesized, that the premotor cortex plays a role in the planning but also in semantic processes and categorization (Fadiga et al., 2000; Martin & Chao, 2001). The dorsolateral prefrontal cortex, frontal operculum and the insula participate in different aspects of language processing, and so forth. A complete list of cortical areas involved in language would encompass areas of even more cortical regions.

Subcortical nuclei, for example, the basal ganglia and the thalamus, participate in language processing as well (Friederici, 2006). Finally, an important aspect of speech control is the neuromuscular control of tongue, lips, larynx, pharynx, the vocal cords and the muscles necessary for breath and so forth, which are controlled, in addition to cortical motor and somatosensory areas, by the cerebellum and cranial nerves and their nuclei.

4.5. INTERHEMISPHERIC CYTOARCHITECTONIC DIFFERENCES

Language dominance and lateralization for speech, favoring the left hemisphere in approximately 95% of subjects, mainly refer to a functional specialization of the human brain. The microstructural underpinnings of this specialization are largely unknown. Visual inspection of histological sections of postmortem brains alone does not enable detecting systematic differences between areas of the dominant hemisphere and their contralateral homologues.

A few studies, however, have demonstrated quantitative interhemispheric differences with respect to the anatomy of language-related areas. In 10 postmortem brains volume of left BA 44 were significantly larger than those of the right hemisphere whereas volumes of BA 45 did not seem to differ significantly between the hemispheres (Amunts et al., 1999). However, for 6 of the 10 subjects (including all females), the volume of BA 45 was greater in the left hemisphere than the right (Amunts et al., 1999; Uylings et al., 2006).

The cytoarchitecture of areas 44 and 45 has been shown to be asymmetric as quantified by density profiles ("density" is estimated by the grey level index, GLI, a measure of the volume fraction of cell bodies; Schleicher et al., 1999). Density profiles reflect laminar changes in the volume fraction of cell bodies from the surface to the white matter border. Feature vectors were collected from these profiles, and a multivariate analysis was performed. The analyses enabled the detection of significant differences between profiles sampled in the left and right hemisphere, thus indicating left–right differences in cytoarchitecture (Amunts & Zilles, 2001). No major asymmetry in total number of neurons in BA 44 and 45 was detected in a recent postmortem study (Uylings et al., 2006). This study was based on the same sample of postmortem brains as employed in the previous study of 10 postmortem brains (Amunts et al., 1999, 2001).

Interestingly, significant interhemispheric differences in cytoarchitecture have already been found in 1 year old infants (Amunts et al., 2003). A significant increase of asymmetry with age was reported in BA 45 but not in BA 44. An adultlike, left-larger-than-right asymmetry in the GLI was reached approximately at age 5 in BA 45, and age 11 in BA 44. It has been hypothesized that the delayed maturation of these language-related areas as compared to motor areas is due to the microstructural changes accompanying the development of language abilities and the influence of language practice on cytoarchitecture during childhood (Amunts et al., 2003).

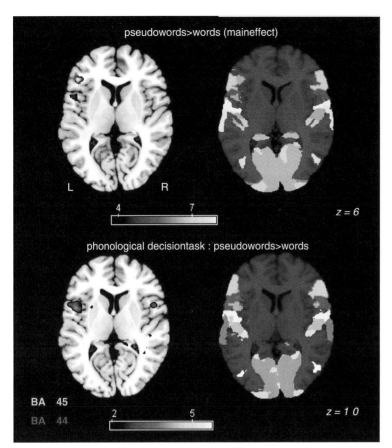

FIGURE 4.3 Superimposition of functional activations obtained during an fMRI study analyzing the activations in Broca's region (left BAs 44 and 45) during the processing of visually presented words and pseudowords and cytoarchitectonic probability maps (Heim *et al.*, 2005). Subjects had to perform either a lexical or a phonological decision task. The upper row shows the activation for the lexical decision task (the contrast pseudowords versus words), the lower row shows the effect for the phonological decision task. The left images display SPM{t} maps rendered on the MNI reference brain; the right shows the same projected on the cytoarchitectonic maximum probability maps (Eickhoff *et al.*, 2005) of BA 44 and 45 (Amunts *et al.*, 2004) (see Plate 3).

4.6. CYTOARCHITECTONIC MAPS FOR THE ANALYSIS OF LANGUAGE

Cytoarchitectonic areas can be compared with functional imaging data more directly when both modalities are registered to a common reference space such as the T1 weighted MR single-subject brain of the Montreal Neurolog-ical Institute (http://www.bic.mni.mcgill.ca/). The registration of postmortem data eliminates all aspects of intersubject variability in their macro-anatomy and makes brains comparable to each other. As a next step, postmortem brains and their delineated areas are superimposed and probabilistic maps calculated. These maps show, for each voxel of the reference space, the frequency with which certain cortical area was present in the sample of postmortem brains in the stereotaxic space of the standard reference brain (Figure 4.3). The list of probabilistic maps available for structural–functional comparisons is rapidly growing, and has recently come to include approximately 100 cortical areas, subcortical nuclei and fibre tracts. Published data are available for the scientific community at http://www.fz-juelich.de/ime.

A combined functional magnetic resonance imaging study on verbal fluency and cytoarchitectonic mapping of BA 44 and 45 served as a first "proof of principle" of the application of the probabilistic maps of areas 44 and 45 with respect to language processing (Amunts *et al.*, 2004). The aim of the study was the analysis of the involvement of BA 44 and 45 in a verbal fluency task. Verbal fluency was investigated using fMRI in 11 healthy volunteers, who covertly produced words from predefined categories. A factorial design was used with factors verbal class (semantic versus overlearned fluency) and switching between categories (no versus yes). Both the *in vivo* fMRI and histological postmortem data were registered to a common reference brain using the above mentioned elastic warping tool. The cytoarchitectonic probability maps showed the involvement of left areas 44 and 45 with verbal fluency relative to the baseline. Semantic relative to overlearned fluency showed greater involvement of left BA 45 than that of 44 (Figure 4.3). Thus, although both areas participate in verbal fluency, they do so differentially. Left BA 45 was more involved in semantic aspects of language processing, while BA 44 was probably involved in high-level aspects of programming speech production *per se*.

Recently, Santi and Grodzinsky (2006) have applied cytoarchitectonic probabilistic maps of BA 44 and 45 in order to distinguish different aspects of syntactic processing from a more general working memory task in an fMRI experiment. Two conditions were investigated – one was syntactic movement, the other was antecedent-reflexive binding. Working memory is required for both tasks to link two non-adjacent positions; but syntactically, they are governed by distinct rule systems. It was found, that the binding-related activation of the left inferior frontal gyrus was more anterior and inferior to the region produced by movement. That is, the binding-related activation overlapped with left BA 45, whereas the movement-related activation overlapped with BA 44. Therefore, the data provided arguments for a functional specialization within Broca's region at the level of cytoarchitectonically defined areas (BA 44 and 45). The authors argued that the distinct activation is inconsistent with either a syntactic working memory or a general verbal working memory account of Broca's region. The study is a nice example illustrating how knowledge from postmortem anatomy and mapping can be combined with information from brain activation under well defined experimental conditions using functional imaging techniques of the living human brain.

4.7. CHALLENGES AND FUTURE DIRECTIONS

Challenges and progress in cytoarchitectonic research can be envisioned as (a) mapping brain regions that have not been charted yet, (b) revealing specific functions and hierarchies of cytoarchitectonic areas and (c) obtaining a better understanding of the synaptic connectivity of language-related areas.

(a) One challenge is to map as of yet uncharted brain regions. Brodmann's parcellation scheme does not hold true in many brain regions, including some of those involved in language. We know from studies in the macaque brain and other non-human primates that the parcellation, for example of the ventral motor and prefrontal cortex is more complex than Brodmann's map suggests (Luppino *et al.*, 1991; Nelissen *et al.*, 2005). Several studies have already been performed resulting in new, more detailed maps of the prefrontal and temporal cortices. It would be major progress if microstructural information were available for the entire human brain, and if this information could be obtained not only in postmortem but also in living human brains.

(b) Another challenge is to uncover the specific function of a cytoarchitectonic area and define hierarchies among cortical areas. Cytoarchitectonic areas seem to be involved in many brain functions such as motor- and language-related functions. If Brodmann's hypothesis is true that a certain cortical area subserves a certain brain function, a common "denominator" of

the different functions should exist. Therefore, cellular and molecular biology of language-related regions have to be investigated in more detail.

(c) Finally, future research should be directed towards an understanding of the synaptic connectivity of language-related areas, the time sequence of their involvement and their functional connectivity. Most knowledge with respect to synaptic connectivity comes from experimental studies in animals; these data, however, cannot directly be transferred to the human brain. Recent methodical progress in diffusion tensor imaging, diffusion spectrum imaging and tractography revealed putative pathways relevant for language control, and may disclose causal relationships between cortical areas. They will be supplemented by electrophysiological and tracing studies in experimental animals unrevealing anatomical fiber tracts of language related regions (Schmahmann *et al.*, 2007).

References

Amunts, K., Schleicher, A., Bürgel, U., Mohlberg, H., Uylings, H.B.M., & Zilles, K. (1999). Broca's region revisited: Cytoarchitecture and intersubject variability. *The Journal of Comparative Neurology, 412*, 319–341.

Amunts, K., Schleicher, A., Ditterich, A., & Zilles, K. (2003). Broca's region: Cytoarchitectonic asymmetry and developmental changes. *The Journal of Comparative Neurology, 465*, 72–89.

Amunts, K., & Zilles, K. (2001). Advances in cytoarchitectonic mapping of the human cerebral cortex. In T.P. Naidich, T.A. Yousry, & V.P. Mathews (Eds.), *Neuroimaging clinics of North America. Anatomic basis of functional MR imaging* (pp. 151–169). Philadelphia: Harcourt.

Amunts, K., Weiss, P.H., Mohlberg, H., Pieperhoff, P., Gurd, J., Shah, J.N., Marshall, C.J., Fink, G.R., & Zilles, K. (2004). Analysis of the neural mechanisms underlying verbal fluency in cytoarchitectonically defined stereotaxic space. The role of Brodmann's areas 44 and 45. *Neuroimage, 22*, 42–56.

Broca, M.P. (1861). Remarques sur le siége de la faculté du langage articulé, suivies d'une observation d'aphemie (Perte de la Parole). *Bulletins et Memoires de la Societe Anatomique de Paris, 36*, 330–357.

Brodmann, K. (1909). *Vergleichende Lokalisationslehre der Großhirnrinde in ihren Prinzipien dargestellt auf Grund des Zellenbaues.* Leipzig: Barth JA.

Calvert, G.A. (2001). Crossmodal processing in the human brain: Insights from functional neuroimaging studies. *Cerebral Cortex, 11*, 1110–1123.

Duffau, H., Capelle, L., Denvil, D., Gatignol, P., Sichez, N., Lopez, M., Sichez, J.-P., & van Effenterre, R. (2003). The role of dominant premotor cortex in language: A study using intraoperative functional mapping in awake patients. *Neuroimage, 20*, 1903–1914.

von Economo, C., & Koskinas, G.N. (1925). Die Cytoarchitektonik der Hirnrinde des erwachsenen Menschen. Berlin: Springer.

Eickhoff, S., Stephan, K.E., Mohlberg, H., Grefkes, C., Fink, G.R., Amunts, K., & Zilles, K. (2005). A new SPM toolbox for combining probabilistic cytoarchitectonic maps and functional imaging data. *Neuroimage, 25*, 1325–1335.

Fadiga, L., Fogassi, L., Gallese, V., & Rizzolatti, G. (2000). Visuomotor neurons: Ambiguity of the discharge or "motor" perception? *International Journal of Psychophysiology, 35*, 165–177.

Friederici, A.D. (2006). What's in control of language? *Nature Neuroscience, 9,* 991–992.

Germann, J., Robbins, S., Halsband, U., & Petrides, M. (2005). Precentral sulcal complex of the human brain: Morphology and statistical probability maps. *The Journal of Comparative Neurology, 393,* 334–356.

Hagoort, P. (2006). On Broca, brain, and binding. In Y. Grodzinsky & K. Amunts (Eds.), *Broca's Region* (pp. 242–253). Oxford, New York: Oxford.

Haug, H. (1979). The evaluation of cell-densities and nerve-cell size distribution by stereological procedures in a layered tissue (cortex cerebri). *Microscopica Acta, 82,* 147–161.

Heim, S., Alter, K., Ischebeck, A.K., Amunts, K., Eickhoff, S.B., Mohlberg, H., Zilles, K., von Cramon, Y.D., & Friederici, A. (2005). The role of the left Brodmann's areas 44 and 45 in reading words and pseudowords. *Cognitive Brain Research, 25,* 982–993.

Luppino, G., Matelli, M., Camarda, R.M., Gallese, V., & Rizzolatti, G. (1991). Multiple representations of body movements in mesial area 6 and the adjacent cingulate cortex: An intracortical microstimulation study in the macaque monkey. *The Journal of Comparative Neurology, 311,* 463–482.

Martin, A., & Chao, L.L. (2001). Semantic memory and the brain: Structure and processes. *Current Opinion in Neurobiology, 11,* 194–201.

Morosan, P., Rademacher, J., Schleicher, A., Amunts, K., Schormann, T., & Zilles, K. (2001). Human primary auditory cortex: Cytoarchitectonic subdivisions and mapping into a spatial reference system. *Neuroimage, 13,* 684–701.

Morosan, P., Schleicher, A., Amunts, K., & Zilles, K. (2005). Multimodal architectonic mapping of human superior temporal gyrus. *Anatomy and Embryology, 210,* 401–406.

Nelissen, K., Luppino, G., Vanduffel, W., Rizzolatti, G., & Orban, G.A. (2005). Observing others: Multiple action representation in the frontal lobe. *Science, 310,* 332–336.

Ojemann, G., Ojemann, J., Lettich, E., & Berger, M. (1989). Cortical language localization in left, dominant hemisphere. *Journal of Neurosurgery, 71,* 316–326.

Ono, M., Kubik, S., & Abernathey, C.D. (1990). *Atlas of the cerebral sulci.* Stuttgart, New York: Thieme.

Riegele, L. (1931). Die Cytoarchitektonik der Felder der Broca'schen Region. *J Psychol Neurol, 42,* 496–514.

Roland, P.E., & Zilles, K. (1994). Brain atlases – a new research tool. *Trends in Neuroscience, 17,* 458–467.

Santi, A., & Grodzinsky, Y. (2006). Taxing working memory with syntax: bihemispheric modulations. *Human Brain Mapping,* November 28 [E-publication ahead of print].

Schleicher, A., Amunts, K., Geyer, S., Morosan, P., & Zilles, K. (1999). Observer-independent method for microstructural parcellation of cerebral cortex: A quantitative approach to cytoarchitectonics. *Neuroimage, 9,* 165–177.

Schmahmann, J.D., Pandya, D.N., Wang, R., Dai, G., D'Arceuil, H.E., de Crespigny, A.J., & Wedeen, V.J. (2007). Association fibre pathways of the brain: Parallel observations from diffusion spectrum imaging and autoradiography. *Brain, 130,* 630–653.

Skipper, J.I., Nusbaum, H.C., & Small, S.L. (2005). Listening to talking faces: Motor cortical activation during speech perception. *Neuroimage, 25,* 76–89.

Talairach, J., & Tournoux, P. (1988). *Coplanar stereotaxic atlas of the human brain.* Stuttgart: Thieme.

Tomaiuolo, F., MacDonald, B., Caramanos, Z., Posner, G., Chiavaras, M., Evans, A.C., & Petrides, M. (1999). Morphology, morphometry and probability mapping of the pars opercularis of the inferior frontal gyrus: An in vivo MRI analysis. *European Journal of Neuroscience, 11,* 3033–3046.

Uylings, H.B.M., Jacobsen, A.M., Zilles, K., & Amunts, K. (2006). Left–right asymmetry in volume and number of neurons in adult Broca's area. *Cortex, 42,* 652–658.

Wilson, S.M., Saygin, A.P., Sereno, M.I., & Iacoboni, M. (2004). Listening to speech activates motor areas involved in speech production. *Nature Neuroscience, 7,* 701–702.

Zilles, K., Palomero-Gallagher, N., Grefkes, C., Scheperjans, F., Boy, C., Amunts, K., & Schleicher, A. (2002a). Architectonics of the human cerebral cortex and transmitter receptor fingerprints: Reconciling functional neuroanatomy and neurochemistry. *European Neuropsychopharmacology, 12,* 587–599.

Zilles, K., Schleicher, A., Palomero-Gallagher, N., & Amunts, K. (2002b). Quantitative analysis of cyto- and receptor architecture of the human brain. In J.C. Mazziotta & A. Toga (Eds.), *Brain mapping: The methods* (pp. 573–602). Amsterdam: Elsevier.

Further Readings

Grodzinsky, Y., & Amunts, K. (Eds.) (2006). *Broca's region*. Oxford: Oxford University Press.

This book is the product of a workshop held in 2004. It brings together contributions from anatomy, linguistics, physiology, clinical neurology, psychology, computational biology and evolutionary biology. It contains state-of-the-art knowledge with respect to Broca's region as well as a collection of historical papers by Broca, Lichtheim, Hughlings-Jackson, Brodmann and others.

Schütz, A., & Miller, R. (Eds.) (2002). *Cortical areas: Unity and diversity*. New York: Taylor and Francis.

The book focuses on homogeneity versus heterogeneity of the cerebral cortex and its relationship to brain function. It introduces different architectonic concepts, explains the connections between cortical areas and connectivity and gives an overview of phylogenetic concepts.

Toga, A., Thompson, P., Mori, S., Amunts, K., & Zilles, K. (2006). Towards multimodal atlases of the human brain. *Nature Neuroscience Reviews, 7,* 952–966.

The paper reviews recent concepts of a human brain atlas. It stresses the advantage of a multimodal approach combining data from cyto-, myelo-, receptorarchitecture and modern fiber tracking methods.

Zilles, K. (2004). Architecture of the human cerebral cortex. Regional and laminar organization. In G. Paxinos & J. Mai (Eds.), *The human nervous system* (pp. 997–1055). Amsterdam: Elsevier.

The chapter provides an overview about the principal subdivisions of the cerebral cortex, the typical cytoarchitecture of main brain regions and its quantitative analysis. It also discusses different mapping concepts.

5

Microgenesis of Language: Vertical Integration of Linguistic Mechanisms Across the Neuraxis

DON M. TUCKER[1,2,4], GWEN FRISHKOFF[3,4] and PHAN LUU[1,2]

[1]*Department of Psychology, University of Oregon Eugene, OR, USA*

[2]*Electrical Geodesics, Inc., Eugene, OR, USA*

[3]*Learning Research and Development Center, University of Pittsburgh, Pittsburgh, PA, USA*

[4]*NeuroInformatics Center, University of Oregon Eugene, OR, USA*

ABSTRACT

Microgenetic theory provides a coherent framework for interpreting the multiple levels of neural organization that contribute to language comprehension. A key proposal is the organization of function across brainstem, diencephalic, and limbic, as well as cortical, networks. We consider evidence that neurolinguistic representations develop from the hemispheric "core" (limbic areas along the medial wall) to the "shell" (specialized regions of lateral neocortex). This anatomical framework suggests how top–down (expectancy-driven) and bottom–up (sense data-driven) directions of control might work together to achieve hierarchically organized conceptual structures of language comprehension. In this way, microgenetic theory provides one account of how language comprehension may emerge from the interaction of domain-general processes (such as memory and attention) and the language-specific networks.

5.1. INTRODUCTION

Using powerful new technologies and well-honed experiment designs, neurolinguistics has achieved rapid progress over the past decade in documenting patterns of brain activity in language. Certain findings – such as left temporal activation in meaning comprehension – have been well replicated, providing confidence in modern methods for brain and language research (for a recent review, see Demonet *et al.*, 2005). Nonetheless, important questions remain about how to interpret these effects: How does language comprehension interact with domain-general processes, such

as memory and attention? Are there language-specific networks that are revealed in measures of brain activity? What is at stake with such questions is not only the species-centered problem of whether language is unique to humans, but also the more fundamental question of how we think about language, and about how the brain helps shape language structure and function. In the present chapter, we address these issues in the context of *microgenetic theory*. Although this theory has received little attention in recent years, it provides a coherent framework for interpreting the multiple levels of neural organization that contribute to language comprehension.

5.1.1. Basic Principles and Plan for Chapter

The idea of microgenesis rests on three principles: orthogenesis, iconicity, and vertical integration. When applied to language, these three principles imply that a careful study of neuroanatomy can lead to new hypotheses about neurolinguistic processing.

The *orthogenetic principle* states that psychological processes undergo a progression from global and integrated toward finely articulated and differentiated patterns of neural structure and function. An example is the articulation of action. Any action, such as the stroke of the cue stick in billiards, is organized first at the proximal core of postural control. From this basis in the axial muscles, the action becomes differentiated into peripheral limb activity, such as the pivotal swing of the arm, leading to final articulation with the distal (wrist and finger) muscles. Microgenetic theory holds that this progression, from global and axial toward focal and distal, is as true of language as it is of any skilled action. Indeed, electrophysiological studies of letter recognition have shown a *global precedence effect* (activation of responses to holistic shapes, prior to detailed representations), suggesting that orthogenesis may serve as a unifying principle in action and perception (Friston, 2005; see Section 5.2).

A second principle of microgenesis emphasizes the similarity, or *iconicity*, of brain structure and function (Section 5.3). Within this view, the core structures and functions of language embody principles that determine how cognitive processes unfold, on a millisecond time scale, within the human brain. The principle of iconicity is particularly important to our proposal that the organization of linguistic acts is embodied in the neural architecture, and thus can be understood by studying brain anatomy.

Following Brown's (2002) analysis of the progressive organization of thought across the phyletic levels of the neuraxis, we have proposed a third microgenetic principle, that of *vertical integration* (Luu & Tucker, 1998; Tucker, 2001, 2002). This principle follows closely from the principle of iconicity, and states that actions and thoughts stem from coordinated processing at multiple levels of brain structure. In this context, we propose that a holistic, limbic-based representation is formed at the onset of each mental process. This representation is then transformed at each stage of processing, through cycles of feedforward and feedback connections between brain regions (Section 5.4).

The implication of microgenetic theory is that cognitive processes such as language comprehension remain integrally linked to more elementary brain functions, such as motivation and emotion (Section 5.5). Thus, microgenetic theory suggests one possible answer to the question of how language may be related to domain-general processes: linguistic and nonlinguistic functions should be tightly integrated, particularly as they reflect common pathways of processing.

In recent years, certain views that are consonant with microgenetic theory – such as the notion of "embodied cognition" (Clark, 2006) – have gained currency in psychology. In Sections 5.2–5.5, we explain how a microgenetic framework can explain the embodied neural substrates of language. We further illustrate how modern methods in neurolinguistics are well suited for testing the key assumptions of microgenetic theory and for revising the theory to incorporate the growing evidence on neural mechanisms of language.

5.2. PRINCIPLES OF BRAIN STRUCTURE AND FUNCTION

5.2.1. Levels of Brain, Levels of Representation

Regions of the mammalian brain are defined by laminar structure (number and thickness of different cellular layers)

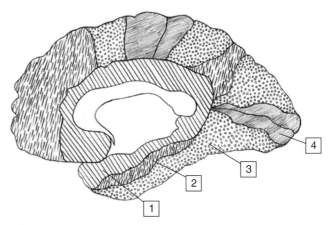

FIGURE 5.1 Medial view of right hemisphere showing major cytoarchitectonic divisions, corresponding to four levels of cortical structure: (1) diagonal stripes = limbic cortex, (2) dashes = heteromodal association cortex, (3) stippled = unimodal sensory or motor cortex, (4) shaded = primary sensory or motor neocortex. *Source*: Modified from Tucker (2007).

and cytoarchitectonics (distribution of cell types). According to these criteria, the brain comprises a hierarchy that ranges from the simplest and least differentiated structures – for example, the reticular formation of the brain stem – to the richly layered cerebral cortex (Mesulam, 2000). Figure 5.1 shows the medial wall of the right brain, with regions shaded according to the degree of laminar differentiation (number of layers).

The oldest and least differentiated cortical regions are known as *limbic cortex*. Limbic regions have close connections with the hypothalamus and brainstem nuclei, which are involved in emotional arousal and regulation of *visceral* (internal) functions, such as body temperature and reproductive functions. Important functions of the limbic cortex include emotion and motivation, but also classically "cognitive" processes such as learning and memory. By contrast, the five-layered primary sensory and motor regions of the *isocortex* support discrete *somatic* functions (e.g., Figure 5.1, division 4). Intermediate levels of cortex – that is, association cortices – are involved in the elaboration of sensory and motor processes and in the integration of sensorimotor and limbic functions (e.g., Figure 5.1, divisions 2 and 3). Finally, *paralimbic* regions, including the cingulate cortex, have been associated with cognitive functions, such as attention and goal-directed behavior, as well as motivation and emotion.

As illustrated in Figure 5.2, cytoarchitectonically distinct regions of the brain are interconnected to form a processing hierarchy that mediates between inner (visceral) and outer (somatic) constraints on the organism (Mesulam, 2000; Tucker, 2001). Representation across these levels may give rise to cognitive representations that are responsive to both sensory input (*external or somatic context*), as well as internal, motivational states (*internal or visceral context*).

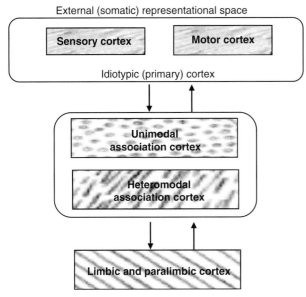

External (somatic) representational space

Idiotypic (primary) cortex

Internal (visceral) representational space

FIGURE 5.2 Levels of brain structure, corresponding to levels of representation. Limbic and paralimbic regions (e.g., cingulate cortex) support representation of the *internal (visceral)* space, while idiotypic (primary sensory and motor) regions provide an interface with the *external (extrapersonal/somatic)* environment. *Source*: After Mesulam (2000).

5.2.2. Language Across Corticolimbic Levels

The question for theories of language then becomes how verbal constructions arise out of this architecture. Recent studies (see Box 5.1) provide new insights into the brain's functional architecture that have implications for hierarchical processing in language. These studies have revealed that isocortex, characterized by six or more cellular layers and a high density of granular neurons, support *feedforward processing* of information and are therefore integral to processing of incoming, sensory information. By contrast, regions characterized by fewer layers and a sparser density of granular neurons participate more commonly in *feedback processing* (see Box 5.2).

According to microgenetic reasoning, damage at each level of the corticolimbic hierarchy results in predictable symptoms. Indeed, early researchers such as Jackson and Luria suggested that neurolinguistic deficits provide an opportunity to observe levels of processing that are implicit in every linguistic act, but are not normally observed in healthy individuals. Following this tradition, Brown (1988) categorized linguistic (speech) errors with respect to stages of normal word production. Different categories of speech error were assumed to reflect damage to a particular level, revealing language that was, so to say, suspended at an intermediate stage of processing (Table 5.1). Brown further linked each deficit to a particular level of neural organization.

Box 5.1　Laminar patterns and regional connectivity of the mammalian cortex

Cortical regions are broadly divided into three types according to their patterns of lamination (number and types of neuronal layers). *Granular* cortex is the most finely differentiated, with six or more layers and a well-developed layer IV (high concentration of granular neurons). Examples include primary sensory areas. *Agranular* cortex has fewer layers (typically two or three) and lacks a layer IV. Limbic regions, including medial temporal structures, are agranular. Finally, *dysgranular* areas are characterized by an intermediate degree of lamination (4–5 layers) and an "incipient" layer IV (Shipp, 2005).

Inter-regional connections are characterized by their patterns of regional connectivity. *Forward* or *ascending* connections (from more to less laminar regions) originate in layers I–III and terminate in layer IV. *Backward or descending* connections terminate in superficial layers (I–III), and projections are more diffuse and less topographical. Neuroanatomists have suggested a correlation between patterns of regional connectivity and the functional nature of information transmission. While forward projections along the visual pathway have been viewed as conveying information along the processing pathway in a hierarchic manner that preserves and refines object representations, consistent with the topographic organization of forward connections, backward projections are viewed as modulatory, providing diffuse control over the intensity or "gain" of activity in a particular region (Friston, 2002).

Studies by Pandya, Barbas, and colleagues have illustrated how differences in patterns of connectivity can be used to predict the directionality of processing. For example, Rempel-Clower and Barbas (2000) observed that projections from lateral prefrontal cortex (BA 46) terminate mostly in upper areas of the ventral temporal cortex in rhesus monkeys, consistent with a feedback pattern of frontal–temporal modulation. This example shows how it is possible to reason from patterns of connectivity to infer the directionality of input from one cortical region to another.

Friston, K. (2002). Functional integration and inference in the brain. *Progress in Neurobiology, 68*(2), 113–143.

Rempel-Clower, N.L., & Barbas, H. (2000). The laminar pattern of connections between prefrontal and anterior temporal cortices in the rhesus monkey is related to cortical structure and function. *Cerebral Cortex, 10*(9), 851–865.

Shipp, S. (2005). The importance of being agranular: A comparative account of visual and motor cortex. *Philosophical Transactions of the Royal Society of London, B Biological Sciences, 360*(1456), 797–814.

Semantic errors were thus attributed to deep (i.e., medial), limbic areas, while more superficial (lexical and phonological) errors were attributed to more focal areas of left lateral cerebral cortex (Table 5.1).

Box 5.2   Anatomy of the medial temporal lobe

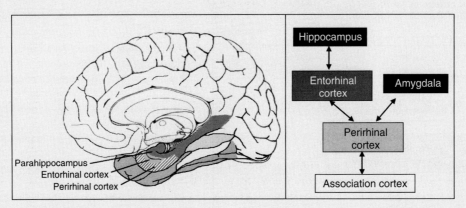

The left panel of the figure shows a medial view of the right hemisphere and its major divisions. The hippocampus and amygdala are nested within the temporal lobe and are not seen. *Entorhinal cortex* lies just below the hippocampus. It is divided into a lateral area, which receives inputs from the perirhinal cortex, and a medial area, which has reciprocal projections to the hippocampus. Adjacent to the entorhinal cortex is the *perirhinal cortex*, which is situated posterior and medial to the temporal pole. Perirhinal cortex is divided into a medial region (BA 35) and a lateral region (BA 36). *Parahippocampal gyrus* lies behind perirhinal cortex and peripheral to the hippocampal–amygdala complex. The collateral sulcus separates the parahippocampal gyrus from occipito–temporal cortex, including the fusiform gyrus (which is lateral to the parahippocampal gyrus) and the lingual gyrus (behind parahippocampal gyrus).

The right panel summarizes patterns of connectivity between areas of rat medial temporal lobe (Kajiwara *et al.*, 2003). We have shaded the divisions to correspond to the laminar subdivisions shown in the left-hand panel (see Figures 5.1 and 5.3 in the main text). Rempel-Clower and Barbas (2000) found

that projections from more granular regions of frontal cortex (e.g., left inferior PFC) terminated in deep layers of medial (agranular) entorhinal cortex. Conversely, projections from agranular or dysgranular regions of frontal cortex (e.g., orbitofrontal cortex) terminated in upper layers of granular cortex within temporal regions (e.g., left inferior temporal cortex), reminiscent of "backward" or limbifugal pathways (*cf.* Figure 5.3 in main text). These pathways may support the kind of frontal–temporal patterns of interaction that have been ascribed to "re-entrant" mechanisms, which have been implicated in selective attention. Given the integral role of attention in language, these pathways are likely to play a central role in language comprehension.

Kajiwara, R., Takashima, I., Mimura, Y., Witter, M.P., & Iijima, T. (2003). Amygdala input promotes spread of excitatory neural activity from perirhinal cortex to the entorhinal–hippocampal circuit. *Journal of Neurophysiology, 89*(4), 2176–2184.
Rempel-Clower, N.L., & Barbas, H. (2000). The laminar pattern of connections between prefrontal and anterior temporal cortices in the rhesus monkey is related to cortical structure and function. *Cerebral Cortex, 10*(9), 851–865.

TABLE 5.1   Substitutions for the Word "Chair" (Example from Brown, 1988)

Subtype	Example	Source
Asemantic	"Wheelbase"	Bilateral paralimbic or limbic
Associative	"Throne"	Bilateral paralimbic
Categorical	"Table"	Left associative cortex
Evocative	"You sit on it ... "it's a..."	Left associative cortex
Phonological	"Shair"	Left focal neocortex

Following the tradition of early microgenetic theorists, Brown further suggested that the temporal unfolding of speech across the neuraxis reflects a progressive articulation of representations at each stage of processing. As illustrated

in Table 5.2, this same global-to-local progression can be conceptualized at multiple time scales: from evolution (*phylogeny*), to organismic development (*ontology*), to dynamic neural function (*microgeny*). In this way, microgenesis points to a continuity in the evolution and development of brain structure and function. The global-to-local principle of microgenesis thus leads to testable predictions about the sequencing and interactivity of neural systems in language.

5.2.3. Archicortical and Paleocortical Bases of Microgenesis

Microgenetic theory was conceived without the benefit of recent evidence on the patterns of neural connectivity (Brown, 2002). Yet modern findings show an organizing principle of cortical connections that provides an interesting and concrete

TABLE 5.2 Mechanisms In Brain Development, Considered at Different Time Scales

Process	Domain	Time scale	Mechanisms
Phylogenesis	Human evolution	Millennia	Laminar structure; encephalization
Ontogenesis	Child development	Decades	Activity-dependent shaping of functional anatomy
Morphogenesis	Embryonic development	Days	Growth of the neuraxis
Microgenesis	Neuronal activity	Milliseconds	Predictive coding; backward resonance

anatomical framework for microgenetic theory. From their extensive studies of monkey cortex, Pandya and Yeterian (1985) proposed that each stage of cortical differentiation during brain evolution is marked by an increase in laminar complexity and an increase in the number of cells in layers IV and III. If this proposal is correct, it suggests that patterns of regional connectivity directly mirror the phylogenetic sequence. A rough analogy can be made in the study of sedimentary layers, or strata, in geology as clues to events of the past. The difference, in the microgenetic framework, is that brain "strata" remain active, and are interconnected in meaningful ways that have enduring consequences for brain and cognitive function. Although Pandya's studies of anatomical connectivity are uncontroversial, his interpretations in relation to Sanides's theory of cortical differentiation have proved difficult to accept for many neuroscientists. Nonetheless, this evolutionary framework provides an elegant way to interpret cognitive processing within the brain's organizational scheme (Deacon, 1989). Specifically, Brown's theory of microgenesis could be related directly to Pandya's findings on the hierarchical organization of brain structure and function (Derryberry & Tucker, 1990; Tucker & Luu, 1998; Tucker, 2007).

Within the corticolimbic connectional pattern of each hemisphere, the limbic networks at the core of the hemisphere are responsible for representing visceral functions, with strong modulation by the hypothalamus (Swanson, 2003). In contrast, the neocortical networks on the lateral surface of the hemisphere are responsible for the sensorimotor interface with the environment. This "core–shell" architecture of each cerebral hemisphere thus arbitrates between motive representations at the core and the requirement for veridical environmental representations at the shell, creating a tension that may motivate the need for re-entrant processing across *limbifugal* (shell-to-core) and *limbipetal* (core-to-shell) directions of the corticolimbic pathways (Tucker, 2001, 2002, 2007).

5.3. FROM ANATOMICAL STRUCTURE TO NEUROLINGUISTIC FUNCTION

The more radical view of microgenesis posits that each mental process unfolds in the context of the brain's evolved structures, traversing levels of the neuraxis in the same direction as those structures were laid down in evolution (Brown, 2002). In particular, Brown has suggested that linguistic processes engage a sequence of brain regions from "depth to surface," beginning in bilateral, limbic regions and terminating in left-lateral cortex (Brown, 1977; Figure 5.3). According to this idea, the progression from medial (limbic) to left-lateral brain activation is reflected in the sequencing of linguistic functions, beginning with activation of a broad semantic category, or concept (ANIMAL), and ending with selection of the phonemic form that corresponds with a specific visual or phonological word form (lion). If this view is correct, language should engage brain areas in an order that is consistent with the evolution of cortical structure. In this section, we critically examine one implication of this idea, namely, that the more general levels of semantic processing should recruit medial, limbic networks.

5.3.1. Anatomy and Functions of the Medial Temporal Lobe

Recent reviews based on neuroimaging of healthy adults have described a widespread network of areas that are activated in semantic tasks, including middle and inferior temporal cortex, cingulate cortex, and inferior prefrontal cortex (PFC) (Demonet *et al.*, 2005). By and large, these studies have shown that meaning comprehension does recruit temporal cortex – broadly defined – consistent with classical neurolinguistic models. However, the precise role of medial temporal structures has been controversial.

Historically, this controversy can be traced to studies of patients with damage to the medial temporal lobe (MTL), including the hippocampus and surrounding structures (Hoenig & Scheef, 2005). Deficits in semantic processing were first reported in studies of patient H.M., who underwent surgical resection of the hippocampus for treatment of epilepsy. However, high-resolution imaging subsequently revealed that H.M.'s lesion was not confined to the hippocampus, but encroached on neighboring regions as well. Later research suggested that semantic deficits may not rely on the hippocampus *per se*, but may involve damage to adjacent medial temporal structures, lateral temporal cortex, or

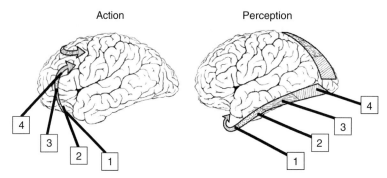

Action   Perception

FIGURE 5.3 Left: Primary direction of corticolimbic traffic for organizing output, starting from limbic integration and progressing toward specific action modules in motor cortex. Right: Primary direction of corticolimbic traffic for integrating perception, with processing typically assumed to begin in specific modules in sensory cortex (in this case the shaded arrows start from the visual area) with increasing elaboration in networks toward limbic cortex. For both input and output, the arrows show separate network pathways for traffic for both the dorsal (top) and ventral (bottom) surface of the hemisphere. Although it is easiest to describe in one direction, the processing is bidirectional. Processing that proceeds toward limbic networks can be described as limbipetal, whereas that proceeding from limbic networks can be described as limbifugal. *Source*: Modified from Tucker (2007).

both (Hoenig & Scheef, 2005). In addition, investigations of hippocampal structure and function led to a consensus that the hippocampus may be essential for encoding of new knowledge, but not for the retrieval of previously encoded semantic representations (Aggleton & Brown, 1999).

Although the hippocampus itself may not be essential for semantic memory in humans, it is important to keep in mind that the MTL in primates comprises multiple structures in addition to the hippocampus (see Box 5.2 for details). Further, in primates the limbic cortex – including ventromedial aspects of the MTL – appears to have taken on many of the memory consolidation functions that are carried out by the amygdala and hippocampus in rodents (Ungerleider & Mishkin, 1982). It would therefore make sense to examine the contributions of other regions of the human MTL in semantic processing tasks, such as the parahippocampal and perirhinal regions. This undertaking can be informed by an understanding of how MTL structures are interconnected with the rest of the brain (see Box 5.2). Recent studies have revealed a dorsal–ventral division of labor within the primate MTL, reflecting tissues of the archicortex and paleocortex that form the subcortical bases for dorsal and ventral pathways of the isocortex. While the hippocampus and parahippocampal regions belong to the dorsal (where or how) regions of archicortex, the perirhinal cortex appears to belong to the ventral (what) processing stream of paleocortical origins (Suzuki & Amaral, 2003). Indeed, perirhinal cortex can be viewed as the end point in the ventral object processing stream (Suzuki & Amaral, 2003). Given these observations, and Brown's prediction that meaning should engage limbic regions, we can predict that perirhinal cortex may be important in semantic representation. The next two sections examine this idea in the context of two forms of evidence: that from patients with discrete MTL lesions and that from neuroimaging studies of MTL function in healthy individuals.

5.3.1.1. Evidence for MTL Involvement In Semantic Processing

In a seminal study, Davies *et al.* (2004) examined medial temporal structures of 10 patients with semantic dementia and 10 patients with Alzheimer's disease who suffered loss of episodic memory with general sparing of semantic processes. Their results showed greater atrophy of the perirhinal cortex in the semantic dementia patients, compared with the Alzheimer's patients and age-matched controls. In addition, patient performance on a range of semantic tests was correlated with perirhinal volume. Davies *et al.* interpreted these data as strong support for the role of medial temporal cortex – and, more specifically, ventral anterior MTL – in semantic processing (but see Levy *et al.*, 2004 for a different interpretation of these data).

Converging evidence for perirhinal involvement in semantic processing has come from neuroimaging studies involving healthy adults. Bartha *et al.* (2003) asked 70 participants to complete a semantic decision task while their inferior and (MTLs) were scanned using functional magnetic resonance imaging (fMRI) methods. More than 90% of the subjects showed MTL activation during the semantic (as compared with the baseline) task. Bartha *et al.* did not report a detailed analysis of rostral (perirhinal) versus caudal (hippocampal) regions. However, it is interesting to note that they observed bilateral, symmetric activation of MTL regions during this task, consistent with Brown's prediction that access to meaning engages bilateral limbic areas.

5.3.1.2. Linguistic Functions of Dorsal and Ventral MTL

It can be argued that lexical (word-level) semantic processing may be viewed as an instance of "item-based" (object) processing and may therefore preferentially recruit the ventral MTL. By contrast, sentence-level semantic processing, which requires contextual semantic integration, may recruit additional areas of the dorsal MTL. This view would be consistent with a current theory of functional differences in ventral and dorsal MTL. To test this idea, Hoenig and Scheef (2005) examined temporal lobe activation in word-level (item-based or lexical) versus sentence-level (relational) semantic processing in a single task. Subjects viewed sentences that ended in either a semantically ambiguous word (e.g., "spade") or an unambiguous near-synonym (e.g., "shovel"):

A. He dug with the spade.
B. He dug with the shovel.

Following each sentence subjects were presented with a probe word (e.g., "ace" or "garden") and asked to determine if the probe fit the prior sentence context (semantic congruity judgment). Condition A allowed for isolation of lexico-semantic processing, through comparison of (1) the brain response to a probe word that was related to the contextually appropriate meaning of the word (e.g., "garden") with (2) the brain response to a probe word that was related to the inappropriate meaning (e.g., "ace"). Consistent with Bartha et al. (2003), the main effect of semantic verification (i.e., sentence-level processing) engaged widespread areas, including bilateral hippocampus. By contrast, lexical meaning activated perirhinal cortex, but not the hippocampus. These findings support the idea that word-level semantic processing may rely on anteroventral regions of the MTL, at the end of the object processing stream.

In a convergent line of work, Tyler, Moss, and associates have used positron emission tomography (PET) and fMRI methods to examine task-dependent recruitment of temporal cortical regions during semantic categorization (Tyler et al., 2004). Their pattern of results shows a frontal–posterior subdivision of MTL activation that depends on the level or depth of semantic processing (coarse versus fine-grained judgments), reminiscent of the orthogenetic (global-to-local) principle in microgenesis.

The work reviewed here raises interesting questions for the neural bases of language, and more particularly, for microgenetic theory. First, anterior ventral MTL appears to support meaning retrieval, consistent with Brown's prediction that paralimbic cortex should be engaged in semantic tasks. Second, perirhinal cortex may specifically support fine-grained semantic discriminations, such as required to denote specific fruits or bird species. This finding does not seem to fit with Brown's prediction that more medial regions should be engaged in activation of holistic or "diffuse" representations. However, it may be that a specific fruit, such as an apple, is actually a simpler, more concrete concept than a superordinate or abstract concept, such as the category "fruit." In fact, substantial evidence suggests this is the case

(see Edelman, 1998 for a review). The perirhinal, limbic cortical representation of basic categories may thus prove consistent with a microgenetic view, in which limbic cortex organizes a more primitive, general meaning, and arbitration with neocortex is required for the more abstract representation of superordinate categories (Tucker, 2007).

5.4. TIME DYNAMICS OF FRONTO–TEMPORAL ACTIVATIONS

As mentioned previously, one challenge for microgenetic theory is to reconcile the observed order of events in perception with the orthogenetic principle, which posits that perception, as well as action, begins in medial, limbic areas. Because our sensory organs are constantly receiving input during the awake state, perception tends to operate on sensory input, and the flow of information in object recognition is thus from primary sensory areas toward association cortex. However, the existence of dreams, mental imagery, and hallucinations suggests that perception does not require sensory input (e.g., Deacon, 1989; Rosenthal, 2003). In these cases, perception appears to be initiated by paralimbic structures and experienced in association and primary cortex.

In this section, we consider evidence for bidirectional corticolimbic processing, mediated by re-entrant pathways, in language comprehension. We focus on the time dynamics of frontal and temporal activity in word and sentence comprehension, and consider evidence for interactions across levels.

5.4.1. Interactionist versus Serial Views of Perception

An important question is precisely when prefrontal regions are engaged, and when and how prefrontal activity influences processing in posterior, perceptual regions. Microgenetic theory would lead one to expect top–down modulation of activity in perceptual areas at very early stages of processing, reflecting a high degree of interactivity. More generally, interactionist theories have posited "predictive coding," in which representations of expectancies are important to shaping the perceptual process (Friston, 2005). According to this view, we should expect to see language processing unfold in a bidirectional manner across levels, with feedback being an integral part of the dynamics of language representation. This interactionist proposal can be contrasted with theories that emphasize serial processing and the "impenetrability" of early sense-perception. Such theories posit a strict hierarchical sequence, with engagement of frontal regions occurring post-perceptually, that is, after object representations have been completely assembled in primary processing regions (e.g., Pylyshyn, 1999). On this latter view, top–down pathways are thought to influence only post-perceptual processing suggesting that feedback from frontal to temporal areas should only be seen at later stages of processing.

5.4.2. Frontal and Temporal Activity in Semantic Processing

In studies of semantic comprehension, event-related potential (ERP) and magnetoencephalography (MEG) studies have revealed a variety of spatiotemporal patterns or "components" that are sensitive to differences in word, sentence, and discourse level meaning (Frishkoff *et al.*, 2004). A well-known pattern is the N400, a scalp-negative deflection in the ERP that is inversely related to ease of semantic processing. Intracranial electroencephalography (EEG) recorded during semantic processing implicate the anterior ventral MTL in producing the N400 (reviewed in Marinkovic *et al.*, 2003). Recent advances in scalp-recorded EEG and MEG methods provide measures with high spatial and temporal resolution that can be used to image neural activity during language processing (e.g., Marinkovic *et al.*, 2003; Frishkoff *et al.*, 2004). These studies have shown that cortical activity in word comprehension begins in primary sensory areas and then rapidly engages a sequence of areas along the ventral, object processing stream, terminating in anterior temporal and left ventral prefrontal regions by around 400–500 ms (Marinkovic *et al.*, 2003; Frishkoff, 2007; Frishkoff *et al.*, 2004).

Electromagnetic studies are beginning to reveal how fronto-temporal dynamics in language may reflect re-entrant processing in posterior association areas after engagement of frontal brain regions. Marinkovic *et al.* (2003) recorded MEG responses during a semantic categorization task in which some words were repeated. Comparison of responses to novel versus repeated words revealed activity in regions associated with lexical (word form) and semantic priming. Using MRI-constrained cortical mapping of activity across time,

they observed priming effects over left inferior prefrontal and left ventral temporal regions at around 350–500 ms for words presented in visual and auditory modalities.

Frishkoff *et al.* (2004) examined the time dynamics of neural activity in both frontal and posterior (temporal and parietal) cortical regions during a sentence comprehension task. The initial detection of semantic incongruity (~250 ms) engaged the left PFC and left anterior cingulate. In the critical N400 interval (300–500 ms), sources in left and right lateral PFC, right temporal cortex, and both anterior and posterior cingulate areas were responsive to the semantic manipulation. Source waveforms (Figure 5.4) showed that left hemisphere activity preceded right hemisphere activity, and semantic effects in frontal regions began earlier and were more sustained than the more transient effects within posterior cortical regions. These patterns may be consistent with neuropsychological models that have emphasized the sustained activity necessary to support top–down feedback provided by frontal areas to modulate activity in posterior cortical networks (Friston, 2005).

Interactions between frontal and posterior regions in language meaning comprehension may be important at the paralimbic as well as the neocortical level. The cingulate cortex, for example, has both anterior and posterior regions that may support differing cognitive and linguistic functions. Intracranial recordings of cingulate activity in rabbits, and ERP and fMRI data from humans, have provided evidence for fast versus slow time dynamics during learning within anterior versus posterior cingulate cortices, respectively (Tucker & Luu, 2006). According to these data, while the posterior cingulate cortex (PCC) may support incremental (slowly adapting, phasic) changes in memory representation, that is, gradual

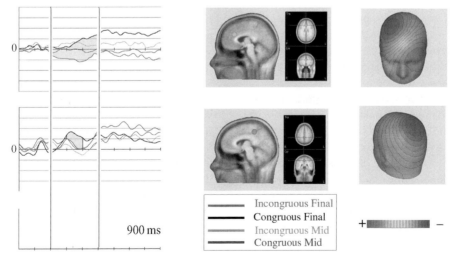

FIGURE 5.4 Top row, time course of electrical activity in ACC. Bottom row, time course of electrical activity in PCC. Left panel, regional source waveforms for ACC (top) and PCC (bottom). Gray areas mark intervals from 200 to 500 ms when semantic effect is statistically significant. Middle panel, source location. Right panel, scalp projection of ACC (top) and PCC (bottom) sources. *Source*: Data from Frishkoff *et al.* (2004) (see Plate 4).

shifts in context, the anterior cingulate cortex (ACC) may be responsive to sudden changes in context that represent larger, or more salient, violations of expectancy. In this context, it is interesting to note that the sentence comprehension task in Frishkoff *et al.* (2004) also evoked different patterns of activity in anterior and PCC (Figure 5.4), consistent with the two mechanisms ascribed to these regions in studies of learning and memory. The ACC responses to semantic violations began earlier (~200–250 ms), and the activity was greater and more sustained (lasting well beyond 500 ms) than that in PCC, which differentiated the task conditions about 100 ms later (~300–350 ms) and was more phasic.

5.5. VERTICAL INTEGRATION: COORDINATION OF INTERNAL AND EXTERNAL CONTEXTS FOR MEANING

Within the core–shell model of hemispheric organization, the process of microgenesis would recruit motivational processes that are subserved by visceral networks of the hemispheric "core," in order to consolidate memory representations across the "core–shell" network hierarchy (Tucker, 2007). In this final section, we describe recent evidence that suggests an integral role for motivational and emotional constraints on semantic processing.

5.5.1. Hemispheric Contributions to Mood and Meaning

For many years, evidence of the right hemisphere's importance to emotional experience and behavior has led to speculations that the right hemisphere's cognition and language processing are strongly modulated by influence from limbic networks (Tucker, 2002). Differential hemispheric contributions in semantics can be understood as a specific instance of the global–local hypothesis of hemispheric asymmetries in visual object processing (Luu & Tucker, 1998). According to this view, the left hemisphere facilitates processing of close (or focal) semantic relationships, whereas the right hemisphere may support a broader range of meaning associations (Frishkoff, 2007 for recent ERP evidence).

Frishkoff (2004) examined how different dimensions of emotional arousal – that is, positive and negative affect – influence left and right hemisphere contributions to the scope of semantic priming. As predicted, they observed that negative arousal selectively enhanced the magnitude of semantic priming seen in the N400 effect for strong associates, particularly over the left hemisphere. These results suggest that emotional states may be important in modulating left and right hemisphere roles in semantic processing. Interestingly, negative affect (but not positive affect) was associated with effects over medial frontal scalp

sites, consistent with prior research by Luu and Tucker (1998), which showed modulation of anterior cingulate cortex activity by negative mood states (e.g., depression and anxiety). By contrast, positive arousal was associated with increased ERP evidence of priming over the right hemisphere, and this was observed for both strong and weak associates (Frishkoff, 2004).

5.5.2. Medial Frontal Networks in Semantic Memory

Given the importance of the cingulate cortex as an interface between visceral (motivational–emotional) and somatic (cognitive) functions, it is not surprising to find the anterior cingulate cortex implicated in the semantics of language as well. In a number of experiments in our laboratory, motivated decisions and actions have been associated with EEG activity in midline frontal cortex that is suggestive of regulation of these decisions by the anterior cingulate cortex (reviewed by Tucker & Luu, 2006). The importance of frontolimbic interaction was also seen in language processing in a study in which subjects evaluated trait words as descriptive of themselves or not (see Box 5.3). The progression of activity, with early signs of engagement of limbic regions and later signs of engaging frontopolar cortex, may be suggestive of a microgenetic process, in which language is first understood at the limbic–visceral (gut) level, and then progressively articulated through re-entrant arbitration with somatic representations of environmental referents.

5.6. CHALLENGES AND FUTURE DIRECTIONS: LEVELS OF BRAIN, LEVELS OF LANGUAGE?

Vertical integration leads one to expect that even complex linguistic acts – such as comprehending the meaning of a sentence – can be related to more fundamental cognitive, motivational, and emotional processes that are subserved at least in part by subcortical brain structures. As reviewed here, research on motivational and emotional influences on language processing provides evidence that is consistent with this hypothesis: limbic and right hemisphere networks are often engaged in the interpretation of meaning. But does this evidence pertain to language generally or just to emotional responses specifically? The microgenetic approach implies that meaning itself is organized around a visceral base, such that language cannot be abstracted from this base. At the same time, if connections imply function, then the language representations of the intermediate "association" cortices (e.g., Broca's and Wernicke's areas) are not only motivated by visceral concepts but operate in concert with sensorimotor cortices through such mechanisms as time-locked retroactivations of the sensorimotor regions.

<table>
<tr><td valign="top">

Box 5.3   Embodied meaning in medial frontal cortex

The meaning of language can be understood at several levels, the most important of which may be the emotional response that both follows and guides understanding. Within a connectionist framework, the dense connectivity among limbic networks, as well as the necessary role of limbic networks in consolidating memory, implies that the integration of meaning requires participation of corebrain networks that are modulated by emotional, adaptive constraints (Tucker, 2002).

In an EEG study of self-evaluation, subjects viewed trait words and responded quickly whether the word described them or not (Tucker *et al.*, 2003). Although the average response time was around 700 ms, there were earlier signs of activity in frontolimbic networks that suggested stages in the understanding of the meaning of the word and in the organization of the response. By about 300 ms, the first significant contrasts were for good (socially desirable) versus bad trait words, regardless of whether subjects endorsed the word as self-descriptive. This contrast was localized to a source in the anterior cingulate cortex. By about 450 ms, there was an endorsement effect, localized to the anterior frontal midline. The patterns of activity observed in this task are consistent with a microgenetic process, in which a consistent sequence of limbic and frontal networks is engaged for subjects in general. Although this process is almost certain to be re-entrant rather than linear, with multiple limbic and neocortical networks converging on the decision, it imples a form of vertical integration, in recruiting both affective and representational constraints on the semantic process.

Tucker, D.M. (2002). Embodied meaning: An evolutionary-developmental analysis of adaptive semantics. In T. Givon & B. Malle (Eds.), *The evolution of language out of pre-language* (53rd edn, pp. 51–82). Amsterdam: J. Benjamins.

Tucker, D.M., Luu, P., Desmond, R.E., Jr., Hartry-Speiser, A., Davey, C., & Flaisch, T. (2003). Corticolimbic mechanisms in emotional decisions. *Emotion*, *3*(2), 127–149.

</td></tr>
</table>

A future challenge for the neuroscience of language will be to test these ideas more explicitly through novel experimental paradigms and with the use of new technologies that enable fine-grained measurement of the timing, as well as the spatial extent, of neural activity during language comprehension. From recent evidence we know that time course is critical to interpreting system-level dynamics at each stage in neurolinguistic processing. Indeed, on a microgenetic view, the anatomy (or structures) of language is intrinsically linked to the time–dynamics (or functions) of different brain regions, reflecting a unified architecture for brain functional mapping.

References

Aggleton, J.R., & Brown, M.W. (1999). Episodic memory, amnesia, and the hippocampal-anterior thalamic axis. *Behavioral and Brain Sciences*, *22*(3), 425–489.

Bartha, L., Brenneis, C., Schocke, M., Trinka, E., Koylu, B., Trieb, T., Kremser, C., Jaschke, W., Bauer, G., Poewe, W., & Benke, T. (2003). Medial temporal lobe activation during semantic language processing: fMRI findings in healthy left- and right-handers. *Brain Research, Cognitive Brain Research*, *17*(2), 339–346.

Brown, J. (1977). *Mind, brain, and consciousness: The neuropsychology of cognition*. New York: Academic Press.

Brown, J.W. (1988). *The life of the mind: Selected papers*. Hillsdale, MI: Lawrence Erlbaum.

Brown, J. (2002). *The self-embodying mind: Process, brain dynamics, and the conscious present*. Barrytown, NY: Station Hill Press.

Clark, A. (2006). Language, embodiment, and the cognitive niche. *Trends in Cognitive Science*, *10*(8), 370–374.

Davies, R.R., Graham, K.S., Xuereb, J.H., Williams, G.B., & Hodges, J.R. (2004). The human perirhinal cortex and semantic memory. *European Journal of Neuroscience*, *20*(9), 2441–2446.

Deacon, T. (1989). Language evolution and neuromechanism. In W. Bechtel & G. Graham (Eds.), *A companion to cognitive science* (Vol. 8, pp. 288–290; discussion 290–281). Malden, MA: Blackwell.

Demonet, J.F., Thierry, G., & Cardebat, D. (2005). Renewal of the neurophysiology of language: Functional neuroimaging. *Physiological Review*, *85*(1), 49–95.

Derryberry, D., & Tucker, D.M. (1990). The adaptive base of the neural hierarchy: Elementary motivational controls of network function. In A. Dienstbier (Ed.), *Nebraska symposium on motivation* (pp. 289–342). Lincoln, NE: University of Nebraska Press.

Edelman, S. (1998). Representation is representation of similarities. *Behavioral and Brain Science*, *21*(4), 449–467. discussion 467–498.

Frishkoff, G. A. (2004). Brain electrical correlates of emotion and attention in lexical semantic priming. PhD Thesis, University of Oregon, Eugene, Oregon.

Frishkoff, G.A. (2007). Hemispheric differences in strong versus weak semantic priming: Evidence from event-related brain potentials. *Brain and Language*, *100*(1), 23–43.

Frishkoff, G.A., Tucker, D.M., Davey, C., & Scherg, M. (2004). Frontal and posterior sources of event-related potentials in semantic comprehension. *Brain Research, Cognitive Brain Research*, *20*(3), 329–354.

Friston, K. (2005). A theory of cortical responses. *Philosophical Transactions of the Royal Society of London B*, *360*, 815–836.

Hoenig, K., & Scheef, L. (2005). Mediotemporal contributions to semantic processing: fMRI evidence from ambiguity processing during semantic context verification. *Hippocampus*, *15*(5), 597–609.

Levy, D.A., Bayley, P.J., & Squire, L.R. (2004). The anatomy of semantic knowledge: Medial vs Lateral temporal lobe. *Proceedings of the National Academy of Sciences, USA*, *101*(17), 6710–6715.

Luu, P., & Tucker, D.M. (1998). Vertical integration of neurolinguistic mechanisms. In B. Stemmer & H.A. Whitaker (Eds.), *Handbook of neurolinguistics* (pp. 55–203). San Diego, CA: Academic Press.

Marinkovic, K., Dhond, R.P., Dale, A.M., Glessner, M., Carr, V., & Halgren, E. (2003). Spatiotemporal dynamics of modality-specific and supramodal word processing. *Neuron*, *38*(3), 487–497.

Mesulam, M.M. (Ed.). (2000). *Principles of behavioral and cognitive neurology* (2nd edn). Oxford: Oxford University Press.

Pandya, D.N., & Yeterian, E.H. (1985). Architecture and connections of cortical association areas. In A. Peters & E.G. Jones (Eds.), *Cerebral cortex. Volume 4. Association and auditory cortices* (pp. 3–61). New York: Plenum Press.

Pylyshyn, Z. (1999). Is vision continuous with cognition? The case for cognitive impenetrability of visual perception. *Behavioral and Brain Science*, *22*(3), 341–365.

Rosenthal, V. (2003). Microgenesis, immediate experience, and visual processes in reading. In A. Carsetti (Eds.), *Seeing, thinking, and knowing* (pp. 221–244). Norwell, MA: Kluwer Academic Publishers.

Suzuki, W.A., & Amaral, D.G. (2003). Where are the perirhinal and parahippocampal cortices? A historical overview of the nomenclature and boundaries applied to the primate medial temporal lobe. *Neuroscience, 120*(4), 893–906.

Swanson, L.A. (2003). *Brain architecture: Understanding the basic plan.* New York: University Press.

Tucker, D.M. (2001). Motivated anatomy: A core-and-shell model of corticolimbic architecture. In G. Gainotti, (Ed.), *Handbook of neuropsychology* (2nd edn, Vol. 5: Emotional behavior and its disorders, pp. 125–160). Amsterdam: Elsevier.

Tucker, D.M. (2002). Embodied meaning: An evolutionary-developmental analysis of adaptive semantics. In T. Givon & B. Malle (Eds.), *The evolution of language out of pre-language* (53rd edn, pp. 51–82). Amsterdam: J. Benjamins.

Tucker, D.M. (2007). *Mind from body: Experience from neural structure.* New York: Oxford University Press.

Tucker, D.M., & Luu, P. (1998). Cathexis revisited. Corticolimbic resonance and the adaptive control of memory. *Annals of the New York Academy of Sciences, 843*, 134–152.

Tucker, D.M., & Luu, P. (2006). Adaptive binding. In H. Zimmer, A. Mecklinger & U. Lindenberger (Eds.), *Binding in human memory* (pp. 84–114). New York: Oxford University Press.

Tyler, L.K., Stamatakis, E.A., Bright, P., Acres, K., Abdallah, S., Rodd, J.M., & Moss, H.E. (2004). Processing objects at different levels of specificity. *Journal of Cognitive Neuroscience, 16*(3), 351–362.

Ungerleider, L.G., & Mishkin, M. (1982). Two cortical visual systems. In D.J. Ingle, R.J.W. Mansfield & M.A. Goodale (Eds.), *The analysis of visual behavior* (pp. 549–586). Cambridge, MA: MIT Press.

Further Readings

Readings related to principles of brain structure and function

Deacon, T. (1989). Holism and associationism in neuropsychology: An anatomical synthesis. In E. Perecman (Ed.), *Integrating theory and practice in clinical neuropsychology* (pp. 1–47). Hillsdale, NJ: Lawrence Erlbaum.
Provides an excellent introduction to microgenetic theory, and offers a clear and unbiased explanation of competing hypotheses about human cognition and neuropsychology from the dual perspectives of "holism" (a more extreme version of microgenetic theory) and "associationism" (the alternative view, which is more commonly assumed in cognitive neuroscience).

Tucker, D.M. (2001). Motivated anatomy: A core-and-shell model of corticolimbic architecture. In G. Gainotti (Ed.), *Handbook of neuropsychology* (2nd edn, Vol. 5: Emotional behavior and its disorders, pp. 125–160). Amsterdam: Elsevier.
Presents a theory of human neuroanatomy that is inspired by microgenetic theory grounded in modern neuroanatomy (particularly evidence from the "Boston" school, including work by Sanides, Pandya, and Barbas). Cortical-subcortical pathways shape cognitive, emotional, and motivational processes.

Readings related to anatomical structure and neurolinguistic function

Hoenig, K., & Scheef, L. (2005). Mediotemporal contributions to semantic processing: fMRI evidence from ambiguity processing during semantic context verification. *Hippocampus, 15*(5), 597–609.
Gives a succinct overview of current views on the involvement of MTL and surrounding cortex (e.g., fusiform gyrus) in semantic processing, and reviews current controversies. Presents an fMRI experiment that was designed to selectively activate temporal regions related to word-level (item-based or lexical) versus sentence-level (relational) semantic processing. Results may support the idea that word-level semantic processing may rely on ventral, anterior regions of the MTL, at the end of the object processing stream.

Tyler, L.K., Stamatakis, E.A., Bright, P., Acres, K., Abdallah, S., Rodd, J.M., & Moss, H.E. (2004). Processing objects at different levels of specificity. *Journal of Cognitive Neuroscience, 16*(3), 351–362.
Provides evidence to support the view that complexity of object representations along the ventral stream increases from caudal to rostral areas, and that anteromedial temporal cortex (including the perirhinal cortex) functions as the endpoint of this pathway.

Readings related to time dynamics of fronto-temporal activations

Friston, K. (2002). Functional integration and inference in the brain. *Progress in Neurobiology, 68*(2), 113–143.
Outlines the idea of perceptual "inference" – that expectancies shape low-level visual perception – and relates this concept to feedback mechanisms in neural processing. There is a good section on functional "integration" in neural processing, and a review of processing pathways within the anatomical framework of Mesulam (1998). Friston presents an intriguing, if controversial, theory of "predictive coding," the neural analoge of perceptual "inference." *Cf.* Pylyshyn (1999) for a contrasting view.

Readings related to vertical integration: coordination of internal and external contexts for meaning.

Gibbs, R.W., Jr. (2003). Embodied experience and linguistic meaning. *Brain and Language, 84*(1), 1–15.
Outlines principles underlying theories of embodied cognition as they apply to language processing, as contrasted with semantic theories that ground meaning in "disembodied" symbols and abstract, propositional structures. The conclusion offers suggestions for how both critics and supporters of embodied cognition might profit from new experimental paradigms that seek to test the extent to which bodily experience influences language and cognitive processing.

Tucker, D.M. (2002). Embodied meaning: An evolutionary-developmental analysis of adaptive semantics. In T. Givon & B. Malle (Eds.), *The evolution of language out of pre-language* (pp. 51–82). Amsterdam: J. Benjamins.
Describes principles underlying human development (ontogeny) and brain evolution (phylogeny), and relates these principles to a theory of neurolinguistic processing that emphasizes vertical integration of language at each level of the brain's neuraxis.

6

A Brief Introduction to Common Neuroimaging Techniques

FRANK A. RODDEN[1] and BRIGITTE STEMMER[2]

[1]Section of Experimental MR of the CNS, Department of Neuroradiology,
Eberhard Karls Universität Tübingen, Tübingen, Germany
[2]Faculty of Arts and Sciences and Faculty of Medicine,
Université de Montréal, Canada

ABSTRACT

This chapter was written for the reader who knows little or nothing about the neuroimaging techniques discussed here but would like to learn enough about them to understand the other chapters in this book. The introduction to this chapter warns the general reader of conceptual hazards associated with all neuroimaging methods. Subsequent sections can be read independently; they describe what sorts of images the technique under discussion produces, a simplified explanation of how that technique works, and its advantages and disadvantages. Finally, the reader is referred to literature in which that method is used in neurolinguistic research and to information sources on details of the technique.

6.1. INTRODUCTION

The interpretation of neuroimaging data associated with linguistic phenomena is harder than it looks. The tendency to reify language into something material is perhaps nowhere more subtly treacherous than when imaging techniques are applied (van Lancker-Sidtis, 2006; Bennett et al., 2007; Sidtis, 2007). The images delivered by the techniques described below are perhaps best considered in close conjunction with their etymological counterpart, "imaginations." In every imaging method to be described, infinitesimally tiny physical forces that occur in the brain (which are presumed to be associated with elements of language) must (1) be tremendously magnified and then (2) radically transformed into something the reader can see. Sometimes,

the rubric for interpreting these images is "the stronger the force, the more intense the image" although this is not always the case. In some methods, it is only the presence of forces that the images represent – with no specific regard being given to the strength of the forces.

When considering the results of each method, the reader should always be aware of the degree of resolution that method allows: both in terms of space and time. In other words, one should pay attention to how "sharp" the image is and "over what period of time" the image was made. For some of the methods, the images are spatially quite "blurred" and correlations between the images and linguistic phenomena can be assigned only to such bulk objects as the cerebral hemispheres or to large regions of the brain (several centimeters); in other methods, the resolution is on the order of millimeters. Likewise, the "exposure time" of the various imaging methods range from milliseconds (electroencephalogram, EEG; magnetoencephalography, MEG) to several seconds (functional magnetic resonance imaging, fMRI, positron emission tomography, PET).

One should also be sensitive as to whether the images are in two or three dimensions. Some images are in only two dimensions corresponding to forces that are derived from the two dimensional surface of the brain; other images represent forces in three dimensions – originating from within the brain itself. All images are susceptible to distortions or artefacts, due to intrinsic properties of the system employed and/or to interactions between the particular system and the various tissues of the brain, cerebrospinal fluid, air filling the sinuses of the skull, the skull and the muscle, skin, fat and hair covering the skull.

Another set of problems associated with imaging data of the brain arises from the simple fact that, no matter what particular phenomenon is under investigation, most of the brain is utterly uninvolved in the object of interest. While, for instance, an experimental subject is concentrating on, say, reading a poem in a foreign language, only a miniscule part of her brain is involved in that task. At the same time, the rates of her breathing and heartbeat are being monitored, her posture is being maintained and the blood concentrations of her hormones are being regulated.

Furthermore, with respect to what she is reading on the written page, her brain is not only occupied with the task of transforming very minute alterations of lightness and darkness into letters and then grouping those letters into words, but her peripheral vision is on the look-out for unexpected changes in the environment; she is simultaneously hearing, feeling, smelling and tasting what is going on around her and within herself. Her emotional status is being sustained; her memory is both recording what is being read – and remembering other poems, perhaps comparing them with the present one. Her brain is associating the poem that she is reading with the rest of her life.

All of these tasks are being carried out in the living brain that is being scrutinized by some imaging method. The assumption is, of course, that, aside from the phenomenon of interest, everything else in her brain is remaining relatively constant and thus forms a background state to, say, reading the poem. This assumption must always be addressed.

And finally, not only is the imaged brain a locus of complex, ongoing changes in local activity with respect to time and space, the reader must also be aware of possible confounding variables conceptually hidden within the events being observed. When interpreting the brain imaging data of, say an experimental subject reading a poem in a foreign language, a whole spectrum of mental states such as surprise, attention, state of arousal and so forth must be considered. Some of these elements are associated with particular patterns of brain activity that may be relevant in interpreting the imaging data on "reading poetry in a foreign language."

The bad news of this section is that even a basic understanding of the technical methods covered below would require the reader to be conversant not only with Maxwellian electromagnetic field theory but also quantum mechanics and atomic chemistry; the good news is that a *working* knowledge that will allow the reader to follow the logic of imaging experiments as they relate to linguistic matters, requires no more than a basic fourth grade knowledge of electricity, magnetism and radioactivity.

Even given an adequate understanding of the imaging methods described below, conceptual pitfalls in interpreting imaging experiments abound, however. Skepticism in science remains a cardinal virtue.

The technical devices described below might best be considered as aids in helping the investigator's *own* imagination as she/he delves deeper into the mysterious correlations between language and the living meat of the brain.

6.2. ELECTROMAGNETIC FUNCTIONAL NEUROIMAGING TECHNIQUES

Historically, the first neuroimaging method used to investigate correlations between language and the living brain was the EEG. The EEG is a graphic representation of electrical brain activity, that is, minute variations of electrical voltage or "electrical potentials" between two points and in relation to some reference point. In the conventional scalp EEG, the recording of these changes in voltage is obtained by placing electrodes (either individually or integrated into a cap) on the scalp with a conductive gel or paste. In neurosurgical operations during which the surface of the brain is exposed, EEG's can also be recorded directly from the leathery outer envelope of the brain (see Chapter 10). Scalp voltages directly reflect neuronal activity. Signals are generated by large clusters of cortical pyramidal neurons. To make these signals "visible," they are amplified and extracted from electrical activity extraneous to brain activity (e.g., of the heart or muscles), or in EEG terminology, the signals are filtered. The filtered EEG signal is then recorded with a pen or, in most modern research instruments, is digitized via an analog-to-digital converter and displayed on a computer screen. A typical adult human EEG signal is about 20–100 μV when measured from the scalp and about 1–2 mV (i.e., around a hundred times stronger) when measured from subdural electrodes (electrodes on the surface of the brain). EEG is used clinically to help diagnose brain disorders such as the presence of epileptic seizure activity. It also provides information about the changes in general states of vigilance such as arousal or consciousness. Changes in such global mental states are associated with changes in the frequency and amplitude distribution of the EEG. For example, alpha brain waves with a relatively high amplitude and 7–12 oscillations per second (i.e., 7–12 Hz) are associated with a relaxed but awake state, while beta waves (13–20 Hz) are of lower amplitude and associated with a more alert or "active" mental state. Very high frequency oscillations (about 30–100 Hz) have been associated with cognitive functions (such as attention and memory). Simply recording a spontaneous, or raw, EEG is relatively unspecific with respect to the more fine-grained changes in mental activity. For example, given only a raw EEG, it would be impossible to determine whether a person was reading or watching a movie or just looking at a landscape. This is where other EEG signals come into play. These signals are very small (from a few microvolts up to around 10 microvolts) and cannot be seen in the raw EEG but need special techniques to be made visible.

6.2.1. The Event-Related Potential Technique

A common technique to make these small signals visible is to present a task (or a so-called event) to a subject while the EEG is being recorded, and then to analyze only the time window relevant to the task – in EEG jargon: one "time locks" the onset of the event to the EEG recording. This allows the identification of brain activity related to the task (or event) and reflected as EEG waveforms, the "event-related potentials (ERPs)" from which this technique derives its name. While the raw EEG reflects all brain activity at a particular point in time, the ERP is that part of the activity associated with the response to a specific event such as deciding whether a sentence is semantically correct or not (see Chapter 9, for language research with ERPs). Compared to the raw EEG, an ERP is of relatively small amplitude and, to make the ERP emerge from the background EEG, one typically repeats a large number of the same or similar events (in language experiments typically 50 or more) and averages the EEG traces that occur in response to these events. In addition, extraneous waves or "noise" not associated with the event are eliminated by specific analysis techniques. ERPs can also arise after repeated optical (visual evoked potentials, VEPs), acoustic (acoustic evoked potentials, AEPs) or tactile stimuli (somatosensory evoked potentials, SSEPs). These potentials occur very early (roughly between 10 and 100 ms after presentation of the stimulus) and are thought to reflect early sensory processes. Long-latency components prior to 200 ms have been associated with late sensory and early perceptual processes while those after 200 ms have been related to higher level cognitive processes such as language or memory. ERP waveforms are broken down into negative- and positive-going fluctuations (components) that are named by their polarity and either their latency (lag time) or ordinal position after stimulus onset. For example, the N400 is a negative-going waveform (N for negative) peaking about 400 ms after stimulus onset. The P3 is the third positive peak (P for positive) that appears after stimulus onset (Box 6.1). (For a discussion of language-related ERPs see Chapter 9.)

Box 6.1   From EEG to ERP: the oddball paradigm

The oddball paradigm is a very commonly used experiment that reliably produced a specific ERP component, the so-called P3. *Panel A*: The subject views the letter X or O presented on a computer screen. Whenever an X or O appears on the screen, a computer program sends this information (via so-called marker codes or triggers) to the program that records the EEG. These can then be "seen" in the EEG recording. The X is presented more often than the O, here a total of 80 times and the O 20 times, in random order and while the EEG is continuously recorded from a midline parietal electrode site. The recorded signals are filtered and amplified, making it possible to see the EEG waveforms. *Panel B*: Based on the marker codes signaling the onset of the stimulus (X or O), the EEG is now segmented into 800 ms time windows (or epochs). This is shown by the rectangles. A great deal of trial-to-trial variability can be seen in the EEG, but experienced viewers can already see a downward (positive) deflection following the infrequent O stimuli. *Panel C*: If one now averages across all waveforms in the O time window and all waveforms in the X time window, the P3 component (on the right side) becomes apparent. The P3 is much stronger for the infrequent O compared to the frequent X stimuli.

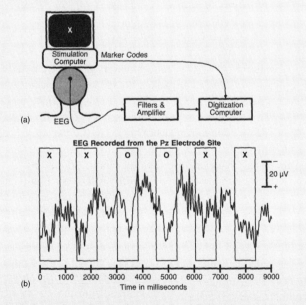

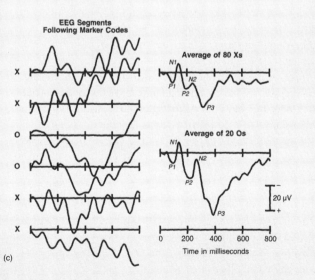

Luck, Steven J., An Introduction to the Event-Related Potential Technique, figure from page 8 "Example of ERP experiment," © 2005 Massachusetts Institute of Technology, by permission of The MIT Press.

How can we infer the sources of the ERPs recorded at the scalp? It has proven impossible to determine from a given pattern of electrical charge observed on the surface of the scalp the many possible locations of underlying neural generators. This is referred to as the *inverse problem*. Some simplifying assumptions and sophisticated modeling techniques are used to determine the location of the activity within the brain. One important assumption is that neural generators can be modeled as electrical dipoles, that is, conductors with one positive and one negative end. Another assumption is that the head is sphere shaped. A model of such a spherical head is created and a dipole placed at some location within the sphere. One then predicts the distribution of voltages that this dipole would create on the surface of the sphere and compares it with the data actually recorded. The model is supported if there is only little difference between the prediction and the actually observed data. If, however, the difference is large, the dipole is shifted to another location and the procedure repeated until the best match is found. Sometimes one needs more than one dipole to obtain a good match but the more dipoles are added the more difficult it gets to identify a unique generator.

6.2.2. Magnetoencephalography

MEG is, in essence, the magnetic equivalent to EEG. While EEG detects the electric potentials generated by neural currents, MEG detects the magnetic fields produced by electrical activity in the brain. Biomagnetic signals are detected by using so-called superconducting quantum interference devices (SQUIDs). Attaching electrodes to the skin is thus not necessary. MEG and EEG provide direct information about the dynamics of evoked and spontaneous neural activity and the location of their sources in the brain. Because the magnetic signals emitted by the brain are extremely weak, shielding from external magnetic signals, including the earth's magnetic field, is necessary and MEG measurements must, therefore, be carried out in an appropriately magnetically shielded room. Magnetic signals arise orthogonally to electric currents flowing along nerve cells close to the surface of the brain. As is the case with EEG, in order for a detectable signal to arise, hundreds of thousands of parallel neurons must be simultaneously transmitting. With EEG, however, the electrical fields "spread out" due to volume conduction as they traverse the tissues between their sources in the brain and the detecting electrodes pasted to the subject's scalp. MEG signals, on the other hand, remain focused. This allows for very accurate estimates of the location (0.1–1 cm) and timing (a few milliseconds) of cerebral activity. For neurolinguistic research this ability to compare millisecond by millisecond changes in the course of brain activation is critical inasmuch as the changes in the brain associated with linguistic phenomena such as syntax and semantics occur on very short time scales (in the milliseconds

range). For exact timing experiments, MEG and EEG are orders of magnitude superior to fMRI or PET. They are, however, inferior to fMRI or PET with respect to inferring the sites of brain activation (spatial resolution).

For the use of MEG in language studies see, for example, Salmelin (2007).

6.2.3. Benefits and Limitations

Both EEG and MEG are direct measures of brain activity; they are completely non-invasive and the measurement environment is silent. Unlike EEG (but similar to fMRI), patients with ferromagnetic implants are excluded as candidates for MEG measurements. Although fMRI or PET (see Section 6.3) yields higher spatial resolution, they are indirect measures of brain activity and are filtered through such intervening variables as nerve-activity-related changes in blood circulation or the uptake of radioactivity. MEG and EEG provide excellent temporal resolution (in the millisecond range). While spatial resolution is rather coarse in EEG due to the distortion of the neuroelectric signals, in MEG neuromagnetic signals traverse the skull and scalp without distortion and spatial resolution for MEG is thus very good. Both techniques reflect bioelectrical activity on the *surface* of the brain (i.e., not primarily *within* the brain). Sources located deep in the brain are thus difficult or impossible to measure with these techniques. MEG can only pick up currents flowing tangential to the scalp (corresponding to sulcal activations), EEG can measure both tangential and radial (corresponding to gyral activations) currents. Finally, compared to EEG (thousands of dollars), MEG is very expensive (millions of dollars) and comparable to an fMRI installation in costs (see Table 6.1 for overall comparison of techniques).

6.3. HEMODYNAMIC FUNCTIONAL NEUROIMAGING TECHNIQUES

For the past several years, popular magazines have featured a plethora of articles purporting to show the brain centers of love, grief, humor and so forth. Such articles usually include images of the brain with areas of activity exhibited as colored areas. Most of these images were derived either from fMRI data or from PET data. Compared with EEG equipment, both fMRI and PET installations are huge and cost millions.

An important property of both fMRI and PET is that the images that are generated are always representations of *differences* between two states of the brain – for example, the brain in the "resting" state compared to the brain while, say, doing arithmetic calculations or humming a song. This means that in the design of every experiment involving fMRI or PET, two different "states" must be defined. The art of the science of deriving valid results from these two methods lies (1) in designing experiments in which two discrete

TABLE 6.1 Characteristics of Imaging Techniques in Language Research

Technique	Method and/or measures	Spatial/temporal resolution	Invasiveness/discomfort to subject	Loudness	Availability	Monitary cost	Unique advantages	Unique disadvantages
EEG, ERP	Bioelectric activity (fluctuating voltage changes detected at scalp)	Centimeters/milliseconds	Non-invasive/very little (more tolerant to movement than, e.g., fMRI)	None	High	Low	Good temporal resolution, ready availability, ease of recording	Low spatial resolution
fMRI	Relative change in deoxyhemoglobin; blood-oxygen-level-dependent (BOLD) effect	Millimeters/1–5 s	Non-invasive/moderate: strict movement restrictions, narrow quarters	Very high	Limited	High	Good spatial resolution	High magnetic field; loud; low temporal resolution
DTI	Application of specific radiofrequency and magnetic field-gradient pulses to track the movement of water molecules in the brain	Millimetres/–	Non-invasive/moderate: strict movement restrictions, narrow quarters	High	Limited	High	Good spatial resolution	Loud; movement restrictions
PET	Intravenously injected radioactive tracer; perfusion; glucose metabolism; oxygen utilization	Millimeters/about 90 s	Invasive (radioactive tracer)/moderate: no head movement, narrow quarters	None	Very limited	Very high	Ability to measure brain metabolism + pharmacodynamics	Poor temporal resolution; exposure to radioactivity; limited to short tasks
MEG/MSI	Biomagnetic activity (magnetic fields measured by superconductive detectors and amplifiers at scalp)	Millimeters at cortex (less precise for deep sources)/milliseconds	Non-invasive/little: strict movement restriction	None	Extremely limited	Very high	Excellent temporal and good cortical spatial resolution	Detects only dipoles oriented tangentially to the skull but not radially
rTMS	Application of rapidly changing magnetic fields to specific areas of the brain leading to excitation or inhibition of neuronal cells	Centimeters/–	Invasive (application of magnetic stimulus)/moderate to high (possible headache)	Low	Limited	Low	Ability to induce "virtual, temporary lesions"	Possible headache; in very rare cases epileptic seizure
NIRS (optical imaging)	Changes in the brain's oxygen absorption based on optical properties of hemoglobin; concentration of deoxyhemoglobin, oxyhemoglobin, total hemoglobin	Millimeters/milliseconds	Non-invasive/very little (more tolerant to movement than fMRI)	None	Limited	Low	Low costs combined with excellent temporal and very good spatial resolution; tolerant to movement	Limited depth penetration (maximum of about 5 cm)
ES (ESM)	Application of electrical current directly to the surface of the brain. Induces a local temporary "lesion" that disrupts normal function	Millimeters/milliseconds	Very invasive: brain penetration	None	Extremely limited	High (surgery costs; technique itself low cost)	"Gold standard" of spatial and temporal resolution	Brain bleeding, infection

DTI: diffusion tensor imaging; EEG: electroencephalography; ERP: event-related potentials; ES: electrocortical stimulation; ESM: electrocortical stimulation mapping; fMRI: functional magnetic resonance imaging; MEG: magnetoencephalography; MSI: magnetic source imaging; NIRS: near-infrared spectroscopy (also referred to as optical imaging); PET: positron emission tomography; rTMS: repetitive transcranial magnetic stimulation.

brain states ("before and after," "with and without" and so forth) are hypothesized to be associated to some (linguistic) phenomenon and then (2) painstakingly excluding as many confounding linguistic, psychological and physiological factors that might be occurring simultaneously with the phenomenon of interest but which are "accidental" and not essential to effects associated with the linguistic phenomenon *per se*.

Two experimental paradigms are common to both fMRI and PET: *block* and *event-related designs*. In the block model, the experimental conditions are separated into distinct blocks and each condition is presented for a specific period of time (usually on a time scale of seconds or fractions of seconds). For example, let us assume we have two experimental conditions involving reading words in Italian. In the first task condition, the participant has to decide whether a word is masculine or feminine (gender condition) and in the second condition whether the word is an animal or an object (animal/object condition). A third condition is used as a control task in which the subject reads pseudowords (pseudowords condition) while alternately pressing buttons. The experimental tasks are organized in blocks with each block containing the stimuli for each experimental condition. For example, for the gender condition, 10 words are presented in block A, for the animal/object condition, 10 words are in block B and for the pseudoword conditions, 10 pseudowords are in block C. In the experiment, the blocks are then presented in alternating (and often randomized) order, for example A–C–B–C–A–C–B–C and so on. Local changes in brain activity are then sought which correlate with changes in these "block states."

In the event-related design, the stimuli for each condition are not organized in blocks but are presented as discrete short-duration events whose timing and order may be randomized. For example, the presentation sequence of the stimuli mentioned above could be: gender word – animal category word – pseudoword – object category word – pseudoword and so on. The basic logic of the event-related design is identical with the block model: local changes of brain activity are sought which correlate with changes in the "event states."

6.3.1. Functional Magnetic Resonance Imaging

The goal of all functional imaging methods is to display areas of altered brain activity as the result of some stimulus or condition; fMRI does this by detecting minute changes in blood oxygenation and flow that occur in response to very localized changes in neural activity. The assumption is that when a certain brain area is momentarily particularly active (e.g., during a mathematical calculation), it consumes more oxygen and, to meet that increased demand, blood flow increases to that active area. fMRI is based on increases in the blood-oxygen level in discrete parts of the brain. The *effect* that fMRI exploits is called the "Blood-Oxygen-Level-Dependent Effect" or, short, the "BOLD effect."

When we engage in a mental task (such as calculating), compared to the resting state, the activity of neurons increases and thus their metabolism increases. The body "overcompensates" for the increased metabolic activity with an increase in regional cerebral blood flow (rCBF). More oxygen is thus supplied to an area of the brain than is consumed. The excess oxygenated blood in the active brain regions flushes the deoxygenated hemoglobin from the capillaries and venules. There is thus an excess of oxyhemoglobin and a relative decrease of deoxyhemoglobin. Deoxyhemoglobin has different magnetic properties compared to oxyhemoglobin. In the presence of a magnetic field (such as the one produced in the magnetic resonance tomography (MRT)) deoxyhemoglobin molecules act like tiny magnets and distort the magnetic field (dephasing). Because there is a relative decrease in deoxygenated blood, proportionally, we have less disturbance of the magnetic field than before activation leading to a stronger signal. It takes time for the fresh blood (and thus oxygen) to reach those areas in the brain with increased demand, and, as a consequence, it takes about 2–3 s for the BOLD response to rise above baseline and 4–6 s to peak.

The cylindrical tube of a magnetic resonance imaging (MRI) scanner houses a very powerful electromagnet. It is within this tube that the subject must lie motionless while the data that produce images from the BOLD effect are being collected. The noise in the tube is so extremely loud (about 93–98 dB in a 1.5 T scanner and more in higher Tesla scanners) that subjects need to wear protective earphones during the experiments. This, of course, can be a distinct disadvantage in a study of language (when using auditory stimuli) as can the requirement of absolute head immobility (speaking is a problem, as it produces movement and susceptibility artifacts).

(For a summary of language studies using the fMRI technique see Chapter 8.)

6.3.2. Diffusion Magnetic Resonance Imaging: Diffusion Tensor Imaging and Diffusion-Weighted Imaging

Diffusion tensor imaging (DTI) is a relatively new *structural* (and not *functional*) MRI technique that provides information on the structural integrity of white matter; it maps white matter fiber orientation and tracks white matter pathways. It provides complementary information to images of cortical function obtained by fMRI. DTI is based on diffusion MRI that measures the diffusion of water molecules in biological tissues. In such an (anisotropic) environment the motions of water molecules are directionally restricted and with DTI it is possible to measure diffusion in multiple directions. This allows researchers to make brain maps of fiber directions and thus investigate how different regions of the brain are connected or, in case of disease, how these connections are disturbed. Caution is, however, warranted.

Conventional DTI-fiber tracking fails to identify the existence or extent of many white matter connections and will also identify connections that do not exist in reality. (For more details and alternate methods see Behrens *et al.*, 2007.)

Another and related modification of regular MRI is diffusion-weighted imaging (DWI) that produces magnetic resonance images of biological tissues *weighted* with the local characteristics of water diffusion. Diffusion-weighted images are very useful to diagnose, for example, vascular strokes in the brain, to study the diseases of the white matter or to infer which part of the cortex is connected to another.

For details on these methods and their applications see Hillis (2005); Behrens *et al.* (2007); Le Bihan (2003), and the special issue on "White matter in Cognitive Neuroscience," edited by Leighton and Ulmer (2005), see also Chapter 40, this volume.

6.3.3. Positron Emission Tomography

PET can be used in several measuring modalities. In one modality, PET can measure the regional utilization of glucose (or other biochemicals). In a second modality, PET can measure pharmacological parameters (uptake in brain tissue, accumulation, release) of molecules of interest (pharmaceuticals, derivatives of pharmaceuticals, neurotransmitters and their analogs, and so forth). In a third modality, it can be used to measure the same nerve-activity-related changes in cerebral blood flow described above for fMRI (with less spatial and temporal resolution but otherwise with all the possibilities and difficulties associated with fMRI). So far, experiments in linguistic research have primarily used PET in this modality.

Box 6.2 explains "how PET works."

(For the use of PET in language research see Chapter 7.)

Box 6.2   How does PET work?

A tracer compound labelled with a positron emitting radionuclide is injected into the participant's circulatory system. She/he is placed within the field of view of an array of detectors capable of registering incident gamma rays. As the radioisotope decays, it emits a positron that subsequently annihilates on contact with an electron. Annihilation leads to liberation of energy in the form of two gamma rays traveling in opposite directions. The gamma rays interact with the scintillating material of the detectors producing a flash of visible light. A photomultiplier tube enhances the scintillations of light so that the signal can be detected. Individual detector units are linked such that two detection events unambiguously occurring within a certain time window may be called "coincident" and thus be determined to have come from the same radioactive annihilation. Using statistics collected from tens-of-thousands of coincidence events, a set of simultaneous equations for the total activity of each parcel of tissue can be solved, and a map of radioactivities as a function of location can be plotted. The resulting map shows where the molecular probe has been. Most PET scans in neurolinguistic studies have been "coregistered" with structural (MRI) scans, to yield the most accurate anatomic information possible.

Radionuclides used in PET scanning are typically isotopes with short half lives such as oxygen-15 (about 2 min). Due to its short half life, the radionuclides must be produced in a cyclotron which must be not too far away from the PET scanner. The radionuclide is incorporated into compounds normally used by the body such as glucose, water or ammonia and then injected into the body in order to visualize traces of where they become distributed. The "neuroimage" is based on an assumption that areas of high radioactivity are associated with brain activity. When oxygen-15 water is used, what is actually measured indirectly is, as is the case with fMRI, the flow of blood – enriched with oxygen-15 containing water – to different parts of the brain.

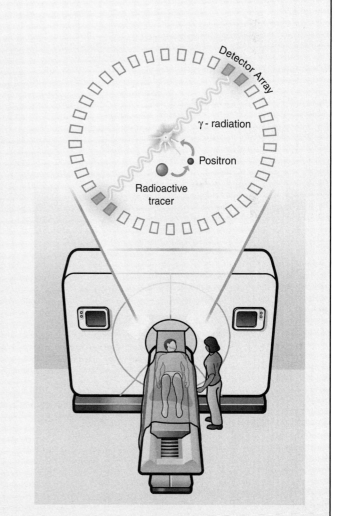

Image courtesy of The Science Creative Quarterly, http: scq.ubc.ca, illustrator Jiang Long. (See Plate 5.)

6.3.4. Benefits and Limitations

PET requires that relatively harmless amounts of non-ionizing radioactivity be induced into the subject and thus cannot be considered non-invasive. Its temporal and spatial resolutions are less precise than fMRI but much better than those yielded by the electromagnetic techniques (see Table 6.1 for overall comparison of techniques). As is the case with fMRI measurements, during PET measurements, the experimental subject's head must be held motionless for the duration of the experiment. In fMRI, speaking induces artifactual signal changes which are a distinct disadvantage in some language experiments. In addition, in fMRI there is signal loss in regions relevant for language processing such as anterior temporal regions due to susceptibility artifacts. A distinctive advantage for PET is that it does not have this problem (see Table 6.2 for a comparison of PET and fMRI).

6.4. REPETITIVE TRANSCRANIAL MAGNETIC STIMULATION

As opposed to the methods discussed thus far, repetitive transcranial magnetic stimulation (rTMS) is a device not to *detect* local brain activity, but rather to *produce* it. rTMS is a neurophysiological method used to activate nerve cells in areas of interest in the brain by applying rapidly changing magnetic fields through the scalp. As can be seen in Box 6.3 the set-up for rTMS is relatively simple. Since rTMS can be performed repeatedly on the same person, every subject can act as her/his own control and information can be derived about what happens when the activities of certain brain regions are disrupted during various linguistic tasks. These virtual, temporary lesions produced by rTMS are, of course, totally reversible although changes in brain activity can last beyond the duration of rTMS application itself. Another advantage of the virtual lesion induced by rTMS is that the confound of the unknown "premorbid" linguistic abilities of patients is absent in rTMS studies.

A major advantage of rTMS is its ability to demonstrate, or at least to suggest, causality. Other non-invasive mapping techniques (EEG, MEG or fMRI) can allow researchers to determine what regions of the brain are *activated* when a subject performs a certain linguistic task, but this is not proof that those regions are actually used for the task; they merely show that a region is *associated* with a task. If cortical activity in that "associated region" is temporarily suppressed via rTMS, however, and the subject's level of performance in the linguistic task is thus reduced during or after the stimulation, the likelihood of the functional necessity of that region to that task obviously increases.

It should be noted that in some subjects (roughly 3%), rTMS produces discomfort or headaches; these side effects do, however, usually disappear within a few minutes, or, at most a few hours. There is also the slight possibility (estimated to be less than 0.1% for any given study) that rTMS can induce an epileptic seizure in a healthy volunteer and thus, appropriate preparations must be made to treat such seizures should they arise. Subjects with a history of seizures are usually excluded as subjects. As is the case with fMRI and MEG, anyone with any sort of metal in their heads (excluding their mouths) and/or nearby regions (such as cardiac pacemakers) should also be excluded from rTMS studies.

TABLE 6.2   Comparison of fMRI and PET (based and modified after Scott & Wise (2003), Speech Communication, **41**, 7–21)

	fMRI	PET
How does the signal come about?	Magnetic characteristics of hydrogen atoms in various organic compounds	Gamma rays after mutual annihilation of positron–electron pair
How is the signal detected?	Coils (that act as a kind of antenna)	Scintillation detectors
How is the signal localized?	Frequency coding	Coincidence detectors
What do the raw data reflect?	Density distribution of hydrogen atoms in various compounds	Distribution of radioactive atoms
What is the relationship (coupling) of the signal to neuronal activation?	Increase of signal is a sign of more blood circulation in the locality due to neuronal activity	Change of radioactivity (rCBF) as indicator of change in neuronal activity
Spatial resolution	BOLD: 3 mm (according to acquisition time, signal-to-noise ratio, field strength)	About 5 mm
Temporal resolution	0.5–5 s	Around 90 s
Safety	Non-invasive as long as the subject has no metal in her/his brain	Slightly invasive (not appropriate for pregnant women)
Noise of scanner	High	Low
Artificial signal loss (also depends on the parameter setting of the MRI)	Posteromedial orbital frontal; medial anterior temporal regions; midportion of ventrolateral temporal lobe (thus signal loss in areas relevant in language studies)	Scatter of gamma rays and weakening of the signal as it traverses brain tissue
Movement	Very sensitive to head and tissue movement	Very sensitive to movement

Box 6.3   How does TMS work?

As can be seen in the figure, a coil of wire, encased in plastic, is held (or mounted) close to the head. When current is allowed to pulse through the coil by the discharge of a *capacitor* (at a rate which the experimenter can adjust, of around 1 cycle per second or 1 Hz) a rapidly changing current flows in its windings. This produces a *magnetic field* oriented *orthogonally* to the plane of the coil. This magnetic field passes unimpeded through the skin and skull, inducing an oppositely directed current in the brain. The strength of the magnetic field decreases with distance and therefore only a few centimeters of the cortex is penetrated (see top right).

In rTMS (*r* for repetitive), stimuli are applied to a given brain area (that can be stereotactically determined) at around 60 pulses per second during several consecutive seconds. This stimulation induces transient "noise" in the information processing and thus amounts to a "virtual, temporary lesion" (from tens of milliseconds up to an hour) in brain function. The number of stimuli per second, the strength of the stimuli, the duration of the train of stimulation, the interval between trains, the total number of trains and the total number of stimuli in a given session or to a given brain position can all be varied. All these aspects of rTMS are referred to as stimulation parameters.

Mark S. George (2003). Stimulating the brain. *Scientific American*, September 2003, 67–73.

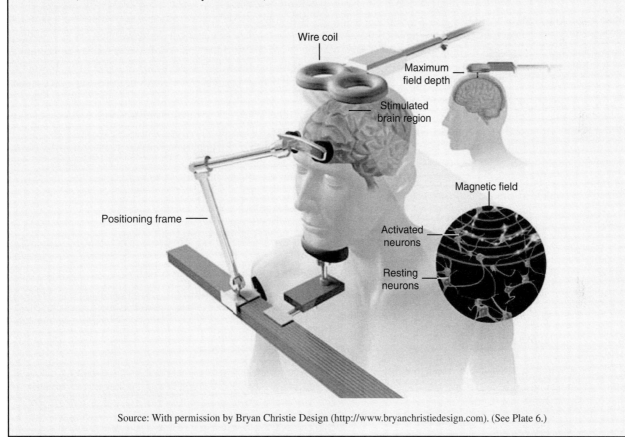

Source: With permission by Bryan Christie Design (http://www.bryanchristiedesign.com). (See Plate 6.)

Limitations of rTMS include its applicability only to those limited areas of the brain close to the skull. Furthermore, the fact that rTMS produces sounds (although not nearly as loud as fMRI) can interfere with linguistic investigations, and, as already mentioned, side effects such as headaches occur in a small percentage of subjects.

Examples for using rTMS in language studies include idiom comprehension (Rizzo *et al.*, 2007), picture naming in Alzheimer disease (Cotelli *et al.*, 2006; Mottaghy *et al.*, 2006) and studies of aphasia (Naeser *et al.*, 2005).

6.5. NEAR-INFRARED SPECTROSCOPY

Although it sounds astonishing, but with a simple light source such as a light bulb, one can illuminate the brain's surface. When the brain is particularly active, for example when doing a calculation task, more blood rushes to the areas of the brain associated with such a task and the brain changes colors, so to speak. The light bulb's spectrum will be more or less absorbed depending on the "color" (corresponding to the activity) of the brain. With appropriate detectors that measure near-infrared light one can watch the brain at work. This is a very simplified account of how near-infrared spectroscopy (NIRS) optical recording technology works. In more sophisticated terms, NIRS uses the wavelengths of light in the near-infrared range (700–1000 nm) to measure changes in the brain's oxygen absorption based on optical properties of hemoglobin. Although it is a technology that has been used for decades in physiological studies (such as cerebral flood flow in premature infants), it is used increasingly in cognitive (including language) studies due to its non-invasiveness, low costs and applicability to infants, children and adults. It has good temporal (milliseconds) and spatial (millimeters) resolution. With increasing depth, however, spatial resolution decreases (about 4 cm of brain penetration) and this limits the use of the technique. The detectors are attached to the scalp or they are integrated into some sort of head cap placed on the participant's head. NIRS technology is thus such that the subject's head need not be confined to a narrow space (such as fMRI or PET). NIRS is also relatively movement tolerant, an important characteristic for speech production studies. Another advantage is its capability to rapidly and simultaneously measure changes in local cerebral blood volume (CBV), and oxy- and deoxyhemoglobin concentrations. The number of publications using NIRS for language studies is currently still very limited although it is anticipated that the near future should see more publications.

For more information on the technology see, for example, Franceschini and Boas (2004) and Huppert et al. (2006). For an example of a language study using NIRS combined with ERPs see Horovitz and Gore (2004).

6.6. CHALLENGES AND FUTURE DIRECTIONS

The challenges that neuroimaging faces are (at least) two fold: (1) purely technical and (2) logical-linguistic-philosophical. The chances that the first set of these challenges will be successfully met seems good: if the past is any indication, each of the methods mentioned above will continue to be refined to yield increasingly fine spatial and temporal resolution of brain events associated with linguistic phenomena. Combining the techniques described above and others still to be developed will allow the investigator to exploit the advantages of one technique while making up for its disadvantages with a supplementary technique. Combinations are currently being developed and tested in many units around the world.

Beside technical developments, another challenge is the utility of neuroimaging techniques in therapeutic settings. For example, it has been shown that people can influence their own EEG brain waves or learn spatial and temporal control over the fMRI BOLD response. These abilities can be used to help people communicate who are unable to do so by conventional means (e.g., Weiskopf et al., 2004). While using EEG as a "communication" device in patients with severe medical conditions seems feasible due to the relatively little costs involved, the exploitation of high cost fMRI in this regard remains to be seen.

The logical-linguistic-philosophical challenges seem greater than the technical ones. Particularly in the cognitive sciences, there is a tendency to measure for the sake of measurement. Teasing apart the cognitive components that might be associated with language, rationality, consciousness or memory, to such an extent that causal inferences can be made between the brain and these phenomena is fraught with hazards. Changes in the brain abound, but a determination of which changes are associated on a one-to-one basis with such subtle linguistic categories as semantics, syntax or pragmatics, may or may not be possible: not because of technical limitations, but because of logical ones. As is emphasized in the book by Bennett et al. (2007), it is the role of the cognitive scientist to determine whether propositions are true or false. It is the role of the cognitive philosopher to determine whether or not a proposition makes sense. Exactly what the phrase "the physiological basis of language" means is a challenge that will have to be met on many fronts.

Acknowledgments

This work was supported by the Canada Research Chair Program awarded to BS. The collaborative effort of the work is reflected in the alphabetical arrangement of the authors' names.

References

Behrens, T.E.J., Johansen Berg, H., Jbabdi, S., Rushworth, M.F.S., & Woolrich, M.W. (2007). Probabilistic diffusion tractography with multiple fibre orientations: What can we gain? Neuroimage, 34, 144–155.

Bennett, M., Dennett, D., Hacker, P., & Searle, J. (2007). Neuroscience and philosophy: Brain, mind and language. New York: Columbia University Press.

Cotelli, M., Manenti, R., Cappa, S.F., Geroldi, C., Zanetti, O., Rossini, P.M., & Miniussi, C. (2006). The effect of transcranial magnetic stimulation on action naming in patients with Alzheimer disease. Archives of Neurology, 63(11), 1602–1604.

Franceschini, M. & Boas, D. (2004). Noninvasive measurement of neuronal activity with near infrared optical imaging. Neuroimage, 21, 372–386.

Hillis, A. (2005). Brain/language relationships identified with diffusion and perfusion MRI. Clinical applications in neurology and neurosurgery. Annals New York Academy of Sciences, 1064, 149–161.

Horovitz, S.G. & Gore, J.C. (2004). Simultaneous event-related potential and near-infrared spectroscopic studies of semantic processing. Human Brain Mapping, 22, 110–115.

Huppert, T.J., Hoge, R.D., Diamond, S.G., Franceschini, M.A., & Boas, D.A. (2006). A temporal comparison of BOLD, ASL, and NIRS hemodynamic responses to motor stimuli in adult humans. Neuroimage, 29, 368–382.

Le Bihan, D. (2003). Looking into the functional architecture of the brain with diffusion MRI. Nature Reviews Neuroscience, 4, 469–480.

Leighton, P.M., & Ulmer, J.L. (Eds.) (2005). White matter in cognitive neuroscience. Special issue. Annals New York Academy of Sciences, 1064.

Mottaghy, F.M., Sparing, R., & Topper, R. (2006). Enhancing picture naming with transcranial magnetic stimulation. *Behavioural Neurology*, *17*(3–4), 177–186.

Naeser, M.A., Martin, P.I., Nicholas, M., Baker, E.H., Seekins, H., Kobayashi, M., Theoret, H., Fregni, F., Maria-Tormos, J., Kurland, J., Doron, K.W., & Pascual-Leone, A. (2005). Improved picture naming in chronic aphasia after TMS to part of right Broca's area: An open-protocol study. *Brain and Language*, *93*(1), 95–105.

Rizzo, S., Sandrini, M., & Papagno, C. (2007). The dorsolateral prefrontal cortex in idiom interpretation: An rTMS study. *Brain Research Bulletin*, *71*(5), 523–528.

Salmelin, R. (2007). Clinical neurophysiology of language: The MEG approach. *Clinical Neurophysiology*, *118*, 237–254.

Sidtis, J.J. (2007). Some problems for representations of brain organization based on activation in functional imaging. *Brain and Language*, *102*, 130–140.

Van Lancker-Sidtis, D. (2006). Does functional neuroimaging solve the questions of neurolinguistics? *Brain and Language*, *98*, 276–290.

Weiskopf, N., Scharnowski, F., Veit, R., Goebel, R., Birbaumer, N., & Mathiak, K. (2004). Self-regulation of local brain activity using real-time functional magnetic resonance imaging (fMRI). *Journal of Physiology Paris*, *98*, 357–373.

Further Readings

Bennett, M., Dennett, D., Hacker, P., & Searle, J. (2007). *Neuroscience and philosophy: Brain, mind and language.* **New York: Columbia University Press.**
This book is a current discussion among cognitive scientists and philosophers on the values and pitfalls of various "language–mind–brain" models.

George, M.S., & Belmaker, R.H. (Eds.) (2007). *Transcranial magnetic stimulation in clinical psychiatry.* **Arlington, VA: American Psychiatric Publishing.**
The book is interesting for those wanting to learn more about TMS and its clinical use. Although the book focuses on focal TMS as a therapeutic tool, it also provides detailed background information on the physics of the procedure, safety standards and how to administer TMS. A review of basic neurophysiological studies with TMS is also given.

Huettel, S.A., Song, A.W., & McCarthy, G. (2004). *Functional magnetic resonance imaging.* **Sunderland, MA: Sinauer.**
This book provides an excellent and detailed introduction to the fMRI technique, including the physics and biology upon which it is based,
experimental design and statistical analysis. Related techniques such as PET or EEG/ERP are also discussed.

Luck, S.J. (2005). *An introduction to the event-related potential technique.* **Cambridge, MA: The MIT Press.**
This is an introductory guide to the ERP technique from a theoretical as well as practical perspective and is recommended to the reader who has little or no background knowledge of the technique. (For further recommendations concerning EEG/ERP see also Chapter 9.)

Mazziotta, J.C., & Toga, A.W. (Eds.) (2000). *Brain mapping: The systems.* **London: Elsevier.**
This volume discusses the various maps and atlases of the brain such as types of maps, maps by anatomical location, by functional systems and by dynamic (developmental, aging) systems.

Mazziotta, J.C., Toga, A.W., & Frackowiak, S.J. (Eds.) (2000). *Brain mapping: The disorders.* **London: Elsevier.**
The book discusses background and technical issues of brain mapping in neurological disorders in adults and pediatric populations, in psychiatric disorders, and in therapy and recovery of function.

Pascual-Leone, A., Davey, N., Rothwell, J., Wasserman, E., & Puri, B.K. (Eds.) (2002). *Handbook of transcranial magnetic stimulation.* **London: Arnold.**
This reference book discusses the basic science, fundamental principles and essential procedures of TMS from a clinical and research perspective.

Toga, A.W., & Mazziotta, J.C. (Eds.) (2002). *Brain mapping: The methods* **(2nd edn). London: Elsevier.**
The volume provides details on the various imaging and non-imaging methods to map the brain.

Toga, A.W., Thompson, P.M., Mori, S., Amunts, K., & Zilles, K. (2006). Towards multimodal atlases of the human brain. *Nature Reviews Neuroscience*, *7*, 952–966.
This article provides an excellent overview and discussion of the creation of modern atlases of the human brain based on imaging techniques, brain mapping methods and analytical strategies.

http://www.neuroguide.com/neuroimg_1.html.
The "Neurosciences on the Internet" site provides a list of interesting and helpful resources related to neuroscience such as neuroanatomy, neuropathology, neuroimaging and so forth.

PART II

NEUROIMAGING OF LANGUAGE

7

PET Research of Language

BARRY HORWITZ[1] and RICHARD J.S. WISE[2]

[1]*Brain Imaging and Modeling Section, National Institute on Deafness and Other Communications Disorders, National Institutes of Health, Bethesda, MD, USA*
[2]*Faculty of Medicine, Division of Neuroscience and Mental Health and Medical Research Council Clinical Sciences Centre, Imperial College London, Hammersmith Hospital, London, UK*

ABSTRACT

This chapter discusses why researchers continue to use positron emission tomography (PET) to study the brain basis of language function, even though functional magnetic resonance (fMRI) has come to dominate functional brain imaging. We contrast the advantages and disadvantages of PET compared with fMRI. We then illustrate a few important PET studies of the past decade that show the kinds of issues for which PET is the most appropriate hemodynamic functional imaging modality – studies of language comprehension and studies of language production. The language comprehension studies focus on hemispheric differences and anterior–posterior differences. For language production, the emphasis is on language production of narratives. We end by commenting on the future directions that language research utilizing functional neuroimaging will likely take.

7.1. INTRODUCTION

In 1988 Petersen and colleagues published an article in *Nature* on single word processing. This important article demonstrated how a relatively new brain imaging technique, positron emission tomography (PET), could enable one to make inferences about some aspects of the neural basis for a number of singularly human cognitive functions. Unlike earlier PET studies that examined sensory processing and whose findings could be related to non-human neurophysiological and neuroanatomical data, the paper by Petersen and colleagues provided a window into the functioning of the brain of healthy human subjects during a high-level cognitive task with a spatio-temporal resolution and area of coverage that other brain imaging methods (e.g., the non-tomographic xenon-133

inhalation method) could not match. This type of data, therefore, offered the promise that the neural underpinnings for uniquely human cognitive functions, especially those related to language processing, could now be explored in living, healthy subjects. Furthermore, this method seemed capable of enabling researchers to explore the functional abnormalities in patients with various kinds of language disorders. Since then, there has been an explosion of research investigations of language using PET, and – since the mid-1990s – an analogous technique called functional magnetic resonance imaging (fMRI) (see Cabeza & Kingstone, 2001, for a comprehensive review).

As the other chapters in this book make clear, a variety of methods are used to understand the neural basis of language processing. Until the advent of functional brain imaging, most knowledge concerning the neurobiological correlates of language processing was derived from neuropsychological investigation of brain damaged patients or by electrical stimulation and recording from individuals undergoing neurosurgery. More recently, techniques like transcranial magnetic stimulation, which can produce "virtual" lesions, have become valuable tools for studying the brain basis of linguistic cognition. However, PET and fMRI (and also electro(magneto)encephalography (EEG/MEG)) have produced an extraordinary wealth of new data that have added considerable information about the functional neuroanatomy of specific cognitive functions (and dysfunctions).

Why are functional brain imaging data so important for understanding neurolinguistics? First, they, of course, permit one to directly have a measure of brain functional activity that can be related to brain structure and to behavior. Second, they can be acquired non-invasively (or in the case of PET, relatively non-invasively, since radioactive iso-topes must be injected into the blood stream; see Chapter 6, for details) from healthy normal subjects, as well as from patients with brain disorders. Third, because these data are obtained simultaneously from much of the brain, they are quite unique for investigating not just what a single brain area does, but also how brain regions work together during

the performance of individual tasks. This latter point is important because the traditional methods used to understand the neural basis of language investigate one "object" at a time (e.g., the ideal brain damaged patient has a single localized brain lesion). The potential to assess how brain regions interact to implement specific neurolinguistic and other cognitive functions has necessitated the development of network analysis methods, and has given rise to a new paradigm in which cognitive functions are conceived as being mediated by distributed interacting neural elements (Mesulam, 1998; Horwitz et al., 1999).

The various functional brain imaging methods are reviewed in Chapter 6 (see Box 7.1 for a brief overview of PET analysis methods). In subsequent sections of this chapter, we will focus on the use of PET to study the neural processes involved in language functioning. We will begin with some remarks contrasting the advantages and disadvantages of PET versus fMRI. We will then discuss a few of the important PET studies of the past decade that will illustrate the kinds of issues for which PET is the most appropriate hemodynamic functional imaging modality. Our last section will address where we see the field going.

Box 7.1   PET analysis methods

There are two primary approaches toward analyzing PET (and fMRI) data. The first, called the subtraction paradigm (Horwitz, 1994), proposes that different brain regions are engaged in different functions, and is implemented by comparing the functional signals between two scans (in its most simple formulation), each representing a different experimental condition. The brain locations of the large signal differences between the two scans are assumed to correspond to the brain regions differentially involved in the two conditions. The second method, the covariance paradigm, rests on the notion that a task is mediated by a network of interacting brain regions, and that different tasks utilize different functional networks (Horwitz, 1994). By examining the covariance in brain activity between different pairs of brain areas (i.e., the functional connectivity), information is obtained about which areas are important nodes in the network under study and how these nodes are functionally connected (see Box 7.2 for an example of functional connectivity analysis of PET data applied to single word reading in normal and dyslexic subjects).

Besides functional connectivity, functional neuroimaging studies can also involve the related concept of effective connectivity. Because two brain regions may show a strong functional connectivity whether or not they are anatomically linked (e.g., the two may receive direct inputs from a third region), methods have been employed to evaluate the effective connectivity, which is the direct effect that activity in one region has on a second (i.e., the functional strength of the anatomical link between the two regions in a given task). The effective connections between a set of brain regions can be obtained by combining the regions' functional connections with a model of the anatomical links between them using techniques such as structural equation modeling (McIntosh et al., 1994) or dynamic causal modeling (Friston et al., 2003).

Friston, K.J., Harrison, L., & Penny, W. (2003). Dynamic causal modelling. *Neuroimage, 19*, 1273–1302.

Horwitz, B. (1994). Data analysis paradigms for metabolic-flow data: Combining neural modeling and functional neuroimaging. *Human Brain Mapping, 2*, 112–122.

McIntosh, A.R., Grady, C.L., Ungerleider, L.G., Haxby, J.V., Rapoport, S.I., & Horwitz, B. (1994). Network analysis of cortical visual pathways mapped with PET. *Journal of Neuroscience, 14*, 655–666.

7.2. PET VERSUS fMRI – SOME METHODOLOGICAL ISSUES

fMRI has a number of advantages over PET, and these advantages have led to fMRI becoming the predominant modality for functional neuroimaging studies of language (see Chapter 8). The three most important advantages are better spatial resolution, better temporal resolution, and not requiring the injection of radioactive substances into the subjects being scanned. This last one allows more scans to be performed in a single subject, and it also means that fMRI can be performed in a non-medical environment, since intravenous lines are not necessary.

However, the use of PET to image regional cerebral blood flow (rCBF) has a number of significant advantages over fMRI, and these advantages are particularly important for addressing key issues in language functioning. First, unlike PET, the gradient coils that are employed in conventional fMRI scanning are quite loud, and can interfere with a subject hearing auditory inputs and with the ability of investigators to record a subject's vocal output. Moreover, the noise produced by the scanner also can reduce the sensitivity of auditory neural responses to auditory inputs even if the subject can hear them. Although fMRI scanning protocols such as sparse sampling (also called clustered acquisition) can overcome some of these problems, they impose limits on the kinds of experimental designs one can employ.

Another significant problem with using fMRI to study language function is that image artifacts occur if the subjects speak. First, speaking results in a susceptibility artifact that is particularly strong in anterior and ventral brain areas. Second, the movement itself can lead to image artifacts, although image preprocessing algorithms can usually deal with these effectively. Although speaking-induced susceptibility artifacts can be distinguished from neurally induced activity (fMRI signal changes due to neural activity are not instantaneous because of the hemodynamic delay, whereas the artifactual signal change arising from speaking is), nevertheless, this problem means that continuous language production cannot be imaged. PET does not have this problem.

Finally, a number of investigators have shown that the left anterior temporal cortex plays a significant role in language processing (Spitsyna *et al.*, 2006; see below for more information), and this area, especially the region around the temporal pole, is particularly sensitive to the fMRI susceptibility artifact. Consequently, in many fMRI studies, data from this area may not be usable. It has been shown that the susceptibility artifact can be reduced by using higher spatial resolution (Devlin *et al.*, 2000), but that in turn often results in limited coverage of other brain areas.

The net effect of these limitations in using fMRI is that imaging studies of speech comprehension and language production, especially beyond the word level, are more readily done employing PET than fMRI, and as will be shown below, important findings continue to be reported.

7.3. CRUCIAL PET FINDINGS

In this section we will discuss PET studies that have addressed questions concerning hemispheric differences and anterior–posterior differences in language comprehension and the brain areas involved in speech and sign language.

7.3.1. Language Comprehension

We will highlight in this subsection recent PET studies that have addressed two fundamental questions concerning the brain organization associated with language comprehension: (1) hemispheric differences and (2) anterior–posterior differences.

7.3.1.1. Hemispheric Differences

It was apparent from the earliest PET studies of speech perception that when subjects heard speech relative to a baseline of silence, activity was almost equally distributed between the left and right superior temporal gyri (e.g., see Petersen *et al.*,1988; Wise *et al.*,1991). This absence of asymmetry indicated that a crude subtractive methodology was not going to replicate the asymmetry evident from clinical observations, namely that impaired comprehension after aphasic stroke was a consequence of left hemisphere lesions, particularly those centered around the lateral (Sylvian) sulcus

(Caplan, 1987). Yet if the early PET studies did not demonstrate anything "special" about the response of the left superior temporal gyrus (STG) (which is the location of unimodal primary and association auditory cortex) to heard words, this was hardly surprising. Spoken language is the most complex sound that we routinely encounter, and over the range of spectral and temporal detail conveyed by speech we can detect phonemes, syllables, stress, and variations in amplitude and pitch. These convey verbal information, in the form of phonetic cues and features, obviously, but also non-linguistic information that both supports comprehension of the verbal message and allows the listener to deduce the affect, sex, age, and individual identity of the speaker. Further, the categorical perception of a sequence of sounds as a word, irrespective of whether the "perceptual unit" is at the level of phonemes or syllables, is remarkably robust, and we can tolerate considerable distortions to speech before it becomes totally incomprehensible. This redundancy in the speech signal suggests that many separate cues and features are processed in parallel, and perception and comprehension is further assisted by top-down processing; we hear next what we expect to hear, given the sense of what has gone before. Moreover, the evidence from neurological cases suggests that although pure word deafness is often only observed after bilateral superior temporal lesions, left-sided lesions alone can result in impaired speech perception, and this impairment does not occur after purely right-sided lesions (Griffiths *et al.*, 1999).

Therefore, the hypothesis was that a refined study design would show with functional imaging that left hemisphere activity predominated over right during speech perception and comprehension. As Scott and Johnsrude (2003) emphasized, the selection of the baseline condition is critical. There have been a series of PET studies which used a variety of non-linguistic acoustic stimuli as the baseline condition; for example, pure tones (Demonet *et al.*, 1992), signal correlated noise – the time-amplitude envelopes of speech filled with white noise, resulting in some temporal but no spectral information – (e.g., Zatorre *et al.*, 1992; Mummery *et al.*, 1999), reversed speech signal (speech played backwards) (Crinion *et al.*, 2003), and spectrally rotated speech (Scott *et al.*, 2000). The advantages and disadvantages of these baseline stimuli have been reviewed in Scott and Wise (2004). One problem with non-linguistic baseline stimuli is that even when they match speech closely in terms of acoustic complexity, they invariably distort or abolish affective prosody and information about the speaker. Therefore, a contrast of speech against one of these baseline stimuli will include responses to both verbal and non-verbal information carried by the speech signal. The review also discusses the use of unfamiliar foreign languages, which, while they might appear to be the best unintelligible baseline to contrast with intelligible native speech, as they will include prosodic and speaker information, nevertheless they also include the confound of unfamiliar phonemes and different rules for combining phonemes; for example, the Japanese word structure is strictly CVCV, whereas English allows a CCCVCCC structure. What influence these confounds will have on observed activity in a functional imaging study is largely unknown. Given such considerations, one can see, why fMRI might compound such subtle problems with its added noise.

Nevertheless, left lateralization of signal in response to speech perception and comprehension has been increasingly observed. One of the first PET studies that demonstrated clear lateralization used a combination of intelligible and unintelligible sentences in a 2 × 2 factorial design (Scott et al., 2000). Sentences presented as clear speech were acoustically matched with the same sentences after spectral rotation (inversion) to render them unintelligible. A further set of sentences was distorted by a technique known as noise-vocoding (Shannon et al., 1995), whereby temporal information is largely preserved but the spectral information is reduced to a few broad frequency bands (six in this study). Perceptually, this distorted speech, which simulates the acoustic information reaching the auditory nerve after a cochlear implant, sounds like a harsh whisper, and it is intelligible after a brief period of familiarization. The "matched" baseline stimulus for the noise-vocoded sentences were made by spectral inversion. The data demonstrated that the left STG responded equally to speech, rotated speech and noise-vocoded speech relative to rotated noise-vocoded speech. This was interpreted as a response to phonetic cues and features, present in both versions of intelligible sentences and also present in the unintelligible rotated speech, but not in the rotated noise-vocoded sentences. Intelligibility, confined to the clear and noise-vocoded speech, activated a left anterior region, centered on the superior temporal sulcus. The main response of the right temporal lobe across contrasts was to clear speech and its spectrally rotated version, stimuli that contained a strong sense of pitch and intonation. Therefore, this study demonstrated a left–right asymmetry in the responses to speech and stimuli that were derived from speech. It also demonstrated a rostral–caudal asymmetry, with intelligibility activating the anterolateral left temporal cortex.

7.3.1.2. Anterior–Posterior Differences

According to the clinical literature on aphasic stroke, one might have predicted that intelligibility would activate Wernicke's area or adjacent cortex in the posterior temporal lobe and inferior parietal lobe – lesions in this region are associated with poor comprehension after aphasic stroke. It remains the view of many that the functional imaging data supports the observations on stroke aphasia, and that lexical semantics is dependent on left temporo-parietal cortex (e.g., Hickok & Poeppel, 2004). However, a growing interest in the functional organization of non-human primate auditory cortex has indicated the existence of an anterior auditory pathway, a "what" pathway (Rauschecker & Tian,

2000; Husain et al., 2004), and thus, by extension, maybe an auditory verbal "what" pathway in the human (Scott & Johnsrude, 2003). The use of magnetic resonance imaging techniques to display human white matter tracts in vivo has led to a reappraisal of the effects of lobar stroke on brain function, when damage to both a cortical area and projection pathways, long and short, contribute to the observed behavioral impairment (Catani & ffytche, 2005). PET has been used to demonstrate that an infarct in Wernicke's area has an impact on the function in the ipsilateral hemisphere, in the intact anterior temporal cortex (Crinion et al., 2006).

The proposal that anterior temporal lobe cortex is involved in word comprehension finds strong support from the clinical observations on patients with the neurodegenerative condition known as semantic dementia where predominantly anterior temporal lobe atrophy is associated with a progressive, and ultimately profound, loss of semantic knowledge (see Chapter 27, for more on semantic dementia). The deficit is not confined only to language comprehension, but it does affect comprehension at the single content word level. This last point is important, because there is a functional imaging literature, mainly from fMRI, that equates anterior temporal lobe activity with comprehension at the sentence and narrative level. For example, Humphries et al. (2006) equates the posterior and anterior temporal lobes with lexical semantics and syntax, respectively, although the authors do indicate that the functional distinction is relative and not absolute. Nevertheless, there is an issue about the context in which the stimuli are presented. To illustrate, the study by Spitsyna et al. (2006) investigated automatic (implicit) comprehension of sentences and/or narratives, without an associated task demand. Under these circumstances, it would appear that there is a relative increase in the strength of the response of the anterior temporal to verbal meaning conveyed within the context of sentences and narratives (see Xu et al., 2005). Left anterior temporal lobe activation can also be seen in response to single words when there is an explicit task demand based on selecting and retrieving word meaning (Sharp et al., 2004).

Figure 7.1 gives an example of the results that can be extracted from a PET study of implicit language comprehension that uses different modalities of stimulus presentation and a range of different baseline conditions – for more details, see Spitsyna et al. (2006). Narratives were presented either as speech or as written text. The modality-specific baseline conditions were rotated speech and text-like arrays of false fonts. A further baseline task, an odd/even decision on serially presented numbers 1–10, was included to unmask language processing activity that is associated with "passive" baseline tasks. The use of an "active" baseline condition reduces the medial and lateral activity associated with what has been termed the default mode of brain function, which includes medial prefrontal activity associated with "self-referential thoughts," the kind of stimulus-independent thoughts that may occupy the

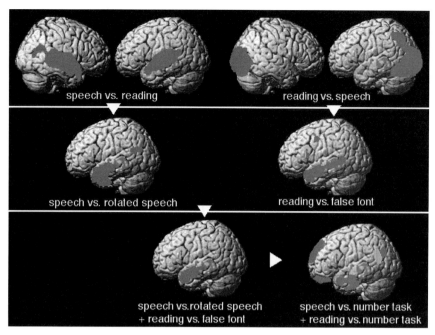

speech vs. reading

reading vs. speech

speech vs. rotated speech

reading vs. false font

speech vs. rotated speech
+ reading vs. false font

speech vs. number task
+ reading vs. number task

FIGURE 7.1   Activity evoked by narratives, both spoken and written, rendered onto the left and right cerebral hemispheres (group averaged data, n = 11 normal subjects). Solid red regions are located over the lateral and inferior surfaces of the hemispheres, hatched red regions are located over the medial surfaces. The contrasts of speech with reading and reading with speech demonstrated bilateral, symmetrical activity in the superior temporal gyri and the occipital lobes, respectively. The asymmetry in posterior parietal cortex (left > right) during reading is the consequence of visual attention and reading saccades being directed to the right in left-to-right readers. Contrasting speech with its modality-specific baseline condition of spectrally inverted (rotated) speech, and reading with its modality-specific baseline condition of text-like arrays of false font, demonstrated activity centered around the superior temporal sulcus, predominantly lateralized to the left. The conjunction of activity for these two contrasts was centered over left anterolateral temporal cortex – a region that responded to intelligible language independent of modality. By using an alternative baseline condition (number task), an explicit task on simple number semantics (an odd/even decision on randomly presented numbers, 1–10), activity was also demonstrated in the anterior fusiform gyrus (the "basal" language area) and just ventral to the angular gyrus. There was also prominent activity in the left superior frontal gyrus, orbito-frontal cortex and in retrosplenial cortex (hatched region). The rationale for using the number task as an alternative baseline condition is described in the text. Data from Spitsyna *et al.* (2006) (see Plate 7).

awake brain unengaged by attention to external stimuli (Gusnard *et al.*, 2001). As stimulus-independent thoughts will include retrieval of both episodic and semantic memories, their intrusion during a baseline task will mask activity associated with declarative memory processes under investigation with the activation task. The effect of using an active baseline task, devoid of meaning (other than simple number semantics) is to considerably increase the extent of activity observed with speech and reading comprehension (Figure 7.1). Such a contrast reveals both anterior and posterior temporal lobe regions, reconciling the separate location of areas associated with access to meaning from the literature on aphasic stroke and semantic dementia. Although the function of these two regions is almost certainly not identical, it would appear probable that their functions are complementary – and the hypothesis is that future connectivity studies will demonstrate integration of their activity during language comprehension.

A third area activated in common was the anterior fusiform gyrus. As this region responded to both heard and written language, its response was polymodal, and its activity may represent that of polymodal perirhinal (paralimbic) cortex. This region is located at the end of a ventral, and strongly left-lateralized, "stream" of activity associated with reading (Spitsyna *et al.*, 2006). Early in this "stream" is the area that has become known as the visual word form area (McCandliss *et al.*, 2003), although the specificity of this region for the processing of written words alone has been challenged (Price & Devlin, 2003). Nevertheless, a lesion of this area, or its connection to left and right primary visual cortex, reliably results in the condition known as pure alexia – poor single word reading, accompanied by laborious covert or overt letter-by-letter reading and a reversed spelling strategy to allow any written comprehension at all.

In summary, the results of these PET comprehension studies lead to the conclusion that verbal comprehension uses unimodal processing streams that converge in both anterior and posterior cortical regions in the left temporal lobe.

7.3.2. Language Production

As mentioned earlier, imaging studies in which the participants are required to produce overt language, especially output beyond a single word, are particularly difficult to do with fMRI, and therefore PET has continued to be used for these investigations. We will discuss some of these studies in this subsection. The focus will be on PET studies requiring continuous language production.

7.3.2.1. Propositional Language Production

Speech of any kind, whether single word repetition, counting (one form of non-propositional speech) or normal narrative speech production (propositional speech), activates the supplementary motor area (SMA), bilateral primary sensorimotor cortex, the left anterior insula/frontal operculum, basal ganglia and thalamus, and bilateral paravermal cerebellum (Wise *et al.*, 1999; Braun *et al.*, 2001; Blank *et al.*, 2002).

Focusing on propositional speech, Braun *et al.* (1997, 2001) used PET measurements of rCBF to evaluate functional brain activity during spontaneous narrative speech and several other tasks involving speech-like features. Each subject performed several tasks, including narrative speech, which consisted of recounting a series of events from memory (e.g., talking about a vacation), and a control task that consisted of producing laryngeal and oral articulatory movements and associated sounds devoid of linguistic content. Compared to a resting state, the orolaryngeal motor control task activated bilaterally dorsal posterior frontal operculum (pars opercularis), pre- and postcentral gyrus and primary and anterior auditory cortex. Compared to the orolaryngeal motor control task, the speech condition activated a large number of regions in the left hemisphere, including regions in the anterior and ventral frontal operculum (pars opercularis, triangularis, orbitalis), pre-SMA, lateral premotor cortex, dorsolateral prefrontal cortex, anterior cingulate and insula. Bilateral activations were observed in posterior superior and middle temporal cortex. Another recent PET study of propositional speech production in normal subjects found a similar set of activated regions (Blank *et al.*, 2002).

Horwitz and Braun (2004) used these data to examine the functional interactions of some of these areas during language production (see Box 7.2 for details about how functional connectivity is evaluated for PET data). The question addressed was the specificity of brain areas traditionally associated with language, since many such regions are located near motor and auditory areas and are often activated in non-language tasks (e.g., for the data discussed in

---

Box 7.2 PET data functional connectivity analysis

The central idea behind functional connectivity analysis is that activities that covary together indicate that the neurons generating the activities may be interacting. As indicated in Box 7.1, two neural entities are said to be functionally connected if their activities are correlated. Functional connectivity does not necessarily imply a causal link, whereas effective connectivity does.

For rCBF PET activation data, some investigators calculate interregional functional connectivity by correlating rCBF data within a task condition and across subjects. The reasoning behind this method starts with the fact that subjects perform tasks with different abilities, as shown by differences in accuracy, reaction time and other measures of performance. This subject-to-subject variability suggests that the activity of the brain network mediating a task also varies from subject to subject.

An example of this approach is Horwitz *et al.* (1998), where PET functional connectivity analysis was applied to normal and dyslexic subjects. The task consisted of pronouncing pseudowords. In agreement with the classic neurologic model for reading, which is based on studies of alexic patients, rCBF in the left angular gyrus showed strong functional connectivity with rCBF in visual association areas in occipital and temporal cortex, and with rCBF in language areas in superior temporal and inferior frontal cortex. In contrast, these strong functional connections were absent in subjects with developmental dyslexia, indicating that dyslexia is characterized by a functional disconnection of the angular gyrus that mirrors the anatomical disconnection seen in alexia, a finding supported by other investigators using fMRI (Pugh *et al.*, 2000) and diffusion tensor imaging of white matter (Klingberg *et al.*, 2000).

Horwitz, B., Rumsey, J.M., & Donohue, B.C. (1998). Functional connectivity of the angular gyrus in normal reading and dyslexia. *Proceedings of the National Academy of Science*, 95, 8939–8944.

Klingberg, T., Hedehus, M., Temple, E., Salz, T., Gabrieli, J.D., Moseley, M.E., & Poldrack, R.A. (2000). Microstructure of temporo-parietal white matter as a basis for reading ability: Evidence from diffusion tensor magnetic resonance imaging. *Neuron*, 25(2), 493–500.

Pugh, K.R., Mencl, W.E., Shaywitz, B.A., Shaywitz, S.E., Fulbright, R.K., Constable, R.T., Skudlarski, P., Marchione, K.E., Jenner, A.R., Fletcher, J.M., Liberman, A.M., Shankweiler, D.P., Katz, L., Lacadie, C., & Gore, J.C. (2000). The angular gyrus in developmental dyslexia: Task-specific differences in functional connectivity within posterior cortex. *Psychological Science*, 11(1), 51–56.

---

the previous paragraph, the pars opercularis was activated by both the speech condition and the orolaryngeal task). Functional connectivity within each condition was calculated using reference voxels corresponding to language areas significantly activated for narrative speech compared to the orolaryngeal motor control task. We chose left hemisphere perisylvian regions in the posterior STG (likely corresponding to what classically has been called Wernicke's area) and in the inferior frontal cortex (near or in what classically has been denoted as Broca's area). The posterior

STG area had strong functional links during speech with left inferior parietal and frontal perisylvian regions, with left middle and inferior temporal gyri, and with the anterior cingulate/supplementary motor area. Most of these strong functional connections were absent in the motor control task, especially the link to frontal perisylvian regions. The left

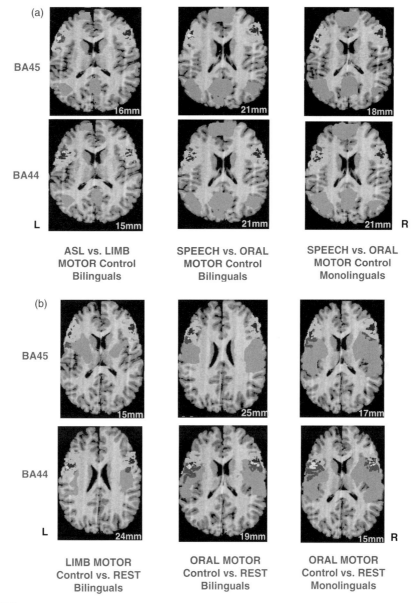

FIGURE 7.2　(a) Activations of BA45 (top row) and BA44 (bottom row) during production of language narratives compared to a motor control task; and (b) activation of BA45 (top) and BA44 (bottom) comparing each motor control task to a resting condition. Shown are representative horizontal slices (left side of each image corresponds to the left side of the brain; the level in mm superior to the AC-PC plane (z-coordinate of Talairach & Tournoux atlas, 1988) is indicated on each slice). Images displayed in the two columns on the left are from the bilingual (English and ASL) subjects, and those in the column on the right are from the monolingual English speakers. Voxels in dark blue correspond to core parts of the specific Brodmann area, those in light blue to peripheral voxels. Voxels significantly more active in one condition compared to a second (Z > 2.33) are shown in green. Voxels in the peripheral part of a Brodmann area that had a significant PET activation are displayed in red, and core voxels that were significantly activated are shown in yellow. From Horwitz *et al.* (2003) [Talairach, J., & Tournoux, P. (1988). *Co-planar stereotaxic atlas of the human brain* (M. Rayport, Trans.). New York: Thieme.] (see Plate 8(a), 8(b)).

frontal opercular region had significantly large functional connections during speech with other language-associated areas in the left hemisphere, including regions in posterior temporal and inferior parietal cortex. Generally, these strong functional connections were absent during the motor control task even though the frontal opercular area was activated in the motor control task when compared to the rest condition. Furthermore, the strong functional interactions between the frontal opercular region in the left hemisphere were to both left and right hemisphere posterior superior temporal gyri regions. This result is consistent with the model proposed by Braun *et al.* (2001) that hypothesizes that language production proceeds from early bilateral posterior cortical stages of lexical access to later left-lateralized anterior stages of articulatory-motor encoding.

7.3.2.2. Broca's Area for Speech and Sign Language

Are the locations in the brain of language regions related to the fact that most human languages employ speech? That is, Wernicke's area is near auditory cortex, and Broca's area is just anterior to the mouth area of motor cortex. Not all linguistic functions, however, use audition as input and mouth movement as output; sign languages, in particular, employ vision and limb movement. To investigate aspects of this issue, a human PET study in hearing adults whose parents were deaf and who were fluently bilingual for speech and America Sign Language (ASL) was performed and showed that narrative production by both languages activated a common set of language areas (Braun *et al.*, 2001). The tasks used were the same type as discussed above: for speech – narrative speech and an orolaryngeal motor control task; for ASL – extemporaneously recounting a story using ASL was the narrative language task along with a limb–facial motor control task devoid of linguistic content. The areas activated by both narrative tasks, which access the modality-independent aspects of language use, included left frontal operculum and bilateral perisylvian cortex in the STG, but common sites of activation also extended to a number of extrasylvian areas presumably involved in paralinguistic functions (declarative memory, attention, visual imagery) that are engaged at the level of discourse.

One of the classical language regions is Broca's area in the inferior frontal gyrus, which consists of two cytoarchitectonically defined regions – Brodmann's areas (BA) 44 and 45. Recently, it has become possible to investigate these subdivisions using probabilistic brain maps (see Chapter 4, for details). These maps were derived by histological analysis that determined the locations and spatial extents of BA44 and BA45 in the left and right hemispheres of 10 individual brains, after which the cytoarchitectonic data were transposed to a common space so as to form a probabilistic atlas, where voxel value indicates the percentage of brains having that location as BA44 or BA45. The above

PET data were used to investigate the role that these two cytoarchitectonic subdivisions of Broca's region play in language production in the bilingual (English-ASL) subjects discussed above (Horwitz *et al.*, 2003). The probabilistic atlas data were applied to the PET data for the two narrative tasks and the two motor control tasks (see Figure 7.2). This allowed us to determine the probability that BA44 or BA45 was activated during each language production task relative to its motor control task. It was found that BA45, not BA44, is activated by both speech and signing. It was BA44, not BA45, that was activated by the two motor control tasks (relative to a resting condition). The same patterns of activation were found for oral language production in a group of English speaking monolingual subjects. These findings thus implicate BA45 as the part of Broca's area that represents the conceptual-language interface that is fundamental to the modality-independent aspects of language generation.

In summary, these studies demonstrate that language production, whether in the form of speech or ASL, share a widespread group of modality-independent brain regions, which include not only classical perisylvian language regions in the left hemisphere, but also areas that extend beyond the traditional regions. Furthermore, speech and ASL production seem to engage the same cytoarchitectonically defined parts of Broca's area. Finally, during speech production, there is a strong functional connectivity between Broca's and Wernicke's areas in the left hemisphere.

7.4. CHALLENGES AND FUTURE DIRECTIONS

In spite of the extensive use of fMRI during the past decade, the above discussion makes it clear that the use of PET to measure rCBF has continued to provide valuable and essentially unique information about the brain organization in humans of language-related processes. However, although we envision that PET will continue to be used in the future, we suspect that more and more there will be a shift away from this technology toward an even more extensive use of fMRI, in spite of the difficulties we discussed in the early sections of this chapter. New hardware and software developments are likely to mitigate some of fMRI's disadvantages in so far as language and auditory studies are concerned.

However, the *real future* of using neuroimaging to assess language functioning lies, we believe, in a totally different direction. It does not lie in just performing more tasks that try to fractionate the linguistic processes into even finer and more subtle detail, although there surely will be a number of studies that do this. It does not lie in just improving the spatial and temporal resolution of neuroimaging technology, although that will surely be accomplished. Rather, we believe that the future of neuroimaging lies in combining multi-modality data acquisition with computational neural modeling.

In neuroscience, as in other areas of biology, there now exists the ability to acquire large quantities of data at multiple spatial and temporal scales of investigation (molecules to the brain as a whole; milliseconds to a lifetime). Because no one method transcends all the different levels, investigators often have trouble interpreting and understanding their particular subset of data in terms of all the other relevant data that others acquired. This makes it difficult to frame a single, unified account relating specific behaviors to their underlying neural mechanisms. We saw a simple example of this problem in the language comprehension section: the bilateral results of the first PET studies seemed at variance with the results of human lesion studies that seemed to demonstrate a strong left hemisphere lateralization of language comprehension. Although more careful experimental design seems to have resolved this issue, it is not clear that experiment alone will always work. What is needed is a formal way to systematically relate multiple data sets (i.e., fMRI, PET, MEG, lesion, behavior, neuroanatomical, perhaps even neuron electrophysiological data), thus bridging the various spatiotemporal levels of neuroscientific investigation. It has been argued that computational neural modeling provides the way to do this (Horwitz, 2004, 2005; Horwitz & Braun, 2004). Already there have been a number of studies in which various types of computational modeling have been combined with human imaging data, including studies of auditory object processing (Husain et al., 2004), mirror neurons (Arbib et al., 2000), and sentence processing (Just et al., 1999). The first two of these used biologically realistic neural modeling to related neural activity to fMRI/PET data, and the last used a cognitive model of sentence comprehension to explain how fMRI activation levels varied as a function of sentence complexity in three brain areas (Broca, Wernicke and dorsolateral prefrontal cortex). These modeling efforts, if successful, result in either a way to understand the neural mechanisms that underlie specific linguistic functions (in the case of using neurobiological models) or to understand the brain locations for specific cognitive components of a linguistic function. Ultimately, the goal will be to provide a unified account of the neural mechanisms supporting language processing.

Acknowledgments

We wish to thank Dr. Allen Braun for useful discussions. This work was supported, in part, by the NIDCD Intramural Research Program.

References

Arbib, M.A., Billard, A., Iacoboni, M., & Oztop, E. (2000). Synthetic brain imaging: Grasping, mirror neurons and imitation. *Neural Networks*, *13*(8–9), 975–997.

Blank, S.C., Scott, S.K., Murphy, K., Warburton, E., & Wise, R.J. (2002). Speech production: Wernicke, Broca and beyond. *Brain*, *125*, 1829–1838.

Braun, A.R., Guillemin, A., Hosey, L., & Varga, M. (2001). The neural organization of discourse: An H215O-PET study of narrative production in English and American sign language. *Brain*, *124*, 2028–2044.

Braun, A.R., Varga, M., Stager, S., Schulz, G., Selbie, S., Maisog, J.M., Carson, R.E., & Ludlow, C.L. (1997). Altered patterns of cerebral activity during speech and language production in developmental stuttering. An H215O positron emission tomography study. *Brain*, *120*, 761–784.

Cabeza, R., & Kingstone, A. (Eds.) (2001). *Handbook of functional neuroimaging of cognition*. Cambridge. MA: MIT Press.

Caplan, D. (1987). *Neurolinguistics and Linguistic Aphasiology: An Introduction*. Cambridge: Cambridge Univ. Press.

Catani, M., & ffytche, D.H. (2005). The rises and falls of disconnection syndromes. *Brain*, *128*, 2224–2239.

Crinion, J.T., Lambon-Ralph, M.A., Warburton, E.A., Howard, D., & Wise, R.J. (2003). Temporal lobe regions engaged during normal speech comprehension. *Brain*, *126*, 1193–1201.

Crinion, J.T., Warburton, E.A., Lambon-Ralph, M.A., Howard, D., & Wise, R.J. (2006). Listening to narrative speech after aphasic stroke: The role of the left anterior temporal lobe. *Cerebral Cortex*, *16*, 1116–1125.

Demonet, J.-F., Chollet, F., Ramsay, S., Cardebat, D., Nespoulous, J.-L., Wise, R., Rascol, A., & Frackowiak, R. (1992). The anatomy of phonological and semantic processing in normal subjects. *Brain*, *115*, 1753–1768.

Devlin, J.T., Russell, R.P., Davis, M.H., Price, C.J., Wilson, J., Moss, H.E., Matthews, P.M., & Tyler, L.K. (2000). Susceptibility-induced loss of signal: Comparing PET and fMRI on a semantic task. *Neuroimage*, *11*, 589–600.

Griffiths, T.D., Rees, A., & Green, G.G.R. (1999). Disorders of human complex sound processing. *Neurocase*, *5*, 365–378.

Gusnard, D.A., Akbudak, E., Shulman, G.L., & Raichle, M.E. (2001). Medial prefrontal cortex and self-referential mental activity: Relation to a default mode of brain function. *Proceedings of the National Academy of Science*, *98*(7), 4259–4264.

Hickok, G., & Poeppel, D. (2004). Dorsal and ventral streams: A framework for understanding aspects of the functional anatomy of language. *Cognition*, *92*, 67–99.

Horwitz, B. (2004). Relating fMRI and PET signals to neural activity by means of large-scale neural models. *Neuroinformatics*, *2*, 251–266.

Horwitz, B. (2005). Integrating neuroscientific data across spatiotemporal scales. *Comptes rendus Biologies*, *328*, 109–118.

Horwitz, B. & Braun, A.R. (2004). Brain network interactions in auditory, visual and linguistic processing. *Brain and Language*, *89*, 377–384.

Horwitz, B., Tagamets, M.-A., & McIntosh, A.R. (1999). Neural modeling, functional brain imaging, and cognition. *Trends in Cognitive Science*, *3*, 91–98.

Horwitz, B., Amunts, K., Bhattacharyya, R., Patkin, D., Jeffries, K., Zilles, K., & Braun, A.R. (2003). Activation of Broca's area during the production of spoken and signed language: A combined cytoarchitectonic mapping and PET analysis. *Neuropsychologia*, *41*(14), 1868–1876.

Humphries, C., Binder, J.R., Medler, D.A., & Liebenthal, E. (2006). Syntactic and semantic modulation of neural activity during auditory sentence comprehension. *Journal of Cognitive Neuroscience*, *18*, 665–679.

Husain, F.T., Tagamets, M.-A., Fromm, S.J., Braun, A.R., & Horwitz, B. (2004). Relating neuronal dynamics for auditory object processing to neuroimaging activity. *Neuroimage*, *21*, 1701–1720.

Just, M.A., Carpenter, P.A., & Varma, S. (1999). Computational modeling of high-level cognition and brain function. *Human Brain Mapping*, *8*, 128–136.

McCandliss, B.D., Cohen, L., & Dehaene, S. (2003). The visual word form area: Expertise for reading in the fusiform gyrus. *Trends in Cognitive Science*, *7*(7), 293–299.

Mesulam, M.-M. (1998). From sensation to cognition. *Brain*, *121*, 1013–1052.

Mummery, C.J., Ashburner, J., Scott, S.K., & Wise, R.J. (1999). Functional neuroimaging of speech perception in six normal and two aphasic subjects. *Journal of the Acoustical Society of America*, *106*, 449–457.

Petersen, S.E., Fox, P.T., Posner, M.I., Mintun, M., & Raichle, M.E. (1988). Positron emission tomographic studies of the cortical anatomy of single-word processing. *Nature*, *331*, 585–589.

Price, C.J., & Devlin, J.T. (2003). The myth of the visual word form area. *Neuroimage*, *19*(3), 473–481.

Rauschecker, J.P., & Tian, B. (2000). Mechanisms and streams for processing of "what" and "where" in auditory cortex. *Proceedings of the National Academy of Science*, *97*(22), 11800–11806.

Scott, S.K.,, Blank, C.C., Rosen, S., & Wise, R.J. (2000). Identification of a pathway for intelligible speech in the left temporal lobe. *Brain*, *123*, 2400–2406.

Scott, S.K., & Johnsrude, I.S. (2003). The neuroanatomical and functional organization of speech perception. *Trends in Neuroscience*, *26*(2), 100–107.

Scott, S.K., & Wise, R.J. (2004). The functional neuroanatomy of prelexical processing in speech perception. *Cognition*, *92*, 13–45.

Shannon, R.V., Zeng, F.G., Kamath, V., Wygonski, J., & Ekelid, M. (1995). Speech recognition with primarily temporal cues. *Science*, *270*, 303–304.

Sharp, D.J., Scott, S.K., & Wise, R.J. (2004). Monitoring and the controlled processing of meaning: Distinct prefrontal systems. *Cerebral Cortex*, *14*, 1–10.

Spitsyna, G., Warren, J.E., Scott, S.K., Turkheimer, F.E., & Wise, R.J. (2006). Converging language streams in the human temporal lobe. *Journal of Neuroscience*, *26*, 7328–7336.

Wise, R., Chollet, F., Hadar, U., Friston, K., Hoffner, E., & Frackowiak, R. (1991). Distribution of cortical neural networks involved in word comprehension and word retrieval. *Brain*, *114*, 1803–1817.

Wise, R.J., Greene, J., Büchel, C., & Scott, S.K. (1999). Brain regions involved in articulation. *Lancet*, *353*(9158), 1057–1061.

Xu, J., Kemeny, S., Park, G., Frattali, C., & Braun, A. (2005). Language in context: Emergent features of word, sentence, and narrative comprehension. *Neuroimage*, *25*, 1002–1015.

Zatorre, R.J., Evans, A.C., Meyer, E., & Gjedde, A. (1992). Lateralization of phonetic and pitch discrimination in speech. *Science*, *256*, 846–849.

Further Readings

Brown, C.M., & Hagoort, P. (Eds.) (1999). *The neurocognition of language*. Oxford: Oxford University Press.
This book presents a series of articles on the cognitive neuroscience of language by experts in the field, including several articles that use functional neuroimaging methods. There are also several articles that discuss functional brain imaging methodology. Thus, this book places neuroimaging results in the context of other studies of language function.

Cabeza, R., & Nyberg, L. (1997). Imaging cognition: An empirical review of PET studies with normal subjects. *Journal of Cognitive Neuroscience*, *9*, 1–26.
This review article provides a very nice overview of PET results obtained from studies of higher-order cognitive processes, including attention, perception, language, and memory. The review shows the set of brain regions that are consistently activated by each cognitive function. Thus, this article places PET studies of language in the context of PET studies of other cognitive functions.

Demonet, J.F., Thierry, G., & Cardebat, D. (2005). Renewal of the neurophysiology of language: Functional neuroimaging. *Physiological Review*, *85*(1), 49–95.
This is a review of functional brain imaging studies of language. Besides having a good section on brain imaging methodology, this review also includes a discussion of language and brain plasticity, thus incorporating information about developmental disorders, post-lesion recovery of function, and language reorganization in neurodegenerative diseases.

Vigneau, M., Beaucousin, V., Herve, P.Y., Duffau, H., Crivello, F., Houde, O., Mazoyer, B., & Tzourio-Mazoyer, N. (2006). Meta-analyzing left hemisphere language areas: Phonology, semantics, and sentence processing. *Neuroimage*, *30*(4), 1414–1432.
The title of this review explains exactly what the article is about. The meta-analysis covers PET and fMRI studies from 1992 to 2004. The authors conclude that their results argue for the notion of large-scale language networks, rather than a modular organization of language in the left cerebral hemisphere.

8

Functional Magnetic Resonance Imaging (fMRI) Research of Language

URI HASSON and STEVEN L. SMALL

Department of Neurology, The University of Chicago, Chicago, IL, USA

ABSTRACT

Prior to the advent of functional imaging, research into the biology of language depended on investigating the anatomy and behavior of large groups of individuals with focal brain injury. The addition of functional imaging, magnetic and electrical recording and cortical and transcranial stimulation methods to the toolkit of the scientist in this field has enabled complementary perspectives on the organization of the brain for language. In this chapter, we survey a subset of this new literature, focusing on data obtained using fMRI and on the interpretation of those data with respect to brain structure and function. A major challenge in this research has been to delineate how different levels of processing are neurally instantiated and the interface between these levels. We specifically address the neural representations and mechanisms underlying sub-lexical speech perception, lexical representation and retrieval, sentential syntax and semantics and higher-level discourse.

8.1. INTRODUCTION

The advancement of non-invasive imaging methods such as positron emission tomography (PET) and functional magnetic resonance imaging (fMRI) has made it possible to study the neurophysiological basis of language. These studies, in areas such as speech recognition, parsing (lexical) semantic memory, sentence and discourse comprehension and production, are informative for both the psychologist and the neuroscientist. On one hand, they constrain cognitive theory, and on the other, they represent the essential constituent of theory development in the neurophysiology of language.

That said, newcomers to fMRI quickly discover that fMRI studies differ in the extent to which their results have implications for cognitive (functional) theories of language. Likewise, studies differ in how informative they are for understanding the anatomy and physiology of the brain. For instance, many neuroscientists focus their research on the role of particular brain structures (e.g., the inferior frontal cortex), and in their work they will define that region as a "region of interest" (ROI) and exclusively report neural activity in that region across different language tasks, ignoring other regions. This type of investigation is useful for the development of neurobiological theories, but may be less important for development of functional theories, as it does not adequately characterize the full network of regions that work together collaboratively to perform tasks. Analogously, some psychologists focus on the degree to which brain activation patterns differ under varying experimental conditions in order to support or refute psycholinguistic theories, but put less emphasis on thorough neurophysiological description of these differences. Although theory development in psychology and linguistics, as well as basic findings in neurophysiology are all valuable uses of functional imaging, much of the field strives for results that are at the intersection of the two and thus can help integrate theories of brain and function. In this review, we focus on four central research themes underlying language processing in the brain. These include (a) sub-lexical speech perception, (b) single-word comprehension, (c) sentence processing (syntactic and semantic) and (d) discourse.

8.2. RECOGNIZING AUDITORY INPUT AS SPEECH

When individuals are presented with spoken, non-meaningful sub-lexical speech stimuli such as consonant-vowel syllables, numerous brain regions demonstrate reliable activity. An important subset of these areas is more activated by such stimuli than by non-speech stimuli. One of the main goals of research in speech perception is to clarify how the acoustic input is represented phonetically and comes to be understood as speech.

Certain brain regions are thought of as specialized for auditory processing. These regions, located bilaterally on the supratemporal plane include (a) transverse temporal gyrus of Heschl; thought to be the site of the primary (or core) auditory cortex in the human, and (b) auditory association cortexes; the planum polare (anterior to Heschl's) and planum temporale (posterior to Heschl's). The core and association regions differ in their functional properties: the primary area is sensitive to pure frequencies, but not complex sounds, whereas the association areas show the opposite pattern.

Drawing on comparative studies with primates, it has been suggested that specialization for speech (as contrasted with generic auditory processing) begins at the level of the auditory association cortex. In primates, the auditory cortex anterior to the core region has been shown to differentiate among diverse monkey calls (Rauschecker & Tian, 2000). A number of studies suggest that humans also have a comparable neural pathway in which acoustic information is understood at a more abstract level, as an "auditory object." These studies have shown that regions on the anterior part of the left superior temporal gyrus (STG) and superior temporal sulcus (STS) of the lateral temporal cortex are more sensitive to auditory speech than to stimuli with similar acoustic complexity (Scott & Johnsrude, 2003). To this extent, auditory processing in the human seems to match that of primates.

However, there are a number of problems with this account. First, while primates show rudimentary higher-level auditory processing in auditory regions anterior to auditory cortex, in humans such processing is also found in the planum temporale posterior to auditory cortex (Griffiths & Warren, 2002). This region may be particularly sensitive to speech input (Vouloumanos et al., 2001), and it seems to process auditory input in a specialized way when it could potentially contain speech (Meyer et al., 2005). The posterior left STG could also be implicated in some of these higher-level functions (Narain et al., 2003). Another problem with this account is interpretive: the argument is based on the finding that speech stimuli evoke more activity than non-speech stimuli in anterior regions of temporal cortex. However, this finding could originate not from specialized mechanisms for perceiving sound categories, but from the fact that speech processing necessitates discriminating among highly similar sound categories. That is, speech comprehension might be a quantitatively more difficult categorization task, but not one that is qualitatively different than other complex auditory discrimination tasks (Belin et al., 2004).

While auditory processing largely activates temporal regions, audiovisual speech evokes different patterns of neural activity. In a pivotal study, Calvert et al. (1997) demonstrated that many brain regions involved in auditory speech perception are also activated during silent lip reading, and further research suggested that the left STS is particularly important for integrating auditory and visual input (Calvert et al., 2000). In that region, neural activity during presentation of auditory input was strongly affected by whether it was accompanied by a matching or mismatching visual stimulus. Finding an interaction between auditory and visual information in STS suggested that integration

---

Box 8.1 Dynamics of auditory and visual integration during audiovisual speech comprehension

How are auditory and visual inputs integrated during audiovisual speech comprehension? Initial findings pointed to the STS as such an integration hub (Calvert et al., 2000). Because this region processes sensory input once it has been analyzed in lower-level, unimodal regions, it was thought that integration takes place relatively late in the processing stream. However, there is some controversy on this issue. First, visual information can affect neural activity in brain areas involved in the earliest stages of acoustic processing (the brainstem; Musacchia et al., 2006) and also in the primary auditory cortex (Pekkola et al., 2003). Second, methods sensitive to the timeline of processing (e.g., magnetoencephalography) indicate that visual information affects auditory processing in the vicinity of the primary auditory cortex before it affects processing in STS (Mottonen et al., 2004). Thus, the stage of processing at which visual information integrates with auditory input during speech processing is still a matter of ongoing research.

Calvert, G. A., Campbell, R., & Brammer, M. J. (2000). Evidence from functional magnetic resonance imaging of crossmodal binding in the human heteromodal cortex. *Current Biology*, *10*(11), 649–657.

Mottonen, R., Schurmann, M., & Sams, M. (2004). Time course of multisensory interactions during audiovisual speech perception in humans: A magnetoencephalographic study. *Neuroscience Letters*, *363*(2), 112–115.

Musacchia, G., Sams, M., Nicol, T., & Kraus, N. (2006). Seeing speech affects acoustic information processing in the human brainstem. *Experimental Brain Research*, *168*(1), 1–10.

Pekkola, J., Ojanen, V., Autti, T., Jaaskelainen, I. P., Mottonen, R., Tarkiainen, A., & Sams, M. (2003). Primary auditory cortex activation by visual speech: An fMRI study at 3 T. *Neuroreport*, *16*(2), 125–128.

of auditory and visual input occurs relatively late in the processing stream, after both auditory and visual input have been independently elaborated in areas involved in lower-level processing (Calvert *et al.*, 2000), but there is some debate on this issue (see Box 8.1). Audiovisual speech perception may also rely on premotor and motor cortexes, as these show greater activity for audiovisual speech than for either auditory or visual speech tracks presented separately (Skipper *et al.*, 2005).

8.3. WORD REPRESENTATION: FORM AND MEANING

In addition to phonological knowledge, word comprehension entails accessing semantic knowledge. There has been considerable work on the organization of lexical semantic knowledge (semantic memory) in the brain, on the basis of both lesion studies and imaging methods. Brain regions have been identified that are relatively selective to living versus non-living categories, or concrete versus abstract nouns (see, Martin & Chao, 2001 for review). Understanding the organization of semantic knowledge in the brain is fundamental for understanding neural activity during word comprehension, as this is the type of information that single words denote.

One of the major goals of imaging research is to understand how semantic knowledge is accessed and manipulated during language comprehension. The main difficulty in answering this question is that word comprehension (whether presented in auditory or visual modality) entails phonological processing of the sort discussed in the previous section and semantic processing having to do with the access and manipulation of information. Consequently, it is not possible to tell which brain regions are specifically involved in semantic processing by observing brain activity during the processing of a single-word. Much of the research of semantic access has been engrossed in developing methods to address this issue.

One solution has been to manipulate the task under which words are processed. In an early study (Demb *et al.*, 1995), words were visually presented to participants, who performed one of two tasks: in the more complex task they judged whether the word was abstract or concrete, and in the simpler task they judged whether the word was presented in upper- or lower-case. Both tasks entail processing of the printed word and making a judgment, but clearly diverge in their semantic processing demands. In that study, a region in left inferior frontal gyrus (IFG) showed more neural activity in the complex task than in the simple task. Furthermore, repeated performance of the more complex task resulted in decreased neural activity in the same IFG region (a phenomenon called *repetition suppression*), but repeated performance of the simpler task did not result in repetition suppression. Because the words were printed in both task conditions, the suppression found only for the more difficult task suggests that this suppression reflects easier semantic processing related to repetition rather than easier phonological processing. A later study (Poldrack *et al.*, 1999) largely replicated these findings, demonstrating greater IFG activity in a semantic task (abstract/concrete judgment for printed words) than in a phonologically oriented task (counting a word's syllables). However, this study further revealed an interesting partitioning between the more anterior and more posterior parts of IFG: whereas the more posterior part (~BA 44) showed above-baseline activity for both phonological and semantic tasks, the more anterior part (~BA 47/45) showed above-baseline activity only for the semantic task. A study by Wagner *et al.*, (2000) made a similar point. In that study, participants made a semantic judgment for visually presented words. Some of these words had been seen before (repeated items) whereas others had not (novel items). The crucial manipulation depended on the context in which the words that were now repeated were previously presented: in one condition, these words were presented in the context of the identical semantic judgment task; in the other condition, they were presented in the context of a different, non-semantic task (upper-case/lower-case judgment). In the anterior portion of left IFG, repeated presentation of words resulted in reduced neural activity, but only when these words were repeated in the context of the same task as they were presented before. In contrast, a different pattern of repetition was evident for posterior IFG: that region demonstrated less neural activity for the repeated words independently of whether they were previously presented in the same or different task. Thus, the posterior part of IFG seemed to benefit from repeated exposure to a word independent of the task in which it was presented – consistent with its role in phonological processing.

There is much debate on the nature of semantic processes subserved by the anterior and posterior parts of IFG. One topic of debate addresses the nature of the semantic processes indexed by activity in anterior left IFG. The details of this debate are outside the scope of the current chapter, and it has yet to be seen whether all the suggested functions play a fundamental role in natural language comprehension. Another point of contention is whether the purported anterior–posterior distinction indeed reflects a dichotomy that is based solely on a semantics versus phonology continuum. With respect to this matter, Gold *et al.* (2005) have shown that the anterior part of left IFG (~BA 45/47) was also reliably active during a phonological task in which participants were asked to "regularize" words that have irregular orthography-to-phonology mappings (e.g., participants were asked to pronounce *pint* in a way that rhymes with *hint*). Gold *et al.*'s findings are consistent with the hypothesis that activity in anterior IFG indexes general types of controlled processing, rather than solely access to semantic knowledge.

Few fMRI studies have examined how access and selection of semantic meanings take place during natural sentence processing. In one such study (Rodd *et al.*, 2005), participants either heard sentences that contained a few ambiguous words (e.g., The *shell* was *fired* towards the *tank*) or sentences with less lexical ambiguity (Her secrets were written in her diary). They conducted two experiments, each with a different task: in one, participants actively listened to the sentences knowing they had to answer a question about them. In the other, they passively listened to the sentences. In the active task experiment, high-ambiguity sentences were associated with more IFG activity (bilaterally) than the low-ambiguity sentences. In contrast, in the passive task experiment, differences were found in left inferior frontal sulcus (IFS) and in left middle temporal gyrus (MTG) but not in IFG. (Furthermore, these results were found using a somewhat less conservative analysis method.)

While the role of left IFG in semantic processing has been extensively studied, it is far from the only region that is involved in semantic processing of single words. Indeed, quite a few other regions have been identified in the literature: Wagner *et al.* (2000) report decreased activity in the left STG and MTG and the left superior and middle frontal gyri (SFG, MFG) when words are repeated in the context of the same task. Poldrack *et al.* (1999) report a greater increase in activity during semantic compared to phonetic tasks bilaterally in SFG, MFG and in the medial frontal gyrus (among other regions; that study imaged only the frontal parts of the brain).

Another paradigm used in studying word comprehension is the semantic priming method. As established by decades of behavioral research, processing a word is easier if a semantically related word has just been previously presented (the target word is then said to be "primed" by the previous one). It is reasonable to assume that on the neural level, accessing a word is more efficient when the word's meaning has been recently primed. This logic has lead to a number of investigations into the neural mechanisms underlying semantic priming: specifically, brain regions showing less activity for primed than unprimed words have been linked to semantic retrieval of words' meanings. In one study, Copland *et al.* (2003) presented participants with words such as *money* or *river* after these words were primed by a word such as *bank*. When neural activity for these words was compared to that of a word semantically unrelated to *bank* (e.g., *sky*), both semantically related words demonstrated neural facilitation in the left MTG and left anterior IFG (BA 47/11). However, not all studies of semantic priming have found facilitation in left IFG, and some have also revealed facilitation in temporal cortex (*cf.* Hasson *et al.*, 2006, for a recent review).

8.4. FROM WORDS TO SENTENCES: SYNTACTIC PROCESSING

Beyond the single-word level, a large body of research has focused on the neural mechanisms underlying semantic and syntactic aspects of sentence comprehension. This is possibly due to the relatively entrenched dichotomy between semantic and syntactic processes in certain linguistic and philosophical theories.

Many studies have attempted to identify neural correlates of syntactic complexity. In an early study, Just *et al.* (1996) presented participants with three sorts of sentences that differed in their structural complexity because they either contained conjoined clauses, subject-relative clauses or object-relative clauses. In brain regions roughly corresponding to Broca's and Wernicke's area (bilaterally), the volume of neural activity increased with sentence complexity. Given that three experimental conditions contained the same content words, the authors argued that the complexity of the sentence was responsible for the increased neural activity, but did not speculate on specific component functions these regions perform.

Expanding on the issue of structural complexity, some researchers have put forward the stronger claim that certain theoretical constructs of syntactic theories are related to neuronal regions (a regular relationship between subcomponents of syntactic theory and brain loci; Grodzinsky & Friederici, 2006, p. 240). Notably, Grodzinsky has argued that there are different types of dependency relations in sentences, and that parsing these relations is associated with a distinct pattern of neural activity. To examine this claim, Ben-Shachar *et al.* (2003) studied neural activity during comprehension of sentences that, from a linguistic perspective, either contained or did not contain a particular syntactic transformation. In that study, sentences that contained transformations were associated with increased neural activity in left IFG and bilaterally in posterior STS, which was taken to indicate their involvement in this very specific type of syntactic processing. Yet, a detailed examination of the results reveals an interesting pattern: in both regions, sentences with transformations showed above-baseline activity, but sentences without transformations showed below-baseline activity. This pattern is intriguing; if these regions were indeed involved in general syntactic processing, we would expect that in both conditions neural activity would be reliably above baseline. Instead, only the more difficult syntactic conditions were associated with above-baseline activity. Similar findings are seen in a study where participants were presented with either subject-relative or more complex object-relative sentences (Cooke *et al.*, 2002). Both types of sentences could contain either few or many words between the antecedent and the gap. In this study, only the most difficult condition, consisting of object-relative sentences with long antecedent-gap linkages, showed above-baseline activity in left IFG. Taken together, these findings may indicate that the increased activity associated with complex transformation indicates a categorically different mode of operation in left IFG, rather than a qualitative increase in activity that is related to syntactic complexity.

Indeed, identifying brain regions that differentiate between sentences of different syntactic complexity can be interpreted in at least two ways: on the more syntax-specific interpretation, this effect could indicate that certain brain regions are specialized in carrying out formal syntactic operations. On a more general interpretation, syntactic difficulty increases the demands on working memory, which in turn results in increased neural activity. A number of studies have tried to differentiate these two components, and some findings suggest that the syntactic effects in left IFG may reflect *maintenance* of dislocated arguments in working memory (Cooke *et al.*, 2002; Fiebach *et al.*, 2005). Thus, there is considerable debate on the explanation of syntactic transformation effects, and future research is needed to address this issue.

A different approach to studying syntactic processing was used by Dapretto and Bookheimer (1999): participants were presented with pairs of statements and determined whether the two statements had the same meaning. In one of the conditions these statements had different syntactic forms (e.g., *the policeman arrested the thief*; *the thief was arrested by the policeman*), and in another condition they were based on word substitutions (*the lawyer questioned the witness*; *the attorney questioned the witness*). When sentence-pairs differed in syntax, there was relatively increased activity in posterior IFG, whereas when they differed in the noun used, there was more activity in anterior IFG. This finding supports the purported dissociation between posterior and anterior aspects of IFG we have discussed.

Temporal regions may also be involved in syntactic processing. Left STG demonstrates less activity during blocks of sentences that share the same syntactic structure compared to blocks where different structures are mixed (Noppeney & Price, 2004). Both left STG and left IFG show increased activity for sentences that are more difficult to parse, independent of whether they are presented in spoken or written form (Constable *et al.*, 2004). When compared to simple correct sentences, sentences that include semantic or syntactic violations are associated with increased activity in posterior STG (for both types of violations) and in anterior STG (for syntactic violations; Friederici *et al.*, 2003). Other findings on the involvement of STG in syntactic processing are reviewed in Grodzinsky and Friederici (2006).

8.5. FROM WORDS TO SENTENCES: SEMANTIC PROCESSING OF SENTENCES

As we have discussed, there is some debate on whether there are brain regions that are specialized for syntactic functions. That is, it is unclear whether the theoretical element referred to as "syntax" has a unique/privileged status in the brain. There is much less debate on whether there are brain regions particularly important for semantic processing of sentences, which in this chapter will subsume processes that underlie the ability to comprehend the meaning of sentences.

In psychology, the study of sentential semantic processing often refers to the online processes by which the cognitive system constructs the meaning of sentences, and to the nature of the end product that results (e.g., is it "image-like," or a-modal/propositional in nature). Neuroscientists tackle questions that overlap to some extent. These include, but are not limited to (a) identifying regions involved in establishing sentence meaning, (b) establishing whether these regions play a language-specific role in a more general role in meaning construction, (c) understanding the processing of literal and non-literal meanings or (d) studying if sentence comprehension activates modality-specific networks related to the content of those sentences (e.g., do sentences speaking of action activate action-related motor regions, see Box 8.2).

One method that may identify brain regions involved in semantic processing is to present sentences in visual and auditory form and characterize the brain regions active for both. Using this method, Constable *et al.* (2004) revealed reliable activity to spoken and printed sentences predominantly in left hemisphere regions, including STG, MTG and IFG. However, this activity is likely to index both semantic and syntactic processing. The anterior left IFG may play a particularly important role in the integration of semantic and syntactic processes. Vandenberghe *et al.* (2002) have shown that activity in this region is sensitive to whether a sentence has a canonical grammatical structure, but *only* when the words in the sentences can be put together to form a meaningful sentence; when they cannot, then this region is not sensitive to grammaticality.

Finding that a brain region is active during the presentation of linguistic information does not necessarily mean that this region performs a function that is uniquely linguistic. Emphasizing this point, Humphries *et al.* (2001) presented participants with narrative information either via sentences (e.g., the sentence *there was a gunshot and then somebody ran away*) or via sound effects (presenting the sound of a gunshot followed by the sound of fading footsteps). Perhaps the most striking result of the study was not the difference between the two conditions, but the fact that both resulted in above-baseline activity in middle (auditory) and posterior regions of the temporal lobe and left IFG. The direct contrast between the conditions did reveal increased neural activity for the sentence condition in the anterior temporal lobes (bilaterally), posterior STS and a few other regions, suggesting these are particularly important for accessing or integrating information presented in linguistic form.

Box 8.2   Meaning and embodied representations

Does the comprehension of action sentences rely on a-modal semantic representations, or on more action-like simulations of the situations described in those sentences? This longstanding question in cognitive science has been recently addressed by neuroimaging studies. Interestingly, these suggest that action sentences systematically activate brain regions associated with observation and execution of physical actions. Sentences referring to actions performed by the mouth, hand and leg evoke greater activity in posterior left IFG than abstract sentences (Tettamanti *et al.*, 2005). In monkeys, this region has been found to contain neurons that fire during both observation and execution of certain goal-directed actions (Rizzolatti & Craighero, 2004). Thus, in humans this region may code actions at a level that is abstract enough to be accessible to language. Furthermore, the sentences referring to leg, mouth and hand differentially activated premotor regions associated with actions of these afferents. Other research shows that brain regions differentially sensitive to observation of mouth, foot and hand actions are also differentially sensitive to sentences mentioning actions performed by these effectors (Aziz-Zadeh *et al.*, 2006). Such studies suggest that language comprehension, at least in the action domain, is supported by motor systems used to perform the actions referred to in those sentences, which is consistent with theoretical approaches in the "embodied cognition" framework.

Aziz-Zadeh, L., Wilson, S. M., Rizzolatti, G., & Iacoboni, M. (2006). Congruent embodied representations for visually presented actions and linguistic phrases describing actions. *Current Biology, 16*(18), 1818–1823.

Rizzolatti, G., & Craighero, L. (2004). The mirror-neuron system. *Annual Review of Neuroscience, 27*(1), 169–192.

Tettamanti, M., Buccino, G., Saccuman, M. C., Gallese, V., Danna, M., Scifo, P., Fazio, F., Rizzolatti, G., Cappa, S. F., & Perani, D. (2005). Listening to action-related sentences activates fronto-parietal motor circuits. *Journal of Cognitive Neuroscience, 17*(2), 273–281.

Semantic integration processes have also been studied by examining neural processing of sentences that contain different types of violations. Kuperberg *et al.* (2000) contrasted brain activity during comprehension of normal sentences, with that seen during comprehension of sentences with different sorts of meaning or syntax violations (e.g., *the man buried/slept/drank the guitar*). In their study, the left IFG was particularly sensitive to the differences between normal- and meaning-violated sentences, but the fusiform gyrus was also sensitive to this difference. In another study (Friederici *et al.*, 2003), semantic violations were associated with increased activity in the middle portion of MTG as well as the Insula (bilaterally).

Several studies have examined sentential processing by identifying brain regions that show decreased activity to sentences when they are presented for a second time. Currently, the results of such studies support a role for temporal regions in sentential semantics. Stowe *et al.* (1999) used a visual presentation method, and compared the initial reading of sentences or mixed-word lists to their repeated presentation. They found decreased activity in cortical areas including the left fusiform gyrus (extending to the inferior parietal lobule and MTG; i.e., relatively posterior regions), left lingual gyrus and right STG/MTG. However, similar repetition effects were found for sentences and for mixed-word lists suggesting that the regions in which the repetition effects were found were not necessarily involved in sentence-level semantic processes (in that study, the lag between the initial and repeated presentations was 44 min, which could have contributed to diminished accessibility of the sentence meaning). Hasson *et al.* (2006) presented auditory sentences twice; these sentences either contained or did not contain subordinate clauses. Repetition of both types of sentences was associated with reduced activity bilaterally in temporal regions. However, in the right lingual gyrus they found decreased activity for repetition of non-subordinate-clause sentences, but repetition enhancement for repetition of subordinate-clause sentences. The authors also found that repetition was associated with decreased activity in left IFG, but only when the sentences were repeated in the context of an active task demanding an explicit sensibility judgment. No repetition effects in IFG were found during passive listening.

8.6. FROM SENTENCES TO DISCOURSE

Comprehension of connected sentences in the context of discourse entails discourse-level processes that are absent during comprehension of single sentences. Experimental cognitive research of discourse comprehension has examined many such processes, for example, those involved in integration of content across sentences. The understanding of the neural underpinnings of these processes is only in its initial stages.

Integration of sentential information with prior knowledge is fundamental to discourse comprehension. How does this occur? St. George *et al.* (1999) presented participants with texts that were difficult to comprehend unless presented with a title that summarized what the text was about. Temporal regions in the right hemisphere showed greater activity for untitled than for titled stories, but temporal regions on the left showed the opposite pattern. However, another study that used a conceptually similar manipulation in which texts were clarified with a picture revealed different results (Maguire *et al.*, 1999). In this study, unusual texts were either clarified by preceding them with a descriptive photo or not. It was found that clarifying a text's meaning resulted in increased activation in medial brain regions. Similarly, in examining comprehension of narratives versus unlinked sentences, Xu *et al.* (2005) found increased activity in the anterior temporal pole, and medial prefrontal cortex (bilaterally) among other regions. One way to summarize these disparate findings is that the comprehension of meaningful narratives is associated with increased neural activity as compared to less meaningful ones, although at present, the nature of the specific manipulation seems to affect the anatomical pattern of results more than does the presence of textual clarification *per se*.

World knowledge is critical for deciphering causal relationships between sentences. Neural regions involved in establishing a causal link between two sentences were first examined by Mason and Just (2004), who presented pairs of sentences varying in causal strength. They found that neural activity in temporal regions in the right hemisphere as well as regions in right IFG was mediated by the strength of the causal link between two sentences. When the sentences were moderately related (as opposed to highly related or distantly related) these regions demonstrated the greatest number of active voxels. In that study, it was the moderate-link condition that most strongly demanded construction of a causal scenario from world knowledge. Yet, a different study using a highly similar methodology (Kuperberg *et al.*, 2006) revealed quite different findings: while the moderate-link condition was associated with the highest activity in certain regions, these did not overlap with those reported by Mason and Just.

A number of studies have attempted to identify brain regions involved in integrating consecutive discourse ideas. Ferstl *et al.* (2005) found that hearing a statement that is inconsistent with prior discourse context was associated with increased neural activity in the vicinity of the right anterior temporal pole and bilaterally in anterior IFG. Individual differences in perceiving inconsistencies were linked to activity in dorso-medial prefrontal cortex. Another study demonstrated that transitions between narrative events during reading are associated with increased activity in midline and right temporal regions (Speer *et al.*, 2007).

The emerging picture from studies of discourse comprehension is that discourse-level relations in text affect activity in brain regions that are not typically involved in the comprehension of single, context-independent sentences. Furthermore, in contrast to the relatively entrenched position that language is mainly left lateralized, discourse relations seem to affect activity bilaterally (Jung-Beeman, 2005).

8.7. CHALLENGES AND FUTURE DIRECTIONS

One of the main challenges facing future research is establishing how and whether language processing capitalizes on more general functions. As reviewed, extracting speech categories from the auditory input may rely on general mechanisms. Similarly, accessing semantic knowledge via words likely depends to some extent on general mechanisms mediating access to semantic knowledge whether triggered verbally or by other means (e.g., pictures or gestures). Compositional processes at the sentence level may share common functions and neural basis with processes mediating music comprehension, comprehension of meaningful sound sequences or even pantomime sequences. On the discourse level, establishing consistency or inconsistency with prior information could rely on neural substrates with a general role in noticing unexpected stimuli in an input stream. Thus, understanding the relation between language comprehension and related domains would be an essential step towards establishing what types of neural processing are more or less specialized for language comprehension.

Another research direction that is developing to be of major interest will address how non-verbal input such as hand gestures or face movements affect the neural processing of language. The presence of gestures can serve to emphasize or add to information conveyed verbally. Thus, the presence of gesture will result in increased activity in certain regions (as more information is processed), but less activity in others, as interpretation *per se* may be easier. Understanding how such integration takes place will enhance understanding of both verbal and non-verbal processing.

Finally, we expect that much research effort will be dedicated to understanding the extent to which language comprehension relies on more basic perceptual or motor systems, and the specific circumstances in which these systems play a greater or lesser role. In particular, it would be important to explicate how activity in such areas serves language comprehension, and whether their activity is sensitive to higher level, a-modal functions expressed by language, such as negation.

To summarize, the neurophysiology of language has benefited greatly from fMRI research to date. The challenges facing this research in the future are interesting, and successfully dealing with those challenges is likely to lead to a much better understating of the human language system.

REFERENCES

Belin, P., Fecteau, S., & Bedard, C. (2004). Thinking the voice: Neural correlates of voice perception. *Trends in Cognitive Science*, 8(3), 129–135.

Ben-Shachar, M., Hendler, T., Kahn, I., Ben-Bashat, D., & Grodzinsky, Y. (2003). The neural reality of syntactic transformations. *Psychological Science*, 14, 433–440.

Calvert, G.A., Bullmore, E.T., Brammer, M.J., Campbell, R., Williams, S.C.R., McGuire, P.K. *et al.* (1997). Activation of auditory cortex during silent lipreading. *Science*, 276(5312), 593–596.

Calvert, G.A., Campbell, R., & Brammer, M.J. (2000). Evidence from functional magnetic resonance imaging of crossmodal binding in the human heteromodal cortex. *Current Biology, 10*(11), 649–657.

Constable, R.T., Pugh, K.R., Berroya, E., Mencl, W.E., Westerveld, M., Ni, W., & Shankweiler, D. (2004). Sentence complexity and input modality effects in sentence comprehension: An fMRI study. *Neuroimage, 22*(1), 11–21.

Cooke, A., Zurif, E.B., DeVita, C., Alsop, D., Koenig, P., Detre, J., Gee, J., Pinãngo, M., Balogh, J., & Grossman, M. (2002). Neural basis for sentence comprehension: Grammatical and short-term memory components. *Human Brain Mapping, 15*(2), 80–94.

Copland, D.A., de Zubicaray, G.I., McMahon, K., Wilson, S.J., Eastburn, M., & Chenery, H.J. (2003). Brain activity during automatic semantic priming revealed by event-related functional magnetic resonance imaging. *Neuroimage, 20*(1), 302–310.

Dapretto, M., & Bookheimer, S.Y. (1999). Form and content: Dissociating syntax and semantics in sentence comprehension. *Neuron, 24*(2), 427–432.

Demb, J.B., Desmond, J.E., Wagner, A.D., Vaidya, C.J., Glover, G.H., & Gabrieli, J.D. (1995). Semantic encoding and retrieval in the left inferior prefrontal cortex: A functional MRI study of task difficulty and process specificity. *Journal of Neuroscience, 15*(9), 5870–5878.

Ferstl, E.C., Rinck, M., & von Cramon, D.Y. (2005). Emotional and temporal aspects of situation model processing during text comprehension: An event-related fMRI study. *Journal of Cognitive Neuroscience, 17*(5), 724–739.

Fiebach, C.J., Schlesewsky, M., Lohmann, G., von Cramon, D.Y., & Friederici, A.D. (2005). Revisiting the role of Broca's area in sentence processing: Syntactic integration versus syntactic working memory. *Human Brain Mapping, 24*(2), 79–91.

Friederici, A.D., Rüschemeyer, S.-A., Hahne, A., & Fiebach, C.J. (2003). The role of left inferior frontal and superior temporal cortex in sentence comprehension: Localizing syntactic and semantic processes. *Cerebral Cortex, 13*(2), 170–177.

Gold, B.T., Balota, D.A., Kirchhoff, B.A., & Buckner, R.L. (2005). Common and dissociable activation patterns associated with controlled semantic and phonological processing: Evidence from fMRI adaptation. *Cerebral Cortex, 15*(9), 1438–1450.

Griffiths, T.D., & Warren, J.D. (2002). The planum temporale as a computational hub. *Trends in Neurosciences, 25*(7), 348–353.

Grodzinsky, Y., & Friederici, A.D. (2006). Neuroimaging of syntax and syntactic processing. *Current Opinion in Neurobiology, 16*(2), 240–246.

Hasson, U., Nusbaum, H.C., & Small, S.L. (2006). Repetition suppression for spoken sentences and the effect of task demands. *Journal of Cognitive Neuroscience, 18*(12), 2013–2029.

Humphries, C., Willard, K., Buchsbaum, B., & Hickok, G. (2001). Role of anterior temporal cortex in auditory sentence comprehension: an fMRI study. *Neuroreport, 12*(8), 1749–1752.

Jung-Beeman, M. (2005). Bilateral brain processes for comprehending natural language. *Trends in Cognitive Sciences, 9*, 512–518.

Just, M.A., Carpenter, P.A., Keller, T.A., Eddy, W.F., & Thulborn, K.R. (1996). Brain activation modulated by sentence comprehension. *Science, 274*(5284), 114–116.

Kuperberg, G.R., McGuire, P.K., Bullmore, E.T., Brammer, M.J., Rabe-Hesketh, S., Wright, I.C., Lythgoe, D.J., Williams, S.R.C., & David, A.S. (2000). Common and distinct neural substrates for pragmatic, semantic, and syntactic processing of spoken sentences: An fMRI study. *Journal of Cognitive Neuroscience, 12*(2), 321–341.

Kuperberg, G.R., Lakshmanan, B.M., Caplan, D.N., & Holcomb, P.J. (2006). Making sense of discourse: An fMRI study of causal inferencing across sentences. *Neuroimage, 33*, 343–361.

Maguire, E.A., Frith, C.D., & Morris, R.G.M. (1999). The functional neuroanatomy of comprehension and memory: The importance of prior knowledge. *Brain, 122*, 1839–1850.

Martin, A., & Chao, L.L. (2001). Semantic memory and the brain: Structure and processes. *Current Opinion in Neurobiology, 11*(2), 194–201.

Mason, R.A., & Just, M.A. (2004). How the brain processes causal inferences in text. *Psychological Science, 15*(1), 1.

Meyer, M., Zaehle, T., Gountouna, V.E., Barron, A., Jancke, L., & Turk, A. (2005). Spectro-temporal processing during speech perception involves left posterior auditory cortex. *Neuroreport, 16*(18), 1985–1989.

Narain, C., Scott, S.K., Wise, R.J.S., Rosen, S., Leff, A., Iversen, S.D., & Matthews, P.M. (2003). Defining a left-lateralized response specific to intelligible speech using fMRI. *Cerebral Cortex, 13*, 1362–1368.

Noppeney, U., & Price, C.J. (2004). An fMRI study of syntactic adaptation. *Journal of Cognitive Neuroscience, 16*, 702–713.

Poldrack, R.A., Wagner, A.D., Prull, M.W., Desmond, J.E., Glover, G.H., & Gabrieli, J.D.E. (1999). Functional specialization for semantic and phonological processing in the left inferior prefrontal cortex. *Neuroimage, 10*, 15–35.

Rauschecker, J.P., & Tian, B. (2000). Mechanisms and streams for processing of "what" and "where" in auditory cortex. *Proceedings of the National Academy of Sciences, 97*(22), 11800–11806.

Rodd, J.M., Davis, M.H., & Johnsrude, I.S. (2005). The neural mechanisms of speech comprehension: fMRI studies of semantic ambiguity. *Cerebral Cortex, 15*(8), 1261–1269.

Scott, S.K., & Johnsrude, I.S. (2003). The neuroanatomical and functional organization of speech perception. *Trends in Neurosciences, 26*(2), 100–107.

Skipper, J.I., Nusbaum, H.C., & Small, S.L. (2005). Listening to talking faces: Motor cortical activation during speech perception. *Neuroimage, 25*, 76–89.

Speer, N. K., Zacks, J. M., & Reynolds, J. R. (2007). Human brain activity time-locked to narrative event boundaries. *Psychological Science, 18*, 449–455.

St. George, M., Kutas, M., Martinez, A., & Sereno, M.I. (1999). Semantic integration in reading: Engagement of the right hemisphere during discourse processing. *Brain, 122*, 1317–1325.

Stowe, L.A., Paans, A.M.J., Wijers, A.A., Zwarts, F., Mulder, G., & Vaalburg, W. (1999). Sentence comprehension and word repetition: A positron emission tomography investigation. *Psychophysiology, 36*, 786–801.

Vandenberghe, R., Nobre, A.C., & Price, C.J. (2002). The response of left temporal cortex to sentences. *Journal of Cognitive Neuroscience, 14*(4), 550–560.

Vouloumanos, A., Kiehl, K.A., Werker, J.F., & Liddle, P.F. (2001). Detection of sounds in the auditory stream: Event-related fMRI evidence for differential activation to speech and nonspeech. *Journal of Cognitive Neuroscience, 13*(7), 994–1005.

Wagner, A.D., Koutstaal, W., Maril, A., Schacter, D.L., & Buckner, R.L. (2000). Task-specific repetition priming in left inferior prefrontal cortex. *Cerebral Cortex, 10*, 1176–1184.

Xu, J., Kemeny, S., Park, G., Frattali, C., & Braun, A. (2005). Language in context: emergent features of word, sentence, and narrative comprehension. *Neuroimage, 25*(3), 1002–1015.

Further Readings

Hasson, U., Nusbaum, H. C., & Small, S. L. (2007). Brain networks subserving the extraction of sentence information and its encoding to memory. *Cerebral Cortex* Advance Access published on March 19, 2007, DOI 10.1093/cercor/bhm016.
An examination of brain regions involved in the integration of discourse contents, and their encoding to memory. Highlights the role of temporal, frontal and midline brain regions in encoding discourse content.

Jung-Beeman, M. (2005). Bilateral brain processes for comprehending natural language. *Trends in Cognitive Sciences*, *9*, 512–518.
A review of higher level language comprehension with emphasis on the role of the right hemisphere in semantic processing at the single word and discourse level.

Vigneau, M., Beaucousin, V., Herve, P.Y., Duffau, H., Crivello, F., Houde, O., Mazoyer, B., & Tzourio-Mazoyer, N. (2006). Meta-analyzing left hemisphere language areas: Phonology, semantics, and sentence processing. *Neuroimage*, *30*(4), 1414–1432.
An analysis of data from 129 scientific reports of neural activity during language processing, broadly defined.

Xu, J., Kemeny, S., Park, G., Frattali, C., & Braun, A. (2005). Language in context: Emergent features of word, sentence, and narrative comprehension. *Neuroimage*, *25*(3), 1002–1015.
A comprehensive fMRI study of the differences between single-word, single-sentence and discourse processing.

CHAPTER

9

Event-Related Potentials in the Study of Language

KARSTEN STEINHAUER[1] and JOHN F. CONNOLLY[2,3]

[1]Centre for Research on Language, Mind and Brain (CRLMB) and School of Communication Sciences and Disorders (SCSD),
Faculty of Medicine, McGill University, Montréal, Québec, Canada

[2]Centre de Recherche, Institut Universitaire de Gériatrie de Montréal (CRIUGM),
Université de Montréal, Montréal, Québec, Canada

[3]Centre de Recherche en Neuropsychologie et Cognition (CERNEC),
Département de Psychologie, Université de Montréal, Montréal, Québec, Canada

ABSTRACT

Why bother recording event-related brain potentials (ERP) in language studies? What is their "added value?" We attempt to answer these questions by providing an overview of recent ERP work investigating word segmentation and phonological analysis during speech processing, semantic integration mechanisms, syntactic processing, and the analysis of prosody in speech. Our view is that the inclusion of ERP has increased our understanding of how the brain accomplishes language. It is equally true that for every question answered there are new questions raised; we address some of these yet unresolved questions as well. The skeptical reader may also ask, "so what?" Are there any practical advantages to such work? The fact is there are practical benefits, such as the newly developed ability to assess cognitive abilities in non-communicative patients including those in so-called vegetative states – a significant advance in health care that is the direct result of basic ERP research in neurolinguistics.

9.1. INTRODUCTION

Investigating how the brain accomplishes language comprehension is a particular challenge compared to other areas of cognitive neuroscience. Animal models are of limited value given that no other species has a comparably complex communication system. On the other hand, areas as distinct as sensory physiology and formal linguistics contribute important details to our understanding of language processes in the brain. Most ERP research discussed here is strongly rooted in psycholinguistic models of word recognition, syntactic parsing, and the online integration of information. This work provides a solid foundation upon which to examine cognitive factors and the brain mechanisms involved in the transformation of a low-level signal into the highly complex symbolic system we know as human language.

Three major goals of using neuroimaging are to understand *where* language is processed in the brain and *when* and *how* the different levels of linguistic processing unfold in time. Thanks to their ability to provide continuous online measures with an excellent temporal resolution, even in the absence of behavioral tasks, ERPs contribute primarily to the second and third goals and can add the valuable perspective of real-time brain dynamics underlying linguistic operations. We will concentrate on these by making reference to a number of established "ERP components," that is, characteristic brain potentials at the scalp assumed to reflect specific neurocognitive processes, and, on occasion, corresponding event-related magnetic field (ERMF) research. We will outline the factors that influence these brain responses, and by inference, the linguistic processes they reflect. As ERP and ERMF measures are not completely without power in addressing *where* in the brain certain language functions may occur, those aspects will briefly be mentioned as well.

ERP research in linguistics has historically had a strong element of component discovery since the first "language" component, the "semantic" N400, was observed (Kutas & Hillyard, 1980). In the intervening years additional ERP components associated with acoustic–phonetic, phonological, orthographic, prosodic, and syntactic processes have been discovered. Those of them concerned with basic operations such as phoneme discrimination or word segmentation tend to be early (100–200 ms), fast, and automatic. Other components reflect integration or revision processes and tend to have larger latencies (up to 1 s). Parallel processes can be distinguished primarily in terms of ERP scalp distributions.

A recent theme that has gained increasing prominence is the degree to which these "linguistic" components are actually domain specific. Most of them have been described initially as being related to language processes only. However, subsequent research has usually weakened the case for domain specificity. For example, the P600 (Osterhout & Holcomb, 1992) had been linked to syntactic processing. More recently, however, it has been proposed that language and music share processing resources and that a functional overlap exists between neural populations that are responsible for structural analyses in both domains (Patel *et al.*, 1998). It will become apparent that most "linguistic" ERP components may also be associated with non-linguistic functions. It might even be proposed that some of them might be more accurately thought of as domain-non-specific responses that reflect basic operations critical for, but not limited to, linguistics processes.

9.2. LANGUAGE-RELATED COMPONENTS AND THEIR FUNCTIONAL SIGNIFICANCE

The following subsections will discuss which ERP components have contributed to our understanding of psycholinguistic processes in phonology, lexical/conceptual semantics, syntax, as well as their respective interactions.

9.2.1. The N100: An Exogenous Component with Linguistic Functions?

The complexity of examining language functions with ERP is captured very well by the first component that we will discuss, the N100: a *Negativity peaking around 100* ms. Long considered an exogenous response sensitive to the physical features (e.g., loudness or brightness) of an auditory, visual, or tactile stimulus, it has more recently been linked to word segmentation processes (Sanders & Neville, 2003). Noting the disagreement in the literature as to whether continuous speech stimuli elicit the early sensory components (including the N100), these studies sought to clarify whether word onsets within a context of continuous speech would elicit the early sensory or "obligatory" components. This work also examined whether the hypothesized word onset responses were related to segmentation and word stress. ERPs to word initial and word medial syllables were obtained within different types of sentence context. It was found that word onset syllables elicited larger anterior N100 responses than word medial syllables across all sentence conditions.

Word onsets in continuous speech can vary in their physical characteristics (e.g., loudness, duration) by virtue of whether the syllable is stressed or unstressed. Thus, word onset effects on the N100 were examined as a function of word stress with the finding that stressed syllables evoked larger N100 responses than unstressed syllables at electrode sites near the midline. Such an effect was expected given the physical differences that exist between stressed and unstressed syllables. However, it was concluded that the N100 was monitoring more than the physical characteristics of the stressed and unstressed syllables because the N100 to these stimuli showed a different scalp distribution compared to that seen for the N100 to word onset and word medial syllables which had been equated for physical characteristics. Further evidence for a language-related role for the N100 was found in an examination of Japanese–English bilinguals who failed to show N100 segmentation effects to English stimuli similar to native English speakers. This observation contrasts with the finding that the Japanese–English bilinguals showed clear N100 responses to sentence onsets thus exhibiting normal acoustic ERPs. The conclusion was made that non-native speakers do not use the acoustic differences as part of a speech comprehension system in the same manner as native speakers. Although these effects require replication as a final confirmation, they nevertheless demonstrate very well the neural flexibility involved in language function as well as the "multipurpose" nature of some ERP components that can reflect simple acoustic processing or complex linguistic segmentation. This work also demonstrates quite well the importance of evaluating all aspects of an ERP component insofar as the primary difference between the N100 as an acoustic or language component was its scalp distribution and thus, by implication, its neural generators.

9.2.2. Prelexical Expectations: The Phonological Mapping Negativity

Terminal words of spoken sentences that violate contextually developed phonological expectations (as in *1b*) elicit this fronto-centrally distributed ERP component that peaks in the late 200 ms range (270–310 ms) and is earlier than and distinct from the N400 reflecting pure semantic anomalies (*1c*) (see also Section 9.2.3). In combined violations of phonological and semantic expectations (*1d*), the Phonological mapping negativity (PMN) precedes the N400 (Figures 9.1 and 9.2(b)).

(1a) *Father carved the turkey with a knife (expected word: knife)*
(1b) *The pigs wallowed in the pen (mud)*
(1c) *The gambler had a streak of bad luggage (luck)*
(1d) *The winter was harsh this allowance (year)*

This component was labeled the phonological mismatch negativity (PMN) (Connolly & Phillips, 1994, from which

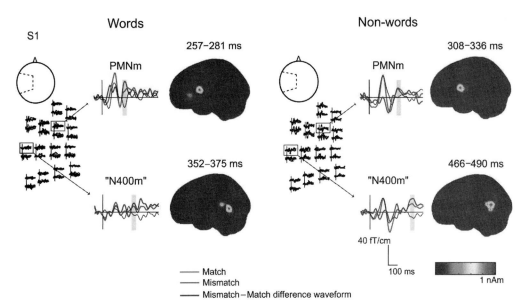

FIGURE 9.1   Phonological mapping negativity (PMN) and semantic N400. MEG responses to words (left) and non-words (right) for one participant for those left-hemisphere channels showing maximum amplitude for the magnetic PMN (PMNm) and the N400-like response. The corresponding estimates of the PMN- and N400m-like response sources (over a 25 ms time window centered at the peak of the response) are depicted in the brain images. The gray vertical bars indicate the 50 ms time periods within which significant PMNm- and N400m-like responses occurred. *Source*: Modified after Kujala *et al.*, 2004 (see Plate 9).

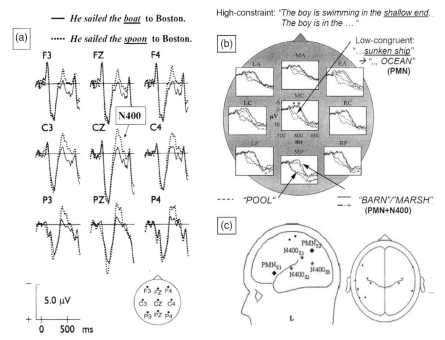

FIGURE 9.2   ERPs to target words illustrating N400 effects. Negative polarity is plotted upwards. (a) Semantic anomaly in sentences. Semantically implausible content words in sentence contexts elicit larger centro-parietal N400s (dotted lines) than plausible words (solid lines) between 300 and 600 ms. *Source*: Modified after Steinhauer *et al.* (2001). (b) PMN and N400 effects in a cross-modal priming study in which written sentences that ended in subordinate words primed superordinate words that ended spoken sentences. Exemplar-subordinate probabilities for congruent endings determined high-constraint primes (shallow end) or a low-constraint primes (sunken ship) which primed auditory target words in the paired sentences (in capitals) that were either congruent (Pool/Ocean) or not (Barn/Marsh). While incongruent targets (e.g., Barn/Marsh) always elicited PMNs and N400s, congruent targets following low-constraint contexts (Ocean) yielded PMNs only (L = left, R = right, M = middle, A = anterior, C = central, P = parietal). *Source*: Modified after D'Arcy, 2004. (c) Source localization revealed distinct neural generators for PMN and N400 components, primarily in the left hemisphere.

Box 9.2   Clinical assessment of language using cognitive ERP

Unlike the history of evoked potentials in assessing sensory function (EP; Chiappa, 1997) cognitive ERP have only recently been employed to examine patients' language abilities and the functional integrity of systems upon which language depends. The failure to employ ERP in clinical settings was partially due to the outdated belief that these components were insufficiently reliable in their occurrence, physical characteristics (such as latency), and functional specificity. Today, however, a high level of specificity and replicability have been established, in some cases by adapting psychometrically valid neuropsychological tests for computer presentation and simultaneous ERP recording. These tests provide a method for neuropsychological assessment of patients who are otherwise impossible to assess due to the severity of their brain injuries, for example, after stroke (D'Arcy *et al.*, 2003). The link between basic research and clinical application is exemplified by an auditory ERP study that tested semantic and phonological sentence processing in a traumatic brain injury patient in a persistent vegetative state (Connolly *et al.*, 1999). The top part of the figure shows axial computerized tomography (CT) scans of patient's head. Left brain is on the right side of scan, the entry wound is left frontal. Three-dimensional figures in the bottom part depict N400 responses in the left (T3–F3) and right (T4–F4) hemisphere to sentences such as *The gambler had a streak of bad luggage* (see Sections 9.2.2 and 9.2.3). Amplitude is on left vertical axis, scalp location on bottom axis, and time (ms) on right axis. Despite his apparent vegetative state, the patient exhibited classic N400 responses to the semantically incongruous sentence endings. Having demonstrated the patient's comprehension abilities were intact, the inference was drawn that he was mentally intact and possessing sufficient mental resources to merit rehabilitation (ultimately highly successful) instead of the scheduled discharge to a long-stay facility and the associated poor prognosis for patients in such conditions.

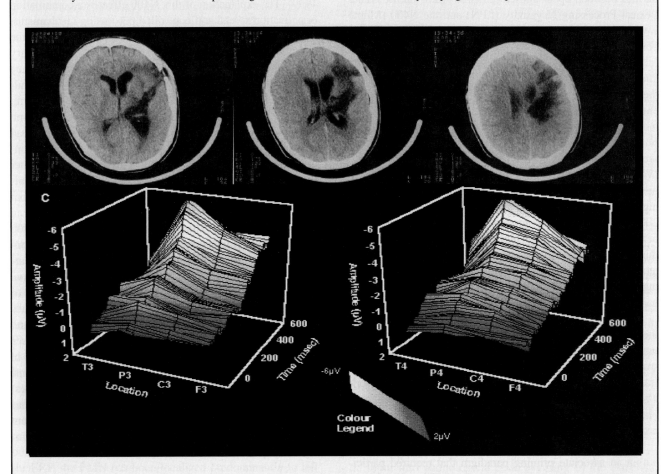

Chiappa, K.H. (1997). *Evoked potentials in clinical medicine* (3rd edn). New York: Lippincott-Raven.

Connolly, J.F., Mate-Kole, C.C., & Joyce, B.M. (1999). Global aphasia: An innovative assessment approach. *Archives of Physical Medicine and Rehabilitation*, 80, 1309–1315.

D'Arcy, R.C.N., Marchand, Y., Eskes, G.A., Harrison, E.R., Phillips, S.J., Major, A., & Connolly, J.F. (2003). Electrophysiological assessment of language function following stroke. *Clinical Neurophysiology*, 114, 662–672.

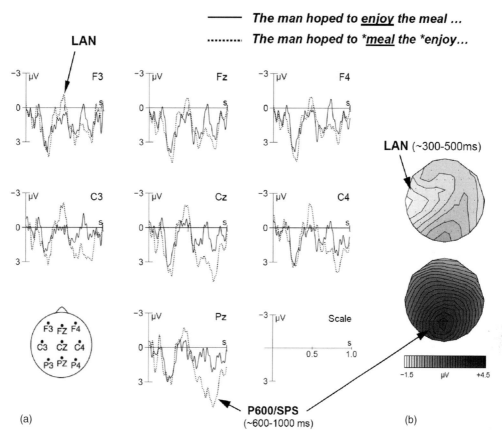

FIGURE 9.3 Biphasic LAN and P600 pattern elicited by syntactic word category violations in an English reading study. Negative polarity is plotted upwards. (a) ERP plots of the ungrammatical target words (dotted lines) show a LAN effect between 300 and 500 ms, that is in a similar time interval as the N400 (see Figure 2(a)), which was followed by a late posterior P600/SPS between 600 and 1000 ms. (b) Voltage maps of the difference waves (violation minus correct control) illustrate the scalp distribution of both ERP components (Steinhauer, unpublished data).

only required to identify the conceptual meaning of content words (*John*, *hit*, *ball*) but it also needs to analyze the grammatical relations between them. This syntactic parsing of the hierarchical structure of utterances will reveal whether *John* is the agent of the action, and whether this action is continuing or has been completed. Parsing takes place incrementally in real time at rates of approximately three words per second and involves (1) the analysis of word order and word category information including function words (*is*, *the*, *was*, *by*) and (2) the checking of certain features that need to be congruent between linked sentence constituents (e.g., subject–verb agreement in English).

The most common way of studying ERP correlates of morpho-syntactic processing has utilized violation paradigms. The rationale is that violations should disrupt or increase the workload of the brain systems underlying the type of processing that is of interest. These changes are intended to elicit a specific differential ERP. Two such ERP components have indeed been identified as markers of two stages of syntactic processing: an early, often left-lateralized anterior negativity (LAN) typically occurring between 100 and 500 ms that has been linked to automatic first pass

parsing, and a late centro-parietal positive component between 500 and 1000 ms (P600), that may reflect rather controlled attempts to reanalyze and fix the anomaly at a later stage (Figure 9.3). The qualitative differences between these two components and the semantic N400 have been taken as evidence that syntactic and semantic information are processed differently in the brain (Osterhout & Holcomb, 1992).

9.2.5. Early and Other Left Anterior Negativities

LANs have primarily been reported for outright syntactic violations and not, for example, for structure ambiguities. Typical conditions eliciting LANs are word category violations (in 2a,b; Neville *et al.*, 1991; Friederici, 2002) and violations of number agreement between the subject and verb (2c). (*Note*: an asterisk "*" marks the ungrammatical word.)

*(2a) He criticized Max's * of proof the theorem*

*(2b) Die Bluse wurde am *ge-bügelt* (The blouse was at the *ironed)

*(2c) The children *plays in the garden*

Among these, only word category violations (*2a,b*) have been found to yield a particularly early LAN (ELAN) between 100 and 300 ms, which appeared more reliable in auditory than visual studies, and has been linked to neural generators in Broca's area and the anterior temporal lobe. Some models suggest that this ELAN is distinct from other later LAN effects and reflects interruptions of highly automatic processes during the very first phase of building up a phrase structural representation that is required in subsequent processing stages. Within this framework, other morpho-syntactic operations (and respective violations) affecting agreement features or verb arguments which already depend on a phrase marker, are processed in parallel to semantic information, and elicit the later LANs between 300 and 500 ms (concurrently with the semantic N400; Rossi *et al.*, 2005). Unlike P600s (see Section 9.2.6), ELANs were not influenced by the relative proportion of violations in an experiment, suggesting their "autonomous" status independent of processing strategies (Hahne & Friederici, 1999).

In psycholinguistics, the short latency of the "syntactic" ELAN component has been of theoretical importance as it lent strong empirical support to so-called syntax first models that claim an initial autonomy phase for the syntactic parsing device, as opposed to more interactive models. However, the short ELAN latency has convincingly been shown to depend on the rapid availability of word category information in the respective experimental paradigms (e.g., the prefix *ge* in the prevailing German paradigm in (*2b*)), rather than the proposed early stage of processing *per se*. In absence of such early phonological markers, word category violations elicit LANs in the typical 300–500 ms time window of other morpho-syntactic violations (Hagoort *et al.*, 2003), even though primacy of syntactic over semantic processes may still hold. The critical impact of phonological markers on the ELAN latency raises yet another issue; since the German prefix *ge* is not restricted to verb forms (and, therefore, is not a reliable word category marker), the ELAN likely reflects violations of phonological expectations related to word category violations rather than an automatic response to syntactic violations as such. This in itself, however, is remarkable as it suggests an extremely early phonological mismatch/detection mechanism based on experimental processing regularities and resulting expectations, modulating the ERP in less than 200 ms. Such an account may also explain the modality differences, that is, the greater robustness of ELANs in auditory experiments. Compatible with this notion are studies that reported a similar early anterior negativity over the right hemisphere (ERAN (early right-anterior negativity)) for certain musical violations (Patel *et al.*, 1998). Hagoort *et al.* (2003) set out to replicate ELAN effects in a Dutch reading study that avoided word initial markings of the word category. As expected, they observed an anterior negativity only between 300 and 500 ms which, moreover, was bilaterally distributed rather than left lateralized. Lau *et al.* (2006) found that

clear LAN-like effects occurred only if local phrase structure imposed high constraints on the target word, whereas less predictable structures resulted in attenuated LAN effects. Predictability and expectations may be crucial to our understanding of LAN-like effects in morpho-syntactic processing more generally. Previous reviews have argued that failure to replicate [E]LANs was due to the failure to create outright syntax violations. However, even the standard paradigm used to successfully elicit ELAN effects in German in (*2b*) does actually not meet this particular criterion (see (*3*)). That is, at the position of the supposed "outright violation" sentences can still be completed such that a syntactically correct (although semantically somewhat odd) sentence results:

(*3*) *Die Bluse wurde am gebügelt noch festlicher wirkenden Jackett mit Nadeln befestigt*
The blouse was to the ironed even more festive seeming jacket with pins fixed (Literal translation)
The blouse was pinned to the jacket which, after being ironed, appeared even more festive (Paraphrase)

The adverbial and adjectival use of participles in German illustrated in (*3*) poses a major problem to the traditional interpretation of ELAN components. Based on word category information alone, the brain has simply no reason to assume an outright syntax violation, unless (a) it either knows in advance that such sentences are not included in the experiment (i.e., a pragmatic constraint) or (b) it uses prosodic cues (or punctuation) to determine that the sentence will not be continued beyond the past participle. In this latter case, however, the word category is entirely irrelevant as *any* single-word completion following *am* would cause a syntax violation.

To summarize, whereas the ELAN appears to be related to violations of expected speech sounds or orthographic patterns in particularly constrained structural environments, the somewhat later anterior negativities between 300 and 500 ms may be more directly linked to structural/syntactic processes proper. Several current models have associated these later LAN components with (the interruption of) proceduralized cognitive operations such as rule-based sequencing or structural unifications, either within the linguistic domain or across cognitive domains (see Chapter 18, this volume; Hoen & Dominey, 2000). In fact, there exist a few reports of LAN-like effects for non-linguistic sequencing (Hoen & Dominey, 2000). A rule-based interpretation of LANs beyond syntax would also be compatible with LAN effects found for over-regularizations in morpho-phonology (e.g., *childs* instead of *children*).

9.2.5.1. Working Memory

In the previous section we discussed rule-based accounts of LAN-like components. Another account explains these components in terms of working memory (WM) load increases.

This appears appropriate for syntactic structures involving long-distance dependencies (such as *wh* questions) and may, in fact, refer to a distinct set of left anterior negativities. LAN effects reflecting WM load usually tend to display broader distributions and longer durations than the focal, transient morpho-syntactic LAN components (Martin-Loeches *et al.*, 2005). Whether a unified WM-based LAN interpretation could appropriately account for LANs elicited by word category or agreement violations remains a controversial issue.

9.2.5.2. Scalp Distribution of LAN Components

Despite their name, the scalp distribution of LAN-like components is not always left lateralized, nor is it always frontal. Factors underlying this variability are not well understood, nor are the reasons explaining why the same paradigm employed for eliciting LANs in some studies fail to do so in others (Lau *et al.*, 2006). The consistency of phonological or orthographic markings along with predictable sentence structures may be important for the elicitation of ELAN components. Recent data indicate that left lateralization of LANs may be modulated by linguistic proficiency levels even in native speakers (Pakulak & Neville, 2004).

9.2.6. P600/Syntactic Positive Shift

The second syntax-related ERP component is a late positivity between 500 and 1000 ms, dubbed the P600 (Osterhout & Holcomb, 1992) or *syntactic positive shift* (SPS), which may be preceded by [E]LAN components (Figure 9.3). The P600/SPS has been linked to more controlled processes during second pass parsing and, unlike LANs, is often found to be modulated by non-syntactic factors including semantic information, processing strategies, and experimental tasks. P600 components have been found across languages for a large variety of linguistic anomalies, such as (a) non-preferred "garden path" sentences that require structural reanalyses due to local ambiguities (Osterhout & Holcomb, 1992; Mecklinger *et al.*, 1995; Steinhauer *et al.*, 1999) and (b) most types of morpho-syntactic violations (such as those in *1a,b,c*). Kaan *et al.* (2000) demonstrated that structurally more complex sentences may evoke a P600 even in the absence of any violation or ambiguity. Taken together, these findings would suggest that the P600/SPS is a rather general marker for structural processing.

The considerable range of (linguistic) phenomena eliciting P600 effects raised the question of whether this response was language-specific at all. A direct comparison of linguistic and musical violations found P600-like waveforms in *both* domains; moreover, their amplitudes displayed parametric modulation as a function of violation strength (Patel *et al.*, 1998) – a finding that clearly questions the P600 as a language-specific response. It was suggested the P600/SPS may rather be viewed as a member of the *P300 family* of

WM-related components, providing a parsimonious domain-general P600 account (Coulson *et al.*, 1998; but see Friederici *et al.*, 2001). Studies examined if the P600 behaved like a P300 and shared its topographical profile, but the overall results were inconclusive. For example, increasing the probability of violations did reduce the P600 amplitude in some studies (pro P300 interpretation) but not in others (contra P300) (Friederici *et al.*, 2001). Patient data showed that basal ganglia lesions affect only the P600 but not the parietal P300 (Kotz *et al.*, 2003), suggesting a dissociation of the components. Current thinking is that the P600 should not be viewed as a monolithic component, but may occasionally comprise P3b-like subcomponents. This hypothesis was strongly supported by a study using temporal–spatial principle component analysis (PCA) to tease apart P600 subcomponents (Friederici *et al.*, 2001). In fact, the authors suggested that P600 subcomponents may reflect the diagnosis of syntactic problems, attempts to fix them, secondary checking processes, and phonological revisions.

9.2.7. Verb Argument Structure Violations and Thematic Roles

Verb argument structure violations seem to elicit more complex patterns than other violations, arguably because they affect thematic role assignments (i.e., "who did what to whom?") in addition to syntactic aspects. These effects also vary across languages. A German study found that, whereas violating the case of an object noun phrase (NP) by swapping dative and accusative case markings elicited a LAN/P600, violating the *number* of arguments by adding a direct object NP to an intransitive verb, as in (*4*), evoked an N400/P600 instead (Friederici & Frisch, 2000).

(4) Sie weiß, dass der Kommissar (NOM) den Banker (ACC) *abreiste (V)
She knows that the inspector (NOM) the banker (ACC) departed (V, intransitive)

Revisions of case assignment elicited N400 effects (sometimes without P600s), while thematic role revisions yielded P600-like positivities (sometimes without preceding negativities) (Bornkessel & Schlesewsky, 2006; Kuperberg *et al.*, 2006). The most striking findings in this respect have been unexpected P600s instead of N400s for sentences in which the thematic roles violated animacy constraints as in (*5*).

(5) For breakfast the eggs would only #eat toast and jam.

Several research groups have suggested that these findings may require a reinterpretation of P600 (sub)components, for example in terms of a re-checking mechanism in cases of conflicts between the syntactic parser and a parallel thematic evaluation heuristics (e.g., van Herten *et al.*, 2005).

9.2.8. Interactions Between Syntax, Semantics, Discourse, and Prosody

Some of the most interesting questions in psycholinguistics concern the integration and interplay of different kinds of information, such as syntax and semantics. Can the syntactic parser be viewed as an autonomous, encapsulated module, as suggested by some "syntax first" models? Or rather is there a continuous multidirectional exchange of all varieties of information as proposed by interactive models? How do N400, LANs, and P600s interact in the case of double violations?

As far as the incoming target word itself is concerned, most available data suggest an early stage of largely parallel semantic/thematic and morpho-syntactic processing followed by a later integration stage (P600 interval) that allows for interaction between different types of information. Thus double violations typically tend to elicit additive effects of a syntactic LAN/P600 pattern and a semantic N400 with possible non-additive modulations of the P600 (Osterhout & Nicol, 1999; Gunter et al., 1997).

One exception, however, is that unlike other syntax violations, word category violations seem to require immediate phrase structural revision and thus temporarily block further semantic processing during this period (thereby preventing or delaying N400 effects of the semantic incongruency). Thus, in a standard grammaticality judgment task, these double violations did not elicit a late N400-like response until a P600 reflecting a structure revision had already occurred (Friederici et al., 1999). This suggests that semantic integration in sentences is at least partly guided by syntactic structure. Typical N400s were observed only if semantic task instructions explicitly required instant semantic integration. Conversely, in order for semantic information to be able to influence syntactic parsing decisions, it must be available much earlier in the sentence, or may not show an effect at all. A study on German "garden path" sentences by Mecklinger et al. (1995) demonstrated that even reliable semantic plausibility information failed to facilitate syntactic reanalysis suggesting that initial syntactic analyses are relatively independent from local semantic information.

However, there are at least two types of contextual information that have been shown to radically change initial parsing preferences: referential support and prosodic cues. First, van Berkum et al. (1999) demonstrated that a discourse providing either one or two potential referents

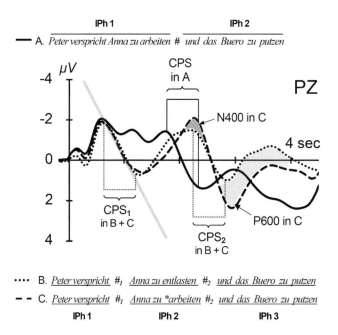

FIGURE 9.4 Illustration of the closure positive shift (CPS) at prosodic boundaries of spoken German sentences, and a prosody-induced syntax violation negative polarity is plotted upwards. Boundary positions and intonational phrases (IPh) are aligned to the time axis. Sentence A (solid line) has only one prosodic boundary (#) after the verb "arbeiten," while B (dotted line) and violation condition C (dashed line) have two such boundaries. At each boundary position a large CPS component was elicited in each condition. Magnitude and slope of the CPS (illustrated by the thick gray line at CPS1) are very similar at all boundary positions. In C, the syntax-prosody mismatch on the verb "arbeiten" additionally elicited an N400/P600 pattern which superimposes the second CPS. Waveforms represent a grand average ERP at PZ across 40 subjects and approximately 5000 trials per condition. Prosodic boundary information is not only important during language learning but also guides the listener's syntactic analysis and sentence comprehension. *Source*: Modified after Steinhauer (2003) and Steinhauer et al. (1999). Translation of sentences: A. Peter promises Anna to work # and to clean the office. B. Peter promises # to support Anna # and to clean the office. C. Peter promises # *[to work Anna] # and to clean the office. (Conditions A and C are lexically identical and differ only prosodically.)

for a NP determined whether readers were biased toward a complement clause or relative clause reading. If a context sentence had introduced "*two* girls" in the discourse, a singular NP "*the* girl" required further specification, thus favoring the (usually non-preferred) relative clause reading (as indicated by an enhanced P600/SPS component). Second, Steinhauer *et al.* (1999) demonstrated that prosodic information in speech can dramatically alter parsing preferences typical for reading. This study showed that the presence or absence of an intonational phrase boundary determined whether the following NP was parsed as the object of either a preceding verb or a subsequent verb. Introducing prosodic boundaries between verbs and their object NPs caused a prosody-induced verb argument structure violation which elicited an N400/P600 pattern to the incompatible verb (*cf.* condition C in Figure 9.4). These two studies demonstrate that at least some kinds of context information can immediately influence the syntactic parsing mechanism.

Box 9.3   Order of emerging ERP components in language development

Language-related ERP components emerge during childhood in a temporal order that nicely corresponds to the development of respective linguistic and cognitive subdomains. As a general pattern, ERP components in childhood are initially larger and more broadly distributed both spatially and temporally and develop the more focused and specialized ERP profiles of adults usually until puberty (Holcomb *et al.*, 1992; Mills *et al.*, 1997; Hahne *et al.*, 2004). The diagram illustrates the timeline of cognitive development and the emergence of corresponding ERP components during the first 3 years of life (adapted from Friederici, 2006). The MMN reflecting one's ability to discriminate sounds is the earliest ERP response and is already present in newborns. During the first months, babies are able to discriminate phonemes of all natural languages. However, at about 10 months a particular specialization for sound distinctions important in their mother tongue is reflected by larger and more robust MMN effects whereas speech sounds that do not

belong to the phonemic inventory of their first language lose the ability to elicit MMNs (categorical perception). The next ERP response found in infants is the CPS reflecting prosodic phrasing. The CPS is present no later than at 8 months, that is, when infants are able to distinguish between adequate and inadequate pausing in speech. As large prosodic boundaries typically coincide with syntactic boundaries and the presence of function words, the presence of the CPS component may indicate the onset of "phonological bootstrapping" in language acquisition. The lexico-semantic N400 component emerges at 12–14 months, just after infants have started to babble. The N400 was observed when infants saw a picture of an animal or simple object (e.g., a dog) and heard a word that did not match (e.g., pencil). Last, LAN and P600 responses to simple syntactic violations develop only 1 year later, at an age of 24 months (P600) and 32 months (LAN).

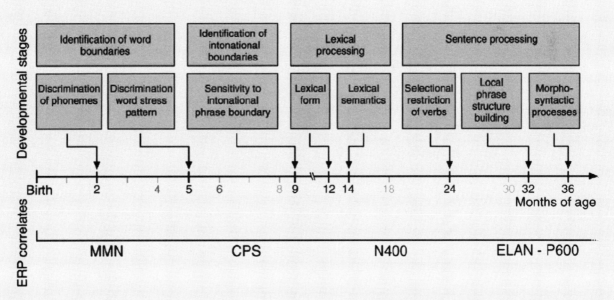

Friederici, A.D. (2006). Neurophysiological markers of early language acquisition: From syllables to sentences. *Trends in Cognitive Science*, 9(10), 481–488.

Hahne, A., Eckstein, K., & Friederici, A.D. (2004). Brain signatures of syntactic and semantic processes during children's language development. *Journal of Cognitive Neuroscience*, 16(7), 1302–1318.

Holcomb, P.J., Coffey, S.A., & Neville, H.J. (1992). Visual and auditory sentence processing: A developmental analysis using event-related brain potentials. *Developmental Neuropsychology*, 8(2/3), 203–241.

Mills, D.L., *et al.* (1997). Language comprehension and cerebral specialization from 13 to 20 months. *Developmental Neuropsychology*, 13, 397–445.

Box 9.4   ERP components in second language: evidence for "critical periods?"

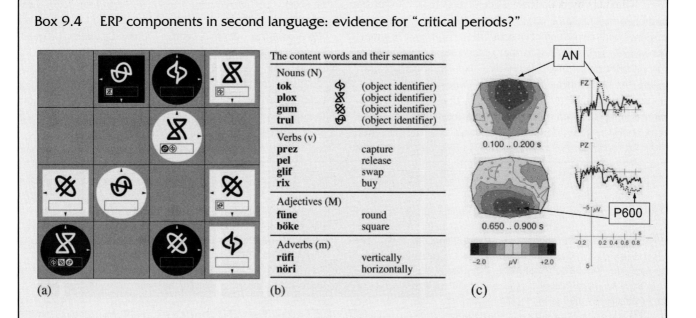

The content words and their semantics		
Nouns (N)		
tok	⊕	(object identifier)
plox	⊗	(object identifier)
gum	⊗	(object identifier)
trul	⊕	(object identifier)
Verbs (v)		
prez		capture
pel		release
glif		swap
rix		buy
Adjectives (M)		
füne		round
böke		square
Adverbs (m)		
rüfi		vertically
nöri		horizontally

(a)   (b)   (c)

Native-like mastery of a language seems almost impossible if this language was not acquired early in childhood. This observation is often explained with an early "critical period" (CP) during which the brain is particularly well prepared to learn the sounds, words, and grammatical rules. An ERP study by Weber-Fox and Neville (1996) tested Chinese subjects who had learned English at different ages, and found support for a CP in syntactic but not semantic processing. Semantic anomalies elicited native-like N400s in all groups, but even short delays in age of exposure to English prevented LANs in syntax conditions. More posterior and right-lateralized negativities, delayed P600/SPS components, or no ERP effects were found instead. Data seemed to indicate that late L2 learners are unable to do early automatic parsing and rely on compensatory brain mechanisms that are distinct from those of native speakers. Alternatively, ERPs might primarily reflect the level of proficiency which was at least partially confounded with age of exposure. To tease these factors apart, Friederici *et al.* (2002) trained adult subjects in the artificial miniature language "Brocanto" to native-like proficiency. A computer-implemented chess-like board game (panel a) was employed to engage subjects in speaking Brocanto: sentences referred to the moves of the game (panel b). After training, high proficient subjects displayed the typical "native-like" ERP patterns of syntactic

processing (panel c): an early anterior negativity (AN) followed by a P600, here shown for a syntactic subcondition that was controlled for transfer effects between first language (German) and second language (Brocanto). Subsequent studies investigating adult L2 learners of *natural* languages found similar but mixed evidence (Clahsen & Felser, 2006; Steinhauer *et al.*, 2006). Overall, while ERP support for a CP in L2 grammar learning appeared unambiguous by 2001, more detailed research and new paradigms have raised new controversies.

Clahsen, H., & Felser, C. (2006). Grammatical processing in language learners. *Applied Psycholinguistics*, 27(1), 3–42.

Friederici, A.D., Steinhauer, K., & Pfeifer, E. (2002). Brain signatures of artificial language processing: Evidence challenging the critical period hypothesis. *Proceedings of the National Academy of Sciences*, 99(1), 529–534.

Steinhauer, K., White, E., King, E., Cornell, S., Genesee, F., & White, L. (2006). The neural dynamics of second language acquisition: Evidence from event-related potentials. *Journal of Cognitive Neuroscience*, Supplement 1, 99.

Weber-Fox, C.M., & Neville, H.J. (1996). Maturational constraints on functional specializations for language processing: ERP and behavioral evidence in bilingual speakers. *Journal of Cognitive Neuroscience*, 8(3), 231–256.

Source: Modified after Friederici *et al.* (2002); Figures 2 and 4.

9.2.9. Prosodic Phrasing: The Closure Positive Shift

The Steinhauer *et al.* (1999) study discussed above also identified a novel ERP correlate of prosodic processing, which was labeled the *closure positive shift* (CPS; Figure 9.4). This component is reliably elicited at prosodic boundaries

and is assumed to reflect prosodic phrasing (closure of intonational phrases) in listeners cross-linguistically. Unlike most other language-related components, it is independent of linguistic violations. In both first and second language acquisition the CPS is among the first brain responses observed (see Boxes 9.3 and 9.4) and may help learners identify syntactic phrase boundaries and even word boundaries. The CPS is also elicited (1) by boundaries in delexicalized and hummed sentence melodies and (2) during silent reading, both at comma positions and when subjects were instructed to reproduce prosodic boundaries at specific positions (Steinhauer & Friederici, 2001). The former finding suggests that the CPS is independent of lexical/syntactic information and may be domain general; the latter one establishes a link between covert prosody and punctuation (in reading and writing). By revealing that, and how, prosody guides language processing, ERPs have addressed longstanding issues in psycholinguistics.

9.3. CHALLENGES AND FUTURE DIRECTIONS

During the last 25 years, electrophysiological investigations have contributed to our understanding of the various processes involved in speech and text comprehension, their roles in language development and clinical applications. In all of these areas there remain many new and interesting challenges to be met.

What different kinds of cognitive subprocesses contribute to the classical ERP components discussed above? Do language-specific ERP (sub)components exist?

What more can ERPs tell us about shared domain space amongst language, music, and other cognitive domains?

How can we move beyond "violation paradigms" in isolated words and sentences towards more ecologically valid paradigms of language processing?

What are the differential effects of explicit (e.g., classroom) versus implicit (e.g., immersion-like) training environments on L2 acquisition? How do they affect which brain systems are involved in language?

References

Bornkessel, I., & Schlesewsky, M. (2006). The extended argument dependency model: A neurocognitive approach to sentence comprehension across languages. *Psychological Review, 113*(4), 787–821.

Connolly, J., Phillips, N., Stewart, S., & Brake, W. (1992). Event-related potential sensitivity to acoustic and semantic properties of terminal words in sentences. *Brain and Language, 43*(1), 1–18.

Connolly, J.F., & Phillips, N.A. (1994). Event-related potential components reflect phonological and semantic processing of the terminal word of spoken sentences. *Journal of Cognitive Neuroscience, 6*(3), 256–266.

Coulson, S., King, J.W., & Kutas, M. (1998). ERPs and domain specificity: Beating a straw horse. *Language and Cognitive Processes, 13*(6), 653–672.

D'Arcy, R.C.N., Connolly, J.F., Service, E., Hawco, C.S., & Houlihan, M.E. (2004). Separating phonological and semantic processing in auditory sentence processing: A high-resolution event-related brain potential study. *Human Brain Mapping, 22*(1), 40–51.

Friederici, A.D. (2002). Towards a neural basis of auditory sentence processing. *Trends in Cognitive Sciences, 6*(2), 78–84.

Friederici, A.D., & Frisch, S. (2000). Verb-argument structure processing: The role of verb-specific and argument-specific information. *Journal of Memory and Language, 43*, 476–507.

Friederici, A.D., Steinhauer, K., & Frisch, S. (1999). Lexical integration: Sequential effects of syntactic and semantic information. *Memory and Cognition, 27*(3), 438–453.

Friederici, A.D., Mecklinger, A., Spencer, K.M., Steinhauer, K., & Donchin, E. (2001). Syntactic parsing preferences and their on-line revisions: A spatio-temporal analysis of event-related brain potentials. *Cognitive Brain Research, 11*(2), 305–323.

Gunter, T.C., Stowe, L.A., & Mulder, G. (1997). When syntax meets semantics. *Psychophysiology, 34*(6), 660–676.

Hagoort, P., Wassenaar, M., & Brown, C. (2003). Syntax-related ERP-effects in Dutch. *Cognitive Brain Research, 16*(1), 38–50.

Hahne, A., & Friederici, A.D. (1999). Electrophysiological evidence for two steps in syntactic analysis: Early automatic and late controlled processes. *Journal of Cognitive Neuroscience, 11*(2), 194–205.

Hoen, M., & Dominey, P.F. (2000). ERP analysis of cognitive sequencing: A left anterior negativity related to structural transformation processing. *Neuroreport, 11*(14), 3187–3191.

Holcomb, P., Grainger, J., & O'Rourke, T. (2002). An electrophysiological study of the effects of orthographic neighborhood size on printed word perception. *Journal of Cognitive Neuroscience, 14*(6), 938–950.

Kaan, E., Harris, A., Gibson, E., & Holcomb, P. (2000). The P600 as an index of syntactic integration difficulty. *Language and Cognitive Processes, 15*(2), 159–201.

Kotz, S.A., Frisch, S., von Cramon, D.Y., & Friederici, A.D. (2003). Syntactic language processing: ERP lesion data on the role of the basal ganglia. *Journal of the International Neuropsychological Society, 9*(7), 1053–1060.

Kujala, A., Alho, K., Service, E., Ilmoniemi, R.J., & Connolly, J.F. (2004). Activation in the anterior left auditory cortex associated with phonological analysis of speech input: Localization of the phonological mismatch negativity response with MEG. *Cognitive Brain Research, 21*(1), 106–113.

Kuperberg, G.R., Caplan, D., Sitnikova, T., Eddy, M., & Holcomb, P.J. (2006). Neural correlates of processing syntactic, semantic, and thematic relationships in sentences. *Language and Cognitive Processes, 21*(5), 489–530.

Kutas, M., & Hillyard, S.A. (1980). Reading senseless sentences: Brain potentials reflect semantic incongruity. *Science, 207*(1), 203–205.

Lau, E., Stroud, C., Plesch, S., & Phillips, C. (2006). The role of structural prediction in rapid syntactic analysis. *Brain and Language, 98*(1), 74–88.

Martin-Loeches, M., Muñoz, F., Casado, P., Melcon, A., & Fernandez-Frias, C. (2005). Are the anterior negativities to grammatical violations indexing working memory? *Psychophysiology, 42*(5), 508–519.

McPherson, W.B., & Holcomb, P.J. (1999). An electrophysiological investigation of semantic priming with pictures of real objects. *Psychophysiology, 36*(1), 53–65.

Mecklinger, A., Schriefers, H., Steinhauer, K., & Friederici, A.D. (1995). Processing relative clauses varying on syntactic and semantic dimensions: An analysis with event-related potentials. *Memory and Cognition, 23*(4), 477–494.

Münte, T.F., Wieringa, B.M., Weyerts, H., Szentkuti, A., Matzke, M., & Johannes, S. (2001). Differences in brain potentials to open and closed class words: Class and frequency effects. *Neuropsychologia, 39*(1), 91–102.

Neville, H., Nicol, J.L., Barss, A., Forster, K.I., & Garrett, M.F. (1991). Syntactically based sentence processing classes: Evidence from event-related brain potentials. *Journal of Cognitive Neuroscience, 3*(2), 151–165.

Newman, R.L., Connolly, J.F., Service, E., & McIvor, K. (2003). Influence of phonological expectations during a phoneme deletion task: Evidence from event-related brain potentials. *Psychophysiology, 40*(4), 640–647.

Niedeggen, M., Rosler, F., & Jost, K. (1999). Processing of incongruous mental calculation problems: Evidence for an arithmetic N400 effect. *Psychophysiology, 36*(3), 307–324.

Osterhout, L., & Holcomb, P.J. (1992). Event-related brain potentials elicited by syntactic anomaly. *Journal of Memory and Language, 31*, 785–806.

Osterhout, L., & Nicol, J. (1999). On the distinctiveness, independence, and time course of the brain responses to syntactic and pragmatic anomalies. *Language and Cognitive Processes, 14*(3), 283–317.

Pakulak, E., & Neville, H. (2004). Individual differences in online syntactic processing in monolingual adults as reflected by ERPs. *17th Annual CUNY Conference on Human Sentence Processing*, College Park (Maryland, USA), 149.

Patel, A.D., Gibson, E., Ratner, J., Besson, M., & Holcomb, P.J. (1998). Processing syntactic relations in language and music: An event-related potential study. *Journal of Cognitive Neuroscience, 10*(6), 717–733.

Rossi, S., Gugler, M.F., Hahne, A., & Friederici, A.D. (2005). When word category information encounters morphosyntax: An ERP study. *Neuroscience Letters, 384*(3), 228–233.

Sanders, L.D., & Neville, H.J. (2003). An ERP study of continuous speech processing. II. Segmentation, semantics, and syntax in non-native speakers. *Cognitive Brain Research, 15*(3), 214–227.

Steinhauer, K. (2003). Electrophysiological correlates of prosody and punctuation. *Brain and Language, 86*, 142–164.

Steinhauer, K., & Friederici, A. (2001). Prosodic boundaries, comma rules, and brain responses: The closure positive shift in ERPs as a universal marker for prosodic phrasing in listeners and readers. *Journal of Psycholinguistic Research, 30*(3), 267–295.

Steinhauer, K., Alter, K., & Friederici, A.D. (1999). Brain potentials indicate immediate use of prosodic cues in natural speech processing. *Nature Neuroscience, 2*(2), 191–196.

Steinhauer, K., Pancheva, R., Newman, A.J., Gennari, S., & Ullman, M.T. (2001). How the mass counts: An electrophysiological approach to the processing of lexical features. *Neuroreport, 12*(5), 999–1005.

van Berkum, J.J., Hagoort, P., & Brown, C. (1999). Semantic integration in sentences and discourse: Evidence from the N400. *Journal of Cognitive Neuroscience, 11*(6), 657–671.

van Herten, M., Kolk, H.H.J., & Chwilla, D.J. (2005). An ERP study of P600 effects elicited by semantic anomalies. *Cognitive Brain Research, 22*(2), 241–255.

Van Petten, C., & Luka, B.J. (2006). Neural localization of semantic context effects in electromagnetic and hemodynamic studies. *Brain and Language, 97*(3), 279–293.

Further Readings

Brown, C.M., & Hagoort, P. (Eds.) (2000). *The neurocognition of language.* **Oxford, UK: Oxford University Press.**
This work provides a critical overview of how neurocognitive methods have been used to address psycholinguistic research questions in listeners, readers, and speakers, focusing in particular on electrophysiological techniques (EEG, ERPs, MEG). The book first introduces the reader to linguistic theory, psycholinguistics, and cognitive neuroscience, and then explains how current theories and models have been tested using brain imaging and ERPs.

Handy, T.C. (Ed.) (2005). *Event-related potentials: A methods handbook.* **Cambridge, MA: The MIT Press.**
Provides a good introduction to the world of ERP recording, analysis, and evaluation.

Nunez, P.L. & Srinivasan, R. (2006). *Electric fields of the brain: The neurophysics of EEG.* **Oxford, UK: Oxford University Press.**
The definitive and comprehensive text on the EEG signal from the relevant principles of physics to the analysis of the response with extensive discussion of issues related to recording the EEG (e.g., choosing the best reference).

Rugg, M.D. & Coles, M.G.H. (1996). *Electrophysiology of mind.* **Oxford, UK: Oxford University Press.**
Despite its age, this classic book is one of the best introductions to ERPs in cognitive neuroscience.

Zani, A., & Proverbio, A. (Eds.) (2003). *The cognitive electrophysiology of mind and brain.* **Amsterdam: Academic Press.**
An introduction to EEG/ERP and MEG measures in perception, language, and memory.

10

Direct Electrical Stimulation of Language Cortex

BARRY GORDON[1,2] and KERRY LEDOUX[1]

[1]Cognitive Neurology/Neuropsychology Group, Department of Neurology,
The Johns Hopkins University School of Medicine, Baltimore, MD, USA
[2]Department of Cognitive Science, The Johns Hopkins University, Baltimore, MD, USA

ABSTRACT

Direct cortical electrical stimulation has been an important tool for helping to understand the neural organization and representation of language functions *in vivo*. Here, we briefly describe the direct cortical electrical stimulation technique, and summarize some of the major results of research using cortical stimulation to study the relationship between language and the brain. Many findings confirm the classical model of language–brain organization derived from studies of patients with chronic lesions; other findings, such as the high degree of individual variability in language localization, and the existence of language centers throughout the dominant hemisphere, warrant modification of that model, and highlight the different perspectives offered by lesion-type studies and by functional correlational studies.

10.1. INTRODUCTION

Direct cortical electrical stimulation (DCES) was originally developed to map the location of sensory, motor, memory, and particularly language functions in the brains of patients being considered for resectional surgery. During the procedure, areas of the brain are stimulated by electrical current, while the patient (who is awake for the mapping procedure) is asked to engage in different behaviors (such as moving a finger or speaking out loud). Under these conditions the stimulation generally produces what is in effect a temporary, localized functional lesion. The effects of the stimulation on the target behaviors are observed to determine which types of behavior are dependent upon the brain region being stimulated. DCES is thus used to try to predict what deficits might result from permanent resection of brain regions, to inform the risk/benefit analysis of any contemplated resections. Individualized mapping, on a subject-by-subject basis, has proven necessary because individuals often prove to have significant deviations from standardized functional-neuroanatomic maps. In addition to this clinical use, DCES as a research tool has led to an increased understanding of the processing components involved in language and of their neuroanatomic associations.

10.2. DESCRIPTION OF THE TECHNIQUE

DCES is applied through electrodes resting directly on the surface of the cortex. The technique thus requires some access to the cortical surface, and is limited to patients undergoing surgical exposure of the cortex for clinical reasons. Electrode placement is determined clinically to best characterize the areas to be resected and their participation in critical functions in that individual. Language mapping is generally indicated when considering resection of the language-dominant hemisphere. Most frequently, electrodes in studies of language localization using DCES have been placed in perisylvian regions: the superior and middle temporal gyri, portions of the inferior temporal gyrus, the angular gyrus, inferior parietal regions, and the posterior inferior frontal lobe. However, more diverse areas (for instance, the basal temporal regions) have also been explored in a number of studies.

During both intraoperative and extraoperative testing procedures (see Box 10.1), electrical current is usually delivered across bipolar electrode contacts, using short-duration square-wave pulses of alternating polarity. The effect of stimulation generally appears to be extremely focal, limited to the cortical tissue immediately subjacent to the electrode pair, and also largely to the cortical gray matter. Additionally, under routine testing conditions, the effects of stimulation appear to have nearly immediate onset as well as nearly immediate offset. Cortical electrical stimulation as generally used is not believed to permanently alter or damage the underlying tissue, based on both animal studies and histopathologic studies of stimulated human brain tissue.

Box 10.1   Intraoperative versus extraoperative testing

Two general methods of DCES have been used, intraoperative and extraoperative. In the intraoperative method, stimulation is applied at the time of the craniotomy. The patient is sedated during the first part of the operation when the skull and dura are opened. Then, the patient is tapered off general anesthesia until awake and able to respond. A second approach, the extraoperative one, has relied on implanted electrode arrays that allow stimulation testing after the patient has recovered from the initial surgery. Such arrays are most commonly implanted to record EEG activity at the surface of the brain to best localize epileptiform activity, but can also be used for stimulation mapping. During an initial surgery, electrode arrays are inserted and are left in place. The patient remains in the hospital for the period during which the electrodes remain in place, generally several days to 2 weeks (depending upon clinical needs). The electrodes (and any cerebral tissue identified for resection) are removed during a second, final, surgery.

Intraoperative testing can provide smaller interelectrode distances than those generally allowed by subdural arrays. Additionally, the surgeon has flexibility in independently selecting electrode sites based on the functional anatomy that has been determined from prior testing, a feature not possible with a chronically implanted array. The mapping offered by the intraoperative procedure may thus be more detailed, in some respects. However, the extraoperative technique, with its grids or strips, can potentially provide coverage of much wider cortical areas (such as occipital areas or inferior temporal areas). The time available for electrical stimulation testing is generally much greater with the extraoperative procedure, and the conditions of testing may be more amenable to experimental control outside of the operating room. The patient's ability and willingness to cooperate may also be improved by extraoperative testing. Finally, extraoperative testing allows repeated stimulation, over time, at the same site(s).

In the most common scenario during language testing, onset of the presentation of the electrical current is timed to coincide with or precede the onset of stimulus presentation for a given trial of a language task. For example, during an object naming task, stimulation might be timed to the presentation of the picture to be named. Alternatively, some researchers have varied the onset of stimulation relative to presentation of the behavioral stimulus in order to assess its effect at different processing stages during the language task (see, e.g., Hart *et al.*, 1998). Stimulation is terminated upon response, or, in cases of a lack of response, after a pre-determined period of time. For maximal safety, the duration of any single train of electrical stimulation is kept relatively brief, generally below 10–15 s, and often less than 5 s. This factor has led to the selection or design of language tasks with trials that can be meaningfully accomplished within this time frame. The most common language task used in conjunction with DCES is visual confrontation naming, in which a picture of an object is presented, and the patient is asked to name it; others include single word reading, sentence reading, naming to definition, writing, auditory comprehension, spontaneous speech, auditory repetition, syllable discrimination, and verbal working memory tasks. Extensive pre-testing is usually done to familiarize the patient with the tasks and to establish baseline levels of performance. Generally, stimulation is applied during some trials, and not during others; behavioral performance is compared across the two conditions.

The effects of stimulation may be positive or negative. *Positive effects* are the elicitation of motor movements, sensations, or, at times, behaviors. An example would be the occurrence of visual phenomena (such as geometric patterns) with stimulation of occipital areas, or clonic and tonic motor activity with stimulation of motor cortex. *Negative effects* involve the temporary disruption of an ongoing behavior upon stimulation. An example is the inhibition of rapid alternating finger movements during stimulation. In many cases, the patient can immediately resume the disrupted behavior upon cessation of stimulation. Negative effects therefore are in essence functional "lesions," extremely proscribed in their location and duration. The detection of negative effects requires the active initiation of different behaviors by the awake patient in order to determine if those behaviors are disrupted by stimulation.

The effect of DCES on language functions is generally a negative one. For example, upon stimulation, the patient may experience a complete inability to continue the language task in which he or she is engaged. At times, this effect may be one of interference more than a complete block, such that response times are slowed, hesitations are observed, paraphasias are produced, or pronunciation is noticeably difficult. These effects will often resolve immediately upon (or very shortly after) cessation of the electrical current.

An example may help to illustrate the procedure and its effects. Suppose a patient undergoing DCES was asked to engage in a confrontation naming task. She would be shown pictures of objects and asked to name each one out loud. She would have been extensively pre-tested on the pictures to be sure that she knew all the objects, and she might have been trained to use a particular phrase in naming them: "This is a ———." During DCES, stimulation would be applied randomly during some (but not all) trials, probably timed to the presentation of the picture. On trials without stimulation, we would expect the patient to be able to respond correctly, and without hesitation or speech error. (These trials serve as important controls, to ensure the patient is capable of performing the task correctly under what are often stressful conditions.) On trials with stimulation, a number of things might occur. The patient might not be able to speak at all, in which case the site of stimulation might be labeled as involved in basic mechanisms of speech production. The patient might be able to say the trained phrase "This is a...", but may then have trouble naming the object; she might even say things like, "I know what it is, I just can't get to it..." This might be scored as a specific deficit in naming (as opposed to general speech arrest). The patient might make phonological errors ("gat" instead of "cat") or semantic errors ("dog" instead of "cat"), indicating the involvement of the stimulation site in specific component processes of naming. She might be delayed significantly in naming the object, but be able finally to produce its name. Or she might show naming behavior that is not in any appreciable way different from her performance on non-stimulation trials, suggesting no role for this particular stimulation site in the naming process.

Our increased understanding of the nature of language representation in the brain from studies of DCES, then, has been based on inferences similar to those made when studying organic brain lesions: that the disruption of language behavior by a lesion in a given brain region strongly suggests the essential involvement of that region in that behavior.

10.3. COMPARISON TO OTHER TECHNIQUES

DCES offers distinct advantages and disadvantages when compared to other functional-neuroanatomic methods usable with human subjects. Probably its most striking advantage over other methods in general is that it is not only a lesional method, which allows direct interpretation of the results, but one in which the "lesion" is relatively small (~1 cm^2), temporary, movable, and repeatable in the same subject. Most discrete lesions in human subjects are otherwise the result of accidental conditions such as ischemic or hemorrhagic stroke, trauma, or encephalitis. Such lesions occur without planned localization, follow their own neuroanatomic distribution, tend to be much larger than the functional lesions induced by DCES, and are irreversible. Also, the effects of natural lesions are typically studied at least days after onset; only the rarest circumstances allow study of their effects within hours. As a result of the time delay, there is an opportunity for functional reorganization with accidental lesions that is less of a concern with DCES.

However, DCES also has a number of disadvantages. Its effects may not be purely "lesional" at the neural level, even though they may appear to be so behaviorally. It may also have distant effects, both positive and negative, which are not easily appreciated during its application. And perhaps most importantly, the opportunity to use DCES is relatively rare, and arises only in special populations: those in whom direct cortical recording of possible seizure activity is felt to be necessary, and those in whom functional mapping is necessary as part of pre-surgical planning. Most such cases have had uncontrollable epilepsy; some have had tumors; some both. It is still unclear how representative the functional neuroanatomic organization of such patients will be of individuals without these conditions. Most research has failed to show a relationship between the extent of an epileptic focus and language localization, as might be expected if epilepsy forced reorganization. But this fundamental limitation of cortical stimulation studies cannot be eliminated.

Two methods that bypass this limitation, and allow studies of normal subjects, are functional neuroimaging (using methods such as fMRI) and transcranial magnetic stimulation (TMS). Functional neuroimaging and DCES result in fundamentally different types of data. Functional neuroimaging shows the areas that can be active during a given task (those that *participate* in the task), while DCES marks the areas without which a given task cannot be accomplished (those that are *necessary* for the task). A more direct analog of DCES is therefore TMS, which has now been widely used in a variety of studies. TMS is similar to DCES in that the effects of TMS can be stimulatory or inhibitory. TMS offers the inestimable advantage over DCES of being usable in healthy participants. It has disadvantages relative to DCES in having presumably larger volumes of current induction and difficulties controlling the exact location of the current pulse in the brain.

10.4. SELECTED FINDINGS WITH CORTICAL STIMULATION TECHNIQUE

In addition to the clinical use, findings from DCES have also contributed to insights about the processing components involved in language and of their neuroanatomic associations.

10.4.1. Correspondence with the Classic Functional-Neuroanatomic Model

What might be called the classic model of neuroanatomic localization of language functions was generally derived from single-case and group studies of patients with language deficits acquired following relatively large, permanent focal lesions due to ischemic infarctions. In this model,

Box 10.2   Language mapping in the non-dominant hemisphere

Most of the electrical cortical stimulation studies discussed in the main text have been limited to the language-dominant (generally left) hemisphere. A handful of studies have explored language localization in the non-dominant hemisphere. Andy and Bhatnagar (1984) found sites throughout the non-dominant cortex at which naming performance was impaired with stimulation. Boatman *et al.* (1998) studied a right-handed patient with seizures. Although cortical stimulation testing was conducted only over the left-hemisphere, bilateral intracarotid sodium amobarbital testing allowed some conclusions to be drawn about speech processing in both hemispheres. Syllable discrimination remained intact following injection to each hemisphere, suggesting that each hemisphere could independently support acoustic–phonetic processing as needed for that task. Subsequently, however, electrical stimulation of the left hemisphere impaired syllable discrimination, leading the researchers to posit preferential processing by left hemisphere mechanisms despite bilateral speech perception capabilities in this patient. Jabbour *et al.* (2005) looked at language localization in the right hemisphere of patients in whom a language-dominant hemisphere had not been demonstrated, that is, they performed DCES in six patients who had been classified as having bilateral language representation by the intracarotid amobarbital procedure. Four of the six patients showed right-hemisphere language areas analogous to those classically described in the left hemisphere (in frontal and temporal regions). One patient showed involvement of widespread right-hemisphere sites in language (at which, however, stimulation did not always evoke errors), and in the final patient, no right-hemisphere language sites were identified (possibly due to the placement of the electrodes). The significance of such right-hemisphere language sites in these patients (with bilateral language representation) is unknown; these sites might duplicate left-hemisphere areas, or language abilities in such individuals might require an interaction of areas across the hemispheres.

Andy, O.J., & Bhatnagar, S. (1984). Right-hemispheric language evidence from cortical stimulation. *Brain and Language, 23,* 159–166.

Boatman, D., Hart, J., Lesser, R.P., Honeycutt, N., Anderson, N.B., Miglioretti, D., & Gordon, B. (1998). Right hemisphere speech perception revealed by amobarbital injection and electrical interference. *Neurology, 51,* 458–464.

Jabbour, R.A., Hempel, A., Gates, J.R., Zhang, W., & Risse, G.L. (2005). Right hemisphere language mapping in patients with bilateral language. *Epilepsy and Behavior, 6,* 587–592.

by the findings from DCES studies. Stimulation-induced interference effects have been demonstrated for most language tasks in most individuals at inferior frontal and posterior temporal sites (e.g., Ojemann *et al.*, 1989; Ojemann, 1991). When assessed on multiple language tasks, a high number of sites that show stimulation deficits across tasks fall within the traditional Broca's and Wernicke's areas (Ojemann, 1991). (See Box 10.2 for language mapping in the non-dominant hemisphere.)

10.4.1.1. Lesion Effects and Aphasia

The behavioral deficits arising from temporary functional lesions created during cortical stimulation often resemble those observed following accidental lesions, and their localization supports interpretations of these disorders within the classical model. For example, deficits in language production and comprehension of the sort demonstrated in many Broca's aphasics have been reported with electrical stimulation of Broca's area (Schaffler *et al.*, 1993). Stimulation of either Broca's area or Wernicke's area has been shown to result in deficits in language comprehension; however, production deficits in those same patients arose only with stimulation of Broca's area (Schaffler *et al.*, 1996), a pattern that is again commensurate with the findings of neuropsychological studies of Broca's and Wernicke's aphasias.

Another example comes from studies of the angular gyrus, both in patients with epilepsy (Morris *et al.*, 1984), and in patients with tumors located near this region (Roux *et al.*, 1999). This area is thought to be the site of damage producing the varied symptoms of Gerstmann syndrome (finger agnosia, right/left disorientation, acalculia, and agraphia). This association was supported by the demonstration of similar deficits with stimulation of the angular gyrus. Additionally, such deficits were elicited with stimulation outside the angular gyrus, and deficits in other behaviors (reading, color naming) were elicited with stimulation of the angular gyrus. These results might help explain why Gerstmann syndrome can result from damage outside the angular gyrus, and why damage to the angular gyrus can result in other behavioral deficits.

In another example, stimulation of sites in and around classical Wernicke's area during multiple language tasks (including auditory comprehension, repetition, and spontaneous speech tasks) was shown to produce patterns of performance similar to those reported in transcortical sensory aphasia (TSA). TSA is characterized by impaired auditory comprehension with intact repetition and overall fluent speech. Interestingly, the precision of electrical stimulation allowed a test of one hypothesis of the neurological basis of TSA. Previous researchers had suggested that the impaired comprehension with intact repetition of TSA arose due to the isolation (by lesion) of an intact Wernicke's area, which

language functions are localized primarily to a region of the inferior frontal gyrus (Broca's area) and a posterior region of the superior temporal lobe (Wernicke's area) in the left (dominant) hemisphere. This model is broadly supported would result in the disconnection of phonology and lexical semantics. That TSA was induced with stimulation of Wernicke's area alone suggests that its sparing is not a necessary condition for this type of aphasia (Boatman et al., 2000).

Finally, the category-specific naming deficits that have been reported in some patients with accidental lesions have also been elicited during DCES; for example, one patient showed category-specific disruption of living objects but not non-living objects upon stimulation of a site in the inferior temporal lobe. (Upon subsequent resection of this region, the patient exhibited similar category-specific deficits in naming; Pouratian et al., 2003.)

In summary, then, electrical stimulation studies have shown broad support for the classic model of localization of language functions. The temporary lesions produced by DCES have frequently been shown to result in behavioral deficits that are similar to those observed in patients with chronic organic lesions, and frequently support the interpretation of aphasic syndrome deficits within the classic neurologic model.

10.4.2. Divergence from the Classic Neurologic Model

10.4.2.1. Discreteness of Representation

The results of cortical stimulation studies of language, however, also demand some refinements to the classic neurological model. The sites at which stimulation interference effects are observed are remarkably localized and discrete. Sites of 1–2 cm^2 have been identified as showing task-specific susceptibility to stimulation; immediately adjacent sites may not show effects of stimulation at all. Virtually no patients have demonstrated stimulation-related errors on language tasks at all sites throughout Broca's and Wernicke's areas as classically defined. Instead, small language-related regions exist in and around these zones, adjacent to other tissue within these areas that does not show interference upon stimulation. This discreteness of localization has led some to suggest a mosaic organization of the sites involved in a given language process (Ojemann et al., 1989; Ojemann, 1991).

An example of the discreteness of localization of function was reported by Boatman and colleagues (Boatman et al., 1997; Boatman, 2004). They performed a series of cortical stimulation studies of speech perception processes in which they specifically targeted acoustic–phonetic processing using an auditory syllable discrimination task. Acoustic–phonetic processing involves the discrimination of speech sounds based on phonetic features such as voicing or manner of articulation. This type of processing is thus one of the earliest, low-level stages of speech perception. Patients were asked to indicate whether syllables that they heard were the same or different; when different, the syllables differed by basic acoustic–phonetic features of consonants. Across studies, performance on this syllable discrimination task

was disrupted by stimulation of a relatively localized region of the middle-posterior superior temporal gyrus in all patients tested. An important component of the perception of speech, then, was localized discretely to a specific cortical area (Boatman et al., 1997; Boatman, 2004).

10.4.2.2. Individual Variability within Anterior and Posterior Language Areas

Another divergence from the classic neurological model comes from the finding of a high degree of individual variability in the localization of essential language sites in cortex. Although most (not all) patients show an area of language localization anteriorly in frontal regions and posteriorly in temporal regions, the exact location of these areas varies widely. For some patients, language areas are identified within traditional Broca's and Wernicke's areas (although, as mentioned, rarely across these entire regions); for others, language regions may be slightly adjacent to these centers, and some patients will not show representation near these centers at all. Ojemann et al. (1989), in a review of cortical stimulation results of 117 patients tested at their center, showed that the area of greatest overlap of interference effects during naming across patients was the inferior posterior frontal cortex (Broca's area). But even at this site, 21% of the 82 patients tested with stimulation during naming did not show interference effects. Of the other 34 zones in which stimulation effects were tested, none was shown to lead to disruption of naming in more than 50% of the patients tested. The prediction of the precise location of language-essential cortex in any individual patient based on anatomical landmarks alone is therefore impossible with current methods.

10.4.2.3. Language Representation Outside of Classical Language Regions

Sites of language interference upon stimulation are also not confined to Broca's and Wernicke's areas. Stimulation of regions surrounding the Sylvian fissure in all directions has been shown to interfere with language processing on a number of different tasks. Outside of Broca's area, stimulation-induced language deficits have been evoked at sites throughout the frontal lobe (including along the superior and middle frontal gyri). Within the temporal lobe, stimulation effects have been demonstrated outside of Wernicke's area, at sites along anterior regions of the superior temporal gyrus and all along the middle temporal gyrus. Inferiorly, stimulation effects have been reported in the basal temporal language area, including inferior temporal gyrus, fusiform gyrus, and parahippocampal gyrus. Cortical stimulation interference effects have been demonstrated all along the supramarginal gyrus and in more superior parietal regions, as well as at regions at the end of the Sylvian fissure, at the posterior temporal–occipital junction.

10.4.2.4. Different Stimulation Sites for Different Language Tasks

A comparison of the sites that, when stimulated, lead to disruption of a given language task with the sites disrupted during another language task will sometimes demonstrate sites of overlap, but also, frequently, sites that are not common to the two (Ojemann *et al.*, 1989; Ojemann, 1991; Ojemann *et al.*, 2002). Presumably, to some extent, this finding reflects the fact that different tasks are the product of a variable mixture of different language processes (phonologic, orthographic, semantic, and so forth), which in turn may be differentially localized in the brain. Naming and reading, for example, are believed to share at least some component processes, including semantic identification (when reading for meaning) and phonological output. Other component processes (such as orthographic identification in the case of reading) may be unique to one or the other task. When naming and reading are tested in the same patients, some sites may show stimulation-induced interference during both tasks, and may thus be sites of localization of some common component processes. Frequently, sites are also tested at which naming deficits are observed in the face of intact reading performance, and vice versa. These sites might be separated by as little as 1 cm, but show precisely differential effects for the two tasks. These sites, then, might best be described as localizing processes that are unique to each task. Other tasks that have been compared within the same patients, and that have demonstrated both overlap and separation in localization, include naming and verb generation; naming and single word comprehension; naming, reading, and writing; and speech perception and production.

If DCES causes interference with basic components of language processes that might or might not be common to different language tasks, we might also expect that more complex tasks (that require the integration of several basic component processes for their execution) would recruit larger areas of cortex than simple tasks that more purely tap one component process. If so, larger areas of cortex should be susceptible to interference effects during more complex tasks than during simpler tasks. An example of such a finding was demonstrated using speech perception tasks of increasing complexity during DCES (Boatman, 2004). As mentioned previously, low-level acoustic–phonetic processing was localized to a circumscribed region of mid-posterior superior temporal gyrus. The more complex phonological processing demanded by phoneme identification tasks (which includes low-level acoustic–phonetic processing plus higher-level speech identification processes) was disrupted by stimulation of a broader region, which included the middle-posterior superior temporal gyrus site critical to acoustic–phonetic processing, but which also included regions within other temporal areas, and within the frontal and parietal lobes. Finally, the still higher-level process of accessing lexical-semantic systems, as assessed

by sentence comprehension tasks, showed the broadest cortical distribution, with stimulation-induced deficits arising at multiple sites throughout the temporal, parietal, and frontal lobes, but again encompassing those disrupted by the lower-level tasks (Boatman, 2004). Thus, component processes that are essential to higher-level processes may have the most precise cortical representation, and complex tasks that require the integration of several component processes may be represented over larger cortical regions.

Differential localization has also been reported when language tasks are presented in two different languages, or in different sensory modalities. Stimulation studies with bilingual patients, tested in both of the languages in which they are proficient, have sometimes shown sites in which naming or reading is disrupted in both languages, but have also shown sites in which language behavior is disrupted in one language but not the other (see, e.g., Lucas *et al.*, 2004; Walker *et al.*, 2004). Overlapping and specific sites have been demonstrated in proficient patients tested with both spoken language and signed language (see, e.g., Corina *et al.*, 1999). Similar findings have been reported when language functions were tested across two different modalities; for example, Roux *et al.* (2006) found sites at which color naming was impaired when stimuli were presented visually, but not when presented auditorily (e.g., a question such as, "What color is grass?"). Hamberger *et al.* (2001) reported finding sites susceptible to stimulation during auditory confrontation naming only in anterior temporal sites, with both auditory and visual naming disrupted at posterior temporal sites.

10.4.2.5. Cognitive Efficiency and Precision of Localization

Several findings suggest that greater efficiency in a cognitive operation may be correlated with greater precision in cortical localization. Women were more likely than men to have language areas only in the frontal lobes; that is, women were more likely to show greater precision of cortical representation of language than were men (Ojemann, 1991). Taken together with the common finding of higher verbal skills in women than in men, this finding may suggest a correlation between cognitive efficiency and precise localization. Ojemann (1991) also found that patients with high verbal IQs (measured pre-operatively) devoted less total surface area to naming than did patients with lower verbal IQs. Additionally, some bilingual patients studied showed greater cortical representation (i.e., a larger number of disrupted sites) for the language in which they were least proficient. Ojemann *et al.* (2002) showed a decrease in the number of sites of interference of a verb generation task as participants become more practiced with the task. These results suggest a positive correlation between proficiency and precision of localization. They are countered, however, by the results of a cortical stimulation study in pediatric patients, in which fewer sites of stimulation-induced errors in naming were identified in children than in adults (and indeed, the youngest children in the study showed fewer sites than the older children; S. Ojemann *et al.*, 2003). The question of whether total cortical area of representation reflects one's behavioral proficiency in a given cognitive realm remains open to future study.

10.4.3. Electrical Stimulation and Subcortical Structures

Increasingly, our understanding of the role of subcortical structures and connecting pathways in language is being informed by studies using electrical stimulation techniques. Already, our understanding of the role of subcortical structures has benefited from research with depth electrodes, placed intraoperatively (for days or weeks) or chronically implanted, usually for the treatment of dyskinesia or chronic pain. A number of studies, for example, have implicated a role for the thalamus in language and verbal memory functions (see, e.g., Bhatnagar & Mandybur, 2005). Another line of work has extended the intraoperative stimulation technique to the white matter tracts lying beneath resected tissue. Stimulation is applied to white matter areas as they are exposed during the resection, with the clinical goal of further mapping essential functions within these tracts while maximizing the resection (especially of tumors). Mapping of the subcallosal fasciculus, the periventricular white matter, and the arcuate fasciculus has provided insight into the role of these subcortical pathways in different types of aphasia (Duffau *et al.*, 2002). Similarly, the mapping of a pathway connecting Broca's area (proposed as an area of articulatory rehearsal) and the supramarginal gyrus (proposed as a potential phonological store) suggests an anatomical basis for the phonological loop (Duffau *et al.*, 2003). Another promising method of studying subcortical pathways was described by Henry *et al.* (2004), who combined intraoperative cortical stimulation and diffusion tensor MRI fiber tracking methodologies to establish connections among essential cortical sites. Matsumoto *et al.* (2004) developed a refined cortico-cortical evoked potential (CCEP) method for studying neuronal connectivity. DCES was used to identify language areas in each patient. Single pulse electrical stimuli were then delivered to these areas, and the resulting electrical activity, time-locked to the stimulus, at other regions throughout the cortex (as detected by the electrocorticogram) was averaged to derive the CCEP. Such measures will increase our understanding of the critical connections among cortical language areas.

10.5. CHALLENGES AND FUTURE DIRECTIONS

Further refinement of the language tasks used during cortical stimulation testing might improve our understanding of the localization of language functions, especially in light of the suggestion of a correlation between task complexity and the number of cortical regions disrupted by stimulation. Psycholinguistic theories of language processing can point researchers in directions of the types of associations and dissociations for which one might seek evidence of neural representation. Using a conjunction of tasks, or stimuli that vary along psychologically meaningful dimensions, might allow a hierarchical investigation to isolate some of the cognitive psychological subprocesses thought to underlie tasks like reading or naming. Some of the stimulation studies reviewed above have adopted this approach to draw refined conclusions about language processing (see, e.g., Hart *et al.*, 1998), but there is much more work that could be done.

Future developments with structural and functional imaging techniques may eventually obviate or at least minimize the need for cortical mapping with DCES. Recently, several groups have explored the possibility of using functional neuroimaging techniques as an alternative to or in conjunction with electrical stimulation mapping for the identification of eloquent areas in surgical patients (see, e.g., Bookheimer *et al.*, 1997; FitzGerald *et al.*, 1997, McDermott *et al.*, 2005; Medina *et al.*, 2005). The use of such technologies in studies of language processing in normal adults has become quite sophisticated, and has greatly expanded our understanding of language lateralization and localization. The application of these techniques to the preoperative determination of essential language areas in surgical patients shows a great deal of promise, and offers a number of advantages over the stimulation method (not least of which is their non- or minimally invasive nature, but also including patient comfort, speed of acquisition, and the ability to map the entire brain, not just those areas exposed by craniotomy). At present, though, functional neuroimaging seems best placed to complement stimulation mapping, by identifying areas that are most likely to be important to language in a given individual and that therefore should be tested by stimulation. Such an approach might reduce the number of electrodes needing to be placed and the amount of stimulation testing to be done.

Finally, increased understanding of cortical plasticity might come from studies of DCES. Gordon *et al.* (1999) described patients who acquired reading deficits following surgical resection, and followed the recovery of these functions during an extended post-operative period. These patients had undergone cortical stimulation testing with several language tasks (including reading) prior to resection. The relationship between the cortical stimulation results and the recovery of reading function in these patients is being analyzed retrospectively in an attempt to better understand the mechanisms by which such recovery might occur. Duffau *et al.* (2002) had the opportunity to perform cortical mapping on a set of three patients twice. Based on the results of the initial mapping, the patients' tumors were not completely resected in order to spare eloquent areas. Because of tumor recurrence, the cortical mapping was performed a second time, 12–24 months after the first. Duffau *et al.* (2002) reported evidence of functional reorganization of the language, sensory, and motor maps in these patients over the 12–24 month intervals, such that areas that had shown essential function on the first mapping were negative on the second. To a certain extent, these mappings were validated when tumor resections were done, guided by the results of the stimulation mappings, and proved to be without neurological sequelae. Although the opportunities for multiple mappings are rare, the information provided from such studies could prove very informative.

Direct electrical stimulation of the human cortex therefore continues to be refined as a useful tool for exploring the functional components of language, and their relationship to brain regions, and for refining our understanding of this uniquely human ability and its brain substrates.

References

Bhatnagar, S.C., & Mandybur, G.T. (2005). Effects of intralaminar thalamic stimulation on language functions. *Brain and Language, 92*, 1–11.

Boatman, D. (2004). Cortical bases of speech perception: Evidence from functional lesion studies. *Cognition, 92*, 47–65.

Boatman, D., Gordon, B., Hart, J., Selnes, O., Miglioretti, D., & Lenz, F. (2000). Transcortical sensory aphasia: Revisited and revised. *Brain, 123*, 1634–1642.

Boatman, D., Hall, C., Goldstein, M.H., Lesser, R., & Gordon, B. (1997). Neuroperceptual differences in consonant and vowel discrimination as revealed by direct cortical electrical interference. *Cortex, 33*, 83–98.

Bookheimer, S.Y., Zeffiro, T.A., Blaxton, T., Malow, B.A., Gaillard, W.D., Sato, S., Jufta, C., Fedio, P., & Theodore, W.H. (1997). A direct comparison of PET activation and electrocortical stimulation mapping for language localization. *Neurology, 48*, 1056–1065.

Corina, D.P., McBurney, S.L., Dodrill, C., Hinshaw, K., Brinkley, J., & Ojemann, G. (1999). Functional roles of Broca's area and SMG: Evidence from cortical stimulation mapping in a deaf signer. *Neuroimage, 10*, 570–581.

Duffau, H., Capelle, L., Sichez, N., Denvil, D., Lopes, M., Sichez, J.P., Bitar, A., & Fohanno, D. (2002). Intraoperative mapping of the subcortical language pathways using direct stimulations: An anatomo-functional study. *Brain, 125*, 199–214.

Duffau, H., Denvil, D., & Capelle, L. (2002). Long term reshaping of language, sensory, and motor maps after glioma resections: A new parameter to integrate in the surgical strategy. *Journal Neurology Neurosurgery Psychiatry, 72*, 511–516.

Duffau, H., Gatignol, P., Denvil, D., Lopes, M., & Capelle, L. (2003). The articulatory loop: Study of the subcortical connectivity by electrostimulation. *Neuroreport, 14*, 2005–2008.

FitzGerald, D.B., Cosgrove, G.R., Ronner, S., Jiang, H., Buchbinder, B.R., Belliveau, J.W., Rosen, B.R., & Benson, R.R. (1997). Location of language in the cortex: A comparison between functional MR imaging and electrocortical stimulation. *American Journal of Neuroradiology, 18*, 1529–1539.

Gordon, B., Boatman, D., Lesser, R.P., & Hart Jr., J. (1999). Recovery from acquired reading and naming deficits following focal cerebral lesions. In S. Broman & J.E.E. Fletcher (Eds.), *The changing nervous system: Neurobehavioral consequences of early brain disorders* (pp. 199–213). New York: Oxford University.

Hamberger, M.J., Goodman, R.R., Perrine, K., & Tamny, T. (2001). Anatomic dissociation of auditory and visual naming in the lateral temporal cortex. *Neurology, 56*, 56–61.

Hart Jr., J., Crone, N.E., Lesser, R.P., Sieracki, J., Miglioretti, D.L., Hall, C., Sherman, D., & Gordon, B. (1998). Temporal dynamics of verbal object comprehension. *Proceedings of the National Academy of Sciences of the United States of America, 95*, 6498–6503.

Henry, R.G., Berman, J.I., Nagarajan, S.S., Mukherjee, P., & Berger, M.S. (2004). Subcortical pathways serving cortical language sites: Initial experience with diffusion tensor imaging fiber tracking combined with intraoperative language mapping. *Neuroimage, 21*, 616–622.

Lucas, T.H., II, McKhann, G.M., II, & Ojemann, G.A. (2004). Functional separation of languages in the bilingual brain: A comparison of electrical stimulation language mapping in 25 bilingual patients and 117 monolingual control patients. *Journal of Neurosurgery, 101*, 449–457.

Matsumoto, R., Nair, D.R., LaPresto, E., Najm, I., Bingaman, W., Shibasaki, H., & Luders, H.O. (2004). Functional connectivity in the human language system: A cortico-cortical evoked potential study. *Brain, 127*, 2316–2330.

McDermott, K.B., Watson, J.M., & Ojemann, J.G. (2005). Presurgical language mapping. *Current Directions in Psychological Science, 14*(6), 291–295.

Medina, L.S., Bernal, B., Dunoyer, C., Cervantes, L., Rodriguez, M., Pacheco, E., Jayakar, P., Morrison, G., Ragheb, J., & Altman, N. (2005). Seizure disorders: Functional MR imaging for diagnostic evaluation and surgical treatment—Prospective study. *Radiology, 236*, 247–253.

Morris, H.H., Luders, H., Lesser, R.P., Dinner, D.S., & Hahn, J. (1984). Transient neuropsychological abnormalities (including Gerstmann's syndrome) during cortical stimulation. *Neurology, 34*(7), 877–883.

Ojemann, G.A. (1991). Cortical organization of language. *Journal of Neuroscience, 11*, 2281–2287.

Ojemann, G.A., Ojemann, J., Lettich, E., & Berger, M. (1989). Cortical language localization in left, dominant hemisphere: An electrical stimulation mapping investigation in 117 patients. *Journal of Neurosurgery, 71*, 316–326.

Ojemann, J.G., Ojemann, G.A., & Lettich, E. (2002). Cortical stimulation mapping of language cortex using a verb generation task: Effects of learning and comparison to mapping based on object naming. *Journal of Neurosurgery, 97*, 33–38.

Ojemann, S.G., Berger, M.S., Lettich, E., & Ojemann, G.A. (2003). Localization of language function in children: Results of electrical stimulation mapping. *Journal of Neurosurgery, 98*, 465–470.

Pouratian, N., Bookheimer, S.Y., Rubino, G., Martin, N.A., & Toga, A.W. (2003). Category-specific naming deficit identified by intraoperative stimulation mapping and postoperative neuropsychological testing: Case report. *Journal of Neurosurgery, 99*(1), 170–176.

Roux, F.E., Boetto, S., Sacko, O., Chollet, F., & Tremoulet, M. (1999). Writing, calculating, and finger recognition in the region of the angular gyrus: A cortical stimulation study of Gerstmann syndrome. *Journal of Neurosurgery, 99*, 716–727.

Roux, F.E., Lubrano, V., Lauwers-Cances, V., Mascott, C.R., & Demonet, J.F. (2006). Category-specific cortical mapping: Color-naming areas. *Journal of Neurosurgery, 104*, 27–37.

Schaffler, L., Luders, H.O., & Beck, G.J. (1996). Quantitative comparison of language deficits produced by extraoperative electrical stimulation of Broca's, Wernicke's, and basal temporal language areas. *Epilepsia, 37*, 463–475.

Schaffler, L., Luders, H.O., Dinner, D.S., Lesser, R.P., & Chelune, G.J. (1993). Comprehension deficits elicited by electrical stimulation of Broca's area. *Brain, 116*, 695–715.

Walker, J.A., Quinones-Hinojosa, A., & Berger, M.S. (2004). Intraoperative speech mapping in 17 bilingual patients undergoing resection of a mass lesion. *Neurosurgery, 54*, 113–118.

Further Readings

Duffau, H., Capelle, L., Sichez, J.P., Faillot, T., Abdennour, L., Law Koune, J.D., Dadoun, S., Bitar, A., Arthuis, F., Van Effenterre, R., & Fohanno, D. (1999). Intra-operative direct electrical stimulations of the central nervous system: The Salpêtrière Experience with 60 patients. *Acta Neurochirurgica, 141*, 1157–1167.
A review of the intraoperative stimulation research done at the Hôpital de la Salpétrière, primarily with patients with tumors and vascular malformations. A nice overview of the clinical implications of the technique.

Gordon, B., Boatman, D., Hart, Jr., J., Miglioretti, D., & Lesser, R.P. (2001). Direct cortical electrical interference (stimulation). In R.S. Berndt (Ed.), *Handbook of neuropsychology* (Vol. 3, 2nd edn, pp. 375–391). Amsterdam: Elsevier.
An overview of the methodology of, and one institution's experience with, extraoperative direct cortical electrical stimulation done through subdural electrode arrays, using relatively extensive language testing protocols.

Ojemann, G.A. (1983). Brain organization for language from the perspective of electrical stimulation mapping. *Behavioral and Brain Sciences, 6*, 189–206.
A detailed account, with commentaries by many other investigators, of the seminal findings of Ojemann and his colleagues with intraoperative direct cortical electric stimulation mapping of language and related functions.

Ojemann, G.A. (2003). The neurobiology of language and verbal memory: Observations from awake neurosurgery. *International Journal of Psychophysiology, 48*, 141–146.
A recent, accessible summary of the pioneering work of one of the most important researchers in the field. A good launching point for those interested in delving further into the study of language processing by electrical stimulation.

11

Transcranial Magnetic Stimulation (TMS) as a Tool for Studying Language

KATE E. WATKINS[1] and JOSEPH T. DEVLIN[2]

[1]Department of Experimental Psychology, University of Oxford, Oxford, UK
[2]Department of Psychology, University College London, London, UK

ABSTRACT

Transcranial magnetic stimulation (TMS) is a powerful non-invasive method for brain stimulation, which allows us to interfere with speech and language processes in the brains of healthy participants. It, thereby, provides new information on the neural and cognitive basis of language organization, studies of which were previously limited to patients with neurological impairment or to correlative functional imaging methods. Here, we provide illustrative examples of recent studies, which serve to demonstrate the variety of ways in which TMS is employed to complement and extend information from imaging and patient studies. We show how the results of TMS studies are sometimes counterintuitive, often facilitating rather than inhibiting responses. Careful control tasks and stimulation of other brain areas are often required before these results can be clearly interpreted. Several classic and new questions about the brain and language have been addressed with TMS. The field is now poised for an explosion of further exploration.

11.1. INTRODUCTION

Until the advent of transcranial magnetic stimulation (TMS), interfering with brain function was limited to studies of patients. The pioneering work of Wilder Penfield and others in the areas of electrocorticography and Wada testing were critical in extending our understanding of the neural basis of language. Limited conclusions could be drawn from these studies, however, as they were based on patients with neurological conditions usually of long standing, which may have altered the brain organization due to recovery or compensatory processes. TMS offers an extension and advancement of these early methods by allowing us to examine language organization in normal healthy individuals. This non-invasive technique tests the *necessity* of an area for a cognitive task unlike others that measure the *correlation* between brain function and a cognitive task either in space, for example, functional magnetic resonance imaging (FMRI) and positron emission tomography (PET), or in time, for example, electroencephalograms (EEG) and magnetoencephalography (MEG). TMS is often referred to as providing a "virtual" lesion, the effects of which are brief and reversible. In addition to interfering with brain function, it can provide new correlates of brain function such as measures of cortical excitability; these measures may be more sensitive to brain function than measures of blood flow and correlated electrical activity.

Here, we summarize the use of TMS in studies of the cognitive and neural basis of language processing. We begin with TMS studies of hemispheric dominance for language and use these to describe the various ways in which TMS is applied (see Box 11.1). This introductory section on language dominance is followed by TMS studies of the classic language circuit, which have examined the role of Wernicke's area in lexical processing, and Broca's area in phonology and semantics. We then describe studies that have explored areas outside this classic circuit examining the role of the motor system in speech perception and differences in processing of nouns and verbs following motor and prefrontal cortex stimulation. Finally, in conjunction with other imaging techniques, we describe how TMS is being used to assess reorganization of function in the brains of patients with aphasia.

Box 11.1 What is transcranial magnetic stimulation?

A transcranial magnetic stimulation (TMS) coil is comprised of windings of metal wire encased in plastic and connected to a capacitor. A brief electric current is discharged through the wires, producing a rapidly changing magnetic field orthogonal to the plane of the coil. When placed over the scalp, the field is unimpeded by the skull, fluid and meninges surrounding the brain and, by simple electromagnetic induction, causes a current to flow in neurons lying parallel to the plane of the coil. In the cortex, this is most likely to cause spiking or depolarization in axons whose orientation is tangential to the cortical surface, that is those lying parallel to the surface, or running perpendicular to the cortical thickness on the crown of a gyrus or bank of a sulcus (Rothwell, 1997).

A single pulse of TMS over the primary motor or visual cortex can be used to elicit positive phenomena such as motor-evoked potentials (MEPs) or phosphenes, respectively (Barker et al., 1985). When delivered to other cortical regions, TMS can disrupt neural activity by introducing noise into the local circuitry and interfering with behavior. These negative effects are typically measured by changes in error rates or response latencies (Walsh & Rushworth, 1999).

With high-frequency repetitive TMS (rTMS), trains of pulses are delivered at fixed inter-pulse intervals of between 20 and 200 ms (50–5 Hz frequency) disrupting function for the duration of the stimulation, on the order of seconds or less. This is useful if we are unsure when precisely to stimulate. Once an area has been shown to be involved in a particular cognitive function or behavior, more precise relationships with time can be determined using single pulses. Low-frequency rTMS (1 Hz or less) delivered for several minutes (typically 10–15), disrupts the function of a brain area and these effects outlast the stimulation itself usually for durations similar to the stimulation time.

Barker, A.T., Jalinous, R., & Freeston, I.L. (1985). Non-invasive magnetic stimulation of human motor cortex. *Lancet*, *I*(8437), 1106–1107.

Rothwell, J.C. (1997). Techniques and mechanisms of action of transcranial stimulation of the human motor cortex. *Journal of Neuroscience Methods*, *74*(2), 113–122.

Walsh, V., & Rushworth, M.F.S. (1999). The use of transcranial magnetic stimulation in neuropsychological testing. *Neuropsychologia*, *37*, 125–135.

11.2. ASSESSING HEMISPHERIC SPECIALIZATION FOR LANGUAGE WITH TMS

Some of the earliest uses of TMS in the study of language assessed hemispheric specialization, lateralization or "dominance." This is unsurprising given the need for a non-invasive replacement for the sodium amobarbital (or Wada) technique used in patients who are about to undergo neurosurgery (see Chapter 3, this volume). These studies provide a useful introduction to the different ways in which TMS can be used both to interfere with brain function and to provide a correlate of the involvement of a brain area in a function (see Box 11.1).

11.2.1. High-Frequency rTMS

In the first TMS studies of language dominance, high-frequency rTMS was applied in 10-s trains of different frequencies over left and right hemisphere language regions (Pascual-Leone et al., 1991). Left inferior frontal gyrus (LIFG) stimulation caused speech interruption in all patients; none had speech problems during right hemisphere stimulation. One patient described the effect of TMS as follows: "I could move my mouth and I knew what I wanted to say, but I could not get the numbers to my mouth" (Pascual-Leone et al., 1991, p. 699). The concordance between the results of TMS and Wada testing in these first six patients gave great confidence to the use of TMS for non-invasive studies of language lateralization. Unfortunately, this early success has not been consistently replicated calling into question the reliability of TMS as an alternative to Wada testing.

For example, recent studies of speech lateralization in healthy participants (Stewart et al., 2001; Aziz-Zadeh et al., 2005) demonstrate that both hemispheres may contribute to the execution of speech with additional areas in the left hemisphere involved in other language production processes. Stimulation over posterior inferior frontal sites in the left or right hemisphere results in speech arrest in association with an electromyographic (EMG) response (i.e., a muscle twitch) in the face muscles. This suggests that the motor representations of the articulators are stimulated. In contrast, stimulation over a more anterior location in the left, but not right, inferior frontal cortex also produces speech arrest but no EMG response. This suggests that stimulation of Broca's area itself results in impaired processes required for language production.

11.2.2. Low-Frequency rTMS

Language lateralization varies across individuals and is particularly important consideration for studies of patients with aphasia. Low-frequency rTMS was used to assess the effects of this variation in a group of normal healthy participants (Knecht et al., 2002). Hemispheric lateralization during word generation was initially established using a blood flow method called transcranial Doppler ultrasonography. One-Hz rTMS stimulation (for 10 min at 110% of the motor threshold (MT); see Box 11.2) over Wernicke's area increased response times on a word-picture matching tasks in individuals with left, but not right, language dominance

Box 11.2 Intensity of stimulation: thresholds

The intensity of stimulation is an important consideration in transcranial magnetic stimulation (TMS) studies, although the choices can be somewhat arbitrary. In the motor and the visual cortex, the effects of stimulus intensity can be directly measured and thresholds established. These represent the percentage of maximum stimulator output required to elicit an motor-evoked potential (MEP) or a phosphene, respectively. The most commonly used threshold is the motor threshold (MT), typically established as the percentage of stimulator output required to elicit an MEP of size > 50 μV in more than 5 out of 10 consecutive pulses. Motor and visual thresholds are most likely related to skull thickness and therefore serve to normalize the stimulation across individuals. To some extent, however, an individual's resting "excitability" will be related to their threshold. When recruiting participants for TMS studies, it is important to consider factors that might raise their cortical excitability and lower threshold, such as a personal or family history of epilepsy, caffeine, alcohol or drug intake, sleep deprivation, hunger or anxiety. Screening for these factors can reduce the likelihood of an adverse event such as a seizure (see Wassermann, 1998 for details). Interestingly, however, motor and visual thresholds appear uncorrelated (Stewart et al., 2001). This may be due to differences in skull thickness overlying these regions, or it may reflect differences in the anatomy of the brain region targeted. For instance, the functional area may lie deeper within the sulcus in one region and on the crown of the gyrus in the other. Also, more often than not we wish to stimulate areas outside the visual and motor systems where it is not possible or practical to determine a threshold. In such cases, arbitrary choices are often made with respect to stimulus intensity, which is worth considering when reviewing the results of TMS studies.

Stewart, L., Walsh, V., & Rothwell, J.C. (2001). Motor and phosphene thresholds: A transcranial magnetic stimulation correlation study. *Neuropsychologia*, *39*(4), 415–419.

Wassermann, E.M. (1998). Risk and safety of repetitive transcranial magnetic stimulation: Report and suggested guidelines from the International Workshop on the Safety of Repetitive Transcranial Magnetic Stimulation, June 5–7, 1996. *Electroencephalography and Clinical Neurophysiology*, *108*, 1–16.

11.2.3. Single-Pulse TMS and EMG

Finally, single-pulse TMS and EMG can be used to measure motor excitability during language tasks. Such a measure may be more sensitive than those obtained with other techniques, which usually assess a proxy of neural activity such as blood flow. In one study, the motor excitability of the hand area of motor cortex was measured under the following conditions: reading aloud, reading silently, spontaneous speech or production of non-speech vocal sounds (Tokimura et al., 1996). The size of the motor-evoked potential (MEP) elicited in the hand muscle by TMS was facilitated (reflecting increased motor excitability) during speech production but was only clearly lateralized for the reading condition. In contrast, the silent reading condition and the non-speech sound production did not result in changes in excitability. This suggests increased motor excitability in primary motor areas during speech production, a perhaps unsurprising result. We used similar methods to examine motor excitability in the representation of speech articulators during speech perception rather than production (Watkins et al., 2003). We found that during listening to speech or viewing speech-related lip movements, the size of the MEP in the lip muscles elicited by single pulses of TMS (120% MT) was facilitated, but only for stimulation over the left hemisphere and not the right (see Figure 11.1(a) and (b)). Also, MEPs measured in the hand muscle in response to single-pulse TMS over the hand representation in the left motor cortex showed no change during any of the speech perception or control conditions (see Figure 11.1(c)). The results of these studies offer an intriguing possibility for non-invasive study of language lateralization using TMS.

11.2.4. Summary

We have described a series of studies that illustrate the variety of ways in which TMS is used to examine hemispheric dominance for speech and language processing. However, it is important to point out the limitations of TMS for such assessments. Unlike sodium amobarbital, which affects the entire middle cerebral artery territory in one hemisphere for several minutes, TMS only affects a small brain area for a very brief time. In addition, not all speech sites are equally accessible to stimulation as they may be located at different depths from the scalp or oriented differently in the two hemispheres. It is difficult, therefore, to make a fair comparison between hemispheres with regard to their involvement or relative involvement in speech production. Also, stimulating the scalp over peri-Sylvian language areas can be painful for the participant when it produces muscle twitches on the head or stimulates cranial nerves such as the trigeminal bundle. All these effects limit the efficacy of TMS for studies of language dominance (see Box 11.3).

as established by transcranial Doppler ultrasonography. Conversely, stimulation over the right hemisphere homolog of Wernicke's area increased response times for those with right, but not left hemisphere dominance. Importantly, the disruptive effect of the TMS interference was related to the degree of language lateralization. TMS had less of an effect in participants with bilateral language organization and a greater effect in those who were strongly lateralized.

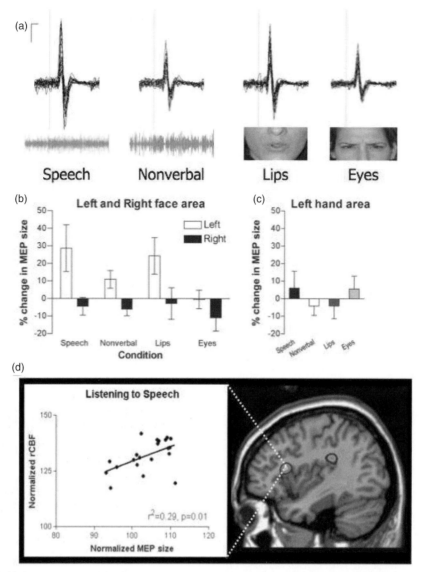

FIGURE 11.1   Motor excitability during speech perception. (a) Data from stimulation of the left primary motor face area in a single subject when listening to speech, listening to non-verbal sounds, viewing speech and viewing eye movements. During the listening conditions, subjects viewed a screen of noise and during viewing conditions, subjects heard auditory white noise. EMG recordings from individual trials are superimposed and the dotted line indicates the time of stimulation. The horizontal bar represents 10 ms and the vertical bar 0.5 mV. (b) Average MEP sizes for the same four stimulus conditions with stimulation to the left and right face area. (c) Average MEP sizes for the four stimulus conditions with stimulation to the hand area of motor cortex. The *x*-axis through the 0% level represents the mean MEP size in the control condition and error bars represent standard error of the mean. Panels (a)–(c) are modified from Watkins *et al.* (2003), with permission. (d) The relation between regional cerebral blood flow in Broca's area and the size of the MEP evoked by single-pulse TMS over the mouth region of primary motor cortex (left panel). On the right, an activation map showing the anatomical location of the significant positive relationship illustrated in the graph.
Source: From Watkins & Paus (2004), with permission.

11.3. DIFFERENT EFFECTS OF TMS ON TASK PERFORMANCE: POSTERIOR TEMPORAL CORTEX AND PICTURE NAMING

The effects of TMS are usually observed in increased reaction times or decreased accuracy of responses, often in comparison with either unstimulated trials or stimulation over control sites. Decreases in reaction times due to stimulation are frequently reported, which emphasizes the importance of baselines and comparison conditions to separate regionally specific facilitation effects from more global ones. These facilitatory effects are difficult to interpret, however,

Box 11.3   Advantages and disadvantages
of transcranial magnetic stimulation

- Advantages
  - Participants act as their own controls, allowing comparisons to be made before, during or after stimulation. This avoids the potential confounds associated with neuropsychological studies of patients with lesions, such as compensatory plasticity, the large and varied extents of naturally occurring lesions, and damage to subjacent fibers-of-passage.
  - TMS can be used to examine the timing of information processing within a brain area with a temporal resolution of tens of milliseconds (Amassian, 1989; Schluter *et al.*, 1998).
  - In conjunction with positron emission tomography (PET), functional magnetic resonance imaging (fMRI) or electroencephalograms (EEG), or by using more than one coil, one can examine functional connectivity between brain regions and within cognitive systems.
  - In patients, one can assess whether brain activity revealed by functional imaging is compensatory or maladaptive.
  - TMS offers the possibility of treatment in patients with speech and language disorders.
- Disadvantages
  - TMS is limited to brain areas at or near the cortical surface because the effect of the stimulation reduces with distance from the coil.
  - The brief passage of current through the coil produces a loud click and a sensation on the scalp, both of which can affect the experimental results and need to be controlled for (Walsh & Rushworth, 1999).
  - When stimulation affects peripheral nerves or muscles, it can cause discomfort or sometimes pain.
  - Areas outside of primary motor or visual cortices can be difficult to locate accurately because there is no obvious physiological response to measure such as MEP or phosphene induction.

Amassian, V.E. (1989). Suppression of visual perception by magnetic coil stimulation of human occipital cortex. *Electroencephalography and Clinical Neurophysiology, 74,* 458–462.

Schluter, N.D., Rushworth, M.F., Passingham, R.E., & Mills, K.R. (1998). Temporary interference in human lateral premotor cortex suggests dominance for the selection of movements. A study using transcranial magnetic stimulation. *Brain, 121* (Pt 5), 785–799.

Walsh, V., & Rushworth, M.F.S. (1999). The use of transcranial magnetic stimulation in neuropsychological testing. *Neuropsychologia, 37,* 125–135.

(Topper *et al.*, 1998; Mottaghy *et al.*, 1999; Andoh *et al.*, 2006). All of these studies suggest involvement of an area in a task but it may be controversial to conclude in all cases that a necessary role has been demonstrated. Below we review a set of studies that have examined lexical retrieval using word-picture matching or picture naming, yielding a set of apparently contradictory results. We use these studies firstly to illustrate the differential effects of TMS on behavior and secondly to demonstrate how different control conditions and control stimulation sites can aid interpretation of some of these behavioral results.

Increased picture-naming latencies were observed during stimulation over the left posterior infero-temporal cortex (BA37) using short (600 ms) trains of rTMS (10 Hz at 75% maximum stimulator output) (Stewart *et al.*, 2001). These increases were found relative to stimulation over the right homologous region and the vertex during performance of the same task. Also, the effect of TMS was selective to this task as it did not affect word and non-word reading or color naming. These results are straightforward and consistent with previous functional imaging findings confirming that the posterior inferior-temporal cortex region is necessary for picture naming.

In another picture-naming study, however, very different behavioral effects of TMS were observed. Stimulation (20-Hz rTMS for 2 s at 55% maximum output) over Wernicke's area in this case decreased naming latencies (Mottaghy *et al.*, 1999). Stimulation over Broca's area and the visual cortex had no affect leading the authors to conclude that stimulation at a low intensity "preactivated" or primed Wernicke's area. The explanation of priming, either of representations in Wernicke's area or of those more remotely located, seems a parsimonious one, but as yet there is no physiological explanation for TMS-induced behavioral facilitation.

In some studies the interpretation of results is even more complicated. For instance, one study (Drager *et al.*, 2004) tested participants on a word-picture verification task before and after a 10-min session of 1-Hz rTMS delivered over one of five locations (Broca's and Wernicke's areas, their right hemisphere homologs, and midline occipital cortex). They also included a sham-TMS control condition, which produces TMS pulses with the coil turned 90° away from the scalp. Reaction times after the 10 min of 1-Hz TMS were consistently faster than those before stimulation regardless of the site, which suggests a non-specific arousal effect of TMS. The authors re-calculated site-specific reaction times relative to the mean across all five sites following stimulation (rather than relative to reaction times before stimulation), to reveal an inhibitory effect of stimulation in Wernicke's area and a facilitatory effect in Broca's area. It is interesting to note that had there been only a single control condition (for example mid-line occipital or sham stimulation), it would have been difficult to separate these regional effects from the overall facilitation effect.

and it may be controversial to conclude from such results that an area is *necessary* for a task. It is striking that many reports of facilitatory effects are associated with Wernicke's area stimulation, suggesting that this may be a special case

11.4. TMS STUDIES OF SEMANTIC AND PHONOLOGICAL PROCESSING IN LIFG

Several recent studies have used TMS to address the role of different portions of the LIFG in semantic and phonological processing (Devlin *et al.*, 2003; Nixon *et al.*, 2004; Gough *et al.*, 2005). Functional imaging studies commonly report activity in anterior portions of the LIFG during semantic processing tasks but the presence of semantic deficits in patients with lesions in this area is not clearly established.

In a simple semantic task such as deciding whether a visually presented word referred to a man-made (e.g., "kennel") or natural object (e.g., "dog"), TMS (10 Hz for 300 ms at 110% MT) over anterior LIFG increased reaction times relative to trials with no TMS; there was no effect of TMS on reaction times for simple judgements concerning the visual properties of the presented words (Devlin *et al.*, 2003). In a similar study, using fMRI-guided rTMS (7 Hz for 600 ms at 100% MT) over anterior LIFG, response times for decisions based on word meanings were significantly slowed relative to trials with no TMS; there were no effects over two other control sites (Kohler *et al.*, 2004). Taken together, these studies support the claim that anterior LIFG is necessary for semantic processing.

A number of studies have also confirmed a necessary role for LIFG in phonological processing. In one study of phonological working memory participants saw a word on a computer screen (e.g., "knees") and then held it in memory during a 1–2 s delay before deciding whether it sounded the same as a subsequently presented non-word (e.g., "neaze") (Nixon *et al.*, 2004). Stimulation (10 Hz for 500 ms at 120% MT) was applied over posterior LIFG during the delay period and selectively increased the error rate during the phonological task, but not during a comparable visual working memory task. Similarly rTMS (5 Hz for 2.4 s at 115% MT) over posterior LIFG increased response times for a phonological working memory task requiring silent reading and syllable counting of a visually presented word (Aziz-Zadeh *et al.*, 2005).

Taken together the results of these TMS studies of LIFG significantly extend the previous neuroimaging results by demonstrating that anterior LIFG is *necessary* for semantic processing while the posterior LIFG is *necessary* for phonological processing. It is possible, however, that the LIFG acts as a single functional region required for both semantic and phonological processing. Alternately, there may be sub-regions within LIFG specialized for semantic and phonological processing but co-activated due to incidental processing. The TMS results described above do not distinguish between these possibilities. Therefore, to test for a double dissociation between semantic and phonological processing within LIFG, Gough *et al.* (2005) designed the following TMS experiment. Two letter strings were presented simultaneously on a computer screen and participants made a semantic (e.g., same meaning? "idea-notion"), phonological (e.g., sound the same? "nose-knows") or visual decision (e.g., look the same? "fwtsp-fwtsp") decision. Compared to no stimulation, TMS of anterior LIFG selectively increased response times when participants made semantic but not phonological decisions. Conversely, stimulation of posterior LIFG selectively interfered with phonological but not semantic decisions (Figure 11.2). Neither site of stimulation affected the response times in the visual control task. In other words, the authors demonstrated a functional double dissociation for semantic and phonological processing within LIFG in sites separated by less than 3 cm. Although this double dissociation was first suggested by functional imaging, it required the spatial precision of TMS to independently disrupt the regions and clarify their distinct contributions to word processing.

11.5. SPEECH PERCEPTION AND THE MOTOR SYSTEM

A recent series of studies used TMS to measure the excitability of the motor system during speech perception. These studies largely arose out of renewed interested in motor theories of perception related to the neurophysiological findings of mirror neurons in the monkey brain (Fadiga *et al.*, 2005; see Chapter 23, this volume). These neurons fire not only when the monkey performs an action but also when the monkey sees or hears an action being performed, thereby providing a functional link between perception and action. Using single pulses of TMS over the cortical motor representation of a muscle of interest (the tongue or the lips), the size of the MEP elicited by TMS at a fixed intensity provides an index of the excitability of the motor system, which is compared across conditions. During passive listening to stimuli with or without consonant sounds that required tongue movements in their production, MEPs were measured from the tongue (Fadiga *et al.*, 2002). MEP size increased when subjects listened to words and non-words containing the labiodental "rr" phoneme as in the Italian word "terra" but not those with the "ff" phoneme as in the Italian word "zaffo." In a similar experiment measuring MEP sizes in the lip muscle (orbicularis oris), MEP size increased when subjects listened to continuous prose passages or viewed lip movements during continuous speech (Watkins *et al.*, 2003), but was unchanged during listening to non-verbal stimuli or watching eye and brow movements.

The two TMS studies described above suggest a functional connection between speech perception and the motor system underlying speech production, but they do not provide any anatomical information to suggest how this link is mediated. By combining TMS with PET, we investigated the brain regions that mediate the change in motor excitability during speech perception (Watkins & Paus, 2004). Motor excitability

(a) **Normalized reaction times**

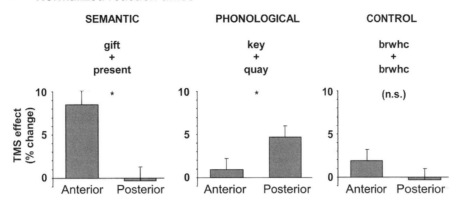

(b) **Stimulation sites**

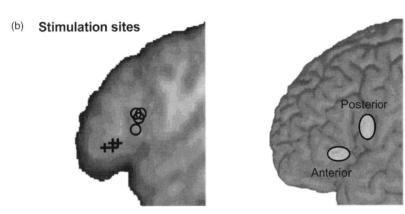

FIGURE 11.2 Effects of stimulation on anterior and posterior LIFG. (a) The bar plots show the mean normalized TMS effects as percent change in reaction times from the non-TMS baseline during synonym judgments (left), homophone judgments (middle) and visual matching (right). Error bars indicate the standard error of the mean and significant differences are indicated with an *($p<0.05$). (b) The bottom panel shows the location of stimulation sites for four participants on their mean structural image with anterior locations marked with crosses and posterior locations marked with circles. Next to it is a 3D rendering with the stimulation sites shown as ovals representing the spatial 85% confidence interval. Stimulation sites were on average 2.5 cm apart on the cortical surface. *Source*: From Gough *et al.* (2005) with permission.

was measured by eliciting MEPs in the orbicularis oris muscle due to TMS over the face area of left motor cortex. As before, MEP size increased during auditory speech perception. The MEP sizes were then regressed against regional cerebral blood flow measures across the whole brain obtained during the TMS. This analysis revealed that increased motor excitability during speech perception correlated with blood flow increases in the posterior part of the LIFG (Broca's area; see Figure. 11.1(d)). In other words, Broca's area plays a central role in linking speech perception with speech production, consistent with theories that emphasize the integration of sensory and motor representations in understanding speech (Liberman & Mattingly, 1985).

The results from these studies show that increased motor excitability of the speech production system is specific to the muscle used in production, that this effect is lateralized to the left hemisphere and that the changes are mediated by neurons in Broca's area. These increases may reflect covert

imitative mechanisms, rapid internal speech or a correlate of the perception process. Whether this in some way aids speech perception or comprehension remains to be tested.

11.6. REPRESENTATION OF ACTIONS AND VERBS IN THE MOTOR SYSTEM AND FRONTAL CORTEX

A series of studies have examined the role of motor areas in the neural representation of actions and verbs. Two such studies have used TMS over the primary motor representations of different effectors to elicit MEPS while subjects processed actions. In the first study, subjects listened to sentences related to hand-actions (e.g., "he sewed the skirt"), foot-actions (e.g., "he jumped the rope") or more abstract actions (e.g., "he forgot the date"), while MEPs were measured from the hand and foot muscles (Buccino *et al.*, 2005).

The size of the MEPs in specific muscles used to produce the actions were specifically modulated when listening to sentences referring to actions involving those muscles rather than actions involving other muscles or abstract actions. In another study, subjects received single pulses of TMS over the left motor cortex representation of the arm or leg while they made lexical decisions (Pulvermüller et al., 2005). Reaction times were decreased to lexical decisions for actions related to arms (e.g., "folding") and legs (e.g. "stepping"), respectively. As noted above, however, the interpretation of facilitatory effects of TMS is not straightforward. The authors suggest that the decrease in response times may be due to partial activation of the representation of actions related to the specific areas being stimulated.

Neuropsychological studies in patients with brain lesions show selective impairments for nouns and verbs suggesting that their neural representations may be spatially distinct. TMS studies offer an opportunity to test this hypothesis in the normal healthy brain. In one study, the authors postulated that verb-specificity may be due to the close relationship between verbs and actions and used TMS to investigate the role of left dorso–lateral prefrontal cortex (DLPFC) in action naming (Cappa et al., 2002). A set of Italian speaking participants were shown pictures of common objects and asked to either name the object (e.g., "telefono" [a telephone]) or the associated action ("telefonare" [to telephone]). rTMS (20 Hz for 500 ms at 90% MT) of left DLPFC decreased naming latencies for verbs relative to right DLPFC and sham stimulation (coil is placed perpendicular rather than tangential to the skull, thereby providing similar sensory effects of TMS but no stimulation). In contrast, the latencies for object naming were unaffected. Based on this condition-specific facilitation effect, the authors suggested that verbs may be preferentially impaired by left frontal lesions because damage to DLPFC affects action observation and representation which are more tightly linked with verbs than nouns.

Other studies, however, call this interpretation into question (Shapiro et al., 2001). In one study, participants were asked to inflect nouns and verbs (e.g., "song"→"songs" or "sing"→"sings") either before or after 10 min of 1-Hz stimulation over left DLPFC. Reaction times were significantly slowed for verbs, but not nouns. In order to determine whether this effect was due to the action-related meaning of the verbs, a second experiment used pseudowords (e.g., "flonk") treated as either nouns or verbs. Because pseudowords do not have any associated meaning, the authors reasoned that TMS would only affect reaction times if the region was important for processing the grammatical class of verbs rather that words with action-related meanings. DLPFC stimulation selectively slowed reaction times only in the verb condition – a finding interpreted as evidence for a neuroanatomical basis for grammatical categories *per se* rather than a by-product of the differences in meaning between nouns and verbs.

It is not entirely clear how to reconcile these findings. In both cases, DLPFC stimulation preferentially affected verbs relative to nouns, although in one case this manifested as facilitation (Cappa et al., 2002) and in the other, inhibition (Shapiro et al., 2001). Further work will be necessary to determine the range of verbs and tasks that engage DLPFC. For instance, are the effects modulated by the amount of "action" inherent in the verb? In other words, do "running" and "throwing" require greater DLPFC involvement than "thinking" or "sleeping"? If so, the functional link between hand or leg motor regions and specific action words may be mediated via DLPFC in much the same way the ventrolateral prefrontal cortex mediates the link between perceiving speech and the mouth region of motor cortex (Watkins & Paus, 2004).

11.7. TMS STUDIES IN PATIENTS WITH APHASIA

Several researchers have investigated the role of the right inferior frontal gyrus (IFG) in recovery from non-fluent aphasia due to left hemisphere stroke (Martin et al., 2004; Naeser et al., 2005a; Naeser et al., 2005b; Winhuisen et al., 2005). Functional imaging studies have frequently shown increased right hemisphere activity in patients with aphasia. The question of interest here is whether the abnormal level of right hemisphere activity reflects a successful compensatory mechanism aiding recovery of function or a maladaptive one, which impedes recovery. An initial study in four chronic non-fluent patients used 1-Hz rTMS (90% MT for 10 mins) to test four different sites in the right hemisphere (Martin et al., 2004). The targeted sites were anterior IFG, posterior IFG, mouth area of primary motor cortex and posterior superior temporal gyrus. Only stimulation over anterior IFG led to an improvement in naming accuracy and reaction times. In a follow-up study, the patients were given 20 min of 1-Hz rTMS at 90% MT daily for 2 weeks (Naeser et al., 2005b). Significant improvements on naming accuracy and reaction times were seen immediately after the 10th treatment session and these effects persisted for at least 2 months in all patients and for 8 months in one of the patients (Naeser et al., 2005a).

In a similar experiment, a group of 14 patients with brain tumors were examined using PET functional imaging and high-frequency rTMS (4 Hz at MT) during verb generation (Thiel et al., 2005). Brain activity revealed by PET was compared in the left and right IFG. The location of maximum activity on the PET scan in these areas was then targeted with TMS while patients attempted to generate verbs. LIFG stimulation resulted in increased latencies for all controls and 10 out of 14 patients; a further three patients did not respond to the presented noun; and one patient produced a verb unrelated to the presented noun. In contrast, right IFG stimulation resulted in increased latencies in only 5 of the 14 patients and none of the controls. Laterality indices calculated for the brain activity in IFG were significantly lower in the patients who showed TMS interference for right IFG stimulation. That is, those with right hemisphere dominance in the PET study were impaired at verb generation when the right IFG was stimulated. The authors concluded that in these five patients, right hemisphere activity was necessary for language function – a conclusion which cannot be drawn from the imaging data alone.

Although TMS provides some support for the idea that homologous regions assume lost language functions following left hemisphere lesions (Thiel *et al.*, 2005; Winhuisen *et al.*, 2005), it also demonstrates that, at the very least, the story is considerably more complicated. For one thing, TMS confirms the importance of residual left hemisphere function, as suggested by previous functional imaging studies. In fact, left hemisphere stimulation interfered with performance more consistently than right hemisphere stimulation, which only affected the subset of patients with the strongest rightward asymmetries. The reason for these asymmetries remains unclear, but one factor likely to play a role is premorbid language organization (Knecht *et al.*, 2002). Another complicating finding is that in some cases, right hemisphere stimulation interfered with performance (Thiel *et al.*, 2005; Winhuisen *et al.*, 2005) while in others, it improved performance (Martin *et al.*, 2004; Naeser *et al.*, 2005a; Naeser *et al.*, 2005b). There are, of course, significant differences between the studies including the types of patients (i.e., chronic non-fluent versus fluent aphasics) and the type of TMS (i.e., long trains of low-frequency versus short bursts of high-frequency TMS). Nonetheless, the findings support the idea that one cannot draw a simple conclusion regarding right hemisphere involvement in recovery. Understanding these differences poses a major challenge for cognitive neuroscience and may require adopting more sophisticated models of recovery that move beyond the simple notions of "homologous transfer of function" and "necessary and sufficient" brain regions.

11.8. CHALLENGES AND FUTURE DIRECTIONS

As illustrated here, TMS has already begun to offer novel and important insights into the cognitive and neural basis of language processing. When used to produce "virtual" lesions, TMS can assess whether an area is necessary for a task by demonstrating inhibitory effects on behavior. In contrast, behavioral facilitation (Cappa *et al.*, 2002; Sakai *et al.*, 2002; Pulvermüller *et al.*, 2005) is not readily explained in terms of externally induced neuronal firing and consequently, a significant challenge will be to elucidate the physiological mechanisms underlying these effects. Without this explanation, interpretation of such results will remain speculative and controversial. TMS can also provide a useful measure of cortical excitability for identifying functional links between language and the motor system, which, for instance, demonstrate a close link between speech perception

and the motor cortex areas responsible for producing speech (Fadiga *et al.*, 2002; Watkins *et al.*, 2003; Watkins & Paus, 2004). A critical test of this relationship will be to determine whether this functional link in some way aids speech perception or comprehension. Finally, recent developments in TMS methodology offer novel opportunities for investigating cortical connectivity (Paus *et al.*, 1997), and suggest that the application of these tools could help to trace the neural circuitry underlying human language processing. In short, the field seems poised to expand enormously in virtually all areas of language research, building on the early successes and developing novel methods capable of answering an even wider range of questions.

References

Andoh, J., Artiges, E., Pallier, C., Riviere, D., Mangin, J.F., Cachia, A., Plaze, M., Paillere-Martinot, M.L., & Martinot, J.L. (2006). Modulation of language areas with functional MR image-guided magnetic stimulation. *Neuroimage*, 29(2), 619–627.

Aziz-Zadeh, L., Cattaneo, L., Rochat, M., & Rizzolatti, G. (2005). Covert speech arrest induced by rTMS over both motor and nonmotor left hemisphere frontal sites. *Journal of Cognitive Neuroscience*, 17(6), 928–938.

Buccino, G., Riggio, L., Melli, G., Binkofski, F., Gallese, V., & Rizzolatti, G. (2005). Listening to action-related sentences modulates the activity of the motor system: A combined TMS and behavioral study. *Brain Research. Cognitive Brain Research*, 24(3), 355–363.

Cappa, S.F., Sandrini, M., Rossini, P.M., Sosta, K., & Miniussi, C. (2002). The role of the left frontal lobe in action naming: rTMS evidence. *Neurology*, 59(5), 720–723.

Devlin, J.T., Matthews, P.M., & Rushworth, M.F. (2003). Semantic processing in the left inferior prefrontal cortex: A combined functional magnetic resonance imaging and transcranial magnetic stimulation study. *Journal of Cognitive Neuroscience*, 15(1), 71–84.

Drager, B., Breitenstein, C., Helmke, U., Kamping, S., & Knecht, S. (2004). Specific and nonspecific effects of transcranial magnetic stimulation on picture-word verification. *European Journal of Neuroscience*, 20(6), 1681–1687.

Fadiga, L., Craighero, L., Buccino, G., & Rizzolatti, G. (2002). Speech listening specifically modulates the excitability of tongue muscles: A TMS study. *European Journal of Neuroscience*, 15(2), 399–402.

Fadiga, L., Craighero, L., & Olivier, E. (2005). Human motor cortex excitability during the perception of others' action. *Current Opinions in Neurobiology*, 15(2), 213–218.

Gough, P.M., Nobre, A.C., & Devlin, J.T. (2005). Dissociating linguistic processes in the left inferior frontal cortex with transcranial magnetic stimulation. *Journal of Neuroscience*, 25(35), 8010–8016.

Knecht, S., Floel, A., Drager, B., Breitenstein, C., Sommer, J., Henningsen, H., Ringelstein, E.B., & Pascual-Leone, A. (2002). Degree of language lateralization determines susceptibility to unilateral brain lesions. *Nature Neuroscience*, 5(7), 695–699.

Kohler, S., Paus, T., Buckner, R.L., & Milner, B. (2004). Effects of left inferior prefrontal stimulation on episodic memory formation: A two-stage fMRI–rTMS study. *Journal of Cognitive Neuroscience*, 16(2), 178–188.

Liberman, A.M., & Mattingly, I.G. (1985). The motor theory of speech perception – revised. *Cognition*, 21, 1–36.

Martin, P.I., Naeser, M.A., Theoret, H., Tormos, J.M., Nicholas, M., Kurland, J., Frengi, F., Seekins, H., Doron, K., & Pascual-Leone, A. (2004). Transcranial magnetic stimulation as a complementary treatment for aphasia. *Seminars in Speech and Language*, 25(2), 181–191.

Mottaghy, F.M., Hungs, M., Brugmann, M., Sparing, R., Boroojerdi, B., Foltys, H., Huber, W., & Topper, R. (1999). Facilitation of picture naming after repetitive transcranial magnetic stimulation. *Neurology, 53*(8), 1806–1812.

Naeser, M.A., Martin, P.I., Nicholas, M., Baker, E.H., Seekins, H., Helm-Estabrooks, N., Cayer-Meade, C., Kobayashi, M., Theoret, H., Fregni, F., Tormos, J.M., Kurland, J., Doron, K.W., & Pascual-Leone, A. (2005a). Improved naming after TMS treatments in a chronic, global aphasia patient – case report. *Neurocase, 11*(3), 182–193.

Naeser, M.A., Martin, P.I., Nicholas, M., Baker, E.H., Seekins, H., Kobayashi, M., Theoret, H., Fregni, F., Maria-Tormos, J., Kurland, J., Doron, K.W., & Pascual-Leone, A. (2005b). Improved picture naming in chronic aphasia after TMS to part of right Broca's area: An open-protocol study. *Brain and Language, 93*(1), 95–105.

Nixon, P., Lazarova, J., Hodinott-Hill, I., Gough, P., & Passingham, R. (2004). The inferior frontal gyrus and phonological processing: An investigation using rTMS. *Journal of Cognitive Neuroscience, 16*(2), 289–300.

Pascual-Leone, A., Gates, J.R., & Dhuna, A. (1991). Induction of speech arrest and counting errors with rapid-rate transcranial magnetic stimulation. *Neurology, 41,* 697–702.

Paus, T., Jech, R., Thompson, C.J., Comeau, R., Peters, T., & Evans, A.C. (1997). Transcranial magnetic stimulation during positron emission tomography: A new method for studying connectivity of the human cerebral cortex. *Journal of Neuroscience, 17*(9), 3178–3184.

Pulvermüller, F., Hauk, O., Nikulin, V.V., & Ilmoniemi, R.J. (2005). Functional links between motor and language systems. *European Journal of Neuroscience, 21*(3), 793–797.

Sakai, K.L., Noguchi, Y., Takeuchi, T., & Watanabe, E. (2002). Selective priming of syntactic processing by event-related transcranial magnetic stimulation of Broca's area. *Neuron, 35*(6), 1177–1182.

Shapiro, K.A., Pascual-Leone, A., Mottaghy, F.M., Gangitano, M., & Caramazza, A. (2001). Grammatical distinctions in the left frontal cortex. *Journal of Cognitive Neuroscience, 13*(6), 713–720.

Stewart, L., Meyer, B.U., Frith, U., & Rothwell, J. (2001). Left posterior BA37 is involved in object recognition: A TMS study. *Neuropsychologia, 39,* 1–6.

Stewart, L., Walsh, V., Frith, U., & Rothwell, J.C. (2001). TMS produces two dissociable types of speech disruption. *Neuroimage, 13,* 472–478.

Thiel, A., Habedank, B., Winhuisen, L., Herholz, K., Kessler, J., Haupt, W.F., & Heiss, W.D. (2005). Essential language function of the right hemisphere in brain tumor patients. *Annals of Neurology, 57*(1), 128–131.

Tokimura, H., Tokimura, Y., Oliviero, A., Asakura, T., & Rothwell, J.C. (1996). Speech-induced changes in corticospinal excitability. *Annals of Neurology, 40*(4), 628–634.

Topper, R., Mottaghy, F.M., Brugmann, M., Noth, J., & Huber, W. (1998). Facilitation of picture naming by focal transcranial magnetic stimulation of Wernicke's area. *Experimental Brain Research, 121,* 371–378.

Watkins, K. & Paus, T. (2004). Modulation of motor excitability during speech perception: The role of Broca's area. *Journal of Cognitive Neuroscience, 16*(6), 978–987.

Watkins, K.E., Strafella, A.P., & Paus, T. (2003). Seeing and hearing speech excites the motor system involved in speech production. *Neuropsychologia, 41*(8), 989–994.

Winhuisen, L., Thiel, A., Schumacher, B., Kessler, J., Rudolf, J., Haupt, W.F., & Heiss, W.D. (2005). Role of the contralateral inferior frontal gyrus in recovery of language function in poststroke aphasia: A combined repetitive transcranial magnetic stimulation and positron emission tomography study. *Stroke, 36*(8), 1759–1763.

Further readings

Devlin, J.T., & Watkins, K.E. (2007). Stimulating language: Insights from TMS. *Brain, 130*(3), 610–622.
The paper provides a comprehensive review of TMS studies of language. It covers the literature published since 1991 when TMS was first introduced as a tool to study language. Further details of many of the studies reviewed here are provided with a specific focus on studies that have extended our understanding of either the cognitive or the neural basis of language.

Robertson, E.M., Theoret, H., & Pascual-Leone, A. (2003). Studies in cognition: The problems solved and created by transcranial magnetic stimulation. *Journal of Cognitive Neuroscience, 15*(7), 948–960.
This paper reviews the potential problems associated with the use of TMS in cognitive studies and provides a critical discussion of the limitations of the method.

Sack, A.T. (2006). Transcranial magnetic stimulation, causal structure-function mapping and networks of functional relevance. *Current Opinion in Neurobiology, 16*(5), 593–599.
This paper provides further detail on recent advances in the combination of TMS with other imaging modalities.

PART III

EXPERIMENTAL NEUROSCIENCE OF LANGUAGE AND COMMUNICATION

12

Disorders of Phonetics and Phonology

HUGH W. BUCKINGHAM[1] and SARAH S. CHRISTMAN[2]

[1]Department of Communication Sciences and Disorders, Linguistics Program, Louisiana State University, LA, USA

[2]Department of Communication Sciences and Disorders, The University of Oklahoma Health Sciences Center, Oklahoma City, OK, USA

ABSTRACT

This chapter highlights some recent research that has significantly impacted our understanding of phonetic and phonologic disorders in adults. In contrast to the popular serial language processing models of the past, exciting new findings from a variety of scientific disciplines now support an interactive and distributional view of phoneme processing in adults with acquired neurolinguistic impairments. We will describe how connectionist computer modeling has combined with innovative assessments of brain function to re-shape our understanding of the origins of sound errors in aphasia and apraxia of speech. Initial discussion will address how disrupted phonological processes yield impairments of phoneme activation and create phonemic paraphasias in words. Subsequent discussion will review how disturbed phonetic processes produce errors in sound and syllable planning, thereby creating the segmental symptoms of apraxia of speech. Lastly, we will assess the different accounts of phonemic paraphasias and consider the production of non-words in the fluent aphasias.

12.1. INTRODUCTION

Twenty-first century investigations into the breakdown of phonological and phonetic processes in adults with aphasia and apraxia of speech typically blend traditional linguistic concepts with ideas from psychology, computer science and the neurosciences. New findings from cortical localization of speech-language functions, for example, are forcing us to accept a mosaic view of the working brain, with fuzzier boundaries of regions highly sensitive to structural/physiological disruption than previously thought. Similarly, the contributions of cognitive impairments in attention and memory to disorders previously considered to be purely linguistic in nature (such

as aphasia) are increasingly appreciated by clinical scientists and have had considerable impact on the structure of modern aphasia rehabilitation programs. Advances in our understanding of neurological and computer-modeled networks now allow us to chart phoneme selection/sequencing and motor speech planning disorders in new ways. We will assess current thinking about how functional connectionist models operate at the higher levels of the language sound system in ways that simulate what the human being does – in health and in disease.

12.2. CHANGING CONCEPTS IN THE NEUROLINGUISTIC ANALYSIS OF PHONETICS AND PHONOLOGY

Phonology is the rule system of language that constrains how sounds may be organized into syllables and words. Phonetics is the science investigating how sounds are produced across the world's many languages. Phonological processing is at work when abstract sound units (phonemes) are selected and sequenced prior to planning the articulatory gestures for speech. An impairment of phoneme processing is considered a linguistic problem and is often a symptom of *aphasia*, an acquired language disorder that frequently follows stroke. Difficulty planning the articulatory movements for speech is considered a non-linguistic (motoric, phonetic) problem and is the central symptom of adult *apraxia of speech*, an acquired motor speech disorder that also frequently follows stroke. There is evidence that there are no clear cut linguistic and neurological divisions in these phonetic/phonemic systems and this in turn forces aphasiologists to consider their interactions in describing phonetic and phonological breakdowns subsequent to brain damage in the dominant (usually, the left) hemisphere. We begin with a brief outline in this section of new psychological models that focus upon connections between three levels of language: semantic/conceptual, words, and sounds.

12.2.1. The Effects of Connectionist Modeling of Phonetics and Phonology

Connectionist models of phonological processing usually operate with three levels of representation and computation: conceptual/semantic, lexical, and phonological (e.g., Dell et al., 1997). At times, a "word shape" level is added to this architecture (Dell & Kim, 2005). Bi-directional connections among all three levels of processing allow these models to simulate semantic and phonological slips-of-the-tongue in normal adults and semantic and phonemic paraphasias in adults with aphasia (e.g., Dell et al., 1997; Martin, 2005). Unfortunately, connectionist models have been generally unconcerned with accounting for finer phonetic detail, and consequently, they have been of little use in simulating speech disruptions that are "apraxic" (clearly motoric) in nature.

Connectionist systems learn by associative principles. They are dynamic (stable but flexible) and they appear to learn simply by virtue of continued active processing. They are responsive to inputs and are constrained by the patterns of the inputs that are fed into their systems. In language, word structures and their phonemic connections comprise most of the formal architecture of the lexicon. The language-specific restrictions on permitted sound sequences (phonotactics and the sonority sequencing principle) limit how phonemic units combine during normal and impaired phonological processing for any one human language. For example, if a language's preferred structure places sounds with least sonority at syllable boundaries, then even non-sensical or neologistic words produced in aphasia (or by a "lesioned" connectionist machine) will tend to reflect that preference. Metaphorically, the language system that the model learns is said to "know" these constraints and this is not overly different from the claim that in the human, the nervous system knows these constraints.

A key question in neurolinguistics concerns how language processing systems in humans and in connectionistic machines come to know their constraints. Researchers who take a nativist stance argue that high-level system constraints are fairly universal and are inherent in the original, internal construction of the system (e.g., they are present at birth). Conversely, those who support an associationist position argue that system constraints are derived from exposure to external environments, therefore varying from community to community. Associative, connectionistic phonological learning systems use mechanisms (i.e., mathematical algorithms) that discover the sound patterns of the word inputs with which they are presented. The statistical probabilities of the input patterns and forms as a whole are learned or abstracted by the model and, in turn, those

probabilities govern or constrain the normal and disordered errors that the system produces. In this sense, by its simulations of human behavior, the model is claimed to be analogous to that of human language users.

A new approach to phonology, optimality theory (Prince & Smolensky, 2004), integrates key concepts from both the nativist and associationist positions on language processing. Optimality theory is a complex linguistic model of constraint-based patterns and markedness hierarchies and it is billed clearly as nativist. It should therefore have "nothing to do with connectionism," although at the same time it is claimed to be "...deeply connected to connectionism" (Prince & Smolensky, 2004, pp. 234–235). Any evaluation of neurolinguistic theory must solve this conflict. We must ask whether and to what extent the tightly constrained paraphasias, which ultimately play into the phonetics, need any other explanation than that they emerge from the language architecture itself. This is troublesome to many neurolinguists who believe that language acquisition requires more than input frequency patterns. This controversy is likely to reappear as more neurolinguistic research considers constraint-based linguistic modeling, and explores how human and machine systems learn by linking input with output. We return to this issue later.

12.2.2. Can Phonetics and Phonology Be Precisely Dissociated?

It has been claimed that (physical) phonetics and (mental) phonology can be dissociated such that there must be some kind of "interface" straddling them. Historical linguistic change, for example, has often shown that physical motor-phonetic and sensory-perceptual processes underlie most abstract (mental) sound changes through time (Ohala, 2005), and, in fact, acoustic-phonetic processing has explained most everything known about synchronic sound patterns in languages (Ohala, 2005). Generative linguists such as Noam Chomsky and Morris Halle claim that articulatory gesture shaping at the phonetic level of description may be "phonologized" ultimately into language knowledge but for Ohala, there is such a marked degree of interaction between the phonetic and phonological aspects of the sound system that the presumed "interface" between them collapses. This mixture can be observed in the brain activation patterns of normal individuals listening to language and in the sound error patterns produced by adults with aphasia and with acquired apraxia of speech.

Classic neurolinguistic theory posits that phonetic (motor speech) errors arise from left inferior frontal lobe disruption whereas phonological (linguistic) errors arise from phoneme selection and ordering disruptions secondary to temporal lobe damage. Twenty-first century research findings are refuting this dichotomy. Phonemic substitution errors have long been observed from frontal lobe damage and subtle phonetic asynchronies have been reported from temporal lobe damage as well (Buckingham & Christman, 2006). Gandour *et al.* (2000) have shown that when speakers of tone languages are presented with rapidly changing pitch patterns that match tonemes in their language, there is a significantly greater degree of metabolic activity in left opercular (motor speech) cortex than in temporal (phonological) cortex. This opercular region does not show metabolic activity, however, with the introduction of rapidly changing pitch patterns that are not contrastive in these same languages, nor is there any unilateral left opercular focal metabolism with, say, Thai tones, delivered to Chinese speakers (who have tonemes, but not those of Thai). English speakers predictably exhibit no opercular area metabolic activity for any set of tonemes whatsoever. Thus, paradoxical patterns of brain activation in healthy adults seem to support the integration, rather than the segregation, of phonetic and phonemic processing across language cortex.

Evidence from speakers with acquired apraxia of speech supports the fuzziness of boundaries in levels of language processing previously assumed to be distinct. Acquired apraxia of speech is a disorder whose phonetic symptoms (especially sound substitutions and omissions) are frequently misinterpreted as phonological (McNeil *et al.*, 2000) even though errors arise from impairment of an intermediate stage of production, subsequent to the phonological selection and sequencing of phonemes, but prior to actual articulatory execution. According to Code (2005) there are four principle features of the syndrome. First, vocalic and consonantal lengthening is observed in syllables, words, and phrases (regardless of whether those vowels and consonants have been correctly selected relative to their phonemic targets). Second, the junctures between segments may have lengthened durations. Third, movement transitions and trajectories from one articulatory posture to another during speech may be flawed, thus creating spectral distortions of perceived phonemes. Finally, distorted sound substitutions may appear. These substitutions are described as sound selection errors and they are not caused by difficulties in the production of properly timed anticipatory co-articulation (a frequent problem in acquired apraxia of speech).

Most of the features of acquired apraxia of speech described by Code are clearly motoric in nature and represent phonetic impairments in speech timing and movement planning. The sound selection errors are problematic, however, in that their source is phonological. Given that patients with pure acquired apraxia of speech typically have brain lesions in and around pre-motor Broca's area (McNeil *et al.*, 2000) or, more controversially, in the anterior gyrus of the left hemisphere Island of Reil (the insula) (Dronkers, 1996; Ogar *et al.*, 2005, but also see Hillis *et al.*, 2004 for discussion of the insula), the evidence for phonological disturbance is even more intriguing. In patients with frontal cortical damage and pure acquired apraxia of speech, it appears that phonetics and phonology are indeed inextricably bound and interwoven. Relevant brain areas appear to have seriously fuzzy boundaries when it comes to language comprehension and production. In sum, it appears that the neat theoretical dissection of brain and language into non-overlapping zones and levels is a paradigm that will need replacement as neurolinguistic science advances.

12.3. ANATOMICAL AND COMPUTATIONAL COMPLEXITY FOR BROCA'S AREA

Two recent studies have re-evaluated the nature and function of the left hemisphere Broca's area (Hagoort, 2005; Thompson-Schill, 2005), and both examinations have consequences for pathological breakdowns of phonetics and phonology. First, Hagoort suggests that Broca's area should really be considered more accurately a Broca's "complex." He argues forcefully and convincingly that the classical designation of Brodmann areas (BA) 44 and 45 for Broca's area is seriously underdetermined and notes that areas 44 and 45 are not particularly well-motivated biologically as comprising one unified zone. Foundas *et al.* (1998), in detailed MRI investigations of left pars opercularis (BA 44) and pars triangularis (BA 45), have demonstrated that these gyral regions differ in gross morphology and cytoarchitecture. Of interest is their finding that patients with small lesions restricted to BA 45 have language problems, while those with small lesions in BA 44 are dysfluent with articulatory disorders. This might explain why patients with damage to the traditional Broca's area may exhibit symptoms of aphasia and/or apraxia of speech but, it does not explain the presence of phonemic selection errors in patients with pure acquired apraxia of speech unless there is more language (phonological) processing occurring in BA 44 than we currently suppose.

Thompson-Schill (2005) has proposed that Broca's area may be involved in selecting from among competing sources of linguistic information (such as those for phoneme selection versus sequencing). This general cognitive capacity may have been recruited for linguistic functions and linked for that purpose to Broca's area. This notion contradicts traditional claims that the left temporal lobe selects phonemes, while the frontal lobe *sequences* them and reinforces our fuzzy boundary argument for phonetics and phonology on both neuroanatomical and linguistic grounds.

It should be noted that any selectional power of Broca's area would clearly go beyond phoneme selection to lexical selection as well. This might complicate our understanding of neologistic jargon aphasia, since abstruse neologisms involve breakdowns in both the selection and sequencing of words and phonemes, and those pathologies typically arise secondary to left *temporal* lesions. In any event, understanding the mechanisms of selection will help us to better explain the overlap of acquired apraxia of speech with aphasia, and thus, the phonetic with the phonological.

12.4. INTRACTABLE PROBLEMS IN THE NEUROLINGUISTICS OF SEGMENTAL PARAPHASIAS

Several issues in neurolinguistics have been investigated for decades, even centuries, without complete success. Although the relationship between the mind and the brain remains one of the largest issues yet to be fully understood, there are several smaller theoretical problems related to the structure of language that have proven remarkably difficult to resolve.

12.4.1. Are There Segmental Rules in Language?

Chomsky's influential generative grammar approach to language proposed that language units (words, for example) were sequenced and moved to form sentences via the application of linguistic rules. Could rules also operate to place sounds (segments) into syllables or is the sound organization process more analogous to filtering, where basic restrictions or constraints on sound sequences allow well-organized sequences to move forward into production but block ill-formed sequences from realization? Could sonority be one such constraint?

12.4.1.1. Sonority

Some aspects of normal and disordered phonological processing are best described as the products of linguistic constraints rather than rules. Buckingham and Christman (2006) have demonstrated that sound errors in fluent aphasia generally abide by the constraints of the sonority sequencing principle (SSP). This principle expresses a universal tendency in language for syllable structures to be organized such that sounds with least sonority (e.g., obstruents, "O") are positioned at syllable peripheries and sounds higher in sonority (nasals, liquids, glides, vowels, "NLGV," respectively) are positioned closer to (or even serve as) the syllable peak. In the least marked case, a sequence of sounds from syllable onset to peak (initial demisyllable) should gradually increase in sonority (ONLGV)

Box 12.1   The dispersion corollary of the principle of sonority

The dispersion corollary of the sonority principle (Clements, 1990) allows us to go beyond mere CV metrics to evaluate crescendo and decrescendo to and from syllabic nuclei. Initial demisyllables, C(C)(C) V and final demisyllables, VC(C)(C) can be measured by this corollary. Clements (1990) claims that the preferred slopes are sharper in the crescendo from the left syllable perimeter to the nucleus and more gradual in the decrescendo from the vowel to the right syllable perimeter. We will concentrate on the initial demisyllable for purposes of explication. Maximally discontinuous would be the CV /pa/, (OV). A preferred slope would be the equally spaced CCV / slow/, (OLV), skipping intermediate N and G.

The stipulated mathematics of the dispersion calculation, which has been a tool in physics for computing forces in potential fields and in acoustics for measuring perceptual distances between the nuclei in vowel systems, is that lower numerical values indicate less marked sequences and spatial configurations in initial demisyllables, with the higher values indicating less marked sequences in the final demisyllables, since there the unmarked slope of the decrescendo is smoother. The dispersion formula is a summation calculation (see Figure) that goes as follows: The dispersion value of an initial demisyllable sequence is the sum of the inverses of the squares of all the distances of all consonant onsets to the nucleus. For example, /baɪ/ would be an OV. From O to V, the distance is 4 (counting from O, you get N L G V = 4). For /naw/, N to V has a distance of 3 (L G V = 3). The $C_1 C_2$ V for /spaɪ/ would have three distances to compute: C_1 to C_2 would have a distance of 1; C_1 to V has a distance of 4, as does C_2 to V. Relative markedness values can be computed:

$$D = \sum_{i=1}^{m} 1/d_i^2$$

The dispersion equation (*Source*: Clements, 1990, p. 304).

/pa/ 1/16 = .06
/leɪ/ 1/4 = .25
/ju/ 1/1 = 1
/snow/ 1/16 + 1/9 + 1/1 = 1.17
/pju/ 1/16 + 1/9 + 1/1 = 1.17
/braw/ 1/16 + 1/4 + 1/4 = .56

Note that the smooth CCV for "brow" has the lower value of .56, indicating the level crescendo, as opposed to the other CCV's, which do not.

Buckingham and Christman (2006) discuss dispersion with regard to studies that use only canonical form metrics as a measure of complexity. Moreover, Christman (1994) demonstrates how phonemic paraphasias and neologisms hold to the sonority principle and its dispersion corollary.

Buckingham, H. W., & Christman, S. S. (2006). Phonological impairments: Sublexical. In K. Brown (Ed.), *Encyclopedia of language and linguistics* (2nd edn, pp. 509–518). Oxford, UK: Elsevier.
Christman, S. S. (1994). Target-related neologism formation in jargon aphasia. *Brain and Language, 46*, 109–128.
Clements, N. (1990). The role of sonority in core syllabification. In J. Kingston & M. E. Beckman (Eds.), *Papers in laboratory phonology: Between the grammar and physics of speech* (pp. 283–333). Cambridge, UK: Cambridge University Press.

whereas from syllable peak to coda (final demisyllable), sounds should gradually decrease in sonority (VGLNO), phonotactics permitting. Various unmarked sonority sequences such as obstruent-vowel initial demisyllables and obstruent-liquid onset clusters appear to outnumber more marked sonority sequences in aphasic language.

Buckingham and Christman (2006) discuss recent studies showing that both bizarre and target-related neologisms abide by the predicted unmarked constraints of sonority. They discuss a corpus of neologisms that resemble target words yet contain phoneme substitutions or movements. These also abide by expected sonority patterning. Until recently, most studies of neologisms were analyzed through language processing models that incorporated phonological mechanisms for sound selection, scanning, copying, checking, ordering, and erasure. These "scan-copier" and "check-off monitors" could break down, however, as a consequence of brain damage and create neologisms via complex phoneme errors. Although adults with aphasia might produce many such errors, these phonemic paraphasias abided by the "rules" of (especially unmarked) phonological structure dictated by phonotactic and sonority constraints. It appears that the rules, actually constraint-based phenomena, must be represented in some multi-level, distributed fashion in the nervous system since they are impervious to even severe forms of brain damage.

Frequency of occurrence is one factor that favors sonority theory over simple descriptions of canonical form (Nickels & Howard, 2004) in explanation of these preserved phonological regularities. For example, although many syllables produced by speakers with aphasia have simpler CV rather than CCV forms, shape cannot account for the greater frequency of OLV productions relative to GV productions unless sonority spacing parameters are also invoked (e.g., a CCV syllable of the OLV type is considered less marked in sonority than is a CV of the GV type) (see Box 12.1). Of course, in English, many marked and unmarked syllables (re: the SSP) may be found. In normal language use, as in phonemic paraphasia and the construction of neologisms, sonority preferences are not inviolate; speakers may construct marked forms and languages may contain marked forms as well. The nature of a phonological constraint (versus a rule) is that it *stretches* (metaphorically), to allow flexibility in form and use while still providing for structure and regularity in the sound system. Future work in neurolinguistics will employ constraint-based theoretical approaches to craft explanations of the phonological regularities in the paraphasic and neologistic utterances produced by speakers with fluent aphasia.

12.4.1.2. Connectionism and Constraint-Based Linguistics

Connectionist modeling (e.g., Dell *et al.*, 1997; Dell & Kim, 2005; Martin, 2005) has forced neurolinguists to consider

that most "phonological rules" may in actuality, be "phonological constraints." Indeed, Prince and Smolensky have characterized the need to explore constraint-based phonology as a kind of "conceptual crisis" within the discipline (2004, p. 2). Recent works from Dell *et al.* (1997), Martin and Dell (2004), Dell and Kim (2005), and Martin (2005) have demonstrated that, when fed large amounts of lexical data, connectionist, interactive-activation systems can learn the patterns of the inputs and faithfully match them with outputs. Once they have abstracted the sound patterns from their input vocabularies and fixed the pattern strengths through frequency of occurrence computations, these systems appear to "know" phonological constraints. They will then control phonotactics, sonority principles, and other patterned formalities (such as syllable constituency, meter, suprasegmentals, stress spacing, and beat movement) in all of their outputs.

When these connectionist (see Box 12.2) systems are subsequently "lesioned" or altered in ways that will cause them to generate phonological errors, they generate patterns of errors similar to those exhibited by human speakers when producing slips-of-the-tongue and phonemic paraphasias. Noise in the machine's connectionist system (scattered weight weakenings) simulates non-pathologically driven slips-of-the-tongue, while other sorts of mathematical alterings can result in the production of outputs that simulate paraphasias. The success of these systems in simulating and predicting human error patterns has forced serious reconsideration of how best to model phonological and phonetic errors in aphasia.

12.4.1.3. Perseveration

Perseveration errors in patients with fluent aphasia have recently been simulated in the work of Martin and Dell (2004), whose connectionistic model produced perseverations with distribution patterns similar to those seen in humans: left-to-right (or carryover) errors occurred more frequently in association with acute and/or profound impairments than with milder disturbances. Martin and Dell's (2004) machine produced progressively fewer perseverative errors as its simulated "lesioning" became less severe; this in turn raised the ratio of right-to-left (anticipatory) errors generated by the model. This finding mirrors behavior in normal slips-of-the-tongue (e.g., more anticipatory errors than perseverative errors), and it suggests that perseveration is a negative indicator for recovery from brain damage. It also offers support for Freud's hypothesis of continuity from normal to pathological behavior in some cases (see Box 12.3) (Dell *et al.*, 1997; Buckingham, 1999; Martin & Dell, 2004; Dell & Kim, 2005; Martin, 2005).

In addition to the analyses of anticipatory and perseverative paraphasias, Buckingham and Christman (2004) have shown that recurrent perseveration and the blending patterns of newly combined word forms both adhere closely to syllable constituent constraints. They also demonstrated

Box 12.2   Stimulus–response learning

Learning theories that are grounded in a complex of stimulus–response co-occurrences and frequencies are said to be associationistic in that their codes of competence are drawn from and dependent upon performance processes in real-time operation. Associative theories (referred to by the arcane term "stochastic") of the acquisition of language knowledge are based on frequency of experience and how often individual items or sets of co-occurring items are presented during learning stages. These frequencies of occurrence, in turn, determine the ultimate statistical probabilities of the production of learned patterns, referred to as the "constraints." Association theory is Aristotelian in origin and weaves its way through the history of association psychology, through the British empiricists, culminating in stimulus–response behaviorism from Sechenov and Pavlov to Watson and Skinner. Behaviorism lost ground to the cognitive revolution of the mid-twentieth century, largely through the influence of Chomskyan generative grammar. More recently, with the rise of neural networks and parallel distributed processing and other forms of connectionist modeling,

we are witnessing a resurgence of Aristotelian metaphysics in linguistics (Clark, 2005). The constraint-based linguistics of optimality theory is being increasingly fastened to the language simulations of connectionist models (Prince & Smolensky, 2004). A number of previously committed generative linguists have joined ranks with the connectionists, some (e.g., Goldsmith, 2005) claiming that associationist linguistics is actually rooted in pre-generative structuralist models, while others (e.g., Newmeyer, 2003) continue to support generative theory. These debates can be understood within the context of centuries-old association psychology.

Clark, B. (2005). On stochastic grammar. Discussion notes. *Language*, *81*, 207–217.
Goldsmith, J. (2005). Review article. [Review of the book *The legacy of Zellig Harris: Language and information into the 21st century, vol 1: Philosophy of science, syntax and semantics.*] *Language, 81,* 719–736.
Newmeyer, F. (2003). Grammar is grammar and usage is usage. *Language, 79,* 682–707.

Box 12.3   The continuity hypothesis

The nineteenth century neurologist, Charles Brown-Sequard and his student, William James, in their early work on epilepsy, reasoned that, "…degree by degree we are led to look on epilepsy as an increased degree of the normal reflex excitability of certain parts of the nervous centers" (Menand, 1998, p. 91). Subsequently, James amplified this view and gave it prominence in his work, coming to argue that there is no great divide between health and disease. This notion accords with the claims made by both John Hughlings-Jackson and Sigmund Freud (Buckingham, 2006) that slips-of-the-tongue in health and phonemic paraphasia in disease share enough structural features to exist on a health-disease continuum, differing quantitatively from less severity and frequency of production in normality to greater severity and frequency of production in pathology. Most analyses and taxonomies of slips and paraphasias in the ensuing century or so have demonstrated this continuity effect in the

data – most recently in connectionist modeling. Additionally, Wijnen and Kolk (2005) have summarized the extensive influence of the continuity claim. They write, "The implication is that language pathology can be understood as the product of a processing architecture that is structurally uncompromised, but in which the dynamics of processing are quantitatively altered, due to a reduction of critical resources" (p. 294).

Buckingham, H. (2006). Was Sigmund Freud the first neogrammarian neurolinguist? *Aphasiology, 20,* 1085–1104.
Menand, L. (1998). William James and the case of the epileptic patient. *New York Review of Books*, December 17, 81–93.
Wijnen, F., & Kolk, H. H. J. (2005). Phonological encoding, monitoring, and language pathology: Conclusions and prospects. In R. J. Hartsuiker, R. Bastiaanse, A. Postma & F. Wijnen (Eds.), *Phonological encoding and monitoring in normal and pathological speech* (pp. 283–303). New York: Psychology Press.

that word onsets perseverate far more often than codas, perhaps because the "onset + rime" binary division of the syllable leaves the onset "less glued" to rime (the vowel plus the coda). This, of course, is yet another constraint-based phenomenon that can be deciphered by a connectionist learning algorithm.

12.4.1.4. Serial Models versus Connectionist Models in Aphasia Description

One of the clearest studies of how rule-oriented psycholinguistic mechanisms can be simplified by constraint-based

connectionist modeling is found in Wheeler and Touretzky (1997). These authors present a syllable "licensing" theory that operates with parallel distributed processing and that generates fully syllabified forms with segments assigned to onset, nucleus, and coda slots in a probabilistic fashion. Word forms are created by associating the features of target phonemes with their appropriate slots (locations) in syllables. The syllable assignment processes run in parallel with a "local order constrainer," which assures that phonemes are properly sequenced. If there are any erroneous associations of sounds with syllable slots, the overall phonological composition of the resulting syllables will still obey constraints on syllable structure. Any severe error of the licenser that the local order constrainer misses will be picked up by late-stage phonotactic editing.

When a target word is fed into this processing model, the output will generally be "faithful" to it (i.e., input and output will match completely). Over thousands of runs, a few errors will be unfaithful. Weakening the featural node strengths of the model will yield even greater "unfaithfulness" of input–output matching since more phonemic errors will be created. In this way, Wheeler and Touretzky showed continuity among between normal slip errors and phonemic paraphasias. They simulated errors that abided by the probabilistic linear patterns for English, whether the errors were minimally or maximally unfaithful to the input target. In each case, however, all output forms simulated human phonology in that they were "possible" words of English.

It appears that a connectionist model, properly programmed, can output forms that match the statistical predictions of a language's phonology and can do so without recourse to psycholinguistic mechanisms like scan-copiers, checkoff monitors, and random generators. With regard to the generation of maximally unfaithful errors like "abstruse" neologisms (Buckingham & Christman, 2006), Wheeler and Touretzky (1997) claim that these, too, may be located on a continuum of faithfulness of phonological input–output matches. This suggestion that neologistic jargon arises from paraphasic rearrangement of correctly accessed semantic forms is one of two approaches to the origin of neologisms and it has been labeled the "conduction theory" of neology (Buckingham & Christman, 2006). This does not allow for failed semantic access prior to phonological processing and yet individuals with aphasia do make complicated (e.g., semantic plus phonological) errors during language production. Consequently, to simulate a full range of possible neologisms, the "lesions" to a connectionist system would have to involve altered associations to all levels, including the conceptual and semantic. A lexical disturbance would then be as likely a component of neology production as a purely phonological disturbance. Theories of neology must be able to account for aberrant forms created in a number of ways.

12.4.1.5. The Problem of Opacity of Form

The presence of phonemic paraphasias in the verbalizations of speakers with aphasia has always created problems for listeners who are trying to determine the target words that speakers may have had in mind. When numerous sound errors distort the phonological forms of target words into bizarre-sounding neologisms, the identity of those targets may no longer be discernable. Such neologisms are usually described as "opaque," "unrecognizable," or "abstruse." Historically, most researchers and clinicians have operationalized their definitions of neologisms, reasoning that if a nonsense word differed from its target word by more

than half of its phonemes, then it was to be considered a neologism. Without specifically saying so, this was the "conduction" account of neology. It implied a continuum of phonemic level error only, in the sense of Wheeler and Touretzky (1997), and it therefore ruled out any possibility of a neologism caused by semantic and then phonological paraphasia. Given the centrality of anomia to most aphasic syndromes, it is likely that the production of abstruse neologisms signals the presence of underlying semantic *and* phonological processing problems with the possibility of both types of errors combining to yield an incomprehensible nonsense word. Ahlsen (2006, p. 58) has described this as the "*anomia theory*" of neology.

It is also possible to form bizarre neologisms when no underlying target at all has been accessed – if there exists a mechanism for creating word structures rather spontaneously.

A "*random syllable generator*" has been proposed to characterize just such a mechanism within the anomia theory of neology. It tacitly assumes that a speaker's phonological competence could create possible syllables and "nonce" words that, while devoid of meaning, nevertheless would abide by phonological constraints (phonotactics, the SSP, and others) and undergo normal contextual accommodation prior to articulation. Many neurolinguists have scoffed at the idea of a "random generator," largely because they misunderstood the metaphorical nature of the suggestion, preferring to argue that brain damage does not create *de novo* mechanisms. Despite the observation that dual routes remain possible for the production of neologisms ("conduction" versus "anomia"), recent connectionist accounts of neologism formation discard an anomia theory of neology (Gagnon and Schwartz, 1996; Hillis *et al.*, 1999).

Hillis *et al.*'s case study (1999) examines the language of a patient with conduction aphasia whose principal trouble rests with phonological activation. This patient reads passively, can match words with pictures, comprehends well, speaks fluently, and in general appears to maintain full connectivity between the semantic/conceptual and lexical strata in a "Dell-like" connectionistic model (e.g., Dell *et al.*, 1997; Dell & Kim, 2005). Accordingly, Hillis *et al.* (1999) could simulate this patient's oral naming errors by focally "lesioning" the connection weights between the lexical and the sub-lexical levels of the model. Activation decay rates were left normal and all connection values between the semantic and word levels were unaltered. In this patient, and in accordance with predictions from Wheeler and Touretzky's (1997) model, phonemic paraphasias and neologisms were frequently produced on oral naming, repetition, and oral reading tasks and they were common in fluent spontaneous speech as well. Hillis *et al.*'s patient had nervous system damage, "...to connections between the posterior superior temporal gyrus in dorsal area 22 and more ventral regions of area 22" (1999, p. 1814); this was a disconnection lesion, in the sense described by Norman Geschwind. Hillis *et al.* claim that dorsal 22 is a site of sub-word phonological processing, while ventral 22 is "likely" to be a region for the representation and processing of whole word forms. The authors argue that neologisms are explained by the conduction theory, and that they arise entirely from altered connection weights locally restricted to word–sub-word linkages. Hillis *et al.* cite other studies, which *have* simulated neology through global lesioning of decay rates as well as of connection weights between the semantic and the word levels of a connectionist model. For some reason, however, Hillis *et al.* could not simulate neology in their patients' paraphasic behavior by globally lesioning their model.

Ultimately, a globally lesioned connectionist model has produced neologisms when connection weights and/or decay rates among all levels were weakened. In this case, lexical substitutions occurring during the creation of neologisms cannot be ruled out. Wheeler and Touretzky (1997) as well as Hillis *et al.* (1999) have simulated neologisms in jargon through phonemic paraphasia, where underlying anomia is not likely. In fact, Wheeler and Touretzky (1997, p. 194) admit that, "We have not yet considered cases where recovery of the lexical phonological representation is itself impaired, but that would be a natural extension of our theory." Accordingly, these two studies account for neology in clear terms of phonemic paraphasia, but they are necessarily restricted to the conduction account of neologisms, and cannot make serious claims that, in other situations, neologisms may well arise from anomia.

To date, the conduction and anomia theories of neologism formation remain (e.g., Kohn & Smith, 1994). Findings from studies of recovery of function in aphasia, such as those of Kohn *et al.* (1996), have provided some of the strongest evidence yet that, for certain patients, the "anomia" theory works best for neologism production, while for certain others, the "conduction" theory works best. In studies where neologisms have been observed in acute aphasias, it has been practically impossible to discern whether they originate from severe phonemic paraphasia or from semantic plus phonological distortion of target word forms. When patients reach chronic stages of aphasia, however, differences in the mechanisms of neologism production can be revealed. Some patients are left with phonemic paraphasias and little anomia, whereas others recover to anomia without much phonemic paraphasia. Kohn *et al.*'s observation of these differences in aphasia outcomes suggested to her that patients who recover to phonemic paraphasia probably generated neologisms via phonological miscalculations only but patients who recover to semantic anomia probably generated neologisms via combinations of semantic and phonological errors. In both cases, phonological processing was observed to improve with recovery but, in clinical

terms, the second patient might have always benefited from treatments to improve access to semantics whereas the first patient might have benefited more from treatment focused on improved access of phonological forms. Although this is a premise to be tested, it illustrates how, ultimately, many arcane-sounding discussions and debates about theories of language processing can lead to knowledge that may inform rehabilitation efforts in clinical settings.

12.5. CHALLENGES AND FUTURE DIRECTIONS

Our chapter has addressed some of the current issues that have profoundly influenced the description and analysis of disorders of phonetics and phonology. New directions in constraint-based generative phonology have smoothed some of the previous rough edges between generative phonology and performance-oriented explanations of phonological processing. Detailed linguistic research has provided new ways to apply markedness theory to the systematic characterization of sound production errors, whether in normal slips-of-the-tongue, in phonemic paraphasia, or in abstruse neology. There are well known probabilistic and statistical frequencies of the sound patterns of any language that arise from the metrical, syllabic, and linear orderings of the sounds in language. Preservation of these patterns has been repeatedly demonstrated in the phonological pathologies associated with aphasia, and they can be simulated by appropriately programmed computer models (connectionistic machines) that seem to learn from their inputs. Normal and disordered phonological processing appears to be governed not necessarily by linguistic rules, but rather by sets of structural constraints (phonotactics, sonority, and others), some of which may be universal. Constraint-based linguistics appears to best describe the structure of sound production performance in normal speech, in slips-of-the-tongue, and in aphasia. We have concentrated on aphasia in this chapter.

There can be no doubt that connectionism has helped to marry the codes of phonology and the processes of production in our current approaches to the study of aphasia. Connectionist models have collapsed the cherished distinction in linguistics of competence and performance by practically turning code and process into one. This blurring of rigid boundaries in linguistic theory is accompanied by the blurring of rigid boundaries in our traditional brain maps. Neuro-imaging studies (primarily using positron emission tomography (PET) and functional magnetic resonance imaging (fMRI)) have warned us that the nervous system does not structurally or regionally carve up phonetics and phonology so easily in the brain. Major sensory centers are now known to have more motor cells than previously thought and major frontal lobe systems have been shown to participate in activities of perception and cognition (such as selecting an information source from among all available). The distributed and redundant nature of brain organization itself seems to argue for a major overlap of phonetics and phonology, which brings it more in line with the observed scarcity of "so-called" pure (motoric) apraxia of speech. It is simply the case that most adults with acquired apraxia of speech exhibit phonemic *and* phonetic errors in their spoken language. Moreover, evidence suggests that patients with lesions in Wernicke's area produce more phonetic errors than previously thought. It is not surprising that some phoneticians reject the idea of an "interface" between the phonetics and phonology, arguing instead for the *integration* of those two aspects of the human sound system.

Finally, we have traced the intractable problem of the origin of the abstruse neologism in jargon aphasia, and have placed the inquiry into the overall framework of optimality theory and connectionism. We showed how most connectionist accounts of the abstruse neologism focus exclusively upon their creation from excessive phonemic paraphasia caused by weakened connection weights between the lexical and phonological level nodes of the model. Certainly, such an approach can create bizarre nonsense words but we have argued here, as elsewhere, that another explanation may also be feasible and it is simply unreasonable to rule out word finding problems as part of the pathology underlying a subset of complex neologisms.

Evidence from the recovery patterns of patients with neologistic jargon aphasia may drive further connectionist modeling that will better capture the range of neologistic jargon aphasia. Future research should investigate the effects of different functional lesionings of connectionist systems, such as the introduction of random noise, the weakening of focal as well as global connections, the lowering of the initial input strengths, and the manipulation of decay rates, both focal and global, with the goal, ultimately, of more completely explaining our understanding of normal and disordered human lexico-phonological processing.

References

Ahlsen, E. (2006). *Introduction to neurolinguistics.* Amsterdam: John Benjamins.

Buckingham, H.W. (1999). Freud's continuity thesis. *Brain and Language, 69,* 76–92.

Buckingham, H.W. & Christman, S.S. (2004). Phonemic carryover perseveration: Word blends. *Seminars in Speech and Language, 25,* 363–373.

Buckingham, H.W. & Christman, S.S. (2006). Phonological impairments: Sublexical. In K. Brown (Ed.), *Encyclopedia of language and linguistics, 2nd edn* (pp. 509–518). Oxford, UK: Elsevier.

Code, C. (2005). Syllables in the brain: Evidence from brain damage. In R.J. Hartsuiker, R. Bastiaanse, A. Postma, & F. Wijnen (Eds.), *Phonological encoding and monitoring in normal and pathological speech* (pp. 119–136). New York: Psychology Press.

Dell, G.S. & Kim, A.E. (2005). Speech errors and word form encoding. In R.J. Hartsuiker, R. Bastiaanse, A. Postma, & F. Wijnen (Eds.), *Phonological encoding and monitoring in normal and pathological speech* (pp. 17–41). New York: Psychology Press.

Dell, G.S., Schwartz, M.F., Martin, N., Saffran, E.M., & Gagnon, D. A. (1997). Lexical access in aphasic and non-aphasic speakers. *Psychological Review, 104,* 801–838.

Dronkers, N.F. (1996). A new brain region for coordinating speech articulation. *Nature, 384,* 159–161.

Foundas, A.L., Eure, K.F., Luevano, L.F., & Weinberger, D.R. (1998). MRI asymmetries of Broca's area: The pars triangularis and pars opercularis. *Brain and Language, 64,* 282–296.

Gagnon, D. & Schwartz, M.F. (1996). The origins of neologisms in picture naming by fluent aphasics. *Brain and Cognition, 32,* 118–120.

Gandour, J.T., Wong, D., Hsieh, L., Weinzapfel, B., Van Lancker, D., & Hutchins, G.D. (2000). A cross-linguistic PET study of tone perception. *Journal of Cognitive Neuroscience, 12,* 207–222.

Hagoort, P. (2005). Broca's complex as the unification space for language. In A. Cutler (Ed.), *Twenty-first century psycholinguistics: Four cornerstones* (pp. 157–172). Mahwah, NJ: Lawrence Erlbaum.

Hillis, A.E., Boatman, D., Hart, J., & Gordon, B. (1999). Making sense out of jargon: A neurolinguistic and computational account of jargon aphasia. *Neurology, 53,* 1813–1824.

Hillis, A.E., Work, M., Barker, P.B., Jacobs, M.A., Breese, E.L., & Maurer, K. (2004). Re-examining the brain regions crucial for orchestrating speech articulation. *Brain, 127,* 1479–1487.

Kohn, S. & Smith, K. (1994). Distinctions between two phonological output disorders. *Applied Psycholinguistics, 15,* 75–95.

Kohn, S., Smith, K., & Alexander, M. (1996). Differential recovery from impairment to the phonological lexicon. *Brain and Language, 52,* 129–149.

Martin, N. (2005). An interactive activation account of aphasic speech errors: Converging influences of locus, type, and severity of processing impairment. In R.J. Hartsuiker, R. Bastiaanse, A. Postma, & F. Wijnen (Eds.), *Phonological encoding and monitoring in normal and pathological speech* (pp. 67–85). New York: Psychology Press.

Martin, N. & Dell, G.S. (2004). Perseverations and anticipations in aphasia: Primed intrusions from the past and future. *Seminars in Speech and Language, 25,* 349–362.

McNeil, M.R., Doyle, P.J., & Wambaugh, J. (2000). Apraxia of speech: A treatable disorder of motor planning and programming. In L.J. Gonzalez Rothi, B. Crosson, & S.E. Nadeau (Eds.), *Aphasia and language: Theory to practice* (pp. 221–265). New York: Guilford Press.

Nickels, L. & Howard, D. (2004). Dissociating effects of number of phonemes, number of syllables, and syllabic complexity on word production in aphasia: It's the number of phonemes that counts. *Cognitive Neuropsychology, 21,* 57–78.

Ogar, J., Slama, H., Dronkers, N., Amici, S., & Gorno-Tempini, M.L. (2005). Apraxia of speech: An overview. *Neurocase, 11,* 427–432.

Ohala, J. (2005). Phonetic explanations for sound patterns: Implications for grammars of competence. In W.J. Hardcastle & J.M. Beck (Eds.), *A figure of speech: A Festschrift for John Laver* (pp. 23–38). Mahwah, NJ: Lawrence Erlbaum.

Prince, A. & Smolensky, P. (2004). *Optimality theory: Constraint interaction in generative grammar.* Oxford, UK: Blackwell Publishing.

Thompson-Schill, S.L. (2005). Dissecting the language organ: A new look at the role of Broca's area in language processing. In A. Cutler (Ed.), *Twenty-first century psycholinguistics: Four cornerstones* (pp. 173–189). Mahwah, NJ: Lawrence Erlbaum.

Wheeler, D.W. & Touretzky, D.S. (1997). A parallel licensing model of normal slips and phonemic paraphasias. *Brain and Language, 59,* 147–201.

Further Readings

Buckingham, H.W. (1977). The conduction theory and neologistic jargon. *Language and Speech, 20,* 174–184.
This paper laid out the argument that neologisms could originate according to the "conduction theory" as well as the "anomia theory." It predicted **two** possible paths of recovery from jargon: (1) mild phonemic paraphasia and (2) a persistent anomia

Butterworth, B. (1979). Hesitation and the production of verbal paraphasias and neologisms in jargon aphasia. *Brain and Language, 8,* 133–161.
This paper proposed for the first time the notion of a "random generator," a metaphor referring to speaker–hearers' extended phonological knowledge, which permits them to appreciate non-words that are nevertheless phonotactically acceptable and to pronounce "nonce" words.

Christman, S.S. (1992). Abstruse neologism formation: Parallel processing revisited. *Clinical Linguistics and Phonetics, 6,* 65–76.
This paper contains the first full-fledged statistical analysis of a large corpus of neologisms, which were demonstrated to abide strictly by the principle of sonority, as long as the phonotactics were in place.

Garrett, M.F. (1984). The organization of processing structures for language production: Applications to aphasic speech. In D. Caplan, Lecours, A.R. & Smith, A. (Eds.), *Biological perspectives on language* (pp. 172–193). Cambridge, MA: The MIT Press.
This paper represents one of Merrill Garrett's first forays into aphasiology, where he demonstrated how his serial model of sentence production could serve to characterize paraphasias and locate the principal production levels involved in their generation.

Lecours, A.R. (1982). On neologisms. In J. Mehler, E. Walker, & M. Garrett (Eds.), *Perspectives on mental representations* (pp. 217–247). Hillsdale, NJ: Lawrence Erlbaum.
This work by Andre Roch Lecours further explored the opaque nature of neologisms that reveal no resemblance whatsoever to any stipulated target word. He distinguished these from the so-called "literal paraphasias," which bore some relation to the target. He, too, distinguished the "abstruse" neologism from the "target related" phonemic error.

Ohala, J. (1990). There is no interface between phonology and phonetics: A personal view. *Journal of Phonetics, 18,* 153–171.
This paper is a well argued work by one of the most prolific contemporary phonologists, who draws heavily from acoustic theory as well as from historical sound change and the mechanisms that underlie those changes. His well taken point is that in reality there is little hope of teasing apart phonetic levels of explanation from phonological structure and function.

13

Impaired Morphological Processing

GONIA JAREMA

Department of Linguistics, University of Montreal and Research Centre,
Institut universitaire de gériatrie de Montréal, Montreal, Quebec, Canada

ABSTRACT

A fundamental ability of human cognition is to comprehend and produce complex words such as *walked*, *greatness*, and *blueberry*. This ability has been repeatedly demonstrated to break down across a variety of neurological disorders. Crucial to our understanding of morphological impairments following brain damage are questions such as: What is the role that grammatical category (e.g., a word being a verb or a noun) or regularity (e.g., a verb being regular or irregular) play in the breakdown of morphology? What is the relationship between a deficit in morphological processing and a deficit in the ability to process syntax or phonology? Can imaging studies offer new insights to this field of inquiry? This chapter will discuss research that has addressed these central questions and conclude with a brief discussion of new research directions and challenges for the future.

13.1. INTRODUCTION

Words are the building blocks of language and, consequently, impaired lexical processing in individuals with neurological disorders can be a devastating experience, as it greatly impedes their ability to construct sentences and communicate successfully. A major reason why words are difficult to process in neurological patients with linguistic disabilities is that in many languages words are complex, rather than simple, entities. Consider, for example, the English word *cat* and its relatives *cats*, *cattish* and *catfish*. It may be difficult to think of the latter as complex, rather than simple, considering the ease with which they are normally used. Yet these words contain two constituents, or morphemes (cat+s, cat+ish; cat+fish), and hence feature

complex word-internal structures. Morphology can thus be defined as the study of word-internal structure and the way in which morphemes combine to form words. Accordingly, patients exhibiting difficulty in understanding or producing complex lexical items can be characterized as having a morphological impairment. The study of morphological impairments is of great interest to neurolinguistics and to neuroscience because it can further elucidate the fractionizing of linguistic function and because it offers new opportunities for investigating the neural substrate underlying the cognition of language, which in turn can inform our theories of language processing. In order to better understand morphological impairments, we will first take a closer look at the functions morphology fulfills.

As illustrated by the examples cited above, morphology serves a variety of functions. Bound morphological markers, or affixes, such as *-s* and *-ish* produce words that are structurally related to their base forms (e.g., *cat* in the previous examples). However, whereas the suffix *-s* only marks, or *inflects*, the base form *cat* for plural, that is, only adds grammatical information to the noun, the suffix *-ish* actually *derives* a new word, resulting in a change both in grammatical category and in meaning. This is captured by the linguistic distinction between inflectional morphology on the one hand, and derivational morphology, on the other. Moreover, words such as *catfish* illustrate that new lexical items can also be formed through compounding, a process through which existing words are combined to form new words. This mechanism of word formation is universal and highly productive in English and in many languages. Productive derivational affixes (e.g., *re-* and *-ness*) and compounding are therefore readily used to coin semantically transparent, that is easily interpretable, novel words such as *retable*, *funkiness*, and *sandboard*. Indeed, knowing the meanings of the parts of such newly coined words as well as the rules governing the legality of these combinations, one can understand the meaning of the resulting complex forms without any difficulty. Thus both derivational morphology and compounding primarily have a word-formation function, while inflectional morphology primarily has a grammatical function. This partitioning of morphological operations is reflected neuropsychologically as there exists ample evidence demonstrating that patients' performance can show dissociations which parallel the linguistic distinction between inflection, derivation, and compounding. This chapter will address the issue of morphological breakdown following brain damage in these distinct domains of linguistic knowledge.

13.1.1. Morphological Breakdown

The performance of individuals with morphological impairments has been studied systematically over the past five decades. Following Goodglass and Hunt's (1958) seminal study that investigated the way in which aphasic patients show increasing difficulty when having to process the English suffix *-s* as a marker of plurality (*hats*) versus tense, number and person (*eats*), versus possession (*hat's*), scores of studies have demonstrated that patients suffering from a variety of neurological disorders can exhibit morphological impairments. While aphasic patients, in particular agrammatic Broca's aphasics, have been studied most widely and across many languages, morphology has also been shown to be impaired in individuals with neurodegenerative disorders such as Dementia of the Alzheimer's Type (DAT), possibly pointing to an underlying impairment of working memory across the two patient populations. The observation that these deficits are found in patients with focal lesions, as is the case in aphasia, as well as in patients with diffuse lesions, as is the case in dementia, may also point to a less localized involvement of brain matter dedicated to the language faculty than is traditionally held. Moreover, patients with right hemisphere (RH) lesions have also been demonstrated to have difficulty with morphological processing, further supporting a more holistic view of brain recruitment and highlighting the importance of inter-hemispheric connectivity and neural circuitry in language function. In the following section we will consider the factors that have been invoked to account for patients' problems in processing word structure.

13.1.2. Current Issues

Among issues that are widely discussed in the literature is the role that grammatical category or phonological regularity might play in the processing difficulties observed. Thus one ongoing debate focuses on the question of impaired verb morphology as compared to better-preserved noun morphology. Another recurring question is whether morphologically

complex words that are morpho-phonologically regular (e.g., *cooked*, *books*) are processed differently from those that are irregular (e.g., *caught*, *children*). Yet another issue that continues to provoke intense discussion among researchers centers on dissociations found in the breakdown of inflectional features such as tense and agreement and the theoretical motivations underlying these dissociations, leading authors to probe the relationship between breakdown in morphological processing and breakdown in the ability to process syntax or phonology. Thus the very notion of a purely morphological deficit has come under scrutiny. These are some of the main issues that are currently being debated in the literature and that will be discussed in this chapter.

13.2. NOUN VERSUS VERB MORPHOLOGY

One of the most robust neuropsychological findings is that verbs are more difficult to process than nouns. With respect to morphological impairments, patients with left frontal damage have repeatedly been shown to have an increased difficulty inflecting verbs as compared to nouns (e.g., Shapiro & Caramazza, 2003; Shapiro *et al.*, 2000; Tsapkini *et al.*, 2002). A selective impairment of verb morphology has long been linked to the observation that agrammatic aphasic patients (i.e., patients with an anterior lesion generally involving Broca's area and presenting with nonfluent, highly reduced speech lacking grammatical features) produce fewer verbs than nouns and that some patients are poorer at naming verbs than nouns. The latter finding has been interpreted as reflecting, for example, a dichotomy at the conceptual level between actions and objects (Marshall *et al.*, 1998) or between object names defined in terms of perceptual features and action names defined in terms of functional features (Bird *et al.*, 2000). However, in a study using both existing homophonous verbs and nouns (*he judges, the judges*) and pseudo-verbs and pseudo-nouns (*he wugs, the wugs*) to probe the production of an agrammatic aphasic patient showing more difficulty producing verbs than nouns, Shapiro and Caramazza (2003) found that words and pseudo-words yielded an identical pattern: Both verbs and pseudo-verbs were produced correctly less frequently than nouns and pseudo-nouns in a sentence-completion task. This study thus crisply demonstrated that grammatical category does indeed play a role, but that the deficit is morphological rather than conceptual–semantic, as pseudo-words are non-existing and therefore not memorized. Interestingly, the opposite pattern, that is better production of verbs and pseudo-verbs than nouns and pseudo-nouns, has also been reported in a case study of a fluent aphasic patient (fluent aphasia is linked to left temporal or temporo-parietal lesions and is mainly characterized by word-finding difficulties and by phonological and semantic paraphasias, or errors) by Shapiro *et al.* (2000), supporting the hypothesis that verb and noun morphology can be spared or impaired differentially across different pathologies.

In summary, the noun–verb dissociation appears to be a pervasive phenomenon in morphological breakdown, despite widely diverging interpretations for its causality. But noun and verb deficits can nevertheless pattern together as both categories may in turn dissociate along the dimension of regularity.

13.3. REGULAR VERSUS IRREGULAR MORPHOLOGY

An important question that arises in the study of morphological disorders is whether words that are morphologically regular are impaired in a manner that is different from words that are morphologically irregular. Regularly inflected nouns and verbs will show alternations that are predictable and rule-based. By contrast, irregular forms are unpredictable and must be stored in memory. For example, in English, regular plural nouns can be formed by affixing the plural marker -s to the singular form of a noun, and regular verbs in the past tense are formed by affixing -ed to their base form. Irregular plural nouns (e.g., *mice*) and past tenses (e.g., *caught*) have to be memorized. The following question then arises: Are neurologically impaired patients better at producing or comprehending inflected words that are morphologically transparent, that is, that can be readily composed or decomposed into their stems and affixes (*paint-ed*, *hat-s*), or at producing irregular forms that are morphologically opaque and that must be retrieved from the mental lexicon? Indeed, morphologically related word pairs such as *mouse* and *mice* or *catch* and *caught* are perceptibly dissimilar and therefore more difficult to associate. As a matter of fact, it appears to be the case that patterns of behavior may vary across languages and patient populations, as will be illustrated below.

A growing trend in neuropsychological research over the last decade or so has been to examine whether specialized domain-specific neural systems subserve distinct cognitive components of language representation and processing. An influential study adopting this approach reported a double dissociation in the processing of English regular versus irregular past-tense forms across different pathologies (Ullman *et al.*, 1997). A group of patients with Parkinson's disease and a patient with anterior (i.e., non-fluent) aphasia were more impaired on regulars, while patients with Alzheimer's disease and patients with posterior (i.e., fluent) aphasia were more impaired on irregulars. Ullman *et al.* (1997) linked regular morphological processing to left anterior neural circuits, hypothesized to subserve the rule-governed procedural system, and irregular morphological processing to left posterior circuits, hypothesized to subserve declarative memory (and lexical representation). If, in keeping with this dual-mechanism model, rule-based computations occur in the anterior regions

of the cortex, then agrammatic Broca's aphasics should systematically exhibit difficulty with regular, but not irregular, morphological processing. This prediction, however, is not born out, even within a single language. Shapiro and Caramazza (2003), for example, found that their English-speaking agrammatic patient R.C., who had shown more difficulty producing verbs than nouns correctly, also exhibited a more severe deficit with irregular than regular verbs. According to Ullman *et al.* (1997), R.C., whose lesion is anterior, should have shown the opposite pattern, that is more difficulty producing regulars. Interestingly, this differential pattern did not obtain for nouns, indicating that only the impaired grammatical category is affected by regularity.

The issue of the neural underpinnings of regular versus irregular morphology was recently reexamined in a study investigating non-fluent and fluent aphasics on regular (rule-based) versus irregular (memorized) past-tense forms (Ullman *et al.*, 2005). Importantly, the study's comprehensive literature review reveals that links between specific brain regions and specific linguistic functions have not yet been clearly established. However, certain trends do seem to emerge. Electrophysiological and neuroimaging studies suggest that posterior regions may subserve the lexicon while frontal regions may subserve the grammar, although unequivocal neuroanatomical dissociations between lexicon and grammar are still lacking. Thus while several event-related potential (ERP) studies point toward LAN (left anterior negativity) effects for regular inflection and central-posterior N400 effects for irregular inflection, other studies have not found such effects. Results from neuroimaging are also mixed, but patterns of activation again suggest that regular and irregular inflection have distinct neural underpinnings linked to left frontal cortex and left temporal cortex, respectively. In their study, Ullman *et al.* specifically targeted double dissociations between regular and irregular inflectional morphology in relation to their neuroanatomical substrates across tasks, modalities and, importantly, using both existing and meaningless novel (i.e., unlisted) regular and irregular forms. Results clearly demonstrated that non-fluent aphasics with left anterior lesions performed more poorly on regular than irregular past tense or plural forms, while the opposite pattern was obtained for fluent aphasics with left posterior lesions, confirming the claim that left frontal regions play a role in grammatical functions, while left posterior cortical regions subserve lexical memory and thus stored irregularly inflected words (see Box 13.1).

A possible confound in studies investigating the regular/irregular dichotomy across grammatical categories in a language such as English is that both nominal and verbal regular forms differ from irregular forms in the way they alternate. Regular forms undergo suffixation, while irregular forms do not. Alternate forms of irregular verbs such as *eat-ate* are thus equal in syllabic length, and so are those of irregular nouns such as *goose-geese*. In contrast, *greet-greeted* and

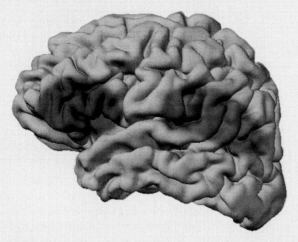
bus-buses are not. Thus both *greet* and *bus* are monosyllabic, while *greeted* and *buses* are bisyllabic. A consequence of this difference is that irregular nouns in the plural and irregular verbs in the past tense are shorter than regular plural nouns and past-tense verbs. In an effort to circumvent this putative confound, Tsapkini *et al.* (2002) investigated morphological processing and the effects of regularity in a highly inflected language, Greek, in which nominal and verbal systems feature regular and irregular forms that are always suffixed. The study's non-fluent patient (SK) presented with a selective deficit in the production of irregular verbal morphology. In contrast, the opposite pattern, with more impaired regular than irregular inflection, was observed in the production of a Greek-speaking fluent aphasic patient (Tsapkini *et al.*, 2001), demonstrating a double dissociation with respect to the regular/irregular distinction in a language where length can be controlled across this divide. Note that the contrasting patterns observed in these two patients are inconsistent with an account that maps rule-governed mechanisms onto anterior neural networks and stored information onto posterior neural networks

(Ullman *et al.*, 1997, 2005), as such a model would predict the exact opposite patterns.

As illustrated above, effects of regularity have been reported across the diagnostic categories of agrammatic aphasia, Parkinson's disease, and Alzheimer's disease. However, it appears to be the case that the dissociations between regular and irregular morphology observed across languages and patient populations do not yield consistent findings, nor converging accounts, possibly because the concept of regularity is not always an-all-or-nothing phenomenon and might best be viewed as a continuum (Tsapkini *et al.*, 2002). As is apparent from the studies reviewed, the debate surrounding effects of regularity has largely focused on impaired inflectional morphology to which we will now turn.

13.4. IMPAIRMENTS OF INFLECTIONAL MORPHOLOGY

Neurologically impaired individuals with a morphological deficit frequently omit or substitute word-final grammatical markers, or inflections. They might, for example, omit the past-tense inflection in a word such as *walked*, and instead produce *walk*, a form that is erroneous when a verb in the past tense is required by the context (e.g., "Yesterday John *walk home."). Or they might substitute *walked* with *walking*, again a form that would be inappropriate here (Yesterday John *walking home.). They might also drop the final *-s* marking plural (e.g., "She reads two *book") or 3rd person singular (e.g., "she *read"). These kinds of errors are characteristic of agrammatic Broca's, or non-fluent, aphasia, a syndrome that has received much attention in research on morphological impairments. Drastically diverging accounts of difficulty with inflectional morphology based on varying patterns of breakdown and on varying theoretical considerations have been proposed in the literature. Below, examples of morphological, syntactic and phonological accounts will be considered to illustrate this divergence. Because investigations of neurological patients showing difficulty with markers that signal tense (*-ed*), subject–verb agreement (*-s*) and gender (e.g., *-a* in the Italian feminine noun *via*) have been pivotal in the debate surrounding the functional locus of inflectional deficits, they will serve here to illustrate these varying perspectives.

13.4.1. Tense versus Agreement

One proposal put forth to account for impaired inflectional morphology in aphasia is that a selective morphological impairment reflects an inability to convert featural information such as tense into morphological material (Thompson *et al.*, 2002). This would signify that although patients "know" that, for example, in the context of a past event a verb must appear in the past-tense form (e.g., *Last year people traveled more*) and that a verb must agree with its subject (e.g., *the*

Box 13.2   Syntactic accounts of inflectional impairments

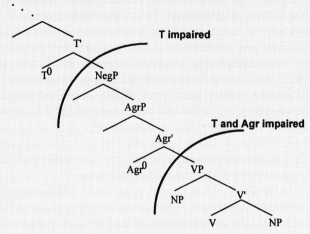

Pruning the syntactic tree (adapted from Friedman and Grodzinsky, 1997).

The Tree Pruning Hypothesis (TPH) stipulates that the degree of severity of agrammatic production is a function of the locus of impairment, or "pruning" (indicated by an arched line), in the syntactic phrase marker, or "tree," and that any node *above* an impaired node is defective, while any node *below* an impaired node is intact. If, for example, the tree is pruned at T, tensed verbs are impaired, but subject–verb agreement is spared as Agr is below T in the tree. If the tree is pruned at Agr, both subject–verb agreement and tensed verbs are impaired as T is above Agr. Friedmann and Grodzinsky's TPH is, however, problematic because theoretical linguists disagree on the exact ordering of inflectional nodes, while the TPH requires a specific ordering of these nodes to adequately accommodate the data. Furthermore, neuropsychological

data show that T is more vulnerable than Agr across many languages, irrespective of the order of T in relation to Agr in a language. In German, for example, Agr is claimed to be positioned higher in the syntactic tree than T, yet in a study investigating German-speaking agrammatic aphasics, tense was found to be more impaired than subject–verb agreement in a sentence completion and a grammaticality judgement task (Wenzlaff & Clahsen, 2004). Adopting Chomsky's Minimalist Program, Wenzlaff and Clahsen rejected an explanation in terms of a hierarchical ordering of T and Agr and proposed the Tense Underspecification Hypothesis (TUH) which posits that only the semantically interpretable T-features are underspecified and therefore unstable in agrammatic aphasia, while the uninterpretable Agr-features are intact. A similar account was suggested by Burchert *et al.* (2005), who investigated the production of tense and subject–verb agreement in nine German agrammatic patients on a sentence-completion task. However, because Burchert *et al.* observed a double dissociation between T and Agr, they extended the TUH to allow for underspecification of the interpretable Agr-features as well. The resulting Tense-Agr Underspecification Hypothesis (TAUH) accounts for the selective impairment of either T or Agr, or an impairment in both, and is thus more in keeping with the patterns evidenced across languages.

Burchert, F., Swoboda-Moll, M., & Bleser, R.D. (2005). Tense and agreement dissociations in German agrammatic speakers: Underspecification versus hierarchy. *Brain and Language, 94*, 188–199.

Friedmann, N., & Grodzinsky, Y., (1997). Tense and agreement in agrammatic production: pruning the syntactic tree. *Brain and Language, 56*, 397–425.

Wenzlaff, M., & Clahsen, H. (2004). Tense and agreement in German agrammatism. *Brain and Language, 89*, 57–68.

boy plays soccer), they are nevertheless unable to access the morphology that marks the features of tense and number and might instead produce "*Last year people *travel more*," or "*the boy *play soccer*". Other researchers view morphological deficits as stemming from an underlying *syntactic* deficit, that is a deficit resulting from a difficulty in encoding sentences. For example, Bastiaanse and van Zonneveld (2004) argued that in Dutch the typical problem with verb inflection in agrammatic aphasia arises in the context of obligatory verb-movement from canonical verb-final position in subordinate sentences to non-canonical verb-second position in matrix sentences. Unlike English, in which the canonical word-order is subject–verb–object, Dutch is a subject–object–verb language (*ik denk dat de jongen een fiets koopt*, "lit. I think that the boy a bike buys." In non-subordinate (matrix) sentences, the verb has to move to the second position (*de jongen koopt een fiets*, "The boy buys a bike"),

an operation that is unavailable to agrammatic patients. Bastiaanse and van Zonneveld suggest that the deficit is located in Levelt's "grammatical encoder" in which sentence patterns are being formulated (Levelt, 1989).

Friedmann and Grodzinsky (1997) also favored a syntactic account, however with reference to the ordering of functional categories such as T(ense) and Agr(eement) in the "syntactic tree," a representational metaphor for the hierarchical organization of syntactic operations. Inflected nouns and verbs are retrieved from the lexicon and move up the syntactic tree in order to check and license their inflectional features against the features present in the tree. The rationale then is that any inflectional node above an impaired (underspecified) node will also be impaired, while any node below an impaired node will be spared (see Box 13.2). This is precisely the pattern that Friedmann and Grodzinsky reported for their Hebrew-speaking patient.

adjective–noun (*blackboard*), and noun–adjective (French *haricot rouge*, "kidney bean," literally "bean red") constructions. Furthermore, compounds can feature word-internal agreement (Italian *alta moda*, "high fashion") and linking elements (Polish *wodospad*, "waterfall"). The following questions might then be asked. Do patients with neurological disorders have difficulty processing compounds? If so, what are the error patterns observed across compound types and language structures? First, compound processing has been shown to dissociate from other types of morphological disorders and to be selectively impaired in aphasia (e.g., Delazer & Semenza, 1998). Second, substitution and omission errors have been very informative in elucidating how patients process compounds and in revealing possible loci of difficulty. For example, in picture naming, target compounds are generally substituted by other compounds or by neologistic compounds such as Italian *raschiavetro, "scrape glass," for *tergicristallo*, "windscreen wiper," literally "clean cristal" (Delazer & Semenza, 1998). Third, aphasic patients tend to omit one of the compound's constituents (e.g., Badecker, 2001). Taken together, these findings from two typologically differing languages (Italian and English) suggest that knowledge of compound-internal structure is preserved in aphasia and that compositional processes are at play during compound processing. Further evidence for this account is provided by Badecker (2001) who observed constituent misorderings (*fire wood → wood fire*) and by Mondini *et al.* (2004) who demonstrated that verbs are omitted more frequently than nouns in noun–verb compounds. Another interesting observation is related to the fact that a compound's left or right constituent can be its "head." Thus *board* is the head of *blackboard* because a blackboard is a type of board and because the head *board* determines the grammatical category of the compound and its gender in gender-marked languages (e.g., German *Luftballon*, "balloon," where the head *ballon*, "ball" and therefore also the entire word is masculine, while *Luft*, "air" is feminine). This would lead to the prediction that since heads play a central role in the parsing and interpretation of compounds, they should be more stable than non-head constituents. The issue is somewhat obscured by the fact that positional and headedness effects cannot be easily teased apart, at least in a language such as English where compounds are always right-headed. However, in Italian, a language that also licenses left-headed compounding, second-constituent substitutions are as frequent as first-constituent substitutions (Delazer & Semenza, 1998). Blanken (2000), who reports similar findings for German, suggests that this may reflect a difficulty at the highest level of compound processing, the conceptual–semantic level, that is before structural processes come into play.

The question whether compound-internal syntactic processes are impaired or spared has also been addressed, the rationale being that if lexicalized objects are "blind" to syntax, then compound-internal agreement should be more stable

than agreement within sentences. This is precisely what Mondini *et al.* (2002) set out to investigate in a study of noun–adjective (*febbre gialla*, "yellow fever") and adjective–noun (*alta moda*, "high fashion") compounds and matched noun phrases (*febbre alta*, "high fever"; *strana moda*, "strange fashion") in Italian-speaking agrammatic aphasics. The patients inflected compound-internal adjectives without difficulty, but were unable to inflect adjectives embedded in noun phrases. Furthermore, no positional effect was found for compounds, while subjects produced more errors in the non-canonically ordered (adjective–noun) noun phrases. These results were interpreted as suggesting that compounds are immune to syntactic impairment. However, difficulty with the production of prepositions required in Italian prepositional compounds does point toward a syntactic deficit, at least for this type of compound (Mondini *et al.*, 2005), although it must be noted that the status of prepositional compounds as true compounds is still being debated by theoretical linguists (Dressler, 2006).

13.7. CHALLENGES AND FUTURE DIRECTIONS

Much remains to be learned about morphological impairments at both the cognitive and neural levels. While a plethora of studies have provided many pieces of the puzzle, the precise locus of breakdown continues to be elusive. Some authors have sought to explain morphological processing difficulties in syntactic terms, others in phonological terms, while yet others maintain that the effects are genuinely morphological in nature. It may be the case that these and other alternative views exist at least in part because researchers rely on different modalities, different tasks, and different stimulus types. One approach, then, might be to undertake cross-population, cross-language studies that employ both receptive and productive tasks, as well as similar experimental paradigms and similar stimulus materials. Moreover, linking behavioral data to neurophysiological and imaging data is proving to be more and more revealing of the nature of morphological processing. The study by Ullman *et al.* (2005) reviewed above is an excellent example of recent efforts to broaden our understanding of the neural correlates of linguistic functions. It is hoped that advancements in imaging methodologies, and the coming together of complementary evidence from a variety of techniques that bring to light both the spatial and temporal dimensions of the neural circuitry subserving language, will continue enhancing our understanding of morphological deficits in neurologically impaired populations.

To conclude, the coming together of increasingly more controlled and sophisticated behavioral and imaging studies across patient populations and across languages offers an extraordinary opportunity to researchers to begin shedding light on many of the grey zones in our understanding of the relationship between cognitive processes of normal and pathological language processing and their neural substrates. Unraveling the mysteries of this relationship is a formidable, but most promising challenge for future research in the neuropsychology of linguistic abilities, including the ability to compute morphology.

Acknowledgments

This research was supported by a Major Collaborative Research Initiative Grant (#412-2001-1009) from the Social Sciences and Humanities Research Council of Canada to Gary Libben (Director), Gonia Jarema, Eva Kehayia, Bruce Derwing, and Lori Buchanan (Co-Investigators).

References

Badecker, W. (2001). Lexical composition and the production of compounds: Evidence from errors in naming. *Language and Cognitive Processes*, *16*, 337–366.

Bastiaanse, R., & van Zonneveld, R. (2004). Broca's aphasia, verbs and the mental lexicon. *Brain and Language*, *90*, 198–202.

Bastiaanse, R., Jonkers, R., Ruigendijk, E., & Van Zonneveld, R. (2003). Gender and case in agrammatic production. *Cortex*, *39*, 405–417.

Bird, H., Howard, D., & Franklin, S. (2000). Why is a verb like an inanimate object? Grammatical category and mnestic deficit. *Brain and Language*, *72*, 246–309.

Bird, H., Lambon-Ralph, M.A., Seidenberg, M.S., McClelland, J.L., & Patterson, K. (2003). Deficits in phonology and past-tense morphology: What's the connection. *Journal of Memory and Language*, *48*, 502–526.

Blanken, G. (2000). The production of nominal compounds in aphasia. *Brain and Language*, *74*, 84–102.

Braber, N., Patterson, K., Ellis, K., & Lambon-Ralph, M.A. (2005). The relationship between phonological and morphological deficits in Broca's aphasia: further evidence from errors in verb inflection. *Brain and Language*, *92*, 278–287.

Delazer, M., & Semenza, C. (1998). The processing of compound words: a study in aphasia. *Brain and Language*, *61*, 54–62.

Dressler, U.W. (2006). Compound types. In: G. Libben & G. Jarema (Eds.), *The representation and processing of compounds words* (pp. 23–44). Oxford: Oxford University Press.

Friedmann, N., & Biran, M. (2003). When is gender accessed? A study of paraphasias in Hebrew anomia. *Cortex*, *39*, 441–463.

Friedmann, N., & Grodzinsky, Y. (1997). Tense and agreement in agrammatic production: pruning the syntactic tree. *Brain and Language*, *56*, 397–425.

Goodglass, H., & Hunt, J. (1958). Grammatical complexity and aphasic speech. *Word*, *14*, 197–207.

Kulke, F., & Blanken, G. (2001). Phonological and syntactic influences on semantic misnamings in aphasia. *Aphasiology*, *15*, 3–15.

Levelt, W.J.M. (1989). Speaking: From intention to articulation. Cambridge: MIT Press.

Marangolo, P., Incoccia, C., Pizzamiglio, L., Sabatini, U., Castriota-Scanderbeg, A., & Burani, C. (2003). The right hemisphere involvement in the processing of morphologically derived words. *Journal of Cognitive Neurosciences*, *15*, 364–371.

Marshall, J., Pring, T., & Chiat, S. (1998). Verb retrieval and sentence production in aphasia. *Brain and Language*, *63*, 159–183.

Miceli, G., Capasso, R., & Caramazza, A. (2004). The relationships between morphological and phonological errors in aphasic speech: data from a word repetition task. *Neuropsychologia*, *42*, 273–287.

Mondini, S., Jarema, G., Luzzatti, C., Burani, C., & Semenza, C. (2002). Why is "Red Cross" different from "Yellow Cross"? A neuropsychological study of noun–adjective agreement within Italian compounds. *Brain and Language*, *81*, 621–634.

Mondini, S., Luzzatti, C., Zonca, G., Pistarini, C., & Semenza, C. (2004). The mental representation of verb–noun compounds in Italian: evidence from a multiple single-case study in aphasia. *Brain and Language*, *90*, 470–477.

Paganelli, F., Vigliocco, G., Vinson, D., Siri, S., & Cappa, S. (2003). An investigation of semantic errors in unimpaired and Alzheimer's speakers of Italian. *Cortex*, *39*, 419–439.

Perlak, D., & Jarema, G. (2003). The recognition of gender-marked nouns and verbs in Polish-speaking aphasic patients. *Cortex*, *39*, 383–403.

Shapiro, K., & Caramazza, A. (2003). The representation of grammatical categories in the brain. *Trends in Cognitive Sciences*, *7*, 201–206.

Shapiro, K., Shelton, J., & Caramazza, A. (2000). Grammatical class in lexical production and morphological processing: Evidence from a case of fluent aphasia. *Cognitive Neuropsychology*, *17*, 665–682.

Semenza, C., Girelli, L., Spacal, M., Kobal, J., & Mesec, A. (2002). Derivation by prefixation in Slovenian: Two aphasia case studies. *Brain and Language*, *81*, 242–249.

Thompson, C.K., Fix, S., & Gitelman, D. (2002). Selective impairment of morphosyntactic production in a neurological patient. *Journal of Neurolinguistics*, *15*, 189–208.

Tsapkini, K., Jarema, G., & Kehayia, E. (2001). Manifestations of morphological impairments in Greek aphasia: A case study. *Journal of Neurolinguistics*, *14*, 281–296.

Tsapkini, K., Jarema, G., & Kehayia, E. (2002). Regularity revisited: Evidence from lexical access of verbs and nouns in Greek. *Brain and Language*, *81*, 103–119.

Ullman, M.T., Corkin, S., Coppola, M., Hickok, G., Growdon, J.H., Koroshetz, W.J., & Pinker, S. (1997). A neural dissociation within language: Evidence that the mental dictionary is part of declarative memory and that grammatical rules are processed by the procedural system. *Journal of Cognitive Neuroscience*, *9*, 266–276.

Ullman, M.T., Pancheva, R., Love, T., Yee, E., Swinney, D., & Hickok, G. (2005). Neural correlates of lexicon and grammar: evidence from the production, reading, and judgment of inflection in aphasia. *Brain and Language*, *93*, 185–238.

Further Readings

Justus, T. (2004). The cerebellum and English grammatical morphology: Evidence from production, comprehension, and grammaticality judgments. *Journal of Cognitive Neuroscience*, *16*, 1115–1130.

This study contributes to our understanding of the neural substrate of morphological processing by providing evidence that damage to the cerebellum results in impairments of subject–verb agreement morphology.

Friedmann, N. (2005). Degrees of severity and recovery in agrammatism: Climbing up the syntactic tree. *Aphasiology*, *19*, 1037–1051.

This is an interesting elaboration of the author's previous studies on impairments of agreement and tense inflection. Framed within the previously proposed Tree Pruning Hypothesis, this study investigates the production abilities of Hebrew- and Arabic-speaking agrammatic aphasics and demonstrates that characterizing agrammatic aphasia using the syntactic tree construct not only accounts for individuals differences between patients and degrees of severity, but also for the stages of spontaneous recovery over time.

Miceli, G. Turriziani, P., Caltagirone, C., Capasso, R., Tomaiuolo, F., & Carmazza, A. (2002). The neural correlates of grammatical gender: An fMRI investigation. *Journal of Cognitive Neuroscience*, *14*, 618–628.

This investigation is one of the first to establish the neural correlates of grammatical gender by showing that the processing of gender is subserved by a fronto-temporal network. Moreover, this study is of particular interest as it demonstrates that neighboring left frontal areas are activated when aphasic patients perform morphological and phonological tasks, thus accounting for the co-occurrence of morphological and phonological errors in aphasia.

Tyler, L.K., de Mornay-Davies, P., Anokhina, R., Longworth, C., Randall, B., & Marslen-Wilson, W. (2002). Dissociations in processing past tense morphology: Neuropathology and behavioral studies. *Journal of Cognitive Neuroscience, 14,* **79–94.**

Tyler, *et al.* bring a new experimental paradigm (primed lexical decision) and a new patient population (patients with a history of herpes simplex encephalitis (HSE)) to the study of dissociations between regular and irregular inflectional morphology following brain damage. They examine the processing of past-tense morphology by non-fluent aphasic patients with damage to the left posterior and inferior frontal lobe and by HSE patients with semantic impairments following bilateral damage to the inferior temporal lobes. The study's behavioral and neuropathological data are in line with the view that distinct functional and neural architectures underlie the processing of past-tense morphology, with frontal neural networks subserving decomposition of regular forms and temporal neural networks subserving the access of stored (full) irregular forms.

14

Disorders of Lexis

GARY LIBBEN

Department of Linguistics, University of Alberta Edmonton, Alberta, Canada

ABSTRACT

Difficulties in the production and comprehension of words accompany virtually all types of language impairment subsequent to brain damage. Moreover, there is often substantial observational continuity between disorders of lexis caused by brain damage and normal word finding difficulties occasionally experienced by non-brain-damaged individuals. This chapter presents an overview of the ways in which lexical production and comprehension can be impaired as well as an overview of recent controversies concerning the organization of words in the brain. We discuss the issue of how true disorders of lexis can be distinguished from impairments that just coincidentally affect words and we present a framework within which these deficits can be categorized. Finally, the issue of how disorders of lexis inform our understanding of the nature of the mental lexicon and word knowledge is discussed.

14.1. INTRODUCTION

We are all familiar with disorders of lexis. We all have them. At times, we feel that we know a concept, but have to talk around it because we do not have the word that would express the concept exactly. We forget words that we once knew. We substitute one word for another. Our representation of a word's meaning is sometimes fuzzy. Sometimes it is just plain wrong.

Most annoyingly, we have all experienced the perception that we know a word but cannot seem to grab it at a particular moment. In this "tip-of-the-tongue" state (Schwartz, 2002), it is rather amazing (and psycholinguistically informative) the extent to which we feel that we know everything about the word (e.g., first letter, number of syllables, what it rhymes with) except what we are looking for.

While most everybody experiences the lexical disturbances above, for unimpaired adults they constitute the exception rather than the rule. Mostly, our lexical systems function in an automatic, effective, and seamless manner so that lexical access occurs in well under half a second, and perfectly accurately. Indeed, it seems an almost effortless and trivial task to obtain the meanings and forms of words as we engage in casual conversation or as we pass signs while driving. This is not the situation subsequent to many types of brain damage. This chapter discusses the manner in which the lexical disorders that we all occasionally experience can constitute the rule rather than the exception following damage to the brain. We focus on the extent to which lexical disorders can be considered a distinct type of language disorder and how the study of such disorders can help answer questions regarding the nature of lexical knowledge and lexical representation and processing in the brain.

14.2. DISORDERS OF LEXIS: A BRIEF OVERVIEW

Lexical disorders accompany almost every type of language disturbance associated with brain damage. Anomia, the impairment of the ability to retrieve words, is extremely common, and can take a variety of forms (Goodglass & Wingfield, 1997). In some cases, the core problem seems to be semantic in nature, in other cases the difficulty seems not so much the ability to access the correct meaning, but difficulty in accessing and producing the correct form. Differences between these two can be revealed by the types of substitution errors (paraphasias) that result when a person with anomia attempts to produce the word that he or she cannot access. Semantic paraphasias are ones in which a semantically related word is produced instead of the target word (e.g., *magazine→book*). Phonemic paraphasias are ones in which the produced word is related to the intended word by sound (e.g., *pill→bill*).

A person suffering from disorders of lexis will likely find that his or her lexical disorder manifests itself predominantly as either difficulty with lexical access or difficulty with lexical selection. The former type is strongly associated with non-fluent Broca-type aphasia and the latter is strongly associated with fluent Wernicke-type aphasia.

Among bilinguals and multilinguals, the most common pattern is one in which lexical difficulties subsequent to language disturbance are roughly in proportion to the person's relative lexical ability in his or her languages prior to the onset of language pathology. In some cases, however, lexical ability in one language is affected much more than another (see Chapter 33), suggesting that the vocabulary associated with individual languages must be represented in the brain in a manner that allows them to be affected selectively in language pathology.

As will be discussed in greater depth below, patterns of lexical disorders that individuals experience within a single language suggest that word types can show differential impairment. Abstract words are typically more vulnerable than concrete words, rare words are typically more vulnerable than common words (with a notable exception being the difficulties that agrammatic patients have with function words, which are typically of very high frequency). Some patients can show more difficulty with particular grammatical classes of words (e.g., nouns or verbs). Dissociations such as these among word types provide a testing ground for whether categories defined in purely linguistic terms can be selectively impaired and the extent to which such impairments can be reduced to impairments defined along more psychologically basic dimensions such as a continuum of lexical abstractness (Bird *et al.*, 2000).

In considering disorders of lexis, it is important to note that most words are not atoms of meaning, but are composed of combinations of meaningful units or morphemes. For speakers of English, words such as *dog, cat, table,* and *chair* seem to constitute the norm, not only because they are high frequency concrete nouns, but also because they are monomorphemic, having no subunits of meaning. In fact, the majority of English words are multimorphemic, as compounds (e.g., *powerboat*), prefixed words (e.g., *empower*), suffixed words (e.g., *powerful*) or, as in the case of *powerboating,* involving a combination of morphological processes (Libben, 2006). Indeed, in many languages of the world, simple monomorphemic words do not exist.

The fact that morphological complexity is a dominant characteristic of words opens an important dimension to the understanding of lexical ability and its impairment. Lexical production involves not only access to words, but also the composition of words from simpler morphological elements (e.g., *power + boat = powerboat*). Comprehension, particularly of uncommon or novel words, can involve the inverse operation of morphological decomposition. Deficits associated with these operations or in processing words that have undergone these operations can be considered to fall under a subcategory of lexical disorders known as morphological disorders (see Chapter 13), which have been at the center of a great deal of controversy concerning the extent to which regular (e.g., *swallowed*) and irregular (e.g., *ate*) verbs are represented in the same manner in the brain (Pinker, 1999). Disorders of lexis also have shown that grammatical suffixes (e.g., the suffix *-ing* in *writing*) are typically more vulnerable to impairment than derivational ones (e.g., the suffix *-er* in *writer*) and that lexical deficits can affect particular types of morphological constructions such as compounds (Semenza & Mondini, 2006).

Difficulty with lexical functioning is also seen outside the strictly defined domains of aphasia. For example, disorders of lexis seen in persons with Alzheimer's disease can very closely approximate those seen in Wernicke's aphasia. Subcortical dementia associated with Parkinson's disease shows, in addition to a general deterioration of control over vocal apparatus, a tendency to omit grammatical suffixes in speech and writing.

A good deal of lesion data and brain imaging data has associated particular types of words and particular lexical functions to specific areas of the brain. Damage to Broca's area severely compromises the ability to produce function words. Inferior frontal lobe damage affects verbs more than nouns. Damage to the left superior temporal gyrus affects auditory lexical processing, and the writing of words is compromised by damage to the left superior parietal lobule and the posterior end of the middle frontal gyrus. In all these cases, however, we are cautioned against jumping to the easy, but probably wrong, conclusion that when damage to a particular brain region results in a diminished ability to perform some lexical function, that area of the brain is, in the absence of pathology, responsible for that function.

The mention of "inhibition" in the preceding example brings us to a final point in this brief overview. In all probability, normal lexical functioning involves the interplay between activation and deactivation of lexical representations. Libben *et al.* (2004) have claimed that lexical processing in both unimpaired and impaired language users lead to over-activation. They claim that lexical processing does not reflect a "neat" underlying system that seeks from the outset to construct parsimonious morphological and semantic representations for incoming stimuli. Rather, it tolerates a great deal of redundancy and over-activation, which is then pruned through a process of deactivation. Thus many disorders of lexis (particularly in domains such as the processing of multimorphemic words or in some cases of deep dyslexia) may not be the result of a failure to activate lexical representations, but rather the failure to deactivate them.

14.3. WORDS IN THE BRAIN

To date, no research team has isolated the representation of a word in the brain, nor do we have agreed upon theories of what such a word representation might look like. One promising step in this direction, however, has been presented by Pulvermüller (1999, 2001), who has proposed a framework by applying a modification of principles of Hebbian cell assemblies to the representations of word forms. In this approach, words have distributed web-like representations that correspond to both their semantics and their use. As shown in Figure 14.1, for example, this framework suggests that vision-related words would be differentiated from action-related words by their spatial distribution in the brain. Action words would show a lexical web that extends anteriorly to the motor strip, whereas vision-related words would show a posterior projection of their lexical webs.

14.3.1. Word Types Distinguished by Abstractness

Pulvermüller claims that the Hebbian approach to lexical representation in the brain can also be extended to explain classic dissociations of word types in aphasia. Under this view, the observed differences between function words and content words in aphasia may reflect the different distributions of these webs. Function words typically do not have perceptual or motor properties and are thus represented in a less diffuse manner in the brain such that they can be said to be localized in the perisylvian region of the left hemisphere in the case of right-handers. Content words, on the other hand, are expected to have webs that extend beyond the perisylvian region. Similar argumentation can be applied to account for differences between abstract and concrete words. Pulvermüller (1999) cites, in support of this claim, a study of event-related brain potentials (ERP) by Kounios and Holcomb (1994) in which it was found that abstract words generated differential ERP responses across the two hemispheres of the brain, whereas concrete words showed similar responses. Thus, in Pulvermüller's framework, abstract words and function words might pattern together in aphasia because they both have less distributed representations in the brain. This is consistent with other proposals such as those of Paivio (1986) and Plaut and Shallice (1993) who have claimed that differences between abstract and concrete words lie in the fact that the latter have richer networks of semantic features or sensory associations.

Work by Crutch (2006) has also addressed the question of the differences between abstract and concrete words in terms of their representations in the brain. His focus, however,

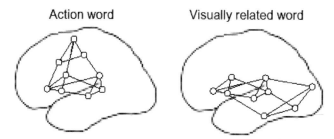

FIGURE 14.1 Hypothesized distributed networks for action words and visually related words in a Hebbian word web (*Source*: From Pulvermüller, 2001, p. 520) (© Elsevier.)

was on qualitative differences, as opposed to quantitative differences in the extent of the lexical web for each word type. He argues that content words can be said to be organized in a categorical network, so that semantic errors shown by patients will typically involve a semantic feature shared by both the target and the erroneous output (e.g., *coast→seashore*). Abstract words, on the other hand are claimed to be stored in an associative network, yielding associative errors by patients (e.g., *evidence→police*).

14.3.2. Word Types Distinguished by Grammatical Category

The question of whether particular lexical categories of words have different types and locations of representation in the brain has been the subject of a great deal of research. Here, increasingly fine-grained analyses of lexical disorders stand to improve our understanding of both the differences among word types and the brain resources that are required for their comprehension and production.

At the center of this discussion has been the distinction between nouns and verbs. In general, noun deficits are associated with anomic aphasia and damage to the temporal/parietal areas of the left hemisphere of the brain. Verb deficits are more common in cases of damage to anterior portions of the left hemisphere and are particularly associated with agrammatic aphasia. This distinction is supported by the fact that, cross-linguistically, verbs are typically more involved in the grammar of a sentence than are nouns. Thus, it seems unsurprising that verbs would show a greater degree of deficit in agrammatic aphasia. But there are many counter examples – cases of anterior damage in which nouns are more affected than verbs and cases of posterior damage in which verbs are more affected than nouns. Rapp and Caramazza (2002) have reported a double dissociation with the same patient. This patient, KSR, showed more difficulty producing nouns as opposed to verbs in speech, but more difficulty producing verbs as opposed to nouns in writing. Rapp and Caramazza argue that evidence such as this points us away from the view that verb/noun dissociations are related to semantic variables such as abstractness (as suggested by Pulvermüller (1999) and Bird *et al.* (2000)). Rather, they claim that at least in the case of KSR, the double dissociation must result from differential impairment in functionally distinguished output lexicons that are specific to particular modalities (i.e., speaking versus writing) and that make reference to the grammatical categories of nouns and verbs.

A case for the relevance of such grammatical categories in the representations of verbs and nouns in the brain has been made by Shapiro and Caramazza (2003). They claim that the processing of nouns and verbs is served by distinct neural circuits that are related to the grammatical properties of these words. This of course does not mean that nouns and verbs could not be differentially represented in the brain as a consequence of their semantic differences as well. But the categorical grammatical properties constitute the focus of their attention. While acknowledging that neuroimaging data have failed to show spatially distinguishable activation for nouns and verbs thus far, they present data from neuropathology that point toward verbs and nouns being represented in different areas of the brain. Moreover, they claim that the distinctions between nouns and verbs as grammatical categories extend to the involvement of different types of morphological processes associated with each. They claim that difficulties with verbal morphology might stem from damage to part of the left mid-frontal gyrus superior to Broca's area, whereas difficulties with nominal morphology might involve more inferior neural structures. In Figure 14.2, which contrasts the performance and lesion types of two patients that they report, we see an example of the type of evidence that leads to these conclusions.

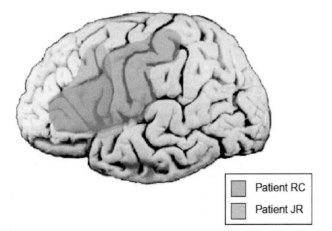

	Patient RC
	Patient JR

FIGURE 14.2   Two patients reported by Shapiro and Caramazza showing different brain regions implicated in naming of nouns and verbs. Patient RC (lesion shown in red) is relatively impaired in naming verbs. Patient JR (lesion shown in blue) is more impaired in noun production. *Source*: From Shapiro and Caramazza, 2003, p. 204 (© Elsevier.) (see Plate 11).

14.3.3. Some Reservations

We have seen in the discussion above examples of rather different approaches to how lexical information can be encoded in the brain. The approach of Pulvermüller concentrates on the relation of words to the actions, perceptions, and functions with which they might be associated. The approach of Caramazza and colleagues concentrates on the relevance of categorical grammatical information. In both cases we should be cautious of reifying our theoretical constructs. When we speak of words in the brain, it is easy to slip into believing that words have some independent reality such that they can be stored in this or that manner. We would be less likely to think about emotions being represented in the brain in such a manner, despite the fact that we can give them names such as "anger" or "joy" and describe them as either positive or negative. Additionally, it should be noted that there are alternatives to the view that nouns and verbs are actually different categories of words, rather than different functions that lexical forms can take. In the linguistic framework of Distributed Morphology (e.g., Marantz, 1997), it is claimed that words do not come with lexical properties, but rather acquire properties such as "noun" and "verb" as consequences of their use in sentence contexts. Using this theoretical framework, Barner and Bale (2002) present arguments for how a brain with no nouns and no verbs could nevertheless generate the patterns of impairment that seem as though they are category-specific effects.

14.4. THE DOMAIN OF LEXICAL DISORDERS, THE NOTION OF A MENTAL LEXICON, AND THE NOTION OF WORD

The overview above brings to our attention the following characteristics associated with lexical deficits:

1. Lexical deficits accompany almost every type of language disturbance.
2. There are many different kinds of words and they are not equally affected by damage to the brain.
3. A person's vocabulary cannot be said to have a particular location in the brain.
4. Different linguistic functions such as speaking, writing, listening, and reading can be differentially impaired and create modality-specific lexical impairments.

The four points above, lead us to the central point of consideration in this section: Are lexical deficits sufficiently homogenous and autonomous such that they should be characterized as a distinct type or level of language disorder? As we have already seen, disorders of lexis are bundled with other characteristics of a syndrome and to problems in modality-specific language processing disorders. But yet, there is also an important way in which disorders of lexis are distinct. A rule of thumb for the identification of distinctly lexical disorders might be the following:

> A disorder of lexis is one in which the explanation of behavior must make reference to properties of the human vocabulary system that we call the mental lexicon and/or to lexical properties of words.

Seen from this perspective, not all impairments of word processing ability fall within the domain of disorders of lexis. An acquired reading disorder that has as its core the inability to process alphabetic scripts would not. A reading disorder in Hebrew might also not qualify as a lexical disorder, if at bottom the deficit involves the inability to reconstruct vowel patterns from syntactic and semantic context (in the Hebrew writing system, only consonants are written – the vowels are inferred by the reader). In the domain of auditory processing, even a syndrome such as *word deafness*, despite its name, might not be appropriately be considered as falling within the domain of lexical disorders. This syndrome, associated with damage to the superior temporal gyrus, is one in which patients can speak, read, and write relatively well and can process environmental sounds. They cannot, however, discriminate words. There is some evidence, however, that word deafness is not truly lexical in nature, but is rather an impairment in the ability to distinguish rapid changes in auditory input (see Price, 2000). If this is indeed the case, then the fact that this deficit targets words might simply fall out from the fact that words require more temporally and acoustically fine-grained auditory processing as compared with environmental sounds. Thus, no appeal to the nature of the mental lexicon or to the lexical properties of words would be required in the explanation of behavior.

The line of reasoning outlined above naturally leads us to questions: What are the properties of the mental lexicon and the properties of words that would need to be referenced in an explanation of disorders of lexis? We will treat them in turn in the two sections below.

14.4.1. The Mental Lexicon

Traditionally, the mental lexicon has been seen as the repository of all lexical knowledge possessed by an individual language user. Coltheart *et al.* (2001) note that the notion of a mental lexicon was first introduced into the psycholinguistic literature by Ann Treisman in her 1961 doctoral dissertation. The notion appeals to our intuition that the human mind must possess a knowledge store which functions as a dictionary in the mind – one which allows easy access to written and oral forms of individual words and associates those forms to meaning. But this notion immediately raises a number of issues. The first is whether we imagine a single individual to be in possession of a single mental lexicon or multiple mental lexicons. From the outset, work on acquired dyslexia pointed our attention to the fact that individuals

with reading disorders (in particular, those labeled as surface dyslexics) can lose the ability to access words on the basis of their stored visual forms but maintain the ability to access the meanings of words, the ability to access them auditorily, and the ability to produce them orally. Cases such as these raise the possibility that the mental lexicon may be composed of modality-specific input lexicons as well as modality-specific output lexicons (see Coltheart *et al.*, 2001 for an excellent review of the evolution of this thinking).

The issue of single versus multiple mental lexicons also comes to the foreground when we consider how words are represented in the minds of bilinguals and multilinguals. Is it best to conceptualize the lexical knowledge of a bilingual as one in which separate lexicons are maintained for each language (with connections between them), or does such a person have a single integrated mental lexicon with individual words identified as belonging to this or that language? Recent research seems to favor the latter view (Libben, 2000; Dijkstra & Van Heuven, 2002; Dijkstra *et al.*, 2005).

The second issue raised by the notion of a mental lexicon is whether there are any passive knowledge stores in the brain that just "sit there" the way a desktop dictionary does. In all probability, there are not. Indeed, it might be better to conceptualize a mental lexicon as a knowledge store that constitutes the ability for lexical *activity* (Jarema & Libben, 2007). All lexical knowledge is, in fact, activity. These activities include not only ones that have external manifestation such as speaking or writing, but also covert activity such as simply recognizing a word. Thus, the mental lexicon is probably not a dictionary, it is probably not a list, and it is definitely not cognitively inert. The mental lexicon is almost certainly not located in a single area of the brain, but it nevertheless serves as a vital construct for understanding the nature of lexical knowledge and lexical impairment. It is the construct under which we can unite those lexical properties that are associated with all words.

14.4.2. The Notion of a Word

Words seem like things, or at least objects of representation. Yet, it is possible that seeing words as things is a consequence of the ways in which we think and talk about words in our cultures. For example, whereas it would seem very natural to ask a learner of English as a second language how many words he or she knows, it would seem much less natural to ask how many phonemes. Similarly, in the domain of language disorders, it seems much more natural to think about phonological disorders in terms of what an individual can no longer *do*, but we often think about a disorder of lexis in terms of what an individual no longer *has*. At bottom, however, the term "word" is probably just a shorthand for a set of lexical activities that are associated with stable meanings and forms.

Box 14.1 How many words?

According to Aitchison (1989), native speakers of a language know about 50,000 words. This would be roughly one tenth of the number of words listed in the Oxford English Dictionary. An educated adult with a substantial technical vocabulary in his or her field of expertise would probably be in possession of about 100,000 words. Multilinguals could have at least twice this number of lexical items in their combined vocabulary.

TABLE 14.1 Properties Relevant to a Characterization of Disorders of Lexis. Properties under each of the Three Categories are Illustrative and not Intended to be Exhaustive

Word properties	Lexicon properties	Lexical use
Class openness	Activation	Speaking
Grammatical category	Facilitation	Listening
Morphological complexity	Deactivation	Reading
Abstractness	Productivity	Writing
		Inflection

14.5. A FRAMEWORK FOR UNDERSTANDING DISORDERS OF LEXIS

We have claimed above that a disorder of lexis must make reference to properties of words or of the lexical system, and have sketched some of those properties. In Table 14.1, we distinguish three categories of properties that constitute a framework within which to understand disorders of lexis. The first category is that of word properties, those properties that distinguish one word from another. The second category refers to the lexicon as a whole. Here, we highlight general features of lexical activation within the lexicon as a whole, the extent to which words can facilitate one another, the extent to which deactivation of inappropriate or competing lexical representations occurs and, finally, the extent to which the system sustains lexical productivity.

In the final column of Table 14.1, we represent the functions of speaking, listening, reading, and writing. The intention here is to focus not on the general properties of these functions, but rather on the distinct aspects of lexical representations that they mobilize when an individual speaks, listens, reads, or writes. From the perspective of lexical processing, these can be seen as a type of repeated measures effect – different aspects or faces of a word. In cases of language pathology in which individuals consistently have difficulty with a particular function, for example reading, the disorder might be properly considered to be a type of lexicon effect, because it affects that aspect of lexical processing in general. If, on the other hand, the reading disorder is restricted to individual words, or individual classes of words, it might be best seen as a disorder of word properties.

An additional point needs to be made regarding the contents of the third column of Table 14.1. It may be the case, that inflectional morphology can also be seen as a repeated-measures lexical factor. Just as the spoken and written forms of a word can be seen as two versions of the same thing, inflectional variants such as *walk*, *walked*, and *walking* may also be different versions of the same thing, in contrast to the derivational pair *walk* and *walker*, which involve distinct but related individual words. As has often been claimed, inflectional variants may not have distinct representations in the

Box 14.2 The lexical decision task

Henderson (1989) has claimed that, in a fundamental way, the modern study of the mental lexicon was created by the advent of the lexical decision task, which is still the task of choice for most research on lexical processing. In the lexical decision task, a participant is presented with a single word, usually visually in the center of a computer screen. The participant's task is to decide, as quickly and as accurately as possible, whether the word is a real word of his or her language. The participant indicates the decision by pressing either a computer key labeled "yes" or on labeled "no." The idea behind this deceptively simple task is that a "yes" response indicates that the presented word has indeed been activated in the mind of the experimental participant, and the time taken to press the "yes" key, reflects the accessibility of the word in the mind of the participant.

mental lexicon as separate words, but be produced "on the fly" so that the form of a word corresponds to the demands of sentence context.

The organization shown in Table 14.1 does not make claims regarding the nature of lexical deficits, but rather provides a framework for the creation of an appropriate inventory of properties and an evaluation of their interactions. The report of Rapp and Caramazza (2002), discussed above, constitutes an example of such an interaction. The fact that their patient KSR had more difficulty with nouns as compared to verbs in speaking, but more difficulty with verbs as compared to nouns in writing can be conceptualized as a claim of an interaction between properties of columns 1 and 3 of Table 14.1.

14.6. CHALLENGES AND FUTURE DIRECTIONS

My goal in this chapter has been to isolate those aspects of lexical processing disorders that can be considered to be truly lexical in nature and to ground disorders of lexis within a view of the mental lexicon and lexical knowledge that has activity as its core. If this turns out to be a profitable approach it carries with it a number of implications for the future.

14.6.1. A Proliferation of Variables

An adequate understanding of disorders of lexis might need to make appeal to a very large number of lexical variables which are unlikely to be linear in nature and less likely to be dichotomous. The good news is that this will align the study of lexical disorders more closely with recent trends in the psycholinguistic investigation of lexical processing in which researchers such as Hay and Baayen (2005) have identified a very large number of interacting variables that shape the lexical processing system and have pointed the way toward modeling and analysis techniques that are able to handle such complexity.

14.6.2. The Incorporation of Semantics

In many ways, the most important variable in the study of lexical processing and its impairment is the one that is least well understood. A remaining challenge is to find a way in which to incorporate the role of meaning in a theory of the mental lexicon in a manner that does justice to the complexity of the problem. Elman's (2004) proposal that meaning is best understood not as a property of a word, but rather in terms of the effect that it has on mental states seems like a promising step in that direction.

14.6.3. No Independent Variables

It may be that meaning is not the only property of words that is illusory. It may be the case that, more generally, the concept of words as having properties "out there" has no place in a theory of the lexicon. Traditionally, we have worked under the assumption that the words that enter an individual's lexical system bring with them a set of properties. This is unlikely to be the truth of the matter. A verb is a verb only if an individual uses it as a verb. A concrete word is concrete only if one has concrete associations to its form. Similarly, it is perhaps only appropriate to consider a word to be morphologically complex if individuals are shown to perform lexical actions that involve access to subunits of meaning within that word. Thus, studies of lexical deficits in the processing of morphologically complex words will need to be based on normative studies of the extent to which speakers activate constituent morphemes within the words in question, not simply whether they contain prefixes or suffixes.

It is my view that the movement of the study of disorders of lexis in the direction outlined above will constitute a further step toward explanatory adequacy and as a by-product will equip us with more ecologically valid models of lexical ability and its impairment.

Acknowledgments

This research was supported by a Major Collaborative Research Initiative Grant (#412-2001-1009) from the Social Sciences and Humanities Research Council of Canada to Gary Libben (Director), Gonia Jarema, Eva Kehayia, Bruce Derwing, and Lori Buchanan (Co-investigators).

References

Barner, D., & Bale, A. (2002). No nouns, no verbs: Psycholinguistic arguments in favor of lexical underspecification. *Lingua, 112*, 771–791.

Bird, H., Howard, D., & Franklin, S. (2000). Why is a verb like an inanimate object? *Brain and Language, 72*, 276–309.

Coltheart, M., Rastle, K., Perry, C., Langdon, R., & Ziegler, J. (2001). DRC: A dual route cascaded model of visual word recognition and reading aloud. *Psychological Review, 108*(1), 204–256.

Crutch, S.J. (2006). Qualitatively different semantic representations for abstract and concrete words: Further evidence from the semantic reading errors of deep dyslexic patients. *Neurocase, 12*, 91–97.

Dijkstra, A., & Van Heuven, W.J.B. (2002). The architecture of the bilingual word recognition system: From identification to decision. *Bilingualism: Language and Cognition, 5*, 175–197.

Dijkstra, T., Moscoso del Prado Martin, F., Schulpen, B., Schreuder, R., & Baayen, R. (2005). A roommate in cream: Morphological family size effects on interlingual homograph recognition. *Language and Cognitive Processes, 20*(1), 7–42.

Elman, J. (2004). An alternative view of the mental lexicon. *Trends in Cognitive Sciences, 8*(7), 301–306.

Goodglass, H., & Wingfield, A. (Eds.) (1997). *Anomia: Neuroanatomical and cognitive correlates*. San Diego, CA: Academic Press.

Hay, J.B., & Baayen, R.H. (2005). Shifting paradigms: Gradient structure in morphology. *Trends in Cognitive Sciences, 9*, 342–348.

Jarema, G., & Libben, G. (2007). Introduction: Matters of definition and core perspectives. In G. Jarema & G. Libben (Eds.), *The Mental Lexicon: Core perspectives* (pp. 1–5). Oxford: Elsevier.

Kounios, J., & Holcomb, P.J. (1994). Concreteness effects in semantic priming: ERP evidence supporting dual-coding theory. *Journal of experimental Psychology: Learning, Memory and Cognition, 20*, 804–823.

Libben, G. (2000). Representation and processing in the second language lexicon: The homogeneity hypothesis. In J.A. Archibald (Ed.), *Second language grammars* (pp. 228–248). London: Blackwell Press.

Libben, G. (2006). Reading complex morphological structures. In S. Andrews (Ed.), *From inkmarks to ideas* (pp. 259–274). New York: Psychology Press.

Libben, G., Buchanan, L., & Colangelo, A. (2004). Morphology, Semantics, and the Mental Lexicon: The failure of deactivation hypothesis. *Logos and Language, 4*(1), 45–53.

Marantz, A. (1997). No escape from syntax: Don't try morphological analysis in the privacy of your own lexicon. *University of Pennsylvania Working Papers in Linguistics, 4*, 201–225.

Paivio, A. (1986). *Mental representations: A dual coding approach*. Oxford: Oxford University Press.

Pinker, S. (1999). *Words and rules*. New York: Basic Books.

Plaut, D.C., & Shallice, T. (1993). Deep dyslexia: A case study of connectionist neuropsychology. *Cognitive Neuropsychology, 10*, 377–500.

Price, C.J. (2000). The anatomy of language: Contributions from functional neuroimaging. *Journal of Anatomy, 197*, 335–359.

Pulvermüller, F. (1999). Words in the brain's language. *Behavioral and Brain Sciences, 22*, 253–336.

Pulvermüller, F. (2001). Brain reflections of words and their meaning. *Trends in Cognitive Sciences, 5*(12), 517–524.

Rapp, B., & Caramazza, A. (2002). Selective difficulties with spoken nouns and written verbs: A single case study. *Journal of Neurolinguistics, 15*, 373–402.

577785

Schwartz, B.L. (2002). *Tip-of-the-tongue states: Phenomenology, mechanism and lexical retrieval*. Mahawah, New Jersey: Lawrence Erlbaum.

Semenza, C., & Mondini, S. (2006). The neuropsychology of compound words. In G. Libben & G. Jarema (Eds.), *The representation and processing of compound words* (pp. 71–95). Oxford: Oxford University Press.

Shapiro, K., & Caramazza, A. (2003). The representation of grammatical categories in the brain. *Trends in Cognitive Sciences, 7*, 201–206.

Further Readings

Aitchison, J. (2003). *Words in the Mind: An Introduction to the Mental Lexicon* (3rd edn). Oxford: Blackwell Publishing.
This is an extremely well written book that offers the reader an accessible introduction to psycholinguistic research on the mental lexicon and the issues that are at the heart of the field of investigation.

Jarema, G., & Libben, G. (Eds.) (2007). *The Mental Lexicon: Core perspectives*. Oxford: Elsevier.
This edited volume assembles a variety of perspectives on the fundamental nature of the mental lexicon. Authors present the key achievements in the field as well as their views on the trajectory of future research.

Pinker, S. (1999). *Words and rules*. New York: Basic Books.
In this book, Steven Pinker explores the consequences of a dual mechanism approach to the representation in the mind. The book has as its core the claim that there is a fundamental difference between regular and irregular morphology.

Pulvermüller, F. (1999). Words in the brain's language. *Behavioral and Brain Sciences, 22*, 253–336.
The article, which is commented upon in this chapter, presents an accessible and provocative perspective on the manner in which words may be represented in the brain.

15

Disorders of Syntax

ALAN BERETTA

Department of Linguistics, Michigan State University, East Lansing, MI, USA

ABSTRACT

In this chapter, I ask what we could learn from studying the syntax of Broca's aphasics. In order to answer this question, I first raise the broader question of what investigation of language in the brain, that is, neurolinguistic inquiry in general, could possibly be informative about. Against that backcloth, I consider some alternative models of the patterns of sparing and loss in Broca's aphasia that have been explored over the last decade or so and assess the contribution they have made to neurolinguistic understanding. I will conclude that to meet the challenges that lie ahead, some fundamental demands will have to be met.

15.1. INTRODUCTION

Someone suffers a stroke, or other neural insult, such that language is impaired. Clinically, of course, we care. But is there any reason for us to care if our interest is, not clinical, but theoretical? That is, if what we wish to understand is something about the relation between language and brain, is there anything that an examination of aphasia can tell us? Given the largely random nature of hemorrhaging or vascular blockage and the attendant tissue damage that may compromise the language faculty (among other functions), it is quite possible that the impairments could be random too, with respect to language, in which case there would be absolutely nothing to say.

However, it turns out that language loss in one kind of aphasia is not random. In a breakthrough study some 30 years ago, it was discovered that Broca's aphasics could understand some kinds of sentences but not others. They could understand actives but not passives, subject relative clauses but not object relative clauses. In other words, the

pattern of sparing and loss was selective. A selective pattern, with different sentence types falling either side of a divide, ought to match a divide in our theory of language. If we can hit upon the right theoretical element such that it naturally yields distinct groups for the understood and non-understood sentences, then a link is established between brain organization of language and our theory of language. The link is tenuous, to be sure, for reasons I will explain, but establishing any theoretical link between language and brain is exceptionally difficult, so much so that some believe that linking aspects of mind (such as language) to brain is a bridge too far for humans ever to accomplish (for example, McGinn, 1993). If we listen to those who think that mind–brain relations are unknowable, then we will be reduced to chanting feint hymns to the cold, fruitless moon. Of course, nobody knows what science will and will not accomplish, but I mention this entirely negative (though by no means irrational) assessment up front in order to temper expectations about what we might already have understood about language–brain relations through an examination of syntactic disorders in Broca's aphasics, and also to achieve a more realistic perspective about the challenges that lie ahead.

I will proceed as follows. First, I will argue for a particular view of the neurolinguistic enterprise that all we can possibly mean by welding *linguistic* to *neuro* is theory unification. It follows from this view that if we are not asking coherent theoretical questions, we are moving no closer to any possible theory unification. I will allege that this view runs counter to the prevailing trend in much of the work on Broca's aphasia, and neurolinguistic inquiry in general, which avoids linguistic theory like the plague, treats brains principally in (possibly theory-free) localizationist terms, and implicitly prefers commonsense description.

Once the stage is set, I will confine discussion to models of comprehension in Broca's aphasia that at least *claim* to be motivated by linguistic theory and assess their contribution to understanding. (I will not consider Wernicke's aphasia because little of a syntactically interesting nature has been uncovered in this population.) In the end, it will emerge that what we have come to learn is of a rather modest nature, but this should not be deflating. The deflation comes long before that, in fact in the very next section, so that by the time we arrive at a consideration of prospects for the future, we will have picked ourselves up and dusted ourselves down sufficiently to view what we *have* learnt via an at least partial commitment to theory as a lot better than the commonsense descriptive alternative and to confront the challenges ahead realistically but coherently.

15.2. THE ENDGAME: THEORY UNIFICATION

If we are going to be in a position to assess the work that has been done on syntactic disorders in aphasia, it is first of all important to consider what such work could possibly achieve, so that we might gauge the extent to which it has contributed to that ultimate goal. So, what could this work possibly achieve? To answer this, it is necessary to consider the broader goal of neurolinguistics.

It would be possible to conceive of neurolinguistics as a domain of inquiry that links language, in its ordinary everyday sense, to brain, in a similarly commonplace, traditional sense. In fact, it is not only possible, but the norm. That is, a glance at some of the major journals that report cognitive neuroscience research reveals that most of the reports of, say, positron emission tomography (PET) or functional magnetic resonance imaging (fMRI) experiments on language resolutely avoid reference to any aspect of linguistic theory and make no contact with any aspect of brain theory. So, what *do* these studies do?

With respect to language, what they do is ask subjects to listen to stories, or think of a word that begins with a letter string such as *cou*, or make rhyme judgments, and so on. It is difficult to see how these tasks tap into anything that linguistic theory suggests is real (and no case to that effect is made). Many studies, it seems, are content to proceed just as we would if all we had was a commonsense notion of language.

With respect to brains, what they do is report locations where activation rises above some threshold, the idea being to link function, in this case language, to neural regions. Isn't location part of brain theory? After all, it is an uncontroversial fact that different functions can occur in different regions of the brain. However, if location is a theoretical postulate, it has to be doing explanatory work, and yet it is hard to find any reference to such explanatory work. If I were to receive instruction to the contrary, I would be

grateful, but lacking any to date, I am inclined to wonder if the fact of function having location might not be merely epiphenomenal, just as linearity is epiphenomenal in linguistic theory. It is an inevitable fact that words have to follow each other in a linear sequence in every sentence ever heard or spoken but linearity has no theoretical status since it explains nothing. Thus, function has to happen somewhere in the brain, but where it happens may not be theoretically informative. (Of course, it is useful to know where something happens if only so that it is clear where inquiry should be pursued. Location is being vigorously investigated, and sensibly so, but this chapter seeks to draw attention to the ultimate goal of neurolinguistics: merging theories.)

To sum up, much neurolinguistic experiment is descriptive and atheoretical, preoccupied with describing the locations of everyday aspects of language. So far as I can tell, this approach to neurolinguistics is based on a full awareness of the enormous challenge of finding out how something as complex as the human brain accomplishes something as complex as language. Confronted by such a daunting task, many take the view that a great deal of description is required in order to establish some basic facts and, the corollary of this view, that concern about theory is premature. This Dragnet approach ("just give us the facts, ma'am, any facts") has the advantage of good common sense, perhaps why many subscribe to it. However, there are reasons to question whether it makes equally good scientific sense.

First, it is a truism that common sense has long been abandoned in the natural sciences. As the physicist Richard Feynman (1985) observed, quantum theory is absurd if we approach it from a common sense perspective; physicists have learned that theory tortures common sense and they have learned not to care. So, if we know that common sense contributes nothing to the growth of understanding, and that theory that agrees with evidence is all that has ever yielded insight, then it seems odd to press the case for a commonsense descriptive approach specifically with respect to language and brain. We would all look askance at anyone who suggested that physics, confronting a particularly difficult question, should shelve theoretical notions of, let us say, energy or light, and replace them with their non-technical everyday senses. And yet, when it comes to examining language in the brain, that is precisely what is so often done, with language being treated pretheoretically. Why should normal scientific criteria be renounced when what is at issue is language?

Let us assume, then, in contradiction to the prevailing norm, that commonsense description is of limited value, and adopt in its place the prevailing norm in naturalistic inquiry, namely, theoretically motivated experimentation. Adopting this approach, neurolinguistics can only mean an attempt to integrate what we know about brain and what we know about language, informally, brain theory and linguistic theory (which is itself, of course, a theory of an aspect of brain, but not a theory whose insights have been due to the traditional sciences). Unification of these two theories is obviously remote; at present, neither constrains the other in the slightest. But however remote, theory unification is the endgame. This approach minimally requires that if we investigate language in the brain, we ask the brain a theoretically coherent question. In addition, the sort of response that we seek from the brain ought to be a kind of response that matters to some aspect of brain theory. Let us now turn to the literature on syntactic disorders and judge its current status in these terms, and then consider its future prospects from the same perspective.

15.3. SYNTACTIC COMPREHENSION IN BROCA'S APHASIA

In studies of aphasia, as in much of the neuroimaging literature, common sense is, as it were, common. Instead of reviewing a massive atheoretical literature in a cursory manner, I will focus on the accounts that make some claim to be motivated by linguistic theory. It will become evident that even some of these are less principled than their proponents appear to think, but the limitations of these models are instructive, I believe. I will contrast models that are heavily reliant on linearity in assigning thematic roles to structural positions with a model that depends on hierarchical structure and is indifferent to linearity. Also, I will attend only to models of comprehension.

I have referred above to the epiphenomenal nature of linearity. Let us pursue it a little further. Fact: in the world's languages, subjects generally precede objects. Fact: subjects are generally assigned thematic roles that are higher in thematic hierarchies than objects. Fact: Agents (the do*ers* of an action) generally occur before Themes (the do*ees* of an action). The order of thematic roles does not follow from any principle in linguistic theory. It is, therefore, just a fact, nothing more. From the point of view of common sense, it is very hard to ignore immediately apparent, brute facts. But that is why we have theories. Theories could not care less about facts that do not follow from principles, so out of countless billions of facts, it is possible to home in on the (often obscure) ones that matter. It may come as something of a surprise, then, that one model after another of syntactic comprehension in Broca's aphasia that presents itself as a model derived from linguistic theory should be so fundamentally beholden to mere linear trends.

15.4. THE LINEAR MODELS

Linear models to account for patterns of sparing and loss in Broca's aphasia have seduced a wide range of researchers over the last 20 years (e.g., Caplan & Futter, 1986; Linebarger, 1995; Grodzinsky, 2000; Piñango, 2000;

Avrutin, 2006). These proposals all differ, sometimes only very slightly indeed, in their implementation, but what is common to them all is the idea that Broca's aphasics have recourse to a strategy in which thematic roles are assigned in a linear order (based, e.g., on how high they are in thematic hierarchies). The Agent role, being the highest, is assigned to the first relevant noun phrase (NP) in a sentence; a role that is lower is assigned to a second relevant NP, and so forth. In what follows, I will assess the empirical status of two of the linear models before considering them with respect to the characterization of neurolinguistic inquiry outlined above.

15.4.1. The Trace Deletion Hypothesis and the Slow Syntax Hypothesis

The TDH (see Box 15.1) is probably the most widely known of the linear accounts. However, it is now an *ex*-model, it is no more, and it has passed away. But it has not gone gently into that good night; and it has not moved on without leaving a trace, so to speak. Its major legacy as an account of comprehension in Broca's aphasia is that it drew attention to the existence of linguistic theory in a field that had a primarily clinical focus, which must be regarded as an achievement. But let us see why rumors of its demise have not been exaggerated.

As the thumbnail sketch in Box 15.1 makes clear, under the TDH, Broca's aphasics turn to a default strategy which has the effect, in passives and object relatives, that their representation involves two Agents, instead of an Agent and a Theme. It is frequently observed that Broca's aphasics perform at chance on these structures when they carry out sentence–picture matching tasks. This is consistent with guessing which participant is actually performing the depicted action, that is, which participant really is the Agent. Note, however, that this claim had been upheld with standard sentence–picture matching tasks in which there are only two participants in an action (see Figure 15.1). One of them must be the Agent and one of them must be the Theme. In this standard two-participant picture task, two Agents could not possibly be performing the same action. Therefore, the aphasic double-Agent representation necessarily conflicts with the world (as presented in the pictures). By modifying the standard task, Beretta & Munn (1998) changed the world so that it matched the supposed double-Agent representation. They devised a sentence–picture task in which there were *three* participants in each action (two Agents and one Theme), rather than the standard *two* (see Figure 15.1). Thus, in an action of, for example, *kicking*, patients still hear a sentence such as (1):

1. The giraffe was kicked by the woman

But now, one of the pictures has *a giraffe* and *a woman* both kicking something else (say, *a dog*). This picture is

FIGURE 15.1   *Left*: The standard pictures with *two* participants *Right*: The double-Agent pictures with *three* participants. *Source*: Adapted from Beretta & Munn 1998, *Brain and Language*; with permission.

not the correct picture to point to, but if aphasics truly had a double-Agent representation of sentences like (1), they should always choose it. Why? Because it presents no conflict with their representation. The world, that is to say, in this picture, is in perfect accord with their supposed double-Agent representation. (Recourse to an argument that patients would know that the verb *kick* must involve an Agent and a Theme cannot be made since the claim that a double-Agent representation is even possible rests on an appeal to informational encapsulation that the TDH has been at pains to maintain in order to protect the hypothesis from the very same charge. Such an argument cuts both ways.)

Broca's aphasics demonstrated that they had no problem with three-participant pictures with active sentences, showing that performance on the passives could not be an artifact of the task. On the passives, these same Broca's aphasics performed in a strikingly uniform manner. They avoided the picture that was consistent with their alleged

double-Agent representation, instead guessing randomly between the remaining two pictures. By shunning the one picture that would permit a double-Agent representation to stand, Broca's aphasics clearly demonstrated that they did *not* resort to the default strategy even when the invitation to do so should have been overwhelming.

This experiment did not just show that the TDH could not explain an anomalous finding, such as, let us say, that subjects in some study performed better on object relatives in English than on subject relatives. Rather, what this experiment did was to plunge a dagger into the very heart of the TDH by demonstrating that the representation of Broca's aphasics cannot contain two Agents, let me stress that, *cannot contain two Agents*, meaning no appeal could ever have been made to the default strategy. Without the linear default strategy, of course, the TDH comes to be nothing at all, explaining not a single finding with respect to aphasic comprehension.

Box 15.1   The trace deletion hypothesis

Many Broca's aphasics perform at chance on passives, such as (1), and above chance on actives, such as (2).

(1) The dog was chased by the cat.
(2) The cat chased the dog.

What is the trace deletion hypothesis (TDH) and how does it account for these basic facts? The TDH (Grodzinsky, 2000) proposes is that in a representation that is otherwise normal, traces (markers of syntactic movement) in Θ-positions are deleted. The loss of trace has serious repercussions: moved referential NPs thereby lose all thematic information borne by the trace. Broca's aphasics do not infer the missing role because the parser is encapsulated; the representation it outputs, even though incomplete, is final. Instead, Broca's aphasics resort to a default strategy. This strategy assigns a role to an NP by its *linear* position (e.g., first NP = Agent), if it does not already have a role. To see how this works, let us consider how it applies to (1) and (2).

In (1), the NP in the *by*-phrase, *the cat*, receives the Agent role. Since no movement is involved, it is not linked to a trace, and is thus assumed to be unproblematic. However, the NP, *the dog*, moves and so involves a trace. Since the trace it is linked to is deleted in a Broca's representation, role assignment is impossible via normal syntactic processes, so the NP receives no role. The default strategy operates on this incomplete output of the parser and assigns the Agent role to *the dog*. The result is that in the aphasic representation, both the *dog* and the *cat* are Agents. Thus, in a sentence–picture matching task, with two NPs competing for Agenthood, guessing is inevitable, and chance performance on passives is explained.

Above-chance performance on active sentences is accounted for by the default strategy compensating for the loss of trace. In (2), the object NP receives the role of Theme normally. The subject NP involves movement (out of the VP), so *the cat* receives no role. However, the default strategy assigns it the Agent role, which happens to be correct. Therefore, on a sentence–picture matching task, aphasic performance is expected to be normal.

Grodzinsky, Y. (2000). The neurology of syntax: Language use without Broca's area. *Behavioral and Brain Sciences, 23*, 1–71.

15.4.2. The TDH, the SSH, and Cross-Linguistic Word Order Variations

English is not the best language to test linear accounts, because word-order is relatively fixed. However, in other languages, word-order is much freer, and phrases can be moved to sentence–initial positions in ways that can reverse the typical order of thematic roles. It is possible, for example, in active sentences (that aphasics are normally good at) to have the Theme role occur first, and in passive sentences (that are

usually hard for aphasics) to have the Agent role occur first. In these circumstances, linear accounts, such as the TDH and the SSH (see Box 15.2), have an opportunity to see if linearity is actually going to do any work for them. Experiments with Broca's aphasics in Korean, Spanish, and Japanese are revealing in this respect, but as we will see, Broca's aphasics did not interpret non-canonically ordered actives or canonically ordered passives as the linear accounts require that they

Box 15.2   The slow syntax hypothesis

Let us consider the same sentences as in Box 15.1, and ask: what is the SSH and how does it explain chance performance on (1) and above chance performance on (2)?

(1) The dog was chased by the cat.
(2) The cat chased the dog.

The slow syntax hypothesis (SSH) (Piñango, 2000) proposes that comprehension failure in Broca's aphasics is due to a lack of alignment between the order of thematic roles in syntactic representation and the linear order of thematic roles in argument structure.

In an unimpaired language faculty, *semantic linking* establishes correspondence between thematic roles and *linear* positions. *Syntactic linking* guarantees that arguments and syntactic functions correspond. Syntactic linking is assumed to be faster. It occurs before and always trumps semantic linking.

Linear semantic linking provides the potential for interference with syntactic linking in Broca's aphasia. It is proposed that syntax, in Broca's aphasia, is slow, so it does not get deployed first and it does not vanquish semantic linking. From the perspective of sentence interpretation, if the output of the two linking mechanisms happens to coincide, interpretation should be normal, but if the output of the two mechanisms conflicts, then the consequence is that Broca's aphasics are forced to guess who is doing what to whom. Now, let us see how it works on (1) and (2).

In (1) and (2), the thematic roles associated with the verb *chase* are {Agent; Theme}. In (1), syntactic linking ensures that the NP, *the cat*, receives the Agent role and that the Theme role is assigned to *the dog*. Allegedly linear semantic linking assigns Agenthood to the first NP, *the dog*, and the Theme role to the second NP, *the cat*. The consequence is that the two linking mechanisms compete. The result is that, in Broca's aphasics, either role could be assigned to either NP. In a sentence–picture matching task, consequently, they resort to guessing who is chasing whom, and chance performance is explained.

In the active sentence (2), the alignment of both semantic and syntactic linking is identical, which explains why Broca's aphasics interpret the sentence correctly.

Piñango, M. (2000). Canonicity in Broca's sentence comprehension: The case of psychological verbs. In Y. Grodzinsky, L. Shapiro & D. Swinney (Eds.), *Language and the Brain* (pp. 327–350). New York: Academic Press.

should. Although the different languages have rather different properties and the specifics of the syntactic analyses vary, considering them all together suffices for present purposes.

Korean permits the object to move linearly before the subject, as in (2).

2.   saja-lul$_i$ key-ka$_k$ t$_k$ mul-eyo t$_i$

     lion-Acc   dog-Nom   bite-COMP

     "The dog bit the lion"

The lion is a left-dislocated object which leaves a trace in object position. The subject (*the dog*) moves from a VP-internal position to IP. Thus there are two movements of referential NPs. Spanish also allows left-dislocation of the object; again, coupled with subject movement, there are also two movements in an active sentence such as (3).

3.   [A la girafa]$_i$ [la mujer]$_k$ t$_k$ t$_i$ la   está empujando

     The giraffe   the woman   it   is   pushing

     "The woman is pushing the giraffe"

In Japanese, also, the same two movements apply in an active:

4.   Hanako$_i$-o   Taro$_k$-ga t$_k$ t$_i$ nagutta

     Hanako-Acc Taro-Nom   hit

     "Taro hit Hanako"

Hagiwara & Caplan (1990) found that Broca's aphasics performed at chance on sentences such as (4) in Japanese, and Beretta *et al.* (2001) found the same for the sentences in Korean (2) and Spanish (3). The SSH agrees with this result since semantic linking is at odds with syntactic linking. The TDH, by contrast, predicts below-chance performance, since both thematic roles have to be assigned by the default strategy according to their linear positions. This would have the consequence that the first NP receives the Agent role and the second NP receives the Theme role, the opposite of correct assignment, and inconsistent with the experimental result. The SSH survives, the TDH does not.

Passives in Korean, Spanish, and Japanese permit the by-phrase to appear in sentence–initial position. This has the effect that both NPs undergo movement in sentences such as (5–7).

5.   key-ekey   saja-ka   mul-hi-eyo

     dog-by   lion-Nom   bite-PASS-COMP

     "The lion is bitten by the dog"

6.   Por la mujer   la girafa está siendo empujada

     By the woman the giraffe is   being   pushed

     "The giraffe is being pushed by the woman"

7.   Taro-ga   Hanako-ni   nagu-rare-ta

     Taro-NOM Hanako-ACC hit-PASS-PAST

     "Taro was hit by Hanoko"

What are the experimental facts in this case? Broca's aphasics performed at chance levels on sentence–picture tasks (Hagiwara & Caplan, 1990; Beretta *et al.*, 2001). However, both the TDH and the SSH predict above-chance performance, contrary to fact. The TDH expects subjects to have no problem understanding sentences like (5), (6), and (7) because the default strategy assigns the Agent role to the first moved NP and Theme role to the second moved NP, which in all three cases happens to be the correct assignment that an unimpaired representation would yield. The strategy should have compensated perfectly for the failure of the syntactic parser in the impaired representation. But it did not.

The SSH fares no better. It also anticipates above-chance performance because syntactic linking establishes that the NP in the by-phrase receives the Agent role, while the subject NP receives the Theme role. Semantic linking requires that the Agent role be assigned to the first NP and the Theme role to the second NP. The two mechanisms thus agree with each other perfectly, so comprehension in Broca's aphasics ought to be fine, again, contrary to fact.

How have the linear models accommodated data regarding above sentences in which there are two dependencies? Far from recognizing that there is a serious problem with a linear strategy, the TDH now actually proposes *yet another strategy* in addition (Grodzinsky, 2006). The problem with the above sentences would go away if one of the dependencies could be eliminated, so the latest version of the TDH simply gets rid of the offending extra dependency relation by decree: a strategy, labeled *theta-bridging*, stipulates that Broca's aphasics do not need to bother with subject movement, thus overcoming an unwelcome intrusion by syntactic theory that subject movement out of VP does indeed occur.

Other than by proliferating strategies, linear models, given a chance to show what linearity can do for them, collapse like a house of cards as soon as they are confronted with cross-linguistic tests which re-order NPs either canonically or non-canonically in ways that go against the statistical tendencies.

15.4.3. Other Linear Models

I have not discussed other linear models here, but no substantive issue is thereby lost. For example, the model proposed by Avrutin (2006) is similar in most relevant respects to the TDH and the SSH, except that the linear order of Agent and Theme in this case is presented as a fact about discourses, rather than sentences, which is neither here nor there from our perspective since the epiphenomenon that provides the competition with syntactic processes is essentially the same.

15.5. A HIERARCHICAL MODEL: THE DOUBLE-DEPENDENCY HYPOTHESIS

Fortunately, not all models that care about linguistically interesting contrasts have adopted the appeal to linear trends to do the heavy lifting for them. Mauner *et al.* (1993) devised the double-dependency hypothesis (DDH) (see Box 15.3) with the expressed aim of dispensing with the TDH's linear default strategy. They succeeded in showing that disruption to certain kinds of referential dependencies can sort the core data (actives, passives, subject and object relatives in English) in the right ways. No strategies, no epiphenomena, just theoretical concepts. The DDH has been shown to be immune to word-order effects cross-linguistically, accounting for all of the above examples, and more, entirely

---

Box 15.3   The double-dependency hypothesis

Again, as in Boxes 15.1 and 15.2, the data that need to be explained are chance performance on (1) and above chance performance on (2).

(1) The dog was chased by the cat.
(2) The cat chased the dog.

What is the double-dependency hypothesis (DDH)? And how does it account for these facts? The DDH (Mauner *et al.* 1993), like the linear models, deals with random performance by positing a deficit that will present the Broca's aphasic with more than one alternative representations where the normal subject has just one. The DDH maintains that in a class of dependencies which are assigned only one thematic role, disruption between the two terms of the dependency occurs. For example, the dependency that obtains between a moved NP and its trace involves only one thematic role; in the normal case, both elements are coindexed; in Broca's aphasia, by contrast, the relation between the NP and the trace is disturbed, that is, the constraint on coindexation is lost. Let us see how it works on (1) and (2).

In (2), the two dependencies are $<[\text{the boy}]_i, t_i>$ and $<\text{ed}_j, [\text{the girl}]_j>$. The referential NP, *the boy* and the foot of its chain, the trace *t*, form one dependency. The other dependency is between the referential NP, *the girl*, and the foot of its chain, the passive morphology *–ed*. In Broca's aphasia, loss of the constraint on coindexation results in a larger grammar and thus an ambiguous interpretation which would permit both the correct coindexation already described and an anomalous coindexation, $<[\text{the boy}]_i, \text{en}_i>$ and $<t_j, [\text{the girl}]_j>$. Chance performance is thus explained.

In sentences where there are *two* relevant dependencies, it is unclear which NP is co-indexed with what. Since the co-indexation is ambiguous and since thematic role assignment is thereby also ambiguous, Broca's aphasics must guess who is doing what to whom. However, where there is only *one* such dependency, as there is in the active sentence (1), no ambiguity is possible, and therefore interpretation should be normal.

Mauner, G., Fromkin, V., & Cornell, T. (1993). Comprehension and acceptability judgments in agrammatism: Disruptions in the syntax of referential dependency. *Brain and Language*, 45, 340–370.

---

naturally. Its empirical coverage is wider than for the linear models (for a review, see Beretta, 2001). In view of this, it is noteworthy that the DDH has been studiously avoided by proponents of the linear models. To adapt Alfred Ayer's pleasingly disingenuous question, why should they mind being wrong when someone has shown them that they are?

That said, the SSH does have one very good reason for disdaining the DDH: it has nothing to say about timing, a clear and present limitation. The SSH, by contrast, has clearly recognized the importance of examining temporal issues (Burkhardt *et al.*, 2003). Unfortunately, in this model, the timing of syntax is slow relative to a supposedly linear semantic mechanism. No evidence, however, has ever been forthcoming that such a linear mechanism is part of an *un*impaired representation, so why should aphasics endorse it? Take away the linear mechanism and the SSH not only lacks a timing component, but ceases to be. Nevertheless, the concern with timing that the SSH represents (see also Avrutin, 2006) is, I believe, a positive development, and one I will return to below.

15.6. WHAT HAVE WE LEARNT?

If the ultimate goal of neurolinguistic inquiry is theory unification, then ignoring theory cannot contribute to that goal. The linear models do not entirely ignore linguistic theory, which is why I am discussing them, but they can make not a single prediction without the agency of a linear strategy that competes with syntactic processes in the sentences Broca's aphasics are observed not to understand but does not compete in those they do understand. There is no theory of strategies and linearity follows from nothing in linguistic theory. Without invoking linearity, the linguistic contrasts by themselves do not have the consequence that correctly and incorrectly interpreted sentences fall out either side of a divide. Without linearity, there is simply no divide in the TDH and SSH.

Independently of worries about their linear or hierarchical nature, has experimental work within *any* of the models yielded insight linking linguistic theory to brain theory? There have been claims that a certain kind of syntactic dependency relation, and perhaps it alone, is housed in the vast tract of brain known as Broca's area. However, claims of this sort have been undermined by a range of neuroimaging studies, which show that damage to Broca's area need not give rise to Broca's aphasia, that Broca's area has been implicated in semantic processing, working memory, and a range of nonlinguistic tasks (Rizzolatti & Arbib 1998; Kaan & Swaab, 2002).

But even if the claim could be sustained, as already discussed, location may not be doing much explanatory work. In my view, about all that can reasonably be claimed about the experimental evidence is that the patterns of sparing and loss in aphasia have shown that the brain makes quite a broad range of subtle distinctions that linguistic theory cares about. Note that this claim says nothing about brain theory. In truth, it says nothing that is not already implicit in linguistic theory, namely, that syntactic properties are brain properties. Studies of aphasia, it might be fair to say, have provided some evidence that helps to make explicit what was already implicit. Brains do syntax somewhere. It's a fact.

15.7. CHALLENGES AND FUTURE DIRECTIONS

Given that this rather measured assessment applies to much of my own work, either I'm having my nineteenth nervous breakdown or the assessment is about right. If the assessment is indeed quite accurate, then it suggests a different approach to the study of syntactic disorders will have to be adopted if, even in principle, we are going to have some hope of meeting the formidable challenges that confront the quest for eventual theory unification.

I mentioned timing in the last section. Let me now return to it. Why should timing be any less epiphenomenal than location might be? Just as function has to occur somewhere in x–y–z space, but where may not be explanatory, so too function must play out in time, but perhaps when may not be explanatory either. But here are some reasons to think that timing may provide a more profitable orientation for inquiry.

First, we have precise and fine-grained timing hypotheses about language processing, but no similarly precise and fine-grained hypotheses about location (Phillips, 2004). Also, timing, unlike location, intersects with theories of how brains compute. For example, timing in neuronal coding proposals, such as the reported synchronicity of firing in assemblies of neurons (Singer, 2000), has for a long time played a major explanatory role, which potentially rescues neurolinguistic inquiry from epiphenomenality – assuming, of course, that synchronicity turns out to be basically correct! Thirdly, we have tools at our disposal that record the synchronous firing of large populations of neurons, for example, magnetoencephalography (or MEG). In addition, MEG has been used to examine linguistically interesting contrasts in normal subjects (see, e.g., Phillips *et al.*, 2001; Beretta *et al.*, 2005; Pylkkänen *et al.*, 2006; Fiorentino & Poeppel, 2007). MEG might prove helpful in tracking the time course of relevant processes in impaired brains, relating them to normal processes.

MEG has not yet been used, so far as I know, to examine disorders of syntax, but EEG has. Friederici, von Cramon, & Kotz (1999) examined the neural responses of patients with different lesion sites to syntactic violations in sentences such as (8):

8.   *Das Eis   wurde im   gegessen
      The ice-cream was   in-the eaten

What they found for patients with lesions in the left anterior region was that they failed to show the normal ELAN (early left anterior negativity) waveform that is associated with early syntactic processing in healthy subjects, though the sentences did elicit a normal P600, a component that has been linked to later aspects of syntactic processing. This is quite intriguing when we consider the finding in the light of proposals, such as the SSH, that syntax is slowed in Broca's aphasia. These proposals may not account for aphasic syntactic comprehension patterns very compellingly, but the idea that syntactic processing is slowed down following brain damage agrees with the EEG finding and may be a good way of interpreting it.

If we could fast forward into the future, we might one day be able to say about the unification of language and brain, "so *that*'s how it turned out!" The answer may be like nothing currently conceived. But here we are on this bank and shoal of time, armed with the theories and tools we have now, trying to figure out how best to proceed. We know how *not* to proceed: ignore theory of language, ignore theories of brain, and posit atheoretical mechanisms in their place. No unification problem seems to have been solved that way. One quite reasonable way we might proceed is to postpone neurolinguistic inquiry until the relevant theories are substantially more advanced. However, given the neuroimaging and electrophysiological tools now at our disposal, another option that may commend itself to us is to attempt neurolinguistic inquiry now. If we choose the latter, the only hope is to remain faithful to the norms of naturalistic inquiry.

What role can disorders of syntax play? The locations of syntactic function revealed by aphasia are anatomically rather coarse. fMRI is a better measure in this respect. But, either way, it would be helpful to know if there is some reason to believe that location is illuminating. Aphasic performance on behavioral tasks may show that brains make the syntactic distinctions that our theory of language (a brain theory in its own right) assumes they must. This is something, perhaps, but not something that is very revealing about the relation between brain theory (as derived from the basic sciences) and linguistic theory. I have suggested that attention to timing may be informative, because we have quite detailed timing hypotheses with respect to language, we have theories of neuronal coding that care about timing, and we have tools with millisecond resolution. This suggests that it would be useful to develop a clearer understanding of the timing of linguistic processes in healthy subjects. Aphasic processes are fast or slow relative only to the normal case, so experiment using aphasic subjects may be envisaged as complementary and auxiliary to experiment with healthy subjects.

References

Avrutin, S. (2006). Weak syntax. In Y. Grodzinsky & K. Amunts (Eds.), *Broca's region* (pp. 49–62). Oxford: Oxford University Press.

Beretta, A. (2001). Linear and structural accounts of theta-role assignment in agrammatic aphasia. *Aphasiology, 15*, 515–531.

Beretta, A., & Munn, A. (1998). Double-Agents and trace-deletion in agrammatism. *Brain and Language, 65,* 404–421.

Beretta, A., Fiorentino, R., & Poeppel, D. (2005). The effects of homonymy and polysemy on lexical access: An MEG study. *Cognitive Brain Research, 24,* 57–65.

Beretta, A., Schmitt, C., Halliwell, J., Munn, A., Cuetos, F., & Kim, S. (2001). The effects of scrambling on Spanish and Korean agrammatic interpretation: Why linear models fail and structural models survive. *Brain and Language, 79,* 407–425.

Burkhardt, P., Piñango, M.M., & Wong, K. (2003). The role of the anterior left hemisphere in real-time sentence comprehension: Evidence from split transitivity. *Brain and Language, 86,* 9–22.

Caplan, D. & Futter, C. (1986). Assignment of thematic roles by an agrammatic aphasic patient. *Brain and Language, 27,* 117–135.

Feynman, R.P. (1985). *QED: The strange theory of light and matter.* Princeton, NJ: Princeton University Press.

Fiorentino, R., & Poeppel, D. (2007). Compound words and structure in the lexicon. *Language and Cognitive Processes, 22.*

Friederici, A.D., von Cramon, D.Y., & Kotz, S.A. (1999). Language related brain potentials in patients with cortical and subcortical left hemisphere lesions. *Brain, 122,* 1033–1047.

Grodzinsky, Y. (2000). The neurology of syntax: Language use without Broca's area. *Behavioral and Brain Sciences, 23,* 1–71.

Grodzinsky, Y. (2006). A blueprint for a brain map of syntax. In Y. Grodzinsky & K. Amunts (Eds.), *Broca's region* (pp. 83–107). Oxford: Oxford University Press.

Hagiwara, H., & Caplan, D. (1990). Syntactic comprehension in Japanese aphasics: Effects of category and thematic role order. *Brain and Language, 38,* 159–170.

Kaan, E., & Swaab, T. (2002). The brain circuitry of syntactic comprehension. *Trends in Cognitive Sciences, 8,* 350–356.

Linebarger, M. (1995). Agrammatism as evidence about grammar. *Brain and Language, 50,* 52–91.

Mauner, G., Fromkin, V., & Cornell, T. (1993). Comprehension and acceptability judgments in agrammatism: Disruptions in the syntax of referential dependency. *Brain and Language, 45,* 340–370.

McGinn, C. (1993). Oxford: Blackwell *Problems in philosophy: The limits of inquiry.*

Piñango, M. (2000). Canonicity in Broca's sentence comprehension: The case of psychological verbs. In Y. Grodzinsky, L. Shapiro, & D. Swinney (Eds.), *Language and the Brain* (pp. 327–350). New York: Academic Press.

Phillips, C. (2004). Linguistics and linking problems. In M. Rice & S. Warren (Eds.), *Developmental language disorders: From phenotypes to etiologies* (pp. 241–287). Mahwah, NJ: Erlbaum.

Phillips, C., Pellathy, T., Marantz, A., Yellin, E., Wexler, K., Poeppel, D., McGinnis, M., & Roberts, T. (2001). Auditory cortex accesses phonological categories: An MEG mismatch study. *Journal of Cognitive Neuroscience, 12,* 1038–1055.

Pylkkänen, L., Llinas, R., & Murphy, G. (2006). Representation of polysemy: MEG evidence. *Journal of Cognitive Neuroscience, 18,* 1–13.

Rizzolatti, G., & Arbib, M.A. (1998). Language within our grasp. *Trends in Neuroscience, 21,* 188–194.

Singer, W. (2000). Response synchronization, a neuronal code for relatedness. In J.J. Bolhuis (Ed.), *Brain, perception, memory: Advances in cognitive neuroscience* (pp. 35–48). Oxford: Oxford University Press.

Further Readings

The following four readings are, in my view, must-reads for those who are interested in neurolinguistics. They are all engagingly written and deeply thoughtful. Any inquiry we may undertake will likely benefit from carefully attending to what they have to say.

Chomsky, N. (1995). Language and nature. *Mind*, 104, 1-61.
This is a classic paper, in two parts. The first part is absolutely fundamental for anyone trying to think seriously about the relation of linguistics and brain science. Chomsky has written about this topic quite often over the years. The dude abides.

Marantz, A. (2005). Generative linguistics within the cognitive neuroscience of language. *The Linguistic Review*, 22, 429–455.
This book does not address the integration of linguistics and brain science, but states the case for generative linguistics taking center stage in any cognitive neuroscience of language. Linguistic theory guides inquiry as to how the brain stores and produces linguistic representations; neurolinguistic studies provide additional sources of data for linguistic theory.

Poeppel, D., & Embick, D. (2005). Defining the relation between linguistics and neuroscience. In: A. Cutler (Eds.), *Twenty-first century psycholinguistics: Four cornerstones*. Mahwah, NJ: Erlbaum.
This book chapter provides a well thought-out discussion of the possible relation between linguistics and brain science. The authors describe the 'standard research program in neurolinguistics' and observe that it is of limited theoretical interest. They go on to outline an approach that draws on computational theories of language and brain.

Priestley, J. (1777). *Disquisitions relating to matter and spiritTail*. Kessinger.
Don't be put off by a reference to an 18th century chemist in a handbook of neurolinguistics that is intended to be cutting edge. Most people, I think, start off thinking about language ad brain from a common sense perspective. So did Priestley. This book follows his journey from that perspective to its total rejection in favor of a view of biology that includes the mental. I liked to think of Priestley riding shotgun for me as I was writing this chapter.

16

The Neural Bases of Text and Discourse Processing

CHARLES A. PERFETTI and GWEN A. FRISHKOFF

Learning Research and Development Center, University of Pittsburgh, Pittsburgh, PA, USA

ABSTRACT

Understanding discourse requires the comprehension of individual words and sentences, as well as integration across sentence representations to form a coherent understanding of the discourse as a whole. The processes that achieve this coherence involve a dynamic interplay between mental representations built on the current sentence, the prior discourse context, and the comprehender's background (world) knowledge. In this chapter, we outline the cognitive and linguistic processes that support discourse comprehension and explore the functional neuroanatomy of text and discourse processing. Our review suggests an emerging picture of the neurocognition of discourse comprehension that involves an extended language processing network, including left dorsal and ventral frontal regions, left temporal cortex, medial frontal cortex, and posterior cingulate. While convergent evidence points to the importance of left frontal and temporal networks in discourse processing, the role of right hemisphere networks is less clear.

16.1. INTRODUCTION

In text and discourse processing, a central idea is that of *coherence* – meaningful links that make a discourse "hang together" between adjacent sentences (*local coherence*) and across larger units (*global coherence*). The goal of this chapter is to describe how the brain supports cognitive and linguistic processes that help establish discourse coherence. We begin by asserting the obvious: text and discourse comprehension engage neural systems that are implicated in language perception (auditory or visual language input), word processes, and sentence comprehension. In addition, when directly compared with word- and sentence-level comprehension, discourse comprehension appears to recruit other areas, including left prefrontal cortex (PFC), anterior temporal regions,

medial frontal cortex, and the posterior cingulate. These regions have been related to general cognitive mechanisms (e.g., attention, memory) that are necessary for the retrieval and maintenance of mental representations across time, as well as to language-specific devices for linking meanings within and across sentences.

While the evidence reviewed here leads to convergent findings, it also suggests some current controversies and areas where further work is needed. One such area concerns hemispheric asymmetries in text and discourse comprehension. While studies of patients with right hemisphere damage have been taken to suggest a unique role for the right hemisphere in discourse comprehension, recent work suggests a need for more refined theories and additional studies, to reconcile the current body of evidence on the left versus right hemisphere contributions to discourse processing.

16.2. COGNITIVE AND LINGUISTIC PRINCIPLES OF DISCOURSE PROCESSING

In this section we review major ideas from the psycholinguistics of text and discourse comprehension. This prior work suggests that comprehenders strive to build coherent representations, or what are called *mental models*, during text and discourse processing. Mental models are built from *propositions* – the "idea units" of language. The challenge for a neural theory of discourse comprehension is to identify the neurocognitive and neurolinguistic mechanisms that serve to link together successive propositions. In what follows we discuss two types of mechanisms, inferential processes and discourse-grammatical cues. Together, these processes help to establish text and discourse coherence. In following sections, we discuss the neural underpinnings of these processes.

16.2.1. Mental Models in Text and Discourse Comprehension

Text researchers use the term proposition to refer to the basic semantic units – the "idea units" – of a text or discourse. In effect, propositions represent the core ideas expressed in a sentence or clause – an action, event, or state of affairs involving one or more participants (e.g., "Jack slept," "John kissed Mary," "The ball is round"). In "Harry let Fido out," the meaning of the verb (let ... out) entails that Harry carried out some action that resulted in Fido changing locations from inside (some place) to outside. The fact that we tend to interpret this sentence as Harry let the dog out of the house illustrates the role of background knowledge: "Fido" is the name of a dog, and "letting out" describes a common event in a household with pets. Thus, a proposition encodes basic relational meanings, partly independent of syntactic expression, while the proposition plus relevant knowledge yields a specific meaning or interpretation.

Establishing coherence across sequences of propositions involves additional processes that extend beyond the single sentence. Words become linked to referents introduced in prior text, or established through cultural transmission of knowledge (e.g., that "Fido" is the name of a dog). As these links are made, the comprehender builds a representation of what the text is about, a mental model.

In fact, according to the influential model developed by Kintsch and van Dijk (see Kintsch and Rawson, 2005), comprehension involves not one mental model, but two. (1) A model of what the text says (the text base, consisting of ordered propositions) and (2) a model of what the text is about (the situation model). The propositional structures of the text base are extracted from sentences, accumulate across successive sentences, and are supplemented by inferences necessary to make the text locally coherent. The situation model is formed from the text base by combining knowledge sources through additional inference processes. A text base thus amounts to a representation of meaning that is close to the language of the text, essentially amodal and propositional. In contrast, a situation model comprises nonpropositional and nonverbal information, and may include modality specific (e.g., visual–spatial), as well as semantic representations (Mellet et al., 2002). Explaining the nature of these representations, how they are formed, and how they are maintained and integrated during online comprehension is central to theories of text and discourse comprehension.

16.2.2. Grammatical Markers of Discourse Coherence

The text processing view treats discourse as linguistic input to be understood by an individual reader. A complementary view, grounded in linguistic insights, emphasizes the socio-pragmatic nature of discourse, and proposes that a key function of grammatical systems is to support alignment of speaker/hearer representations during communication (Givón, 2005). According to this framework, the linguistic structures that support communication operate as socio-pragmatic cues. For example, a pronoun ("he," "she," "it," and so forth) cues the comprehender to link a previously mentioned referent (John, Mary, the ball). To use these cues appropriately, a speaker (or writer) must consider not only the propositional information to be encoded, but also the knowledge and intentional states of the comprehender. For example, referring to John's daughter has a different cueing effect depending on whether the comprehender knows John and, in particular, whether he knows that John has a daughter. If not, then referring to John's daughter out of the blue can lead to a breakdown in communication – from the comprehender's viewpoint, a break in coherence.

The text perspective and the discourse-grammatical perspective converge to identify coherence as a key issue in language comprehension. Functionalist accounts of discourse-grammatical structures describe the linguistic mechanisms that serve communication through coherence. Many of these mechanisms operate at the level of local coherence, preserving stretches of conversation (and text reading) from coherence breakdowns. These mechanisms must operate in close concert with cognitive (attentional, working memory) and socio-emotional processes that are relevant for communication (see Box 16.1). Theories of text comprehension, in turn, provide complementary insights on how inferences can function to help establish global, as well as local, coherence.

16.2.3. Inferencing and Coherence

Prior work (e.g., Gernsbacher & Robertson, 2002) has identified various types of coherence links, including those that establish continuity of the discourse topic or theme (referential coherence), event time and location (temporal and spatial coherence), and causal or intentional relationships between events. Here we focus on referential coherence to illustrate some general principles.

Building on our previous example, consider the following text sample (from Sanford & Garrod, 2005):

1. In the morning Harry$_1$ let out his dog Fido$_2$.
2. In the evening he$_1$ returned to find a starving beast$_2$.

Note the use of a pronoun "he" in sentence (2): this expression is typically understood to refer to the same real-world entity as the name "Harry" in sentence (1), a phenomenon called coreference. A variety of anaphoric devices (definite articles, pronouns) and deictic expressions ("this" or "that") signal coreference in English. These devices are part of a grammatical system that provides instructions

Box 16.1   Individual differences in working memory and discourse: evidence from ERPs

Reading researchers have long suspected that discourse comprehension is tied to working memory (WM). The challenge in comprehending text is precisely that of activating relevant information at the appropriate time, storing information in short-term memory, and reactivating information as needed to support referential links across clause and sentence boundaries. In support of this view, Ericsson and Kintsch (1995) cite studies that suggest "reading span" (a measure developed by Daneman & Carpenter) predicts text comprehension skill, even after controlling for other reading and language skills.

In ERP studies, researchers have identified a pattern known as the "left anterior negativity" (LAN) that is active during sentence and discourse comprehension and that varies with WM skill (King & Kutas, 1995). The LAN occurs relatively early (t150–300 ms) and is strongest over left anterior electrodes, consistent with early activation of the left PFC. Interestingly, the LAN responds not only to variations in syntax that may affect WM, but also to cues that can affect memory strategies, even when syntactic structure is held constant. For example, Münte *et al.* (1998) examined variations in LAN in two conditions where the sentence structures had identical syntax, but differed in the temporal (referential) links between successive clauses.

1. *After* we submitted the article, the journal changed its policy.
2. *Before* we submitted the article, the journal changed its policy.

In (1), the initial word (*after*) cues the comprehender that the order of the two events is the same as the order of the two clauses that encode these events. In (2), the intial word (*before*) indicates that the temporal order will be different from the surface order. Interestingly, subjects with high WM scores showed a greater LAN in (2). This evidence may point to more effective memory strategies among high comprehenders.

Ericsson, K.A., & Kintsch, W. (1995). Long-term working memory. *Psychological Review, 102*(2), 211–245.

King, J., & Kutas, M. (1995). Who did what and when? Using word- and clause-level ERPs to monitor working memory usage in reading. *Journal of Cognitive Neuroscience, 7*, 376–395.

Münte, T.F., Schiltz, K., & Kutas, M. (1998). When temporal terms belie conceptual order. *Nature, 395*(6697), 71–73.

for how to make a discourse locally coherent. In the words of Givón (2005), the key function of grammar is to provide an "*automated* discourse processing strategy" (italics added).

A second instance of coreference in this same example illustrates *backward, or bridging, inference*. The first noun phrase (NP), "a starving beast," in sentence (2) is understood to refer to the same real-world entity as "his dog Fido" in sentence (1). Note that the second NP is marked by an indefinite article (a). This contrasts with the usual practice of using the definite article to signal "old" or "given" (i.e., previously mentioned) information. The nonstandard use of "a" in this context cues the comprehender to draw an inference: Fido was not starving in the morning when Harry left. This further allows the inference that Fido had no food during the time that Harry was gone. In this case, the coherence device is the use of two different NPs that must be made coreferential for the text to be coherent.

Finally, whereas backward inferences are often obligatory, *forward or predictive inferences* are strictly optional and can be costly to processing resources. They may not be made except when compelled by a need for either textual or causal coherence (for reviews, see Beeman *et al.*, 2000; Perfetti *et al.*, 2005).

16.2.4. Summary

Readers strive to develop some degree of coherence in the meaning they derive from a text. To do this, they establish links within and across sentences, using grammatical cues and drawing various kinds of inferences. Grammatical devices cue relatively automatic processes that help to establish coherence links, but such links also can be established through inferences, which engage additional processes that depend on the comprehender's standard for coherence, cognitive capacity, and language skills.

Understanding how these multiple processes are coordinated in real time during text and discourse comprehension requires an explicit theory of cognitive and neural mechanisms – the focus of remaining sections in this chapter.

16.3. THE NEUROSCIENCE OF TEXT AND DISCOURSE COMPREHENSION

Recent reviews (Mar, 2004; Ferstl, 2007) attest to the growing interest in the neural basis of text and discourse comprehension. Our discussion, which benefits from these prior reviews, will conclude that language comprehension involves a left-lateralized network of brain areas with limited and task-specific involvement of the more anterior, dorsal and ventral, and prefrontal areas. The controversial role of right versus left hemisphere contributions will be addressed in Section 16.4.

16.3.1. The Role of the Temporal Lobes in Discourse Comprehension

To discover what is special about text and discourse processing, it is important to consider direct comparisons of brain activation elicited by discourse with activation to isolated words, sequences of (unconnected) words, and (unconnected) sentences. A number of imaging studies have included such comparisons. When connected discourse or isolated (unconnected) sentences are compared with word lists, the anterior temporal lobes show greater activation (Mazoyer *et al.*, 1993). Given prior research linking anterior temporal lobes to semantic comprehension (see Chapter 5), increased activity in anterior temporal lobes may reflect added demands for semantic processing in comprehending connected text.

Additional evidence for the role of anterior temporal lobes in discourse-level semantic processing has come from electromagnetic (event-related potential, ERP and magnetoencephalography, MEG) studies. The N400 component (and its MEG counterpart, the mN400), has consistently been linked to neural sources in the anterior temporal lobe (see Van Petten & Luka, 2006 for a recent review). Interestingly, the N400 response has been found to vary with demands on sentence- and text-level integration, as well as word-level semantic comprehension. For example, in a recent study, Hagoort *et al.* (2004) presented sentences, such as *The Dutch trains are yellow/white/sour and very crowded*. Dutch subjects know very well the famous yellow trains of the Netherlands, so when the word *white* appears they know the sentence is false. The N400 elicited by the pragmatic anomaly (*white*) was indistinguishable in latency and distribution from the N400 to the semantically anomalous ending, *sour*. When they presented the same materials in an functional magnetic resonance imaging (fMRI) task, Hagoort *et al.* (2004) found that *sour* and *white* both produced increased activation in the left inferior PFC and near areas associated with semantic processing, including the left temporal lobe (for an illustration see also Box 17.1). Although more work is needed on whether the brain honors the distinction between semantics and pragmatics, this study at least suggests that it is the comprehender's knowledge, whether based on what is true or what is sensible, that is reflected in N400 measures of semantic integration. Likewise, this may suggest that the increased temporal lobe activity in processing connected discourse, versus unconnected sentences, reflects a difference in degree, rather than one of kind.

By combining results from recent fMRI and ERP studies, we can conclude that the anterior temporal lobes are important in sentence- and text-level semantic integration. This leads back to our original question: What, if anything, is unique to text-level processing?

16.3.2. The Role of PFC in Discourse Comprehension

The difference between text and sentence processing comes to this: in reading text, information must be integrated across sentence boundaries to maintain coherence.

Studies that compare connected discourse with sentences that lack global coherence (Mazoyer *et al.*, 1993) suggest that one locus for routine integration processes that are supported by coreference (e.g., based on argument overlap) is the superior dorsomedial prefrontal region (BA 8–9). This region also appears to be involved when integration demands long reaches for knowledge. For example, Ferstl and von Cramon (2001) had subjects read sentence pairs that lacked explicit overlap to support integration:

1. *The lights* have been on since last night. *The car* doesn't start.

2. Sometimes *a truck* drives by the house. *The car* doesn't start.

When sentences could be linked through a backward inference, as in (1), activation was greater in the superior dorsomedial prefrontal region and posterior cingulate cortex. When cohesive ties were added to the second sentence to suggest a link to the first, for example, *that's why the car doesn't start*, increased activation was observed in left PFC for the unrelated case (2) but not the related cases (1). Ferstl and von Cramon (2001) suggest the activation in the unrelated case with the "why" phrase added reflects additional processing required to reconcile the linguistic information in favor of integration (this is why) with the pragmatic understanding that the car's starting and the truck's passing are unrelated.

An fMRI study reported in Schmalhofer and Perfetti (2007) provides additional support for a frontal-medial response during inferencing. Adapting the materials and procedure of the ERP study by Yang *et al.* (2007) (see Box 16.2), this study had subjects read two-sentence passages that varied the ease of integration between a target word and the information from a preceding sentence. When integration was possible only by making a predictive inference in the first sentence, higher activation was observed in superior dorsomedial PFC (BA 8–9). Additional activation in left (but not right) ventral PFC was linked to verification judgments that were performed later in the task. In this part of the task, subjects were asked to decide whether an event (e.g., wine spilled) was or was not implied by the previous two sentences. For example, when the first sentence referred to turbulence during a flight while wine was being served, *wine spilled* typically elicited a "yes" response. During the verification judgment, the inference condition produced additional activation in left (but not right) inferior frontal gyrus, suggesting additional processing during the judgment, and implying that the predictive inference (that wine spilled) was not explicitly made during the reading of the first sentence. These results replicate findings from the behavioral literature,

Box 16.2   Individual differences in inferencing: evidence from ERPs

Studies of text comprehension have consistently found individual differences in readers' ability to make coherence links. A recent ERP study by Yang *et al.* (2007) examined different types of word-to-text integration among strong and weak readers. The main interest was in the N400 response to the second mention of a referent (e.g., *explosion*) in relation to the first mention in a previous sentence. There were three conditions which differed in the extent to which the first sentence established a referent that could be accessed at the word *explosion* in sentence 2 (see Table).

The amplitude of the N400 to the critical word (*explosion*) was reduced when the previous sentence had referred to an explosion (referentially explicit condition). It was also reduced when the previous sentence had referred to explosion using different words (referentially paraphrased condition). Importantly, there was no reliable N400 reduction when the critical word was related to the prior sentence only by inference (inference condition). There was, however, substantial variability among subjects in this condition, consistent with individual differences in the tendency to make forward inferences (see Box 16.3). Thus, explicit and meaning-based paraphrase relations patterned together in reducing the N400, whereas a process that depended on the situation model to generate an inference did not. Adults classified as low skill in comprehension showed weaker and slower (sluggish) integration effects, especially for the paraphrase condition. Thus, while these results do not suggest that making inferences is necessarily a problem for low

comprehenders (but see Box 16.3), they suggest that making word-based meaning connections is likely to contribute to difficulty in text comprehension.

Yang, C.-L., Perfetti, C.A., & Schmalhofer, F. (2007). Event-related potential indicators of text integration across sentence boundaries. *Journal of Experimental Psychology: Learning, Memory, and Cognition, 33*(1), 55–89.

Experiment Conditions in Yang *et al.* (2007)

Condition	Sample passage
Explicit	After being dropped from the plane, the bomb hit the ground and *exploded*. The *explosion* was quickly reported to the commander.
Paraphrased	After being dropped from the plane, the bomb hit the ground and *blew up*. The *explosion* was quickly reported to the commander.
Inference	After being dropped from the plane, the bomb hit the ground. The *explosion* was quickly reported to the commander.
Unrelated	Once the bomb was stored safely on the ground, the plane dropped off its passengers and left. The *explosion* was quickly reported to the commander.

Box 16.3   The role of working memory in forward inferencing

Working memory plays an integral role in text and discourse comprehension, and several influential theories have been advanced to explain how inferential processes may be linked to working memory capacity (see Sanford & Garrod, 2005, for a recent review). In an EEG study, St. George *et al.* (1997) found that individual differences in working memory skill affect EEG measures under conditions that promote forward inferencing (see Box 16.2). Probe sentences were presented (e.g., *the turkey burned*) representing forward inferences that could be drawn from a previous text (e.g., *she forgot about the turkey in the oven*). Whereas readers with high working memory capacity showed N400 effects in this forward inference condition, readers with low working memory capacity did not. These results are

consistent with the idea that forward inferences are optional and are made only under certain circumstances that vary with texts and reader dispositions. These findings support prior work that has emphasized individual differences in text and comprehension (see Perfetti *et al.*, 2005).

Perfetti, C., Landi, N., & Oakhill, J. (2005). The acquisition of reading comprehension skill. In M.J. Snowling & C. Hulme (Eds.), *The science of reading: A handbook* (pp. 227–247). Oxford: Blackwell.
Sanford, A.J., & Garrod, S.C. (2005). Memory-based approaches and beyond. *Discourse Processes, 39*(2–3), 205–224.
St. George, M., Mannes, S., & Hoffman, J.E. (1997). Individual differences in inference generation: An ERP analysis. *Journal of Cognitive Neuroscience, 9*(6), 776–787.

and suggest that left ventral and dorsomedial PFC may be important in certain inferencing processes.

In reviewing the fMRI research on inferencing, Ferstl (2007) concludes that there is clear evidence for a contribution of dorsal medial PFC to the process of establishing coherence from sentences; across studies, however, the results are variable, and further work is needed to map this

activity to specific cognitive and linguistic processes that support discourse coherence.

ERP research has provided additional evidence for engagement of prefrontal processes in discourse comprehension. In addition to the N400, which is associated with semantic processes, an earlier negativity, more frontal in its distribution, may be associated with referential processes in both written and spoken language comprehension (Van Berkum *et al.*, 2003). Van Berkum *et al.* (2003) found that when the referent of a NP was ambiguous (e.g., "the girl" when the discourse had previously introduced two different girls), there was an early negativity (peak at 300–400 ms) with a frontal distribution that was distinct from the N400, where central and parietal recording areas tend to be most affected by meaning congruence. An open question is how this response may be related to the left anterior negativity (or LAN), which has been linked to early syntactic processing (see Box 16.1; Chapter 9) and to the medial frontal negativity, identified in prior work on word- and sentence-level semantic comprehension (see Chapter 5).

In summary, what seems remarkable (but probably should not be) is that text processing shows the pattern of activation observed in sentence processing. Both tasks involve temporal lobe and prefrontal (inferior frontal gyrus) activation. In addition, both tasks recruit dorsal PFC in response to task demands (Ferstl, 2007). An apt conclusion from Ferstl is that "… in the absence of an overt, demanding comprehension task, language processing in context proceeds with surprisingly little brain power" (Ferstl, 2007, p. 66).

16.4. RIGHT HEMISPHERE CONTRIBUTIONS TO DISCOURSE COMPREHENSION

Research based on lesion studies has implicated right-cortical regions in certain discourse-pragmatic comprehension tasks (for a review see Chapters 17 and 28). By contrast, our review suggests that both sentence and text processing are generally bilateral and left dominant. On the other hand, a few studies have reported larger right lateralized in discourse comprehension (Robertson *et al.*, 2000). In this section, we review evidence for and against enhanced right hemisphere activity in discourse processing.

16.4.1. Evidence on Right Hemisphere Contributions to Inferencing

Although both the right hemisphere discourse functions and their neural anatomy remain somewhat unclear, some studies find evidence for right hemisphere involvement in inference making. Beeman *et al.* (2000), for example, used a divided visual field paradigm and had subjects name words related to possible inferences. The authors observed priming of words related to forward (predictive) inferences that was restricted to the left visual field and thus the right hemisphere. By contrast, activation of backward (bridging) information immediately after the "coherence break" – when new information was presented that required a bridging inference – was greater in the right visual field (left hemisphere).

Robertson *et al.* (2000) provide additional support for right hemisphere contributions to discourse coherence. In an fMRI study, they presented lists of sentences that contained exclusively either indefinite or definite NPs (a/the child played in the backyard; a/the mother talked on the telephone). The indefinite NP condition produced more activation in anterior cingulate cortex and the left inferior frontal gyrus. The definite NP case produced more activation in right inferior frontal gyrus. This result was interpreted as consistent with the hypothesis that the right hemisphere contributes specifically to integration processes. However, note that lists of sentences with definite articles and no argument (NP) overlap are still lists, and thus require backward integration on every sentence to establish local coherence. Thus, if backward inferences were made in the definite NP condition, then activation of the right hemisphere in this study would be inconsistent with the findings of Beeman *et al.* (2000), who observed priming of backward inferences only in the right visual field (left hemisphere).

Processes that work on either coherent or incoherent texts are serving the construction of a coherent situation model, and right hemisphere temporal and prefrontal areas may be involved in some circumstances (e.g., Ferstl, 2007). On the other hand, there is evidence that posterior cingulate cortex, left anterior temporal, and prefrontal areas are involved in different aspects of the construction job. Maguire *et al.* (1999) used pictures to guide coherence for otherwise incoherent stories and compared the results with easily comprehensible stories. They concluded that the posterior cingulate cortex was especially involved in the successful construction of a situation model. Cingulate activation was higher during the second presentation of a story (which should facilitate the situation model) and when a picture had been present for an incoherent story. Further, cingulate activation was correlated with comprehensibility ratings. In addition, Maguire *et al.* (1999) identified the left anterior temporal lobe with processing of incoherent versus coherent stories; activation was also correlated with recall. They also identified two prefrontal areas: the anterior lateral PFC was associated with recall and with the second presentation of a story, while a ventral medial PFC area (BA 11) was more active in coherent than incoherent (and no-picture) stories and was associated with comprehensibility ratings. This study provides a fuller picture of the integrative and memory processes that must occur in text comprehension as a function of the obstacles imposed on building coherence. However, it reveals no special right hemisphere involvement.

16.4.2. A Special Role for Right Hemisphere Processing in Global Coherence?

As discussed in Section 16.2, text research has been concerned not only with the processes that establish local coherence, but also with the processes that sustain global inferences and support the situation model. According to one view, it may be precisely at the level of global integration that we should expect to see evidence for enhanced right hemisphere processing in discourse comprehension (Beeman *et al.*, 2000; Robertson *et al.*, 2000).

Causal inferences provide one means of effecting global coherence in text comprehension. Mason and Just (2004) correlated behavioral results with imaging data, varying the causal distance between two sentences. For example, the target sentence *The next day his body was covered with bruises* followed one of these three sentences:

1. Joey's brother punched him.
2. Joey's mother got angry at him.
3. Joey went to a friend's house.

Sentence (1) provides a close causal connection for the target sentence, whereas (3) can be causally linked to the target only through a chain of inferences. Sentence (2) is intermediate, requiring a plausible causal inference. Consistent with Ferstl and von Cramon (2001), Mason and Just found that lateral prefrontal activation increased slightly with reduced coherence, but they also found that right frontal–temporal regions were most active when the target was preceded by the intermediately related sentence (2). This increased activation may reflect the memory demands of the successful bridging inference made in the intermediate condition, as suggested by Mason and Just (2004); but see Ferstl (2007) for a different interpretation based on the coarse coding hypothesis.

Although global coherence depends on inferences of various kinds, it depends more fundamentally on referential and coreferential binding, a process that inferences support. Syntactically well-formed sentences can be linked together so as to resist referential binding, producing the effect of a vague and hard to comprehend text. A title for such a text can help support global coherence. St. George *et al.* (1999) found that vague texts presented with titles produced left-lateralized temporal activation, whereas untitled texts produced bilateral temporal activation. This right hemisphere involvement in reading incoherent texts differed from the results of Robertson *et al.* (2000). Not only are the right hemisphere areas different in the two studies (inferior and middle temporal versus lateral PFC) – the direction of the coherence effect was different. The prefrontal area identified by Robertson *et al.* (2000) was associated with easy coherence (sentences linked through definite articles), whereas the right temporal areas of St. George *et al.* were associated with incoherence. Thus, right temporal activation may reflect support for more difficult processing, rather than reflecting a routine role in building coherence. If the reader is working to integrate information across sentences, additional memory resources may be required when this task is possible but difficult (Mason & Just, 2004). It is therefore important to control the difficulty across task conditions in order to link right hemisphere activations specifically with processes related to coherence making.

16.4.3. Right Hemisphere Involvement in Processing Nonliteral and Emotive Discourse

Stories often reflect multiple levels of meaning, for example a literal and metaphorical level or a literal and a moral of the story level. A general conclusion from imaging results is that acquiring these nonliteral meanings involves right hemisphere functions. An often-cited study using Aesop's fables (Nichelli *et al.*, 1995) concluded that the right hemisphere is "where the brain appreciates the moral of a story." The authors observed that the right inferior frontal gyrus and the right anterior temporal lobe were more activated when subjects judged whether an animal character represented a story moral, than whether the character had some specific physical feature.

However, there is reason to question the broader generalization that comprehension of nonliteral meaning generally relies on right hemisphere processing. Rapp *et al.* (2004) presented literal and metaphorical sentences for judgments of emotional valence. For example, the lover's words are harp sounds is a metaphor that would produce a positive valence rating. In comparisons between literal and metaphoric sentences, Rapp *et al.* (2004) found only three areas of greater activation for metaphors, and all were left lateralized: inferior frontal gyrus, and both anterior and inferior temporal lobe areas. Task differences may explain the discrepancy between this study and those that find right hemisphere involvement in nonliteral meaning. However, an interesting theoretical possibility suggested by the Rapp *et al.* (2004) results is that similar left hemisphere semantic processes are engaged across literal and nonliteral meanings and that metaphors require more of those semantic processes in the absence of supporting context.

Furthermore, recent studies have reported activation of the dorsal medial PFC during moral judgments (Greene *et al.*, 2001; Moll *et al.*, 2002) and ethical judgments (Heekeren *et al.*, 2003). Similarly, in research on understanding humor, early conclusions that the right hemisphere has a special role for jokes (Brownell & Martino, 1998) are qualified by later studies. Goel and Dolan (2001), for example, report increased activation in the left inferior frontal gyrus and in bilateral posterior temporal cortex (both inferior and middle) during the reading of jokes relative to humorless continuations of the same sentence.

Given such results, a tentative generalization is that the processing of nonliteral themes is supported by an extended left-lateralized language network that includes frontal areas. This network is also engaged in other more literal language tasks, including those that involve inference and evaluation, and may be supplemented by right hemisphere temporal and frontal areas under some circumstances. However, the

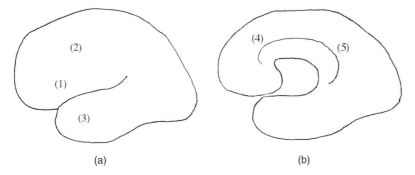

(a)   (b)

FIGURE 16.1 (a) Left lateral and (b) medial views of the brain, showing major areas that are differentially active during discourse versus word and sentence comprehension. (1) ventral PFC (inferior frontal gyrus); (2) dorsolateral PFC; (3) anterior temporal cortex; (4) anterior dorsomedial cortex (including anterior cingulate); and (5) posterior medial cortex, including posterior cingulate and precuneus.

precise role of the right hemisphere in inferencing and evaluation remains to be determined.

16.5. SUMMARY AND CONCLUSIONS

We began by observing that progress in mapping brain functions to processes in discourse comprehension requires attention to what has been learned in cognitive and psycholinguistic research. To date, this body of research suggests that text understanding, rich in implied as well as literal meaning, does not replace the comprehension of words and sentences, but builds on it. According to this view, there is a distinction between a text and the situation it describes; at the same time construction of a situation model still relies on basic comprehension of words and sentences.

Similarly, evidence from the cognitive neuroscience of text and discourse processing generally supports the assumption that areas of the brain that are active during sentence comprehension also support the comprehension of connected text. When listeners encode sentences, left hemisphere language mechanisms are involved in perceiving words, encoding their meanings, parsing the sentence, and integrating the meanings across sentences. The resulting integration of information is realized at two levels: (1) coherent semantic representations of successive clauses and sentences that are subject to verbatim memory loss at clause and sentence boundaries and (2) a situation model based on the updating of information as the text proceeds. What do we know about how the brain represents these two levels of comprehension?

Ferstl (2007) suggests a broad extended language network that supports text comprehension. It involves the lateral PFC, including the inferior frontal gyrus and the dorsolateral PFC, the anterior temporal lobes, including the temporal pole, the dorsomedial PFC, including the anterior cingulate cortex, and the posterior cingulate cortex (Figure 16.1).

Within ventral PFC, the inferior frontal gyrus routinely supports phonological and syntactic functions that are present in language processing generally. The triangular structure within the inferior frontal gyrus supports semantic integration with context. Other areas in ventral PFC, especially the posterior lateral PFC, are recruited when the task demands, or the comprehender's goals require, attentional resources. The anterior temporal lobe functions range over basic sentence comprehension, including processes that correspond to assembling propositions from sentences and mechanisms that link propositional information across sentence boundaries. Underlying these text functions are basic memory functions. For the dorsomedial PFC, the function is so far less precise in its correspondence to a cognitively specified comprehension process. The variety of tasks that elicit dorsomedial PFC activity makes it unlikely that the function of the dorsomedial PFC will be identified in discourse-specific terms, outside of general cognitive processing. Furthermore, within each of these broad areas, the specific neural structures and their exact functions are subject to different interpretations and can be identified more specifically only with considerable hedging.

If we apply this tentative characterization of a language comprehension network extended to handle texts, we do not see a simple mapping of anatomy onto the two levels of text comprehension theory. The text base, the understanding of the words and sentences, relies heavily on areas in ventral PFC, but also requires dorsolateral PFC and anterior temporal cortex. The construction of the situation model is distributed across this network, as far as we can tell. Of course, there should be no surprise to learn that the components of text processing are distributed rather than localized. They depend fundamentally on processes of information encoding, memory, and retrieval, along with basic left hemisphere language processes. The first set – those that support the general cognitive resources that must be part of comprehension – are distributed. The traditional localization of language functions, particularly syntax within the inferior frontal gyrus, provides a structure that is inherited by text-level processing. Thus, the ironic conclusion is that the only special structures for text processing are those that are special for language processing. Beyond these, text processing requires broadly distributed brain resources for the various cognitive, social, and affective processes that are integral to language and communication, including domain-general processes for memory updating that involve the (anterior and posterior) cingulate cortex.

16.6. CHALLENGES AND FUTURE DIRECTIONS

One key challenge for future work will be to reconcile contradictory evidence on the role of the right hemisphere in establishing coherence in discourse comprehension. The work reviewed here suggests several ways in which the right hemisphere may contribute to discourse processing: in supporting backward (versus forward) inferences, in helping to establish global (versus local) coherence, in recruiting additional memory resources, and in processing nonliteral and social/affective meaning. However, results are by no means consistent across studies, and detailed methodological and task analyses will be required to understand these differences.

Another topic that is ripe for future research concerns the nature of syntactic processes, and their interactions with communicative processes. According to functionalist views of syntax, morphosyntactic markers (e.g., pronouns, articles) can be considered overt, automated, and obligatory cues to perform certain discourse-pragmatic functions (e.g., linking referents across clause and sentence boundaries). The question remains whether regions that have previously been linked to syntactic processing (e.g., the anterior region of left inferior frontal gyrus) are also implicated in the kind of communicative inferences that may be required for two individuals to effectively track not only the current theme of a conversation, but also the attentional focus and knowledge states of one another.

Finally, neuroimaging studies of discourse comprehension may add to our understanding of individual differences (Boxes 16.1–16.3). As these studies reveal mechanisms that help explain differences in linguistic and communicative competence, they may lead to practical, as well as theoretical results, with implications for clinical and educational research, as well as for the neuroscience of language.

References

Beeman, M.J., Bowden, E.M., & Gernsbacher, M.A. (2000). Right and left hemisphere cooperation for drawing predictive and coherence inferences during normal story comprehension. *Brain and Language, 71*(2), 310–336.

Brownell, H.H., & Martino, G. (1998). Deficits in inference and social cognition: The effects of right hemisphere brain damage on discourse. In M. Beeman & C. Chiarello (Eds.), *Right hemisphere language comprehension: Perspectives from cognitive neuroscience* (pp. 309–328). Mahwah, NJ: Lawrence Erlbaum.

Ferstl, E.C. (2007). The functional neuroanatomy of text comprehension: What's the story so far? In F. Schmalhofer & C.A. Perfetti (Eds.), *Higher level language processes in the brain: Inference and comprehension processes* (pp. 53–102). Mahwah, NJ: Erlbaum.

Ferstl, E.C., & von Cramon, D.Y. (2001). The role of coherence and cohesion in text comprehension: An event-related fMRI study. *Cognitive Brain Research, 11*, 325–340.

Gernsbacher, M.A., & Robertson, R.W. (2002). The definite article *the* as a cue to map thematic information. In M. Louwerse & W. Van Peer (Eds.), *Thematics: Interdisciplinary studies* (pp. 119–136). Amsterdam: J. Benjamins.

Givón, T. (2005). Grammar as an adaptive evolutionary product. In T. Givón (Ed.), *Context as other minds: The pragmatics of sociality, cognition and communication* (Chapter 4, pp. 91–123). Amsterdam: J. Benjamins.

Goel, V., & Dolan, R.J. (2001). Reciprocal neural response within lateral and ventral medial PFC during hot and cold reasoning. *Neuroimage, 20*, 2314–2321.

Greene, J.D., Sommerville, R.B., Nystrom, L.E., Darley, J.M., & Cohen, J.D. (2001). An fMRI investigation of emotional engagement in moral judgment. *Science, 293*, 2105–2108.

Hagoort, P., Hald, L., Bastiaansen, M., & Petersson, K.M. (2004). Integration of word meaning and world knowledge in language comprehension. *Science, 304*(5669), 438–441.

Heekeren, H.R., Wartenburger, I., Schmidt, H., Schwintowski, H.-P., & Villringer, A. (2003). An fMRI study of simple ethical decision-making. *Neuroreport, 14*, 1215–1219.

Kintsch, W., & Rawson, K.A. (2005). Comprehension. In M.J. Snowling & C. Hulme (Eds.), *The science of reading: A handbook* (pp. 209–226). Oxford: Blackwell.

Maguire, E.A., Frith, C.D., & Morris, R.G.M. (1999). The functional neuroanatomy of comprehension and memory: The importance of prior knowledge. *Brain, 122*, 1839–1850.

Mar, R.A. (2004). The neuropsychology of narrative: Story comprehension, story production and their interrelation. *Neuropsychologia, 42*, 1414–1434.

Mason, R.A. & Just, M.A. (2004). How the brain processes causal inferences in text. *Psychological Science, 15*(1), 1–7.

Mazoyer, B.M., Tzourio, N., Frak, V., Syrota, A., Murayama, N., Levrier, O., Salamon, G., Dehaene, S., Cohen, L., & Mehler, J. (1993). The cortical representation of speech. *Journal of Cognitive Neuroscience, 5*, 467–479.

Mellet, E., Bricogne, S., Crivello, F., Mazoyer, B., Denis, M., & Tzourio-Mazoyer, N. (2002). Neural basis of mental scanning of a topographic representation built from a text. *Cerebral Cortex, 12*(12), 1322–1330.

Moll, J., de Oliveira-Souza, R., Eslinger, P.J., Bramati, I.E., Mourao-Miranda, J., Andreiuolo, P.A., & Pessoa, L. (2002). The neural correlates of moral sensitivity: A functional magnetic resonance imaging investigation of basic and moral emotions. *The Journal of Neuroscience, 22*, 2730–2736.

Nichelli, P., Grafman, J., Pietrini, P., Clark, K., Lee, K.Y., & Miletich, R. (1995). Where the brain appreciates the moral of a story. *Neuroreport, 6*, 2309–2313.

Perfetti, C., Landi, N., & Oakhill, J. (2005). The acquisition of reading comprehension skill. In M.J. Snowling & C. Hulme (Eds.), *The science of reading: A handbook* (pp. 227–247). Oxford: Blackwell.

Rapp, A.M., Leube, D.T., Erb, M., Grodd, W., & Kircher, T.T.J. (2004). Neural correlates of metaphor processing. *Cognitive Brain Research, 20*, 395–402.

Robertson, D.A., Gernsbacher, M.A., Guidotti, S.J., Robertson, R.R., Irwin, W., Mock, B.J., & Campana, M.E. (2000). Functional neuroanatomy of the cognitive process of mapping during discourse comprehension. *Psychological Science, 11*(3), 255–260.

Sanford, A.J., & Garrod, S.C. (2005). Memory-based approaches and beyond. *Discourse Processes, 39*(2–3), 205–224.

Schmalhofer, F., & Perfetti (2007). Neural and behavioral indicators of integration processes across sentence boundaries. In F. Schmalhofer & C.A. Perfetti (Eds.), *Higher level language processes in the* brain: *Inference and comprehension processes* (pp. 161–188). Mahwah, NJ: Erlbaum.

St. George, M., Kutas, M., Martinez, A., & Sereno, M.I. (1999). Semantic integration in reading: Engagement of the right hemisphere during discourse processing. *Brain, 122,* 1317–1325.

Van Berkum, J.J., Zwitserlood, P., Hagoort, P., & Brown, C.M. (2003). When and how do listeners relate a sentence to the wider discourse? Evidence from the N400 effect. *Cognitive Brain Research, 17*(3), 701–718.

Van Petten, C., & Luka, B.J. (2006). Neural localization of semantic context effects in electromagnetic and hemodynamic studies. *Brain and Language, 97*(3), 279–293.

Yang, C.L., Perfetti, C.A., & Schmalhofer, F. (2007). Event-related potential indicators of text integration across sentence boundaries. *Journal of Experimental Psychology: Learning, Memory, and Cognition, 33*(1), 55–89.

Further Readings

Reading related to Section 16.2: Cognitive and Linguistic Principles of Discourse Processing

Givón, T. (2005). Grammar as an adaptive evolutionary product. In T. Givón (Ed.), *Context as other minds: The pragmatics of sociality, cognition and communication* (Chapter 4, pp. 91–123). Amsterdam: J. Benjamins.
This chapter considers how grammatical structures may be related to a variety of cognitive and communicative functions. Givón proposes that major subsystems in language are linked to different memory systems. These cognitively based processes provide basic tools for establishing discourse coherence. Givón also provides numerous examples of what he terms "discourse-pragmatic" markers in language, including morphosyntactic devices for establishing discourse coherence. The perspective represented here is one that shows how linguistic markers can be linked to specific socio-pragmatic goals that motivate human discourse, and communication.

Perfetti, C., Landi, N., & Oakhill, J. (2005). The acquisition of reading comprehension skill. In M.J. Snowling & C. Hulme (Eds.), *The science of reading: A handbook* (pp. 227–247). Oxford: Blackwell.
This chapter provides a thorough introduction to Kintsch's Construction Integration Theory, and discusses the cognitive and linguistic mechanisms that are implicated in reading comprehension. Individual differences in cognitive and linguistic skills are also linked to individual differences in discourse comprehension, and to a theory of reading development.

Readings related to Section 16.3: The Neuroscience of Text and Discourse Comprehension

Ferstl, E.C. (2007). The functional neuroanatomy of text comprehension: What's the story so far? In F. Schmalhofer & C.A. Perfetti (Eds.), *Higher level language processes in the brain: Inference and comprehension processes* (pp. 53–102). Mahwah, NJ: Erlbaum.
This is a rich and highly informative chapter that reviews the state-of-the art in neuroimaging of discourse functions. Note that the studies cited involve primarily healthy adult subjects, and the discussion is focused on studies that have specifically attempted to isolate particular cognitive and brain functions in discourse comprehension.

Mason, R.A., & Just, M.A. (2004). How the brain processes causal inferences in text. *Psychological Science, 15*(1), 1–7.
Replicating previous findings from behavioral studies in an fMRI study, the authors show that (1) subjects are slower to encode inferences that are more distantly related to a particular outcome and (2) subsequent recall is strongest for causal inferences that are moderately related to the outcome sentence. The authors discuss these results in the context of a two-stage theory of inferencing, which posits separate processes related to inference generation and memory integration.

Readings related to Section 16.4: Right Hemisphere Contributions to Discourse Comprehension

Beeman, M.J., Bowden, E.M., & Gernsbacher, M.A. (2000). Right and left hemisphere cooperation for drawing predictive and coherence inferences during normal story comprehension. *Brain and Language, 71*(2), 310–336.
This study provides important evidence on right versus left hemisphere contributions to discourse processing, using the divided visual field method. Results suggest a right hemisphere advantage in priming of forward (predictive) inferences, and a left hemisphere advantage in priming of backward (bridging) inferences. The authors relate these patterns to Beeman's right hemisphere "coarse coding" hypothesis.

Robertson, D.A., Gernsbacher, M.A., Guidotti, S.J., Robertson, R.R., Irwin, W., Mock, B.J., & Campana, M.E. (2000). Functional neuroanatomy of the cognitive process of mapping during discourse comprehension. *Psychological Science, 11*(3), 255–260.
In an fMRI study the authors found greater right hemisphere BOLD activation for lists of sentences that used definite articles rather than indefinite articles to anaphorically relate the nouns in a text. These findings are particularly impressive, given that subjects were not instructed to try to make coherence links across the individual sentences that were presented on each trial.

17

Neuropragmatics: Disorders and Neural Systems

BRIGITTE STEMMER

Faculty of Arts and Science and Faculty of Medicine,
Université de Montréal, Montreal, Canada

ABSTRACT

How do we compute meaning from something that is not said? This is quite a challenging task even for the healthy brain, so how can people whose brain has been afflicted by trauma or disease such as stroke, dementia or psychosis deal with such issues? Research has demonstrated that specific aspects of linguistic pragmatics can be impaired in some individuals with acquired brain damage while in others these aspects are intact. Whereas pragmatic disorders can help to characterize specific patient populations, such disorders are not specific to a certain disease, nor do similar pragmatic "symptoms" have similar causes. Basic neural systems in the prefrontal lobe make necessary contributions to the interpretation and generation of specific aspects of pragmatic behavior; by themselves, however, these systems are not sufficient as they are intertwined with attention, memory, monitoring and affective systems. The nature and extent of the recruitment of these various systems and their interaction seem to depend on the requirements set by the experimental task and the characteristics of the individual. Although more and more details of these systems are emerging, the puzzle is still far from being completed.

17.1. INTRODUCTION

When we hear or utter something like "That was very smart of you" we may interpret this to mean *Well done, you are a clever person* on one occasion, or, *You are an idiot* on another. The way we interpret or phrase these utterances goes beyond the traditional players in linguistics, that is, phonetics, morphology, syntax and semantics, and also involves the *unsaid*. The way language is used and interpreted while taking into consideration the characteristics of the speaker and hearer and the effects of contextual and situational variables has

been subsumed under the notion linguistic pragmatics. It also includes aspects of discourse, inasmuch as the production and interpretation of discourse also relies on pragmatics. The study of how the brain comprehends and produces linguistic pragmatic behavior has become known as neuropragmatics.

Frequently discussed issues in the study of neuropragmatics relate to which aspects of pragmatics are preserved or impaired in individuals with brain pathology. Findings from such research are generally used in several ways such as specifying the functional (pragmatic) abilities of the patients (e.g., for diagnostic or therapeutic purposes), testing psycholinguistic models of processing or gaining insights into the neural substrates underlying the pragmatic impairment. Advances in neuroimaging technologies have made it possible to study the latter two aspects also in healthy populations.

This chapter will introduce the reader to research in neuropragmatics with a focus on pragmatic disorders in *acquired* brain pathologies and the neural systems involved (see also Chapter 28 for a focus on traumatic brain injury (TBI) patients; see Chapter 37 for a focus on autism and Asperger syndrome).

17.2. PRAGMATIC DISORDERS IN ADULT CLINICAL POPULATIONS

Affective prosody, aspects of discourse (e.g., conversational and narrative style, cohesion and coherence), non-literal language (such as "indirect" speech acts or humor) or figurative language (e.g., metaphor, idioms, irony and so forth) are the most frequently investigated pragmatic aspects of linguistic communication in patients with acquired brain pathologies. A broad range of patients have been investigated that shows problems with these different aspects of pragmatics.

17.2.1. Discourse

Before sophisticated brain imaging methods were yet available, individuals with lesions in the right hemisphere (usually due to a stroke) were the most investigated patient population. What made this patient group so interesting to researchers was an observed dissociation between their impaired communicative abilities and well-preserved linguistic abilities. For example, difficulties with topic maintenance, identification and extraction of relevant themes, structure organization and discourse cohesion and coherence as well as the use of prosody to interpret the emotional content of discourse have been characterized as reflecting the conversational and narrative discourse of some right hemisphere damaged (RHD) patients (for summaries see Brownell & Martino, 1998; Myers, 2005). Not all studies report these impairments, however (e.g., Brady *et al.*, 2005). And similar discourse problems have also been described in other patient populations although the underlying cause was explained differently. The discourse problems of RHD patients have mainly been attributed to difficulties with integrating information or drawing appropriate inferences. At times, the question was also raised whether attention or affective impairments observed in some RHD patients were the underlying cause, without, however, clear answers. In some patients with Alzheimer's disease impaired attention and memory functions were more clearly associated with problems in drawing inferences between textual content and real world knowledge (Chapman *et al.*, 1998). Executive functions and social behavior, that is, abilities closely mediated by the frontal lobe, have been suggested to underlie the inability of patients with frontotemporal dementia to establish global coherence (Ash *et al.*, 2006). The discourse problems of patients with a lesion in the left hemisphere have been mainly ascribed to their linguistic impairment (for a summary see Wright & Newhoff, 2005).

Another disorder that produces impairments similar to RHD patients is schizophrenia. Schizophrenia patients have been shown to display difficulties with topic maintenance, distinction of relevant from non-relevant content in narrative discourse and with turn taking and decoding implied meaning in conversation (for a summary see Meilijson *et al.*, 2004; Mitchell & Crow, 2005). Early work proposed that the discourse difficulties of schizophrenic patients rest in their inability to share reality with their interlocutor, to use conventional social norms to guide their speech or to mindread. Some authors suggest that the underlying cause for these difficulties is based on their abnormal lateralization of language – with what are normally left hemisphere language functions being lateralized to the right hemisphere or being more equally distributed between the hemispheres (Mitchell & Crow, 2005; see Chapter 29, this volume).

To sum up, despite different disease etiologies and differences in the underlying pathophysiologies several patient groups display similar problems in discourse production and comprehension. This would seem to indicate that the discourse "symptoms" are not specific to a certain disease but rather that different causes can underlie similar pragmatic impairments. The explanations for the observed discourse impairments vary widely depending on the pathological population (for a more detailed discussion see Section 17.3). Although it is not always clear which neural systems are impaired in these different patient populations, what they seem to have in common (in very general terms) is an implication of the frontal lobes (see also Chapter 28).

17.2.2. Non-literal Language and Figurative Language

Aside from discourse phenomena, another aspect of pragmatics that has received particular attention concerns the interpretation of "the unsaid", as reflected in non-literal language such as "indirect" speech acts (requests) or humor, or in figurative language such as metaphor, sarcasm and irony, or in idioms. Classical models have assumed that the comprehension of literal utterances required just one processing step while non-literal or figurative language involved at least two (or more) processing steps, that is computing the literal meaning first, rejecting it as contextually inappropriate and finally re-interpreting the utterance and arrive at the intended meaning.

17.2.2.1. Indirect Requests

The concept of indirectness has been a key feature of investigations in several neuropsychological studies with RHD patients. The underlying assumption in early studies has been that "indirect" requests (e.g., uttering *I am cold* as a request to close the window) is more abstract and requires more complex or different inferencing processes and thus more processing steps than "literal" or "direct" language (*Close the window, please*). While early studies claimed that RHD patients had problems understanding "indirect" requests, this claim had to be modified as subsequent studies using different designs and theoretical frameworks only partially supported the earlier findings (for a review see Stemmer, 2008). RHD patients were able to produce or comprehend indirect requests but showed difficulties to establish a relationship between request types (non-conventional indirect requests) and the supporting material. These studies also pointed out the importance of distinguishing subtypes of the target phenomena (e.g., direct requests, conventional indirect requests, non-conventional indirect requests) as these may be processed differently and thus may also implicate different neural systems.

Other patient groups that have been shown to be impaired in their abilities to appropriately interpret non-conventional indirect requests (*Awfully dry air in here* meaning *I am thirsty*) are patients with schizophrenia (Corcoran, 2003) and patients with TBI, especially such patients with frontal lobes damage and executive dysfunctions (see Chapter 28).

17.2.2.2. Other Non-literal and Figurative Language

The comprehension of figurative language or other non-literal language is similar to understanding non-conventional indirect requests inasmuch as such figurative or non-literal language also implies comprehending "the unsaid". Complex meta-representational abilities are required that involve making inferences at various levels of complexity and integrating a wide range of information sources. It is thus not surprising that impairments have been described in several patient populations. These include difficulties in the interpretation of metaphors in patients with RHD and in schizophrenia patients (for summaries see Champagne Lavau *et al.*, 2006; Mitchell & Crow, 2005) but not patients with Alzheimer's disease in the early stages (Papagno, 2001). Problems recognizing and appropriately interpreting sarcasm and irony have also been shown in TBI patients with frontal lobe damage and with executive impairment despite at least partially intact mindreading abilities (e.g., McDonald, 1999). Although RHD patients have been able to correctly identify sarcastic remarks in the context of a multiple choice task, they have shown difficulties with tasks involving lies and ironic joke stories, that is, tasks that require second-order meta-representational judgments (Winner *et al.*, 1998).

Similar to figurative language, joke comprehension also involves the generation of inferences drawing on contextual, knowledge and experiential factors. In addition to these more cognitive elements, apprehension, unexpectedness, surprise and appreciation are other elements in joke comprehension. Early behavioral studies reported problems with joke interpretations in RHD and left hemisphere damage (LHD) patients (for a summary see Stemmer, 2008). A later study that investigated patients with right, left or bilateral frontal lesions reported impairments in specific aspects of the humor task with such impairment being evident only in a subgroup of patients with frontal lobe lesions (Shammi & Stuss, 1999). Patients with schizophrenia have also displayed difficulties with joke comprehension. Schizophrenia patients in remission, however, performed much better than those with the active disease suggesting that medication may have interfered with joke comprehension in these patients (Corcoran *et al.*, 1997; Drury *et al.*, 1998).

17.2.2.1. Mentalizing or Mindreading

The ability to infer other peoples' mental states, thoughts and feelings has been referred to as mentalizing or mindreading,

that is, the person is said to have a theory of mind (ToM) (for details see Chapters 28 and 37, this volume). ToM is typically evaluated using first-order and second-order belief tasks that involve reasoning about mental states such as assessing a person's ability to infer that someone can have a mistaken belief that is different from one's own true belief (e.g., see Chapters 28 and 37). In this context, it is important to mention that ToM has been used both as a *descriptive* tool, and, at the same time, served as an *explanatory* model. We will first focus on the descriptive level before taking up the topic again in Section 17.3.

A variety of tasks have been used to test patients with dementia, RHD, schizophrenia and other pathologies. The similarity of findings is striking, in the sense that those patient groups that have been shown to be impaired in non-literal and figurative language production and comprehension also display problems with high-order ToM tasks.

ToM and Dementia

Patients with the frontal variant of frontotemporal dementia (Pick's disease) show changes in personality and behavior such as lack of empathy, socially inappropriate behavior, lack of personal awareness and insight while their memory capacities are relatively spared in the early stages. This is in contrast to patients with Alzheimer's disease who typically present with impairment of episodic memory and of semantic and attentional processing while, in the early stages, personality and social behavior is intact. Comparing these two patients groups on mindreading tasks can thus provide insights into the dependency of mindreading on these skills. Frontotemporal dementia patients show poor performance on a broad spectrum of mentalizing and classical ToM tasks despite good general comprehension and memory abilities (Gregory *et al.*, 2002; Lough *et al.*, 2006). Such patients are also impaired in moral reasoning, emotion processing and empathy (Lough *et al.*, 2006). Patients with Alzheimer's disease also perform poorly on mentalizing and ToM tasks, although in a more selective manner. For example, they showed difficulties only with the second-order false belief task and faux pas task (Gregory *et al.*, 2002). Furthermore, while non-verbal psychological reasoning was impaired, physical reasoning was intact (Verdon *et al.*, 2007). It thus seems that both frontal lobe and memory skills affect mindreading tasks, *albeit* to different extents. Frontal lobe impairment has a more detrimental and a variable effect, possibly due to the broad range of skills mediated by the frontal lobes whereas memory impairments particularly affect those mindreading tasks that generate a high processing load.

Another question concerns the relationship of executive functions and mindreading. There is some indication that executive functions contributed to the poor performance of patients with the frontal variant of frontotemporal dementia although they did not seem sufficient to explain all the problems observed (Snowden *et al.*, 2003; Lough *et al.*, 2006).

The nature and extent of differently involved neural systems seems to be equally important.

ToM and Schizophrenia

An impairment in social functioning is one of the most disabling clinical features in schizophrenia and the more severe and acute the symptoms the more severe the mentalizing impairment (e.g., Corcoran *et al.*, 1997; for a review see Lee *et al.*, 2004). There is currently an inconsistency as to whether specific items of schizophrenia symptomatology (or different types of schizophrenia) relate to impaired mentalizing abilities. Impairment in mentalizing and empathy in schizophrenia seems to be independent of generalized cognitive deficits although no final evaluation of this relationship can yet be given.

ToM and RHD

Some studies have investigated ToM behavior in RHD patients with, however, ambiguous findings or interpretations (for a summary see Siegal & Varley, 2002). For example, in one study some RHD patients demonstrated problems in attributing second-order beliefs; this performance varied, however, in the sense that the RHD patients sometimes responded correctly to the second-order belief questions and failed at other times (Winner *et al.*, 1998). In addition, some of the non-brain-damaged control participants also performed poorly on the second-order belief tasks. Another study investigated the hypothesis that people with RHD show a deficit in ToM in the context of otherwise intact reasoning skills (Happé *et al.*, 1999). Compared to the LHD patients with Broca type aphasia and a healthy control group, the RHD group performed less well on stories and cartoons that required mentalizing abilities but similarly well on stories and cartoons that did not require mentalizing abilities. Although the authors suggested a dedicated cognitive system for ToM and a role for the right hemisphere in *adult* ToM, like most other studies, the results are difficult to interpret due to the high heterogeneity of the patient groups. One of the few studies that compared patient groups according to lesion localization was that by Stuss and colleagues (Stuss *et al.*, 2001). These authors investigated stroke patients with lesions in right, left and bilateral focal frontal and non-frontal regions using visual perspective-taking and a deception task, both implicating first-order attributions. In the visual perspective-taking task, first level direct inferences did not pose problems for any patient group. However, inferences in a transfer condition showed significant differences for the right frontal group compared to all other groups. For the deception task, the right frontal group did not show any problems, although combined right and left frontal patients displayed significant difficulties with the task. In addition, the right medial (inferior and superior) frontal areas and the right anterior cingulate correlated with the number of errors. The authors concluded that the frontal lobes are implicated in some aspects of ToM and that different neural systems may be implicated in different ToM tasks.

17.2.3. Summary

Difficulties with the production and/or interpretation of discourse, non-literal and figurative language and mentalizing have been described in several pathological populations. When interpreting these empirical findings, it is important to realize not only the benefits of patient studies but also their limitations. The number of subjects investigated is usually relatively small but the most important problem is the great heterogeneity of the patients in terms of age, education, time post-onset and disease severity. Equally problematic is the poor information on the neuropsychological tasks, inadequate information on the lesion site and functional severity exhibited by the patients. All these negative factors are then combined with analysis of the patients as group (see also Stuss *et al.*, 2001). As a consequence, general conclusions need to be made with caution and studies need to be replicated. With these cautionary words in mind, what seems to emerge from the patient studies is that the frontal lobes do play an important role in pragmatic processing. Although many studies with RHD patients do not carefully distinguish between frontal and non-frontal lesions, the majority of included RHD included in these studies have frontal lesions and the studies that directly address this issue show the role/involvement of the frontal lobes. Findings from TBI patients with lesioned frontal lobes support the notion that the frontal lobes are critical for the production and interpretation of higher language functions (see Chapter 28) as do research results from patients with frontotemporal dementia and schizophrenia (for a discussion of hemispheric asymmetry and impaired neural substrates in schizophrenia see also Chapter 29, this volume; Mitchell & Crow, 2005).

Although research based on lesion studies has shown the relevance of the frontal lobes in linguistic pragmatics, the functional subdivision of the frontal lobes has rarely been considered. The different pathologies that produce overlapping pragmatic disorders suggest that pragmatic disorders are not disease specific. It seems more likely that the extent and nature of lesions to the neural systems and the resulting (faulty or modified) processing will affect functional systems that contribute in a variety of ways to aspects of pragmatic behavior. Various pathologies will then produce at times similar and at other times different impairments.

17.3. EXPLAINING LINGUISTIC PRAGMATIC IMPAIRMENTS IN CLINICAL POPULATIONS

Numerous hypotheses and theories have been advanced to explain linguistic pragmatic difficulties; there has, however, been limited interest in establishing an appropriate theoretical framework. The earliest explanations suggested that RHD patients had problems integrating pieces of information into a coherent whole or difficulties in generating inferences. More recent explanations have suggested that a faulty ToM underlies their problems. In previous work, we have argued that inference generation *per se* is not a theory and that *the mental model hypothesis* as advanced by Johnson-Laird (for a summary see Stemmer & Cohen, 2002) encompasses other suggested hypotheses or theories. The failure of RHD patients to ignore the implausibility of absurd logical syllogisms has, however, been interpreted as contradiction of the mental model hypothesis (McDonald, 2000). This seems a rather limited view of the mental model hypothesis and it does not seem justified to refute the hypothesis on these grounds alone.

The empirical validity of three theories that attempt to explain pragmatic language impairment in autism, RHD patients and TBI patients was discussed by Martin and McDonald (2003), that is *ToM* as a component of *social inference theory*, the *weak central coherence hypothesis* (WCCH) and the *executive dysfunction account*. These authors show that none of the three positions sufficiently accounts for the pragmatic impairments observed across different patient populations. For ToM it is pointed out that there is a close relationship between the concept of ToM and pragmatic understanding – and this makes it difficult to define the causal relationship between the two. Others have criticized that the ToM framework inappropriately intellectualizes everyday social activities, as experiments are conducted to test the ability to represent the intentions of others instead of observing and analyzing the intentionality in spontaneous social interactions (Leudar *et al.*, 2004). For example, discrepancies have been found between patients' poor performance on ToM tasks in the laboratory and good performance in managing intentionality in interactions outside of the laboratory (McCabe, 2004). Finally, ToM is but one domain of social cognition and its application to patients has mostly been limited to processes used to perceive other people. Other components of social cognition that also need consideration are the perception of self and social knowledge that enables people to manage everyday life. (For a summary see Beer & Ochsner, 2006.)

The WCCH refers to the inability to use context to derive meaning; this inability reflects the failure of a central system to integrate small pieces of information with a globally coherent pattern of information. A problem already mentioned for ToM also pertains to the WCCH: it is used to describe and, at the same time, explain pragmatic deficits. Another shortcoming of the WCCH is that it cannot be applied to explain pragmatic behavior across populations (Martin & Crow, 2003).

The executive dysfunction account is based on the observation that disturbances in executive functions often reflect impairments of the frontal lobe. Executive functions have been conceptualized as the central executive of the information-processing system and as such include the control of attention, goal setting (initiating, planning, problem solving, strategic behavior) as well as that of cognitive flexibility (attention shifting, working memory, self-monitoring, self-regulation) (Stuss & Alexander, 2000). The question is thus whether these specific cognitive abilities underlie pragmatic impairments, or – as some have suggested – especially ToM abilities. While some studies report a relationship between executive dysfunctions and pragmatic impairment, or ToM in particular, others do not – and yet again others have not investigated this issue (for a summary see Martin & Crow, 2003; see also Mitchell & Crow, 2005; Champagne Lavau *et al.*, 2006).

In sum, despite numerous studies of pragmatic disorders, there is currently no agreement on how to best explain the impairments. Several factors contribute to this situation: an important and inherent problem in the majority of studies across patient populations is the grouping of patients – despite great heterogeneity. Another problem relates to the ambiguity of the stimuli used and yet other problems refer to the lack of adequate consideration of testing different theoretical positions and failing to cross-reference findings in other clinical populations. In addition, various researchers use different methods of assessment which thus target different aspects of pragmatics. The same holds for neuropsychological assessment. Finally, as already mentioned, target concepts (e.g., inferencing, ToM, executive functions) are frequently used to describe as well as explain the phenomenon and this leads to circularity of the arguments advanced.

17.4. NEURAL SYSTEMS UNDERLYING PRAGMATIC ABILITIES

With the technological advancement in neuroimaging, it is now possible to correlate pragmatic task performance with brain activation in healthy as well as in pathological populations and specify in more detail the neural systems involved. The majority of studies have used the functional magnetic resonance imaging (fMRI) technique, as it is particularly suitable to investigate the neural systems implicated (see Box 17.1 for the relevance of electroencephalogram/event-related potential (EEG/ERP) studies).

17.4.1. Neural Systems in Discourse Processing

Based on lesion studies, text and discourse processing, and especially inferencing and integration processes, are often associated with the right hemisphere, although the findings are not always clear cut (see Section 17.2). So far,

Box 17.1   One or two steps to interpret language?

When someone tells us that "Dutch trains are white" we can either take the sentence at face value, or, if we are Dutch or an observant traveler, world knowledge tells us that this is not true as Dutch trains are yellow. How do we get to the insight that the sentence is not true and the speaker possibly joking or a liar? By first computing a local context-independent meaning of the sentence from its phonology, syntax and semantics, and then, in a next step, integrating this meaning with our world knowledge? Or do we do this – as some researchers claim – in one step? Peter Hagoort and his colleagues have performed a series of experiments tackling the question whether we use two steps to arrive at the meaning of a sentence or whether we combine in one step the linguistic information with the information from prior discourse,

the speaker, the situational and our world knowledge. For example, in one study they presented violated sentences where the violation could be recognized on linguistic bases alone or only when world knowledge was applied (see the following figure). Using the event-related potential and fMRI technique they showed that the N400 effects obtained for both conditions were identical in onset and peak latency suggesting that lexical–semantic knowledge and general world knowledge were both integrated in the same time-frame during sentence interpretation [panel (a) in the figure]. In addition, the fMRI data identified a common activation are in the left inferior frontal gyrus for both conditions [panel (b) in the figure].

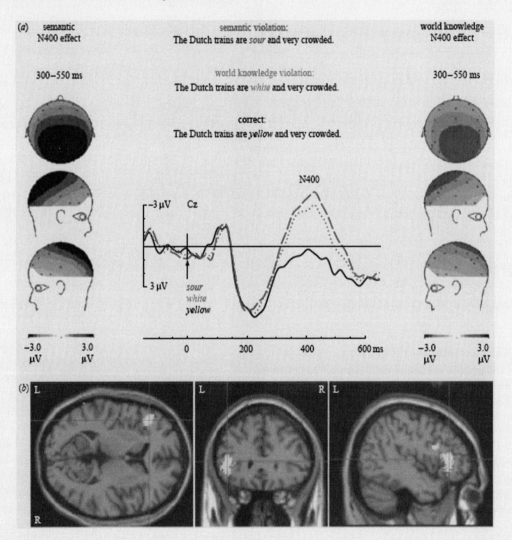

Source: Figure taken from Hagoort & van Berkum (2007) (Fig. 4, p. 805). Printed by permission of The Royal Society. (See Plate 12.)

Hagoort, P., & van Berkum, J. (2007). Beyond the sentence given. *Philosophical Transactions of the Royal Society B*, *362*, 801–811.

Hagoort, P., Hald, L., Bastiaansen, M., & Petersson, K.M. (2004). Integration of word meaning and world knowledge in language comprehension. *Science*, *304*, 438–441.

neuroimaging studies with healthy populations have not provided much evidence for this assumption. Text processing has been associated with left hemisphere dominance or the involvement of both hemispheres, and studies on coherence and inferencing have also produced controversial results (for a detailed review and discussion the reader is referred to Chapter 16; see Box 17.2 for a summary). Similar to sentence processing, activation of inferior frontal and temporal regions have been described in most studies indicating (the intuitively plausible observation) that similar neural systems that underlie sentence processes are also recruited during text and discourse processes. Other neural systems have also been activated; some of them may reflect the recruitment of cognitive processes (such as attention, memory or control processes) involved in the task. So far it has been difficult to determine the neural systems that are specific to text or discourse, or, even more basic, whether such specific systems exist.

17.4.2. Neural Systems in Non-literal and Figurative Language

Different processing mechanisms have been suggested for different subtypes of non-literal and figurative items (e.g., conventional versus novel metaphors) and it is thus possible that these may also implicate different neural systems. A behavioral study using the divided visual field technique showed the involvement of the right hemisphere in the comprehension of novel but not of conventional metaphors (Faust & Mashal, 2007). Further support comes from a neuroimaging study with fMRI that contrasted conventional and novel metaphors (Mashal *et al.*, 2005). The right homolog of Wernicke's area was viewed as playing a special role in the processing of novel metaphors conjointly with the activation of other regions, suggesting a basic network for the processes of reading and comprehension (see Box 17.3). Another study reported an overlapping network in the classical left language areas and specific (mostly left) frontal and temporal networks for conventional metaphoric and specifically activated right temporal areas for ironic statements (Eviatar & Just, 2006; see also Box 17.3). In disagreement with the Mashal *et al.* (2005) study, only left inferior frontal activation for simple novel metaphors was reported by Rapp *et al.*, 2004. Yet another study using the ERP and divided visual field techniques found that both hemispheres were associated with metaphor processing (Coulson & van Petten, 2007).

Using the repetitive transmagnetic stimulation (rTMS) technique only left temporal and left frontal regions have been associated with opaque idiom comprehension, (Oliveri *et al.*, 2004). No disruption, however, was observed with right frontal stimulation. While this study speaks to the importance of these regions in opaque idiom comprehension,

it cannot exclude the possibility that other left or right hemisphere regions contribute to the comprehension process inasmuch as no other areas were stimulated.

Taken together, these studies suggest an activation of classical language areas conjointly with specific regions in the prefrontal and temporal cortices that are differently activated for various types of figurative language. Whether these regions or neural systems are specific for figurative language remains currently unanswered. It seems likely that, depending on task and stimuli demand, and on individual characteristics, attention, memory and control networks are initiated to various degrees.

Behavioral lesion studies have frequently been interpreted to point to an involvement of the right hemisphere in non-literal and figurative language, although findings have not always been straightforward. Based on current neuroimaging studies it is difficult to provide an evaluation – at least at this point they do not seem to support a special role for the right hemisphere.

Another aspect that deserves attention is the affective component involved in pragmatic processing. At least one characteristic that distinguishes humor and joke comprehension from figurative language is its affective component. It is thus not surprising that a cognitive and an affective system were identified in an fMRI study on joke processing. The cognitive system was implicated in "getting the joke" and the affective system in joke appreciation (Goel & Dolan, 2001; see also Box 17.3). It is thus plausible to assume that when affect is involved (and not necessarily only in the form of a joke) in pragmatic processing, the neural systems associated with affect will be recruited. Subsequent humor and joke studies confirmed and further specified these distinctive systems and identified yet another network suggestive of the implication of the reward system (ventral tegmental area, medial ventral prefrontal cortex (PFC), Ncl. accumbens) in joke and humor comprehension (for a summary see Box 17.3).

17.4.3. Neural Systems in Mindreading (ToM)

In Section 17.2, it was suggested that patients with difficulties in mindreading show some involvement of the frontal lobes. Difficulties with mindreading may result from an abnormal interaction between the frontal lobe and their functionally connected cortical and subcortical areas. Here we will now focus on studies directly addressing the issue of neural systems underlying mindreading and, more generally, social cognition (see Box 17.4 for mindreading and mirror neurons).

A series of neuroimaging studies have implicated the medial frontal cortex in mindreading although – similar to behavioral studies – evidence relating to laterality remains controversial (for summaries see Siegal & Varley, 2002; Lee

Box 17.2 Neuroimaging of discourse

Generally, left or bilateral activation has been observed much more frequently than specific right hemisphere activation. In addition to the activation of classical language areas in the left hemisphere (inferior frontal gyrus, superior, middle, inferior temporal gyri and angular gyrus) in most studies, other areas are also activated. Whether these are specific to discourse is currently not clear. The figure summarizes the areas of brain activation that have been identified in different neuroimaging studies on aspects of discourse. The localization is approximate and symbols close to each other may, in fact, overlap. For exact localization the reader is referred to the original studies. Note that the studies are not directly comparable due to different methodological details.

(a) Left lateral view

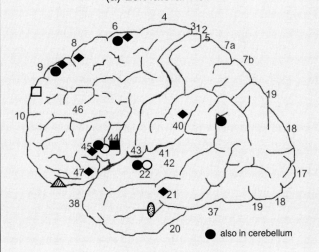

(b) Right lateral view

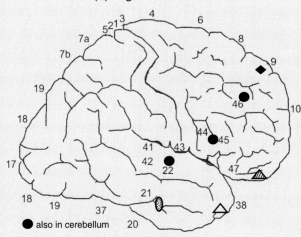

(c) Left medial view

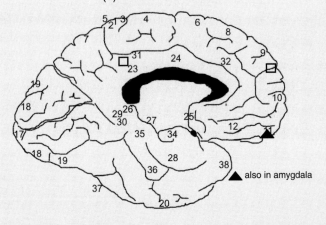

(d) Right medial view

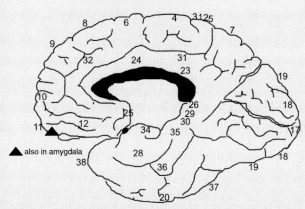

Symbols:

Caplan, R., & Dapretto, M. (2001). Making sense during conversation: an fMRI study. *Neuroreport, 12*, 3625–3632: ○ reasoning (L-IFG BA44,45; L-STG BA22) ● topic maintenance (bilateral with RH bias: BA44/45, BA22; R-DLPFC BA46,9; L-PL BA39; L-SMA BA6)

Ferstl, E.C., & von Cramon, Y. (2001). The role of coherence and cohesion in text comprehension: an event-related fMRI study. *Cognitive Brain Research, 11*, 325–340: □ coherence building L-PCC/IPC BA23/31; L-frontomedian wall/SFG BA9,10), ■ task difficulty: L-IFS/IPCS BA44

Ferstl, E.C., Rinck, M., & von Cramon, D.Y. (2005). Emotional and temporal aspects of situation model processing during text comprehension: An event-related fMRI study. *Journal of Cognitive Neuroscience, 17*(5), 724–739: △ local detection of inconsistencies (R-ATL), ◬ integration of inconsistencies (bilateral vlPFC BA47/1; ▲ emotional involvement (bilateral vmPFC/supraorbital sulcus + amygdaloid complex)

Kuperberg, G.R., Lakshmanan, B.M., Caplan, D.N., & Holcomb, P.J. (2006). Making sense of discourse: An fMRI study of causal inferencing across sentences. *Neuroimage, 33*(1), 343–361: ◆ generation and integration of causal inferences: L-IFG BA45,47, L-MFG BA6/9, L-IPL BA40; L-MTG BA21; L-SFG BA9; R-SFG BA6/8/9 St. George, Kutas, Martinez, & Sereno, 1999: ◖ untitled condition (bilateral ITS; R-MTS)

L, R: left, right; BA: Brodman area; ATL: anterior temporal lobe; DLPF: dorsolateral prefrontal cortex; IFS: inferior frontal sulcus; IFG: inferior frontal gyrus; IPC: inferior precuneus; IPCS: inferior precental sulcus; IPL: inferior parietal lobule; ITS: inferior temporal sulcus; MFG: middle frontal gyrus; MTG: middle temporal gyrus; MTS: middle temporal sulcus; PCC: posterior cingulate cortex; SFG: superior frontal gyrus; SMA: supplementary motor area; STG: superior temporal gyrus; vlPFC: ventrolateral prefrontal cortex; vmPFC: ventromedial prefrontal cortex.

© B. Stemmer

Box 17.3   Neuroimaging of non-literal and figurative language

Similar to the discourse tasks (see Box 17.2), non-literal and figurative language tasks activate the classical language areas. There is currently no agreement to which extent the other activated brain areas are specific to metaphor processing and how they interact with other cortical or subcortical systems. The latter aspect also applies to discourse processing although some proposals have recently been made. Humorous stimuli but not metaphoric stimuli also activated brain regions associated with affective processing. The reader should be aware that the studies are not directly comparable due to different methodological details. Note that localization is approximate and symbols close to each other may, in fact, overlap. For exact localization see original studies.

(a) Left lateral view

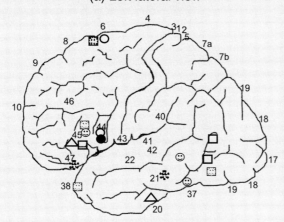

(b) Right lateral view

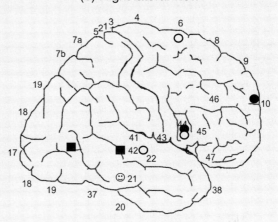

(c) Left medial view

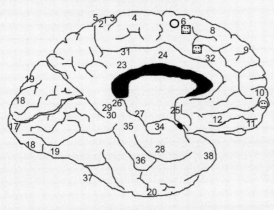

(d) Right medial view

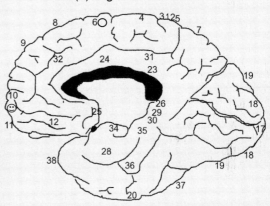

Symbols for figurative language studies:

Eviatar, Z., & Just, M.A. (2006). Brain correlates of discourse processing: An fMRI investigation of irony and conventional metaphor comprehension. Neuropsychologia, 44, 2348–2359: □ conventional metaphors (L-IFG, L-ITG, R-ITG, L-IESC) ■ ironic statements (R-STG, R-MTG)

Mashal, N., Faust, M., & Hendler, T. (2005). The role of the right hemisphere in processing nonsalient metaphorical meanings: Application of principal components analysis to fMRI data. Neuropsychologia, 43, 2084–2100: ○ novel metaphors (bilateral Broca BA44; R-Wernicke (BA22/44), bilateral insula BA13; bilateral SMA BA6) ● conventional metaphors (bilateral Broca BA44; R-SFG BA10)

Rapp, A.M., Leube, D.T., Erb, M., Grodd, W., & Kircher, T.T.J. (2004). Neural correlates of metaphor processing. Cognitive Brain Research, 20, 395–402: △ metaphors (L-IFG BA47/45)

Symbols for joke/humor studies:

Goel, V., & Dolan, R.J. (2001). The functional anatomy of humor: segregating cognitive and affective components. Nature Neuroscience, 4(3), 237–238: ☺ joke (mvPFC BA10/11, bilateral cerebellum); cognition: (L-posterior MTG BA21/37, L-posterior ITG BA37, R-posterior MTG BA21, L-IFG BA44/45)

Mobbs, D., Greicius, M.D., Abdel-Azim, E., Menon, V., & Reiss, A.L. (2003). Humor modulates the mesolimbic reward centers. Neuron, 40, 1041–1048: ▦ bilateral ventral tegmental area, Ncl. accumbens, amygdala; L-IFG BA44/45, L-TOJ BA37, L-temporal pole BA38

Moran, J.M., Gagan, S.W., Adamas, R.B.Jr., Janata, P., & Kelley, W. (2004). Neural correlates of humor detection and appreciation. Neuroimage, 21, 1055–1060: ▦ humor appreciation: bilateral insular cortex, bilateral amygdala); humor detection: L-posterior MTG BA21, L-IFG BA47)

Watson, K.K., Matthews, B.J., & Allman, J.M. (2007). Brain activation during sight gags and language-dependent humor. Cerebral Cortex, 17, 314–324: ☺ L-midbrain, bilateral amygdala (common network for both types of humor)

(The numbers indicate the approximate site of Brodman areas).

L, R: left, right; BA: Brodman area; IESC: inferior extrastriate cortex; IFG: inferior frontal gyrus; ITG: inferior temporal gyrus; MTG: middle temporal gyrus; mvPFC: medial ventral prefrontal cortex; SFG: superior frontal gyrus; SMA: supplementary motor area; STG: superior temporal gyrus; TOJ: temporo-occipital junction

© B. Stemmer

Box 17.4   Mirror neurons and mindreading

A mirror neuron is a neuron that fires both when an individual performs an action and when the individual observes the same action performed by another individual. In other words, the neuron "mirrors" or imitates the action of another person, as though the observer were performing the action herself. Thus, mirror neurons have been associated with imitative behavior and action. As can be seen in the following figure, a core circuit for imitation has been identified: the posterior superior temporal sulcus is associated with a high-order visual description of the action to be imitated, the parietal component with the motor aspects of the imitated action and the frontal component with the goal of the imitated action (for a review see Iacoboni & Dapretto, 2006). Researchers have observed that the mirror neuron system not only codes for action but also for the intention associated with it, and a link was thus established between the mirror neuron system and social cognition (but see Jacob & Jeannerod., 2005 for a critique). It has even been proposed that mirror neurons

are key elements in the understanding of the intentions of others associated with everyday actions. Some disorders, like autism, have been explained with a disruption of the activity of the mirror neuron system. Whether the mirror neuron hypothesis can explain other pathologies that implicate disorders of social cognition remains to be seen.

Does the mirror neuron system give us some indication on the question whether language related aspects of social cognition are lateralized? As a special high-order motor system the mirror neuron system tends to be bilaterally represented. However, empirical research suggests that the left hemisphere – with its specialization for specific language components – has a multimodal (visual, auditory) mirror neuron system, whereas the right hemisphere has only a visual mirror neuron system.

IPL: inferior parietal lobe; PF: prefrontal; PFG: prefrontal gyrus; PMC: posterior medial cortex; IFG: inferior frontal gyrus; STS: superior temporal sulcus; MNS: mirror neuron system.

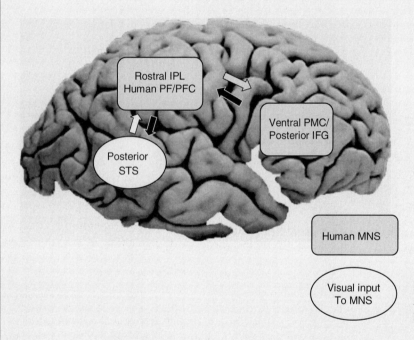

Source: Figure adapted by permission from Iacoboni, & Dapretto. (2006). *Nature Reviews Neuroscience*. Macmillan Publishers Ltd., © 2007.

Iacoboni, M., & Dapretto, M. (2006). The mirror neuron system and the consequences of its dysfunction. *Nature Reviews Neuroscience*, 7, 942–951.

Jacob, P., & Jeannerod, M. (2005). The motor theory of social cognition: A critique. *Trends in Cognitive Sciences*, 9, 21–25.

et al., 2004; Amodio & Frith, 2006; see also Figure 17.1. When other aspects of social cognition such as visual, cognitive and affective domains are considered, neural systems partially overlap – but also differ within the frontal region, as has been shown in stroke patients with damage to right, left or bilateral frontal or non-frontal regions.

The findings suggest that those social cognition tasks that preferentially implicate cognitive processes such as visual perspective-taking, recruit lateral and superior medial frontal regions, while the ability to infer mental states implicates affective connections of the ventral medial frontal cortex to the amygdala and other limbic regions (Stuss *et al.*, 2001).

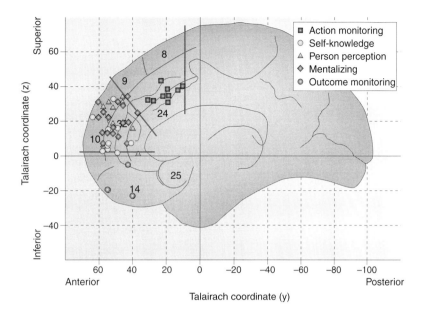

FIGURE 17.1 The figure shows activations in the medial frontal cortex during action monitoring, social cognition and outcome monitoring. The meta-analysis suggests that social cognition tasks (including self-knowledge, person perception, mentalising) activate areas in the anterior rostral medial frontal cortex (arMFC). Monitoring of actions activate the posterior rostral region of the MFC (prMFC) while monitoring of outcomes involves the orbital MFC (oMFC). Figure reprinted by permission from Macmillan Publishers Ltd: *Nature Review Neuroscience*, D.M. Amodio & C.D. Frith (2006). © 2007. (See Plate 13.)

The observation that patients with schizophrenia (with negative symptoms) and patients with ventromedial PFC damage were impaired on second-order affective ToM tasks but not in cognitive ToM conditions would support such a view (Shamay-Tsoory *et al.*, 2007). The extent to which these neural systems overlap is currently not clear. There is also evidence that the neural substrates of mindreading and empathy (i.e., the attribution of emotion) share the same frontal–temporal network – *albeit* with different weighting of subcomponents. While the empathy network relies more on temporal/amygdala and orbitofrontal areas, the mindreading network depends on medial frontal cortex (Lee *et al.*, 2004). Based on a review of neuropsychological and neuroimaging studies on aspects of social cognition, it has been suggested that the anterior regions of the medial frontal cortex are associated with metacognitive representations that enable us to "think about thinking", such as reflecting on what others think of us or on the values that people attach to actions and outcomes (for a review see Amodio & Frith, 2006; see Figure 17.1). In more caudal regions of the medial frontal cortex value is associated with actions while in the more orbital region value is associated with outcomes.

17.4.4. Summary

The nature and extent of the neural systems implicated in pragmatic processing largely depend on which aspects of pragmatics are in focus, the demands the experimental tasks make on the brain system, and, last but not least, the individual characteristics of the individual. It seems unlikely that there is a *single* neural system responsible for linguistic pragmatic behavior. Instead, pragmatic behavior evolves from an interaction of neural systems necessary for the task at hand – in both a bottom–up and top–down fashion. The classical language areas are involved and so are attention and memory (fronto-temporal-parietal systems), monitoring (cingulate cortex) and emotion networks (fronto-limbic-systems). The medial PFC is particularly involved if aspects of social cognition are implicated in pragmatic behavior. The special role of the PFC is generally undisputed. However, the contributing role of its subsystems is still unclear and present research findings are limited and difficult to reconcile. Some tasks that were hypothesized (based on behavioral studies) to involve the right hemisphere have been shown in neuroimaging studies to rely on left or bilateral hemisphere systems. What seems to evolve is that if the brain detects inconsistencies and/or task demands increase, then the right hemisphere is recruited. Generally, however, based on current neuroimaging studies, the specific contribution of the right hemisphere to pragmatic behavior still remains elusive.

We have seen the importance of including psycholinguistic theories to ensure finer graded stimulus construction (e.g., not all metaphors are alike and thus need not necessarily implicate the same processing mechanisms) and its impact on (different) neural systems. Individual differences and variability (e.g., in arousal, affective and cognitive systems) have only rarely been considered although it can be assumed that they also play an important role. Experimental designs that are based on carefully constructed stimuli and carefully analyzed tasks (instead of undifferentiated assumptions on stimulus and task similarity) might be one way to achieve better predictions and test hypotheses on the sufficiency and necessity of the recruited neural systems. And finally, one question that remains unanswered is the very basic one of whether what we test in experimental conditions also applies outside the laboratory.

17.5. CHALLENGES AND FUTURE DIRECTIONS

Linguistic pragmatic behavior is a complex phenomenon with many different aspects. Understanding its processing mechanisms and underlying neural systems requires a cross-disciplinary endeavor – an approach often espoused but in practice still less than optimal. Although research questions from (psycho-)linguistics, philosophical, social cognition and neuroscience research increasingly converge in an effort to unravel how the brain deals with linguistic pragmatic behavior, theory and models discussed in each of these disciplines are less easily reconciled. Hence, one challenge lies in the effort to develop theories and models that satisfy cross-disciplinary standards, which would then serve to test more specific predictions and hypotheses. Besides the theoretical framework, we also need to be guided by methodological, technical and paradigm standards when using neuroimaging techniques. Currently, these are as different and numerous as the number of neuroimaging laboratories that exist. Without some minimal consensus, comparison of studies will remain vague or impossible. Related to this is the requirement to replicate studies – a common and necessary practice in research – an activity which, however, rarely occurs in pragmatic research using hemodynamic techniques. Finally, we need to consider individual variability and differences which have been mostly neglected.

Another dimension that needs development is a more optimal combination of imaging studies to make best use of the strength and limitations of each method. While such a combination of methods has become more and more common in other research areas, it is still rare in pragmatic research. Complementary information could also be gained by investigating pragmatic behavior in patient populations using neuroimaging techniques.

Many neuroimaging studies report areas of activation without much consideration of possible interactions with other brain regions. The relationship between identified cortical and subcortical regions and their organization in interrelated neural systems might be equally important.

Advances in diffusion and fiber tracking imaging might help to elucidate these issues.

A final issue concerns the ecological validity of our findings. We need to verify whether what we test in the laboratory also applies to "real life". This implies, on the one hand, the development of laboratory tasks that better simulate real world behavior (e.g., by using virtual reality techniques), while, on the other hand, applying real world findings to laboratory tests.

Acknowledgments

This work was supported by the Canada Research Chair Program. Thanks go to F.A. Rodden and H.A. Whitaker for helpful suggestions to earlier drafts.

References

Amodio, D.M., & Frith, C.D. (2006). Meeting of minds: The medial frontal cortex and social cognition. *Nature Reviews Neuroscience, 7*, 268–277.

Ash, S., Moore, P., Antani, B.S., McCawley, G., Work, M., & Grossman, M. (2006). Trying to tell a tale. Discourse impairments in progressive aphasia and frontotemporal dementia. *Neurology, 66*, 1405–1413.

Beer, J.S., & Ochsner, K.N. (2006). Social cognition: A multi level analysis. *Brain Research, 1079*, 98–105.

Brady, M., Armstrong, L., & Mackenzie, C. (2005). Further evidence on topic use following right hemisphere brain damage: Procedural and descriptive discourse. *Aphasiology, 19*(8), 731–747.

Brownell, H.H., & Martino, G. (1998). Deficits in inference and social cognition. The effects of right hemisphere brain damage on discourse. In M. Beeman, & C. Chiarello (Eds.), *Right hemisphere language comprehension: Perspectives from cognitive neuroscience* (pp. 309–328). Mahwah, NJ: Lawrence Erlbaum.

Champagne Lavau, M., Stip, E., & Joanette, Y. (2006). Social cognition deficit in schizophrenia: Accounting for pragmatic deficits in communication abilities? *Current Psychiatry Reviews, 2*, 309–315.

Chapman, S.B., Highley, A.P., & Thompson, J.L. (1998). Discourse in fluent aphasia and Alzheimer's disease. *Journal of Neurolinguistics, 11*, 55–78.

Corcoran, R. (2003). Inductive reasoning and the understanding of intention in schizophrenia. *Cognitive Neuropsychiatry, 8*(3), 223–235.

Corcoran, R., Cahill, C., & Frith, C.D. (1997). The appreciation of visual jokes in people with schizophrenia: A study of "mentalizing" ability. *Schizophrenia Research, 24*, 319–327.

Coulson, S., & van Petten, C. (2007). A special role for the right hemisphere in metaphor comprehension? ERP evidence from hemifield presentation. *Brain Research, 1146*, 128–145.

Drury, V.M., Robinson, E.J., & Birchwood, M. (1998). "Theory of mind" skills during an acute episode of psychosis and following recovery. *Psychological Medicine, 28*, 1101–1112.

Eviatar, Z., & Just, M.A. (2006). Brain correlates of discourse processing: An fMRI investigation of irony and conventional metaphor comprehension. *Neuropsychologia, 44*, 2348–2359.

Faust, M., & Mashal, N. (2007). The role of the right cerebral hemisphere in processing novel metaphoric expressions taken from poetry: A divided visual field study. *Neuropsychologia, 45*, 860–870.

Goel, V., & Dolan, R.J. (2001). The functional anatomy of humor: Segregating cognitive and affective components. *Nature Neuroscience, 4*(3), 237–238.

Gregory, C., Lough, S., Stone, V., Erzinclioglu, S., Martin, L., Baron-Cohen, S., & Hodges, J.R. (2002). Theory of mind in patients with frontal variant frontotemporal dementia and Alzheimer's disease: Theoretical and practical implications. *Brain, 125*(4), 752–764.

Happé, F., Brownell, H., & Winner, E. (1999). Acquired "theory of mind" impairments following stroke. *Cognition, 70,* 211–240.

Lee, K.H., Farrow, T.F.D., Spence, S.A., & Woodruff, P.W.R. (2004). Social cognition, brain networks and schizophrenia. *Psychological Medicine, 34,* 391–400.

Leudar, I., Costall, A., & Francis, D. (2004). Theory of mind: A critical assessment. *Theory and Psychology, 14*(5), 571–578.

Lough, S., Kipps, C.M., Treise, C., Watson, P., Blair, J.R., & Hodges, J.R. (2006). Social reasoning, emotion and empathy in frontotemporal dementia. *Neuropsychologia, 44,* 950–958.

Martin, I., & McDonald, S. (2003). Weak coherence, no theory of mind, or executive dysfunction? Solving the puzzle of pragmatic language disorders. *Brain and Language, 85*(3), 451–466.

Mashal, N., Faust, M., & Hendler, T. (2005). The role of the right hemisphere in processing nonsalient metaphorical meanings: Application of principal components analysis to fMRI data. *Neuropsychologia, 43,* 2084–2100.

McCabe, R. (2004). On the inadequacies of theory of mind explanations of schizophrenia. Alternative accounts of alternative problems. *Theory and Psychology, 14*(5), 738–752.

McDonald, S. (1999). Exploring the process of inference generation in sarcasm: A review of normal and clinical studies. *Brain and Language, 68,* 486–506.

McDonald, S. (2000). Exploring the cognitive basis of right-hemisphere pragmatic language disorders. *Brain and Language, 75,* 82–107.

Meilijson, S.R., Kasher, A., & Elizur, A. (2004). Language performance in chronic schizophrenia: A pragmatic approach. *Journal of Speech, Language and Hearing Research, 47,* 695–713.

Mitchell, R.L.C., & Crow, T.J. (2005). Right hemisphere language functions and schizophrenia: The forgotten hemisphere? *Brain, 128,* 963–978.

Myers, P. (2005). Profiles of communication deficits in patients with right cerebral hemisphere damage: Implications for diagnosis and treatment. *Aphasiology, 19,* 1147–1160.

Oliveri, M., Romero, L., & Papagno, C. (2004). Left but not right temporal involvement in opaque idiom comprehension: A repetitive transcranial magnetic stimulation study. *Journal of Cognitive Neuroscience, 16*(5), 848–855.

Papagno, C. (2001). Comprehension of metaphors and idioms in patients with Alzheimer's disease. A longitudinal study. *Brain, 124,* 1450–1460.

Rapp, A.M., Leube, D.T., Erb, M., Grodd, W., & Kircher, T.T.J. (2004). Neural correlates of metaphor processing. *Cognitive Brain Research, 20,* 395–402.

Shamay-Tsoory, S.G., Aharon-Peretz, J., & Levkovitz, Y. (2007). The neuroanatomical basis of affective mentalizing in schizophrenia: Comparison of patients with schizophrenia and patients with localized prefrontal lesions. *Schizophrenia Research, 90,* 274–283.

Shammi, P., & Stuss, D. (1999). Humour appreciation: A role of the right frontal lobe. *Brain, 122,* 657–666.

Siegal, M., & Varley, R. (2002). Neural systems involved in "theory of mind". *Nature Reviews Neuroscience, 3,* 463–471.

Snowden, J.S., Gibbons, Z.C., Blackshaw, A., Doubleday, E., Thompson, J., Craufurd, D., Foster, J., Happé, F., & Nearly, D. (2003). Social cognition in frontotemporal dementia and Huntington's disease. *Neuropsychologia, 41,* 688–701.

Stemmer, B. (2008). Neuropragmatics. In M.J. Ball, M. Perkins, N. Mueller, & S. Howard (Ed.), *Handbook of clinical linguistics* (pp. 61–78). Oxford: Blackwell.

Stemmer, B., & Cohen, H. (2002). Neuropragmatique et lésions de l'hémisphère droit. *Psychologie de l'Interaction, 13*(14), 15–46.

Stuss, D.T., & Alexander, M.P. (2000). Executive functions and the frontal lobes: A conceptual view. *Psychological Research, 63,* 289–298.

Stuss, D.T., Gallup, G.G., & Alexander, M.P. (2001). The frontal lobes are necessary for "theory of mind". *Brain, 124,* 279–286.

Verdon, C.-M., Fossati, P., Verny, M., Dieudonné, B., Teillet, L., & Nadel, J. (2007). Social cognition: An early impairment in dementia of the Alzheimer type. *Alzheimer Disease and Associated Disorders, 21*(1), 25–30.

Winner, E., Brownell, H., Happé, F., Blum, A., & Pincus, D. (1998). Distinguishing lies from jokes: Theory of mind deficits and discourse interpretation in right hemisphere brain-damaged patients. *Brain and Language, 62,* 89–106.

Wright, H.H., & Newhoff, M. (2005). Pragmatics. In L.L. LaPointe (Ed.), *Aphasia and related neurogenic language disorders* (pp. 237–248). New York: Thieme.

Further readings

Mason, R.A., & Just, M.A. (2006). Neuroimaging contributions to the understanding of discourse processes. In M. Traxler & M. A. Gernsbacher (Eds.), *Handbook of Psycholinguistics* (pp. 765–799). Amsterdam: Elsevier.

Ferstl, E.C. (2007). The functional neuroanatomy of text comprehension: What's the story so far? In F. Schmalhofer & C.A. Perfetti (Eds.), *Higher level language processes in the brain: Inference and comprehension processes* (pp. 53–102). Mahwah, NJ: Erlbaum.
Both chapters discuss the neural systems implicated in discourse processing. Although the interpretation of the correlation between behavioral and neuroimaging do not always coincide and are, at times, disputable, the chapters provides a basis for further evaluation, verification and discussion.

Stuss, D., & Knight, R. (2002). *Principles of frontal lobe function.* Oxford: Oxford University Press.
The collection of articles provides an excellent summary on the different aspects of frontal lobe functions – from a theoretical as well as clinical perspective.

The Role of Memory Systems in Disorders of Language

MICHAEL T. ULLMAN

Brain and Language Laboratory, Departments of Neuroscience, Linguistics, Psychology and Neurology
Georgetown University, Washington, DC, USA

ABSTRACT

Language is often assumed to rely on domain-specific neurocognitive substrates. However, this human capacity seems to crucially depend on general-purpose brain memory systems. Evidence suggests that lexical memory relies on declarative memory, which is specialized for arbitrary associations and is rooted in temporal lobe structures. The mental grammar instead depends on procedural memory, a system specialized for rules and sequences and rooted in frontal/basal-ganglia structures. Developmental and adult-onset disorders such as Specific Language Impairment, autism, Tourette syndrome, Parkinson's disease, Huntington's disease and non-fluent aphasia each involve particular grammatical deficits and analogous non-linguistic procedural memory deficits, as well as abnormalities of procedural memory brain structures. Lexical and declarative memory are spared in most of these disorders, and play a compensatory role. In contrast, Alzheimer's disease, semantic dementia, fluent aphasia and amnesia each similarly affect lexical and declarative memory functions and involve abnormalities of declarative memory brain structures, while leaving grammar and procedural memory largely intact.

18.1. INTRODUCTION

Language is often claimed or assumed to rely on dedicated neural, psychological, and computational – that is, neurocognitive – substrates. However, increasing evidence suggests that in fact this uniquely human capacity largely depends on memory systems that also subserve a range of non-language functions, and are moreover found in animals as well as humans (Ullman, 2004). Here we focus on two long-term brain memory systems, declarative and procedural memory, and explore their relations to language in a range of developmental and adult-onset disorders.

18.2. THE DECLARATIVE AND PROCEDURAL MEMORY SYSTEMS

Evidence suggests the existence of multiple distinct memory systems in the brain, including declarative and procedural memory (Eichenbaum & Cohen, 2001; Poldrack & Packard, 2003; Squire *et al.*, 2004; Ullman, 2004). The *declarative memory system* subserves the learning, representation, and use of knowledge about facts (semantic knowledge) and personally experienced events (episodic knowledge), such as the fact that the permafrost in Alaska is melting, or that you fell asleep last night with the light on. The system seems to be specialized for learning arbitrary pieces of information and the associations between them. Knowledge is rapidly learned, with as little as a single exposure being necessary for retention. The learned information can be generalized and used flexibly across different contexts. The acquired knowledge is at least partly, but not completely (Chun, 2000), explicit – that is, available to conscious awareness.

The *procedural memory system* underlies the learning of new and the processing of established perceptual-motor and cognitive "skills" and "habits." The system may be specialized for learning rules and sequences. Learning requires repeated exposure to stimuli, or practice with the skill or habit. Neither the learning nor the remembering of the skills or knowledge is accessible to conscious memory. Thus the system is referred to as an "implicit memory" system. Note, however, that the term "procedural memory" is used here to refer only to one type of implicit "non-declarative" memory system, not to all such

and the protein BDNF (Box 18.1). Inferior frontal cortex, in particular the region of Brodmann's areas (BA) 45/47, is expected to underlie retrieval.

In contrast, the procedural memory system should subserve the gradual implicit learning of knowledge that underlies what is often thought of as the mental grammar – that is, the knowledge subserving the rule-governed sequential and hierarchical combination of complex linguistic representations. The system may be expected to subserve rule-governed knowledge and computations across linguistic domains, including in syntax, morphology (e.g., in regularly inflected forms) and phonology (e.g., in novel word forms, whose phonological elements must somehow be combined). Portions of frontal/basal-ganglia circuits, including BA 44 and the caudate nucleus, should be critical in these linguistic functions (Box 18.2). The caudate nucleus is predicted to play a critical role in acquiring the knowledge, which should be modulated by dopamine, while BA 44 and premotor cortex may be more important in the computation or processing of that knowledge. Given the existence of parallel functionally segregated frontal/basal-ganglia channels (Box 18.2), there is no reason to assume that all grammatical domains or functions should depend on the same channels, or that language-subserving channels necessarily also underlie non-language functions. Rather it is an empirical question as to which segregated channels subserve which linguistic and/or non-linguistic functions (Ullman, 2004, 2006).

The two memory systems are predicted to interact both cooperatively and competitively. Complex linguistic representations are expected not only to be computed by procedural memory, but also to be memorized in declarative memory. The likelihood of such memorization should depend on factors that affect learning or processing information in this system, including item-related variables such as the frequency of complex forms, and subject-related differences in the functionality of the system due to factors such as sex, estrogen levels, and genetic variability (Box 18.3). Additionally, the dysfunction of procedural memory should encourage a compensatory reliance on declarative memory. It is expected that the knowledge learned in declarative memory can be used flexibly, and generalized from existing representations (e.g., *kicked, nicked, licked*) to new representations (*blick-blicked*), allowing for at least some degree of rule-like behavior. Thus at least to some extent the two memory systems should constitute redundant mechanisms for aspects of language (Ullman, 2007). Finally, learning in one system may inhibit learning analogous knowledge in the other, while a dysfunction in one system may enhance the other, potentially facilitating a compensatory role for declarative memory in the use of complex linguistic representations following a dysfunction of the procedural memory system.

These and other predictions regarding the relations between the two memory systems and language have been investigated using a wide range of methods, in developmental, psycholinguistic, neurological, electrophysiological, and neuroimaging studies (Ullman *et al.*, 1997; Ullman, 2004). Across methodologies, the basic approach for testing the predictions laid out above has been to examine whether the expected language and non-language functions both depend on one or the other memory system and that they do so in a similar manner. For example, tasks involving lexical and non-linguistic conceptual–semantic stimuli should elicit analogous functional magnetic resonance imaging (fMRI) activation patterns and event-related potential (ERP) components, and should be similarly modulated by estrogen or BDNF. Moreover, similar tasks with the two types of stimuli should lead to analogous neurocognitive patterns. For example, learning both types of stimuli should engage medial temporal lobe structures, while later recall of both should activate BA 45/47.

Here we focus on evidence related to developmental and adult-onset disorders. As we will see, the evidence suggests the following. Disorders known to affect grammar similarly affect non-linguistic functions of procedural memory, and involve the dysfunction of the neural substrates of this system. Conversely, disorders of procedural memory result in analogous problems for grammar. Thus impairments of grammar and procedural memory co-occur, independently of whether the underlying disorder is traditionally thought of as affecting language or non-linguistic domains. Moreover, these disorders often leave lexical and declarative memory relatively intact. Indeed, in some disorders lexical/declarative memory seems to play a compensatory role for grammatical functions. In contrast, disorders generally thought of as affecting lexical memory similarly affect declarative memory, and vice versa. These disorders often leave grammar and procedural memory and their neural underpinnings largely intact. Thus across a number of disorders one finds double dissociations between declarative and procedural memory across both linguistic and non-linguistic functions.

18.4. DISORDERS OF GRAMMAR AND PROCEDURAL MEMORY

18.4.1. Developmental Disorders

A number of developmental disorders seem to be associated with (different sorts of) procedural memory system dysfunctions and grammatical abnormalities, accompanied by relatively spared lexical and declarative memory. These include Specific Language Impairment (SLI), autism, Tourette syndrome, dyslexia, and Attention Deficit Hyperactivity Disorder (ADHD) (Ullman, 2004; Ullman & Pierpont, 2005; Walenski *et al.*, 2007; Walenski *et al.*, 2006). Here we focus on SLI, autism, and Tourette syndrome.

18.4.1.1. Specific Language Impairment

SLI is usually defined as a developmental disorder of language in the absence of frank neurological damage, hearing deficits, severe environmental deprivation, and mental retardation (Leonard, 1998). The disorder has generally been explained either as an impairment specific to grammar or as a processing deficit, for example, of working memory or of briefly presented stimuli and rapidly presented sequences. However, both classes of theoretical accounts have trouble explaining the pattern of language and non-language deficits, and the heterogeneity across individuals with the disorder (Ullman & Pierpont, 2005).

Evidence suggests that SLI may instead be largely explained by abnormalities of procedural memory system brain structures, in particular of Broca's area (BA 44 and 45) within frontal cortex and the caudate nucleus within the basal ganglia (Ullman & Pierpont, 2005). First, such abnormalities seem to be consistently found in SLI and other developmental language disorders with similar phenotypes, such as disorders of the FOXP2 gene. Second, the pattern of both language and non-language deficits is consistent with such abnormalities. Grammatical impairments are typical of the disorder – not only deficits of receptive and expressive syntax, but also of morphology and phonology. The procedural system dysfunction also clearly extends beyond language. Motor deficits are widely observed in children and adults with SLI. Individuals with SLI have particular difficulty on motor tasks involving complex sequences of movements, such as moving pegs, sequential finger opposition, and stringing beads. The disorder also results in deficits of other functions that depend on the brain structures underlying procedural memory, including working memory, processing rapidly presented sequences, and mental rotation (Leonard, 1998; Ullman & Pierpont, 2005).

In contrast, lexical and declarative memories are relatively spared in the disorder, as evidenced by largely intact word recognition and comprehension, normal word learning and lexical–semantic organization, and spared learning in declarative memory (Ullman & Pierpont, 2005). However, the *retrieval* of lexical knowledge (word finding) is difficult for individuals with SLI, as might be expected if the frontal and basal ganglia structures underlying retrieval (e.g., BA 45) are dysfunctional.

Children and adults with SLI can compensate for their deficit with lexical and declarative memory (Ullman & Pierpont, 2005). For example, unlike typically developing control subjects, individuals with SLI show consistent "frequency effects" on regularly inflected past-tense and plural forms – that is, correlations between the frequency of these forms and performance at producing them. This suggests that individuals with SLI, unlike typically developing subjects, consistently retrieve regular past-tense forms from memory rather than combining them in the mental grammar.

Additionally, these individuals sometimes learn explicit rules such as "add –ed to make a past-tense form." For example, one child reported that "at school, learn it at school. In the past tense put -e-d on it. If it's today it's-i-n-g. Like swimming: 'I went swimming today' and 'Yesterday I swammed'" (Ullman & Pierpont, 2005).

It is important to emphasize that it is *not* being claimed that all individuals identified as SLI are afflicted with a dysfunction of the procedural memory system (Ullman & Pierpont, 2005). Given the broad definition of SLI this would clearly be too strong a claim. Nevertheless, it is predicted that many if not most individuals diagnosed with SLI do have such a dysfunction. Moreover, much of the heterogeneity in the disorder can be explained by the particular combination of frontal/basal-ganglia channels or other procedural system structures that are affected (with the likelihood of SLI presumably being higher with greater or multiple dysfunctions; e.g., see Bishop, 2006), as well as by the compensatory abilities of other systems, in particular of declarative memory (see Box 18.3).

18.4.1.2. Autism

Autism, also referred to as Autism Spectrum Disorder (ASD), is a developmental disorder associated with deficits of language and communication, as well as of social interaction, motor function, and certain other domains. Roughly 20% of children with autism are essentially non-verbal, using fewer than five words per day (Lord *et al.*, 2004). Others acquire functional language to varying degrees, although the exact profile of language and communicative abilities is heterogeneous. Most research on language deficits in autism has focused on the pragmatic difficulties found in the disorder – that is, impairments in using and interpreting language appropriately for the social and real-world contexts in which utterances are made.

Evidence suggests that autism is also associated with deficits of grammar and of non-linguistic functions that depend on the procedural memory system (Ullman, 2004; Walenski *et al.*, 2006). High-functioning individuals with autism may show syntactic impairments in both receptive and expressive language. Inflectional morphology and regular inflection in particular, has been found to be abnormal in both elicited and spontaneous speech. Whereas no deficits are observed in the processing of individual phonemes, impairments are often reported in processing sound combinations in non-words. Both the acquisition and processing of both motor and non-motor sequences have been reported to be abnormal. Complex sequences seem especially problematic. Impairments of other functions that depend on the brain structures of the procedural memory system, such as rapid temporal processing and working memory, have also been observed. Although studies of the neurobiology of procedural and declarative memory brain structures in autism have produced a number of inconsistent results, some patterns are beginning to emerge. Of particular interest here, abnormalities of left frontal cortex, especially Broca's area, have consistently been found in studies that have examined this region.

In contrast, lexical and conceptual knowledge appear to remain largely intact in high-functioning autistics. In fact, in a recent study object naming was found to be *enhanced* in ASD children, as compared to typically developing children (Walenski, Mostofsky, Larson, & Ullman, in press). Additionally, tasks probing learning in declarative memory suggest normal rote learning of individual items such as telephone numbers, as well as intact associative learning, such as remembering word pairs. However, learning personally experienced episodes seems to be consistently impaired, perhaps due to the particular dependence of episodic memory on frontal structures.

Individuals with ASD may compensate for their grammatical deficits with lexical and declarative memory (Walenski *et al.*, 2006). Children with autism rely much more than typically developing children on "formulaic speech," that is speech with prefabricated sequences of words that appear to be stored whole in memory. For example, ASD speech is marked by repetitive and stereotyped utterances such as *thank you* or *you're welcome*.

Moreover, in two functional neuroimaging studies of receptive syntax, ASD subjects showed greater activation, as compared to healthy control subjects, in temporal/temporo-parietal cortex, but less activation in premotor cortex, Broca's region, and the basal ganglia, among other structures (Müller *et al.*, 1999; Just *et al.*, 2004). Similarly, a structural magnetic resonance imaging (MRI) study reported that language-impaired ASD (and SLI) subjects showed a volumetric decrease in left frontal regions, especially in Broca's area, but an increase in a left temporal lobe region (De Fossé *et al.*, 2004). These patterns may also reflect an increased compensatory reliance on declarative memory in the face of dysfunctional grammatical and procedural functions (Walenski *et al.*, 2006).

18.4.1.3. Tourette Syndrome

Tourette syndrome (TS) is a developmental disorder most obviously characterized by the presence of verbal and motor tics, which are both fast and involuntary. The tics can be explained by abnormal dopamine levels and structural abnormalities of the basal ganglia, in particular of the caudate nucleus, resulting in decreased inhibition of frontal activity, a hyperkinetic behavioral profile, and an inability to suppress the tics (Albin & Mink, 2006).

Given this frontal/basal-ganglia dysfunction, it is not surprising that procedural memory and related functions have also been reported to be abnormal in the disorder (Walenski *et al.*, 2007). For example, the acquisition of implicit probabilistic rules (in the "weather prediction task"), which depends at least in part on the caudate nucleus, has been found to be impaired in TS (Marsh *et al.*, 2004). However, deficits have not been found in all tasks traditionally used to probe procedural learning, such as the serial reaction time task, perhaps because of compensatory learning in declarative memory.

Indeed, lexical and declarative memory are largely spared in TS (Walenski *et al.*, 2007). TS children have shown normal performance at both picture naming and stem completion, suggesting that lexical representations remain intact. Acquiring new information in declarative memory, such as list learning and remembering the location of objects, also seems unproblematic (Marsh *et al.*, 2004). And whereas the implicit learning of procedural knowledge in the weather prediction task was found to be impaired in TS, normal performance was observed in a separate test of explicit knowledge in the same subjects (Marsh *et al.*, 2004).

A recent study examined grammatical and lexical processing in a past-tense production task of regular and irregular forms, as well as the processing of previously learned procedural and declarative knowledge in a picture naming task of objects that either involve procedural motor-skill knowledge (tools and other manipulated objects; e.g., *hammer*) or do not involve such knowledge (e.g., *elephant*) (Walenski *et al.*, 2007). TS children were significantly faster than typically developing control children at producing rule-governed past tenses (*slip-slipped, plim-plimmed*) but not irregular and other unpredictable forms (*bring-brought, splim-splam*). They were also faster than controls at naming pictures of manipulated (*hammer*) but not non-manipulated (*elephant*) items. These data were not explained by a wide range of potentially confounding subject- and item-level factors, such as the age and IQ of the subjects, and the frequency and phonological complexity of the items. The results suggest that the processing of procedurally based knowledge, both of grammar and of manipulated objects, is particularly *speeded* in TS. Thus the frontal/basal-ganglia abnormalities in the disorder may lead not only to tics, but to a wider range of abnormally rapid behaviors, including in the processing of procedural knowledge both in the naming of manipulated objects and the production of rule-governed forms in language.

18.4.2. Adult-Onset Disorders

Various neurodegenerative and other adult-onset disorders affect grammar, non-linguistic aspects of procedural memory, and the neural substrates of this system. The different disorders are associated with damage to different portions of procedural memory system brain structures. These lead to different types of dysfunctions, each of which seem to be similar across language and non-language domains.

18.4.2.1. Parkinson's Disease

Parkinson's disease (PD) is associated with the degeneration of dopamine-producing neurons, particularly in the basal ganglia (substantia nigra). This degeneration, which results in high levels of inhibition in the motor and other frontal cortical areas to which the basal ganglia project, is thought to explain why PD patients show suppression of motor activity (hypokinesia) and have difficulty expressing motor sequences (Jankovic & Tolosa, 2007).

The degeneration has also been implicated in PD patients' impairments at procedural learning in a number of perceptual-motor and cognitive tasks, such as sequence learning in the serial reaction time task and probabilistic rule learning in the weather prediction task. In contrast, temporal-lobe regions remain largely intact and lexical and declarative memory relatively spared in low- and non-demented PD patients (Knowlton *et al.*, 1996; Ullman *et al.*, 1997), although these patients often have word finding difficulties, consistent with a role for frontal/basal-ganglia circuits in retrieval.

Grammatical deficits are also found in PD. Even non-demented PD patients show impairments at both expressive and receptive syntax (Grossman *et al.*, 2000; Ullman, 2004). Non-demented severely hypokinetic PD patients show impairments at the production of *–ed*-affixed past-tense forms (e.g., *walked, plagged*), relative to irregulars (Ullman *et al.*, 1997). A similar though weaker pattern is found in patients with lower levels of hypokinesia (Longworth *et al.*, 2005). However, it remains unclear whether the various grammatical deficits observed in PD can be attributed directly to the basal ganglia (perhaps even due to problems with non-procedural functions, such as syntactic integration; Friederici *et al.*, 2003), or whether they can instead or additionally be explained by excessive inhibition from the basal ganglia to frontal cortex, which itself may responsible for rule-governed computation (Ullman, 2006).

18.4.2.2. Huntington's Disease

Like Parkinson's disease, Huntington's disease (HD) results in basal ganglia degeneration, though the disease rapidly progresses to cortical regions as well. However, it affects different basal ganglia circuits than PD, resulting in the disinhibition of frontal areas receiving basal ganglia projections (Jankovic & Tolosa, 2007). This is thought to explain the insuppressible movements – chorea, a type of hyperkinesia – found in patients with HD. These patients have also been reported to show procedural learning deficits, as well as impairments in both expressive and receptive syntax (Murray, 2000).

HD patients have been found to show the opposite pattern of PD patients not only in the type of movement impairment (the suppressed movements of hypokinesia versus the unsuppressed movements of hyperkinesia), but also in the type of errors on *–ed*-suffixed forms. Ullman *et al.* (1997) observed that, unlike normal control subjects, HD patients produced many forms like *walkeded, plaggeded, dugged*, and *digged*. The patients did not produce analogous errors on irregulars like *dugug* or *keptet*, suggesting that the affixed error forms are not explained by articulatory deficits. Rather the data suggest unsuppressed *–ed*-suffixation. This conclusion is strengthened by the finding that the production rate of these over-suffixed forms correlated with the degree of chorea, across patients. Another study, which also found increased rates of over-regularization in HD patients (Longworth *et al.*, 2005), did not examine the correlation between chorea and the production of these forms.

The contrasting patterns in PD and HD, linking movement and *–ed*-suffixation in two distinct types of impairments related to two types of basal ganglia damage, strongly implicate frontal/basal-ganglia circuits in *–ed*-suffixation. They support the hypothesis that these circuits underlie the processing of grammatical rules as well as movement, and suggest that they play similar roles in the two domains. Moreover, given that disinhibition of frontal activity is implicated in the unsuppressed movements of HD, such disinhibition also seems likely to account for the unsuppressed affixation also observed in the disorder.

18.4.2.3. Non-fluent Aphasia

The term "aphasia" generally refers to language impairments resulting from one or more focal lesions in the brain. Clusters of symptoms tend to co-occur in types (syndromes) of aphasia. Although there are a number of different adult-onset aphasia syndromes, several of these can be grouped into either of two larger categories, which are often referred to as non-fluent and fluent aphasia (Feinberg & Farah, 1997).

Non-fluent aphasia seems to reflect, at least in part, damage to brain structures of the procedural memory system. It is associated with lesions of left inferior frontal regions, in particular Broca's area and nearby cortex, as well as of the basal ganglia, inferior parietal cortex, and anterior superior temporal cortex (Feinberg & Farah, 1997). It is also associated with agrammatism, in both expressive and receptive language. Agrammatic speech is characterized by abnormalities in the use of free and bound grammatical morphemes such as auxiliaries, determiners, and affixes. In receptive language patients have particular problems using the syntactic structure of sentences to understand their meaning. Non-fluent aphasics have also been found to have greater difficulty with regular than irregular morphology in both expressive and receptive language tasks, even holding constant word frequency, word length, and various other factors (Ullman *et al.*, 1997; Ullman *et al.*, 2005).

Non-fluent aphasia is strongly associated with impairments of non-linguistic functions that depend on the procedural memory system. These aphasics typically have a range of motor impairments, from articulation to the execution of complex learned motor skills, particularly those involving sequences (ideomotor apraxia) (Feinberg & Farah, 1997). Interestingly, they have also been found be impaired at learning new sequences, in particular sequences containing abstract structure (Goschke et al., 2001; Dominey et al., 2003). However, because the patients in these studies may also have had basal ganglia lesions, it is premature to conclude that the frontal regions alone are critical for learning sequences, even sequences containing abstract structure.

In contrast, non-fluent aphasics are relatively spared in their recognition and comprehension of content words, such as nouns and adjectives. Nevertheless, as would be expected with damage to Broca's area and the basal ganglia, these aphasics generally have lexical retrieval difficulties. Interestingly, evidence also suggests that non-fluent aphasics can compensate for their grammatical impairments by memorizing complex forms in lexical memory, as evidenced by frequency effects for regular past-tense forms (Drury & Ullman, 2002).

18.5. DISORDERS OF LEXICON AND DECLARATIVE MEMORY

Several adult-onset disorders affect lexical and declarative memory, leaving grammar and procedural memory largely intact. Importantly, the particular type of lexical and declarative memory impairments vary across the disorders, depending on the nature of their underlying neuropathology. Whereas disorders with neocortical temporal lobe lesions affect the use of established lexical and conceptual–semantic knowledge (in Alzheimer's disease, semantic dementia, fluent aphasia), those with severe medial temporal damage instead or additionally impair the learning of new lexical and non-linguistic declarative knowledge (in Alzheimer's disease and amnesia).

18.5.1. Alzheimer's Disease

Alzheimer's disease (AD) affects medial as well as neocortical temporal-lobe structures, leaving the basal ganglia and frontal cortex, particularly Broca's area and motor regions, relatively intact (Feinberg & Farah, 1997). Consistent with this degeneration, both the learning of new and the use of established lexical and conceptual–semantic knowledge is impaired in AD. In contrast, AD patients are relatively spared at acquiring and processing motor and cognitive skills, and at both expressive and receptive syntax. Additionally, patients with severe deficits at object naming or fact retrieval make more errors at producing irregular than –ed-affixed past-tense forms (Ullman et al., 1997; Cortese et al., 2006). And across AD patients, error rates at

object naming and fact retrieval correlate with error rates at producing irregular but not –ed-affixed past tenses.

18.5.2. Semantic Dementia

Semantic dementia is associated with the progressive degeneration of inferior and lateral temporal lobe regions, and to a lesser extent medial temporal lobe structures, leaving inferior frontal, premotor, and basal ganglia structures largely intact (Grossman & Ash, 2004). The disorder results in impairments using established lexical and non-linguistic conceptual knowledge, with spared motor, syntactic, and phonological abilities. Additionally, these patients do not have particular difficulty acquiring new knowledge in declarative memory, consistent with a relative sparing of medial temporal structures. Like AD patients, semantic dementia patients have more trouble producing and recognizing irregular than –ed-suffixed past-tenses, and the degree of their impairment on irregulars has been found to correlate with their performance on lexical/semantic memory tasks (Patterson et al., 2001; Cortese et al., 2006).

18.5.3. Fluent Aphasia

Fluent aphasia is at least partly associated with damage to declarative memory brain structures, in particular left temporal and temporo-parietal regions, though the lesions often extend further into inferior parietal structures. Fluent aphasics have impairments in the production, reading, and recognition of the sounds and meanings of content words, as well as of conceptual knowledge (Feinberg & Farah, 1997; Ullman, 2004). In contrast, these aphasics tend to produce syntactically well-structured sentences, and to not omit morphological affixes. However, damage in and around inferior parietal cortex in fluent aphasia can lead to grammatical impairments, supporting a role for this region in the mental grammar and procedural memory. In direct comparisons of regular and irregular morphology, fluent aphasics have been found to show the opposite pattern to that of non-fluent aphasics, with worse performance at irregular than regular forms (Ullman et al., 1997; Ullman et al., 2005).

18.5.4. Anterograde Amnesia

Bilateral lesions of medial temporal lobe structures can lead to an inability to learn new information about facts, events, and words (Eichenbaum & Cohen, 2001; Squire et al., 2004). Neither phonological nor semantic lexical knowledge is acquired following such damage, supporting the hypothesis that these structures underlie the learning of word forms as well as word meanings (Ullman, 2004). This "anterograde amnesia" is typically accompanied by the loss of information for a period preceding the damage – that is temporally graded retrograde amnesia. However, knowledge acquired substantially before lesion onset is largely spared. Thus even though medial temporal lobe structures seem to underlie the learning of new lexical information, adult-onset amnesics should remember words learned during childhood. As expected, the well-studied amnesic H.M. did not differ from normal control subjects at syntactic processing tasks, or at the production of regular or irregular forms in past-tense, plural and derivational morphology (Ullman, 2004).

18.6. CHALLENGES AND FUTURE DIRECTIONS

In sum, evidence from both developmental and adult-onset disorders suggests that the declarative and procedural memory systems play important roles in language, and that disorders of these systems similarly affect language and non-language functions. However, open questions remain. First, some existing evidence is inconsistent. For example, procedural learning deficits have been found in some but not other studies in autism, TS, PD, and HD. These inconsistent findings are not yet understood. Second, various important experiments have not yet been carried out. Thus at the time of writing apparently no published studies have tested SLI individuals on procedural learning, which is predicted to be impaired in this disorder. Third, certain important issues have yet to be resolved. For example, the functional roles played by the basal ganglia, and the precise nature of language deficits in basal ganglia disorders, are still not entirely clear. Nevertheless, the extant evidence clearly supports the view that the two memory systems subserve language, and that many disorders that affect language can be profitably characterized as disorders of these systems. Importantly, the study of the two memory systems at many levels, in both humans and animals, should lead not only to a deeper understanding of language and the disorders that affect it, but also to advances in the diagnosis and therapy of these disorders (Ullman, 2004; Ullman & Pierpont, 2005).

Acknowledgments

The author thanks Alexis Allen, Cristina Dye, Matthew Walenski, and Brigitte Stemmer for helpful comments. This chapter was written with support for the author from NIH R01 HD049347, NIH R03 HD050671, and research grants from the National Alliance for Autism Research and the Mabel Flory Trust.

References

Albin, R.L., & Mink, J.W. (2006). Recent advances in Tourette syndrome research. *Trends in Neurosciences, 29*(3), 175–183.

Bishop, D.V.M. (2006). What causes specific language impairment in children? *Current Directions in Psychological Science, 15*(5), 217–221.

Chun, M.M. (2000). Contextual cueing of visual attention. *Trends in Cognitive Sciences, 4*(5), 170–178.

Cortese, M.J., Balota, D.A., Sergent-Marshall, S.D., Buckner, R.L., & Gold, B.T. (2006). Consistency and regularity in past tense verb generation in healthy aging, alzheimer's disease, and semantic dementia. *Cognitive Neuropsychology, 23*(6), 856–876.

De Fossé, L., Hodge, S., Makris, N., Kennedy, D., Caviness, V., McGrath, L., Steele, S., Ziegler, D., Herbert, M., Frazier, J., Tager-Flusberg, H., & Harris, G. (2004). Language association cortex asymmetry in autism and specific language impairment. *Annals of Neurology, 56,* 757–766.

Dominey, P.F., Hoen, M., Blanc, J.-M., & Lelekov-Boissard, T. (2003). Neurological basis of language and sequential cognition: Evidence from simulation, aphasia, and ERP studies. *Brain and Language, 83*(2), 207–225.

Drury, J.E., & Ullman, M.T. (2002). The memorization of complex forms in aphasia: Implications for recovery. *Brain and Language, 83,* 139–141.

Eichenbaum, H., & Cohen, N.J. (2001). *From conditioning to conscious recollection: Memory systems of the brain.* New York: Oxford University Press.

Feinberg, T.E., & Farah, M.J. (Eds.) (1997). *Behavioral neurology and neuropsychology.* New York: McGraw-Hill.

Friederici, A.D., Kotz, S.A., Werheid, K., Hein, G., & von Cramon, D.Y. (2003). Syntactic comprehension in Parkinson's disease: Investigating early automatic and late integrational processes using event-related brain potentials. *Neuropsychology, 17*(1), 133–142.

Goschke, T., Friederici, A., Kotz, S.A., & van Kampen, A. (2001). Procedural learning in Broca's aphasia: Dissociation between the implicit acquisition of spatio-motor and phoneme sequences. *Journal of Cognitive Neuroscience, 13*(3), 370–388.

Grossman, M., & Ash, S. (2004). Primary progressive aphasia: A Review. *Neurocase, 10*(1), 3–18.

Grossman, M., Kalmanson, J., Bernhardt, N., Morris, J., Stern, M.B., & Hurtig, H.I. (2000). Cognitive resource limitations during sentence comprehension in Parkinson's disease. *Brain and Language, 73*(1), 1–16.

Jankovic, J.J., & Tolosa, E. (2007). *Parkinson's disease and movement disorders* (5th edn). Philadelphia, PA: Lippincott Williams & Wilkins.

Just, M., Cherkassky, V., Keller, T., & Minshew, N. (2004). Cortical activation and synchronization during sentence comprehension in high-functioning autism: Evidence of underconnectivity. *Brain, 127,* 1811–1821.

Knowlton, B.J., Mangels, J.A., & Squire, L.R. (1996). A neostriatal habit learning system in humans. *Science, 273,* 1399–1402.

Leonard, L.B. (1998). *Children with specific language impairment.* Cambridge, MA: MIT Press.

Longworth, C.E., Keenan, S.E., Barker, R.A., Marslen-Wilson, W.D., & Tyler, L.K. (2005). The basal ganglia and rule-governed language use: Evidence from vascular and degenerative conditions. *Brain, 128*(3), 584–596.

Lord, C., Risi, S., & Pickles, A. (2004). Trajectory of language development. In M.L. Rice & S.F. Warren (Eds.), *Developmental language disorders* (pp. 7–29). Mahwah, NJ: Lawrence Erlbaum Associates.

Marsh, R., Alexander, G.M., Packard, M.G., Zhu, H., Wingard, J.C., Quackenbush, G., & Peterson, B.S. (2004). Habit learning in Tourette syndrome: A translational neuroscience approach to developmental psychopathology. *Archives of General Psychiatry, 61,* 1259–1268.

Müller, R.A., Behen, M.E., Rothermel, R.D., Chugani, D.C., Muzik, O., Mangner, T.J., & Chugani, H.T. (1999). Brain mapping of language and auditory perception in high-functioning autistic adults: A PET study. *Journal of Autism and Developmental Disorders, 29*(1), 19–31.

Murray, L.L. (2000). Spoken language production in Huntington's and Parkinson's diseases. *Journal of Speech, Language, and Hearing Research, 43*(6), 1350–1366.

Patterson, K., Lambon Ralph, M.A., Hodges, J.R., & McClelland, J.L. (2001). Deficits in irregular past-tense verb morphology associated with degraded semantic knowledge. *Neuropsychologia, 39,* 709–724.

Poldrack, R.A., & Packard, M.G. (2003). Competition among multiple memory systems: Converging evidence from animal and human brain studies. *Neuropsychologia, 41*(3), 245–251.

Squire, L.R., Stark, C.E., & Clark, R.E. (2004). The medial temporal lobe. *Annual Review of Neuroscience, 27,* 279–306.

Ullman, M.T. (2004). Contributions of memory circuits to language: The declarative/procedural model. *Cognition, 92*(1–2), 231–270.

Ullman, M.T. (2006). Is Broca's area part of a basal ganglia thalamocortical circuit? *Cortex, 42*(4), 480–485.

Ullman, M.T. (2007). The biocognition of the mental lexicon. In M.G. Gaskell (Ed.), *The Oxford handbook of psycholinguistics.* Oxford, UK: Oxford University Press, 267–286.

Ullman, M.T., & Pierpont, E.I. (2005). Specific language impairment is not specific to language: The procedural deficit hypothesis. *Cortex, 41*(3), 399–433.

Ullman, M.T., Miranda, R.A., & Travers, M.L. (2008). Sex differences in the neurocognition of language. In J.B. Becker, K.J. Berkley, N. Geary, E. Hampson, J. Herman & E. Young (Eds.), *Sex on the brain: From genes to behavior.* New York, NY: Oxford University Press, 291–309.

Ullman, M.T., Pancheva, R., Love, T., Yee, E., Swinney, D., & Hickok, G. (2005). Neural correlates of lexicon and grammar: Evidence from the production, reading, and judgment of inflection in aphasia. *Brain and Language, 93*(2), 185–238.

Ullman, M.T., Corkin, S., Coppola, M., Hickok, G., Growdon, J.H., Koroshetz, W.J., & Pinker, S. (1997). A Neural Dissociation within Language: Evidence that the mental dictionary is part of declarative memory, and that grammatical rules are processed by the procedural system. *Journal of Cognitive Neuroscience, 9*(2), 266–276.

Walenski, M., Mostofsky, S.H., & Ullman, M.T. (2007). Speeded processing of grammar and tool knowledge in Tourette's syndrome. *Neuropsychologia, 45,* 2447–2460.

Walenski, M., Tager-Flusberg, H., & Ullman, M.T. (2006). Language in autism. In S.O. Moldin & J.L.R. Rubenstein (Eds.), *Understanding autism: From basic neuroscience to treatment* (pp. 175–203). Boca Raton, FL: Taylor and Francis Books.

Walenski, M., Mostofsky, S.H., Larson, J.C.G., & Ullman, M.T. (in press). Enhanced picture naming in autism. *Journal of Autism and Developmental Disorders.*

Further Readings

Eichenbaum, H., & Cohen, N.J. (2001). *From conditioning to conscious recollection: Memory systems of the brain*. New York: Oxford University Press.
Wide ranging examination of brain memory systems with discussions of the brain systems that mediate the different memory systems, the information they process and the principles by which they operate.

Squire, L.R., Stark, C.E., & Clark, R.E. (2004). The medial temporal lobe. *Annual Review of Neuroscience, 27,* 279–306.
Review of the role of medial temporal lobe structures (hippocampal regions, perirhinal, entorhinal and parahippocampal cortices) in declarative memory. The authors argue that medial temporal lobe structures are implicated in establishing and maintaining long-term memories, but become independent of these memories through a process of consolidation.

Ullman, M.T. (2004). Contributions of memory circuits to language: The declarative/procedural model. *Cognition, 92*(1–2), 231–270.
Theoretical and empirical overview of the declarative/procedural model of language. The author argues that certain language disorders can be viewed as disorders that primarily affect either the declarative or procedural memory system.

19

The Relation of Human Language to Human Emotion

DIANA VAN LANCKER SIDTIS

New York University Department of Speech Pathology, New York, NY, USA

ABSTRACT

Emotions, thought, and language form distinct but overlapping cognitive domains. Thought and language are mutually reliant. An intimacy between thought and emotion is also acknowledged. Although language competence is independent of emotion, most linguistic performance is tinged by emotional content. Affective content in linguistic expression belongs to "paralanguage," which lies outside linguistic structures, but emotional content resides in lexical and phrasal units as well. A prominent paralinguistic component is prosody. Depending on what aspects of prosodic expression and comprehension are studied, right and left hemispheric as well as subcortical structures have been implicated. Brain structures for emotional behaviors and linguistic–affective expression are disparate, but interactional. Current interest in the relations of emotion to language in brain processing has intensified, posing new questions and utilizing advanced neuroscience technologies.

19.1. INTRODUCTION

Language, thought, and emotion form a basic triad in human psychology. Few have doubted the interdependence between the first two, language and thought. In the last century, cognitive psychology proceeded for many years without mention of emotion. But currently the importance of emotion in thinking is well recognized, giving rise to the concept of emotional intelligence, and constituting a new field, emotion psychology. In contrast, only recently has the interface of human language and human emotion received serious consideration. In linguistics, the role played by emotions in language structure and use has been examined indirectly in metaphor analysis, rating studies, and discourse analysis. In psycholinguistics, priming studies have expanded to examine the role of emotion in word storage and retrieval. Functional brain imaging in its many forms aims to identify

the interfaces between emotion and language processing. There is now considerable focus on the subtle and intimate interactions of language and emotion, which appear to operate independently and rely on different brain structures. New neuroscientific approaches to behavior can be used to address such questions. The purpose of this article is to revisit and explore the relationship between language and emotion from the perspective of these currents of interest.

19.2. LANGUAGE AND EMOTION: THEORETICAL PERSPECTIVES

When language is viewed as a structural system – as a vehicle that can be used to discuss or describe anything, its relationship to emotional experiencing is tangential. This is the view that language and emotion have no necessary relationship at all. The primary structural levels of language, phonology, syntax, and the semantic lexicon, provide tools, equally capable of verbalizing about the cosmos, an engineer's diagram, or a personal feeling. In this view, language and emotions differ intrinsically, like bicycles and fish. Evolutionarily, it has been suggested that two signaling systems underlie human communication, one with a limbic basis, corresponding to animal vocalization, and the other cortically represented and unique to humans. In this view, emotive considerations belong to paralanguage, which, by definition, lies outside the language system. To understand the evolution of human language, the debate must focus on the emergence of symbol systems and grammar.

Another perspective on the relationship between human emotion and human language holds that language has roots in emotional expression. Cries, songs, and shouts are seen as motivating linguistic development. Emotional language as well as "automatic" or formulaic expressions have their origin in earlier evolutionary development (Code, 2005), and continue to hold an important place in linguistic competence. In this view, human language is not independent of or separate from emotional and formulaic speech, but is made up of two highly integrated processing modes, novel and holistic (Van Lancker Sidtis, 2004).

Attempts at discovering structure in any one component of the "triad" have been the most successful in the case of language, which is made up of units (e.g., phonemes, morphemes, words, sentences) and rules for combining or organizing them (grammar or syntax). Some structure also has been discovered in thinking, or cognition, through various types of psychological modeling. Emotional phenomena have been the least amenable to codification. Language and emotion have an uneasy coexistence, being disparate entities as well as coworkers in the business of communication. However, disparate language and emotion may be, though, emotional nuances tinge all but the most carefully constructed linguistic expression.

19.3. UNIVERSALITY OF EMOTION EXPRESSION IN LANGUAGE

Universals of emotional expression across languages have been examined in nonverbal aspects of language – prosody and gesture. In humans, cultural differences exist not only in language itself, but also in extra speech sounds, gestures, and facial expressions. Several researchers have found that judgments of affective expression in speech were more difficult across cultural boundaries than within a cultural group (Breitenstein et al., 2001; Scherer et al., 2001; Chen et al., 2004). Processing of emotional adjectives can be investigated using electrophysiological measures, such as electromyography and evoked responses. Studies of emotional words in multilingual speakers reveal interesting findings (Dewaele, 2004). Intensity of emotional content in words has been shown to differ in the first as compared to the second language (Ayçiçegi & Harris, 2004). Evidence that bilingual speakers perceive emotional meanings in their native language more intensely than in the second language comes from greater autonomic reactivity, as measured by galvanic skin response (GSR), to emotional words in the native language. Similarly, recall of autobiographical memories, which often involve use of emotional terms, varies with the language status, first or later learned, used by bilingual speakers (Marian & Neisser, 2000; Matsumoto & Stanny, 2006).

Neuropsychological disorders of emotion affect communicative competence, and conversely, speech and language disorders interfere with efficient communication of emotional and attitudinal information. Therefore, rather than viewing language and human emotion as autonomous and independent, there is ample evidence that emotions, moods, affect, and attitudes underlie and inform nearly every normal spoken expression; that human language contains components that serve to convey this information; that listeners naturally interpret such information from spoken language. The lack of normal sounding paralinguistic material in speech synthesized by computer is a major cause of ratings of unpleasantness or dissatisfaction by listeners. Computer synthesis technology is looking toward enhancing intonational naturalness of speech signal, but this a particularly challenging goal, given the complexity of this domain of speech.

Emotional states may be viewed in terms of their bodily manifestations (e.g., a smile) or the subjective experience of the emotion (e.g., feeling happy). Some knowledge of important cerebral structures and circuits associated with emotional experiencing and emotional behaviors has been derived from animal lesion experiments and from studies of humans with brain damage or psychiatric disabilities. These structures and circuits are heterogeneous and are widely distributed in the brain. Physiological manifestations of emotional states include changes in the autonomic nervous system and the release of neurotransmitters, neuroendocrine secretions, and hormones. Production of emotional words can also be evaluated using these techniques. In this domain of study, verbal signs are examined – distinctive word choices (e.g., strong adjectives), selection of conventional speech formulas with special emotive impact (e.g., *Shut up!*), cursing, or exclamations that may have begun as a reflexive vocalization but have evolved to onomatopoeic words (*ow!*; *wow!*; *ouch!*; *jeez!*). Nonverbal vocal gestures such as sighing, meaningful pausing, or emitting more or less conventionalized cries may also occur. Study of a left-hemispherectomized adult, who developed normally until onset of a brain tumor in his left hemisphere, indicated that a range of vocal-emotive forms, verbal and nonverbal, are supported by the right hemisphere. Following neurosurgery, this patient was profoundly aphasic, but freely utilized emotive vocalizations (see Table 19.1).

19.4. THE LINGUISTIC COMMUNICATION OF EMOTION

Language expresses emotion both indirectly and directly, implicitly or explicitly, intentionally or not, using different "levels" of linguistic form: phonetic, syntactic, semantic, and different "domains," including pragmatics. Infants quickly learn much about communicative interaction from prelinguistic vocal information. Infants recognize their mother's voices shortly after birth. A range of attitudes and emotions is identifiable in the melodies of mother's speech to their infants, which appears across cultures. Developmental studies comparing normal and language delayed children document maturational schedules in abilities to recognize emotions in speech.

Most obvious to the role of emotions in language, the set of words – the language lexicon – provides a rich source of terms to express emotional information about oneself, one's reaction to the world, one's attitudinal judgments, and one's feelings. A prime example of implicit emotional information lies in connotative meanings; connotations convey attitudes, adding to the referential meaning. The lexical items "slim,"

TABLE 19.1 Emotional and Pragmatic Vocal Expressions

Time (min)	Expletives	Pause-fillers	Nonverbal vocalizations
0	Goddammit	Uh	
	Goddammit	boy, well	umh mmm
		well, no	duhh
		Uh	umh, duh
1	Goddammit	no, eh	
		Ah	neah
		Uh	
2	God –	Nah	ugh
	Goddammit	Ah	(laugh)
		Um	(sigh)
		Mm	mm(sigh)
3		Uh	tsk
		oh, yes	whaa
		No	nah
		well, yes	wha
4	Shit	oh, yes	
	Goddammit		(laugh)
		Ah	
			ahh
5	Goddammit	Oh	
		Ah	
	Goddammit		

Emotional and pragmatic vocal expressions emitted during 5 min of interview in a profoundly aphasic, right handed, normally developed adult male following left hemispherectomy for treatment of cancer. Verbal utterances and vocal sounds were produced with normal articulation and prosody, in dramatic contrast to severely impaired naming, speaking, and repetition. In many cases, the vocalizations were appropriate to the context. These observations implicate a right-hemisphere-subcortical circuit in competence for these kinds of utterances.

"slender," "skinny," and "scrawny" all refer to physical thinness, but the first two connote approval, while the latter two are derogatory. Emotive factors have been described in diachronic language change. Associations and prejudice contribute to pejorative sense-change. Use of hyperbole in lexical items can result in loss of unpleasant connotations: English adjectives "awful" and "frightful," are now merely adjectival intensifiers, and have lost their semantic content (awe, fear). Hyperbolic expression can occur in the phonology, with expressions such as "jillion," "kajillion," "umptillion," and so on (to mean a very large number). Slang recycles hyperbole continuously, using such terms as "He's wild" and "That's hot." An extreme use of lexical items to convey strong emotive meaning appears in cursing. The neurological substrates for swearing appear to originate in a limbic-subcortical circuit, as is seen in residual, preserved swearing in aphasia associated with left-hemisphere damage,

and hyperactivated production in Tourette's syndrome (Van Lancker & Cummings, 1999).

The speaker's intentions and attitude are important components of language and inform the listener as to the emotional stance taken. The term attitude in social psychology refers to a stable mental position, or "stance," consistently held by a person toward some idea, or object, or another person, involving both affect and cognition. In language, word choice and intonation join to convey the attitude of the speaker: irony, incredulity, contempt, disdain, approval, and sarcasm are examples of attitudes conveyed in speech. Voice information is also a rich source of impressions about personality and mood which include a wide array of attributes. As familiar voice recognition is likely processed primarily in the right hemisphere, it follows that many such judgments, heard in the speech signal, are performed by the nonlanguage hemisphere.

19.5. PROSODIC COMMUNICATION

Prosody, or the melody of speech, takes first place in the list of media for linguistically expressing emotion and attitude. Through intonation, feelings are naturally revealed in the prosodic information of each person's individual utterances. The information may be concordant or discordant with the propositional, lexical content. Its authority will usually override contradictory verbal content, as in the utterance "I'm not angry!" spoken with high amplitude (loudness), high fundamental frequency (pitch), and accelerating tempo. Very brief utterances can convey distinctive information depending on prosodic content: the word "right" can imply a broad range of meanings, from casual agreement to condemning rejection. "Right" said with rapidly rising, relatively high pitch, means agreement and reinforcement; spoken with low, falling pitch, longer vowel and creaky voice, the same word indicates sarcastic repudiation. Similarly, "no" and "yes," "well," "sure," "hello," and most other brief speech formulas inhabit a phenomenally rich domain of possible attitudinal and affective meanings, depending on their prosodic content.

19.6. SYNTACTIC STRUCTURES AS REFLECTIVE OF ATTITUDINAL MEANING

While syntax is probably the most emotion-sanitized component of language, some aspects may indeed contribute subtle attitudinal attributes – a topic yet to be systematically studied in ordinary language use, although elaborated in literary studies of stylistics, which analyze the matching of such grammatical features as sentence complexity, length, and syntactic structure to content. Here "register," referring to a stylistic range from informal to formal, is relevant. Similarly, in speech, usage of simple versus grammatically complex sentence structures may communicate immediacy versus distance, and may elicit

Box 19.1   Written language

If it is acknowledged that the largest portion of affective expression is present in the prosodic component of speech, written language is relatively impoverished in its ability to communicate emotion, being restricted to words and syntactic structures, with only sparse aid from punctuation (exclamation marks, italics, underscoring, and hyphens) in the formal style. Informally, iconic distortions of letters have a considerable range of expression. Font style or form as well as various diacritics may graphically convey amplitude or category of emotion or attitude. Casual or formal styles, with all their contextual nuances, can be communicated by letter shape choices (see cartoon). For example, large and/or bold face type versus small, light font style may successfully symbolize amplitude range, often correlated with anger or fear; font type may represent category of emotion or attitude: italic or other slanted graphemes may suggest uncertainty, confusion, or bewilderment; font changes and reduplicated punctuation marks (e.g., "You're *leaving*??!") may indicate surprise or shock. These devices, often used in cartoons by graphic artists, are dependent on context for interpretation and have little systematic structure, in comparison to the phonemes, lexemes, and syntactic rules of language.

Cartoon illustrating the use of font type to represent a change in pragmatic style. Pooch Cafe' (©) (2006) Paul Gilligan. Reprinted by permission of Universal Press Syndicate. All rights reserved.

different responses from different listeners (ease versus discomfort). Persons using complex syntactic structures may be perceived as "pedantic," while the terse subject-verb structure may elicit a judgment that the speaker is "sincere." Word order can be used to convey attitude and nuance; importance is given to an element by placing it at the beginning or end of the sentence. Thus placement of connotative lexical items in an utterance can contribute to its impact on the listener.

19.7. LANGUAGE AFFECTS EMOTIONS

Verbal expression can affect the emotions of both speaker and listener. The perpetrator of an unintentional pun or a "Freudian slip" may be embarrassed to hear, in retrospect, what was actually said. In the psychodynamic interpretation of speech errors, puns, and malapropisms (a wrong but somehow related word), suppressed feelings about sex or other culturally sensitive topics slip out accidentally into fluent speech, but most speech errors have a more plausible linguistic explanation. Persons expressing anger can fuel their emotional state by their own talk. Producing words about emotions can shape and direct, reduce, or enhance those emotions in the speaker. The "talking therapy" is a standard technique in clinical psychology for accessing and managing emotions. As for the listener, words can intensely affect the emotions. We hardly need elaborate on the ability of words to incite strong emotional responses. Even in a free society, some words are considered so powerful that it is an unacceptable practice to print them in the newspaper or to say them on radio or television and in some social settings, fines are still incurred. The repertory of such words varies considerably across cultures; numerous societies fear and respect taboo expressions in many different ways. A Michigan court recently revoked a 105-year-old state law that banned the use of "indecent, immoral, obscene, vulgar, or insulting language" in the presence of women and children, when the conviction of a canoeist, who swore after falling into a river, was overturned. A cross linguistic study of the vocal tics produced in Tourette's syndrome revealed that taboo items constituted the "coprolalic" (swearing) utterances across all cultures reported, including those in Europe, South America, and India (Van Lancker & Cummings, 1999). These observations on swearing implicate basal ganglia nuclei of the brain, the site of dysfunction in Tourette's syndrome.

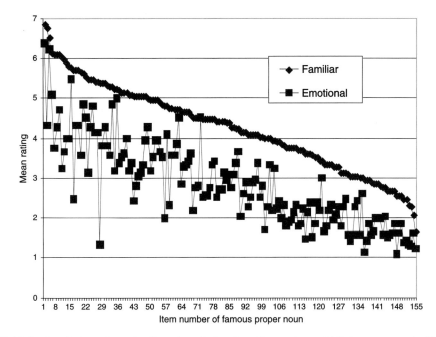

FIGURE 19.1 Familiarity and emotionality ratings. Ratings by college students of 157 famous names, familiar landmarks, and known brand names on two scales: familiarity and emotionality. Ratings were from 1–7 (least emotional, most emotional, least familiar, most familiar).

19.8. PRAGMATICS – LANGUAGE USE – AND EMOTIONAL EXPRESSION

Nearly all speech formulas, idioms, and proverbs evoke auras of feeling across a broad range of emotional and attitudinal perspectives: the valued, the frightening, the courageous, the hateful, or the noble. For example, "She has him eating out of her hand" connotes attitudes of submissiveness, dependence, haughtiness, and dominance. Other nonpropositional expressions – slang, clichés, exclamations, and expletives, also express and evoke strong emotion. In fact, formulaic expressions have as a primary communicative function the ability to convey emotional and attitudinal perspectives in a manner less direct – and less confrontive – than literal utterances.

Personal relevance – another rich source of emotional experience – influences and determines communication in various ways. The ability to establish, maintain, and recognize personally relevant stimuli (e.g., persons, places) is distinguishable from the processing of unfamiliar stimuli and has an influence on one's affective state. Studies in brain-damaged subjects and in normal subjects reveal a significant role of the right hemisphere in recognizing personally familiar proper nouns, such as names of famous persons and landmarks (Ohnesorge & Van Lancker, 2001). In matters of personal relevance, familiarity ratings are correlated with but not identical to emotional ratings (Figure 19.1). Deficits of the familiarity sense, as in Capgras syndrome (i.e., delusional misidentification of people, places or objects) and agnosias for places, voices and faces, are associated with right-hemisphere damage. Production

of social speech formulas is also impoverished following right sided cerebral dysfunction (Van Lancker Sidtis & Postman, 2006). Because formulaic expressions indirectly convey much in the way of attitude and affect in conversation, these patients may have a communication handicap. In contrast, persons with left-hemisphere damage and aphasia have a higher proportion of formulaic expressions, allowing for some successful communication, despite loss of syntactic function.

19.9. BRAIN STRUCTURES UNDERLYING EMOTIONAL LANGUAGE

Brain damage can affect the experiencing of emotional and attitudinal states, having consequences for linguistic expression. In addition, brain damage can specifically affect expression and perception of emotional and attitudinal information in language and speech, despite presumably intact emotional behavioral function. It is often difficult to determine the source of deficient affective–linguistic function – whether the underlying emotional system or communicative competence itself. Or cast in another manner, it may be controversial as to whether a communicative deficit, for example, dysprosodic (flat) speech, is a secondary result of a primary disturbance, such as abulia (deficit in motivation – see Box 19.2), or is a primary symptom in a behavioral constellation, which includes an amotivational state, reduced intonation variations, and other features (Van Lancker Sidtis et al., 2006). A residual "emotional" component in aphasic speech, including swearing, exclamations, and other emotive expressions, has long been identified.

Box 19.2   Abulia, a neurobehavioral diagnosis, refers to reduced motivation. "Flat" speech prosody in these patients mimics emotional disturbance

Abulia is a motivational deficit that is associated with apathy, loss of will, and lack of initiating behaviors. Mood disturbances may not be primary in this condition, in that the patients claim not to be especially depressed or anxious. They just do not feel like doing anything. A salient linguistic feature of abulia is monotonic or "flat" verbal expression. Damage to nuclei of the basal ganglia is correlated with abulia and the accompanying dysprosody (Bhatia & Marsden, 1994). Patients with dysprosody resulting from abulia present with dysprosodic speech when speaking spontaneously, but are able to successfully imitate utterances produced with emotional intonation (Van Lancker Sidtis *et al.*, 2006). Patients with previous musical training may retain the ability to sing in tune. This contrast in speech mode (and singing ability) can be diagnostic of this condition. In evaluating prosodic competence in patients presenting with dysprosodic

speech, it is important to obtain speech samples in different task conditions. These cases are also to be distinguished from deficient motor control of pitch production, which follows lesions in the right hemisphere, and which affects spontaneous speech as well as repetition and singing. Abulia may be mistaken for depression or sadness, or primary dysprosody, but in these cases, dysprosodic expression is secondary to the neurobehavioral disease.

Bhatia, K.P., & Marsden, C.D. (1994). The behavioural and motor consequences of focal lesions of the basal ganglia in man. *Brain, 117*, 859–876.

Van Lancker Sidtis, D., Pachana, N., Cummings, J., & Sidtis, J. (2006). Dysprosodic speech following basal ganglia insult: Toward a conceptual framework for the study of the cerebral representation of prosody. *Brain and Language, 97*, 135–153.

Psychiatric disturbances such as schizophrenia, psychosis, and mania may give rise to anomalous lexical selections, unusual grammatical structures, and prosodic deficits including monotone or hypermelodic speech and altered rate and rhythm. Abulia (see Box 19.2), mood disorders, and executive planning disturbances all can manifest themselves in dysprosody (defective melody of speech). Executive planning disturbances are associated mainly with frontal lobe dysfunction; other neuropsychiatric and neurobehavioral disorders presenting with deficient emotional expression in speech occur in association with subcortical damage. Subcortical functional disturbance in Parkinson's disease leads to deficient perception and production of prosodic contrasts signaling emotions in speech (Pell & Leonard, 2003). Hypophonic (soft) and slowed speech is characteristic of the depressed patient. The speech timing deficits resulting from ataxic disturbances in cerebellar disease may also – *albeit* erroneously – convey an impression of affective–prosodic disturbance, or difficulty expressing emotional states.

The right hemisphere has been associated with processing of affective–prosodic contrasts in speech (Schirmer *et al.*, 2001). Results from functional imaging support the lesion studies in indicating right sided activation (Buchanan *et al.*, 2000; Elliott *et al.*, 2000; Kotz *et al.*, 2003; Mitchell *et al.*, 2003; Wildgruber *et al.*, 2004). Difficulties in unequivocal interpretation of functional imaging signals have been identified (Sidtis, 2007); therefore, this and related findings for cerebral localization of emotional prosodies must await confirmation. Clinical studies suggest that both hemispheres participate in recognition of affective–prosodic contrasts, as shown by errors made by brain-damaged subjects (Sidtis & Van Lancker Sidtis, 2003). The left-hemisphere group misidentified timing information in emotional stimuli, while

right-hemisphere individuals misidentified pitch cues. These observations led to the notion that acoustic cues must be taken into account to explain hemispheric specialization for recognizing emotion contrasts in speech.

Dementing disorders may severely impact the comprehension and production of meaningful emotional communication. In the cortical dementias, speech, while well articulated, becomes relatively empty of semantic content, affective as well as informational. Here, the affective–communicative deficits are likely secondary to primary affective–cognitive disturbance. Efforts to test the abilities of Alzheimer's patients to produce and recognize emotional meanings in speech are hampered by the difficulties presented by their cognitive disorder. In one such attempt, a researcher demonstrated in a lengthy training and practice session through gesture and sound how different emotions have different intonations, and attempted to show an Alzheimer patient how to point to one of four faces, representing happy, sad, angry, and surprised, on hearing the utterance played on the tape. When it seemed certain that the subject understood the task, the utterance "Johnny is walking his dog" spoken with a happy intonation was presented, and the subject was encouraged to point to one of four faces on the response sheet. After carefully scrutinizing the four facial drawings on the response sheet, the individual looked up and said "But there's no dog here." Assessing prosodic information in speech, which is "backgrounded" in the speech signal compared to the linguistic information (verbal content), requires the subject to perform a meta-analytic task, which incurs significant cognitive demands.

In Alzheimer's disease, familiar speech formulas, such as greetings and leave-taking, are often used fluently, even in later stages of the disease, when cognitive impairment is great. These expressions have complex social meanings, but they are sufficiently routinized such that context alone may trigger their use. It is interesting to note that speech formulas often occur inappropriately in these patients; for example, emitting a warm "It's nice to see you again" to a stranger in the clinic, or stating to the examiner, who was performing an evaluation, "Let's do this at a time that is more convenient for you." Demented subjects may use but not correctly understand social speech formulas. This interpretation is supported by the observation that comprehension of idioms and proverbs, which naturally contain complex affective and ideational material, is impaired very early in the course of the disease.

19.10. AFFECT LEXICON

Deficits in processing emotional lexical content in production and comprehension following focal brain damage have been described in various neurological and psychiatric conditions. When recounting stories with and without affective content, patients with right-hemisphere damage utilize fewer affectively laden words than normal control subjects. Autobiographical reports generated by right-hemisphere damaged subjects are impoverished in emotional words, compared reports of their nonbrain-damaged peers. Several studies report greater electrical responses in the brain to emotional words in the right hemisphere than the left, and split visual studies show similar results. Emotional words, in a number of paradigms, generally elicit greater amplitudes in evoked potential studies than neutral words. Increased activity in the amygdala, a major structure in emotional behaviors, appears following administration of emotional words (Hamann & Mao, 2002; Ferstl *et al.*, 2005; Landis, 2006).

19.11. DYSPROSODIC DISTURBANCES

While impaired "melody of speech" was long associated with left-hemisphere damage, there are various sorts of evidence for a role of the right hemisphere in processing stimuli containing affective information (Pell, 1998; Baum & Pell, 1999; Berckmoes & Vingerhoets, 2004). Dysprosody of speech has numerous etiologies. In some cases, patients deny significant mood disorder, and yet they speak with low, monotonic pitch. So far, no overall model accounts for the many disparate reports: questions arise about validity of assessments, the meaning of prosodic measures chosen, task, and modality variations. Subcortical involvement (Karow *et al.*, 2001) affecting neurobehavioral function may account for affective dysprosody in speech. Motor disabilities in producing pitch contrasts may be accountable, or perceptual deficits in recognizing key acoustic cues to emotional utterances in speech may affect performance.

With recognition of a role of the basal ganglia in affective–prosodic behaviors, a new neurobehavioral model of prosody has been proposed. The importance of subcortical structures to intact prosodic production and comprehension is seen in dysarthrias arising from basal ganglia deficits. Further, basal ganglia regulate the facial and prosodic expression of motivation and mood. Postmorbid dysprosody in production may be due to mood changes, cognitive-programming failure, motor dysfunction, or motivational deficits, while problems in recognizing emotional speech may be attributable either to deficient pitch analysis or to impairment in comprehension of verbal–emotional content.

19.12. PRAGMATIC DEFICITS FOLLOWING BRAIN DAMAGE

Communicative functioning, called "pragmatics" – the use of language in everyday situations – has significant representation in the right hemisphere. Inability to perceive – and sometimes produce – figurative meanings, inferences, indirect requests, and verbal humor contribute to an impoverished emotional communicative function and is often associated with right-hemisphere dysfunction. Certain properties that have been attributed to the right hemisphere converge to provide a general picture of its role in pragmatic function. Language in everyday use involves many facets of paralinguistic meaning, which require a longer time window than do phonetic sequences for speech perception. Attitudes, emotions, perspectives, mood, and personal characteristics are communicated via the intonation contour by pitch, loudness, temporal variables, and voice quality. The longer processing times are better suited to right-hemisphere abilities. The right hemisphere is also superior at discerning patterns and profiles, and at integrating detail into a holistic Gestalt. This type of processing is also well suited to pragmatic functions, which involve recognition of social and verbal context, and involve larger units of discourse carried in longer stretches of speech. Innuendo, connotation, and thematic material are properties of discourse units; performance on these patterned aspects of communication has been found to be deficient following right-hemisphere damage. To further facilitate processing of paralinguistic meanings, the right hemisphere is superior at complex pitch perception, and it is well known that fundamental frequency (the acoustic counterpart of pitch) mean and variation are key cues to emotional and attitudinal contrasts in speech, as well as signaling phrase boundaries and conversation turns. Given these facts – a longer processing window, pattern recognition, and complex pitch perception, it is not surprising that most elements of the pragmatics of communication, including recognition of paralinguistic material such as emotions, sarcasm, irony, and humor; response to conversational cues;

and discernment of nonliteral and inferential meanings in speech are often impaired in right-hemisphere damage.

19.13. ASSESSMENT OF COMMUNICATIVE COMPETENCE FOR EMOTIONAL EXPRESSION

Specific communicative disorders often disturb the affective content of speech, language, or pragmatics. Dysprosody refers to failed signaling or identification of affective and attitudinal cues in the physical speech signal. For language, alexithymia involves defective retrieval of lexical items for emotion, and constrained syntactic choices restrict range of affective expression. In communication, pragmatic deficits involve discourse, including nonliteral meanings, inference, theme, and humor. These are all by nature difficult to assess and quantify, and at present, only research-level protocols are available. One standardized protocol is "The Neuropsychology Behavior & Affect Profile" (NBAP), which was created to assess affective change in brain-impaired individuals; it includes a "pragnosia" (deficient pragmatic functions) dimension as well as dimensions for emotional and attitudinal disturbance.

Tests focused on disturbances in speech and language do not traditionally assess comprehension and expression of emotions and the use of affective language. The Boston Diagnostic Aphasia Examination (BDAE) includes as a measurement the "melody of speech," without reference to its role in affective–prosodic expression. The Mini Inventory of Right Brain Injury includes in its language processing section a few questions probing the ability to express emotional tone of voice. The inclusion of objective rating items of affect and prosody as part of communication assessment batteries is rare – indeed, reliably assessing prosodic production is difficult. A special competence for determining intonational detail is required. Informal clinical observations and judgments and use of unpublished protocols are the norm in clinically gauging affective expression and comprehension. New evaluation and treatment instruments for prosodic function are under development.

19.14. SUMMARY

Although emotion can proceed without language, verbal communication is ordinarily and normally imbued with affective and attitudinal nuances. If a listener is not sensitive to the emotional nuances of the speaker, he or she can miss much of the meaning. Similarly, when the speaker has lost the ability to project the pattern of his/her attitudes and emotions onto the linguistic expression, the listener will understand the words without being able to discern their emotional pattern. Although language and emotion are deeply intertwined, one is not dependent on the other. Each operates from different brain systems. One of the greatest challenges in neurolinguistic research lies in understanding

how brain structures integrate to permit verbal comprehension and production of attitude and emotion in speech.

19.15. CHALLENGES AND FUTURE DIRECTIONS

Goals of understanding the relationship of language to emotion enjoy the benefit that one of these two domains, language, is well described and has heuristically established structure. This provides an advantage over trying to map relations between the other two domains of the triad: thought and emotion, which lend themselves much less easily to structural descriptions. Viewing the well known levels and units of linguistic structure, it is possible to examine how emotional attributes imbue each and all of them in different ways. The great challenge lies in mapping these processes in the brain. Cerebral areas of language representation are fairly well established, and certain brain correlates of emotional processing are known, but interrelating these two neuropsychological domains is at a beginning state. In addition, questions remain about the mental reality of posited linguistic structures. That is, does the brain "know" and utilize linguistic structures, as presented in the textbooks, in a straightforward way? Or is some other "currency" or strategy in use? Attempts to correlate linguistic components and processes posited in performance models with identifiable brain structures or networks have not been consistently successful. This uncertainty will be amplified for aspects of language related to emotion. Further, it is currently held that brain structures for language representation are relatively discrete and localized, while brain structures for emotional processing are heterogeneous and extensive. With the advent of functional brain imaging and advanced therapies such as deep brain stimulation, models of brain processing of language and emotion may rapidly evolve, leading to new approaches to these questions.

References

Ayçiçegi, A., & Harris, C.L. (2004). Bilinguals' recall and recognition of emotion words. *Cognition and Emotion, 18*, 977–987.

Baum, S.R., & Pell, M.D. (1999). The neural bases of prosody: Insights from lesion studies and neuroimaging. *Aphasiology, 13*(8), 581–608.

Berckmoes, C. & Vingerhoets, G. (2004). Current directions in psychological processing: Neural foundations of emotional speech. *Science, 13*(5), 182.

Breitenstein, C., Van Lancker, D., & Daum, I. (2001). The contribution of speech rate and pitch variation to the perception of vocal emotions in a German and an American sample. *Cognition and Emotion, 15*, 57–79.

Buchanan, T.W., Lutz, K., Mirzazade, S., Specht, K., Shah, N.J., Zilles, K., & Jancke, L. (2000). Recognition of emotional prosody and verbal components of spoken language: An fMRI study. *Cognitive Brain Research, 9*(3), 227–238.

Chen, A., Gussenhoven, C., & Rietveld, T. (2004). Language-specificity in the perception of paralinguistic intonational meaning. *Language and Speech, 47*(4), 311–349.

Code, C. (2005). First in, last out? The evolution of aphasic lexical speech automatisms to agrammatism and the evolution of human communication. *Interaction Studies, 6*, 311–334.

Dewaele, J.-M. (2004). The emotional force of swearwords and taboo words in the speech of multilinguals. *Journal of Multilingual and Multicultural Development, 25,* 204–222.

Elliott, R., Rubenstein, J.S., Sahakian, B.J., & Dolan, R.J. (2000). Selective attention to emotional stimuli in a verbal go/no-go task: An fMRI study. *Neuroreport, 11*(8), 1739–1744.

Ferstl, E., Rinck, M., & von Cramon, D.Y. (2005). Emotional and temporal aspects of situation model processing during text comprehension: An event-related fMRI study. *Journal of Cognitive Neuroscience, 17*(5), 724–739.

Hamann, S. & Mao, H. (2002). Positive and negative emotional verbal stimuli elicit activity in the left amygdala. *Neuroreport, 13*(1), 15–19.

Karow, C.M., Marquardt, T.P., & Marshall, R.C. (2001). Affective processing in left and right hemisphere brain-damaged subjects with and without subcortical involvement. *Aphasiology, 15,* 715–729.

Kotz, S.A., Meyer, M., Alter, K., Besson, M., von Cramon, D.Y., & Friederici, A.D. (2003). On the lateralization of emotional prosody: An event-related functional MR investigation. *Brain and Language, 86*(3), 66–76.

Landis, T. (2006). Emotional words: What's so different from just words? *Cortex, 42*(6), 823–830.

Marian, V. & Neisser, U. (2000). Language-dependent recall of autobiographical memories. *Journal of Experimental Psychology: General, 129,* 361–368.

Matsumoto, A. & Stanny, C.J. (2006). Language-dependent access to autobiographical memory in Japanese–English bilinguals and IS monolinguals. *Memory, 14*(3), 378–390.

Mitchell, R.L., Elliott, R., Barry, M., Cruttenden, A., & Woodruff, P.W. (2003). The neural response to emotional prosody, as revealed by functional magnetic resonance imaging. *Neuropsychologia, 41*(10), 1410–1421.

Ohnesorge, C. & Van Lancker, D. (2001). Cerebral laterality for famous proper nouns: Visual recognition by normal subjects. *Brain and Language, 77,* 135–165.

Pell, M.D. (1998). Recognition of prosody following unilateral brain lesion: Influence of functional and structural attribute of prosodic contours. *Neuropsychologia, 36*(8), 701–715.

Pell, M.D. & Leonard, C.L. (2003). Processing emotional tone from speech in Parkinson's disease: A role for the basal ganglia. *Cognitive, Affective and Behavioral Neuroscience, 3,* 275–288.

Scherer, K.R., Banse, R., & Wallbott, H.G. (2001). Emotion inferences from vocal expression correlate across languages and cultures. *Journal of Cross-Cultural Psychology, 32*(1), 76–92.

Schirmer, A., Alter, K., Kotz, S.A., & Friederici, A.D. (2001). Lateralization of prosody during language production: A lesion study. *Brain and Language, 76,* 1–17.

Sidtis, J.J. (2007). Some problems for representations of brain organization based on "activation." *Brain and Language, 102*(2), 130–140.

Sidtis, J.J. & Van Lancker Sidtis, D. (2003). A neurobehavioral approach to dysprosody. *Seminars in Speech and Language, 24*(2), 93–105.

Van Lancker Sidtis, D. (2004). When novel sentences spoken or heard for the first time in the history of the universe are not enough: Toward a dual-process model of language. *International Journal of Language and Communication Disorders, 39*(1), 1–44.

Van Lancker, D. & Cummings, J. (1999). Expletives: Neurolinguistic and neurobehavioral perspectives on swearing. *Brain Research Reviews, 31,* 83–104.

Van Lancker Sidtis, D. & Postman, W.A. (2006). Formulaic expressions in spontaneous speech of left- and right-hemisphere damaged subjects. *Aphasiology, 20*(5), 411–426.

Van Lancker Sidtis, D., Pachana, N., Cummings, J., & Sidtis, J. (2006). Dysprosodic speech following basal ganglia insult: Toward a conceptual framework for the study of the cerebral representation of prosody. *Brain and Language, 97,* 135–153.

Wildgruber, D., Hertrich, I., Riecker, A., Erb, M., Anders, S., Grodd, W., & Ackermann, H. (2004). Distinct frontal regions subserve evaluation of linguistic and emotional aspects of speech intonation. *Cerebral Cortex, 14*(12), 1384–1389.

Further Readings

LeDoux, J.E. (2000). Emotion circuits in the brain. *Annual Review of Neuroscience, 23,* 155–184.
LeDoux reviews current information from neuroscience research on cerebral representation of emotion.

Phan,, K.L., Wager, T., Taylor, S.F., & Liberzon, I. (2002). Functional neuroanatomy of emotion: A meta-analysis of emotion activation studies in PET and fMRI. *Neuroimage, 16*(2), 331–348.
These authors review results of brain neuroimaging studies on emotion.

Scherer, K.R. (2003). Vocal communication of emotion: A review of research paradigms. *Speech Communication, 40,* 227–256.
This article summarizes the various approaches to the study of emotional communication in speech.

Turner, J.H. & Stets, J.E. (2005). *The sociology of emotions.* Cambridge, England: Turner and Stets.
Emotional behaviors in cultural and sociological settings are described and explained.

20

Acquired Reading and Writing Disorders

CLAUDIO LUZZATTI

Department of Psychology, University of Milano-Bicocca, Milano, Italy

ABSTRACT

Empirical evidence indicates that while the acquisition of oral language is based on neuroanatomical and functional units that are genetically predetermined, it is difficult to apply these assumptions to written language. This chapter pursues questions such as how reading and writing are represented at a mental level and how and where is it implemented in the human brain. In the first section the classical clinical neuropsychological taxonomy of reading and writing disorders is introduced and its limits discussed, while the second section deals with the contemporary models that describe normal written language processing and analyzes the implications of a more cognitively sound description of acquired dyslexia and dysgraphia. The third section provides an overview of the major cross-linguistic differences between languages, using alphabetic and logographic scripts and the impact these differences have on reading and spelling disorders. Finally, the neuroanatomical foundations of written language arising from both the anatomo-clinical correlative approach and functional neuroimaging studies are discussed.

20.1. INTRODUCTION

The invention of the written language is one of the most important events in the history of mankind; this extraordinary instrument enabled those populations who developed it to completely transform major aspects of their lives. Through written language traders recorded their commercial transactions with other populations, laws were drawn up, passed and published, and artistic, cultural and sacred texts (lyrics, poems and chronicles) were consigned to posterity. In other words, written language is the landmark of the shift from prehistory to history.

But how is this ability represented at the mental level and how and where is it implemented in the human brain? Furthermore, to what extent do mental representations underlying written differ from those underlying oral words? The purpose of this chapter is to provide answers to these questions, outlining the mental and neural foundations of written language and describing the principle aspects of acquired reading and spelling disorders.

The processes that underlie written language acquisition differ substantially both philogenetically and ontogenetically from those underlying oral language. First of all, the natural evolution from apes to humans is primarily characterized by the emergence of oral language and, although linguistic aspects (phonological, lexical and morpho-syntactic parameters) differ greatly from one culture to another, the major principles of language and speech are the same. Furthermore, the acquisition of oral language is an early and fully automatized phenomenon following well-defined developmental steps. All these observations clearly indicate that the acquisition of oral language is based on neuroanatomical and functional units that are genetically predetermined. However, it is difficult to apply these assumptions to written language, since its earliest documented appearance dates back approximately 6000 years (an extremely short time in terms of natural evolution). Much more importantly, many human populations achieved literacy only in the last few generations (this is true also for Western European cultures) and several are completely illiterate still today. Moreover, the acquisition of written language is not automatic but slow and arduous. On the other hand, children from cultures that have not been exposed to written language do learn how to read and write as proficiently as children coming from cultures with a deeply rooted history of literacy. Therefore, it is very unlikely that the acquisition of written language, unlike oral language, is genetically determined and that the brain contains built-in mechanisms specifically devised for reading and writing. On the contrary, it would appear to depend on (nonautomatized) voluntary and effortful learning and to be based on the development of cognitive abilities that are not originally designed to process orthographic information.

Different cultures introduced different types of scripts to transcribe their oral communication into written language. Overall, writing systems can be divided into *alphabetic* (sound-based) and nonalphabetic (*logographic*) scripts. In *alphabetic scripts*, symbols correspond to sounds (phonemes or syllables). This procedure allows the transcription of phonological strings (including high- and low-frequency words, new words and nonwords) by the use of combinations of symbols (usually 20–50). However, in certain languages whose orthography was defined several centuries ago (like French or English), the sound-to-letter correspondence may no longer be completely shallow (transparent). In nonalphabetic (*logographic*) scripts such as Chinese, the symbols do not reflect the phonological word-form; they provide the underlying meaning and can thus be shared by individuals speaking different dialects or languages. The major disadvantage of *logographic* scripts is the vast quantity of symbols (up to 10,000) that must be learned to achieve proficiency in the language.

Finally, there are languages whose writing system is based on a combination of different scripts: for instance, Japanese combines the logographic Chinese (*kanji*) and two sub-word-level (SWL) syllabic (*kana*) scripts. *Kanji* characters are used for nouns and the stems of verbs and adjectives; *hiragana* characters are used both to represent grammatical information (endings of verbs, endings of adjectives and grammatical functional particles) that cannot be expressed with *kanji* and to write words whose kanji might be obscure or presumably unknown to the reader; *katakana* characters are used to transcribe foreign proper names, loan words and foreign words.

20.2. DÉJERINE'S CLASSICAL ANATOMO-FUNCTIONAL DIAGRAM OF READING (1891, 1892)

A literate adult may lose the ability to read and write after acquired brain damage. The impairment may either disrupt these abilities in isolation or in association with a more general aphasic language disorder. For over a century now, neuroscientists have observed patients suffering from reading or spelling impairments after brain damage, drawing information from these cases to describe the mental substrate of written language and its neuroanatomical foundation.

The French neurologist Joseph-Jules Déjerine (1891, 1892) described two patients suffering from reading disorders in the absence of any further verbal deficits and provided an important clinical and theoretical contribution. A brief review

of these cases is mandatory to understand the subsequent developments in the field.

20.2.1. Pure Alexia and Alexia with Agraphia

In the first case, the reading disorder was associated with a spelling impairment (*cécité verbale avec agraphie,* i.e., verbal blindness with agraphia) and was caused by cerebral damage to the left angular gyrus (Déjerine, 1891). In the second case, the reading impairment was pure (*cecité verbale pure*) and followed an occipital and inferior temporal lesion extending to the retroventricular white matter and the callosal splenium. Starting from these observations, Déjerine developed a model of written language based on the assumption of a *written word-form area* in the left angular gyrus (see Box 20.1 and 20.2). According to his model, the right hemisphere would be word-blind, and the visual images of words would have to reach the left angular gyrus to be identified. A lesion of the left angular gyrus (patient 1) would result in the inability to name words and also cause a parallel spelling impairment, since according to the model, visual word-forms have to be retrieved prior to activation of the graphic motor patterns of letters (or prior to oral spelling). Conversely, the left occipital and inferior temporal area would not be specifically involved in reading, but *alexia*, that is, a deficiency in the ability to read, would simply be due to left hemianopia and to the fact that the callosal lesion would hinder visual word representations from reaching the left angular gyrus and the remaining left-hemisphere language areas (patient 2).

Déjerine's anatomo-functional account of written language remained undisputed for the next 70 years, until the second half of the twentieth century, and continues to be the major frame of reference for the clinical description of reading and writing disorders after brain damage.

20.3. CLASSICAL NEUROLINGUISTIC CLASSIFICATION OF ACQUIRED SPELLING DISORDERS

Until the end of the 1960s, the classification of acquired writing impairments, like that of acquired reading impairments, was based on an anatomo-functional taxonomy: *aphasic agraphia*, alexia with agraphia, *pure agraphia, apraxic agraphia* and *callosal agraphia*. This classification is often still applied in a clinical neurological context, and is therefore worthy of description.

20.3.1. Aphasic Agraphia

In agraphia accompanying aphasia, the spelling deficit mirrors the characteristics of the oral language disorder. Patients suffer from a widespread spelling deficit with letter omissions, substitutions and perseverations. In fluent aphasia the deficit often assumes the aspect of jargon agraphia;

Box 20.1   Déjerine's anatomical diagram of written language

The French neurologist J-J. Déjerine developed a model of reading based on his observation of two patients suffering from reading impairments in the absence of any further verbal deficits. CVM: center of visual memories; OVC: occipital visual center; CAM: center of auditory memories (i.e., Wernicke's area); MCA: motor center of word articulation (i.e., Broca's area); HMC: hand motor center.

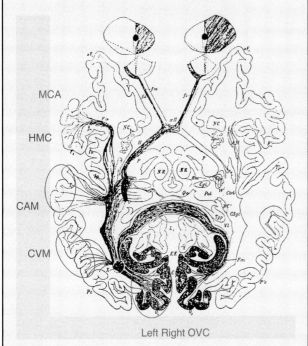

Left Right OVC

Alexia with agraphia would be caused by damage to CVM, while *pure alexia* requires damage to the left OVC and isolation of CVM from the controlateral right OVC. (*Source*: Modified from Déjerine, 1914.)

Déjerine, J.-J. (1914). *Sémiologie des affections du système nerveux.* Paris: Masson.

Box 20.2   Déjerine's diagram of written language drawn in the form of an information-processing flowchart model

The adaptation of the model simplifies the comparison with contemporary models of written language (see below). The most obvious differences with respect to cognitive models are (1) the anatomical frame; (2) the lack of a distinction between lexical and lexical–semantic levels of processing (i.e., between words and underlying meanings) (*Source*: Modified from Luzzatti, 2003).

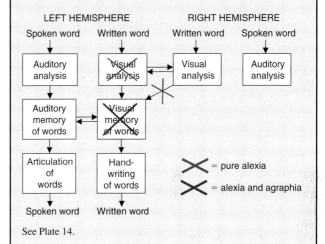

See Plate 14.

Luzzatti, C. (2003). Optic aphasia and pure alexia: contribution of callosal disconnection syndromes to the study of lexical and semantic representation in the right hemisphere. In E. Zaidel & A. Iacoboni (Eds.), *The parallel brain: The cognitive neuroscience of the corpus callosum* (pp. 479–496). Cambridge, MA: The MIT Press.

nonfluent aphasic patients are often impaired in achieving single letter strokes.

20.3.2. Alexia with Agraphia

See Section 20.2.1 for a detailed discussion of alexia with agraphia.

20.3.3. Pure Agraphia

This type of spelling impairment presents without any other language disturbance and is caused by damage to the orthographic representations of letters and words and therefore involves handwriting as well as oral spelling or typing on a keyboard. It usually follows left parietal lesions, though it has been found, in some rare cases, following left frontal lesions.

20.3.4. Apraxic Agraphia

This disorder derives from an inability to produce letter symbols. It may follow either a failure to retrieve letter forms from memory or difficulty in achieving these forms graphically and combining individual letter strokes appropriately. The deficit cannot be traced back to ideomotor or constructional apraxia, since both may appear in the absence of any writing disorder and, vice versa, apraxic agraphia may be found in the absence of constructional or ideomotor apraxia. In this type of writing impairment oral spelling and typing on a keyboard are usually spared.

20.3.5. Callosal Agraphia

This is a rare form of agraphia following damage to the antero-medial part of the corpus callousm that connects the left-hand motor center (in the right hemisphere) to left-hemisphere spelling representations, generating a peripheral spelling disorder for the left hand only. It is mostly associated with left-hand apraxia and left-hand tactile anomia.

20.4. DUAL-ROUTE MODELS OF READING AND COGNITIVE ACCOUNTS OF ACQUIRED DYSLEXIA

The cognitive psychological development of neurolinguistics that emerged at the end of the 1960s provided the basis for a more cognitively sound approach to impaired reading performance after brain damage. A few seminal studies developed in this framework (Marshall & Newcombe, 1966, 1973) showed in fact that Déjerine's model and his account of dyslexia could not explain some aphasic phenomena, in particular the emergence of *semantic errors* (e.g., WOOD ⇒*tree*; NEGATIVE ⇒*minus*; ELEVATOR ⇒ *lift*) and *grammatical class* effects (*nouns* are read better than *verbs* or function *words*), *concreteness* effects (*concrete* nouns are read better than *abstract* nouns) and *word frequency* effects (*high-frequency* words are read better than *low-frequency* words). Furthermore, Déjerine's model could not explain some aspects of dyslexic performance such as the disproportionate inability to read irregular words or nonwords (orthographically plausible but nonlexical letter strings). These observations were consistent with the *cognitive models* developed in these same years to describe the functional processes underlying normal reading performance (Morton, 1969, 1980). Actually, healthy participants are usually able to read words with irregular spelling like PINT and COLONEL as well as nonwords, thus suggesting the need of two complementary reading routes, that is a *lexical* and a *SWL* procedure.

The SWL procedure implies the application of *orthographic-to-phonological conversion* rules, which allow literate individuals to convert letter strings into corresponding strings of phonemes. This route allows a normal speaker to read regular words (i.e., words with shallow orthography like CAT or TOWN) and regular nonwords, like MABLE. In contrast, the *lexical route* assumes that the processing of a lexical string of letters is based on the retrieval of stored knowledge from the *orthographic input lexicon,* the *cognitive system* and the *phonological output lexicon*. The strings of phonemes that are generated along either reading route are then conveyed to the *phonemic buffer,* a short-term store interfacing the phonological output to the corresponding articulatory processor.

With respect to the SWL routine, the *lexical* procedure provides literate individuals with processing that is quicker, consumes fewer cognitive resources and activates the underlying conceptual knowledge automatically. However, this procedure can only be applied to words whose orthography has already been learned; it cannot process nonwords and is the only route available when reading words with irregular orthography.

Figure 20.1 shows the dual-route model of reading and the reading impairments predicted and tested on the model.

20.4.1. Phonological and Surface Dyslexia

The dual-route processing model described so far predicts that either the SWL or the lexical reading procedure may be impaired disproportionately after brain damage. In fact, these contrasting patterns of damage have been fully confirmed. Impairment of the ability to apply the grapheme-to-phoneme conversion rules causes a reading disorder that is known as *phonological dyslexia* and is characterized by impaired reading of nonwords and preserved naming of regular and irregular words. Impaired processing along the lexical route causes a reading disorder that goes under the label of *surface dyslexia* and is characterized by preserved reading of nonwords and regular words and by poor reading of irregular words.

20.4.2. The Direct Lexical Route

A few years after the dual-route model was first described, Schwartz *et al.* (1980) reported the case of a patient (WLP) who suffered from a degenerative brain disease with predominant semantic impairment (almost 10 years later this disorder was given the name *semantic dementia*). WLP had preserved reading ability for irregular words (e.g., *tortoise*), but no understanding of their meaning or ability to sort pictures into categories. WLP could read easily and without effort, often adding comments that emphasized the dissociation between her ability to pronounce the target words and her comprehension impairment (e.g., "*hyena* … what the heck is that?") This observation called for a revision of the dual-route model and for the introduction of a *direct reading route* connecting the *orthographic input lexicon* to the *phonological output lexicon* (by-passing the underlying conceptual knowledge).

20.4.3. Deep Dyslexia

Some dyslexic patients produce semantic errors in reading words (Marshall & Newcombe, 1966, 1973). Their responses are related to the target in meaning but may be totally unrelated to the graphemic and phonological word-form. For example, HOUND may be read as *dog* and WOOD as *tree*. *Semantic errors* is the distinctive feature of deep dyslexia but other types of errors also occur (see Coltheart, 1980, 2000 for a review). These patients are usually unable to read nonwords, whereas their reading of words with irregular orthography does not differ from that of regular words. This pattern of symptoms is similar to that of phonological dyslexia and indicates disproportionate damage to the SWL reading routine.

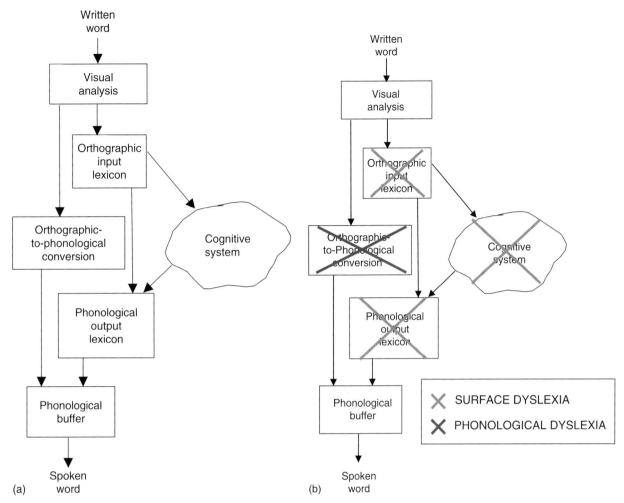

FIGURE 20.1 Dual-route model of reading: (a) lexical and SWL reading routes; (b) functional lesions causing phonological and surface dyslexia (see Plate 15).

Furthermore, word reading in deep dyslexia is characterized by a *part of speech effect* (*nouns* are read better than *verbs* and function *words*) and by a *concreteness effect* (*concrete* nouns are read better than *abstract* nouns).

Semantic errors suggest that deep dyslexic patients try to read via a damaged semantic system. An alternative explanation for the entire pattern of phenomena typifying deep dyslexia is the emergence of right-hemisphere lexical–semantic knowledge after extensive damage to the verbal abilities of the left hemisphere (Coltheart, 1980, 2000). This is obviously in contrast with Déjerine's view of full right-hemisphere verbal blindness. Indeed, it has been shown that the right hemisphere has linguistic knowledge, which is, however, limited to high-frequency concrete nouns. Furthermore, large left-hemisphere perisylvian lesions causing extensive damage to the language areas and leading to deep dyslexia also result in severe *agrammatic* aphasia (an oral language impairment which, among other lexical and syntactic aspects, largely replicates the grammatical class effect found in deep dyslexia).

20.4.4. Letter-by-Letter Reading

Patients suffering from this type of reading impairment are unable to name target words through either the lexical or the SWL route. Although they may still be able to name the individual letters and retrieve the target phonological word-form through a reverse oral spelling procedure, the routine is slow, laborious and often inefficient. There is a strong word length effect, but no word frequency or word class effect (Coslett & Saffran, 1989).

20.4.5. Neglect Dyslexia

This reading disorder usually accompanies a left *unilateral neglect* syndrome. Unilateral neglect typically involves a patient's inattention to his/her left visual field, left body side and the left extrapersonal hemispace. In a word reading task, for example, patients will neglect the left side of the written stimulus, making either omissions (e.g., reading *right* instead of *bright*) or substitutions (reading *summer* instead of *hammer*) (see Haywood & Coltheart, 2001 for a review). When dealing with a sentence, neglect patients usually start reading from the middle of the word string, omitting its left side. Patients are mostly unaware of their impairment (*anosognosia*). Not only reading but also spelling may be impaired in association with left hemineglect. Patients may produce errors both in handwriting and in oral spelling on the left side of words (Baxter & Warrington, 1983). Spelling errors found in neglect dysgraphia have been interpreted as a left-sided neglect of mental orthographic representations and have been used as evidence for a representational account of unilateral neglect.

20.5. DUAL-ROUTE MODELS OF SPELLING AND COGNITIVE ACCOUNTS OF ACQUIRED DYSGRAPHIA

As in the cognitive model of reading described in the previous section, two independent procedures – a *lexical* and a *SWL route* – are thought to be deployed by adult literate subjects for spelling. In the *lexical* procedure words are processed as a unitary entry, whereas *SWL* processing implies the application of phoneme-to-grapheme conversion rules. Dual-route models of spelling can account for the ability shown by literate adults to spell irregular words as well as legal nonwords and for the impaired spelling performance of aphasic brain damaged patients (Patterson, 1986; Shallice, 1981).

The *lexical route for spelling* requires retrieval of stored phonological and orthographic lexical knowledge. Input words activate phonological lexical knowledge (*phonological input lexicon*), which then spreads to the underlying conceptual knowledge, which in turn activates stored orthographic lexical representations (*orthographic output lexicon*).

The *SWL procedure* converts strings of sounds into the corresponding letter sequences through the application of phoneme-to-grapheme conversion rules. This routine enables spelling of regular words and nonwords (nonlexical phonotactically legal phonemic strings). The procedure comprises a succession of steps: first, continuous chains of sounds are broken down into their corresponding strings of phonemes (*auditory-to-phonological conversion*). Next, phonemes are temporarily stored in a *phonemic buffer*, and eventually the phoneme strings are sequentially converted into the corresponding letter strings (phoneme-to-grapheme conversion rules).

Both routes feed the *graphemic buffer*, a short-term memory store interfacing the abstract lexical as well as the SWL orthographic representation with different peripheral spelling procedures (handwriting, writing on a keyboard or oral spelling).

Generally, application of *lexical* procedures enables the spelling of regular and irregular words whose orthography has been learned previously, but is ineffective for spelling new words or nonwords. The SWL procedure, in contrast, enables the spelling of regular words and legal nonwords, but is ineffective when spelling words with irregular orthography. In other words, irregular words can only be spelt along the lexical route and nonwords along the SWL route, whereas regular words may be spelt along both writing procedures. Figure 20.2 shows the dual-route model of spelling and the spelling impairments predicted by this model.

Traditionally, cognitive taxonomy distinguishes *central* from *peripheral* dysgraphic disorders: in the former, the spelling deficit is caused by disproportionate damage to the lexical or to the SWL route (i.e., *surface* and *phonological* dysgraphia, respectively) or by damage to the graphemic output buffer; in the latter type of dysgraphia, the spelling deficit arises downstream from the phonological buffer. Therefore, the impairment does not affect the ability to obtain an abstract orthographic representation, but does impinge on specific output procedures such as written or oral spelling and the ability to make appropriate use of the alternative codes (allographs) that can be employed in writing a word (cursive versus print script; uppercase versus lowercase characters).

20.5.1. Phonological and Surface Dysgraphia

The dual-route model described so far predicts two independent spelling disorders after brain damage, which are caused by disproportionate impairment either of the *SWL* or of the *lexical* spelling route.

Damage to phoneme-to-grapheme conversion causes predominant impairment in the spelling of nonwords, a disorder called *phonological dysgraphia.* Impairment to the lexical route causes a spelling advantage for regular over irregular words with a majority of phonologically plausible errors (e.g., *yacht* is spelled JOT), a type of error that clearly indicates that patients are relying on the SWL procedure. This latter spelling impairment is called *surface dysgraphia* (or *lexical agraphia*).

20.5.2. Graphemic Buffer Dysgraphia

Spelling may be impaired at the graphemic buffer level. Damage to this processing level will lead to a pattern of impairment in which either graphemic transpositions (e.g., *deinal* instead of *denial*) or substitutions predominate. In this latter type of buffer dysgraphia, the target syllabic structure is preserved – vowels substitute for other vowels, (e.g., *ralish* instead of *relish*), consonants for other consonants (*moor* instead of *moon*), clusters for other clusters (*tentle* instead of *temple*) and double consonants for other double consonants (*harrow* instead of *ladder*). A comparable degree of impairment emerges for both words and nonwords, but presuming that the demand for buffer resources is higher for nonwords than for words, the former may be more affected.

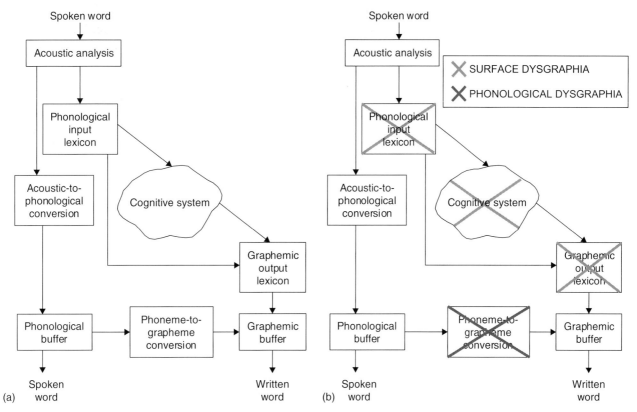

FIGURE 20.2 Dual-route model of spelling: (a) lexical and SWL spelling routes; (b) functional lesions causing phonological and surface dysgraphia (see Plate 16).

20.5.3. Peripheral Dysgraphia (Allographic Dysgraphia)

This is a writing disorder that may originate at a processing stage *peripheral* to the graphemic output buffer. This pattern of damage is usually restricted to handwriting, while oral spelling is spared. Some patients cannot keep apart different sets of characters (allographs), for example lower- and upper-case, and so forth, continuing to mix them while writing by hand. In contrast to *apraxic dysgraphia* (see above), patients suffering from *allographic dysgraphia* may be fluent in writing and usually form their letter shapes well.

20.6. PRINCIPLES FOR THE ASSESSMENT OF READING AND SPELLING IMPAIRMENTS

A sound cognitive assessment of reading and spelling disorders requires tasks that tackle each level of the lexical and the SWL procedures separately. Assessment material has been developed for several languages. The most well-known battery developed for the English language is the Psycholinguistic Assessment of Language Processing in Aphasia (PALPA; Kay *et al.*, 1992).

20.6.1. Reading Impairments

In reading, four types of tasks should be considered: *lexical decision*, *semantic judgment*, *word naming* and *reading comprehension*. Visual lexical decision (i.e., distinguishing real words from nonlexical letter strings) tests the integrity of the orthographic input lexicon. Semantic judgment tasks (i.e., making decisions about the meaning of written words) and reading comprehension assess access to conceptual knowledge. Reading words with regular and irregular letter-to-sound correspondence from different grammatical classes (nouns, verbs, function words) and with different word frequency verifies the status of the lexical route, while reading nonwords aloud tests the integrity of the SWL routine. The *National Adult Reading Test* (NART: Nelson, 1982) assesses the ability to read words with irregular letter-to-sound correspondence. It has been standardized on a large sample of healthy English-speaking adults. As expected, the NART score correlates highly with the education level.

20.6.2. Spelling Impairments

As with reading assessment tasks, writing tasks should also include several sets of items tapping the lexical and the SWL procedures and the various underlying processing units. Tasks should include writing to dictation of words with regular spelling, words with irregular spelling and nonwords. Words should vary by grammatical class, word length and word frequency. Oral and written spelling should be compared to detect patients suffering from allographic dyslexia.

20.7. READING AND SPELLING DISORDERS IN LANGUAGES WITH DIFFERENT SCRIPTS

Much of the progress in developing models of written language and in understanding dyslexia and dysgraphia is the result of studies in English. However, not all alphabetic scripts are as irregular as English. Italian and Spanish, for instance, have a much shallower orthography, which makes it possible to read and spell most words through SWL print-to-sound and sound-to-print conversion routines. However, even these languages offer some irregularities. For instance, in Italian it is not possible to predict whether the stress in tri-syllabic or longer words falls on the penultimate or on the antepenultimate position: for example, the stress in GONDOLA falls on the first syllable (['gondola] and not *[gon'dola]), whereas in MENTOLO it falls on the second syllable ([men'tolo] and not *['mentolo]). There is some degree of irregularity also in spelling. Therefore, dual-route reading and spelling appears to be mandatory even in languages with predominantly shallow orthography. This hypothesis has been confirmed by observing phonological and surface reading and spelling impairments in Italian and Spanish aphasic patients. However, the predominance of phonological dyslexia and the comparative rarity of deep dyslexia in shallow orthography languages like Spanish or Italian (e.g., Ardila, 1991; Toraldo *et al.*, 2006) and the predominance of surface dyslexia in opaque orthographies such as English and Hebrew suggest that the two routes have a different relevance in different languages.

Interesting results emerge from studies on speakers of languages like Japanese or Chinese that use logographic scripts. The reading deficit pattern in Japanese patients strongly resembles that of English-speaking patients, with surface dyslexia following impaired ability to read and understand *kanji* characters and spared SWL processing of *kana* characters, while the opposite pattern of damage emerges in phonological dyslexia (Sasanuma, 1980).

Chinese, on the other hand seems to constitute the symmetrical condition to Italian and Spanish, which is to be attributed to the fact that its script is based on lexical logographic characters. However, notwithstanding its logographic aspects, Chinese too has some SWL processing: Chinese characters frequently contain phonetic radicals. However, the system is quite irregular as these radicals may be positioned to the left or to the right (or the top or the bottom) of the semantic radicals and the same element can act as either a semantic or a phonetic radical in a number of different characters. This implies that an over-reliance

on SWL phonetic information would cause a pattern of reading impairment that corresponds to surface dyslexia in alphabetic scripts. In fact, this pattern of dyslexia has been reported as *legitimate alternative reading of components* (LARC) (see Yin *et al.*, 2005 for a review). In contrast, the absence of a rule-based SWL reading route makes it practically impossible for the opposite pattern of damage to arise, that is a reading impairment such as phonological dyslexia in alphabetic languages.

Similar results have been obtained from Chinese dysgraphic patients whose performance on writing to dictation was markedly better than their written picture naming, which suggests that writing ability along a nonsemantic pathway has been spared. In this case the impairment exhibits intermediate aspects between surface and direct dysgraphia. As in reading, phonological (lexical) dysgraphia cannot be detected in Chinese, since there is no way to spell nonwords in this language.

In conclusion, contrastive studies on acquired reading and writing disorders in languages with different scripts and/or different degrees of regularity indicate that dual-route reading and writing models may be generalized across cultures, but that the relative impact of both routes varies substantially from one script to the other.

20.8. NEUROANATOMY OF WRITTEN LANGUAGE

The neural correlates of reading and spelling disorders and – more importantly – of the lexical and SWL reading and spelling pathways have still not been fully clarified.

20.8.1. The Anatomy of Reading

Phonological dyslexia is usually caused by large left-hemisphere fronto–temporo–parietal perisylvian lesions and a similar anatomical basis has also been found for *deep dyslexia* (e.g., Luzzatti *et al.*, 2001). Extensive damage to left-hemisphere language areas is consistent with Coltheart's account of *deep dyslexia*, namely the emergence of right-hemisphere lexical knowledge (Coltheart, 1980, 2000). *Surface dyslexia* is associated with antero–lateral temporal damage or with temporal lobe atrophy (Patterson & Hodges, 1992). Finally, *letter-by-letter dyslexia* is associated with left occipital and inferior temporal damage. Déjerine considered this lesion to be responsible for pure alexia (without agraphia), not because the damaged structure was the site of specific visual word-forms but rather because of a disconnection between right-hemisphere vision and the left-hemisphere angular gyrus visual word-form area.

However, recent neuroimaging studies suggest that the left inferior temporal occipital cortex (and fusiform gyrus) might be a better candidate for the store of learned lexical orthographic knowledge (Cohen & Dehaene, 2004).

Furthermore, attempts to localize the neural basis of the lexical and SWL routes of reading through neuroimaging studies indicate that written words do not activate a specific lexical pathway but rather modulate activation in areas that are associated with semantic processing. In a similar way, pseudowords were not found to activate any specific SWL pathway, but to modulate activation in areas that are associated with phonological processing (Binder *et al.*, 2003).

Generally, results from neuroimaging studies that have tried to disentangle the neural pathways involved (1) in reading words by means of the lexical route, (2) in reading nonwords by means of the SWL route and (3) in analyzing objects or non-orthographic visual stimuli, are still contradictory. The data obtained so far do not support a clear-cut anatomical separation of these functional pathways (Devlin *et al.*, 2006).

20.8.2. The Anatomy of Spelling

The results obtained from anatomo-correlative and neuroimaging studies are still relatively inconclusive. Some neuropsychological studies showed association between surface dysgraphia and damage to the left angular gyrus (e.g., Rapcsak & Beeson, 2002), providing support for Déjerine's localization of the visual word-form area. However, the emergence of spelling impairments after left inferior temporal and occipital lesions suggests that this area also plays a role in storing orthographic lexical representations to be accessed during reading and spelling (Rapcsak & Beeson, 2004). Converging evidence on a role of the left posterior inferior temporal cortex (fusiform and inferior temporal gyri) in lexical spelling has arisen from functional imaging studies on healthy participants (Beeson *et al.*, 2003). Similar findings have also emerged during mental recall of *Kanji* characters in Japanese subjects (Nakamura *et al.*, 2000).

Altogether, imaging data on reading call into question Déjerine's hypothesis that the left angular gyrus plays a crucial role in storing orthographic representations of familiar words. The empirical evidence in fact appears to suggest a localization of the written word-form in the inferior temporal occipital cortex. Finally, results from neuroimaging studies of spelling are contradictory concerning the neural pathways involved. They do not supply either a clear distinction of the neural pathways involved in the spelling performance compared to those involved in oral language processing (phonology and lexical-semantics) or a distinction of the lexical and SWL spelling routes (e.g., Norton *et al.*, 2007).

20.9. CHALLENGES AND FUTURE DIRECTIONS

Dual-route models of reading and spelling can account for the ability shown by normal literate subjects to process regular words, irregular words and nonword. However, stronger

evidence of a dual-route mental organization of written language derives from the reading and spelling performance of brain damaged patients. Although these models were originally based on the performance of English-speaking dyslexic and dysgraphic patients, several studies have meanwhile been conducted on other languages with much more shallow alphabetic scripts (e.g., German, Italian or Spanish), with consonantal scripts (Hebrew), syllabic scripts (Japanese *Kana*) and ideographic scripts (Chinese, Japanese *Kanji*). Although some differences across types of orthography were documented, the performance of dyslexic and dysgraphic patients points toward the universality of dual-route models both in alphabetic and in nonalphabetic scripts. Selective damage of either the lexical or the SWL routes of processing also suggests the independent neural implementation of these pathways. However, the cerebral correlates of the two functional routes still need clarification. Brain lesions identified in patients correspond only partly with the areas of activation obtained from neuroimaging studies of healthy individuals. Future research shall focus on the neural foundations of these discrepancies.

References

Ardila, A. (1991). Errors resembling semantic paralexias in Spanish-speaking aphasics. *Brain and Language*, *41*, 437–445.

Baxter, D.M., & Warrington, K.E. (1983). Neglect dysgraphia. *Journal of Neurology, Neurosurgery and Psychiatry*, *46*, 1073–1078.

Beeson, P.M., Rapcsak, S.Z., Plante, E., Chargualaf, J., Chung, A., Johnson, S.C., & Trouard, T.P. (2003). The neural substrates of writing: A functional magnetic resonance imaging study. *Aphasiology*, *17*, 647–665.

Binder, J.R., McKiernan, K.A., Parsons, M.E., Westbury, C.F., Possing, E.T., Kaufman, J.N., & Buchanan, L. (2003). Neural correlates of lexical access during visual word recognition. *Journal of Cognitive Neuroscience*, *15*, 372–393.

Cohen, L., & Dehaene, S. (2004). Specialization within the ventral stream: The case for the visual word form area. *Neuroimage*, *22*, 466–476.

Coltheart, M. (1980). Deep dyslexia: A right hemisphere hypothesis. In M. Coltheart, K.E. Patterson, & J.C. Marshall (Eds.), *Deep dyslexia* (pp. 326–380). London: Routledge and Kegan Paul.

Coltheart, M. (2000). Deep dyslexia is right-hemisphere reading. *Brain and Language*, *71*, 299–309.

Coslett, H.B., & Saffran, E.M. (1989). Evidence for preserved reading in 'pure alexia'. *Brain*, *112*, 327–359.

Déjerine, J.J. (1891). Sur un cas de cécité verbale avec agraphie suivi d'autopsie. *Mémoires de la Société de Biologie*, *3*, 197–201.

Déjerine, J.J. (1892). Contribution à l'étude anatomo-pathologique et clinique des différentes variétés de cécité verbale. *Comptes Rendus Hebdomadaires des Séances et Mémoires de la Société de Biologie, 4*, 61–90.

Devlin, J.T., Jamison, J.L., Gonnerman, L.M., & Matthews, P.M. (2006). The role of the posterior fusiform gyrus in reading. *Journal of Cognitive Neuroscience*, *18*, 911–922.

Haywood, M., & Coltheart, M. (2001). Neglect dyslexia with a stimulus-centred deficit and without visuospatial neglect. *Cognitive Neuropsychology*, *18*, 577–615.

Kay, J., Lesser, R., & Coltheart, M. (1992). *PALPA: Psycholinguistic assessment of language processing in aphasia*. Hove, UK: Lawrence Erlbaum.

Luzzatti, C., Mondini, S., & Semenza, C. (2001). Lexical representation and processing of morphologically complex words: Evidence from the reading performance of an Italian agrammatic patient. *Brain and Language, 79,* 345–359.

Marshall, J.C., & Newcombe, F. (1966). Syntactic and semantic errors in paralexia. *Neuropsychologia, 4,* 169–176.

Marshall, J.C., & Newcombe, F. (1973). Patterns of paralexia: A psycholinguistic approach. *Journal of Psycholinguistic Research, 2,* 175–199.

Morton, J. (1969). Interaction of information in word recognition. *Psychological Review, 76,* 165–178.

Morton, J. (1980). The Logogen Model and Orthographic Structure. In U. Frith (Ed.), *Cognitive approaches in spelling.* London: Academic Press.

Nakamura, K., Honda, M., Okada, T., Hanakawa, T., Toma, K., Fukuyama, H., Konishi, J., & Shibasaki, H. (2000). Participation of the left posterior inferior temporal cortex in writing and mental recall of kanji orthography: A functional MRI study. *Brain, 123,* 954–967.

Nelson, H.E. (1982). *The National Adult Reading Test (NART): Test manual.* Windsor, Berk, UK: NFER-Nelson.

Norton, E.K., Kovelman, I., & Petitto, L.A. (2007). Are there separate neural systems for spelling? New insights into the role of rules and memory in spelling from functional magnetic resonance imaging. *Mind, Brain and Education, 1,* 48–59.

Patterson, K.E. (1986). Lexical but nonsemantic spelling. *Cognitive Neuropsychology, 3,* 341–367.

Patterson, K.E., & Hodges, J.R. (1992). Deterioration of word meaning: Implications for reading. *Neuropsychologia, 30,* 1025–1040.

Rapcsak, S.Z., & Beeson, P.M. (2002). Neuroanatomical correlates of spelling and writing. In A.E. Hillis (Ed.), *Handbook of adult language disorders* (pp. 71–99). Philadelphia, PA: Psychology Press.

Rapcsak, S.Z., & Beeson, P.M. (2004). The role of left posterior inferior temporal cortex in spelling. *Neurology, 62,* 2221–2229.

Sasanuma S. (1980). Acquired dyslexia in Japanese: Clinical features and underlying mechanisms. In M. Coltheart, K.E. Patterson & J.C. Marshall (Eds.), *Deep dyslexia* (pp. 48–90). London: Routledge and Kegan Paul.

Schwartz, M.F., Saffran, E.M., & Marin, O.S.M. (1980). Fractionating the reading process in dementia: Evidence from word-specific print-to-sound associations. In M. Coltheart, K. Patterson & J.C. Marshall (Eds.), *Deep dyslexia* (pp.259–269). London: Routledge and Kegan Paul.

Shallice, T. (1981). Phonological agraphia and the lexical route in writing. *Brain, 104,* 413–429.

Toraldo, A., Cattani, B., Zonca, G., Saletta, P., & Luzzatti, C. (2006). Reading disorders in a language with shallow orthography: A multiple single-case study in Italian. *Aphasiology, 20,* 823–850.

Yin, W., He, S., & Weekes, B.S. (2005). Acquired dyslexia and dysgraphia in Chinese. *Behavioural Neurology, 16,* 159–167.

Further Readings

Coltheart, M. (1996). Phonological dyslexia: Past and future issues. *Cognitive Neuropsychology, 13,* **749–762.**
This is an introduction to a special issue on phonological dyslexia. The entire issue is recommended for readers interested in the cognitive neuropsychology of dyslexia.

Coltheart, M. (2000). Deep dyslexia is right-hemisphere reading. *Brain and Language, 71,* **299–309.**
The author discusses the contemporary interpretations of acquired deep dyslexia and compares the results from the performance of dyslexic patients to results from brain-imaging studies. Coltheart concludes that the RH interpretation remains the best explanation of deep dyslexia.

Coltheart, M. (2006). Acquired dyslexias and the computational modeling of reading. *Cognitive Neuropsychology, 23,* **96–109.**
The author compares and tests two approaches to the computational modeling of reading: the connectionist triangle model and the "dual-route cascaded" (DRC) model. Both computational approaches were tested by damaging the model and testing how well the impaired reading obtained from these lesions simulates the reading impairment seen in acquired surface and phonological dyslexia. The results favor the DRC model. The paper is recommended for readers interested in *computational cognitive neuropsychology.*

Price, C.J., & Mechelli, A. (2005). Reading and reading disturbance. *Current Opinion in Neurobiology, 15,* **231–238.**
This is a review of the recent functional neuroimaging studies on reading and reading impairments. Results are analyzed in a cognitive neuropsychological frame and in relation to contemporary cognitive dual-route models of reading.

Yin, W.G., He, S.X., & Weekes, B.S. (2005). Acquired dyslexia and dysgraphia in Chinese. *Behavioural Neurology, 16,* **159–167.**
The authors review major reports of patients who have acquired dyslexia and dysgraphia in Chinese and describe the implications of these data for the functional architecture of reading and writing. The authors' conclusion is that the unique features of Chinese script determine the symptoms of acquired dyslexia and dysgraphia in Chinese. It is an excellent review of the acquired impairments of written language in a cognitive neuropsychological frame. The paper (and the entire special issue of which the paper is part) is recommended for those readers who are interested in acquired dyslexia and dysgraphia across scripts.

21

Number Processing

CARLO SEMENZA

Department of Psychology, University of Trieste, Trieste, Italy

ABSTRACT

Neuropsychological studies on patients with brain lesions have contributed to contemporary understanding of numerical processing on equal footing with studies of healthy populations. Having carved the cognitive chicken at its joints via the traditional method of studying neuropsychological dissociations has thus paid handsomely. Indeed, also thanks to fast-improving neuroimaging techniques, but mainly thanks to increasingly refined conceptual tools newly applied in clinical research, we are witnessing an unprecedented progress in the interdisciplinary domain of mathematical cognition. Fascinating new data are emerging on how and where the brain represents numbers and performs calculation. The relation that numerical cognition has with language is much clearer than it was just a decade ago.

21.1. INTRODUCTION

Numbers affect many aspects of contemporary life. It would be a mistake, however, to think that numbers are just a product of advanced civilization. Even life in primitive conditions could not and cannot dispense with numbers. Only one language (Box 21.1) out of the thousands known, Pirãha, spoken in a limited Amazon region, has no words for numbers! Indeed, written representations of numerals can be found in bone engravings and on cave walls by Neolithic men, dating about 30,000 years ago. The first numerate societies, like the Sumerians and the Babylonians, started using numbers around 10,000 BC, and by 3000 BC they had developed a wide variety of number-specific cultural tools.

Box 21.1 Numbers and cultures

Different cultures use different numerical systems. There is only one language that we know of without number words (Pirãha, Amazon), but several languages are known that have very restricted systems. Examples of this kind come from Australian aboriginal languages, like Mangarayi (that has numbers up to 3), Yidini (up to 5), Hixkaryana (up to 5, plus 10: the word for 4 means "his brother twice over;" 5 is "half our hands," 10 is "our hands completely"). Haraui (New Guinea) has an up to 4 system with addition: thus $3 = 2 + 1$ and $4 = 2 + 2$.

While several cultures still keep track in counting using body parts (up to 74, examples are given in the main text), most cultures use bases, that is collections of numbers from which other numbers can be derived by multiplication and addition. Thus, in base systems, numbers can be expressed as a multiple of a base plus any number that is smaller than the base. For example, in a language with base 60, the number 361 would be expressed as $6 \times 60 + 1$.

Base 10 is the most popular, but there are several others and many cultures use a mixed base system. Thus, there are systems with base 20, 12, 6, 60, 32 and so forth. However, most languages, for historical reasons, contain irregularities. One very regular number word system with base 10 is found in Mandarin Chinese. But look at English: the next number after "ten" is "eleven," not "ten-one" as full regularity would dictate! A very endangered language, Resian Slovene, regularly alternates base 10 with base 20. In French base 20 intrudes in the base 10 system ("quatre-vingt," $80 = 4 \times 20$) and there are mixed numbers like "soixante-dix-sept" ($77 = 60 + 10 + 7$). All the first 100 numbers in Hindi are irregular.

Descriptions of number systems in languages worldwide may be found in the vast work of Bernard Comrie, the leading expert in the field.

Elementary number processing is demonstrable in infants. For example, they show signs of renewed interest when three dots appear after they have been shown two for some time, and they seem to notice the difference if two puppets disappear behind a screen and then only one reappears when the screen is removed. Even animals of several species, not only those more proximate to homo sapiens, are now thought to register numbers in some representational form and enter them in simple mental computations. It is therefore evident that numbers and their manipulation reflect a basic ability developed in living organisms with natural evolution, and that culturally transmitted, number related, abilities are just the latest result of a long and complex evolutionary process.

Number-specific tools represent numbers in various formats: numerals (Arabic, Roman and so forth), number words, dots on dice and so forth. More complex tools are meant for use in calculation: arithmetical facts (e.g., $6 \times 3 = 18$), arithmetical procedures (e.g., addition, subtraction, multiplication, borrowing, carrying and so forth), and arithmetical laws (e.g., commutability: $5 \times 4 = 4 \times 5$).

The use of these tools involves the use of different skills in order to perform several different mathematical tasks. A huge theoretical and empirical effort has been made in order to understand how the cognitive system acquires these skills and deals with these tasks. Significant contributions have come from animal psychology, human experimental psychology in adults, developmental psychology, behavioral genetics and neuroscience. This chapter will summarize the results of this effort, emphasizing how traditional neuropsychology has handsomely contributed to models of numerical processing, on equal footing with experimental studies on typical, non-brain-damaged populations. The observation of counterintuitive clinical phenomena, moreover, prompted research in unforeseen directions.

21.2. THE REPRESENTATION OF NUMBERS

Knowledge of numbers seems somehow independent from other types of knowledge stored in our brain. After left parietal lobe damage, CG, a lady studied by Cipolotti *et al.* (1991), was still proficient in language tasks and reasoning and retained a normal IQ. Yet she could not deal at all, verbally or otherwise, with numbers above four! In contrast, the case reported by Cappelletti, *et al.*, (2001), affected by semantic dementia, was severely impaired in naming, reading and writing of non-number words as well as in several other verbal and pictorial tasks, while showing no comparable deficit with numerical material. Similarly, another patient was reported whose reasoning abilities were almost nil, while his calculation was excellent! In classic amnesia, involving disorders of episodic memory, number skills are instead preserved. Patients with short-term memory impairment may often retain the ability to carry out mental calculation.

Neuropsychological cases thus show how numerical representations and calculation abilities are stored in long term, semantic, memory in a highly modular way and can be selectively impaired.

21.2.1. Number Meaning in the Brain

How are, then, numbers represented in the brain? Why do they differ from other domains of knowledge? The way quantities are appreciated is crucial in understanding how the brain represents number knowledge and needs to be described in some detail.

Capturing the number of elements in a given collection and accessing the appropriate mental representation is performed via three basic tasks: subitizing, counting and estimation. Each of these tasks is independent from other tasks in the brain. Suppose one has to tell how many dots appear in a given array. Reaction times in this task appear to be about the same for each quantity inferior to five dots and increase exponentially thereafter. The ability to correctly appreciate such small quantities without serial processing is referred to as "subitizing." The abilities of patient CG, who could not deal with numbers above four, were thus restricted to the subitizing range. For appreciating quantities of five and more, one has to resort to counting or estimation, abilities that CG did not retain after her brain injury. "Counting" basically consists of pairing each element of a collection to a number word. The last assigned word is the label for the entire collection. This latter property is called "cardinality." The concept of "ordinality," instead, refers to the ability to judge which of two different quantities is larger.

According to Gelman and Gallistel (1992), cardinality is part of innate mathematical knowledge. Such knowledge would also include realising that: (a) each element must correspond to one verbal label only; (b) a conventional order rules verbal labels; (c) counting applies to any collection of objects; (d) changing the labeling order does not modify the numerosity of the collection. A separate ability involves separating already counted objects from those that remain to be counted. Not all authors believe that counting rests on innate knowledge: the alternative view is that children deduct principles of counting from experience (Wynn, 1998).

Ordinality and cardinality are really different aspects of number representation and were shown to dissociate in neuropsychological patients. A selective impairment with number cardinality only has in fact been reported (Delazer & Butterworth, 1997): the patient could correctly order any sequence of numbers, while unable to add 1 to a given number! He thus could say what number came after 6, but, rather paradoxically, he did not know the answer to $6 + 1$!

According to the position of Dehaene and co-authors (e.g., Dehaene & Cohen, 1995) patients with left hemisphere lesions may not be able to appreciate the exact meaning of numbers and may not perform exact calculation. Some of them, however, may nonetheless be able to provide good number estimates and perform approximate calculation. "Estimation" is a quick but less accurate process than counting. Several variables, besides objective numerosity, have been shown to influence estimation: the physical properties of the stimuli, their spatial disposition, practice in the tasks, visual clarity and so forth. Many researchers think that the internal representation of a number is analogical and its precision decreases as the number increases. But what does "analogical" mean in the case of number representation? The understanding of this concept started with Galton in the nineteenth century.

21.2.2. The Number Line

Since Galton (1880), in fact, scientists were made aware that some people are conscious of an internal spatial representation of numbers that mostly takes the form of an imaginary line. This "number line" is used in understanding numbers and, sometimes, in calculation. In modern psychology, the nature of the semantic representation of numbers is indeed controversial. While some people (e.g., McCloskey, 1992) assume a precise representation even for large numbers, others (e.g., Dehaene & Cohen, 1995) hypothesise an analog magnitude representation for approximate quantity, that is number meanings are represented in a continuous, spatially extended, scale. Vis-à-vis such controversy, the concept of "number line" is gaining new popularity. A modern addition to Galton's observation is, in fact, that even people who are not consciously aware of a mental number line may nonetheless use it. Dehaene et al. (1993) showed that within a given interval, people who read left to right are generally faster at making judgements (e.g., odd/even judgements) about smaller numbers with the left hand but faster with their right hand for larger numbers. This effect (known as "SNARC"—Spatial-Numerical Association of Response Codes) may be weaker or inverted in speakers of languages with right-to-left writing! This suggests that an unconscious number line is there in most people and that its orientation depends on the direction of reading. In the case of left-to-right reading, smaller numbers are thus on the left and larger numbers on the right of the line. Interestingly, further evidence for this conclusion comes from patients with left sided neglect. Given the task of telling which is the midpoint of a numerical interval, they show the same displacement effect to the right they would show in line bisection (thus answering 6 instead of 5 when asked the midpoint between 3 and 7).

Order rather than magnitude may however be the critical property of the number line. A SNARC-like effect is in fact observed for other ordered sequences, like the letters of the alphabet or the days of the week.

21.2.3. Fingers Count (and So Do Other Body Parts)

A spatially organized line is not the only way we help ourselves to mentally represent numbers. Our body helps us

as well. Neuropsychologists proved indeed long time ago that dealing with numbers and dealing with fingers are linked abilities. Damage, or temporary inactivation (Rusconi et al., 2005), in the left parietal lobe leads to what is known as "Gerstmann's syndrome," a condition whereby patients cannot calculate and cannot recognize fingers.

Why does our brain process numbers nearby the representation of fingers? This may be because mental representations of numbers are influenced by strategies used to keep track on how many items one has counted up to a point. Several developmental and cross-cultural studies show that spontaneous finger-counting strategies are developed and used by children in almost all human cultures (Butterworth, 1999). However, while fingers are invariably used in body counting, other body parts and body locations (e.g., between-fingers intervals) sometimes with repeated passages, up to 74, are used! The sequential order is not intuitive. For example, Yupno males (with respect to our own culture, political correctness may have different implications in this tribe of Papua New Guinea: female do not count or count less) count up to 33: they start from the little finger of the left hand, then they count fingers on the right hand, the left foot, the right foot, then the ears, the eyes, the nostrils and the nose, the nipples, the navel, the left testicle, the right testicle and the penis. South of them, natives of the Torres Straight also count up to 33, but start from the little finger of the right hand, then count the wrist, the elbow, the shoulder, go on with a mid-chest point and start on the other side with the shoulder, elbow and wrist down to the left hand fingers; then they go to the fingers of the left foot, then the ankle, the knee, the left hip, the right hip and down to the knee and the ankle to end on the little finger on the right foot.

Use of body parts, typically finger counting, appears thus to be a universal strategy to deal with numbers. Using and practising prototypical finger counting may lead to long-term associations between fingers and digits. Long-established finger-counting strategies have been recently shown to influence the way numerical information is projected into physical space and perhaps mentally represented (Di Luca et al., 2006).

The existence of a neural network supporting finger movement representation has been advocated on the basis of neuroimaging observations (Pesenti et al., 2001). Joint activation of left pre-central and parietal areas would sustain finger counting and numerosity quantification. It would support early representations in childhood and, by extension, become the substrate of numerical knowledge and processes in adults.

21.2.4. Number Words Are Special: The Number Lexicon

Cardinal (and ordinal) numbers are also represented in a speaker's language. Neuropsychological studies show that number words can be damaged in a very selective fashion (Dohmas et al., 2006; Semenza et al., 2007). For example, one patient made, in oral production, lexical substitutions of number words but not of other words. Conversely cases were found with phonological impairments in production that spared numerals. In these cases the problem seems to be located in the activation of the lexical system from the semantic system. While leaving many unanswered questions, these cases are nonetheless important because they show how number words are in several respects different from other collections of words.

21.2.5. On Knowing About Nothing: The Elusive Number Zero

The most mysterious number word, and elusive number-related concept is zero. How does our mind appreciate and manipulate the concept of nothingness in the numerical domain? There is good reason to think that this is not an easy task. While adding a null quantity or subtracting it from a given quantity can be easily represented (e.g., by visual imaging), it is less obvious which representation may be invoked in multiplication or division by zero. Zero as an operand, unlike any other operand, makes any quantity disappear!

The idea of a collection with no members or, in mathematical terms, of an empty set, is not difficult to grasp. Yet it took a long time to efficiently represent zero within a numerical system. Zero was not part of the ancient formal number systems. Mathematicians in Babylon, classic Greece (including Archimedes!) and Rome, ignored zero altogether! Working independent from Europe, the Maya are now believed to have preceded everybody by developing a system representing zero in the first centuries AD. Only in the early seventh century, in the Indus valley, did zero develop a meaning in its own right. The notion took about another 150 years to be adopted by Arabs, whose traders passed it to the Europeans only in the twelfth century! This historical delay, even more surprising when confronted with the extraordinary mastery of other mathematical principles, reflects the intrinsic difficulty of manipulating a null quantity in arithmetic.

The same story holds within individual development. Young children, once they have learned to deal with positive integers, and have learned both the word "zero" and the corresponding Arabic symbol, take a considerable time to appreciate that zero is a numerical value corresponding to nothingness.

Neuropsychologists, on their part, showed that the notion of zero remains somehow segregated in the brain. Patients have been described, for example, showing how even the same type of knowledge, that is, the 0-rule in multiplication, may be correctly retrieved or not depending upon the arithmetic context. In describing these phenomena, Semenza *et al.* (2006b) suggested that processing of zero in arithmetic is likely to be mediated by shallow, context-bound rules and procedures, whose binding with conceptual knowledge is rather marginal. Indeed, though automatic and routinized procedures may efficiently support calculation in full-functioning

individuals, they may become isolated and neglected pieces of knowledge in the context of disturbed cognitive systems.

21.3. NUMBER MANIPULATION: TRANSCODING

Written and spoken words are obviously not the only way to represent numbers. Various codes for numbers are used and are indeed independently represented and processed in our brain. Patients have been reported who could not use the alphabetical code but could flawlessly use numbers in the Arabic code. The reversal case has also been documented.

Processes dealing with numbers are often "transcoding" processes, in that they transform a number from a given representational format into another format. Thus, in the contemporary Western culture (other codes, e.g., the Roman code, have been used or are used in other cultures), a task like reading aloud presupposes transcoding Arabic or alphabetically written numbers into spoken number words; in writing on dictation the reverse transcoding process is at work. Other common transcoding tasks are, for example, number repetition or number transformation from one code into the other (e.g., "two" into "2" or vice-versa).

How does the cognitive system perform such tasks? Patients committing revealing patterns of errors in these tasks have been described. Naturally, errors mostly occur when transcoding complex numbers. Complex numbers are built according to a sort of lexical syntax. "Number syntax" has been shown – on the basis of neuropsychological dissociations – to be independent from language syntax and defines the set of acceptable structures formed by combining the basic lexical elements. The semantic component in the system allows attribution of meaning to each sequence.

The most interesting findings thus concern two main types of transcoding errors: "lexical" and "syntactic" errors (Deloche & Seron, 1982). These errors often co-occur and are sometimes observed in relative isolation from each other, reflecting independent "lexical" and "syntactic" processing mechanisms.

"Lexical errors" consist of the incorrect production of one or more of the individual elements in a number (e.g., 4 instead of 8, 57 instead of 58, 2506 instead of 2406). "Syntactic errors" are instead violations of the order of magnitude that spare the correct lexical elements: they are made when the power of ten is wrong with respect to the target. Syntax errors may thus result in longer numbers as in "510,028" instead of "528" or "50,028" instead of "528," or in shorter numbers (when deletions of one or more zeros occur, as in "25" instead of "205"). Different cultures adopt different number syntax systems. Thus the nature of number syntax errors may vary among patients who speak different languages. For example in French, whereby base 20 intrudes in a base 10 system (Box 21.1), patients make syntax errors that cannot be observed in English.

However the distinction is not that simple. Granà *et al.* (2003) showed that there is an important difference between "lexical" and "syntactic" zeros. One patient made errors exclusively on this latter type. While lexical zeros are semantically derived zeros, like those in tens (e.g., the zero in "30," and the first zero in "20,104") that originate from a numerical concept, syntactic zeros are syntactically produced as the result of syntactic rules (e.g., the zeros in "13,004" that have just place value). Syntactic zeros are thus more difficult to manipulate.

One still debated point is whether or not number transcoding and calculation necessarily imply semantic mediation via a central abstract semantic representation and whether asemantic, code-specific routes exist besides the semantic one (different models have been proposed on this subject: for reviews see Noël, 2001, and Cipolotti & van Harskamp, 2001; examples are illustrated in Box 21.2).

21.4. CALCULATION

Calculation needs more than just numbers. Its different components, shown to dissociate in neuropsychological cases, include arithmetical signs, arithmetical number facts and rules, calculation procedures, approximate calculation and conceptual knowledge.

21.4.1. Signs, Facts and Rules

Some very rare patients were described who could not understand and name arithmetical signs (+, −, ×, : and so on), but perform otherwise correct calculation according to their misidentification!

Patients do also exist who can no longer tell the result of single-digit, over-learned addition (e.g., 4 + 2) or multiplication problems (e.g., 5 × 6) listed in the so called arithmetical

---

Box 21.2   Number processing models

Figure below integrates McCloskey's (1992) and Cipolotti and Butterworth's (1995) number processing models. In McCloskey's version, number transcoding happens exclusively via abstract, modality neutral, semantic representations that are used for all calculations; transcoding implies understanding. In Cipolotti and Butterworth's version, transcoding may be performed via non-semantic direct associations: thus "3" in the Arabic format may directly become "three" in the alphabetic format and understanding the meaning of the number three may not be necessary to perform the task. Arithmetical operations are stored separately. Number size is represented as base 10 units.

The triple-code model by Dehaene and Cohen (1995) (see second Figure) specifies the anatomical location of its

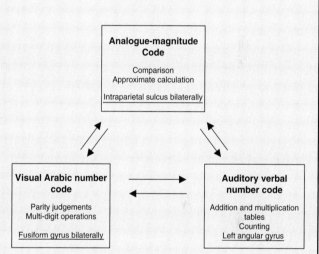

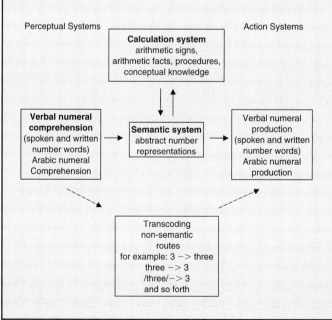

components. Number size is represented in a logarithmically compressed form. There are no separate stores for arithmetical operations. Transcoding and calculation may be performed without semantics. For instance, one digit multiplications may be retrieved directly as over-learned "facts." Multi-digit operations may be accomplished with the aid of visual images.

Cipolotti, L., & Butterworth, B. (1995). Toward a multiroute model of number processing: Impaired number transcoding with preserved calculation skills. *Journal of Experimental Psychology: General, 124*(4), 375–390.

Dehaene, S., & Cohen, L. (1995). Towards an anatomical and functional model of number processing. *Mathematical Cognition, 1,* 83–120.

McCloskey, M. (1992). Cognitive mechanisms in numerical processing: Evidence from acquired dyscalculia. *Cognition, 44,* 107–157.

tables. These problems and their solutions, called arithmetical facts, are thought to constitute an independently stored system of notions (Ashcraft, 1992). Neuropsychological cases show that arithmetical facts are indeed segregated in the brain according to the type of operation and independently retrieved.

Arithmetical rules are distinct from facts and include problems where one of the elements (e.g., the multiplier or the addend) can be substituted by any number (e.g., n + 1, n − 1, n × 1, n : 1, n + 0, n − 0, n × 0). Rules involving zero are, as already mentioned, particularly difficult to learn and can be selectively disturbed by brain damage.

21.4.2. Procedures: Bugs in the Brain?

Arithmetical procedures, that is the algorithms necessary for each type of operation, are also stored independently from each other. Acquired disturbances of procedures thus constitute a rather heterogeneous picture: for example, only selective deficits of multiplication or subtraction have been described (Cipolotti & Van Harskamp, 2001).

Studies with healthy learners highlighted systematic, consistent errors known as *bugs* at certain points of the procedure. Based on selective cases, Semenza *et al.* (1997) also distinguished "memory" problems for the procedural algorithm (typically resulting in systematic misapplication of some specific rule) and "monitoring" problems, deriving from the inability to devote attention and keep track of each specific step in the procedure. Thus, while "memory" patients (Girelli & Delazer, 1996), commit, just like healthy participants, errors of the "bug" type, "monitoring" patients, while always starting the operation in the correct way, would then make an increasing and inconsistent number of errors of all sorts, have little awareness about the precision of the performance and a difficulty in ending the operation. Spatial components of procedures, we will later see, are also thought to be stored in the right hemisphere.

21.4.3. Conceptual Knowledge

The use of procedures does not necessarily require conceptual understanding of each step of the procedure and is applied only in already familiar tasks. What has been called "conceptual knowledge," instead, implies understanding of arithmetical operations, laws and principles pertaining. Also labeled "adaptive expertise," it may or may not be explicitly known as such to the calculator, but allows us to make inferences and can be flexibly adapted to new tasks. Examples of this sort of knowledge are potentially infinite as demonstrated by calculators that answer the question "given X, what is the result of Y?" The "commutability" principle (e.g., X: 22 × 31 = 682; Y: 31 × 22 = ?), "repeated addition" (11 × 4 = 44; 11 + 11 + 11 + 11 = ?), "10a × 10b" (45 × 8 = 360; 450 × 80 = ?), "a − 1 × b" (94 × 5 = 470; 93 × 5 = ?), and "multiplication/division invariance" (71 × 9 = 639; 639: 9 = ?) are just some examples.

Conceptual knowledge is independent from other mathematical notions and abilities. A patient unable to retrieve number facts (Hittmair-Delazer *et al.*, 1994) could brightly use his residual knowledge to overcome his difficulties with tables (e.g., 7 × 8 = ?; 7 × 10 = 70; 7 × 2 = 14; 70 − 14 = 56)! Although unable to tell facts as simple as 4 + 3 = 7, he had no difficulty with algebraic expressions such as (ac + bc): c = a + b! Other patients, in contrast, flawlessly retrieve arithmetical facts while showing no understanding of mathematical concepts whatsoever.

21.5. NUMBERS AND CALCULATION IN THE BRAIN

Neuropsychological data, indicating patterns of dissociation, suggest how the number and calculation system works. What about the brain? A fundamental question arises first: Did a biologically determined mechanism predisposed to represent the semantic domain of knowledge of numbers and arithmetic evolve in the brain? This idea has been advocated on the grounds of observations made in anatomo–clinical studies and brain imaging as well as in different fields of human, adult and child, psychology and animal psychology. However, the notion of a single mechanism must be used with caution. Most mathematical abilities have been indeed shown to dissociate from each other and from linguistic capacities. Arithmetic processing, just like language, is complex and may rely on independent sub-modules, including basic abilities (such as recognition of quantity), higher arithmetic abilities (such as multiplication or addition), as well as a set of non-dedicated abilities. In right handed people, higher arithmetical abilities are located in the left hemisphere. Some are linked to language, some are not. Traditional classifications of acalculia, in fact, distinguished between primary, "pure anarithmetia" and acalculia secondary to language disorders. Right hemisphere or "spatial acalculia" was also included in the traditional classification, although only sparse attempts to specify the functions missing in this particular syndrome were until recently made, mostly on the understanding that some calculation disorders are a consequence of more generic spatial disorders like, for instance neglect.

21.5.1. Core Abilities: The Left Parietal Lobe

Brain localization of number processing has made some progress from the original, still basically correct, idea, that, with the exception of some spatial components located in the right hemisphere, most math skills are located in the left parietal lobe. One cognitive model, the "triple-code" model (Dehaene & Cohen, 1995), is based on recently discovered anatomical correlations (Box 21.2). The model, while assuming that Arabic and magnitude representations of numbers are available to both hemispheres in the vicinity of the parieto–occipito–temporal junction (but extending to the occipito–ventral on the left), assigns the verbal representation underlying the retrieval of arithmetical facts exclusively to the left hemisphere. Arithmetical facts refer to single digit problems, like tables, that are, in learned calculators, retrieved from long-term memory rather than calculated. Only the left hemisphere would possess the representation of the sequence of words corresponding to verbal numerals and lexical mechanisms for identifying and producing spoken numerals. These lexical mechanisms, that would not be specific for numbers, would be implemented in the classic language (Broca's and Wernicke's) areas, including the inferior frontal and the superior and middle temporal gyri, as well as basal ganglia and thalamic nuclei.

Within the left parietal lobe, three major circuits for number processing were recently proposed, that is a core system in the intra-parietal sulcus, a left angular gyrus system and a bilateral posterior superior parietal system (Dehaene *et al.*, 2003). Each of these circuits is said to play a distinct functional role in arithmetic:

(1) A core system, called into action whenever numbers are manipulated and quantity processing is required, is located in the horizontal segment of the intra-parietal sulcus (HIPS). This circuit is systematically activated whenever numbers are processed, independently of the notation used for numbers, and the activation increases when the task requires the processing of quantities. Distinct portions may sustain the appreciation of exact numerosity and analog magnitude (Castelli *et al.*, 2006). This area seems to be activated in calculation: more in approximate than in exact calculation and more in subtraction than in multiplication.

While this circuit is specific to the number domain, the remaining circuits are instead thought to rely on representations and processes that are not specific to the number domain.

(2) A left angular gyrus area, in connection with other left hemisphere areas, supports the retrieval of arithmetic facts. In particular, the left angular gyrus supports the verbal aspects of numerical processing. It is thought to mediate the retrieval of facts stored in verbal memory, but not other numerical tasks, like subtraction, number comparison or complex calculation, that are related to quantity processing (Box 21.3).

(3) A bilateral posterior superior parietal system would support visuo-spatial processes, attention and spatial working memory related to numerical processing. It also represents the number line. It is active in number comparison, approximation or counting. It also appears to increase in activation when subjects carry out two operations instead of one.

21.5.2. Complex Calculation and the Right Hemisphere

The above described circuits are working on simple math tasks. What happens in our brain when we are engaged in complex calculation? Indeed, beyond a certain degree of difficulty, perhaps with the exception of some prodigious calculator (often an idiot-savant), when working memory

---

Box 21.3   Numerical tasks

An example of a comprehensive battery for the evaluation of number and calculation skills:

Counting: including *Verbal counting* (e.g., backward from 15; onward two by two, starting from 3), *Written Counting* (e.g.,: backward from 17; onward three by three, starting from 2) and *Dot Counting* (determining numerosity of given dot patterns).

Comprehension of numbers: including *Number Comparison* (choose the larger of two multi-digit numerals in three different modalities: Arabic numerals, written number words and spoken numerals); *Parity judgements* (on written Arabic numerals); *Analog number scale* (choose among three alternatives the position corresponding to a given Arabic numeral on an analog number scale extending in half cases from 0 to 100 and in the other half from 0 to 50).

Numerical Transcoding: including *Transcoding from Arabic numerals to tokens* (select tokens representing a given Arabic numeral from a set of 3–7 tokens of three different types representing 100, 10 and 1); *Reading Arabic numerals*; *Writing Arabic numerals to dictation*; *Reading number words*; *Transcoding from written number words to Arabic numerals*; *Transcoding from tokens to Arabic numerals*.

Calculation skills and arithmetic principles: including *Arithmetic facts and rules* (simple arithmetic problems, presented in Arabic format and answered verbally; facts include one digit additions, one-two digits subtractions, tables for multiplication and reverse tables for division; rules include problems of the type $n + 0$, $n - n$, $n \times 0$, $n \times 1$, $n : 1$, $n : n$, etc.); *Multiplication facts, multiple choice*; *Mental calculation* (additions, subtractions, multiplications and divisions to be computed mentally and answered orally); *Written calculation* (additions, subtractions and multiplications); *Approximation* (multi-digit additions, subtractions, multiplications and divisions presented with four false answers with the instruction of choosing the one closest to correct); *Text problems* (problems in textual format requiring the computation of a single addition, subtraction, multiplication, division, for example: "A woman weighed 70 kg; meanwhile she lost 4.5 kg; How much does she weigh now?"); *Arithmetic principles* ($24 + 37 = 61$, $37 + 24 = ?$; $34 \times 6 = 204$, $204 : 6 = ?$).

A comprehensive battery of this sort, with normative data, can be found in:

Delazer, M., Girelli, L., Granà, A., & Domahs, F. (2003). Number processing and calculation. Normative data from healthy adults. *Clinical Neuropsychology*, *17*, 331–350.

---

demands exceed its capacity, calculating people must resort to the written modality. What happens at this point?

In written calculation, one must use stored memory for arithmetical procedures and this memory must include a learned spatial layout schema. This schema would help the non-prodigious calculator along each sub-step of the arithmetical operation by representing where exactly each sub-step would be placed. In other words, it helps to keep track of the steps in the operation not only knowing *what* has to be written, but also *where* it has to be written. "Start from the rightmost element,""Add the carry to the next column on the left: none of us remembers and uses these instructions explicitly, in a verbal format, for each step of the operation!" The use of a spatial schema, maintained on-line during calculation, instead, would render the process easier for us and almost automatic. Granà *et al.* (2006), starting from these considerations, suggested that an isolated deficit of this function was at the basis of the deficit in a patient who could not perform complex multiplication any more. This patient had a posterior right hemisphere lesion, involving the polar planum, Heschl gyrus, the temporal planum, posterior insula, the supramarginal gyrus and the angular gyrus, with dilatation of the right ventricle. No left hemisphere lesion was detected!

21.5.3. Where in the Brain Did Calculating Abilities Evolve from?

With the exception of some space-specific learned abilities (such as the case reported above), most of higher arithmetical abilities are located, in right handers, in the left hemisphere. Why? This is, indeed, not a trivial question. Is it because of language lateralization? The answer may only partially depend on whether number and calculation processing are independent abilities, as we tend to believe now (and most data show), or just a special aspect of language as some used to believe. It may instead depend on whether a primitive common capacity evolved in the same hemisphere.

Semenza *et al.* (2006a) sought an answer to this question by looking at (very rare) patients whose laterality patterns are reversed. In six right handed patients affected by aphasia following a lesion to their non-dominant hemisphere (crossed aphasia) and in two left handed aphasic patients with a right sided lesion, acalculia was always found although in different degrees. The type of acalculia depended on the type of aphasia, following patterns that have been previously observed in the most common aphasias resulting from left hemisphere lesions. No sign of *right hemisphere* or *spatial* acalculia (acalculia in left lateralised right handed subjects) was detected. These results suggest that, as a rule, language and calculation share the same hemisphere.

A more cautious interpretation is that these data may just reveal that in cases of right brain lateralisation of language, a number of math functions or math sub-processes migrate with the language functions, but again, why? The reason for this anatomical proximity may lie in the fact that, as recently suggested (Hauser *et al.*, 2002), a primitive computational mechanism capable of "recursion," a combinatory capacity at the basis of both language and calculation faculties, has evolved in the dominant hemisphere for

reasons independent of both functions. This mechanism, by providing an open-ended and limitless system of communication, would have thus allowed the evolution of language and, independently, of skills recruited for calculation.

21.6. CHALLENGES AND FUTURE DIRECTIONS

Different sub-disciplines are all meant to ultimately converge in a theory explaining mental mechanisms involved in math skills. This should happen in terms of a relatively small set of rules, possibly framed in the precise language of mathematics. These rules should be ideally grounded at a neurological level of explanation through a series of additional bridging laws linking the molecular, synaptic and cellular levels with psychological representation and computation. Is the time for such theory near at all? Dehaene's (in press) recent endeavor, for example, may sound as an affirmative answer to this question. His starting point consists in postulating a neuronal code whereby the cardinal of a set of objects is represented approximately by the firing of a population of neurons, each of which firing to a preferred numerosity. Such neurons, indeed, have been recently discovered in monkeys. Dehaene then shows how one can capture the properties of such system with equations that tightly fit behavioral and neural data for some simple numerical tasks like number comparison or same-different judgements. As Galileo's motto goes: "provando e riprovando" (trying and retrying).

References

Ashcraft, M.H. (1995). Cognitive psychology and simple arithmetic: A review and summary of new directions. *Mathematical Cognition*, *1*(1), 3–34.

Butterworth, B. (1999). *The mathematical brain*. London: Macmillan.

Castelli, F., Glaser, D.A., & Butterworth, B. (2006). Discrete and analogue quantity processing in the parietal lobe: A functional MRI study. *Proceedings of the National Academy of Sciences*, *103*, 4693–4698.

Cappelletti, M., Butterworth, B., & Kopelman, M. (2001). Spared numerical abilities in a case of semantic dementia. *Neuropsychologia*, *39*, 1224–1239.

Cipolotti, L., Butterworth, B., & Denes, G. (1991). A specific deficit of numbers in a case of dense acalculia. *Brain*, *114*, 2619–2637.

Cipolotti, L., & van Harskamp, N. (2001). Disturbances of number processing and calculation. In R.S. Berndt (Ed.), *Handbook of Neuropsychology* (pp. 305–334). Amsterdam: Elsevier.

Dehaene, S. (2007). Symbols and quantities in parietal cortex: Elements of a mathematical theory of number representation and manipulation. In P. Haggard, Y. Rossetti, & M. Kawato (Eds.). *Attention and Performance: Sensory-motor foundations of higher cognition* (pp. 527–574). New York: Oxford University Press.

Dehaene, S., & Cohen, L. (1995). Towards and Anatomical and functional model of number processing. *Mathematical Cognition*, *1*, 83–120.

Dehaene, S., Bossini, S., & Giraux, P. (1993). The mental representation of parity and numerical magnitude. *Journal of Experimental Psychology: General*, *122*(39), 371–396.

Dehaene, S., Piazza, M., Pinel, P., & Cohen, L. (2003). Three parietal circuits for number processing. *Cognitive Neuropsychology, 20*(3–6), 487–506.

Delazer, M., & Butterworth, B. (1997). A dissociation of number meanings. *Cognitive Neuropsychology, 14*, 613–636.

Deloche, G., & Seron, X. (1982). From one to 1: An analysis of a transcoding process by means of neuropsychological data. *Cognition, 12*(2), 119–149.

Di Luca, S., Granà, A., Semenza, C., Seron, X., & Pesenti, M. (2006). Finger-digit compatibility in Arabic numeral processing. *Quarterly Journal of Experimental Psychology, 59*(9), 1648–1663.

Dohmas, F., Bartha, L., Lochy, A., Benke, T., & Delazer, M. (2006). Number words are special: Evidence from a case of primary progressive aphasia. *Journal of Neurolinguistics, 19*(1), 1–37.

Galton, F. (1880). Visualized numerals. *Nature, 21*, 252–256.

Gelman, R., & Gallistel, C.R. (1992). Preverbal and verbal counting and computation. *Cognition, 44*, 43–74.

Girelli, L., & Delazer, M. (1996). Subtraction bugs in an acalculic patient. *Cortex, 32*(3), 547–555.

Granà, A., Hofer, R., & Semenza, C. (2006). Acalculia from a right hemisphere lesion. Dealing with "where" in multiplication procedures.. *Neuropsychologia,, 44*, 2972–2986.

Granà, A., Lochy, A., Girelli, L., Seron, X., & Semenza, C. (2003). Transcoding zeros within complex numerals. *Neuropsychologia, 41*, 1611–1618.

Hauser, M.D., Chomsky, N., & Fitch, W.T. (2002). The faculty of language: What is it, who has it and how did it evolve? *Science, 298*, 1569–1579.

Hittmair-Delazer, M., Semenza, C., & Denes, G. (1994). Concepts and facts in calculation. *Brain, 117*(4), 715–728.

McCloskey, M. (1992). Cognitive mechanisms in numerical processing: Evidence from acquired dyscalculia. *Cognition, 44*, 107–157.

Noël, M.P. (2001). Numerical cognition. In B. Rapp (Ed.), *The handbook of cognitive neuropsychology* (pp. 495–518). Philadelphia: Psychology Press.

Pesenti, M., Zago, L., Crivello, F., Mellet, E., Samson, D., Duroux, B., Seron, X., Mazoyer, B., & Tzourio-Mazoyer, N. (2001). Mental calculation expertise in a prodigy is sustained by right prefrontal and medial–temporal areas. *Nature Neuroscience, 4*(1), 103–107.

Rusconi, E., Walsh, V., & Butterworth, B. (2005). Dexterity with numbers: rTMS over left angular gyrus disrupts finger gnosis and number processing. *Neuropsychologia, 43*, 1609–1624.

Semenza, C., Miceli, L., & Girelli, L. (1997). A deficit for arithmetical procedures: Lack of knowledge or a lack of monitoring? *Cortex, 33*, 483–498.

Semenza, C., Delazer, M., Bertella, L., Granà, A., Mori, I., Conti, F., Pignatti, R., Bartha, L., Dohmas, F., Benke, T., & Mauro, A. (2006a). Is math lateralized on the same side as language? Right hemisphere aphasia and mathematical abilities. *Neuroscience Letters, 406*, 285–288.

Semenza, C., Granà, A., & Girelli, L. (2006b). On knowing about nothing. The processing of zero in single and multi-digit multiplication. *Aphasiology, 20*, 1105–1111.

Semenza, C., Bencini, G., Bertella, L., Mori, I., Pignatti, R., Ceriani, C., Cherrick, D., & Magno Caldognetto, E. (2007). A Dedicated neural mechanism for vowel selection: A case of relative vowel deficit sparing the number lexicon. *Neuropsychologia, 45*, 425–430.

Wynn, K. (1998). Numerical competence in infants. In C. Donlan (Ed.), *Development of mathematical skills* (pp. 3–25). Hove: Psychology Press/Taylor & Francis.

Further Readings

Ashcraft, M.H. (1992). Cognitive arithmetic: A review of data and theory. *Cognition, 44*, 75–106.
Cognitive science was meeting neuropsychology of mathematics those days. The specialists may want to know how the meeting started from the point of view of cognitive psychologists.

Butterworth, B. (1999). *The mathematical brain.* **London: Macmillan.**
Human beings are born to count! Brian Butterworth argues that our genes contain a set of instructions for building a "mathematical brain". A wonderful book for anybody interested in the (neuro)psychology of mathematics. Fun and easy to read even for non-scientists. It nonetheless gets very deep in the matter.

Campbell, J.I.D. (2005). *The handbook of mathematical cognition.* **Hove: Psychology Press.**
A comprehensive handbook for advanced reading. Contains a wide variety of papers written by different experts in the field.

Dehaene, S. (1997). *The number sense: How the mind creates mathematics.* **New York: Oxford University Press.**
A mathematician turned into a top neuroscientist explores the new science which brings numbers and neurons together. Sophisticated experiments are made easy to understand even to the layman.

Deloche, G., & Seron, X. (1987). *Mathematical disabilities.* **Erlbaum: Hillsdale.**
This is an old reference book. But the authors can be counted among the few founders of modern neuropsychology of math. It clearly summarizes everything that was known at the time. Still inspirational.

Donlan, C. (1998). *Development of mathematical skills.* **Hove: Psychology Press/Taylor & Francis.**
This is a reference book. It contains a collection of essays about the development of mathematical abilities. Of special interest are the chapters devoted to individual differences.

Updated information on the progress in the field can be obtained by visiting the website of NUMBRA, a Marie Curie funded European network devoted to the study of (neuro)psychology of math: http:// math.nmi.jyu.fi/numbra/

22

Neurolinguistic Computational Models

BRIAN MACWHINNEY[1] and PING LI[2]

[1]Department of Psychology, Carnegie Mellon University Pittsburgh, PA, USA
[2]Department of Psychology, University of Richmond, Richmond, VA, USA

ABSTRACT

It is tempting to think of the brain as functioning very much like a computer. Like the digital computer, the brain takes in data and outputs decisions and conclusions. However, unlike the computer, the brain does not store precise memories at specific locations. Instead, the brain reaches decisions through the dynamic interaction of diverse areas operating in functional neural circuits. The role of specific local areas in these functional neural circuits appears to be highly flexible and dynamic. Recent work has begun to provide detailed accounts of both the overall circuits supporting language and the detailed computations provided in smaller neural areas. These accounts take the shape of both structured and emergent models.

22.1. INTRODUCTION

Recent decades have seen enormous advances in linguistics, psycholinguistics, and neuroscience. Piecing these advances together, cognitive scientists have begun to formulate mechanistic accounts of how language is processed in the brain. Although these models are still very preliminary, they allow us to integrate information derived from a wide variety of studies and methodologies. They also yield predictions that can drive the search for specific neurolinguistic processing mechanisms.

22.2. THE COMPUTER AND THE BRAIN

Current neurocomputational models build on a core set of ideas deriving from the field of artificial intelligence, as it matured in the 1960s. At that time, researchers believed that one could view the brain as a type of digital computer. The four crucial design features of the digital computer are: binary logic, seriality of computation, a fixed memory address space, and modularity of program design. Given this, we can ask whether the human brain also relies on binary logic, seriality, a fixed address space, and modularity.

Neuroscience has provided fairly clear answers to this question. First, we know that individual neurons do indeed fire in an on–off binary fashion. So this feature shows a close match. However, unlike the serial computer, the brain operates in a massively parallel fashion. Imaging studies have shown that, at a given moment in time, neurons are active throughout the brain. Because of this massive parallelism, the binary functioning of individual neurons takes on a very different role in the brain, serving to modulate decisions and activations, rather than making simple yes–no choices in a serial fashion. But it is not the case that the full parallel activations in the brain are equally in consciousness at a given moment. Although the brain has no central processing unit (CPU), there is a system of interrelated executive control processes localized in the frontal cortex that operates with additional support from posterior memory areas. We can think of this frontal system as a neural CPU. This system imposes its own form of seriality on thinking, allowing only a limited number of ideas or percepts to be active in working memory or focal attention at a given time. So, although the brain is massively parallel, it achieves a certain limited form of seriality for processes in focal attention.

Modularity is a crucial feature of program design in the digital computer. Modularity is not hard wired into the computer. Rather, it is enforced by the structure of computer languages and the methods used by particular programs. Some form of modularity is also clearly present in the brain. During neuroembryonic development, cells that are initially undifferentiated migrate from the germinal matrix to specific cortical and sub-cortical areas. These migrating cells maintain their connections to other areas as they migrate but also begin to differentiate as they move to particular cortical areas. Thus, some cortical differentiation is present already at birth. The brain is initially highly plastic. Over time, as we will see in more detail later, areas become increasingly committed to particular computational functions. However, as Bates and colleagues (Elman *et al.*, 1996) have emphasized, "neural modules are made not born." In this sense, modularity is an emergent fact in the brain, just as it is in computer programming. This insight can also help us understand the organization of processing modules in bilingualism (Hernandez *et al.*, 2005).

The biggest difference between the brain and the computer is the fact that the neural CPU cannot access memory via the systematic method used by the computer CPU. When a program is loaded onto a computer, it can reliably search in a particular memory location for each important piece of information. The neural CPU cannot rely on this scheme. In the late 1970s, there was an attempt to identify a system for memory addressing that might indeed parallel the system using by the digital computer (John, 1967). One idea was that individual neurons might be identified on the basis of codes embedded in the expressed portion of the DNA or RNA. If this were true, one might expect that memories could be encoded directly in the cells. To test this, biologists taught small planaria worms to navigate a series of turns in a maze to get some food. Once the worms had been trained, they ground them up and fed them to other untrained planaria. The hope was that the memories stored in the DNA of the trained planaria would be passed on to the untrained worms. At first, the results were promising. But later it appeared impossible to replicate the experiments outside of the original laboratories. The results of these experiments were chronicled in a series of papers called "The Worm Runner's Digest." Looking back, it seems remarkable that scientists could have thought that memories would be encoded this way. However, at that time, the strength of the analogy between the brain and the computer was so strong that the experimental hypothesis seemed perfectly reasonable.

The problem of reliably accessing stored memories is part a more general problem in neural computation. Because individual neurons do not have addresses, they cannot be given unique names. We cannot imagine that a particular neuron represents "a yellow Volkswagen" or that another neuron represents "the past tense suffix." It is clearly impossible to pass symbolic information down neuronal axons. Instead, neurons must acquire a functional significance that arises from their role as participants in connected neural networks. In part, this is because neurons are not as reliable as silicon. Neurons may die and, in some areas of the brain, new neurons may be born. Neural firing is subject to a variety of disruptions caused by conditions varying from fatigue to epilepsy. In

extreme cases, victims of stroke or other injuries may lose large portions of their brain, but maintain the ability to talk and think. However, if a computer has faults in even a few silicon gates in memory, it will be unable to function at all. Thus, the neural CPU must use a very different, more flexible, method for addressing memory. It was the realization of this fundamental fact that led in the 1980s to the rise of neural network models (Grossberg, 1987) as the major method for modeling the brain. All current work in neurocomputational models is illuminated by this basic insight.

22.3. STRUCTURED MODELS

There are two ways in which we can allow facts about the brain to constrain our neurolinguistic models. One method relies on structured modeling and the other on unstructured or emergent modeling. Within the framework of structured modeling, we can distinguish between module-level models and neuron-level models.

22.3.1. Module-Level Structured Models

In this section, we will examine work on module-level models. These models attempt to localize processing in particular neural modules. On an anatomical level, it is clear that the brain is rich in structure. For example, there are at least 54 separate processing areas in visual cortex (Van Essen *et al.*, 1990). But it is not clear whether these areas function as encapsulated modules or rather as interactive pieces of functional networks.

Evidence for neurolinguistic modules has come from three sources: aphasiology, brain imaging, and developmental disorders. The oldest of these sources is the evidence from differing patterns of language deficit in aphasia. One can study patients with lesions of different types in the hope of identifying double disassociations between information-processing skills and lesion types. For example, some patients will have damaged prosodic structure, but normal segmental phonology. Other patients will have damaged segmental structure, but normal prosody. This pattern of results would provide strong support for the notion that there is a localized cognitive module for the processing of prosody. In practice, however, evidence for such double dissociations is difficult to obtain without post hoc partitioning of subject groups. But this partitioning itself casts doubt on the underlying assumptions regarding modularity and dissociability.

A familiar example of this type of model for language is the Geschwind (1979) model of connected language modules. This model is designed to account for how we can listen to a sentence and then imitate it or reply to it. According to this model, language comprehension begins with the receipt of a linguistic signal by auditory cortex in the temporal lobe. This information is then passed on to Wernicke's area for lexical processing. From here, information is passed over the arcuate fasciculus to Broca's area, where the reply or imitation is planned. Finally, the output signal is sent to motor cortex for articulation. This model treats processing as the passing of information between modules. According to this model, damage in a given area will predict loss of the related ability. Thus, damage to Broca's area should lead to Broca's aphasia. Unfortunately, the neurological assumptions of this model have proven problematic. Originally, Wernicke's area was thought to be an association area at the juncture of the temporal and parietal lobes. However, there is little evidence that this area functions as association cortex with any specific linkage to lexical or linguistic processing. Some other components of the Geschwind model are less problematic. In particular, it is clear that sounds are controlled by temporal auditory cortex, and that the final stages of speech output are controlled by motor cortex.

The role of Broca's area in the Geschwind model is also problematic. Although Broca's area is well defined anatomically, it has not been possible to locate specific perisylvian areas that are associated with specific aphasic symptoms or with specific patterns of disruption of naming in direct cortical stimulation (Ojemann *et al.*, 1989). However, recent work using functional magnetic resonance imaging (fMRI) methods has begun to clarify this issue. It has been shown that processing in inferior frontal gyrus (IFG) involves three clusters: (1) tasks emphasizing semantic processing with activation in anterior IFG in the *pars orbitalis*; (2) phonological tasks with activation in the posterior superior IFG; and (3) syntactic tasks with activation between the other two areas in middle IFG or *pars triangularis*, Brodman's area 44/45 (BA) (Bookheimer, 2002; Hagoort, 2005). These separations are not sharp and absolute, but they do seem to represent interesting differentiations in IFG that correspond with traditional linguistic distinctions. However, we must remember that the subtractive methodology used to analyze fMRI data tends to underestimate the contribution of other areas of the brain that are also involved in a particular task.

A final type of evidence for neurolinguistic modules comes from studies of children with developmental disorders. Here, researchers have applied the same logic of double dissociations used in the study of aphasia. In particular, it is often argued that children with Williams syndrome show an intense cognitive deficit with no serious disruption of language functioning. In contrast, children with Specific Language Impairment (SLI) are said to have intact cognitive functioning with marked impairment in language. However, this supposed double dissociation is not so clear in practice. Children with Williams syndrome do have effective control of language, but they achieve this control in ways that are far from normal (Karmiloff-Smith *et al.*, 1997; see Chapter 36, this volume). Moreover, many children with SLI also have problems with related areas of conceptual functioning (e.g., in temporal sequence processing) and there have not

yet been any successful attempts to link the module that is supposedly damaged in SLI to any particular brain region.

A somewhat different approach to localization views alternative cortical areas as participating in functional neural networks. In this framework, a particular cortical area may participate in a variety of functional networks. Within each of these various networks, the area would basically serve a similar computational role. However, because processing demands vary across networks, the specific products of this processing will vary depending on the network involved. Mason and Just (2006) argue that fMRI studies have provided evidence for five functional neural networks for discourse processing:

1. A right hemisphere network involving middle and superior temporal cortex that computes a coarse semantic representation.
2. A bilateral network in dorsolateral prefrontal cortex (DLPFC) that monitors conceptual coherence.
3. A left hemisphere network involving IFG and left anterior temporal for text integration.
4. A network involving medial frontal areas bilaterally and right temporal/parietal areas for perspective taking.
5. A bilateral, but left-dominant, network involving the intraparietal sulcus for spatial imagery processing.

In practice, it is difficult to understand the exact separation between these various processes. For example, there seems to be a conceptual overlap between the second and third of these networks. Also, it is not clear whether the additional activation recorded in right hemisphere sites indicates basic discourse processing or a spillover of processing from the left hemisphere. And it is not clear whether these areas would be involved in different ways for comprehending written versus spoken discourse. Furthermore, networks involved in broader conceptual tasks such as perspective taking probably involve more than just a few cortical areas. To the degree that perspective taking triggers empathy and body mapping (MacWhinney, 2005), it will also rely on additional frontal areas, basal ganglion, amygdala, and mirror neuron systems in both frontal and parietal cortex. Despite these various issues, it seems profitable to continue exploring the interpretation of fMRI results in terms of interlocking functional neural networks. These models allow researchers to study functional localization without forcing them to think of local areas as encapsulated neural modules.

22.3.2. Neuron-Level Structured Models

Structured modeling can also be conducted on a level that avoids direct reference to modules. Morton's (1970) logogen model was a particularly successful model of this type. At the center of each logogen network was a master neuron devoted to the particular word. Activation of this central unit could then trigger further activation of units encoding its phonology, orthography, meaning, and syntax. Pushing the notion of spreading neural activation still further, McClelland and Rumelhart (1981) developed an interactive activation (IA) account of context effects in letter perception. IA models have succeeded in providing clear accounts for a wide variety of phenomena in speech production, reading, auditory perception, lexical semantics, and second language learning. Because each unit in these models can be clearly mapped to a particular linguistic construct, such as a word, phoneme, or semantic feature, the generation of predictions from IA models is very straightforward (see Box 22.1). Because these models do well at representing the processes of activation and competition, they usually provide good models of experimental

---

Box 22.1   Interactive-activation account of speech production

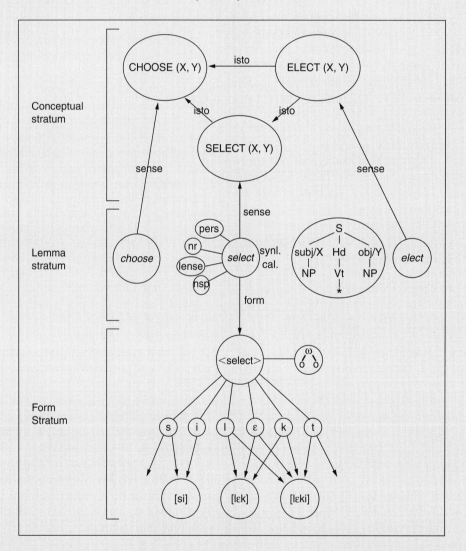

The IA account from Levelt (2004) illustrates the activation of the word "select" and its surrounding syntactic context. The diagram shows how, on the level of form, the word "select" is composed of phonemes that activate specific syllables. Here [lɛk] competes for activation with [lɛkt]. Also, the word has a weak–strong stress pattern. On the level of the lemma, the word has the syntactic feature of present tense and specifies a subject and an object. At this level, it competes with forms like "choose" and "elect." Finally on the conceptual level, notions of *selecting*, *choosing*, and *electing* are all activated, but only *selecting* is chosen in this case.

Levelt, W.J.M. (2004). Models of word production. *Trends in Cognitive Sciences, 4*, 223–232.

data with both healthy and aphasic individuals. Moreover, the core idea of IA maps up well with what we know about how neurons interact in terms of excitatory and inhibitory synaptic connections.

IA models suffer from a fundamental weakness. It is difficult to imagine how real neurons could achieve the local conceptual labeling required by these models. How could the learner manage to tag one neuron as "fork" and another neuron as "spoon?" How could the learner manage to connect up exactly these specific cells to the correct phonemic components? Moreover, should we really imagine that each word or concept is represented by one and only one central neuron? Given the fact that neurons are subject to death and replacement, would we not expect to find even normal speakers continually losing words or phonemes when their controlling cells die? If there were some learning method associated with this architecture, these various problems with IA models could be solved. However, it is not clear how one could formulate a learning algorithm for this type of localist model.

22.4. EMERGENT MODELS

To address this problem, Rumelhart *et al.* (1986) developed a neural network learning algorithm called back propagation. This algorithm makes no assumptions regarding "labels" on neurons. Instead, the functioning of individual neurons emerges through a competitive training process in which connections between neurons are adjusted in proportion to the mismatch (error) in the input-to-output mapping. Because there are no labels on units, there is a tendency for information controlling a given pattern to become distributed across the network. This distribution of information can be characterized as parallel distributed processing (PDP). Because PDP models based on back propagation algorithm are easy to develop and train, this framework has generated a proliferation of models of language learning. PDP models have been successfully applied to a wide variety of language learning areas including the English past tense, Dutch word stress, universal metrical features, German participle acquisition, German plurals, Italian articles, Spanish articles, English derivation for reversives, lexical learning from perceptual input, deictic reference, personal pronouns, polysemic patterns in word meaning, vowel harmony, historical change, early auditory processing, the phonological loop, early phonological output processes, ambiguity resolution, relative clause processing, word class learning, speech errors, bilingualism, and the vocabulary spurt.

22.4.1. Self-Organizing Maps

Although PDP models succeed in capturing the distributed nature of memory in the brain, they do this by relying on reciprocal connections between neurons. In fact, there is

little evidence that individual neurons are connected in this way. Also, PDP relies on back propagation of an explicit error correction signal. Again, evidence that this is present is fairly weak. MacWhinney (2000a) argued that distributed PDP models fail to provide an emergent localist representation of the word, making further morphological and syntactic processing difficult. As we noted earlier, one of the great strengths of IA models is their ability to provide localist representations for words and other linguistic constructs. However, because these forms cannot be learned from the input, researchers have turned to back propagation and PDP to provide learning mechanisms. A more ideal solution to the problem would be to rely on a learning algorithm that managed to induce flexible, localist representations that correspond in a statistical sense to concepts and words.

The self-organizing feature map (SOFM) algorithm of Kohonen (2001) provides one way of tackling these problems. This framework has been used to account in great detail for the development and organization of the visual system (Miikkulainen *et al.*, 2005). Applications of the model to language are more difficult, but are also showing continual progress. To illustrate the operation of SOFM for language, let us consider the DevLex-II word learning model of Li *et al.* (2007). This model uses three local maps to account for word learning in children: an auditory map, a concept map, and an articulatory map (see Box 22.2). The actual word representations are computed from the contextual features derived from realistic child–parent interactions in the CHILDES database (MacWhinney, 2000b). In effect, this meant that words that occurred in similar sentence contexts are represented similarly, and consequently organized by SOFM in nearby neighborhoods in the map. For example, nouns were organized together because they appeared consistently in slots such as "my X" or "the X." These representations can also be coupled with perceptual features to capture the child's early perceptual experiences (Li *et al.*, 2004).

Self-organizing maps offer a promising method for achieving emergent localist encodings. There are also several ways in which the capacity of maps can be expanded through the addition of neurons or overlays with new coding features. Which of these methods is actually used in the brain remains unclear. We know that the brain has an enormous capacity for the storage of memories. However, efficient use of this storage space may rely on hippocampal storage mechanisms to organize memories for efficient retrieval (Miikkulainen *et al.*, 2005). We will discuss this issue further in the final section.

22.4.2. Syntactic Emergence

Although these emergentist models have succeeded in modeling a variety of features in word learning and phonology, it has been more difficult to apply them to the task of modeling syntactic processing. Elman's (1990) simple recurrent network (SRN) model uses recurrent back-propagation

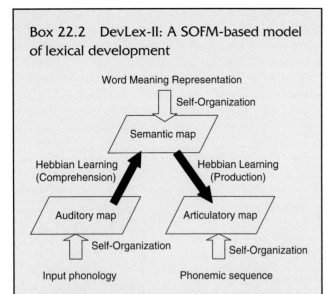

Box 22.2   DevLex-II: A SOFM-based model of lexical development

The DevLex-II model is a neural network model of early lexical development based on SOFM (Li *et al.*, 2007). The model has three sub-networks (maps) to process and organize linguistic information: Auditory map that takes the phonological information of words as input, semantic map that organizes the meaning representations based on lexical co-occurrence statistics, and articulatory map that outputs phonemic sequences of words. The semantic map is connected to both the auditory map and the articulatory map through associative links that are trained by Hebbian learning. Such connections allow for the effective modeling of comprehension (auditory to semantic) and production (semantic to articulatory) processes in language acquisition.

Li, P., Zhao, X., & MacWhinney, B. (2007). Dynamic self-organization and early lexical development in children. *Cognitive Science, 31,* 581–612.

connections to update the network's memory after it reads each word. The network's task is to predict the next word. This framework views language comprehension as a highly constructive process in which the major goal is trying to predict what will come next. An alternative to the predictive framework relies on the mechanisms of spreading activation and competition. For example, MacDonald *et al.* (1994) have presented a model of ambiguity resolution in sentence processing that is grounded on competition between lexical items. Models of this type do an excellent job of modeling the temporal properties of sentence processing. Such models assume that the problem of lexical learning in neural networks has been solved. They then proceed to use localist representations to control IA during sentence processing.

Another lexicalist approach uses a linguistic framework known as Construction Grammar. This framework emphasizes the role of individual lexical items in early grammatical learning (MacWhinney, 1987). Early on, children learn

to use simple frames such as *my + X* or *his + X* to indicate possession. As development progresses, these frames are merged into general constructions, such as the possessive construction. In effect, each construction emerges from a lexical gang. Sentence processing then relies on the child's ability to combine constructions online. When two alternative constructions compete, errors appear. An example would be *say me that story*, instead of *tell me that story*. In this error, the child has treated "say" as a member of the group of verbs that forms the dative construction. In the classic theory of generative grammar, recovery from this error is supposed to trigger a learnability problem, since such errors are seldom overtly corrected and, when they are, children tend to ignore the feedback. Neural network implementations of Construction Grammar address this problem by emphasizing the direct competition between *say* and *tell* during production. The child can rely on positive data to strengthen the verb *tell* and its link to the dative construction, thereby eliminating this error without corrective feedback. In this way, models that implement competition provide solutions to the logical problem of language acquisition.

These various approaches to syntactic learning must eventually find a way of dealing with the compositional nature of syntax. A noun phrase such as "my big dog and his ball" can be further decomposed into two segments conjoined by the "and." Each of the segments is further composed of a head noun and its modifiers. Our ability to recursively combine words into larger phrases stands as a major challenge to connectionist modeling. One likely solution would use predicate constructions to activate arguments that are then combined in a short-term memory buffer during sentence planning and interpretation. To build a model of this type, we need to develop a clearer mechanistic link between constructions as lexical items and constructions as controllers of the on-the-fly process of syntactic combination.

22.4.3. Lesioning Emergent Models

Once one has constructed networks that are capable of learning basic syntactic processes, it is an easy matter to test the ability of these models to simulate aphasic symptoms by subjecting the model to lesions, just as the brain of the real aphasic person was subjected to real lesions. Lesioning can be done by removing hidden units, removing input units, removing connections, rescaling weights, or simply adding noise to the system. It has been shown that these various methods of lesioning networks all produce similar effects (Bullinaria & Chater, 1995). However, work on lesioning of models can also capture patterns of dissociation. For example, in lesioned models, high-frequency items are typically preserved better than low-frequency items. Similarly, patterns with many members will be preserved better than patterns with fewer members, and items with associations to many neighbors better than items with fewer neighbors.

A particularly interesting application of the network lesioning technique is the model of deep dyslexia developed by Plaut and Shallice (1993). The model had connections from orthography to semantics and from semantics to phonology, as well as hidden unit layers and layers of clean-up units for semantics and phonology. Lesions to the orthographic layer led to loss of the ability to read abstract words. However, lesions to the semantic clean-up units led to problem in reading concrete words. This work illustrates how one could use the analysis of lesions to emergent models to elaborate ideas about area-wide patterns of connectivity in the brain.

Like models based on back propagation, SOFM-based models can also be used to study lesion and recovery from lesion. Li et al. (2007) showed how DevLex-II, when lesioned with noise in the semantic and phonological representations and their associative links, can display early plasticity in the recovery from lesion induced at different time points of learning. Their lesion data matched up with empirical findings from children who suffer from early focal brain injury with respect to lexical learning and U-shaped behavior (MacWhinney et al., 2000).

22.5. CHALLENGES AND FUTURE DIRECTIONS

Despite the various successes of both structured and emergent models, neurolinguistic computational models continue to suffer from a core limitation. This is the failure of these models to decode the basic addressing system of the brain. However, recent work in the area of embodied cognition may be pointing toward a resolution of this problem. This work emphasizes the ways in which the shape of human thought emerges from the fact that the brain is situated inside the body and the body is embedded in concrete physical and social interactions. In neural terms, embodied cognition relies heavily on a perception-action cycle that works to interpret new perceptions in terms of the actions needed to produce them. For example, when watching an experimenter grab a nut, a monkey will activate neurons in motor cortex that correspond to the areas that the monkey itself uses when grabbing a nut. These so-called mirror neurons (see Chapter 23) form the backbone of a rich system for mirroring the actions of others by interpreting them in terms of our own actions.

A variety of recent computational models have tried to capture aspects of these perceptual-motor linkages. One approach emphasizes the ways in which distal learning processes can train action patterns such as speech production on the basis of their perceptual products (Westermann & Miranda, 2004). A very different approach, developed in MacWhinney (2005), works out the consequences of the online construction of an embodied mental model of a sentence for linguistic structures. This account views the frontal lobes as working to encode a virtual homunculus that can be used to enact the various actions, stances, and perspective shifts involved in linguistic discourse.

By itself the notion of an embodied perceptual-motor cycle will not solve the core problem of memory addressing in the brain. However, it seems to point us in a very interesting direction. Starting with the core observation that the brain is located in the body and fully connected to all the senses and motor effectors, we could imagine that the primary code used in the brain might be the code of the body itself. When we look at the tonotopic and retinotopic organization of sensory areas or the multiple representations of effectors in motor areas, this view of the brain as an encoder of the body seems transparently true. However, it is less clear how this code might extend beyond these primary areas for use throughout the brain. One radical possibility is that the brain may rely on embodied codes throughout and that the body itself could provide the fundamental language of neural computation.

References

Bookheimer, S. (2002). Functional MRI of language: New approaches to understanding the cortical organization of semantic processing. *Annual Review of Neuroscience, 25*, 151–188.

Bullinaria, J.A., & Chater, N. (1995). Connectionist modelling: Implications for cognitive neuropsychology. *Language and Cognitive Processes, 10*, 227–264.

Elman, J.L. (1990). Finding structure in time. *Cognitive Science, 14*, 179–212.

Elman, J.L., Bates, E., Johnson, M., Karmiloff-Smith, A., Parisi, D., & Plunkett, K. (1996). *Rethinking innateness.* Cambridge, MA: MIT Press.

Geschwind, N. (1979). Specializations of the human brain. *Scientific American, 241*, 180–199.

Grossberg, S. (1987). Competitive learning: From interactive activation to adaptive resonance. *Cognitive Science, 11*, 23–63.

Hagoort, P. (2005). On Broca, brain, and binding: A new framework. *Trends in Cognitive Sciences, 9*, 416–422.

Hernandez, A., Li, P., & MacWhinney, B. (2005). The emergence of competing modules in bilingualism. *Trends in Cognitive Sciences, 9*, 220–225.

John, E.R. (1967). *Memory.* New York: Academic Press.

Karmiloff-Smith, A., Grant, J., Berthoud, I., Davies, M., Howlin, P., & Udwin, O. (1997). *Language and Williams syndrome: How intact is "intact?"Child Development, 68*, 246–262.

Kohonen, T. (2001). *Self-organizing maps* (3rd edn). Berlin: Springer.

Li, P., Farkas, I., & MacWhinney, B. (2004). Early lexical development in a self-organizing neural network. *Neural Networks, 17*, 1345–1362.

Li, P., Zhao, X., & MacWhinney, B. (2007). Dynamic self-organization and early lexical development in children. *Cognitive Science, 31*, 581–612.

MacDonald, M.C., Pearlmutter, N.J., & Seidenberg, M.S. (1994). Lexical nature of syntactic ambiguity resolution. *Psychological Review, 101*(4), 676–703.

MacWhinney, B. (1987). The competition model. In B. MacWhinney (Ed.), *Mechanisms of language acquisition* (pp. 249–308). Hillsdale, NJ: Lawrence Erlbaum.

MacWhinney, B. (2000a). Lexicalist connectionism. In P. Broeder & J. Murre (Eds.), *Models of language acquisition: Inductive and deductive approaches* (pp. 9–32). Cambridge, MA: MIT Press.

MacWhinney, B. (2000b). *The CHILDES project: Tools for analyzing talk.* Mahwah, NJ: Lawrence Erlbaum Associates.

MacWhinney, B. (2005). The emergence of grammar from perspective. In D. Pecher & R.A. Zwaan (Eds.), *The grounding of cognition: The role of perception and action in memory, language, and thinking* (pp. 198–223). Mahwah, NJ: Lawrence Erlbaum Associates.

MacWhinney, B., Feldman, H.M., Sacco, K., & Valdes-Perez, R. (2000). Online measures of basic language skills in children with early focal brain lesions. *Brain and Language, 71*, 400–431.

Mason, R.. & Just, M. (2006). Neuroimaging contributions to the understanding of discourse processes. In M. Traxler & M. Gernsbacher (Eds.), *Handbook of psycholinguistics* (pp. 765–799). Mahwah, NJ: Lawrence Erlbaum Associates.

McClelland, J.L., & Rumelhart, D.E. (1981). An interactive activation model of context effects in letter perception: Part 1. An account of the basic findings.. *Psychological Review, 88*, 375–402.

Miikkulainen, R., Bednar, J.A., Choe, Y., & Sirosh, J. (2005). *Computational maps in the visual cortex.* New York: Springer.

Moll, M., & Miikkulainen, R. (1997). Convergence-zone epsodic memory: Analysis and simulations. *Neural Networks, 10*, 1017–1036.

Morton, J. (1970). A functional model for memory. In D.A. Norman (Ed.), *Models of human memory* (pp. 203–248). New York: Academic Press.

Ojemann, G., Ojemann, J., Lettich, E., & Berger, M. (1989). Cortical language localization in left, dominant hemisphere: An electrical simulation mapping investigation in 117 patients. *Journal of Neurosurgery, 71*, 316–326.

Plaut, D., & Shallice, T. (1993). Deep dyslexia: A case study of connectionist neuropsychology. *Cognitive Neuropsychology, 10*, 377–500.

Rumelhart, D., Hinton, G., & Williams, R. (1986). Learning internal representations by back propagation. In D. Rumelhart & J. McClelland (Eds.), *Parallel distributed processing: Explorations in the microstructure of cognition* (pp. 318–362). Cambridge, MA: MIT Press.

Van Essen, D.C., Felleman, D.F., DeYoe, E.A., Olavarria, J.F., & Knierim, J.J. (1990). Modular and hierarchical organization of extrastriate visual cortex in the macaque monkey. *Cold Spring Harbor Symposium on Quantitative Biology, 55*, 679–696.

Westermann, G., & Miranda, E.R. (2004). A new model of sensorimotor coupling in the development of speech. *Brain and Language, 89*, 393–400.

Further Readings

Just, M.A., & Varma, S. (2007). The organization of thinking: what functional brain imaging reveals about the neuroarchitecture of complex cognition. *Cognitive, Affective, and Behavioral Neuroscience, 7,* **153–191.**
This article compares three recent neurocomputational approaches that provide overall maps of high-level functional neural circuits for much of cognition.

Miikkulainen, R., Bednar, J.A., Choe, Y., & Sirosh, J. (2005). *Computational maps in the visual cortex.* **New York: Springer.**
This book shows how self-organizing feature maps can be used to describe many of the detailed results from the neuroscience of vision.

Ullman, M. (2004). Contributions of memory circuits to language: the declarative/procedural model. *Cognition,* **92, 231–270.**
An interesting, albeit speculative, application of a neurocomputational model to issues in second language learning.

Mirror Neurons and Language

MICHAEL A. ARBIB

*Computer Science Department, Neuroscience Program, and USC Brain Project,
University of Southern California, Los Angeles, CA, USA*

ABSTRACT

Mirror neurons are neurons that fire both when an agent performs an action and when the agent observes the same action performed by someone else. The Mirror System Hypothesis claims that brain mechanisms for language evolved atop a mirror system for grasping through the successive emergence of systems for imitation, pantomime, and protosign (a system of conventionalized gestures). To focus the discussion of neurolinguistics, we relate praxis and language to the action-oriented perception of scenes. Computational models of the canonical and mirror systems for grasping introduce a variety of macaque brain regions of use in determining homologs in the human brain that can ground a processing approach to neurolinguistics.

23.1. INTRODUCTION

A key property of language is *parity*. The hearer can recognize (more-or-less) what the speaker intends to say. What is the neurology of this? Human brain imaging shows that Broca's area is one of several brain regions strongly associated with a system involved in both the generation and recognition of various manual actions – the *mirror system* for grasping. Moreover, having a mirror system for grasping is a key property we share with other primates. Indeed, in monkeys, the "mirror property" can be observed even at the single neuron level. We argue that the study of the monkey brain anchors the comparative neurobiology of mirror systems to yield new insights into both the evolution of the language-ready brain and the human brain mechanisms which support language.

But why might a mirror system for grasping be associated with an area traditionally associated with speech production?

The answer starts with the observation that for deaf people, language may take the form of a signed language. We thus now associate Broca's area with language production as a multimodal performance (possibly involving hands, face, and voice) rather than speech alone.

23.1.1. The Mirror System Hypothesis

Arbib and Rizzolatti (1997; Rizzolatti and Arbib, 1998) advanced the Mirror System Hypothesis which claims that the parity requirement for human language is met because Broca's area evolved atop the mirror system for grasping with its capacity to generate and recognize a set of manual actions. In particular, speech is rooted in communication based on manual gesture. The core evidence for the hypothesis is as follows: First, a neuron is called a *mirror neuron* if its firing correlates with both execution of a specific action and observation of more-or-less related actions. Mirror neurons for grasping (e.g., distinguishing a precision pinch from a power grasp), were first discovered in an area of premotor cortex of the macaque monkey called F5 (Gallese *et al.*, 1996) and later found in parietal cortex. Second, anatomical analysis shows that macaque area F5 is homologous to Brodmann's area 44, which is part of human Broca's area, an area traditionally associated with the production of speech. And third, human brain imaging studies (e.g., Grafton *et al.*, 1996) show activation for both grasping and observation of grasping in or near Broca's area, though we do have little data on activity of individual neurons in this mirror system.

The present chapter reviews the Mirror System Hypothesis as an evolutionary hypothesis, while also showing that its current version posits a set of mirror systems of relevance to the neurolinguistic analysis of the modern human brain.

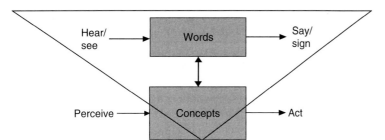

FIGURE 23.1 The figure highlights the distinction between words (or, more generally, constructions) and concepts and indicates the many-to-many relation which links them.

Kemmerer, 2006). Here, rather than combining words according to a few highly general and purely syntactic rules, we employ a large number of *constructions* which may have broad or limited applicability, but each of which clarifies how syntactic combination goes hand in hand with the derivation of new meanings from the meanings of the constituents. The meaning of idiomatic expressions like *kick the bucket* or *take the bull by the horns* cannot be inferred from the meaning of the individual words and highly general rules of grammar. The specialized constructions needed to define the meaning of such idioms blur the division between grammar and lexicon.

As noted in Figure 23.1, each word (whose phonological form we posit is associated with mirror neurons for generation of the word and recognition of its form) gains its meaning by its linkage to concepts (in the form of perceptual, motor, and coordinating schemas) which, with few exceptions have a neural representation which does not involve mirror neurons. To see how to extend Figure 23.1 by generalizing words to constructions in a way that links to conceptual relations, we consider VISIONS, a computer system that can recognize objects and relations in a visually presented scene (Box 23.3). VISIONS employs a wide variety of basic "subscene schemas" such as those relating *sky* and *roof* (one way to recognize a roof is as a form of a certain geometry and size immediately below a region that has been recognized as sky) and *roof, house* and *wall* (if you see a roof, check whether the region below it can be recognized as a wall to confirm that together they constitute views of parts of a house). This suggests that an approach to visual scene interpretation with a large but finite inventory of subscene schemas (*from the visual end*) may provide the linkage with constructions (*from the language end*) rich enough to encompass an exemplary set of sentences.

Extended attention to a given component of a scene allows one to elaborate the scene's description, extending the complexity of the constituents in the parse tree of the sentence. One may have varied reasons for choosing what to describe in more detail. Sentence production can then be viewed as a goal-directed act, akin to reaching to grasp an object – but now the goal is communicative and covert rather than praxic and overt. We claim that "praxic constructions" viewed as converters from cognitive form to

---

Box 23.3 Integrating perceptual and motor schemas

We view concepts as abstracted from the perceptual and motor schemas of an agent interacting with its world. Knowledge is mediated through activity in a network of interacting schema instances. The *activity level* of an instance of a perceptual schema represents a "confidence level" that the object represented by the schema is indeed present while that of a motor schema signals a "degree of readiness" to control some course of action.

In the VISIONS computer system (Draper *et al.*, 1989), *low-level vision* takes an image and extracts such data as regions and boundaries tagged with various features. The knowledge required for *high-level vision* is provided by *perceptual schemas*; the state of interpretation of a scene unfolds in working memory (WM) as a network of schema instances. When a schema instance is activated, it is with an associated area of the image. Moreover, schemas may invoke other schemas – for example, a schema for recognizing houses may invoke a schema for recognizing a wall below where the roof has been recognized. Initial activations may die out or be strengthened by competition and cooperation with other schema instances – after some time, the network of most active instances constitutes the interpretation of the scene.

To model animals and humans autonomously pursuing current goals through the *action–perception cycle*, we may extend the analysis so that motor schemas as well as perceptual schemas are instantiated in WM, with goals changing the perceptual demands as action proceeds. Planning will be intertwined with execution, with patterns of schema activation modified and thus the "plan" updated as action proceeds and new sensory stimulation is obtained (Arbib & Liaw, 1995).

Arbib, M.A., & Liaw, J.-S. (1995). Sensorimotor transformations in the worlds of frogs and robots. *Artificial Intelligence, 72,* 53–79.

Draper, B.A., Collins, R.T., Brolio, J., Hanson, A.R., & Riseman, E.M. (1989). The schema system. *International Journal of Computer Vision, 2,* 209–250.

temporal behavior provides the evolutionary substrate for constructions for communication.

23.4. MODELING THE CANONICAL AND MIRROR SYSTEMS FOR GRASPING

We see the language-ready-brain as a system that integrates a mirror system for the phonological form (which may include manual and facial gestures) of words and constructions and a network of schemas which provide words and sentences with their meaning – and claim that these interacting systems can be illuminated within an evolutionary framework rooted in the mirror system for grasping. We thus "rewind the clock" to consider macaque brain mechanisms for generation and recognition of manual actions, presumed to be akin to that of the macaque-human common ancestor.

Discharge in most grasp-related macaque F5 neurons correlates with an action rather than with the individual movements that form it. One may thus relate F5 neurons to various *motor schemas* corresponding to the actions associated with their discharge. Mirror neurons are complemented by *canonical neurons* which fire when the monkey performs a specific action but not when it observes a similar action. The parietal area AIP (Anterior Intra-Parietal sulcus) and area F5 together anchor the canonical cortical circuit in macaque which transforms visual information on intrinsic properties of an object into hand movements for grasping it. AIP processes visual information to extract grasp parameters relevant to the control of hand movements and is reciprocally connected with the F5 canonical neurons. Primary motor cortex (F1) formulates the neural instructions for lower motor areas and motor neurons. The AIP extracts *affordances* of an object (Gibson, 1966), parameters relevant for action, whereas area IT (inferotemporal cortex) recognizes its category.

The FARS model (named for Fagg, Arbib, Rizzolatti and Sakata) provides a computational account of the system centered on the AIP → F5$_{canonical}$ pathway (Fagg & Arbib 1998): The dorsal stream (that is, that which passes through AIP) does not know "what" the object is, it can only see the object as a set of possible affordances, whereas the ventral stream (from primary visual cortex to inferotemporal cortex, IT) is able to categorize the object. This information is passed to prefrontal cortex (PFC) which can then, on the basis of the current goals of the organism and the recognition of the nature of the object, bias the affordance appropriate to the current task. For example, consider the different affordances exploited to move a mug out of the way versus to lift it to drink from it. The original FARS model suggested that the PFC bias was applied to F5; subsequent neuroanatomical data suggested that PFC and IT may modulate action selection at the level of parietal cortex rather than premotor cortex (Rizzolatti & Luppino, 2003). Figure 23.2a gives a partial view of the FARS model updated to show this modified pathway. AIP may represent several affordances initially, but only one of these is selected to influence F5. This affordance then activates the F5 neurons to command, via primary motor cortex F1, the appropriate grip once it receive a "go signal" from another region, F6, of PFC.

The Mirror Neuron System (MNS) model of Oztop & Arbib (2002) focuses on the learning capacities of mirror neurons. Here, the task is to determine whether the shape of the hand and its trajectory are "on track" to grasp an observed affordance of an object using a known action. In this model, the AIP → F5$_{canonical}$ pathway emphasized in Figure 23.2a is complemented by PF → F5$_{mirror}$. Moving down the left of Figure 23.2b, we see that the model has four parts:

1. Recognizing the location of the object provides parameters to the motor programming area F4 which computes the reach.
2. As in FARS, object features are processed by AIP to extract grasp affordances which are sent on to the canonical neurons of F5 that choose a particular grasp.

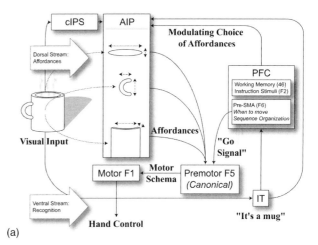

(a)

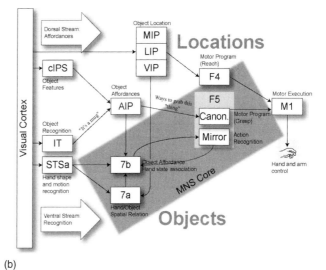

(b)

FIGURE 23.2 (a) In the upper diagram, AIP does not "know" the identity of the object, but can only extract affordances. PFC uses the identification of the object carried out by IT, in concert with task analysis and WM, to help AIP select the appropriate action from its "menu". (b) The center of the diagram shows how mirror neurons may learn what visually observable trajectories of the hand relative to an object correspond to a grasp as encoded by canonical neurons. (See text for further details.)

3. The information about the reach and the grasp is used by the motor cortex F1 to control the hand and arm.
4. The last part comprises components that can learn and apply key criteria for activating a mirror neuron, recognizing that the preshape that the monkey is seeing corresponds to the grasp that the mirror neuron encodes; the preshape that the observed hand is executing is appropriate to the object that the monkey can see (or remember); and that the hand is moving on a trajectory that will bring it to grasp the object.

In the MNS model, the *hand state* was defined as a vector whose components represented the movement of the wrist relative to the location of the object and of the hand shape relative to the affordances of the object. An artificial neural network corresponding to PF and $F5_{mirror}$ was trained to recognize the grasp type from the *hand-state trajectory*, with correct classification often being achieved well before the hand reached the object. The modeling assumed that the neural equivalent of a grasp being in the monkey's repertoire is that there is a pattern of activity in the F5 canonical neurons that command that grasp. During training, the output of the F5 canonical neurons, acting as a code for the grasp being executed by the monkey at that time, was used as the training signal for the F5 mirror neurons to enable them to learn which hand-object trajectories corresponded to the canonically encoded grasps. As a result of this training, the appropriate mirror neurons come to fire in response to viewing the appropriate trajectories even when the trajectory is not accompanied by F5 canonical firing. Bonaiuto *et al.* (2007) have improved the learning architecture of the MNS model and extended the capacity of the model to address data on audio–visual mirror neurons, that is, mirror neurons that respond to the sight and sound of actions with characteristic sounds such as paper tearing and nut cracking, and on response of mirror neurons when the target object was recently visible but is currently hidden. Modeling audio-visual mirror neurons rests on a Hebbian association between auditory neurons and mirror neurons, whereas recognition of hidden grasps employs circuits for working memory (WM) (encoding data on the currently obscured object) and dynamic remapping (to update the estimate of the location of the hand).

The key point is that F5 mirror neuron input encodes the trajectory of the relation of parts of the hand to the object. This prepares the F5 mirror neurons to respond to hand-object relational trajectories even when the observed hand belongs to another monkey or a human. What makes the modeling worthwhile is that the trained network responded not only to hand-state trajectories from the training set, but also exhibited interesting responses to novel hand-object relationships. Such results confirm that mirror neurons can be recruited to recognize and encode an expanding repertoire of novel actions.

23.5. IMITATION AND MIRROR NEURONS: IN SEARCH OF SUBTLETY

The evolutionary sequence of extensions of the mirror system – grasping, simple and then complex imitation, pantomime, protosign and, finally, protosign & protospeech – involves genuine changes at each stage. Many papers on human brain imaging literature seem to imply that having mirror neurons is a sufficient base for imitation. However, macaques do not use imitation to any significant degree. Rather, for many actions within the monkey's repertoire, observation of that action when executed by another will activate the appropriate mirror neurons. Imitation, by contrast, goes in some sense in the "opposite direction", since it may involve using observation to acquire an action not already in the repertoire. Arbib & Rizzolatti (1997) link this to inverse and direct models in motor control (Oztop *et al.*, 2006 give a recent review). Thus the mirror system for grasping must have been augmented by evolutionarily novel mechanisms to yield the overall human capacity for imitation.

23.5.1. A Dual-Route Model of Praxis and Language

Building on this discussion of imitation, we briefly look at the dissociations seen in apraxia which suggest a dual-route model for imitation in humans. In discussing apraxia, De Renzi (1989) notes that an apraxic subject exhibits a *semantic deficit* which involves difficulty both in classifying gestures and in performing familiar gestures on command yet may be able to execute these same gestures quite well in the sense of copying the pattern of a movement without "getting the meaning of the action" of which it is part. Let me call this *low-level imitation* to distinguish it from imitation based on recognition and "replay" of a goal-directed action. Such considerations motivate the *dual-route* conceptual model of the praxis system of Rothi *et al.* (1991). They postulate both (1) a direct route whereby a visual representation of the limbs is converted into intermediate postures for execution, mediating *low-level* imitation; and (2) an indirect route in which an action is recognized as a known goal-directed action and reconstructed thereby.

Their model was based on the analysis of auditory/verbal input by Patterson & Shewell (1987), and used the term "praxicon" for the store of motor schemas for known actions to parallel the term "lexicon" for the store of words in the language system. They provide a bidirectional link between the phonological lexicon (e.g., the action of saying or signing for each word) and its semantics and another bidirectional link between the semantics and the praxicon (Figure 23.3).

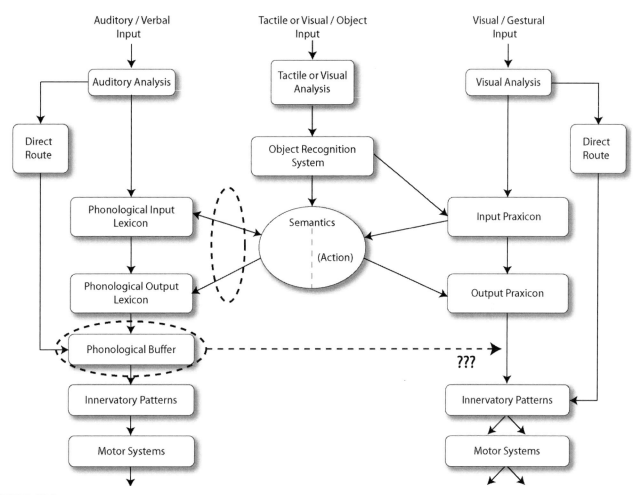

FIGURE 23.3 A dual-route imitation learning model balancing language and praxis. The diagram distinguishes a direct route for perception of meaningless sound sequences and meaningless manual gestures from the recognition of elements in the lexicon and praxicon of the language and praxis systems, respectively. We stress that the right-hand side should be augmented by an "action buffer", and note the relation of the bidirectional link between lexicon and semantics to the Saussurean sign linking the signifier to the signified. (*Source*: Adapted from Rothi *et al.*, 1991.)

Our concern is to understand the implications of this figure when the semantics is extended to encompass "scenes" structured by actions. Rothi *et al.* used the language system as a template for the design of a separate praxis system, linked only by the semantics of action. For us, the challenge is to suggest how the language system evolved "atop" the praxis system. Rothi *et al.* include a phonological buffer but not an action-structure buffer. By contrast, a full account of the praxis system needs a detailed analysis of structured actions, especially if it is to support complex imitation.

23.5.2. The Dorsal and Ventral Streams and Language

Consider the following parallel parieto-frontal interactions:

1. object → AIP → $F5_{canonical}$ praxis
2. action → PF → $F5_{mirror}$ action understanding
3. scene → Wernicke's → Broca's   language production

To see the problem with this, we review two lesion studies on humans. Goodale *et al.* (1991) studied a patient DF with a lesion to the inferotemporal (ventral) pathway. She was unable to verbalize or pantomime size or orientation, yet could preshape her hand to the appropriate size or orientation when performing a manual task. Conversely, Jeannerod *et al.* (1994) studied a patient AT whose lesion was in the parietal (dorsal) pathway. She could communicate the size and orientation of an object, but was unable to preshape her hand when reaching for the same object. Thus what at first sight appear to be the same motor parameters have two quite different representations – one for determining "How" to set the parameters for an action, the other expressing "What" an object and what its describable properties might be. However, although she could not preshape for a cylinder, for which there is no associated size range, AT was able to preshape to grasp an object like a lipstick whose "semantics" included a fairly narrow size range. We thus suggest that the IT "size-signal" has a diffuse effect on grasp programming – enough to bias a choice between two alternatives, or provide a default value when the parietal cortex cannot offer a value itself, but not enough to perturb a single sharply defined value established in parietal cortex by other means. Note the corresponding path from IT to AIP – both directly and via PFC – in linking dorsal and ventral streams in visually guided grasping in the macaque (the FARS model, Figure 23.2a).

Thus, the view that 3 above is a simple extension of 1 and 2 runs into an apparent paradox since the data of patients AT and DF suggest the following scheme:

4. Parietal "affordances" →   preshape
5. IT "perception of object" →   pantomime or verbal
                                  description of size

This means, one cannot pantomime or verbalize an affordance but rather one needs a "recognition of the object" (IT) to which attributes can be attributed before one can express them.

We have now gathered evidence that allows us to revisit Figure 23.1 in a way which makes contact with all that has gone before.

The lower two boxes of Figure 23.4 correspond to the words and concepts of Figure 23.1, but we now make explicit the hypothesis that a mirror system for phonological expression (words) evolved atop the mirror system for grasping to serve communication integrating hand, face, and voice. We also postulate that the concepts – for diverse actions, objects, attributes, and abstractions – are now seen to be represented by a network of concepts stored in long-term memory (LTM), with our current "conceptual content" formed as an assemblage of schema instances in WM.

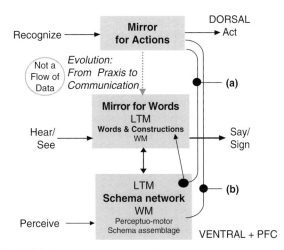

FIGURE 23.4   The mirror system for actions links to the mirror system for words (its "evolutionary cousin") only indirectly, through the schema network: (a) Words link to schemas, not directly to the dorsal path for actions. (b) As in FARS planning in PFC can affect the pattern of dorsal control of action.

Analogously, the "Mirror System for Words" contains a network of word forms in LTM and keeps track of the current utterance in its own WM. An alternative model of how the brain evolved for language stresses the role of WM while downplaying the role of the mirror system for grasping (Aboitiz et al. 2006). It proposes that cortical auditory-vocal networks of the monkey brain can be partly homologized with language networks that participate in the phonological loop (Baddeley, 2003). The model further suggests that semantic and syntactic processing also rely on active memory networks, but with networks for short-term cortical memory becoming increasingly complex in hominid evolution.

The perhaps surprising aspect of Figure 23.4 is that the arrow linking the "Mirror System for Actions" to the "Mirror System for Words" expresses an evolutionary relationship, not a flow of data. Rather than a direct linkage of the dorsal representation of an action to the dorsal representation of the phonological form, we have two relationships between the dorsal pathway for the Mirror for Actions (top) and the schema networks and assemblages of the ventral pathway and PFC (bottom). The rightmost path in Figure 23.4 corresponds to the paths in the FARS model whereby IT and PFC can affect the pattern of dorsal control of action. The path just to the left of this shows that the dorsal representation of actions can only be linked to verbs via schemas.

Supporting evidence for the view sketched above comes from Hickok & Poeppel (2004) who observe that early cortical stages of speech perception involve auditory fields in the superior temporal gyrus bilaterally (although asymmetrically). The authors offer evidence that this cortical processing system diverges into two streams:

1. A *dorsal stream* which maps sound onto articulatory-based representations and projects dorso-posteriorly. It involves a region in the posterior Sylvian fissure at the parietal–temporal boundary, and ultimately projects to frontal regions. This network provides a mechanism for the development and maintenance of "parity" between auditory and motor representations of speech.
2. A *ventral stream* which maps sound onto meaning and projects ventro-laterally toward inferior posterior temporal cortex (posterior middle temporal gyrus) which serves as an interface between sound-based representations of speech in the superior temporal gyrus (again bilaterally) and widely distributed conceptual representations.

We close by noting compatible data on the macaque auditory system. There is evidence for a dorsal and a ventral stream which process auditory spatial and auditory pattern/object information, respectively (e.g., Romanski et al., 1999). Both streams eventually project to the frontal cortex, which integrates both auditory spatial and object information with visual and other modalities. In human and non-human primates, auditory objects, including speech sounds, are identified in anterior superior temporal cortex which projects directly to inferior frontal regions. This same anatomical substrate may support both the decoding of vocalizations in nonhuman primates and the decoding of human speech.

23.6. CHALLENGES AND FUTURE DIRECTIONS

Embodying neurolinguistics requires embedding multiple mirror systems in the broader context of the brain. Modeling to date has shown how to analyze the basic macaque-like mechanisms for grasping, and Oztop *et al.* (2006) analyze beginning attempts to extend this capability to the imitation of single actions. Among the challenges for future work are the following:

1. Develop models of complex imitation, addressing the way in which complex actions can be structured by the recognition of the subgoals they serve to achieve.
2. Investigate to what extent the hierarchical structures that emerge from this process are comparable to the hierarchical structures employed in language construction, and make sense of the transition from praxic imitation to pantomime.
3. We have suggested that construction grammar may provide a good framework for neurolinguistics. However, this choice is but one of many. To take just four examples, Grodzinsky (2000) use generative grammar to explain Broca's aphasia by the Trace-Deletion Hypothesis, but stresses that other regions are also involved in syntax; Hagoort (2003) relates his event-related potential (ERP) findings to a parser based on a lexicalist grammar, lateral inhibition and "unification links" (Vosse & Kempen, 2000); while Schlesewsky and Bornkessel (2004) relate their ERP findings to role and reference grammar (Van Valin, 2005). Building a framework rich enough to incorporate all their findings seems crucial to the future health of neurolinguistics.
4. Fogassi *et al.* (2005) provide data on parietal mirror neurons for grasp which are sorted for overall goal (e.g., a mirror neuron for grasping may fire only when the grasp precedes bring-food-to-mouth, not putting-food-in-container, and vice versa). This relates to the issue of sequencing actions, and in some cases encapsulating the complex actions. We need to integrate such findings with the investigations in 2. to further develop neural mechanisms which support parallels between "motor constructions" and language constructions.
5. The list can be extended indefinitely, but two issues to close this list are: (a) how do the basal ganglia interact with cortical regions in hierarchical language processing; and (b) to what extent can developmental mechanisms be

modeled in sufficient detail to help us better understand what innate mechanisms are required for a brain to be language-ready.

References

Aboitiz, F., García, R.R., Bosman, C., & Brunetti, E. (2006). Cortical memory mechanisms and language origins. *Brain and Language*, *98*(1), 40–56.

Arbib, M., & Rizzolatti, G. (1997). Neural expectations: A possible evolutionary path from manual skills to language. *Communication and Cognition*, *29*, 393–424.

Arbib, M.A. (2005). From monkey-like action recognition to human language: An evolutionary framework for neurolinguistics. *Behavioral and Brain Sciences*, *28*, 105–167.

Baddeley, A. (2003). Working memory: Looking back and looking forward. *Nature Reviews Neuroscience*, *4*, 829–839.

Bonaiuto, B., Rosta, E., & Arbib, M.A. (2007). Extending the mirror neuron system model, I, audible actions and invisible grasps. *Biological Cybernetics*, *96*, 9–38.

Cheney, D.L., & Seyfarth, R.M. (2005). Constraints and preadaptations in the earliest stages of language evolution. *The Linguistic Review*, *22*, 135–159.

De Renzi, E. (1989). Apraxia lk. In: Boller, F., & Grafman, J. (Eds.), *Handbook of neuropsychology*. Vol. 2. (pp. 245–263), Amsterdam: Elsevier.

Fagg, A.H., & Arbib, M.A. (1998). Modeling parietal-premotor interactions in primate control of grasping. *Neural Networks*, *11*, 1277–1303.

Fillmore, C.J., Kay, P., & O'Connor, M.C. (1988). Regularity and idiomaticity in grammatical constructions: The case of let alone. *Language*, *64*(3), 501–538.

Fogassi, L., Ferrari, P.F., Gesierich, B., Rozzi, S., Chersi, F., & Rizzolatti, G. (2005). Parietal lobe: From action organization to intention understanding. *Science*, *308*(5722), 662–667.

Gallese, V., & Lakoff, G. (2005). The brain's concepts: The role of the sensory-motor system in conceptual knowledge. *Cognitive Neuropsychology*, *22*, 455–479.

Gallese, V., Fadiga, L., Fogassi, L., & Rizzolatti, G. (1996). Action recognition in the premotor cortex. *Brain*, *119*, 593–609.

Gibson, J.J. (1966). *The senses considered as perceptual systems*. London: Allen & Unwin.

Goodale, M.A., Milner, A.D., Jakobson, L.S., & Carey, D.P. (1991). A neurological dissociation between perceiving objects and grasping them. *Nature*, *349*, 154–156.

Grafton, S.T., Arbib, M.A., Fadiga, L., & Rizzolatti, G. (1996). Localization of grasp representations in humans by PET: 2, Observation compared with imagination. *Experimental Brain Research*, *112*, 103–111.

Grodzinsky, Y. (2000). The neurology of syntax: Language use without Broca's area. *Behavioral Brain Science*, *23*(1), 1–71.

Hagoort, P. (2003). How the brain solves the binding problem for language: A neurocomputational model of syntactic processing. *Neuroimage*, *20*, 18–29.

Hickok, G., & Poeppel, D. (2004). Dorsal and ventral streams: A framework for understanding aspects of the functional anatomy of language. *Cognition*, *92*, 67–99.

Jeannerod, M., Decety, J., & Michel, F. (1994). Impairment of grasping movements following a bilateral posterior parietal lesion. *Neuropsychologia*, *32*(4), 369–380.

Kemmerer, D. (2006). Action verbs, argument structure constructions, and the mirror neuron system. In M.A. Arbib (Ed.), *Action to language via the mirror neuron system* (pp. 347–373). Cambridge, UK: Cambridge University Press.

Oztop, E., & Arbib, M.A. (2002). Schema Design and Implementation of the Grasp-Related Mirror Neuron System. *Biological Cybernetics*, *87*, 116–140.

Oztop, E., Kawato, M., & Arbib, M.A. (2006). Mirror neurons and imitation: A computationally guided review. *Neural Networks*, *19*, 254–271.

Patterson, K.E., & Shewell, C. (1987). Speak and spell: Dissociations and word class effects. In M. Coltheart, G. Sartori, & R. Job (Eds.), *The cognitive neuropsychology of language* (pp. 273–294). London: Erlbaum.

Rizzolatti, G. & Arbib, M.A. (1998). Language within our grasp. *Trends in Neuroscience*, *21*, 188–194.

Rizzolatti, G., & Luppino, G. (2003). Grasping movements: Visuomotor transformations. In M.A. Arbib (Ed.), *The handbook of rear brain theory and neural networks* (2nd edn, pp. 501–504). Cambridge, MA: The MIT Press.

Romanski, L.M., Tian, B., Fritz, J., Mishkin, M., Goldman-Rakic, P.S., & Rauschecker, J.P. (1999). Dual streams of auditory afferents target multiple domains in the primate prefrontal cortex. *Nature Neuroscience*, *2*(12), 1131–1136.

Rothi, L.J.G., Ochipa, C., & Heilman, K.M. (1991). A cognitive neuropsychological model of limb praxis. *Cognitive Neuropsychology*, *8*, 443–458.

Schlesewsky, M., & Bornkessel, I. (2004). On incremental interpretation: Degrees of meaning accessed during sentence comprehension. *Lingua*, *114*, 1213–1234.

Van Valin, R.D., Jr. (2005). *Exploring the syntax-semantics interface*. Cambridge: Cambridge University Press.

Vosse, T., & Kempen, G. (2000). Syntactic structure assembly in human parsing: A computational model based on competitive inhibition and a lexicalist grammar. *Cognition*, *75*(2), 105–143.

Further Readings

Arbib, M.A. (2005). From monkey-like action recognition to human language: An evolutionary framework for neurolinguistics. *Behavioral and Brain Sciences*, 28, 105–167.

This gives a good sense of the state of debate over hypotheses A1 through A4 for the Mirror System Hypothesis since it is accompanied by numerous commentaries pro and con, together with my response. Supplemental commentaries and my "electronic response" are at http://www.bbsonline.org/Preprints/Arbib-05012002/Supplemental/Arbib.E-Response_Supplemental.pdf.

Arbib, M.A. (Ed.). (2006). Action to language via the mirror neuron system. Cambridge: Cambridge University Press.

In this volume, written to be accessible to a wide audience, experts from child development, computer science, linguistics, neuroscience, primatology, and robotics present and analyze the mirror system and show how studies of action and language can illuminate each other. Topics discussed in the fifteen chapters include: What do chimpanzees and humans have in common? Does the human capability for language rest on brain mechanisms shared with other animals? How do human infants acquire language? What can be learned from imaging the human brain? How are sign- and spoken-language related? Will robots learn to act and speak like humans?

Emmorey, K. (2002). Language, cognition, and the brain: Insights from sign language research. Mahwah, NJ: Lawrence Erlbaum and Associates.

An excellent resource for thinking about neurolinguistics in multi-modal terms. Signed languages depend upon high-level vision and motion-processing systems for perception, and the integration of motor systems involving the hands and face for production. These differences at the periphery make all the more valuable the analysis of neurolinguistic commonalities for speech and sign.

Grodzinsky, Y., & Amunts, K. (Eds.) (2006). Broca's region. Oxford: Oxford University Press.

This book contains a collection of classic papers on Broca's region, including the first two papers written by Paul Broca. These provide a valuable historical background for the studies by anatomists, brain imagers, neurolinguists, and neurophysiologists who present the state of the art in Broca's-region research.

Rizzolatti, G. & Craighero, L. (2004). The mirror-neuron system. *Annual Reviews Neuroscience*, 27, 169–192.

Giacomo Rizzolatti led the team that first discovered mirror neurons, and was my colleague in formulating the Mirror System Hypothesis. This paper provides an excellent review of recent research on the mirror system (though new findings continue to appear at an impressive rate). The authors first describe the functional properties of mirror neurons in monkeys then review the characteristics of the human mirror system. Though noting that social organization is impossible without action understanding, they stress those properties specific to the human mirror system that might explain the human capacity to learn by imitation, as well as its implications for the study of language mechanisms.

24

Lateralization of Language across the Life Span

MERRILL HISCOCK[1] and MARCEL KINSBOURNE[2]

[1]*Department of Psychology and Center for Neuro-Engineering and Cognitive Science,*
University of Houston, Houston, TX, USA
[2]*Department of Psychology, New School for Social Research, New York, NY, USA*

ABSTRACT

The cerebral hemispheres of the adult brain, though they look alike, are dissimilar in function. Some of the most striking functional differences are found in the realm of language. We enumerate various facets of brain and behavioral asymmetry that are observed early in human development and are potential precursors of language lateralization. We also summarize evidence that addresses the question of change in language lateralization during the lengthy period from early childhood to senescence. The upshot of these various kinds of evidence is that the brain is functionally asymmetric from birth (or earlier) and that lateralization of language remains relatively stable across the life span. However, several problems and complexities preclude a simple and definitive answer to the question of life-span changes. We discuss those problems and complexities in the final part of the chapter.

24.1. INTRODUCTION

This chapter concerns the time course of language lateralization. It addresses primarily "when" questions. When are harbingers of language lateralization first observed in human development? When does lateralized language emerge? When is the lateralization of language complete? At what later stage of life, if any, does lateralization begin to change?

Different answers to these "when" questions may be represented on a graph in which degree of language lateralization is plotted against the years of a person's life. The most parsimonious explanation, and the explanation that we have favored in the past is known as "invariant lateralization." Invariant lateralization can be depicted as a horizontal line, for example, as line "a" in Figure 24.1. According to this viewpoint, signs of lateralization are already present at the

time of birth, and the magnitude of differences between left and right sides of the brain remains constant throughout the life span.

The alternative point of view, of course, is that lateralization changes across the life span. There are two variants of this "age-dependent lateralization" position. The first, known as progressive lateralization, could be represented graphically as a monotonic increase in lateralization from a symmetric base state at birth to the mature state of left-sided language. Traditionally, progressive lateralization has been claimed to continue from the beginning of language development until pubescence (Lenneberg, 1967). However, some clinical evidence indicates that language lateralization is completed by the age of 4 or 5 years. It has also been suggested that the process continues throughout the life span. Although the increase in lateralization over time need not be linear, we represent the three hypothetical patterns of progressive lateralization as straight lines (lines "b," "c," and "d") in Figure 24.1.

The other variant of age-dependent lateralization is what we might call lateralized degeneration. This variant concerns only senescence. The central idea is that the functions of one cerebral hemisphere deteriorate before the functions of the other hemisphere. Typically, because certain nonverbal abilities appear to decline at an earlier age than most verbal abilities, it is claimed that the right hemisphere is the first to deteriorate. By virtue of its later or more subtle deterioration, the left hemisphere progressively becomes more "dominant" during the period of cognitive decline. This possibility is depicted as line "e" in Figure 24.1. Asymmetry of deterioration is logically independent of asymmetric development and may be combined with any previous pattern of lateralization.

We begin with a brief summary of the empirical evidence regarding the genesis of hemispheric specialization. We

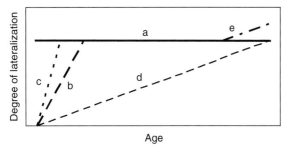

FIGURE 24.1 Curve "a" depicts the developmental invariance model, which posits that the left hemisphere is specialized for language at the time of birth or even prior to birth. The model predicts early emergence of lateralized precursors of language, and no increase or decrease in the degree of lateralization during development. The lateralization-by-puberty model is represented by curve "b", and the lateralization-by-five-years model is represented by curve "c". Both models exemplify progressive lateralization, in which the bilateral symmetry of infant's brain gradually develops into the lateralized adult state. Curve "d" represents a form of progressive lateralization in which the brain continues to become more strongly lateralized during maturity and even into senescence. The progressive degeneration model is represented by curve "e". According to this model, the functions of one hemisphere (usually the right) deteriorate more rapidly than the functions of the other, thus accentuating any differences between hemispheres.

consider (1) brain anatomy, (2) electrophysiological findings, (3) behavioral asymmetries, and (4) clinical evidence. We emphasize language lateralization but the coverage will by necessity be broader. We conclude the first part of the chapter by summarizing evidence that concerns differential decline.

In the second part of the chapter, we discuss complexities that make it difficult to interpret the available evidence. The emphasis then will shift from "when" questions to "what" questions. What do the early asymmetries mean? Exactly what is lateralized? What is the relationship between language lateralization and the skill with which language is used?

24.2. BRAIN DEVELOPMENT

Assertions about invariant or progressive lateralization are based on evidence of brain asymmetries during different stages of development. The evidence comprises anatomical, electrophysiological, behavioral, and clinical data.

24.2.1. Brain Anatomy

Asymmetries in the immature brain. Several anatomical asymmetries have been found in the neonatal brain (Spreen *et al.*, 1995). These findings contradict the previously held belief that the two cerebral hemispheres in infancy are morphologically identical or nearly so. Asymmetries in some cerebral regions can be attributed to more widely spaced cortical columns on the larger side (Buxhoeveden & Casanova,

Box 24.1   Will brain imaging resolve questions about lateralization?

Do methods drive science? The history of lateralization studies suggests that this may be so. From the early nineteenth to mid-twentieth century, functional differences between the human hemispheres were inferred primarily from lesion studies. Though enormously informative, this method addressed mainly the binary question: is a function affected by left- or by right-sided damage? Twentieth century technologies such as electroencephalography, commissurotomy, and Wada test enabled more refined questions.

Functional imaging techniques herald a new era for lateralization studies. Most functional imaging studies thus far only confirm what is already known through other methods, but even this corroboration serves to solidify and elaborate earlier findings and also to validate the imaging methods themselves. Nonetheless, the new imaging technologies will permit researchers to examine more complex and refined propositions about lateralization. Two studies illustrate this prospect.

Liégeois *et al.* (2004) used functional magnetic resonance imaging (fMRI) to monitor the re-organization of language in children following early left-hemisphere damage. They found that children with lesions in or near the anterior language region continued to use the left hemisphere for expressive language. Unexpectedly, language was more likely to shift to the right hemisphere if the lesion in the left hemisphere was remote from classical language areas.

Reuter-Lorenz *et al.* (2000) used positron emission tomography (PET) to illuminate age-related changes in the lateralization of verbal and nonverbal working memory. Instead of the expected differential left (verbal) versus right (spatial) frontal lateralization as seen in young adults, the PET evidence for older participants indicated bilateral frontal activation for both spatial and verbal rehearsal.

The rapidly growing corpus of functional imaging studies points to progressive refinement in the answers that can be obtained, and in the models of cortical functioning that will result. It becomes possible to study differences between hemispheres simultaneously with differences within each hemisphere.

Liégeois, F., Connelly, A., Cross, J. H., Boyd, S.G., Gadian, D.G., Vargha-Khadem, F., & Baldeweg, T. (2004). Language reorganization in children with early-onset lesions of the left hemisphere: An fMRI study. *Brain, 127*, 1229–1236.
Reuter-Lorenz, P.A., Jonides, J., Smith, E.E., Hartley, A., Miller, A., Marshuetz, C., & Koeppe, R.A. (2000). Age differences in the frontal lateralization of verbal and spatial working memory revealed by PET. *Journal of Cognitive* Neuroscience, *12*, 174–187.

2000), but whether cortical volume is an index of any functional characteristic that differs between hemispheres remains to be established. One can only conclude that some of the anatomical asymmetries that characterize the adult human brain are neither absent nor less marked in the infant brain. A similar conclusion applies to the anticlockwise torque that is manifested as a wider frontal lobe on the right side and a greater anterior extension of the right frontal lobe, coupled with a wider posterior left hemisphere and a greater posterior extension of the left hemisphere. Irrespective of its functional significance, if any, in the adult human brain, a similar torque is observed in the immature human brain as well as in the ape brain.

Maturational gradients. Perhaps anatomical disparities between the hemispheres reflect differences in the rate of growth. However, findings are not unanimous as to which hemisphere grows more rapidly. The putative left-to-right gradient of an earlier era was superseded by subsequent evidence that the right cerebral cortex in fact develops more rapidly than the left, which is now commonly believed.

A difference between the maturation rates of the two hemispheres is but one aspect of brain development, and it may not be more important than other maturational gradients. In an attempt to consolidate all available data on the rate at which different brain regions mature, Best (1988) proposed a growth vector representing the resultant of four developmental dimensions: right-to-left, anterior-to-posterior, primary-to-secondary-to-tertiary, and basal-to-cortical. In addition, within each cerebral hemisphere, there are regional patterns in the development of myelination, in neuronal, dendritic, and axonal density, and in the width of different cortical layers.

24.2.2. Electrophysiological Evidence

A substantial body of evidence indicates that electrophysiological asymmetries are present in infancy. Much of this evidence has been reviewed by Molfese and Betz (1988) and by Segalowitz and Berge (1994). Numerous studies have shown consistent differences between the infant's hemispheres in electrophysiological responses to specific speech cues. Event-related potentials (ERPs) indicate that differences in voice onset time (e.g., the difference between "ba" and "pa") elicit an initial response from both hemispheres, followed by a response from the right hemisphere alone. Differences in place of articulation (e.g., the difference between "ba" and "ga") elicit a left-hemisphere response that is followed by a bilateral response. The ERP findings for infants are similar – but not identical – to those for adults.

24.2.3. Behavioral Evidence

Head turning and postural asymmetry. Whether turning spontaneously or in response to stimulation, most infants turn their heads to the right more often than to the left. This rightward bias is one of the earliest behavioral asymmetries to be manifested by the neonate, and it has been linked both to parental handedness and to the infant's subsequent hand preference.

When the infant is placed on his or her back with the head turned to one side, the arm and foot on that side often are extended, and the contralateral arm and foot are flexed. This so-called asymmetric tonic neck reflex (ATNR) occurs in the newborn infant and persists for at least the first 3 months of life, but certain aspects change during that time (Liederman, 1987). For instance, the ATNR is more evident in the legs of newborns than in their arms, whereas the opposite pattern is observed in infants older than 3 weeks. In infants between the ages of 3 and 10 weeks, head-turns to the nonpreferred side are more likely to elicit the ATNR than are head-turns to the preferred side, but this may not apply to neonates. Whether the predominant direction of the ATNR predicts subsequent handedness is a matter of dispute although, as noted previously, head turning by itself seems to bear a relationship to subsequent manual asymmetries.

Motor activity. Early indications of hand preference are of special interest because of theoretical and empirical links between handedness and language lateralization in the adult (e.g., Knecht *et al.*, 2000; Annett, 2002). Unfortunately, the literature is replete with unresolved conceptual and methodological issues and seemingly inconsistent findings. Much of the evidence supports Liederman's (1983) contention that most infant behavior is dominated by the left hemisphere and right hand. For instance, most infants hold objects for a longer time with the right hand than with the left hand, and most infants prefer the right hand for a variety of target-related actions that are performed during the first few months of life. A hand preference for unimanual manipulation of objects develops between the ages of 5 and 7 months, and a hand preference in tasks requiring bimanual manipulation develops by the age of 1 year.

Perception. Studies of auditory perception provide some of the most convincing evidence of early functional asymmetries. This evidence has been summarized by Best (1988). Using dishabituation paradigms to determine infants' ability to discriminate between two speech sounds, investigators have demonstrated an early right-ear advantage (REA) for detection of transitions between consonants (e.g., /ma/ to /da/) and a left-ear advantage (LEA) for transitions in musical timbre (for example, cello to bassoon). For instance, Best *et al.* observed an REA for speech syllables and an LEA for musical stimuli in infants 3 months of age and older. Although an LEA for musical stimuli was found in 2-month-old infants, a corresponding REA for speech perception has not been reported below the age of 3 months.

Box 24.2   Measuring asymmetries and age differences in asymmetries: How hard can it be?

Kolb and Whishaw (2003) summarize attempts by different researchers to measure differences between the size of people's left and right feet. The upshot is that something as seemingly straightforward as measuring foot length may lead to disparate results. If so, how difficult might it be to measure differences between the structure or the functions of left and right regions of the brain?

The situation often becomes even more challenging when one compares groups. Consider the challenge of comparing the performance of younger and older children on a verbal dichotic listening test. Maybe the average 5-year-old reports 30% of left-ear words and 40% of right-ear words, whereas the average 10-year-old reports 60% of left-ear words and 80% of right-ear words. The REA is 10 percentage points for the 5-year-olds and 20 percentage points for the 10-year-olds. Yet, the ear differences are identical in magnitude if calculated as proportions of the overall score. The 10-year-olds are more asymmetric by one calculation but not by the other.

This scaling problem is sometimes compounded by floor and ceiling effects. Many laterality researchers have responded by using mathematical transformations (laterality indices) that adjust difference scores for the overall level of performance.

Regrettably, no transformation is entirely neutral or applicable to all data (Kinsbourne & Byrd, 1985). Some investigators eschew parametric data analyses in favor of less informative alternatives such as tabulating the proportion of individuals in each age group who show an REA. We advocate a third approach that entails selecting two transformations that would be expected to bias the results in opposite directions (Kinsbourne & Hiscock, 1983). If both transformations lead to congruent results, the data can be interpreted accordingly. If not, the findings are considered inconclusive with regard to age-group differences.

Kinsbourne, M., & Byrd, M. (1985). Word load and visual hemifield shape recognition: Priming and interference effects. In M.I. Posner & O.S.M. Marin (Eds.), *Mechanisms of attention: Attention and performance XI* (pp. 529–543). Hillsdale, NJ: Erlbaum.

Kinsbourne, M., & Hiscock, M. (1983). The normal and deviant development of functional lateralization of the brain. In M.M. Haith & J.J. Campos (Eds.), *Handbook of Child Psychology, Infancy and Developmental Psychobiology* (Vol. 2, 4th edn, pp. 157–280). New York: Wiley.

Kolb, B., & Whishaw, I.Q. (2003). *Fundamentals of human neuropsychology* (5th edn). New York: Worth.

A study based on a novel behavioral method suggests that a speech-related brain asymmetry is present even in short-gestation infants. Using limb movements as a measure of immaturity, Segalowitz and Chapman (1980) found that repeated exposure to speech, but not to music, caused a disproportionate reduction of right-arm tremor in infants with an average gestational age of 36 weeks. This was taken as evidence that speech affected the left side of the brain more than the right side.

Other investigators have observed that neonates turn more often to the right than to the left when exposed to speech sounds. MacKain *et al.* (1983) reported that 6-month-old infants detect the synchronization of visual (articulatory) and aural components of adults' speech, but only when the adult is positioned to the infant's right. These findings suggest that speech sounds bias orientation to the right side of space, presumably because the left side of the infant's brain is more responsive than the right side to speech-specific activation. This language-specific asymmetry appears to complement a more pervasive left-hemisphere prepotency that biases orientation to the right.

Childhood laterality. In a review of auditory, visual, tactual, and dual-task laterality studies involving children between the ages of 2 and 12 years, Hiscock (1988) found no consistent evidence of age-related increases in laterality. Irrespective of the modality tested or the method employed, cross-sectional studies typically reveal the expected asymmetries

in the youngest children tested, and those asymmetries are comparable in magnitude to the asymmetries found in older children.

Though few longitudinal studies of the REA for linguistic stimuli have been published, the results are similar to results from cross-sectional studies. When an age-related increase (or decrease) in laterality is observed, the change seems to reflect either the noisiness of the data or extraneous factors that covary with age. For example, a large-scale longitudinal study by Morris *et al.* (1984) yielded different developmental patterns of ear asymmetry for different sub-samples. The authors attributed this variability to the lack of experimental control that is inherent in the free-report dichotic listening method.

Even if laterality remains invariant across the childhood years, a quantitative difference conceivably could exist between the asymmetry of children and of adults. However, direct comparisons of children and adults on dichotic listening tasks have not revealed any consistent difference in the magnitude of the REA.

24.2.4. Clinical Evidence

The literature on childhood aphasia, hemispherectomy, and recovery of function has been reviewed by a number of authors (e.g., Spreen *et al.*, 1995; Bates *et al.*, 1999; Vargha-Khadem, 2001). Even though the hypothesis of progressive lateralization drew much of its support from cases of aphasia in children following right-hemispheric lesions, the preponderance of evidence now suggests that persistent aphasia consequent to unilateral right-hemisphere damage is as infrequent in children as in adults.

Studies of cognitive functioning following unilateral lesions that use quantitative measures are limited by a reliance on IQ and academic achievement tests, which constitute neither sensitive nor comprehensive measures of linguistic and visuo-perceptual functioning. Such measures are not optimal for differentiating between left- and right-sided lesions. Despite this handicap, several studies do show associations between left-sided damage and verbal impairments, and between right-sided damage and nonverbal impairments. Some studies suggest that the effects of unilateral lesions are less selective when the damage is prenatal or perinatal than when it occurs later in development (Vargha-Khadem, 2001). Nonetheless, even unilateral brain lesions of prenatal or perinatal origin can lead to differential impairment of verbal and nonverbal functions. The development of expressive language, in particular, appears to be vulnerable to early left-sided brain damage (Bates et al., 1999).

The implications of hemispherectomy performed at different ages are difficult to specify, mainly because of extraneous factors that confound comparisons between hemispherectomy in children and in adults, but also because of the inferential limitations of small-sample and single-case studies. If any conclusion is justified by the available hemispherectomy evidence, it is that the right hemisphere of children exhibits an impressive ability to support "aspects of everyday verbal communication" when the left hemisphere is compromised (Devlin et al., 2003).

24.3. CHANGES ASSOCIATED WITH AGING

Does the seemingly stable lateralization of the brain in early and middle adulthood change in senescence? The answer to this question is informed by studies of normal aging as well as by clinical studies.

24.3.1. Normal Aging

Physiological changes. Autopsy studies reveal no obvious predominance of neuropathology in either the left or the right hemisphere of the elderly (Esiri, 1994). Studies based on PET scanning report nearly identical levels of glucose metabolism in the left and right cerebral cortices and in the left and right basal ganglia of healthy elderly subjects. A study of 100 postmortem brains revealed comparable age-related decreases in cerebral volume within the left and right hemispheres (Witelson et al., 2006).

Behavioral changes. Various psychometric data, typically cross-sectional, establish that certain kinds of tasks are more prone than others to show age-related declines in average performance. The selectivity of decline is often interpreted as evidence that the right hemisphere deteriorates sooner or more precipitously than the left. For instance, scores on the performance (nonverbal) subtests of the Wechsler Adult Intelligence Scale decline more rapidly with increasing age than do scores on the verbal subtests. Much of this differential decline can be attributed to a cohort effect that affects nonverbal tests more strongly than verbal tests (Flynn, 2006). Nevertheless, a differential decrement of verbal and nonverbal skills has been confirmed by longitudinal data, which are not susceptible to cohort effects. Similarly, tests of "fluid" intelligence – the ability to solve problems requiring novel information and strategies – are more likely to show age-related changes than are tests of "crystallized" intelligence – the ability to utilize previously acquired knowledge and strategies. Again, this pattern is frequently attributed to a greater deterioration of right-hemisphere functioning. Age-related decline in average performance on some tests of fluid intelligence is quite marked. On Raven's standard progressive matrices, an untimed test that contains seemingly novel problems, a raw score at the 50th percentile for 18-year-olds falls at the 95th percentile for 65-year-olds. Much of that difference, however, can be attributed to a cohort effect rather than to cognitive deterioration over time, and even the portion attributable to deterioration does not necessarily reflect asymmetrical deterioration.

Two of the most salient cognitive deficits observed among the elderly are reduced speed of processing (van Gorp et al., 1990) and reduced ability to perform complex or difficult tasks (Crossley & Hiscock, 1992). It is uncertain whether these deficits represent separate limitations or different manifestations of the same limitation. Furthermore, it remains to be established that either deficit implicates the deterioration of one hemisphere more than the other.

24.3.2. Clinical Evidence

Dementia. Alzheimer's disease (AD), especially early-onset AD, often affects object naming, spontaneous speech, and praxis. Although attributable to left-hemisphere disease, these deficits simply may be more noticeable than deficits associated with right-hemisphere disease. Right-hemispheric symptoms such as spatial confusion are also seen frequently. AD patients do not constitute a homogeneous group for laterality studies. Even though some individuals initially may exhibit cognitive profiles suggestive of asymmetric dysfunction, either hemisphere may be the more severely affected. Ultimately the pathology becomes diffuse and bilateral.

Aphasia. The relative incidence of different aphasia types appears to shift as a function of age. Expressive disorders and mixed aphasias (i.e., nonfluent speech with moderately impaired comprehension) predominate throughout the life span, but particularly during the first three decades of life.

The incidence of global aphasia begins to rise in the fourth decade, and sensory (fluent) aphasias become relatively common during the seventh decade. These age-related shifts might be attributable to a gradually increasing degree of lateralization and focal brain organization: An initially bilateral and diffuse language substrate becomes unilateral and focal over time, with expressive functions lateralizing before receptive functions. Alternatively, the changing incidence rates might reflect regional changes in vulnerability to cerebral disease, changes in compensatory ability (i.e., in plasticity), or other variables.

24.4. INTERPRETIVE COMPLEXITIES

The presence or absence of age-related changes in language lateralization is an old question that has attracted a large amount of research and clinical observation. Accordingly, a large corpus of relevant evidence has been accumulated. The evidence, as we have already seen, is often ambiguous. Some of the sources of ambiguity are discussed below.

24.4.1. General Problems in Studying Life-Span Development

Developmental patterns must be constructed. The ideal study of change across the life span would begin at conception and then continue longitudinally through successive stages of life. As a rule, the scientific examination of ontogenetic change in humans does not conform to this ideal but instead begins with the mature organism. As interesting characteristics are observed and investigated in the adult, certain development questions invariably arise. At what stage of development does this attribute become manifest? What are the mechanisms responsible for the emergence the characteristic? To what extent is the attribute influenced by exogenous or endogenous environmental factors? What evolutionary advantage, if any, is bestowed by the characteristic? How does the attribute change as the organism approaches the end of its life trajectory? Studies of young or middle-aged adults are then extended downward to infants and children, and upward to the elderly.

Life-span studies are constrained by methodological limitations. Unfortunately, the methods used with young or middle-aged adults often are not feasible for use with children or elderly people. Visual (tachistoscopic) experiments in which letters or words are presented to one visual half-field cannot be used with infants or young children who cannot read. Most elderly people can read, of course, but their ability to perform a visual task may be compromised by impairments of visual acuity. Dichotic listening is also problematic. Infant researchers have used habituation–dishabituation paradigms to study asymmetries of auditory perception in infants, and some of the infant studies have

shown an REA for language sounds. Infant studies of this kind are difficult to do, however, and results will vary with the physiological and psychological state of the infant at the time of testing. It is especially difficult to test 1- and 2-year-olds, for whom infant methods are inappropriate and methods designed for older children may be unsatisfactory. Dichotic studies have been done successfully with 3-, 4-, and 5-year-olds, but even children in this age range do not respond well to the standard dichotic listening tests that are used with school-age children and adults. The elderly are difficult to assess because hearing impairment is common in older adults. fMRI is difficult to use with young children, but other functional imaging techniques (e.g., magnetic source imaging and functional transcranial Doppler sonography) seem to be more feasible.

24.4.2. Problems Related to the Study of Language Lateralization

Brain *lateralization is an abstraction.* The majority of adults have speech-related regions of cerebral cortex that are larger on the left hemisphere than on the right hemisphere (Dorsaint-Pierre *et al.*, 2006). The majority of humans also are more skilled with the right hand than with the left and more likely to report linguistic signals from the right ear than from the left ear when the signals are presented in dichotic competition. Although there is reason to believe that different asymmetries may be correlated, associations among them are not strong nor are they universal (e.g., Knecht *et al.*, 2000; Dorsaint-Pierre *et al.*, 2006; Fernandes *et al.*, 2006). What asymmetry best represents lateralization? Does any specific asymmetry adequately represent lateralization?

The loose association among different asymmetries illustrates two fundamental points. First, one must distinguish lateralization of the brain from specific manifestations of lateralization. Neuroanatomical asymmetry, handedness, and ear asymmetry are manifestations of a lateralized nervous system, but none is synonymous with lateralization. Lateralization is a construct, which is to say, an artifact. Nature may reveal concrete and specific markers of lateralization, such as handedness, but it is researchers who infer the existence of lateralization. The second truth is that none of the markers is definitive. Some may be better proxies than others, perhaps because they lend themselves to more precise measurement, but none is ideal. The implication of this distinction between construct and manifestations is that lateralized language may mean different things to different people.

It is not easy to specify what is lateralized. As Kinsbourne (1984) has pointed out, neuropsychologists have been more successful in specifying where functions are localized than in specifying the functions that are localized. The dichotomy between "what" and "where" underlies many of the most difficult questions concerning life-span development of hemispheric specialization. A misplaced emphasis on the question of "where" has led to disagreement about the lateralization of functions that are only vaguely defined. For instance, some dichotic listening tasks tap processes that are primarily auditory or phonetic, whereas others involve semantic and mnemonic processing. As a rule, however, investigators have assumed that a particular category of stimuli, such as digit names, represents the full spectrum of linguistic material.

Luria (1973) addressed this problem when he emphasized that the development of a new perceptual or motor skill entails first consolidating isolated elements of the function into an integrated and automatized series of elements and then linking the integrated elements into a network of superordinate functions. Luria made it clear that lesion studies cannot reveal the localization of a mental process unless the structure of the process is understood. Much of the research on age-related change in functional lateralization violates this basic Lurian caveat: Investigators have attempted to lateralize a function or set of functions without any independent knowledge as to how the functions are organized.

Processing asymmetries must be distinguished from activation asymmetries. Only in the ideal case would the pattern of regional brain activation match precisely an individual's cortical localization scheme. Once laterality of activation is distinguished from lateralized representation of various cortical functions (Segalowitz & Berge, 1994), it follows that age-related changes in one aspect of lateralization may be dissociated from changes in the other aspect. For example, markedly asymmetric activation in a patient with AD need not imply a corresponding shift of cortical functions to the more activated side. The asymmetry of activation may instead indicate that the locus of activation and locus of processing are not the same.

Change in lateralization is different from change in the microstructure of an activity. New components of a task may materialize as a consequence of maturation or practice, or existing components may become automatized. In addition, increasing proficiency might entail restructuring a task, as when material-appropriate strategies enhance reading. Other changes in task structure – compensatory as well as deleterious – presumably occur in senescence. Whenever there is an age-related change in the lateralization of an activity, it will be important to establish whether that change reflects a shifting brain basis for a fixed set of task components or a shifting set of task components.

Specialization versus *plasticity*. It is essential to distinguish between decreasing bilaterality of function (evidence of which might be found in normal children) and decreasing equipotentiality (which would be manifested only as differential outcomes following focal brain injury sustained at different ages). Failures to make that distinction have added confusion to the literature on childhood lateralization. The

same problem is encountered when behavioral changes at the other end of the life span are being interpreted. A developmental shift toward bilateralization of a previously unilateral function is different in principle from deterioration of the unilateral function.

Individual- and population-level asymmetries. If evidence continues to support the existence of hemisphere-selective deficits in some patients with AD, it will be important to know whether the disease process itself is asymmetrical in those patients (at least, in the early stages of the disease), or whether the behavioral consequences of an invariably bilateral degenerative process depend on certain premorbid characteristics of the patient. Either answer would contribute significantly to our understanding of AD, but neither answer would support a claim that the left and right hemispheres – at the level of the population – are differentially vulnerable.

The implications of asymmetry are unclear. Failure to develop normal language skills often has been attributed either to an inadequate degree of lateralization or a deficiency in left-hemispheric function. Apparently neither explanation is correct. In a longitudinal study of children from birth until the age of 5 years, Molfese and Molfese (1997) found that individual differences in ERP recorded in the neonate could predict performance on language tests 5 years later. However, prediction did not depend entirely on left-hemisphere ERP. Electrical responses from both hemispheres contributed to the prediction of later language functioning. In another longitudinal study, Guttorm *et al.* (2005) similarly found a significant correlation between ERP to linguistic stimuli in neonates and measures of receptive language skill at the age of 2.5 years. In this study, it was right-hemisphere ERP that predicted receptive language ability. Left-hemisphere ERP predicted verbal memory at the age of 5 years.

Other evidence calls into question the notion of a specific association between brain lateralization and language. Indications of lateralization sometimes are found in the absence of language, and language sometimes is accompanied by bilateral brain activity. Vallortigara (2006) points out that, in various nonhuman species, the left side of the brain is specialized for processing complex auditory signals. Lateralized ERPs to speech cues have been recorded in dogs and rhesus monkeys, prompting Molfese and Betz (1988) to suggest that the critical difference between humans and nonhumans with respect to speech discrimination is reflected not by lateralized ERP components but instead by bilateral components that have been not been found in monkeys and dogs. The right hemisphere is implicated further in language processing by fMRI evidence indicating that bilingual adults have more bilateral and diffuse activation patterns when listening to their second language than when listening to their primary language (e.g., Dehaene *et al.*, 1997).

Differential loss of skills does not necessarily mean differential deterioration of the cerebral hemispheres. The available evidence concerning cognitive changes during aging establishes only that certain measures – especially reaction time and dual-task performance – are more prone than others to show age-related declines in average performance. Even the widely accepted principle that fluid intelligence deteriorates more rapidly than crystallized intelligence remains open to question (van Gorp *et al.*, 1990). Longitudinal data reveal that the rate of cognitive decline varies markedly among individuals.

Witelson *et al.* (2006) studied the postmortem brains of 100 individuals of known IQ. The correlation between postmortem cerebral volume and verbal IQ was +.51 for right-handed women and +.62 for right-handed men. The correlations were comparable irrespective of hemisphere. Moreover, there was a striking sex difference in the relationship between age and brain volume. Irrespective of hemisphere, brain volume decreased significantly with age in men but not in women. When considered together, the data present a peculiar pattern. Regardless of whether brain volume decreases with age (as in men) or does not change with age (as in women), verbal IQ – which is related to brain volume in both men and women – remains constant. Findings such as these are difficult to reconcile with any simple principle for relating the decline of cognitive ability to differential left- and right-hemisphere deterioration or even to concurrent deterioration of both hemispheres.

24.5. CHALLENGES AND FUTURE DIRECTIONS

Brain functions change over the life span as a consequence of maturation, experience, and disease, and those functional changes sometimes are associated with detectable re-organization of the neural substrate. Language is not exempt from this general principle. Only a dynamic substrate would support the acquisition of various culturally specified languages (sometimes in the same brain) and the idiosyncratic modification of those languages during the course of a person's life. Language representation is not hard-wired to the degree that destruction of the relevant left-hemispheric circuitry would always preclude the development or recovery of language. On the contrary, clinical evidence from various sources attests to the resilience of language in the aftermath of left-hemisphere damage, especially when the damage occurs pre- or perinatally.

Yet, the preponderance of evidence from studies of normal brains suggests that the lateralization of language ordinarily does not change during one's life. In the absence of pathology, language representation remains primarily left-hemispheric in the great majority of humans. Lateralized language is the default state, but insults to the brain under some circumstances may cause some or all components of language to be shifted to the opposite side. Neural plasticity allows the default state to be altered.

An obvious challenge, therefore, is to understand plasticity with respect to different facets of language, how plasticity itself changes across the life span, and the circumstances under which it results in interhemispheric (rather than merely intrahemispheric) re-organization of the neural substrate. Three contrasting models of life-span changes in plasticity are depicted in Figure 24.2. Curve "a" represents the capacity for new learning, or fluid intelligence, which may begin its gentle decline before the end of the third decade. Curve "b" depicts a gradual decline in plasticity with respect to various aspects of language that could account for the clinical observations on which the Lenneberg (1967) hypothesis rested. The decline in plasticity between language onset and pubescence commonly was misinterpreted as progressive lateralization. Curve "c" depicts a rapid decrease in plasticity between the perinatal period and the first birthday. This pattern is consistent with evidence indicating that the right hemisphere can support relatively normal development of expressive language only if the left hemisphere is damaged during the prenatal or perinatal period.

Developmental invariance of lateralization, coupled with developmentally decreasing plasticity, has served as a broad heuristic device over the past few decades. Findings from a variety of experimental and clinical studies support this combination of concepts. To an increasing extent, structural and functional imaging techniques will enable researchers to ask more detailed and sophisticated questions about plasticity with respect to different components of language and the factors that determine whether some or all of those components will move to the opposite hemisphere following damage to the language-dominant hemisphere. Imaging studies will also be useful in addressing two other important questions that we have raised in this chapter: (1) the potential incongruence

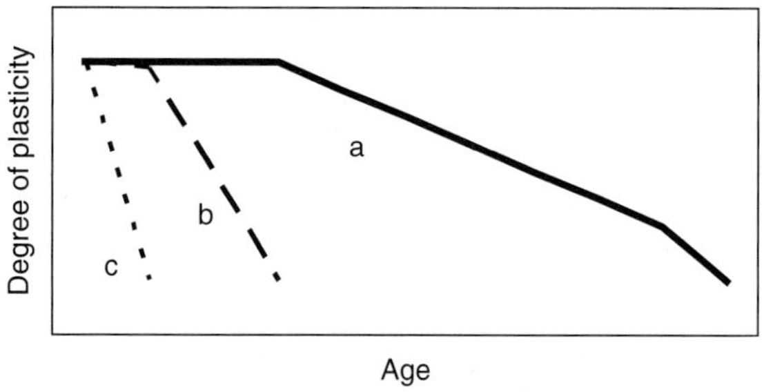

FIGURE 24.2 Curve "a" depicts a conventional model of plasticity, in which overall capacity for learning remains at a maximum for a number of years (perhaps until the end of the third decade), after which it begins to decline gradually. Curve "b", which may pertain to certain automatized phonological aspects of language, indicates a shorter period of maximal learning capacity (ending by pubescence), which is followed by a rapid decline in plasticity with respect to those aspects of language. Curve "c" depicts a decline in plasticity that begins prior to birth, and perhaps as soon as the neurons of the cerebral cortex begin to establish connections. This model corresponds to evidence of impaired expressive speech development in individuals who sustained left-hemisphere damage during the prenatal or perinatal stages.

between the cortical substrate of language and the subcortical mechanism that differentially activates the left and right hemispheres; and (2) the causes and consequences of individual differences in language lateralization and plasticity.

References

Annett, M. (2002). *Handedness and brain asymmetry: The right shift theory.* Hove, East Sussex, UK: Psychology Press.

Bates, E., Vicari, S., & Trauner, D. (1999). Neural mediation of language development: Perspectives from lesion studies of infants and children. In H. Tager-Flusberg (Ed.), *Neurodevelopmental disorders* (pp. 533–581). Cambridge, MA: MIT Press.

Best, C.T. (1988). The emergence of cerebral asymmetries in early human development: A literature review and a neuroembryological model. In D.L. Molfese & S.J. Segalowitz (Eds.), *Brain lateralization in children: Developmental implications* (pp. 5–34). New York: Guilford.

Buxhoeveden, D., & Casanova, M. (2000). Comparative lateralisation patterns in the language area of human, chimpanzee, and rhesus monkey brains. *Laterality, 5,* 315–330.

Crossley, M., & Hiscock, M. (1992). Age-related differences in concurrent-task performance of normal adults: Evidence for a decline in processing resources. *Psychology and Aging, 7,* 499–506.

Dehaene, S., Dupoux, E., Mehler, J., Cohen, L., Paulesu, E., & Perani, D. et al. (1997). Anatomical variability in the cortical representation of first and second language. *Neuroreport, 8,* 3809–3815.

Devlin, A.M., Cross, J.H., Harkness, W., Chong, W.K., Harding, B., Vargha-Khadem, F., & Neville, B.G.R. (2003). Clinical outcomes of hemispherectomy for epilepsy in childhood and adolescence. *Brain, 126,* 556–566.

Dorsaint-Pierre, R., Penhune, V.B., Watkins, K.E., Neelin, P., Lerch, J.P., Bouffard, M., & Zatorre, R.J. (2006). Asymmetries of the planum temporale and Heschl's gyrus: Relationship to language lateralization. *Brain, 129,* 1164–1176.

Esiri, M.M. (1994). Dementia and normal aging: Neuropathology. In F.A. Huppert, C. Brayne & D.W. O'Connor (Eds.), *Dementia and normal aging* (pp. 385–436). Cambridge, UK: Cambridge University Press.

Fernandes, M.A., Smith, M.L., Logan, W., Crawley, A., & McAndrews, M.P. (2006). Comparing language lateralization determined by dichotic listening and fMRI activation in frontal and temporal lobes in children with epilepsy. *Brain and Language, 96,* 106–114.

Flynn, J.R. (2006). Tethering the elephant: Capital cases, IQ, and the Flynn effect. *Psychology, Public Policy, and Law, 12,* 170–189.

Guttorm, T.K., Leppänen, P.H.T., Poikkeus, A.-M., Eklund, K.M., Lyytinen, P., & Lyytinen, H. (2005). Brain event-related potentials (ERPs) measured at birth predict later language development in children with and without familial risk for dyslexia. *Cortex, 41,* 291–303.

Hiscock, M. (1988). Behavioral asymmetries in normal children. In D.L. Molfese & S.J. Segalowitz (Eds.), *Brain lateralization in children: Developmental implications* (pp. 85–169). New York: Guilford.

Kinsbourne, M. (1984). . *Why is neuropsychology progressing so slowly? Contemporary Psychology, 29,* 793–794.

Knecht, S., Drager, B., Deppe, M., Bobe, L., Lohmann, H., Floel, A., Ringelstein, E.B., & Henningsen, H. (2000). Handedness and hemispheric language dominance in healthy humans. *Brain, 123,* 2512–2518.

Lenneberg, E. (1967). *Biological foundations of language.* New York: Wiley.

Liederman, J. (1983). Mechanisms underlying instability in the development of hand preference. In G. Young, S. Segalowitz, C. Corter & S. Trehub (Eds.), *Manual specialization and the developing brain* (pp. 71–92). New York: Academic Press.

Liederman, J. (1987). Neonates show an asymmetric degree of head rotation but lack an asymmetric tonic neck reflex asymmetry: Neuropsychological implications. *Developmental Neuropsychology, 3,* 101–112.

Luria, A.R. (1973). *The working brain: An introduction to neuropsychology.* New York: Basic Books. (B. Haigh, Trans.)

MacKain, K., Studdert-Kennedy, M., Spieker, S., & Stern, D. (1983). Infant intermodal speech perception is a left hemisphere function. *Science, 214,* 1347–1349.

Molfese, D.L., & Betz, J.C. (1988). Electrophysiological indices of the early development of lateralization for language and cognition, and their implications for predicting later development. In D.L. Molfese & S.J. Segalowitz (Eds.), *Brain lateralization in children* (pp. 171–190). New York: Guilford.

Molfese, D.L., & Molfese, V.J. (1997). Discrimination of language skills at five years of age using event-related potentials recorded at birth. *Developmental Neuropsychology, 13,* 135–156.

Morris, R., Bakker, D., Satz, P., & Van der Vlugt, H. (1984). Dichotic listening ear asymmetry: Patterns of longitudinal development. *Brain and Language, 22,* 49–66.

Segalowitz, S.J., & Berge, B.E.M. (1994). Functional asymmetries in infancy and early childhood: A review of electrophysiologic studies and their implications. In R.J. Davidson & K. Hugdahl (Eds.), *Brain asymmetry* (pp. 579–615). Cambridge, MA: MIT Press.

Segalowitz, S.J., & Chapman, J.S. (1980). Cerebral asymmetry for speech in neonates: A behavioral measure. *Brain and Language, 9,* 281–288.

Spreen, O., Risser, A.H., & Edgell, D. (1995). *Developmental neuropsychology.* New York: Oxford University Press.

Vallortigara, G. (2006). Cerebral lateralization: A common theme in the organization of the vertebrate brain. *Cortex, 42,* 5–7.

Van Gorp, W.G., Satz, P., & Mitrushina, M. (1990). Neuropsychological processes associated with normal aging. *Developmental Neuropsychology, 6,* 279–290.

Vargha-Khadem, F. (2001). Generalized versus selective cognitive impairments resulting from brain damage sustained in childhood. *Epilepsia, 42*(Suppl. 1), 37–40.

Witelson, S.F., Beresh, H., & Kigar, D.L. (2006). Intelligence and brain size in 100 postmortem brains: Sex, lateralization and age factors. *Brain, 129,* 386–398.

Further Readings

Bates, E., Vicari, S., & Trauner, D. (1999). Neural mediation of language development: Perspectives from lesion studies of infants and children. In H. Tager-Flusberg (Ed.), *Neurodevelopmental disorders* (pp. 533–581). Cambridge, MA: MIT Press.
A scholarly review and interpretation of the extensive literature concerning effects of early brain lesions on different facets of language development.

Hiscock, M., & Kinsbourne, M. (1995). Phylogeny and ontogeny of cerebral lateralization. In R.J. Davidson & K. Hugdahl (Eds.), *Brain asymmetry* (pp. 535–578). Cambridge, MA: MIT Press.
A detailed critique of the claim that the lateralized brain of the adult human represents the culmination of phylogenetic and ontogenetic progressions from simple and symmetric to complex and asymmetric nervous systems.

Martins, I.P. (1997). Childhood aphasias. *Clinical Neuroscience, 4,* 773–777.
A succinct review of the literature on childhood aphasia.

Van Gorp, W.G., Satz, P., & Mitrushina, M. (1990). Neuropsychological processes associated with normal aging. *Developmental Neuropsychology, 6,* 279–290.
This cross-sectional study of 57- to 85-year-olds indicates that age-related decline in neuropsychological test performance is related to demand for processing speed rather than to specific task content.

Interhemispheric Interaction in the Lateralized Brain

JOSEPH B. HELLIGE

Psychology Department, Loyola Marymount University, Los Angeles, CA, USA

ABSTRACT

The cortex of the human brain is divided into left and right hemispheres, each constituting a somewhat separate information processing system with its own abilities, propensities and biases. The two hemispheres typically coordinate their various activities without effort, leading to a sense of unity in language processing and other domains. After a brief overview of the nature of functional hemispheric asymmetry, with special emphasis on processes related to language, the present chapter considers the special challenges that laterality poses and the mechanisms of interhemispheric interaction that address those challenges. Special consideration is given to the costs and benefits of distributing processing across both hemispheres as tasks become more or less demanding of the resources of a single hemisphere. The chapter ends by reviewing individual differences in the efficiency of interhemispheric interaction and the relationship to language and other activities.

25.1. INTRODUCTION

A prominent characteristic of human brains is division of the cerebral cortex into left and right hemispheres, each constituting a somewhat separate information processing system with its own abilities, propensities and biases. Among the most dramatic hemispheric differences are those related to language. For example, language disorders are far more frequent and more serious after damage to certain areas of the left hemisphere than after damage to corresponding areas of the right hemisphere. In particular, left-hemisphere dominance for such things as speaking, reading words, using grammar and understanding word

meaning has been established for some time. Despite this general picture of left-hemisphere superiority for language, there is also evidence of contributions from the right hemisphere. For example, functional neuroimaging studies typically show activation in many areas of both hemispheres as individuals perform a variety of language tasks. Indeed, when "language" is viewed more broadly for the purpose of communication, aspects of left-hemisphere dominance may even be complemented by right-hemisphere dominance for such things as intonation, emotional tone and building coherent meaning across sentences. Remarkably, the left and right hemispheres typically coordinate their various activities without effort, leading to a sense of unity in language processing and other domains. The present chapter outlines how this coordination takes place and considers more generally the variety of interhemispheric interactions in our lateralized brain.

The chapter begins with a brief overview of the nature of functional hemispheric asymmetry, with special emphasis on processes related to language. This is followed by a discussion of the advantages of having a lateralized brain and of the special challenges that laterality creates for interhemispheric interaction. Against this backdrop, I then review mechanisms of interhemispheric interaction, including discussion of the costs and benefits of distributing processing across both hemispheres as tasks become more or less demanding. Though there is a good deal of consistency from person to person, there is also sufficient individual variation in the efficiency of interhemispheric interaction to have important consequences for language and other activities. Thus, the chapter ends by considering several dimensions of individual variation and looking toward future issues related to interhemispheric interaction.

25.2. FUNCTIONAL HEMISPHERIC ASYMMETRY

Hemispheric asymmetry or laterality is ubiquitous across information processing domains and across contemporary species. For example, within humans, functional hemispheric asymmetries are found in such varied domains as motor control, perception, memory, emotion and language. During the last 30 years, laterality studies have demonstrated that functional asymmetries are also ubiquitous across other species, with some of those asymmetries being similar to those found in humans. In fact, it has been hypothesized that all contemporary vertebrate groups have inherited a basic pattern of laterality from a common chordate ancestor.

In order to understand issues related to interhemispheric interaction, it is useful to consider certain general properties of contemporary hemispheric asymmetry. In addition to being ubiquitous across information processing domains, most hemispheric asymmetries in humans and other species are subtle rather than being all or none. That is, both sides of the brain typically have at least some competence to perform a task or to utilize a specific process, though one side or the other may be superior, preferred or dominant in some other measurable way. For example, for many people the right hand is better at a variety of fine-grained motor activities, but the left hand is not completely without competence for those same activities. In the visual domain, the left hemisphere is superior for processing small local details and the right hemisphere is superior for processing global configuration, but each hemisphere can handle both local and global information to a reasonable extent (see Figure 25.1). One exception to this property of subtlety may be overt speech, which is produced exclusively by the left hemisphere in most people. Even in this case, however, the right hemisphere is capable of producing speech in individuals who are born without a left hemisphere or if the left hemisphere is removed at a sufficiently young age.

Evidence from individuals with unilateral brain injury, from split-brain patients and from neurologically intact individuals has identified additional aspects of left-hemisphere dominance for language, though in most of these other cases the right hemisphere is not completely without ability. The left hemisphere is typically better than the right for speech perception and other phonetic tasks such as rhyming and naming printed words and non-words. The left hemisphere is also dominant for such things as determining whether a string of letters spells a word and for dealing with syntactic (grammatical) information in both the production and understanding of language.

In something of a complementary manner, the right hemisphere is thought to be superior to the left for a variety of additional communication-related factors. For example, studies with both brain-injured and neurologically intact individuals demonstrate right-hemisphere superiority for the production and perception of prosody or intonation in speech. Thus, the speech of patients with right-hemisphere injury is often described as monotonic, lacking vocal inflection and emotion. Such patients also have a difficult time identifying such things as emotion from the vocal inflections of others. Right-hemisphere damage also interferes with the ability to understand certain kinds of jokes that depend on the ability to build a context across sentences and to process the metaphoric meanings of words such as warm and cold (which can refer to feelings as well as to temperature). (For further discussion of language-related laterality, see Hellige, 2001; Banich, 2004.)

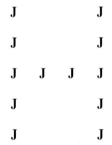

FIGURE 25.1   A large letter H (the global level) made up of small letter J's (the local level). When patients with left-hemisphere damage try to reproduce such figures, the intact right hemisphere correctly produces the global pattern (the large H), but ignores the small local details (the small J's). In contrast, in patients with right-hemisphere damage, the intact left hemisphere correctly produces the local detail (several small J's) but does not arrange them into the correct global pattern. In neurologically intact individuals, global and local levels are processed better when flashed briefly to the left visual field (right hemisphere) and right visual field (left hemisphere), respectively, and functional neuroimaging studies show that attending to global versus local levels produce greater activation in right and left hemispheres, respectively.

25.3. ADVANTAGES AND CHALLENGES OF HEMISPHERIC ASYMMETRY

The ubiquity of laterality suggests that it confers sufficient advantages to have evolved early and to have been sustained and exploited for millennia in many widely differing species. A long-standing idea is that hemispheric asymmetry is advantageous as a way of packing more abilities into a particular amount of neural tissue than would be the case if the two hemispheres were functionally identical. Thus, as one hemisphere becomes specialized for language, the other is freed to become specialized for other activities like localizing stimuli in space. Something akin to this idea of increased processing capacity can be extended to asymmetries that are subtle in the way discussed earlier.

Indications of hemispheric asymmetry in a number of domains suggest that left- and right-hemisphere superiorities complement each other, like two pieces of a puzzle (Hellige, 2001). In terms of speech perception, for example, left-hemisphere dominance for identifying such things as stop consonants (b, d, g, p, t, k) may arise because the left hemisphere is well adapted for processing the rapid acoustic changes that differentiate one such consonant from others. By way of contrast, right-hemisphere dominance for identifying emotional tone may arise because the right hemisphere is well adapted for processing acoustic changes over longer temporal intervals (Ivry & Robertson, 1998). Another language-related example is provided by hemispheric differences in semantic processing. When a word is presented, the left hemisphere very quickly restricts access to one possible meaning (either the dominant meaning or the meaning most consistent with the preceding words), whereas the right hemisphere maintains activation of many possible meanings for a longer period of time (Chiarello, 2003; see also Box 25.1).

Box 25.1   When is a bank not a bank? Or, what's right and what's left

"He did as he was told, put all of the money into a plain brown paper bag, and took the bag to the right bank, only to find the abductors waiting across the Seine river on the south side. 'Never again,' he mused 'will I forget that right and left refer to the north and south sides of the river rather than the other way round.' You just gotta love Paris." If you are like most people, when you read the word "bank" you immediately thought of an institution that houses money. That is, in fact, the most common meaning of the word "bank." Moreover, prior references to a bag of money created a context that was also biased toward that meaning. However, by the time you reached the end of the passage, you probably realized that the reference was actually to the bank alongside a river – the Right Bank of the Seine in Paris to be specific. There is good reason to believe that both brain hemispheres contributed to your performance and likely interacted with each other at various points along the way. When encountering an ambiguous word, the left hemisphere quickly restricts activation to the dominant meaning or the meaning most consistent with the context that has come before. This is typically very useful in quickly deciphering the intended meaning. In contrast, when encountering the same word, the right hemisphere maintains activation of multiple meanings, distant semantic relations and such things as metaphoric interpretation. This is also very useful in quickly resolving the confusion that occurs when information that comes in later is in conflict with the first meaning considered. By the way, chances are you also switched at some point from interpreting "right" as the opposite of "wrong" to "right" as the opposite of "left!" (see Chiarello, 2003).

Chiarello, C. (2003). Parallel systems for processing language: Hemispheric complementarity in the normal brain. In M.T. Banch & M. Mack (Eds.), *Mind, brain, and language: Multidisciplinary perspectives* (pp. 229–250). Lawrence Erlbaum Associates: Mahwah, NJ.

The computational network that is best adapted to one of two complementary processes may be different in subtle ways from the computational network that is best adapted to the other. An instructive example comes from studies of localizing visual stimuli in space. Specifically, the right hemisphere is dominant for processing information about metric distance or what is often referred to as coordinate spatial relations (e.g., determining the distance between a line and a dot), whereas the left hemisphere is dominant for processing information about categorical spatial relations (e.g., whether a dot is above or below a line) (Hellige, 2001, 2006; Kosslyn, 2006). In computer modeling studies, neural network simulations that were constructed to have relatively large overlapping "receptive fields" computed coordinate spatial information better than did neural network simulations that were constructed to have relatively small, non-overlapping "receptive fields," whereas exactly the reverse was found for the computation of categorical spatial relationships. Furthermore, these same simulation studies found that both categorical and coordinate performance was better when the networks were split so that some units contributed only to categorical processing and others contributed only to coordinate processing than when all units contributed to both types of spatial processing. Results like these illustrate how it can be advantageous for complementary processes to be segregated from each other so as to reduce maladaptive crosstalk. There is considerable evidence that two tasks performed simultaneously interfere with each other more when they require specialized abilities of the same hemisphere than when the processing load can be spread more evenly across both hemispheres. Thus, it is likely to be advantageous to have mutually inconsistent or complementary processes at least partially segregated into different hemispheres. Consistent with this possibility, greater cerebral lateralization for two complementary tasks in the domestic chick is associated with enhanced ability to perform those tasks simultaneously: finding food (for which the left hemisphere is typically dominant) and being vigilant for predators (for which the right hemisphere is typically dominant) (Rogers et al., 2004).

The foregoing considerations suggest that laterality may have emerged and been sustained as a way of packing more abilities into a finite amount of neural tissue and as a way of reducing maladaptive interaction between partially inconsistent, complementary neural computations. Along with these advantages, laterality poses challenges associated with the need for communication and collaboration between the two hemispheres. To some extent, the challenges are similar to those involved in connecting cortical areas within a single hemisphere. However, there are important differences. For one thing, each hemisphere is a relatively complete "brain" unto itself, with a broad range of abilities. Without appropriate interhemispheric interaction, this could lead to conflicts, with each hemisphere trying to seize control of processing to direct action toward different goals. Interhemispheric interaction is also demanded by many naturally occurring circumstances in which each hemisphere receives only a portion of the information that is required to identify a stimulus or perform a task. For example, in vision, the left half of centrally fixated objects and words projects directly to the right hemisphere and the right half projects directly to the left hemisphere so that integrating information across the visual midline must involve interaction of the two hemispheres. Interaction is also required when different processing components of a task are accomplished primarily or exclusively by different hemispheres, as when listening to your favorite comedienne tell her version of the "aristocrats" joke, during which the left hemisphere is likely to take the lead for phonetic, syntactic and certain semantic aspects of processing and the right hemisphere is likely to take the lead for processing intonation, prosodic cues to emotion and contextual clues to meaning. In addition, the fact that many functional asymmetries are subtle in the manner discussed earlier also raises the possibility that, when needed, tasks or processes that are typically performed by only one hemisphere can recruit help from the other hemisphere and thereby improve performance.

25.4. MECHANISMS OF INTERHEMISPHERIC INTERACTION

The largest fiber tract connecting the left and right hemispheres is the corpus callosum, consisting of between 200 and 800 million axon fibers. In general, anterior portions of the corpus callosum connect frontal and premotor regions of the two hemispheres, middle portions of the corpus callosum connect motor and somatosensory regions of the two hemispheres and posterior regions of the corpus callosum connect temporal, postparietal and peristriate regions of the two hemispheres (see Figure 25.2). Integrity of the corpus callosum is particularly important for the transfer of explicit, conscious information about the identity of stimuli. The corpus callosum has also been hypothesized to help regulate the state of asymmetric arousal between the two hemispheres and to serve as a kind of barrier between the hemispheres, minimizing potentially maladaptive cross-talk between the processes for which each hemisphere is dominant, though the specific mechanisms by which the corpus callosum accomplishes these different things are not yet fully understood (for more on the corpus callosum, see Hellige, 2001; Banich, 2004; Bloom & Hynd, 2005, and suggestions for further readings).

In addition to the corpus callosum, the two hemispheres are connected via the anterior commissure and via a number of subcortical pathways. Studies with split-brain patients, whose hemispheres can no longer communicate via the surgically severed corpus callosum, indicate that implicit (unconscious) information about stimulus identity, information about

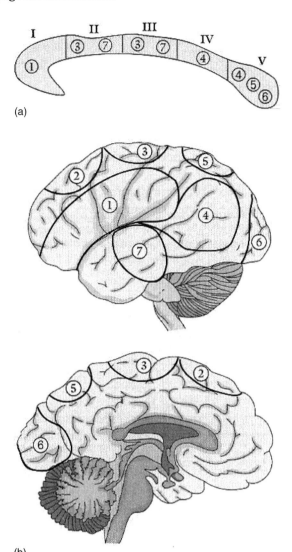

(a)

(b)

FIGURE 25.2 (a) Diagram of the corpus callosum. The number shown in each section indicates the brain region depicted in (b) (top, left hemisphere, lateral view; bottom, left hemisphere, midsagittal view), which are connected through that section of the corpus callosum. The connections occur in topographic manner: anterior sections (I) of the corpus callosum connect anterior sections of the brain (Region 1, which is frontal), middle sections of the callosum (II and III) connect brain regions that are more central (Regions 3 and 7) and posterior sections of the callosum (IV and V) connect posterior sections of the brain (Regions 4, 5 and 6). Some brain regions, such as the frontal region labeled 2, have few, if any, callosal connections. (Banich, M.T. (2004). *Cognitive neuroscience and neuropsychology* (2nd edn). Boston: Houghton Mifflin, © 2004 by Houghton Mifflin Company. Used with permission.)

the location of objects in space, information about categories to which a stimulus belongs and contextual information about an object can be transmitted subcortically. Subcortical structures may also play a role in permitting each hemisphere to receive information about decisions made by the other hemisphere (for more, see Banich, 2004, and further readings).

This extensive system of callosal and subcortical connections allows the two hemispheres to bind the various aspects of language into a coherent experience, even though different aspects depend primarily on processing in one hemisphere. However, although the two hemispheres are capable of sharing many types of information, ranging from sensory input to complex decisions, cooperation at all levels does not necessarily take place all of the time. Thus, it is important to consider when the benefits of distributing processing across both hemispheres might be outweighed by the costs of interhemispheric transfer.

25.4.1. Costs and Benefits of Interhemispheric Interaction

In order to examine the costs and benefits of interhemispheric interaction, it is instructive to compare performance of the same task under conditions that demand interhemispheric collaboration and conditions that do not. Many experiments of this sort take advantage of the fact that the visual projection pathways of humans are such that information from the right visual half-field projects directly to the left hemisphere and information from the left visual half-field projects directly to the right hemisphere. Thus, simultaneous lateralization of two stimuli to the same or opposite visual half-fields can determine whether interhemispheric collaboration is needed to determine whether the two stimuli match according to some criterion (see Figure 25.3). An important conclusion to emerge is that distributing information becomes more beneficial as tasks become more complex or demanding. That is, for very simple tasks (e.g., determining whether two uppercase letters are physically identical), performance is typically better when both matching items are presented to the same visual field and hemisphere than when one is presented to each visual field and hemisphere (a within-hemisphere advantage), suggesting that the costs of interhemispheric transfer outweigh the potential benefits of interhemispheric interaction. However, when processing demands become a bit greater (e.g., indicating whether two letters of different case have the same name), performance is typically better when one matching item is presented to each visual field and hemisphere than when they are both presented to the same visual field and hemisphere (an across-hemisphere advantage), suggesting that the benefits of spreading the processing load across both hemispheres now outweigh the costs of interhemispheric transfer. Dividing relevant input between the two hemispheres is also advantageous if it permits the hemispheres to engage in mutually inconsistent perceptual processes (e.g., one hemisphere restricts processing to upright letters while the other hemisphere restricts processing to inverted letters). Results such as these suggest why it may be advantageous to lateralize complementary or mutually

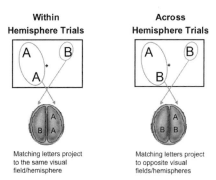

Within Hemisphere Trials

Across Hemisphere Trials

Matching letters project to the same visual field/hemisphere

Matching letters project to opposite visual fields/hemispheres

FIGURE 25.3 Example of an experimental paradigm used to investigate the costs and benefits of interhemispheric interaction. The task is to indicate whether the letter on the bottom matches either of the top two letters, which are always different from each other. The three letters are flashed on the viewing screen for a fraction of a second while the observer's eyes are fixated on the center cross. Thus, letters from each side of the screen project directly to only one brain hemisphere. Two types of matching trials are illustrated above. On within-hemisphere trials, the two letters that match are presented to the same side of the fixation cross and, thus, to the same brain hemisphere. Consequently, the match may be made without interhemispheric interaction. On across-hemisphere trials, the two letters that match are presented to opposite sides of the fixation cross and, thus, to different brain hemispheres. Consequently, interhemispheric interaction is required to make a match. If the costs of interhemispheric interaction outweigh the benefits of distributing processing across both hemispheres, performance should be better on within-hemisphere trials than on across-hemisphere trials. Exactly the opposite result should be obtained if the benefits of distributing processing across both hemispheres outweigh the costs of interhemispheric interaction.

exclusive aspects of language and communication to opposite hemispheres. (For more discussion and review, see Hellige, 2001; Banich, 2004; Patel & Hellige, 2007.)

A key idea to emerge from studies of interhemispheric interaction is that it becomes advantageous to spread processing across both cerebral hemispheres to the extent that a specific task or set of tasks overloads the processing resources of a single hemisphere. This possibility receives additional support from a set of computerized models that simulate two interconnected processing systems or "hemispheres" performing physical identity and name identity letter-matching tasks. In these models, an across-hemisphere advantage emerged spontaneously for the simulated name-identity task but not for the simpler physical-identity task, for which there tended to be a within-hemisphere advantage. Moreover, the complex task tended to involve both "hemispheres" even when the two matching stimuli were projected to the same side, a result similar to that found in an functional magnetic resonance imaging (fMRI) investigation of letter-matching tasks (Monaghan & Pollmann, 2003).

25.4.2. Mixing Stimuli That Are Processed in Different Cortical Areas

Though within-hemisphere overload becomes more likely with increases in task complexity and task difficulty, neither difficulty nor complexity *per se* provides a sufficient explanation of the extent to which a task sufficiently overloads the resources of a single hemisphere so as to produce an across-hemisphere advantage. Consider experiments which mix stimulus formats that are processed via different cortical routes, even though they may eventually lead to a common abstract code. In one such experiment, observers were required to indicate whether two stimuli represented the same numeric quantity. Matching stimuli consisted of two digits, two dice-like dot patterns or a digit and a dice-like pattern (mixed-format condition). Despite added difficulty and complexity, the mixed-format condition produced a significant within-hemisphere advantage rather than the across-hemisphere advantage that would be expected for a task that, like matching different-case letters on the basis of name, cannot be performed on the basis of physical identity (Patel & Hellige, 2007). This counterintuitive within-hemisphere advantage persisted for the mixed-format condition even when the task required a more difficult and complex decision based on whether the two matching stimuli fell into the same magnitude category (small (1,2), medium (3,4), large (5,6)). Similar within-hemisphere advantages have been obtained in experiments that required observers to indicate whether items represented by a word and a picture came from the same semantic category (e.g., the word "coat" and a picture of a shoe) (Koivisto & Revensuo, 2003) and in experiments that required observers to indicate whether a word and a cartoon face depicted the same emotion (e.g., the word "sad" and a cartoon face with a sad expression; see Box 25.2). What all of these experiments have in common is the mixing of stimulus formats that are processed via different cortical routes, even within one hemisphere. Thus, a critical factor in determining the relative costs and benefits of interhemispheric interaction appears to be the extent to which multiple stimuli presented to the same hemisphere must compete for the same cortical route. In a sense, just as the division of labor between the two hemispheres can increase the overall processing capacity of the brain, the division of labor among more areas within a hemisphere can increase the processing capacity of that hemisphere. This may explain why such things as phonetic analysis, processing of syntax and extraction of word meaning, which depend on distinctive cortical areas, can all take place without interference within the normal left hemisphere.

25.4.3. Bihemispheric Redundancy Gain

Another experimental paradigm that has proven useful in the study of interhemispheric interaction compares performance on unilateral visual half-field trials with performance on redundant bilateral trials, on which exactly the same information is presented to both hemispheres. In a number of experiments, performance is found to be better on redundant

---

Box 25.2 Mixing stimulus formats: when are two hemispheres better than one?

Do the two stimuli shown below in column A refer to the same emotion? How about the two stimuli shown in column B? Column C? In each case, the answer is "yes." My colleagues and I have used stimuli like these to examine how mixing different stimulus formats influences the costs versus benefits of distributing processing across both brain hemispheres. When both items were of the same type (columns A and B), performance was better if one of the two matching stimuli was presented to each hemisphere than if both matching stimuli were presented to the same hemisphere. Despite being even more complex, when one item was a word and the other was a cartoon face (column C), the results were quite different. In fact, performance was better if both matching stimuli were presented to the same hemisphere than if one was presented to each hemisphere. Why? Uppercase and lowercase words are processed in common cortical areas, as are different cartoon faces. Thus, simultaneous processing of two words or two faces presented to the same hemisphere may overload that hemisphere, so that the benefits of spreading processing across both hemispheres outweigh the costs of interhemispheric interaction. In contrast, even though they may refer to the same emotion, words and faces are processed along different cortical routes, even when they are presented to the same hemisphere. Thus, within-hemisphere competition is minimized, so that any benefits of spreading processing across both hemispheres are overshadowed by the costs of interhemispheric interaction. Just as the division of labor between the two hemispheres can increase the overall processing capacity of the brain, the division of labor among more areas within a single hemisphere can increase the processing capacity of that hemisphere.

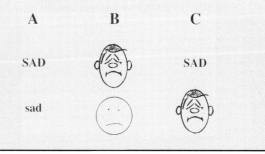

---

bilateral trials (on which both hemispheres can contribute) than on either left or right visual field unilateral trials (a phenomenon referred to as bilateral or bihemispheric redundancy gain). With respect to language, such bihemispheric redundancy gain has been found for lexical decision tasks and for the identification of printed words and pronounceable non-words, for handwritten cursive, and for a variety of languages including English, German, Hebrew, Arabic and Urdu (Adamson & Hellige, 2006; Hellige & Adamson, 2006; Hellige & Adamson, 2007). Factors that influence the balance of costs and benefits associated with interhemispheric interaction also seem to influence the magnitude of bihemispheric redundancy gain. For example, mixing stimulus formats on redundant bilateral trials (e.g., digits to one visual field and corresponding dice-like dot patterns to the other) reduces the amount of bilateral gain, even though the stimuli on both sides represent exactly the same abstract numeric information. In view of the fact that many if not most callosal fibers connect homologous regions of the two hemispheres, such results suggest that some of the interhemispheric interaction responsible for bilateral redundancy gain is greatest for homologous areas of the two hemispheres (see Marks & Hellige, 2003).

25.5. INDIVIDUAL VARIATION

Individuals differ in such things as the size and microstructure of the corpus callosum, the speed with which information about even simple stimuli can be transmitted through the corpus callosum and the efficiency of interhemispheric coordination for more complex tasks. These aspects of interhemispheric interaction appear to be related in important ways. For example, an increase in the size of the corpus callosum is thought to reflect greater potential for interhemispheric interaction. Faster transmission speed through the corpus callosum may also facilitate interhemispheric collaboration (e.g., Cherbuin & Brinkman, 2006). Further, individual differences in interhemispheric connectivity (as reflected by the microstructure of the corpus callosum) are related to language lateralization (as reflected by word generation) such that interconnectivity is stronger or faster in strongly left-dominant individuals compared to moderately left-dominant, bilateral or moderately right-dominant individuals (see Westerhausen et al., 2006).

25.5.1. Variation Across the Life Span

In considering individual variation in interhemispheric interaction, it is useful to begin with changes across the life span. The corpus callosum begins to become myelinated between the ages of 3 and 7 years, with myelination not being completed until about the age of puberty. (Myelin is a fatty sheath around neurons that greatly increases conduction speed.) Thus, at very young ages, the corpus callosum is not fully functional, making it difficult for young children to perform tasks that require information to be shared across the hemispheres. Above the age of 6 or 7 years, however, children, like adults, show an across-hemisphere advantage for name-identity letter matching but not for simpler physical-identity letter matching (see Banich et al., 2000). At the same time, there is variation among children in such

things as the rate of myelination and in other aspects of callosal connectivity. Efficient callosal function may be particularly important during early stages of learning to read, as certain visual processes for which the right hemisphere is dominant must be coordinated with language processes for which the left hemisphere is dominant (see Banich, 2004).

At the other end of the life span, there is anatomical evidence of age-related changes in the corpus callosum, including some evidence of a reduction in size. Of course, there are also age-related changes within the hemispheres, which complicates behavioral predictions. For example, if the decline in callosal connectivity is great relative to within-hemisphere decline, then older individuals should show a larger within-hemisphere advantage than young adults for a simple physical-identity-matching task and a smaller across-hemisphere advantage than young adults for a more complex name-identity-matching task. On the other hand, if age-related declines within the hemispheres are large relative to the decline in callosal connectivity, then exactly the opposite might be expected. Empirical results have been mixed, and suggest that task complexity may be an important moderating factor. For example, older individuals do seem to show a larger within-hemisphere advantage than young adults for very simple physical-identity-matching tasks (e.g., Reuter-Lorenz et al., 1999; Cherry et al., 2005; Reuter-Lorenz & Mikels, 2005), consistent with an age-related decline in callosal connectivity. For a more difficult name-identity-matching task, however, the results are less consistent. Reuter-Lorenz et al. found a larger across-hemisphere advantage for older individuals than for young adults. Cherry et al. found a similar difference for response accuracy, but exactly the opposite difference for reaction time. It may be that, as tasks become sufficiently complex, age-related declines within the hemispheres make it more effective to spread the processing load across both hemispheres, despite some age-related decline in callosal connectivity.

25.5.2. Cognition and Memory

There are a number of indications that interhemispheric interaction may be associated with individual differences in cognition and memory. For example, enhanced interhemispheric interaction characterizes the brains of mathematically gifted adolescents compared with adolescents and college students of average mathematical ability (Singh & O'Boyle, 2004). Whether similar findings characterize youth who are precocious in other domains (including language) remains to be determined.

Increased interhemispheric interaction also appears to be associated with better retrieval of episodic memories. Episodic memory refers to memory for personally experienced events occurring at specific times and places. Studies of brain activity indicate that encoding (i.e., storing) of episodic memories is associated with increased prefrontal activity in the left hemisphere whereas retrieving of episodic memories is associated with prefrontal activity in the right hemisphere. The fact that opposite hemispheres are associated with encoding and retrieval makes it likely that right-hemisphere retrieval mechanisms interact with memory traces encoded by the left hemisphere. Studies of individual variation have provided an important way of studying the association between interhemispheric interaction and episodic memory. The logic of such studies is based on findings that mixed-handedness (compared to strong right-handedness) is associated with increased size of the corpus callosum and with evidence suggesting that a larger corpus callosum is associated with greater interhemispheric interaction in neurologically normal individuals. With this in mind, it has

---

Box 25.3   Interhemispheric interaction and childhood amnesia

What is your earliest childhood memory? Not an event in your childhood that others have told you about, but one you actually remember. No matter how hard we try, we cannot remember events from the earliest years of our lives, a phenomenon referred to as childhood amnesia. Memory for personally experienced events (referred to as episodic memory) is hypothesized to rely on interhemispheric interaction, as right-hemisphere retrieval mechanisms operate on memory traces encoded primarily by the left hemisphere. Thus, individuals with more efficient interhemispheric interaction might be expected to show an earlier offset of childhood amnesia; that is, to recall events from a younger age than individuals with less efficient interhemispheric interaction. To test this possibility, Christman *et al.* (2006) elicited early childhood memories from college students who were either strongly right-handed or mixed-handed. They chose this classification because non-right-handedness is thought to be associated with increases in the size of the corpus callosum and with increased interhemispheric interaction. In fact, Christman *et al.* found that non-right-handers reported earlier episodic memories than did strong right-handers (the accuracy of those memories and the age at which the events occurred was verified by the participants' parents). In a second experiment, making bilateral saccadic eye movements (moving the eyes back and forth) for 30 s, which has been shown to momentarily enhance interhemispheric interaction, also led to reports of earlier episodic memories. Because eye movements made in young adulthood could not have affected the original encoding of childhood memories, interhemispheric interaction seems to have its effect by increasing the ability to retrieve those memories.

Christman, S.D., Propper, R.E., & Brown, T.J. (2006).Increased interhemispheric interaction is associated with earlier offset of childhood amnesia. *Neurospychology, 20,* 336–354.

---

been shown that mixed-handers, relative to strong right-handers, have better memory for both real-world and laboratory events, are less prone to false memories and show an earlier offset of childhood amnesia (i.e., they are able to recall events from younger ages; see Box 25.3). (For more review and discussion, see Christman *et al.*, 2004; Propper *et al.*, 2005; Christman *et al.*, 2006.)

25.5.3. Cognitive Deficits and Emotional Disorders

Less efficient interhemispheric interaction is also hypothesized to be associated with a variety of cognitive deficits and emotional disorders. For example, a number of studies have reported areas of the corpus callosum to be smaller in individuals with developmental dyslexia, but there are also results to the contrary (for discussion, see Bloom & Hynd, 2005). Though it is not yet possible to reconcile the discrepant results, it may be that callosal connectivity is more important during the early stages of learning to read than it is for experienced readers or that only a subset of dyslexics has problems that relate to interhemispheric interaction. Reduced interhemispheric interaction, as reflected in corpus callosum morphology and in behavioral studies, is also thought to be associated with such disorders as schizophrenia (see Mohr *et al.*, 2000; Caligiuri *et al.*, 2005) and Tourette's syndrome (see Plessen *et al.*, 2004), both of which are also associated with problems related to verbalization and language. Though there is evidence of corpus callosum involvement in patients with Alzheimer's disease, behavioral studies show comparable decline on within- and across-hemisphere conditions, perhaps because Alzheimer's disease produces a similar breakdown of cortical connections within and across hemispheres (Reuter-Lorenz & Mikels, 2005).

25.5.4. Gender

Though there are some indications of corpus callosum differences in men and women, many of the results are difficult to interpret in view of sex differences in overall brain size and in view of what may be complex and inconsistent interactions between sex and handedness. Behavioral studies of functional interhemispheric interaction have not produced consistent differences between men and women, though there is some indication that the efficiency of callosal transfer in women varies as the level of estradiol and progesterone vary over the menstrual cycle (Hausmann *et al.*, 2006).

The foregoing examples illustrate the range of individual variation in efficiency of interhemispheric interaction. To be sure, such studies shed light on the biological and behavioral mechanisms that make one person different from another or one group different from another. At the same time, the study of individual variation is an important converging technique that provides a way of testing theories about the nature and extent of interhemispheric interaction more generally. As such, investigations of individual variation will continue to be an important tool for learning more about brain-behavior relations.

25.6. CHALLENGES AND FUTURE DIRECTIONS

A great deal of progress has been made in understanding the neural mechanisms that underlie language and communication. Indeed, cognitive neuroscience has done a very good job of taking the brain apart by identifying specific processing modules. The rate of progress has accelerated with interdisciplinary approaches that attempt to combine different levels of analysis, including more sophisticated *in vivo* measures of brain structure, more clever behavioral manipulations, more sophisticated computational models and an increasing ability to put all of this information together. Modern functional imaging techniques make it clear that most tasks involve neural networks dispersed across both brain hemispheres. A significant challenge concerns the manner in which the different elements of these neural networks fit together and coordinate their activity – at both biological and behavioral levels. As outlined in this chapter, the need to understand integration across areas is particularly acute with respect to the two very general information processing systems characterized by the left and right hemispheres. Though understanding interhemispheric interaction poses continued challenges, the expanding array of tools holds promise for resolving significant issues. The next several years are likely to see the development of imaging techniques that provide even more detail about the microstructure of the corpus callosum and about the flow of activation across different regions of the corpus callosum as individuals perform tasks in real time. This, along with clever behavioral manipulations and modeling techniques will permit an even more detailed analysis of individual variation, addressing such important questions as the relationship between laterality and interhemispheric interaction and about the advantages of having an appropriately lateralized brain. The existence of laterality in so many animal species, some of which have good interhemispheric communication and some of which have virtually none, will also provide important converging tests among theoretical alternatives. All things considered, advances in these various areas provide reason for optimism as we put the brain back together again.

References

Adamson, M.M., & Hellige, J.B. (2006). Hemispheric differences for identification of words and nonwords in Urdu-English bilinguals. *Neuropsychology, 20,* 232–248.

Banich, M.T. (2004). Cognitive neuroscience and neuropsychology (2nd edn). Boston, MA: Houghton Mifflin.

Banich, M.T., Passarotti, A.M., & Janes, D. (2000). Interhemispheric interaction during childhood: I. Neurologically intact children. *Developmental Neuropsychology, 18,* 33–51.

Bloom, J.S., & Hynd, G.W. (2005). The role of the corpus callosum in interhemispheric transfer of information: Excitation or inhibition? *Neuropsychology Review, 15,* 59–71.

Caligiuri, M.P., Hellige, J.B., Cherry, B.J., Kwok, W., Lulow, L.L., & Lohr, J.B. (2005). Lateralized cognitive dysfunction and psychotic symptoms in schizophrenia. *Schizophrenia Research, 80,* 151–161.

Cherry, B.J., Adamson, M., Duclos, A., & Hellige, J.B. (2005). Aging and individual variation in interhemispheric interaction and hemispheric asymmetry. *Aging, Neuropsychology, and Cognition, 12,* 316–339.

Cherbuin, N. & Brinkman, C. (2006). Efficiency of callosal transfer and hemispheric interaction. *Neuropsychology, 20,* 178–184.

Chiarello, C. (2003). Parallel systems for processing language: Hemispheric complementarity in the normal brain. In M.T. Banch & M. Mack (Eds.), *Mind, brain, and language: Multidisciplinary perspectives* (pp. 229–250). Lawrence Erlbaum Associates: Mahwah, NJ.

Christman, S.D., Propper, R.E., & Dion, A. (2004). Increased interhemispheric interaction is associated with decreased false memories in a verbal converging semantic associates paradigm. *Brain and Cognition, 56,* 313–319.

Christman, S.D., Propper, R.E., & Brown, T.J. (2006). Increased interhemispheric interaction is associated with earlier offset of childhood amnesia. *Neurospychology, 20,* 336–354.

Hausmann, M., Tegenthoff, M., Sanger, J., Janssen, F., Gunturkun, O., & Schwenkreis, P. (2006). Transcallosal inhibition across the menstrual cycle: A TMS study. *Clinical Neurophysiology, 117,* 26–32.

Hellige, J.B., (2001). Hemispheric asymmetry: What's right and what's left (Paperback edition). Cambridge, MA: Harvard University Press.

Hellige, J.B. (2006). Evolution of brain lateralization in humans. *Cognition, Brain, Behavior, 10,* 211–234.

Hellige, J.B., & Adamson, M.M. (2006). Laterality across the world's languages. In K. Brown (Ed.), *Encyclopedia of language and linguistics* (2nd edn, pp. 709–718). Elsevier: Oxford.

Hellige, J.B., & Adamson, M.M. (2007). Hemispheric differences in processing handwritten cursive. *Brain and Language, 102,* 215–227.

Ivry, R.B., & Robertson, L.C. (1998). *The two sides of perception.* Cambridge, MA: MIT Press.

Koivisto, M., & Revensuo, A. (2003). Interhemispheric categorization of pictures and words. *Brain and Cognition, 52,* 181–191.

Kosslyn, S.M. (2006). You can play 20 questions with nature and win: Categorical versus coordinate spatial relations as a case study. *Neuropsychologia, 44,* 1519–1523.

Marks, N.L., & Hellige, J.B. (2003). Interhemispheric interaction in bilateral redundancy gain: Effects of stimulus format. *Neuropsychology, 17,* 578–593.

Mohr, B., Pulvermüller, F., Cohen, R., & Rockstroh, B. (2000). Interhemispheric cooperation during word processing: Evidence for callosal transfer of dysfunction in schizophrenic patients. *Schizophrenia Research, 46,* 231–239.

Monaghan, P., & Pollmann, S. (2003). Division of labor between the hemispheres for complex but not simple tasks: An implemented connectionist model. *Journal of Experimental Psychology: General, 132,* 379–399.

Patel, U., & Hellige, J.B. (2007). Benefits of interhemispheric collaboration can be eliminated by mixing stimulus formats that involve different cortical access routes. *Brain and Cognition, 63,* 114–127.

Plessen, K.J., Wentzel-Larsen, T., Hugdahl, K., Feineigle, P., Klein, J., Staib, L.H., Leckman, J.F., Bansal, R., & Peterson, B.S. (2004). Altered interhemispheric connectivity in individuals with Tourette's disorder. *American Journal of Psychiatry, 161,* 2028–2037.

Propper, R.E., Christman, S.D., & Phaneuf, K.A. (2005). A mixed-handed advantage in episodic memory: A possible role of interhemispheric interaction. *Memory and Cognition, 33,* 751–757.

Reuter-Lorenz, P.A., & Mikels, J.A. (2005). A split-brain model of Alzheimer's disease? Behavioral evidence for comparable intra and interhemispheric decline. *Neuropsychologia, 43,* 1307–1317.

Reuter-Lorenz, P.A., Stanczak, L., & Miller, A.C. (1999). Neural recruitment and cognitive aging: Two hemispheres are better than one, especially as you age. *Psychological Science, 10,* 494–500.

Rogers, L.J., Zucca, P., & Vallortigara, G. (2004). Advantages of having a lateralized brain. *Proceedings of the Royal Society B, Biology Letters, 271,* 420–422.

Singh, H., & O'Boyle, M.W. (2004). Interhemispheric interaction during global–local processing in mathematically gifted adolescents, average-ability youth, and college students. *Neuropsychology, 18,* 371–377.

Westerhausen, R., Kreuder, F., Sequeria, S.D.S., Walter, C., Woener, W., Wittling, R.A., Schweiger, E., & Wittling, W. (2006). The association of macro- and microstructure of the corpus callosum and language lateralisation. *Brain and Language, 97,* 80–90.

Further Readings

Rogers, L.J., & Andrew, R.J. (Eds.) (2002). *Comparative vertebrate lateralization.* **Cambridge University Press: Cambridge, UK.**

Vallortigara, G. & Regolin, L. (2006). Animal brain lateralization: Cognitive, ontogenetic and phylogenetic aspects. *Cognition, Brain, Behavior, 10,* **187–210.**

Vallortigara, G. & Rogers, L.J. (2005). Survival with an asymmetrical brain: Advantages and disadvantages of cerebral lateralization. *Behavioral and Brain Sciences, 28,* **575–589.**
These books and articles provide thoughtful review and integration of research on laterality and interhemispheric interaction in non-human species.

Cutica, I., Bucciarelli, M., & Bara, B.G. (2006). Neuropragmatics: Extralinguistic pragmatic ability is better preserved in left-hemisphere-damaged patients than in right-hemisphere-damaged patients. *Brain and Language, 98,* **12–25.**

Ross, E.D., Thompson, R.D., & Yenkosky, J. (1997). Lateralization of affective prosody in brain and the callosal integration of hemispheric language functions. *Brain and Language, 56,* **27–54.**

Soroker, N., Kasher, A., Giora, R., Batori, G., Corn, C., Gil, M., & Zaidel, E. (2005). Processing of basic speech acts following localized brain damage: A new light on the neuroanatomy of language. *Brain and Cognition, 57,* **214–217.**
These articles provide more on possible hemispheric differences and interhemispheric interaction for prosodic and pragmatic aspects of language.

Aboitiz, F., & Montiel, J. (2003). One hundred million years of interhemispheric communication: The history of the corpus callosum. *Brazilian Journal of Medical and Biological Research, 36,* **409–420.**

Zaidel, E., & Iacoboni, M. (Eds.) (2003). *The parallel brain: The cognitive neuroscience of the corpus callosum.* **Cambridge, MA: MIT Press.**
These articles and books provide more information on the corpus callosum and other biological mechanisms of interhemispheric interaction.

Bergert, S., Windmann, S., & Güntürkün, O. (2006). Is interhemispheric communication disturbed when the two hemispheres perform on separate tasks? *Neuropsychologia, 44,* **1457–1467.**

Cherbuin, N., & Brinkman, C. (2006). Hemispheric interactions are different in left-handed individuals. *Neuropsychology, 20,* **700–707.**

Compton, R.J., Wilson, K., & Wolf, K. (2004). Mind the gap: Interhemispheric communication about emotional faces. *Emotion, 4,* **219–232.**

Ratinckx, E., & Brysbaert, M. (2002). Interhemispheric Stroop-like interference in number comparison: Evidence for strong interhemispheric integration of semantic number information. *Neuropsychology, 16,* **217–229.**

Weems, S.A., & Zaidel, E. (2005). Repetition priming within and between the two cerebral hemispheres. *Brain and Language, 93,* **298–307.**
These articles provide additional examples of behavioral investigations of interhemispheric interaction in language and other communication-relevant domains.

CLINICAL NEUROSCIENCE OF LANGUAGE

A. Language in Special Populations and in
Various Disease Processes

26

Acute Aphasias

CLAUS-W. WALLESCH and CLAUDIUS BARTELS

Department of Neurology, University of Magdeburg, Magdeburg, Germany

ABSTRACT

Acute aphasias present themselves in the first hours or days up to a few weeks after stroke or other damage. Their main etiology is stroke. The syndromes of acute aphasia are contaminated by other symptoms of the underlying pathology, that is clouding of consciousness, major impairment of attentional functions, and aspects of acute pathophysiology such as insufficient blood supply (ischemic penumbra), edema, local inflammation, and the effects of the disruption of intracerebral connections that disturb the balance of neurotransmission (diaschisis). On the other hand, compensatory mechanisms, output monitoring and control, and new learning of linguistic contents and rules have yet to set in. Besides the obvious (i.e., large lesions tend to result in greater damage, mild aphasia will rather remain mild, transitory dysfunctions leave no residue), their course is hardly predictable from data obtained during the first days after onset.

26.1. INTRODUCTION

In western societies, the incidence of stroke is about 25 per 10,000/year, and more than half of these patients suffer from acute communication disorders. Stroke is by far the most common cause of aphasia. Acute aphasia is an epiphenomenon of acute regional dysfunction in structures subserving language processing, in most instances caused by ischemia, in others in decreasing order of frequency by hemorrhage, focal epileptic seizure, traumatic brain injury, encephalitis, migraine, and multiple sclerosis. At its very onset, mutism and apraxia of speech, a disorder of speech motor control, with or without severe comprehension deficit are the most prominent clinical presentations; more rarely the onset is characterized by paraphasia, jargon, or anomia.

In the acute stage and beyond, the symptoms of aphasia depend on clinical, pathophysiological and neuropsychological variables.

26.1.1. Clinical Variables

Clinically, the course and permanence of functional impairment differ according to etiology. Whereas ischemic stroke and intracerebral hemorrhage result in permanent structural damage, the functional lesions of transitory ischemia, focal seizure, and migraine are transient. Even with persisting lesions, the acute stage is characterized by additional dysfunction of brain regions that will eventually recover.

26.1.2. Pathophysiological Variables

Pathophysiologically, this includes phenomena such as the resolution of diaschisis (i.e., dysfunction of brain structures in other vascular territories that are functionally linked with the lesioned region, which is caused by the imbalance of inhibitory or excitatory neurotransmission) and the reperfusion of the ischemic penumbra. Loss of function occurs at perfusion rates (i.e., regional cerebral blood flow) twice as high as those that lead to neuronal death. This implies that each ischemic stroke has a border zone (penumbra) in which neuronal tissue is salvageable, for example, by medical treatment to dissolve a blood clot (thrombolysis) but also by reflex dilatation of blood vessels or by capillary sprouting.

The size and development of the penumbra can be determined by MR perfusion imaging. In acute vascular aphasia it has been shown that error rates in naming were more closely linked to the amount of hypoperfused tissue than to the amount of structural damage (Hillis *et al.*, 2000). This indicates that in the acute stage dysfunction but not necessarily permanent damage determines the clinical picture.

On the other hand, persisting ischemia in the penumbra initiates a number of pathophysiological mechanisms that may cause further neuronal damage (for more detail, see Zukin *et al.*, 2004). Reperfusion (renewed blood circulation into an ischemic area resulting from thrombolytic or spontaneous revascularization) may cause additional transient or permanent damage by neuronal swelling and edema or even hemorrhage. With recovery, the synaptic organization of damaged functional brain systems may change over time, enabling compensation. Training may even lead to the development of new neurons (van Praag *et al.*, 1999).

26.1.3. Neuropsychological Variables

Additional neuropsychological impairments such as disorders of attention and consciousness that occur in the acute stage compromise language performance. These impairments resolve with time.

A recent functional magnetic resonance imaging (MRI) study investigated 14 patients with aphasia due to a left middle cerebral artery infarction with an auditory comprehension task 2 days, 12 days, and 1 year post stroke (Saur *et al.*, 2006). The aphasia of the patients recovered considerably over the period of observation; 6 patients recovered completely. In the subacute stage, that is 12 days after stroke, a large increase of activation in the right homolog of Broca's area was observed that correlated with improved language function in the experimental task. In the chronic phase, the peak of activation shifted back to the left hemisphere, associated with further language improvement. These data indicate that the early course of acute aphasias is linked to individually different activations of right-hemisphere language resources, whereas language performance in the chronic stage is based on functions of the damaged but still structurally more competent left hemisphere. The authors speculate that persistence of right frontal activation may indicate a chronic disturbance of interhemispheric interaction, which might be disadvantageous for language processing.

In view of these factors, it seems obvious that the evolution of the underlying pathology and its functional consequences are rather unpredictable, even in stroke (see Box 26.1). One has to remember that the chronic aphasic syndromes reflect stable pathology in an individual with intact personality, nonverbal intelligence, and attention, in the context of individual compensations. As an example of compensation consider Broca's aphasia, which is a common consequence of initially severe language deficit from a large left middle cerebral artery stroke in a younger person. Lexical access to content words is re-established through effort, secondary lexical representations being widespread and thus retrievable from cortical areas unaffected by pathology (Cao *et al.*, 1999). Automatic grammaticalization, on the other hand, is a function of the damaged perisylvian cortex. Through therapy, agrammatic patients may learn to apply grammatical rules, but in many cases this is no longer an automatized process. This explains both the occurrence of paragrammatisms and the spontaneous use of agrammatic clauses in Broca's aphasia (Hofstede & Kolk, 1994); grammaticalization by the application of re-learned rules places demands on attentional resources, which are required for the sake of communication.

Furthermore, especially in cases with a milder degree of linguistic deficit, the acute disruption of the highly automatized process of speech generation and language production often results in error types that are characterized by failing attempts at voluntary control, such as parapraxia (errors of articulatory movements or their sequence) with apraxia of speech or sequences of phonemic paraphasias (i.e., errors in the production of speech sounds) aiming at but missing the intended target. The latter are termed "phonemic approximations" (e.g., "tru – turrel – tollet – …" for "turret"). More severe cases are mute initially.

Box 26.1 Pathology in acute aphasia

In acute aphasia, there is no stable pathology, intellectual and attention functions are compromised, consciousness may be clouded, and compensatory attempts have not yet been initiated. The clinical course in the first few weeks may be determined by the activation of individually different language processing resources of the right hemisphere.

TABLE 26.1   Differences Between Acute and Chronic Vascular Aphasias

	Acute aphasias	Chronic aphasias
Time since onset	Hours to weeks	Months to years
Pathophysiology	Penumbra and dysfunction due to diaschisis (see text) increase functionally damaged area beyond structural damage	Compensation and reorganization decrease dysfunction
Neurobiology	Activation of right-hemisphere resources	Reorganization of left-hemisphere mechanisms
Neuropsychology	Additional disorders of attention, impaired perception of deficits	Compensation and reorganization of cognitive processes
Handling of illness	Despair	Coping
Social role	"Being ill" – receiving social attention and support	"Being disabled" – social exclusion

Table 26.1 summarizes differences between acute (still unstable) and chronic (stable) vascular aphasias. It describes that for pathophysiological reasons (i.e., penumbra, diaschisis), the functional pathology underlying acute aphasia is more extensive than the structural pathology, and that the performance of acutely aphasic patients is further hampered by disorders of attention and awareness. In the chronic stage, effects of compensation and reorganization reduce the functional deficit. It should also be noted that individuals with chronic aphasia are perceived as handicapped which is a less positive social role than that of the acutely ill.

26.2. APPROACHES TO CLASSIFYING ACUTE APHASIAS

As patients suffering from acute aphasia differ greatly in their language symptoms, numerous attempts have been made over more than a century to classify them into categories, starting with the labels of "motor" and "sensory." The information given in Table 26.1 indicates that the classification used for chronic vascular aphasias is misleading when applied to syndromes of the acute stage. In fact, because of the rapid pathophysiological changes outlined above, Kertesz (1979) saw little use at all in syndrome classification during the first 14 days after onset (Box 26.2).

Box 26.2   The term syndrome

The term "syndrome" denotes a cluster of symptoms, the association of which is more frequent than chance. It may or may not indicate the underlying pathological or functional–anatomic basis, and thus is not necessarily of heuristic value. "Syndromes" are derived by three different approaches:

1. Test batteries that use defined cut-off scores in various modalities to assign patients (e.g., the Western Aphasia Battery, Kertesz, 1982).
2. Test batteries that categorize by obligatory symptoms (e.g., the Aachen Aphasia Test, Huber et al., 1984).
3. Statistical means of determining similarity, that is cluster analysis (e.g., Wallesch et al., 1992).

Nowadays, syndromes are seen as prototypes with most patients exhibiting individual deviations from their typical characteristics. With acute aphasias, certain constellations of symptoms indicate a good prognosis despite considerable impairment.

Kertesz, A. (1982). *Western aphasia battery*. New York: The Psychological Cooperation.
Huber, W., Poeck, K., & Willmes, K. (1984). The Aachen Aphasia Test. In F.C. Rose (Ed.), *Progress in aphasiology* (pp. 291–303). New York: Raven.
Wallesch, C.W., Bak, T., & Schulte-Mönting, J. (1992). Acute aphasia – patterns and prognosis. *Aphasiology*, 6, 373–385.

There are only five large clinical studies in the literature that attempt to assess patients suffering from acute aphasia within the first month post onset in a standardized fashion and perform either a cluster analysis in order to identify common patterns of impairment or use a forced-choice syndromatic classification. Because these studies are the main contributors to the available database, they will be presented here in some detail, although two of them have already been conducted in the 1970s and 1980s. However, the methods of language assessment used at that time are still valid and have changed little since.

26.2.1. Study 1: Kertesz (1979)

Sixty-four right-handed aphasic patients were investigated with the Western Aphasia Battery (WAB) 14–45 days after stroke. The WAB assesses fluency and information content of spontaneous speech, comprehension, naming, repetition, reading, and writing as well as "performance" tasks such as praxis, calculation, drawing, block design, and Raven's colored progressive matrices. Cluster analysis, a statistical tool to group multidimensional observations in a way to minimize variance between groups of cases, resulted in the nine following patient groups (in parentheses: "forced" classification into aphasic syndrome based on WAB using categories that are normally used for individuals with chronic aphasia):

1. 11 patients with low language and performance scores (9 global, 2 Wernicke's aphasia).
2. 7 patients with low language scores but better non-verbal performance (6 global aphasia, 1 "isolation of the speech area syndrome," that is severe aphasia with preserved repetition).
3. 7 patients with good comprehension subtests and also relatively good performance scores (6 Broca's, 1 conduction aphasia).
4. 5 patients with high fluency and high repetition scores but poor performance scores (transcortical motor, transcortical sensory aphasia, and isolation of the speech area syndrome – all these syndromes are characterized by preserved repetition).
5. 6 patients with high fluency scores but lower comprehension and performance scores (3 Wernicke's, 2 conduction, 1 transcortical sensory aphasia – fluent aphasia syndromes characterized by paraphasia).
6. 5 patients with relatively good performance, preserved fluency, but otherwise poorer language scores (all Wernicke's aphasia).
7. 9 patients with low fluency scores but high comprehension and performance scores (4 Broca's, 4 conduction, 1 anomic aphasia).
8. 5 patients with low writing, reading and performance scores, with high repetition and fluency scores, and some

comprehension deficits ("alexia and agraphia with mild speech disturbance," 3 anomic, 1 transcortical sensory, 1 Wernicke's aphasia – patients with pronounced written language deficit).

9. 9 patients with uniformly high scores in both language and performance areas (8 anomic, 1 conduction aphasia – mild aphasia).

The identification of 9 patterns among 64 patients as revealed by cluster analysis supports Kertesz' claim of the futility of classificatory attempts in acute aphasia.

26.2.2. Study 2: Wallesch *et al.* (1992)

Ninety-seven patients from a university department of neurology were investigated within the first 14 days after onset of aphasia. Sixty-seven patients could be assessed with a non-standardized form (using every second subtest item) of the Aachen Aphasia Test (Huber *et al.*, 1984). These included 57 patients with ischemic infarction, 4 with primary hemorrhage, 2 after trauma, 2 postoperatively,

and 1 with herpes encephalitis. In addition to the aphasia assessment, verbal fluency, presence of verbal perseveration, echolalia, dysarthria (sensorimotor speech disorder), ideomotor apraxia (a disorder of the execution of movement sequences), hemineglect (an impairment of attending to the right side of space and body), hemiparesis, and clouded consciousness were scored.

Table 26.2 provides an overview of the neuropsychological and neurological symptoms. It demonstrates that a majority of aphasic patients studied within the first 2 weeks post onset suffer from symptoms of global cerebral dysfunction (clouded consciousness). It must be assumed that these interact with linguistic and communicative performance.

The data from the AAT subtests were analysed by a centroid link cluster analysis program. The analysis was terminated when more than 80% of the subjects were assigned to clusters of more than 5% of the sample. Sixty-five of 67 subjects were assigned to one of seven clusters on the basis of their AAT subtest performance. Table 26.3 summarizes these clusters.

The Token Test errors indicate that aphasia severity was the dominant parameter for cluster generation. Cluster analysis revealed two groups of mildly and fluent aphasic patients (clusters 1 and 2) differing mainly in auditory comprehension, and one cluster (3) of fluent but more markedly impaired, paraphasic patients. Clusters 4–7 comprised of mainly non-fluent aphasia, in 4 with partially preserved naming and in 5 with largely preserved repetition ("acute transcortical motor aphasia (TMA)"). Clusters 6 and 7 contained patients with non-fluent aphasia with severe deficits in all modalities. One obvious shortcoming of the applied methodology is that it excluded symptom constellations that occurred in less than 5% of the sample. Thus, one patient with massive phonemic paraphasia in the form of phonemic approximations resulting in very low repetition performance ("acute conduction aphasia") was not assigned to a cluster.

TABLE 26.2 Neuropsychological and Neurological Symptoms Accompanying Acute Aphasia Within 14 days from Onset (Percentages)

Verbal perseveration	41
Echolalia	25
Dysarthria	26
Ideomotor apraxia	23
Hemiparesis	55
Hemiplegia	12
Clouding of consciousness	Mild 42, marked 10

Source: From Wallesch *et al.* (1992).

TABLE 26.3 Clusters Derived from AAT Subtest Data in 65 Acute Aphasic Patients Assigned to One of 7 Clusters. Numbers Give AAT Scores, Except for Token Test (Errors)

Cluster	N	Non-infarct etiology	Repetition Median (max = 75)	Range	Naming Median (max = 60)	Range	Auditory comprehension Median (max = 30)	Range	Token Test errors Median (max = 25)	Range
1	4	0	67	64–71	51	37–55	28	26–30	1	0–2
2	6	0	71	57–73	52	48–57	22	19–23	3	0–5
3	13	2	65	47–74	45	32–53	22	16–27	11	8–17
4	8	2	60	38–63	24	19–36	15	13–18	17	11–23
5	5	1	60	56–74	6	0–20	9	8–14	21	18–24
6	13	3	27	3–40	6	0–26	16	11–18	22	19–25
7	6	0	6	0–13	2	0–13	11	9–13	25	23–25

Source: Data from Wallesch *et al.* (1992).

Outcome data on 36 of the subjects were obtained more than a year later (12 patients had passed away). One third of the surviving patients had recovered completely within the first year according to their AAT performance, especially those who had only been mildly aphasic initially. The instability of acute aphasia syndromes is emphasized when patterns of syndrome development are inspected (Table 26.4). We would like to emphasize that 29 out of 36 patients were not or only mildly aphasic after 1 year. A caveat of the study is that only those patients were included with whom language assessment could be undertaken for at least 20 min within 14 days post onset, thus excluding patients with severe neurological impairment.

All clusters of acute aphasia with the exception of cluster 7 were compatible with complete recovery of language (as measured by the Aachen Aphasia Test). The clusters of acute aphasia were also not predictive of the chronic aphasia syndrome. The authors concluded that there was little profit to gain from a detailed neurolinguistic assessment within the first 2 weeks post aphasia onset, as had been suggested earlier by Kertesz (1979).

26.2.3. Study 3: Laska *et al.* (2001)

These authors presented a Swedish test similar to the WAB to 119 aphasic stroke patients within the first 30 days (in 90% within 11 days) after stroke. Hundred and four patients could be followed-up for 18 months. At 18 months, 27% had completely recovered, 49% still suffered from aphasia, and 24% had died. Three quarters of the deaths were attributed to cardiovascular disease. Both in-hospital and long-term mortality were higher with aphasic than with non-aphasic stroke patients. Forcing a classification by using cut-off scores, in the Swedish adaptation of the WAB, resulted in large proportions of about 20% each for global and Wernicke's aphasia, as well as a large number of

non-standard aphasia syndromes: 15% conduction aphasia, 10% transcortical sensory, and some 30% for mixed fluent and non-fluent aphasia. The authors reported good recovery for patients initially classified as Wernicke's or conduction aphasia. This study also emphasized the syndromatic instability of acute aphasia as well as the potential of substantial improvement over time regardless of severity of the initial symptomatology.

26.2.4. Study 4: Godefroy *et al.* (2002)

This study included only those patients admitted to the Lille acute stroke unit who were examined by a speech therapist with the Montreal–Toulouse battery "usually within the first month after stroke." Of 207 patients diagnosed as aphasic, 52 were classified as global, 30 as Wernicke's, 22 as Broca's aphasia and 5 as anomic. Forty-one were assigned to non-standard syndromes: 17 transcortical motor, 10 "subcortical," 7 conduction, and 7 transcortical sensory aphasia. Fifty-seven were non-classifiable (Godefroy *et al.*, 2002).

26.2.5. Study 5: Pedersen *et al.* (2004)

This study included 488 patients who were consecutively admitted to three Copenhagen stroke units, who were clinically aphasic at admission. They were assessed within the first week after stroke onset with parts of the WAB (Kertesz, 1982), including fluency and information content of speech, naming, comprehension, and repetition. These are sufficient to determine the aphasia syndrome according to the Boston classification and allow the computation of the "Aphasia Quotient" as a measure of severity. Of these 488 patients, 92 had been clinically aphasic at admission but were in full remission when the test was carried out. Another 126 patients had to be excluded because of impaired consciousness, pure

TABLE 26.4 One Year Outcome From Acute Aphasia Based on Previously Established Clusters. Severity Classification According to AAT Evaluation

	Outcome				
Cluster	No aphasia	Mild aphasia	Moderate aphasia	Severe aphasia	Death/unknown
1	1	2	0	0	1
2	3	3	0	0	0
3	5	3	1	0	4
4	2	2	1	0	3
5	1	0	1	0	2
6	1	4	1	1	6
7	0	2	0	2	2
Sum	13	16	4	3	16

Source: Data from Wallesch *et al.* (1992).

dysarthria, refusal or lack of cooperation, interfering somatic or sensory impairment, death, transfer, or discharge.

Of the remaining 270, the WAB classified 32% as globally, 12% as Broca's, 15% as Wernicke's, 27% as anomic, 6% as transcortical sensory, 5% as conduction, 2% as transcortical motor, and 1% as "isolation of the speech area" aphasia. The subgroup of 203 patients with first strokes revealed a similar distribution. Table 26.5 shows how the WAB-syndromes changed in survivors from the first week to one year follow-up.

Whereas there was no significant interaction between syndromes and age in the acute phase, at follow-up the age difference between persons with Broca's aphasia and those with Wernicke's aphasia narrowly missed significance, the latter being on average a decade older. As age is associated with diffuse brain damage which hampers output control and thus impairs control of paraphasia, this may indicate that younger patients with initially fluent aphasia are more likely to achieve output control (transition Wernicke's aphasia to anomic aphasia) and that younger non-fluent patients may be able to activate lexical representation outside the language area by properly allowing for agrammatic communication (transition global to Broca's aphasia). Each syndrome of acute aphasia in the Pedersen et al. study was compatible with complete restitution at 1-year follow-up. Among those who remained aphasic, certain frequent patterns of transition could be observed: global to Broca, Broca to anomic, Wernicke to conduction or anomic. The results of Pedersen et al. once more underline the difficulty, if not futility of syndrome prognosis in the first week after stroke.

26.2.6. Synopsis of Group Studies

The studies agree that classification of acute aphasias both by cluster analysis and by forced assignment, does not result in meaningful predictions of the chronic syndrome. All syndromes of the acute stage were found compatible with complete language recovery (according to the tests used).

With respect to outcome prediction, all studies could establish that mild acute aphasia had an especially good prognosis, and that severe acute aphasia was still compatible with complete recovery. Laska et al. (2001) reported good recovery for syndromes classified as acute Wernicke's and conduction aphasia, Pedersen et al. for acute transcortical sensory (fluent aphasia with comprehension deficit and preserved repetition), conduction (acute aphasia with focally impaired repetition) and anomic aphasia.

26.3. NON-STABLE SYNDROMES OF ACUTE APHASIA

Clinicians occasionally encounter non-stable syndromes of acute aphasia. These are acute conduction aphasia, TMA, acute paraphasia, and "pure" motor aphasia.

26.3.1. Acute Conduction Aphasia

Of 14 patients with acute conduction aphasia, in the study of Pedersen et al. (2004), 10 recovered, 2 converted to anomic aphasia, and 2 remained as conduction aphasics. This is in accordance with our and others' (Kertesz & McCabe, 1977) experience, but differs from data presented by Bartha and Benke (2003). These authors included 20 patients with a fluent aphasia syndrome including a profound deficit in repetition and displaying phonemic paraphasias and self-corrections in the absence of a severe language comprehension disorder. Only 11 of these were assessed within the first 2 weeks post onset. Three of those who were investigated at a later point in time had evolved from Wernicke's aphasia. Despite mild comprehension deficit as an inclusion criterion, Token Test performance was highly variable. Verbal-auditory short-term memory was reduced in all patients but one. Twelve patients could be reassessed more than 100 days later. Of these 7 exhibited chronic conduction aphasia and only 2 had recovered completely. For the investigation of

TABLE 26.5   Change In Type of Aphasia from First week to One Year Follow-Up in Survivors

	At follow-up						
	Global aphasia	Broca's aphasia	Wernicke's aphasia	Transcortical motor aphasia	Conduction aphasia	Anomia	No aphasia
Within first week							
Global aphasia	22%	35%	7%			22%	15%
Broca's aphasia		18%		9%		36%	36%
Wernicke's aphasia			18%		36%	27%	18%
Transcortical sensory aphasia						17%	83%
Conduction aphasia					14%	14%	71%
Anomia						38%	62%

Source: Data from Pedersen *et al.* (2004).

the course and prognosis of acute conduction aphasia, two aspects appear crucial: the time span between onset and diagnostic assessment, and the definition of the syndrome.

In our experience, acute conduction aphasia is more often non-fluent (except for non-propositional comments) than fluent with prominent and often cumbersome attempts at phonemic approximation. These can hamper communication more than uncorrected paraphasia would do. However, their presence indicates awareness of deficits, output monitoring and attempts at correction. This explains why the deficit is more marked in situations where attention is focused on output, such as repetition, and less marked with naming, and least prominent in propositional speech. Although most authors on the subject would agree that repetition impairment is the most prominent feature of the syndrome, there is disagreement about its functional basis. Explanations focus on a phonological control or a phonological buffer deficit. Probably, each explanation is valid for certain patients. With acute aphasia, the control deficit is much more prominent. The syndrome reflects the attempt to gain control over the deficits, which is successful in most cases.

26.3.2. Acute Transcortical Motor Aphasia

The fronal lobe contains a medial system including the anterior cingulate cortex and the supplementary motor area that conveys motivation and initiation of actions and a dorsolateral system that generates action programs based on sensory information and the situational context (e.g., Joseph, 1996). Both systems include cortical and subcortical structures. Acute transient TMA occurs with lesions of the connections between these systems. Typical lesions affect the medial frontal cortex with infarcts of the anterior cerebral artery or connections between the left basal ganglia and the thalamus. The essential clinical feature of TMA is a pronounced dissociation between a marked reduction in the amount and complexity of spontaneous speech and a retained ability to produce speech in response to external stimulation in repetition, reading aloud, and confrontation naming. In the acute phase, speech production may be reduced to the point of mutism. Patients with TMA generally do not speak spontaneously. Answers to questions are usually produced only after a long delay. Comprehension may be only mildly impaired (Rapcsak & Rubens, 1994). The disorder has been termed "aphasia sine (without) aphasia," as the crucial deficits seems to be utterance initiation under multiple degrees of freedom.

26.3.3. Acute Paraphasia

Acute phonemic paraphasia with sometimes largely preserved comprehension is a rare but not unusual finding with subcortical infarctions involving the ventrolateral (Wallesch, 1997) or anteromedian (Carrera et al., 2004) thalamus. Language disorders resulting from thalamic lesions are often confounded by perseveration and fluctuating impairments of attention and consciousness that are particularly frequent with lesions of the non-specific thalamic nuclei. Acute semantic paraphasia with variable degrees of comprehension deficit have been described with lesions of the left head of caudate and anterior limb of the internal capsule (for a review, see Wallesch, 1997). The underlying pathophysiology has been related to the lesion of thalamocortical projections involved in gating the processing of multiple lexical choices.

Figure 26.1 shows T2-MR (enhancing water-containing lesions) and DWI-MR (indicating regions of permanent

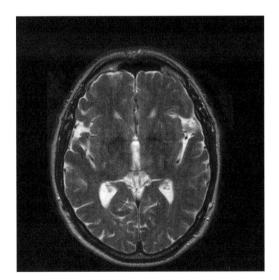
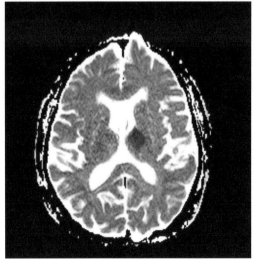

FIGURE 26.1   T2- and DWI-MR scans of a 56-year-old male patient with an anterior thalamic infarct 10 h after symptom onset. At admission, the patient presented with fluctuating somnolence, non-fluent aphasia with only minimal production and largely intact comprehension were noted. For further information, see text. (The scans are oriented to neuroradiological convention: left is right, you look upon the patient from his feet.)

damage) scans of a patient with a left anterior thalamic infarct 10 h after symptom onset. At admission, a fluctuating somnolence, and non-fluent aphasia with only minimal production were noted. Comprehension appeared clinically adequate. Twelve days later, spontaneous speech was still non-fluent with marked word finding difficulties, semantic, and phonemic paraphasia, but no discernible syntactic deficit. Repetition, writing to dictation, and reading aloud were preserved. Behaviourally, he exhibited fluctuations of attention, perseverations, iteration, and discrete echolalia. Naming, comprehension, and Token Test performance were close to the 50th percentile of the Aachen Aphasia Test normative sample.

Box 26.3   Motor aphasia

Although aphasia denotes a linguistic deficit, the term "motor aphasia" has remained. It reflects the common experience made in acute stroke services of patients with a severe, obviously motor expressive deficit accompanied by no obvious comprehension impairment. The underlying functional pathology is that of apraxia of speech, which may accompany both severe and mild aphasia, and may even occur in the absence of aphasia (Ogar et al., 2006).

Ogar, J., Willock, S., Baldo, J., Wilkins, D., Ludy, C., & Dronkers, N. (2006). Clinical and anatomical correlates of apraxia of speech. *Brain and Language*, *97*, 343–350.

26.3.4. "Pure Motor Aphasia": Apraxia of Speech

In the last two decades, there has been considerable debate concerning the theoretical status and underlying pathophysiology as well as the clinical symptoms of apraxia of speech. A recent review enumerated as core symptoms the effortful "groping" for articulatory postures and therefore difficulty forming the correct orofacial position to produce the correct sound, the occurrence of more consonant than vowel errors (as the consonant part of the word is more difficult to produce) inconsistent or variable errors, the production of words or speech sounds or both that approximate the target word, the difficulty of producing adjacent consonants, for example inserting additional vowel sounds, and, finally, the awareness of errors (West et al., 2005).

These symptoms are difficult to establish in a patient who is almost mute or inattentive. However, the list indicates which phenomena should be looked for: searching and groping, awkward and clumsy articulatory movements, trial and error phonations and articulations, vowel insertions, and subject's apparent discontent with his attempts at production.

Apraxia of speech has been defined and theoretically explained by McNeil et al. (1997) as a disorder of speech production caused by inefficiencies in the translation of a well-formed and filled phonological frame to previously learned kinematic parameters assembled for carrying out the intended movement, resulting in intra- and inter-articulator temporal and spatial segmental and prosodic distortions.

The presence of apraxia of speech explains the clinical picture of "motor aphasia," frequently encountered in acute stroke services. It is easily established that the linguistic disorder is insufficient to cause the expressive deficit, by using tests of comprehension or the Token Test. Sometimes, even written propositions are possible (see Box 26.3).

Figure 26.2(a) shows computerized tomography (CT) scans of a 53-year-old teacher who had suffered a cerebral hemorrhage in the frontal operculum 7 days earlier. At admission, he was mute and communicated by hand gestures. At day 7, speech production was characterized by severe parapraxia, especially of initial consonants and only mild anomia, agrammatism, and comprehension disorder. He made 10/50 errors in the Token Test. The patient recovered rapidly with hemorrhage resorption. The residual state is shown in the MRI-scans in Figure 26.2(b) demonstrating hemosiderin (a blood degradation product) in the left frontal operculum at the border between premotor and motor cortex.

26.4. CHALLENGES AND FUTURE DIRECTIONS

New MRI techniques, including diffusion-weighted imaging (DWI) in clinical routine and functional imaging in research settings, will greatly contribute to our understanding of the mechanisms of the acute disruption of linguistic integrity by brain pathology. They may shed some more light into the still confusing syndromatology of the acute aphasias. Of these new techniques, DWI will outline the damaged brain areas at an earlier stage than is possible now. Functional imaging allows to analyze which brain areas are being activated by specific stimuli. It will also demonstrate regions affected by functional pathology, such as diaschisis. Finally, diffusion tensor imaging will outline the intracerebral connections and pathways of an individual aphasic person and their affection or sparing by pathology.

Acknowledgment

We thank Prof. M. Skalej, MD, Institute of Neuroradiology, University Clinic Magdeburg, for permission to reproduce the scans.

(a)

(b)

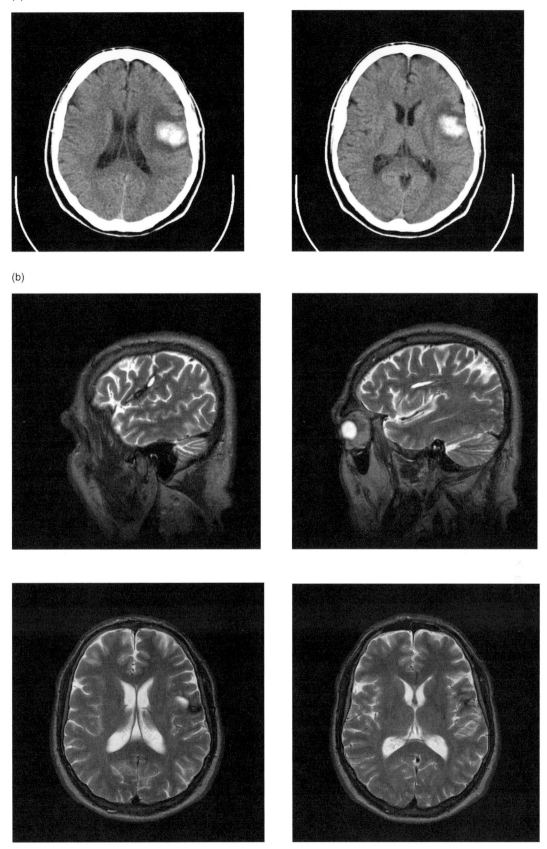

FIGURE 26.2 (a) CT scans of a 53-year-old teacher with marked apraxia of speech 7 days after a cerebral hemorrhage in the frontal operculum. (b) MRI scans of the same patient 14 weeks after the hemorrhage showing degradation products of blood in the area of lesion and a small slit-like defect. See text for further information. (The horizontal scans are oriented to neuroradiological convention: left is right, you look upon the patient from his feet.)

References

Bartha, L., & Benke, T. (2003). Acute conduction aphasia: An analysis of 20 cases. *Brain and Language, 85*, 93–108.

Cao, Y., Vikingstad, E.M., George, K.P., Johnson, A.F., & Welch, K.M.A. (1999). Cortical language activation in stroke patients recovering from aphasia with functional MRI. *Stroke, 30*, 2331–2340.

Carrera, E., Michel, P., & Bogousslavsky, J. (2004). Anteromedian, central, and posterolateral infarcts of the thalamus. Three variant types. *Stroke, 35*, 2826–2832.

Godefroy, O., Dubois, C., Debachy, B., Leclerc, M., & Kreisler, A. (2002). Vascular aphasia: Main characteristics of patients hospitalized in acute stroke units. *Stroke, 33*, 702–705.

Hillis, A.E., Barker, P.B., Beauchamp, N.J., Gordon, B., & Wityk, R.J. (2000). MR perfusion imaging reveals regions of hypoperfusion associated with aphasia and neglect. *Neurology, 55*, 782–788.

Hofstede, B.T., & Kolk, H.H. (1994). The effects of task variation on the production of grammatical morphology in Broca's aphasia: A multiple case study. *Brain and Language, 46*, 278–328.

Huber, W., Poeck, K., & Willmes, K. (1984). The Aachen Aphasia Test. In F.C. Rose (Ed.), *Progress in aphasiology* (pp. 291–303). New York: Raven.

Joseph, R. (1996). *Neuropsychiatry, neuropsychology, and clinical neuroscience* (2nd edn). Baltimore, MD: Williams & Wilkins.

Kertesz, A. (1979). *Aphasia and associated disorders*. New York: Grune & Stratton.

Kertesz, A. (1982). *Western aphasia battery*. New York: The Psychological Cooperation.

Kertesz, A., & McCabe, P. (1977). Recovery patterns and prognosis in conduction aphasia. *Brain, 100*, 1–18.

Laska, A.C., Hellblom, A., Murray, V., Kahan, T., & von Arbin, M. (2001). Aphasia in acute stroke and relation to outcome. *Journal of Internal Medicine, 249*, 413–422.

McNeil, M.R., Robin, D.A., & Schmidt, R.A. (1997). Apraxia of speech: Definition, differentiation, and treatment. In M.R. McNeil (Ed.), *Clinical management of sensorimotor speech disorders* (pp. 311–344). New York: Thieme.

Pedersen, P.M., Vinter, K., & Olsen, T.S. (2004). Aphasia after stroke: Type, severity and prognosis. *Cerebrovascular Diseases, 17*, 35–43.

Rapcsak, S.Z., & Rubens, A.B. (1994). Localization of lesions in transcortical aphasia. In A. Kertesz (Ed.), *Localization and neuroimaging in neuropsychology* (pp. 297–329). San Diego, CA: Academic Press.

Saur, D., Lange, R., Baumgaertner, A., Schraknepper, V., Willmes, K., Rijntjes, M., & Weiller, C. (2006). Dynamics of language reorganization after stroke. *Brain, 129*, 1371–1384.

Van Praag, H., Christie, B.R., Sejnowski, T.J., & Gage, F.H. (1999). Running enhances neurogenesis, learning, and long-term potentiation in mice. *Proceedings of the National Academy of Science, 96*, 13427–13431.

Wallesch, C.W. (1997). Symptomatology of subcortical aphasia. *Journal of Neurolinguistics, 10*, 267–275.

Wallesch, C.W., Bak, T., & Schulte-Mönting, J. (1992). Acute aphasia – patterns and prognosis. *Aphasiology, 6*, 373–385.

West, C., Hesketh, A., Vail, A., & Bowen, A. (2005). Interventions for apraxia of speech following stroke. *The Cochrane database of systematic reviews 2005, Issue 4*. Art. No. 004298.pub2. DOI: 10.1002/14651858.CD00428.pub2.

Zukin, R.S., Jover, T., Yokota, H., Calderone, A., Simionescu, M., & Lau, C.G. (2004). Molecular and cellular mechanisms of ischemia-induced neuronal death. In J.P. Mohr, D.W. Choi, J.C. Grotta, B. Weir & P.A. Wolf (Eds.), *Stroke – pathophysiology, diagnosis, and management* (pp. 829–854). Philadelphia, PA: Churchill-Livingstone.

27

Language in Dementia

MURRAY GROSSMAN

Department of Neurology, University of Pennsylvania School of Medicine, Philadelphia, PA, USA

ABSTRACT

Is language functioning subject to the ravages of neurodegenerative disease the way that memory is compromised in Alzheimer's disease? In this review, we show that progressive neurodegenerative disease in fact has the same devastating effect on language as Alzheimer's disease (AD) has on memory. Patients with frontotemporal dementia (FTD) and other dementing conditions are shown to have different forms of primary progressive aphasia that selectively interfere with specific aspects of language. This includes an impairment of grammatical processing in progressive non-fluent aphasia, difficulty with the concepts underlying word meaning in semantic dementia, a disorder of lexical retrieval in several neurodegenerative conditions, and compromised discourse in non-aphasic patients with impaired social and executive functioning due to FTD. Taken together, these findings emphasize the profound ways in which neurodegenerative diseases can compromise language.

27.1. INTRODUCTION

Can patients with dementia have aphasia? This question provoked considerable debate in the recent past. Comprehensive assessments with validated and reliable aphasia batteries demonstrated language problems in demented patients (Cummings *et al.*, 1985). However, observations such as these were dismissed all too often as a general decline of cognitive functioning. They were attributed to a deficit in memory, attention, or some other non-linguistic factor. This stems from the historical view that "dementia" refers to a disorder of episodic memory caused by diffuse cortical disease. Much of our modern thinking about dementia is due to Mesulam's important challenge to this approach in his modern

description of primary progressive aphasia (PPA). PPA is a progressive disorder of language impairment that is present for 2 years prior to clinical evidence of difficulty in other cognitive domains. Rather than attributing PPA to diffuse disease, moreover, this language disorder was associated with focal cortical dysfunction affecting a circumscribed brain region.

In the following review, I discuss disorders of language that are associated with dementia. This not only includes impairments of language in common dementias such as Alzheimer's disease (Alzheimer's dementia), but also language disorders in less common neurodegenerative diseases such as PPA. I focus on several aspects of language representation and processing, including phonologic, lexical–semantic, grammatical, and discourse levels of language functioning. Each of these language components appears to be disproportionately compromised in particular neurodegenerative diseases. This outcome is observed only because the neuroanatomic basis of these dementias is not diffuse, but instead each of these diseases compromises cerebral functioning in a specific anatomic distribution. Observations such as these are consistent with the claim that specific language processes depend on the integrity of large-scale neural networks, and that focal neurodegenerative diseases selectively compromise such a neural network to cause aphasia.

27.2. PHONOLOGY AND SPEECH ERRORS IN DEMENTIA

Speech errors are very common in specific neurodegenerative diseases. However, the qualitative nature of disrupted speech varies depending on the particular condition. Garbled pronunciation, known as dysarthria, is a frequent manifestation of neurodegenerative diseases with a motor component such as Parkinson's disease, motor neuron disease and small vessel ischemic disease. Experimental tasks may be used to elicit speech errors in Alzheimer's dementia, but the dominant historical view is that phonemic paraphasic errors in speech – substituting a letter sound within a word, such as "fetter" instead of the target word "letter" – occur no more commonly in Alzheimer's dementia than in age-matched healthy seniors (Bayles & Tomoeda, 1983). A detailed analysis of conversational speech, single word production and highly familiar speech sequences was performed in a series of 10 patients with clinically or pathologically defined Alzheimer's dementia presenting with a progressive form of aphasia (Croot et al., 2000). This work shows frequent phonemic paraphasias amounting to 40.6% of the errors occurring in spontaneous speech. These errors are frequently evident in naming, repetition, and reading. Imaging and pathologic analyses implicate perisylvian disease of the left hemisphere in the phonemic errors of these patients.

Clinical observations associate progressive non-fluent aphasia (PNFA) with frequent speech errors and apraxia of speech, although quantitative analyses are rare. In a semi-structured speech sample of four PNFA patients, speech errors seem to occur frequently (Thompson et al., 1997). A detailed analysis of speech in two patients with PNFA shows more difficulty on single word production tasks as the phonological information available in the task decreases (Croot et al., 1998). Naming thus yields more speech errors than reading, which in turn results in more errors than repetition. More errors are also associated with longer words. A careful clinical study of speech errors associated apraxia of speech with PNFA (Josephs et al., 2006).

A large semi-structured speech sample in eight PNFA patients finds phonemic paraphasic error in 5% of their words (Ash et al., 2004). This is significantly more common than the speech errors in Alzheimer's dementia, in patients with a fluent form of progressive aphasia known as semantic dementia, and in healthy, age-matched seniors. The errors are complex, involving numerous deviations from the intended target. These include insertions and deletions of target sounds, exchanges or metatheses, and incorrect vowel targets, occurring alone and in various combinations. Moreover, phonemic paraphasic errors are doubly dissociated with semantic paraphasic errors – the substitution of a word for the intended target, such as "sentence" instead of "letter" – in patients with Alzheimer's dementia or semantic dementia. PNFA patients thus produce fewer semantic paraphasic errors than phonemic errors, and fewer semantic paraphasic errors than patients with Alzheimer's dementia and semantic dementia. By comparison, Alzheimer's dementia and semantic dementia patients produce more semantic paraphasic errors than their own rate of phonemic errors, and more semantic errors than PNFA patients. The observation of multiple word-sound impairments is consistent with a large-scale network for word processing that consists of several dissociable components, and focal disease can result

in different kinds of speech errors depending on the precise location of the impairment within this network.

27.3. LEXICAL RETRIEVAL AND NAMING DIFFICULTY IN DEMENTIA

Am I becoming demented because I have such difficulty finding words? Word-finding pauses (often during word searches), circumlocutions (talking around a topic and describing a target word when it cannot be retrieved), and naming difficulty are very common as we age, but they are even more common in neurodegenerative diseases such as Alzheimer's dementia, PNFA, and semantic dementia (Grossman et al., 2004). However, the basis for impaired naming varies depending on the nature of the condition. Cognitive models of naming (Levelt, 1992) suggest components such as retrieving a lexical representation from the mental lexicon that corresponds to a target concept, and translating the lexical representation to a specific modality of representation so that the word can be produced. In Alzheimer's dementia, naming difficulty has been associated with impairments in lexical retrieval, semantic difficulty, and visual limitations (Lambon Ralph et al., 1997). Likewise, the naming difficulty in semantic dementia appears to involve an interruption of lexical retrieval and degraded semantic knowledge (Lambon Ralph et al., 1998).

Recent work focuses on specific aspects of naming to bring out particular attributes of the naming process. Naming difficulty thus may vary depending on the semantic category being queried. We turn to this source of impairment in more detail below when we consider semantic memory difficulty. Naming deficits also appear to be sensitive to the major grammatical subcategory being probed. For example, verbs are more difficult than nouns in Alzheimer's dementia and frontotemporal dementia (FTD) (Yi et al., 2007). In PNFA, difficulty with verb naming may be associated with the disruption of the motor component of action knowledge that contributes to verb meaning (Bak et al., 2006). This may be related to disease in motor association cortex where knowledge of action features may be represented. Alternately, verb naming difficulty may be due in part to an executive resource impairment that limits processing of verbs because of their complexity (Rhee et al., 2001). Verb naming difficulty in PNFA also may vary depending on the oral or written modality of production (Hillis et al., 2002). In semantic dementia, by comparison, naming may be more difficult for verbs of action than verbs of cognition because of degradation of the visual–perceptual feature knowledge represented in visual association cortex (Yi et al., 2007). Multiple levels of naming difficulty in these patients appear to reflect a large-scale neural network that can be interrupted in several distinct ways in these focal dementias, causing different kinds of naming impairments (Box 27.1).

Box 27.1   The neural basis for naming difficulty in dementia

Support for a "single deficit" approach to anomia comes from MRI studies suggesting shared patterns of cortical atrophy in Alzheimer's dementia and frontotemporal dementia (FTD). Areas of atrophy common to these patients appear to include temporal neocortex, frontal neocortex, and the hippocampus (Baron *et al.*, 2001). Assuming a single source of naming difficulty, a correlative MRI study related impaired lexical retrieval on confrontation naming and category naming fluency tests in a combined group of semantic dementia patients and patients with Alzheimer's dementia to atrophy in a single region – left anterior temporal cortex (Galton *et al.*, 2001).

Support for a "large-scale network" approach comes from MRI studies suggesting partially distinct patterns of cortical atrophy in Alzheimer's dementia and FTD. While direct comparisons of the anatomic distribution of disease in Alzheimer's dementia and FTD have been rare, distributions of atrophy in

the temporal lobe appear to be somewhat different in these conditions (Grossman *et al.*, 2004). Each FTD subgroup also appears to have a partially distinct anatomic distribution of disease, according to *in vivo* measures such as single photon emission computed tomography (SPECT), positron emission tomography (PET), and quantitative analyses of atrophy in high resolution MRI studies. Patients with semantic dementia thus seem to have left anterior temporal disease, inferior frontal regions of the left hemisphere appear to be compromised in patients with progressive non-fluent aphasia (PNFA), and disease in non-aphasic patients is said to be focused in frontal regions of the right hemisphere. This is illustrated in a voxel-based morphometry (VBM) analysis of FTD subgroups (see figure below). A correlative study related naming difficulty in these patient subgroups to partially distinct anatomic distributions of disease (Grossman *et al.*, 2004).

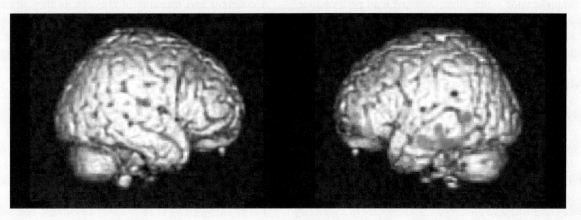

Significant cortical atrophy in PNFA (*n* = 7, green), semantic dementia (*n* = 8, red), and FTD patients with a disorder of social comportment and executive functioning (*n* = 14, blue) using voxel-based morphometry relative to healthy seniors (*n* = 11) (see Plate 17).

Baron, J.-C., Chetelat, G., Desgranges, B., Perchey, G., Landeau, B., de la Sayette, V., & Eustache, F. (2001). *In vivo* mapping of gray matter loss with voxel-based morphometry in mild Alzheimer's disease. *Neuroimage, 14*, 298–309.

Galton, C.J., Patterson, K., Graham, K.S., Lambon Ralph, M.A., Williams, G., Antoun, N., Sahakian, B.J., & Hodges, J.R. (2001). Differing patterns of temporal atrophy in Alzheimer's disease and semantic dementia. *Neurology, 57*, 216–225.

Grossman, M., McMillan, C., Moore, P., Ding, L., Glosser, G., Work, M., & Gee, J.C. (2004). What's in a name: Voxel-based morphometric analyses of MRI and naming difficulty in Alzheimer's disease, frontotemporal dementia, and corticobasal degeneration. *Brain, 127*, 628–649.

27.4. SEMANTIC MEMORY DIFFICULTY IN DEMENTIA

So, you hear a word, but what does it mean? Semantic memory is the long-term representation and processing of the concepts underlying objects, actions, abstract words, and the like. Semantic memory plays a crucial role in many cognitive measures such as naming and word-picture matching. Frequently observed difficulty on tasks such as

these suggests that semantic memory is widely impaired in patients suffering from a variety of neurodegenerative diseases, including Alzheimer's dementia and FTD. However, the basis for the semantic memory impairment in these patient groups is unclear. Our model of semantic memory includes at least two components: The long-term representation of knowledge, such as the features contributing to objects; and the semantic processes making use of this knowledge to integrate these features into a coherent whole

27.5. GRAMMATICAL DEFICITS
IN DEMENTIA

Grammatical processes are thought by many to be quintessentially linguistic. While speech errors occur commonly throughout our lifespan, and while difficulties with naming clearly increase as we age, grammatical deficits are often thought to be a narrow property of a linguistic system that is uniquely human. Indeed, this "impenetrable" quality of grammar was thought to be a model for analyses of modular cognitive functioning.

More recently, two views have begun to modify this classic approach. One perspective asserts that grammar is an emergent property of the lexical semantic and phonologic systems, modified by statistical properties of language use such as frequency. This work derives largely from assessments of the inflectional morphology system. The dual mechanism account of English verb past tense formation thus holds that a rule-based procedure generates regular past tense forms by addition of a dental suffix. The particular allomorph used in speech is phonologically determined, and the suffix added in writing is -ed. Irregular past tense forms such as *caught* are listed in the mental lexicon. In a landmark study, Ullman and his co-workers found that patients with Parkinson's disease have selective difficulty with rule-based past tense formation, and this was associated with their limited procedural memory. They also showed that patients with Alzheimer's dementia have difficulty with irregular past tense formation, and associated this with their declarative memory difficulty recalling individual words (Ullman *et al.*, 1997). An alternate model of past tense formation invokes a single complex mechanism for the production of both regular and irregular forms of the past tense. According to this model, a distributed constraint-satisfaction process depends on the phonological representations of words, and semantic information is recruited additionally to help retrieve the past tense forms of infrequent irregular verbs. Advocates of the single mechanism approach point to computer simulations demonstrating that one complex mechanism can explain past tense formation for both regular and irregular verbs. Evidence from patients with semantic dementia has been marshaled to support this claim (Patterson *et al.*, 2001). Their intact past tense formation of regular verbs is attributed to their preserved phonologic processing. However, their difficulty generating past tense forms of irregular verbs appears to be modulated by a frequency effect, and performance on irregular verbs correlates with performance on a measure of semantic memory.

A second modification of the modular approach to grammar forwards the idea that executive resources support grammatical processing. Grammatical processes also are crucial for sentence comprehension and production. Effortful and agrammatic speech are among the hallmarks of PNFA (Grossman *et al.*, 1996). These patients are also thought to have impaired grammatical comprehension: They

have difficulty with the subset of sentences featuring a complex grammatical structure. Moreover, it appears that PNFA patients have limited working memory (Grossman *et al.*, 1996), and this too contributes to their sentence comprehension difficulty. Sentences with a center-embedded subordinate clause (e.g., "The friendly boy$_i$ with short brown hair that t_i chased the girl lives in Boston") require maintaining the head noun phrase "the friendly boy") in working memory throughout the subordinate clause, for example, so that the noun phrase can be linked appropriately (indicated by the subscript "i") with the information following the subordinate clause (in this case, the silent trace t representing the subject of the verb "chased"). In an on-line word detection study, information maintained in working memory apparently becomes degraded as PNFA patients attempt to understand sentences like this, thus showing how working memory difficulty can contribute to a grammatical comprehension impairment (Grossman *et al.*, 2005). Observations such as these underline that a large-scale neural network helps support complex language processes such as sentence comprehension, and that disease in various portions of this network compromise language functioning (Box 27.3).

27.6. DISCOURSE DEFICIT
IN DEMENTIA

Self-expression in fluent, spontaneous speech is a critical feature in defining the humanness of our species. Impairment of connected speech is devastating to an individual's ability to relate to those around him or her. While most studies of language difficulty in dementia focus on the linguistic levels of phonology, syntax, and semantics, less attention has been directed toward the higher level of discourse and multi-sentence utterances. One distinguishing feature of discourse is that it places significant demands on executive resources beyond those required for other aspects of language processing. This resource-related component appears to play a substantial role in the discourse impairments of dementia patients.

In one report, Alzheimer's dementia patients appear to be impaired on what is termed "macro-level processing" but not on the "microlinguistic" levels of syntax and lexicon (Glosser & Deser, 1990). Another study finds that Alzheimer's dementia patients are impaired at both the levels of "gist" (overall interpretation and inferencing) and "detail" (accuracy of content) (Chapman *et al.*, 2002).

We elicited a semi-structured speech sample from FTD patients by having them provide a narrative to accompany a wordless picture story (Ash *et al.*, 2006). We find narrative difficulty in FTD due in part to the symptoms of PPA. PNFA patients thus have the sparsest output, producing narratives with the fewest words per minute compared to other subjects. Semantic dementia patients have considerable difficulty retrieving the words needed to tell coherent

Box 27.3   The neural basis for grammatical and working memory components of sentence processing in progressive non-fluent aphasia

Several studies describe left inferior frontal atrophy in PNFA (Gorno-Tempini *et al.*, 2004). While this anatomic locus of disease is associated with the PNFA syndrome, no published work tests a direct correlation between cortical atrophy in PNFA and sentence comprehension. We administered a measure of sentence comprehension to PNFA patients. This required subjects to answer a simple probe question about an orally presented sentence, where some sentences contained a subordinate clause. Using voxel-based morphometry (VBM), we correlated grammatical aspects of sentence comprehension with cortical atrophy in PNFA. The figure shows the anatomic distribution of significant cortical atrophy in these patients (green and yellow areas). Within this anatomic distribution, several areas correlated significantly with grammatical comprehension difficulty (yellow areas), including two areas in left inferior frontal cortex. The ventral frontal area in Brodmann area 47 is associated with grammatical comprehension in some fMRI studies of sentence processing in healthy adults, while the more dorsal frontal area in Brodmann areas 44 and 6 is often associated with working memory during sentence processing (Cooke *et al.*, 2005). Additional support for these two components of sentence processing comes from an fMRI activation study in FTD that monitors brain activity during comprehension of grammatically complex sentences that also feature a working memory component. In PNFA, poor comprehension of the grammatical component of these sentences is associated with little recruitment of the inferior frontal area. By comparison, non-aphasic patients with a disorder of social comportment and executive functioning showed

a sentence comprehension deficit that reflected their working memory difficulty, and this was associated with little activation of the dorsal frontal area.

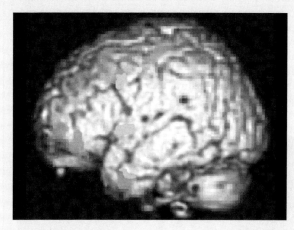

VBM correlation of grammatical comprehension in sentences with cortical atrophy in PNFA (see Plate 19).

Cooke, A., DeVita, C., Gonzalez-Atavales, J., Moore, P., Chen, W., Gee, J.C., & Grossman, M. (2005). Large-scale neural network for sentence processing. *Brain and Language, 96*, 14–36.

Gorno-Tempini, M., Dronkers, N.F., Rankin, K.P., Ogar, J.M., Phengrasamy, L., Rosen, H.J., Johnson, J.K., Weiner, M., & Miller, B.L. (2004). Cognition and anatomy in three variants of primary progressive aphasia. *Annals of Neurology, 55*, 335–346.

---

narratives. Though not aphasic, patients with a social and executive disorder due to FTD have profound difficulty organizing their narratives. They are able to describe the content of a picture, but they cannot link this picture to the preceding picture or to the one that follows, and they cannot effectively express the overall point of the story. This deficit correlates with poor performance on a measure of executive functioning that requires an organized mental search. Impaired day-to-day communication in non-aphasic FTD patients with a disorder of social and executive functioning thus appears to be due in part to a striking deficit in discourse organization (Box 27.4).

27.7. CHALLENGES AND FUTURE DIRECTIONS

This brief review emphasizes the variety of impairments that can be seen at several levels of language processing.

For speech, naming, comprehension, and discourse, it appears that patients with neurodegenerative diseases have distinct impairment profiles. This is unlikely to be due to diffuse disease that compromises brain functioning in a non-specific manner. If so, similar patterns of difficulty would be observed across these various conditions. Likewise, the observed complexity of impairment is unlikely to be due to the destruction of an isolated naming or comprehension faculty. If true, this would have resulted in a single impairment profile within each language domain. We observe instead a rich variety of language deficits that differ in appearance across the various neurodegenerative diseases I addressed.

This work has important theoretical implications, since it emphasizes the large-scale neural networks that support language processing. I emphasize that these patterns of impairment do not replicate the disorders seen in stroke patients. While direct comparisons are rare, Broca's aphasics differ from patients with PNFA; and Wernicke's aphasics do not

28

Frontal Lobes and Language

SKYE McDONALD

School of Psychology, University of New South Wales,
Sydney, Australia

ABSTRACT

People who sustain damage to the frontal lobes of the brain due to focal lesions, traumatic brain injury, frontotemporal dementia or other conditions have impaired communication skills. They can be ineffective and disorganized in both their ability to communicate and in their capacity to understand innuendo and inference. The 1980s and 1990s saw an upsurge in the use of sophisticated techniques to characterize these difficulties. Since that time there has been growing interest in examining the underpinnings of communication impairment focusing upon neuropsychological abilities mediated by the frontal lobes. In this chapter we examine how conventional constructs associated with executive function such as working memory, inhibition, flexibility and general problem solving impact upon communicative competence. In addition, very recent advances in social neurosciences are examined, including theory of mind, emotion and social knowledge, and how these might impact upon the ability to communicate effectively.

28.1. INTRODUCTION

The ability to communicate requires more than simply having access to word forms, meanings and syntax. Communication is a process in which one speaker makes his or her thoughts and intentions known to another in an effective and socially acceptable manner. While language is an important tool in this process, there are clearly other cognitive and affective processes involved. Nowhere is this more apparent than in communication patterns following frontal lobe damage (FLD). People with frontal lobe injuries can have perfectly intact language skills, that is to say, they are able to name objects, construct grammatically acceptable propositional speech and understand the literal meaning of conversations in which they engage. But despite this, their communication skills are frequently aberrant. When attempting to converse their narratives may peter out or lapse into perseverative comments and stereotyped phrases (Alexander *et al.*, 1989). Alternatively they may be over-talkative, disorganized and frankly confabulatory in what they say with apparent indifference to their conversational partner. Similar patterns of impairment are seen in adults with frontal dementia and traumatic brain injuries (TBI) both of which result in frontal systems impairment.

It has been a conundrum as to how to measure these problems and how to explain their occurrence. Fortunately, contemporary linguistic theories, especially those with a social focus have been used to document communication breakdown and the 1980s and 1990s saw a great deal of work in this regard. In particular, an examination of pragmatic inferences, that is the inferred meanings that arise when language is used in context, has been revealing. Theories that address how discourse unfolds have also been useful to reveal problems in language production. With the advent of sensitive measures of these deficits we are in a better position to be able to study and explain them. Added to this, recent advances in neurosciences provide some exciting new possibilities. In this chapter we will provide an overview of how problems in inference and discourse management manifest after frontal lobe lesions and some possible explanations for this.

28.2. PRAGMATIC INFERENCE

Every time we speak we choose our words to be both effective and socially appropriate. Often we prefer to state our objectives indirectly because we wish to be polite or otherwise thought well of by the listener (Brown & Levinson, 1987). When we speak indirectly, we expect the listener to recognize the inferred meaning (the pragmatic inference), for example, *It is hot in here!* as a request for the window to be opened. The "social dysdecorum" that frequently ensues after FLD includes not only socially inappropriate language but also lack of insight, blunted social awareness, tactlessness, poor reasoning and behavioral dyscontrol (Alexander *et al.*, 1989). These collective features suggest that people with frontal lobe lesions may have problems understanding and using pragmatic inference. And indeed, this seems to be the case. In numerous studies over the past 10 years, it has been demonstrated that FLD, due to either focal lesions or traumatic brain injury (TBI), lead to disorders in the understanding of sarcasm in which a speaker says one thing, for example *What a great haircut* to suggest the reverse. Mostly this is because the patient with FLD erroneously interprets the salient literal meaning as the meaning intended. Having established this, the question arises as to why this occurs?

According to traditional speech act theory, the process of decoding such inferences is sequential: (1) the literal meaning is comprehended; (2) some cue indicates this is not sufficient (e.g., the literal meaning is contradicted by the context) and (3) inferential rules are implemented in order to extract the intended meaning. Subsequent theorists have challenged the sequential nature of the comprehension process but evidence continues to suggest that both the literal and inferred meanings are normally processed. Given that FLD does not result in basic language impairment, failure to understand sarcasm is unlikely to reflect a problem with step (1). Consequently deficits must reside in other processes, either (2) or (3).

28.2.1. Prefrontal Functions and Inference

While the prefrontal lobes do not appear to have a primary role in perception or movement, they are critical to the goal directed control of all cognitive processes, that is, executive function. Experimental work in both animals and humans has demonstrated that while cells in the occipital cortex may fire uniformly to identical sensory information, cells in the prefrontal cortex fire differentially to the same stimuli when presented in different contexts (Mesulam, 2002). This suggests that the prefrontal cortex has a role in determining how context alters meaning. Further, adults with FLD frequently respond to superficial characteristics of the environment and are unable to think conceptually. They have poor inferential reasoning on neuropsychological tests and also fail to pick up inferences linking one sentence to the next (Ferstl *et al.*, 2002). Poor general inference making has also been specifically linked to poor understanding of sarcasm (Martin & McDonald, 2005; McDonald *et al.*, 2006).

Executive function is a complex notion and FLD produces a wide variety of deficits. This makes it difficult to develop theoretical models to account for the range of phenomena

seen. In general, it is agreed that executive function is the process by which we regulate our thoughts and behavior in order to meet desired goals and solve problems. Several constructs that are frequently discussed as integral to this process are working memory, inhibition and flexible, attentional focus. There are good reasons why deficits in each of these areas may interfere with inferential reasoning.

Working memory is the process by which information from long term memory stores and/or more immediate perceptual processes is manipulated to enable effortful cognitive processing to occur. Working memory is pivotal to the computation of inferences in written text (Calvo, 2005). Adults and children with poor working memory as a result of TBI are also the poorest at generating inferences (Barnes & Dennis, 2001) and understanding sarcasm (Martin & McDonald, 2005; McDonald *et al.*, 2006). Furthermore, by reducing working memory demands their ability to understand inferences significantly improves (Barnes & Dennis, 2001). Information processing speed is integrally associated with working memory and has also been associated with less efficient inference making in normal readers (Calvo, 2005) and poor pragmatic understanding in adults with TBI (McDonald *et al.*, 2006).

Poor inhibitory control prevents the suppression of irrelevant associations, thus adding to working memory demands. It also increases the likelihood that listeners will react to initial meanings of conversational comments. The ability to shift one's focus of attention flexibly between relevant attributes of the environment is also fundamental to efficient executive processing. Inflexible, perseverative thinking is a common consequence of FLD and results in rigid fixation upon superficial attributes. While disorders of inhibition and flexibility are both likely to disrupt conversational inference, evidence linking them to pragmatic understanding is limited. Only one study to date has examined the relation between problems with inhibition and poor understanding of sarcasm and found this to be significant, if modest (Channon & Watts, 2003). Perseveration, also, has been shown to share a small amount of variance with scores on a sarcasm task (Martin & McDonald, 2005) although this is not a robust finding across studies (McDonald *et al.*, 2006).

The generally small amount of shared variance found between inhibition, flexibility and conversational inference may reflect the lack of reliability of the measures used to date. But it is also possible that it reflects fundamental differences in the types of information used for the two types of tasks. Specifically, social information focuses upon different attributes than that which characterizes the words, numbers and simple figures that so often make up standard neuropsychological tests. For example, social information emphasizes individual rather than categorical differences and is dynamic, changing in response to the observer. Thus, while working memory may be fundamental to all cognitively demanding tasks, the ability to flexibly focus upon (and conversely, ignore) attributes of information may engage different processes and different brain systems depending on whether the task is essentially social or non-social. We will explore this issue further below.

28.2.2. The Social Nature of Conversational Inference

Conversational inference requires making judgments that are unique to social tasks. Sarcasm and the related device, irony, for example, have less to do with making statements about a state of affairs and more to do with the speaker's attitude toward that state of affairs. Scorn and derision are often inextricably linked. This suggests that, in order to interpret such utterances, the hearer must have the means to infer the speaker's attitude and level of sincerity. Other kinds of conversational inference (e.g., hints) may also allude to issues that are on the speaker's mind rather than in the patent context.

Linguistic theorists have grappled with the unique, social facets of pragmatic inference for some time. Although neurosciences have been slow to catch up there has been growing interest in social information processing. Indeed, there is accruing evidence that social information processing is functionally and even neuroanatomically distinct from non-social processing. This provides a major impetus to re-examine the contributions that frontal systems make to understanding conversational inference.

28.3. SOCIAL COGNITION

Given that humans are social animals, it makes sense that social intelligence (i.e., the ability to interact with others in a complex and flexible manner) would have a special status in natural selection. This premise underpins arguments that social information processing may have evolved as a separate processing system within the human brain with different processing demands to that involved in non-social problem solving (Adolphs, 2001). Certainly, dissociations between social and non-social problem solving appear across different developmental disorders. Thus, adults with Asperger's syndrome can excel at non-social problem solving but have poor social functioning. On the other hand, children with Williams syndrome have abnormally low intellectual skills but high levels of social competence. In addition, there are numerous published reports of adults with brain damage who perform normally on tests of executive function, but whose ability to function socially is dramatically reduced (Saver & Damasio, 1991).

There are currently three major constructs that characterize social cognition: (1) "theory of mind" (ToM), that is the ability to make judgments concerning another's mental states; (2) emotion processing, that is the ability to understand and distinguish different affective states and (3) social knowledge schemas that guide judgments and behavior. Evidence concerning the neural substrates of these different components consistently highlights overlapping systems that include the

prefrontal cortex. Furthermore, whereas the dorsolateral surfaces of the prefrontal lobes are engaged in many conventional executive tests, it is the orbitomedial and ventral surfaces of the frontal cortices, in association with limbic structures such as the amygdale and cingulate gyrus that are associated with social cognitive tasks (Adolphs, 2001).

28.3.1. Theory of Mind

ToM is a construct originally evoked to explain autistic behavior. ToM is the capacity to attribute mental states, such as thoughts, beliefs, desires and intentions to others. Given that all human behavior is motivated by beliefs, intentions and so forth, it stands to reason that ToM is essential to understand the communicative efforts of others. Non-literal language devices rely particularly upon ToM because one must understand the mind of the speaker in order to appreciate the inference. For example, metaphor requires understanding that someone is making an implicit comparison between the actual state of affairs and some similar set of circumstances. Children with autism who cannot understand first order ToM (what another person believes) cannot understand metaphors. Sarcasm, on the other hand, relies upon more sophisticated understanding of the mental states of others, for in order to appreciate irony, one must understand that the speaker not only has a particular view, but wants the listener to share that view, despite the literal meaning of his or her utterance. In order to understand this we must understand what one person thinks about another person's thoughts – second order ToM. While children who have first order ToM may be able to appreciate metaphor, they fail to understand sarcasm and irony if they cannot make such second order ToM judgments (Happé, 1993).

There are general similarities between communicative deficits reported in Asperger's disorder (mild autism) (see Chapter 37) and those described following focal, frontal lesions, frontotemporal dementia and TBI. Of the many anecdotal descriptions of FLD available, a general picture emerges of the individual who has difficulty taking into account the thoughts and feelings of others. Descriptors such as "egocentric" and "self-focused" are common and communication style is described as blunt, overly familiar and inappropriate (Alexander et al., 1989). For these reasons, recent work has addressed the question as to whether FLD leads to disorders of ToM and whether, in turn, these are related to problems with communication. Typically, ToM is tested using "false belief" stories where one protagonist acts upon a false belief about the actions of another (such as one fellow drinking from a can of Coke on the assumption that it is full of Coke and not milk, not knowing that his practical joking brother has exchanged the contents). Understanding of ToM in such stories is assessed via comprehension questions. Using such tasks it has been found that people with FLD, for example after TBI, have deficits in ToM reasoning (Milders et al., 2003).

Research examining the role of ToM in communication has concentrated upon an examination of its role in processing sarcasm and deceit. Recognition of deceit, like sarcasm, requires understanding what the speaker intends the listener to believe or feel as a result of the interaction. The comment *No of course you don't look fat!* may be spoken sarcastically wherein the speaker wants the listener to believe the reverse, or may be spoken kindly, in order to make the listener feel better about himself than reality might warrant. In order to determine the difference, second order ToM is required so as to understand the speaker's intention. Children and adults with TBI who have poor second order ToM are also those

most likely to experience problems understanding the meaning of sarcastic exchanges and/or lies (Dennis *et al.*, 2001; McDonald & Flanagan, 2004).

28.3.2. Emotion Processing

The neural processes associated with emotion have engendered an upsurge of interest in the last decade. Whereas earlier research emphasized cerebral laterality, more recent research has emphasized the role of the prefrontal cortex, in concert with the amygdala, anterior cingulate gyrus, ventral striatum and other structures associated with the limbic system (Phillips *et al.*, 2003). Deficits in the ability to recognize emotional expressions in others are found following focal lesions (e.g., the frontal lobes and amygdala) (Hornak *et al.*, 2003), and more diffuse conditions including, once again, TBI (Milders *et al.*, 2003) and frontotemporal dementia (Keane *et al.*, 2002).

Failure to recognize the emotions of others will clearly affect social functioning and communicative competence. It is difficult to imagine how any normal conversation can proceed if one speaker is unable to gauge the emotional responses of their conversational partner. For example, Shimokawa found that people with Alzheimer's disease who had problems with emotion recognition were more likely to be thought of as indifferent and socially awkward (Shimokawa *et al.*, 2001). In adults with TBI, poor emotion recognition has been associated with poor use of humor when conversing (McDonald *et al.*, 2004).

Consideration of the role of emotion recognition in communication feeds into related arguments regarding ToM and intentionality judgments. After all, how do speakers make judgments concerning speaker intentions? While some of this information may arise from shared knowledge concerning the current context and prior events, an important source of information is the emotional demeanor of the speaker. Very little research has examined the relation between emotion and pragmatic inference. That which has been conducted has focused upon sarcasm and results have been ambiguous (McDonald & Flanagan, 2004). Because sarcasm is usually conveyed via a statement that blatantly contradicts the context, it pivots specifically upon the ability to *disregard* the literal interpretation of the (sarcastic) comment. This may place an exaggerated emphasis upon flexibility and inhibition to the exclusion of other abilities that might normally contribute. The role of emotion in other, more subtle kinds of pragmatic inference (such as hints) has not been explored.

28.3.3. Social Knowledge

Within social psychology it has long been recognized that we hold stereotypes about others and that these influence

---

Box 28.1   Is ToM a unique ability?

ToM is a controversial notion. One major area of contention with regards to ToM is whether or not this represents a modular ability that can be differentially impaired or preserved independent of other cognitive abilities. The evidence for its modularity is not compelling but this may partially reflect the fact that ToM is loosely operationalized and tasks used to measure it vary from judgments concerning eye gaze and mood through to comprehending written and spoken stories. Doubly recursive questions (such as "Does John believe that Sally believes ...") typically used to assess comprehension of ToM in written stories make heavy demands upon language, working memory and general inferential reasoning skills. Indeed, some studies have reported a significant relation between ToM and non-ToM inferential reasoning although others have not (Channon *et al.*, 2005). Working memory has been found to make a significant contribution to scores on false belief tasks (Bibby & McDonald, 2004). These findings suggest that generic executive abilities underlie ToM (in stories).

But other kinds of ToM tasks do not place as much emphasis upon working memory or complex reasoning. The "Mind in the Eyes" test, for example, requires participants to choose a label to describe what a person is thinking or feeling, based on a photograph of their eyes alone. Frontal and amygdala damage disrupts the ability to make these kinds of judgments. Despite the fact that this task is less effortful than story comprehension, a relation between ToM thus measured and conventional executive tests of flexibility and self-regulation has been reported (Henry *et al.*, 2006). So the question as to whether these represent functional distinct abilities that coincidently share common neuroanatomical pathways or whether they are related skills with common underpinning processes remains unresolved.

Bibby, H., & McDonald, S. (2004). Theory of mind after traumatic brain injury. *Neuropsychologia, 43*, 99–104.

Channon, S., Pellijeff, A., & Rule, A. (2005). Social cognition after head injury: Sarcasm and theory of mind. *Brain and Language, 93*, 123–134.

Henry, J.D., Phillips, L.H., Crawford, J.R., Ietswaart, M., & Summers, F. (2006). Theory of mind following traumatic brain injury: The role of emotion and executive functioning. *Neuropsychologia, 44*, 1623–1628.

our social behavior. The notion that people with FLD have reduced access to such social knowledge has barely been considered to date. Indeed, it is widely believed that people with circumscribed deficits in either memory (learning) or executive function have relatively preserved access to knowledge and skills that were acquired premorbidly. However, recent research challenges this belief with respect to social knowledge. Functional imaging of normal individuals suggests that ventromedial portions of the frontal lobes, in conjunction with the amygdale are involved when judging "trustworthiness" based on facial characteristics. Conversely, such judgments are impaired with amygdala damage (Adolphs et al., 1998). Personality judgments of extroversion, warmth, neuroticism and adventurousness are impaired in people with FLD (Heberlein et al., 2004). In a related vein, the prefrontal regions, amygdalae and associated structures are engaged when evaluating faces according to gender, race, dominance and general attractiveness.

According to Milne and Grafman (2001) social schema are stored in the prefrontal lobes and are used to make initial social judgments. These schemas are activated rapidly and automatically, influenced by somatic responses via the amygdalae. The evidence for this proposal arose from the finding that seven adults with medial frontal damage were less regulated by implicit gender stereotypes on a reaction time task than non-brain-injured controls (Milne & Grafman, 2001). To date, there has been no research conducted that examines the relationship between deficits in social knowledge and communicative competence.

28.4. DISCOURSE PRODUCTION

The majority of this chapter has focused upon the mechanisms that underpin the ability to understand the communication of others, specifically, the ability to understand pragmatic inference. The measurement of deficits in discourse production is arguably more difficult, and the most innovative work to date has been primarily in the area of TBI. Discourse production has been assessed using a number of theoretical approaches focusing upon a variety of communicative tasks. The research in this area has been reviewed extensively in McDonald et al. (1999) and more recently in McDonald (2007). This research will be touched on briefly here focusing upon polite utterances, narratives and conversations. We will then turn to a discussion regarding potential explanations for such deficits within the framework of frontal lobe functions already discussed.

28.4.1. Polite Utterances

In Brown and Levinson's (1987) seminal work, the pivotal role of politeness in everyday communication was emphasized. Across cultures it was demonstrated how

speakers modulate their utterances in order to maximize the polite effect, by either deferring to the listener's desire to be unimpeded (their negative face) using indirectness as a means to create distance or, otherwise, their desire to be thought well of (their positive face) by the use of complements, in-group language and the like. Essentially the attitude of the speaker toward the listener is tacitly conveyed via their choice of polite device. In work conducted during the 1990s it was established that although speakers with TBI may have an intact repertoire of politeness strategies, they are less flexible than matched control speakers in their ability to use these across different contexts (Togher & Hand, 1998). When faced with diplomatically challenging communication tasks such as hinting, or making requests to overcome reluctance on the part of the listener, adults with TBI had difficulty refraining from stating their desires directly and making comments that were counter-productive to their desired outcome. These studies of politeness suggest that adults with TBI have difficulties using pragmatic inference to moderate the social impact of their language.

28.4.2. Narratives

Analysis of narratives such as story telling or explaining a procedure involves close examination of the text (either written by the participant or transcribed from their speech). This has been useful to document anecdotal accounts of the slow, hesitant or conversely, over-talkative communication patterns seen in many speakers with TBI. Formal analysis of narratives has confirmed that many adult and adolescent TBI speakers speak more slowly than their non-injured peers with more incomplete, ambiguous or uninterpretable utterances and less information.

Not only have researchers been interested to document the efficiency of narratives, but also whether these are coherent, that is whether they have semantic continuity. At a "local" level coherence can be measured by whether there are links that tie information in one proposition to others throughout the text, for example whether a pronoun in one sentence has a clear reference in another. Global coherence, on the other hand refers to whether the sequence of ideas (propositions) conforms to conventional narrative patterns. For example, there is an inherent structure to the way in which stories unfold. Similarly, there are essential elements and important sequences to follow when explaining a procedure.

Despite the sophistication of procedures used to examine both local and global coherence, the emerging picture has been mixed with evidence to suggest that there are difficulties with coherence as well as the reverse. It is possible that the ambiguity arises from the very heterogenous nature of the TBI population with respect to both the nature and severity of brain injury. This often includes FLD but may extend into other areas. Emerging from this rather complex area, one type of task that has revealed relatively consistent deficits is the production of procedural narratives. People with TBI, both adult and adolescent, have been reported to omit essential information, provide a disrupted sequence of explanation and include irrelevant and ambiguous material when explaining how to play a simple game. Raters' impressions of procedural texts produced by TBI speakers support notions of disrupted global coherence by suggesting they are confused and disorganized.

28.4.3. Conversations

Conversational exchanges are even more complex than narratives to quantify but some approaches such as functional

systemic linguistics as developed by Michael A.K. Halliday (for example as outlined in his 1985 text) have proven useful. Using such an approach it has been found that TBI speakers elicit a greater number of questions and prompts from their partners than their non-injured peers. They have poor topic maintenance and poor initiation, and they provide information in an inefficient and disorganized manner. They have also been found to provide more information, including inappropriate information to their conversational partner.

28.4.4. Neuropsychological Underpinnings of Disordered Discourse Production

The same executive function constructs as already described will impact upon the ability to produce effective communication. For example, the ability to produce coherent discourse requires continuous monitoring of output in order to ensure that each pronoun used has an unambiguous source and that ideas are inserted logically and effectively into the language flow. This is an on-line processing task that will depend on good working memory capacity. Consistent with this impaired working memory has been linked to low efficiency and local cohesion as well as to the complexity of clauses (Youse & Coelho, 2005).

The notion of inhibition is clearly relevant in socially appropriate language production. Often what is *not* said can be as important as what is. In line with this there is a modest correlation between poor ability to produce effective requests and indices of disinhibition on formal neuropsychological tests. Although inflexibility is also relevant when producing communication and has been documented with respect to politeness (Togher & Hand, 1998) there has been no independent research linking politeness in discourse to other measures of rigidity in people with TBI.

In general, the ability to produce language to meet socially defined goals can be thought of as a (social) problem-solving task, and consequently many of the same processes and control mechanisms as engaged in non-social tasks may be relevant. Thus, the global coherence of text may be affected by executive disorders that impair the ability to plan and sequence behavior to meet specified goals. In studies of adults and children with TBI (including mild and severe injuries) the amount of essential information provided was associated with measures of conceptual thought, generativity and problem solving. Likewise, inability to transform the gist of written text was correlated to poor problem solving (Chapman et al., 2004). These results strongly suggest that general deficits in the ability to reason flexibly and readily, in general, interfere with the orderly transmission of information. Furthermore, certain kinds of tasks may be more taxing of executive abilities than others. Story generation tasks, for example, wherein the speaker has no template for the structure and flow of information, are likely to be more

---

Box 28.2   It takes two to tango: the role of partners in creating disordered conversation

Claims concerning the aberrant nature of conversational skills in people with frontal, or indeed any clinical condition, need to be considered carefully. After all, conversation is a two way process and the behavior of one co-conversant can have a dramatic effect upon the nature of the conversation, regardless of the skills of the other. Togher and colleagues have applied exchange structure analysis, a form of discourse analysis based on M.A.K. Halliday's work to examine the conversations of speakers with TBI when conversing with their mothers, with therapists, with members of the police force and with other service providers. In these they demonstrated that compared to normal adults without brain injuries, the people with TBI were asked more questions for which the speaker already knew the answer. In addition, they were subjected to repeated checking of the veracity of their responses, were given more information and were asked for less (Togher et al., 1997a, b).

This analysis demonstrated how people who were interacting with speakers they identified as disabled were disempowering them and, in doing so, depriving them of normal opportunities to use any preserved communication skills. In a further powerful demonstration of this, Togher and colleagues showed that speakers with TBI – when placed in a more empowering situation, such as educating school children on the circumstances of their injury – conversed in a manner that was indistinguishable from control speakers (with spinal injuries) (Togher, 2000).

Togher, L. (2000). Giving information: The importance of context on communicative opportunity for people with traumatic brain injury. *Aphasiology, 14*(4), 365–390.
Togher, L., Hand, L., & Code, C. (1997a). Analysing discourse in the traumatic brain injury population: Telephone interactions with different communication partners. *Brain Injury, 11*(3), 169–189.
Togher, L., Hand, L., & Code, C. (1997b). Measuring service encounters with the traumatic brain injury population. *Aphasiology, 11*(4–5), 491–504.

challenging for those with executive disorders than story recall based on exposure to an oral or pictorial story. In general, it appears that story generation tasks are more demanding for adult speakers, be they normal or suffering from TBI.

Social cognition may well be relevant to these particular dimensions of discourse production but has not, as of yet, been examined. Interestingly, however, it seems that emotionally salient experiences are related by adolescents with and without TBI with better informational structure and coherence than less engaging tasks (Leer & Turkstra, 1999). This raises the question as to whether emotional engagement facilitates coherent discourse and, if so, how. This is another fascinating avenue of exploration that awaits empirical research.

28.5. CHALLENGES AND FUTURE DIRECTIONS

Linguistic theory has helped document communication disorders after FLD and there have been huge advances in the past few decades. This has provided an exciting impetus to discovering how the frontal lobes mediate communication. With better measures of communication we are able to examine the relationship between them and well-established "cognitive constructs" associated with frontal lobe function such as working memory, flexibility and inhibition, as well as some more recently conceptualized constructs in "social cognition" such as ToM, emotion and social knowledge. But there are many challenges yet to be met.

In particular, our understanding of the role that social cognition plays in communication has a long way to go. ToM is the construct that has attracted the most attention to date but its role in communication has been limited to an examination of irony and deceit. It could be argued that all communication revolves around understanding the speaker's beliefs and intentions. Indeed, theorists such as Dan Sperber and Deidre Wilson regard these kinds of inferences as central to the communication process. The field is open for far more research into this area. Similarly, there has been no research to date that links ToM to discourse production, either the ability to produce socially appropriate speech acts, such as requests, or extended discourse such as narratives and conversations.

The role of emotion judgments in communication is another field that is wide open for examination. Apart from a few studies examining emotion recognition with respect to sarcasm, there has been little systematic investigation of the role of emotion recognition in the ability to both understand conversational meaning and the ability to temper one's discourse output appropriately. Furthermore, emotion research *per se* is undergoing rapid development with interesting ramifications for research into communication. A fascinating new direction is the idea that emotional experience is mediated by the orbitofrontal system and that this may play an important role in cognition. For example, emotional recognition and emotional experience are intimately associated

such that the making of facial expression leads to emotional changes (Levenson *et al.*, 1990). Adults with focal frontal lesions and TBI can experience reduced emotional responsivity. Furthermore, there is a correlation between this and impaired recognition of emotional expressions (Hornak *et al.*, 2003) suggesting that one may assist the other. Certainly, it appears the ventral frontal system mediates both the early rapid appraisal of emotionally significant stimuli, and affective responses to those stimuli (Phillips *et al.*, 2003).

Interestingly, the role that somatic cues play in cognition may not be limited to appraisal of emotional states. The role of somatic markers in guiding decision making is a new area of research lead by Bechara and colleagues. Their main thesis is that humans anticipate complex response contingencies and can have a somatic response to these even prior to conscious awareness (Bechara, 2004). The majority of work has used the Iowa Gambling Task. This task requires the participant to choose from "high risk" and "low risk" card decks. The risky decks lead to high gains and losses but, if played continuously, will gradually result in loss. The low risk decks have lower gains and losses but lead to gains in the longer term. Normal, healthy adults learn over trials to play with the "low risk" decks but adults with ventromedial damage do not. Both normal adults and those with ventromedial lesions have a change in their skin conductance when provided with feedback regarding their choices. Importantly, normal adults start to demonstrate these changes in anticipation of their choice, even before they can verbalize the rules. People with ventromedial damage do not. As a result of this work it has been argued that somatic responses may assist decision making. This may be especially true in personal and social decision-making situations, where precise calculations of the outcome of any given course of action is not possible (Bechara, 2004). Further work on rare patients with focal damage restricted to either the right or left hemisphere has found that right ventromedial lesions are more likely to cause such deficits than left. These patients are also the ones to experience poor social and interpersonal function, impaired emotion processing and unemployment. Whether or not such somatic responses play a role in communication is unknown although it would be expected that such highly social, interactive behavior as communication would be influenced by somatic responses.

As indicated in our earlier discussion, the role of social appraisals in communication is yet to be examined. The notion that people make social judgments fairly rapidly upon brief encounters with strangers is not new. The idea that judgments about the social status of others influences how we speak to them is not new either. Brown and Levinsons's seminal work on politeness established this many years ago. But the idea that initial social judgments may be mediated by frontal systems and therefore impaired with frontal damage is new. Further, the possibility that frontal lobe injury may impact upon both these social judgments and the ability to moderate polite communication accordingly is as yet unexplored.

Finally, although other constructs associated with "executive function" are well established, their role in communication is still not well documented or understood. There seems to be unequivocal evidence that working memory is required for effortful communication tasks. Discourse production, in general, seems to tap many of the same skills as engaged in other problem-solving tasks. But evidence for the role of specific constructs such as inhibition and flexibility is, as yet, limited. In part, this may reflect the ongoing challenge of documenting discourse characteristics in a manner that is reliable and consistent across tasks and research groups. Applications of discourse analyses that focus upon cohesion, for example, are potentially of great value but produce variable results from one study to another. This makes the next step of determining cognitive correlates frustratingly difficult. But it is also possible that these constructs need to be measured as they apply to specifically social information. In addition, the notion that social and non-social information processing may tap different constructs altogether raises a whole new range of questions as to whether conventional measures of executive function are, in fact, tapping abilities that are less critical to communication than more recently conceptualized social measures. The next decade should provide some exciting answers to these questions.

References

Adolphs, R. (2001). The neurobiology of social cognition. *Current Opinion in Neurobiology, 11*, 231–239.

Adolphs, R., Tranel, D., & Damasio, A.R. (1998). The human amygdala in social judgement. *Nature, 393*, 470–474.

Alexander, M.P., Benson, D.F., & Stuss, D.T. (1989). Frontal lobes and language. *Brain and Language, 37*, 656–691.

Barnes, M.A., & Dennis, M. (2001). Knowledge-based inferencing after childhood head injury. *Brain and Language, 76*(3), 253–265.

Bechara, A. (2004). The role of emotion in decision-making: Evidence from neurological patients with orbitofrontal damage. *Brain and Cognition, 55*(1), 30–40.

Brown, P., & Levinson, S. (1987). Politeness: Some universals in language use. *Studies in interactional sociolinguistics*, Vol 4. New York: Cambridge University Press.

Calvo, M.G. (2005). Relative contribution of vocabulary knowledge and working memory span to elaborative inferences in reading. *Learning and Individual Differences, 15*(1), 53–65.

Channon, S., & Watts, M. (2003). Pragmatic language interpretation after closed head injury: Relationship to executive functioning. *Cognitive Neuropsychiatry, 8*(4), 243–260.

Chapman, S.B., Sparks, G., Levin, H.S., Dennis, M., Roncadin, C., & Zhang, L. et al. (2004). Discourse macrolevel processing after severe paediatric traumatic brain injury. *Developmental Neuropsychology, 25*(1–2), 37–60.

Dennis, M., Purvis, K., Barnes, M.A., Wilkinson, M., & Winner, E. (2001). Understanding of literal truth, ironic criticism, and deceptive praise following childhood head injury. *Brain and Language, 78*, 1–16.

Ferstl, E.C., Guthke, T., & von Cramon, D.Y. (2002). Text comprehension after brain injury: Left prefrontal lesions affect inference processes. *Neuropsychology, 16*(3), 292–308.

Happé, F.G.E. (1993). Communicative competence and theory of mind in autism: A test of relevance theory. *Cognition, 48*, 101–119.

Halliday, M.A.K., (1985). *An Introduction to Functional Grammar.* London: Arnold.

Heberlein, A.S., Adolphs, R., Tranel, D., & Damasio, H. (2004). Cortical regions for judgments of emotions and personality traits from point-light walkers. *Journal of Cognitive Neuroscience, 16*, 1143–1158.

Hornak, J., Bramham, J., Rolls, E.T., Morris, R.G., O'Doherty, J., & Bullock, P.R. et al. (2003). Changes in emotion after circumscribed surgical lesions of the orbitofrontal and cingulate cortices. *Brain, 126*, 1691–1712.

Keane, J., Calder, A.J., Hodges, J.R., & Young, A.W. (2002). Face and emotion processing in frontal variant frontotemporal dementia. *Neuropsychologia, 40*(6), 655–665.

Leer, E.V., & Turkstra, L. (1999). The effect of elicitation task on discourse coherence and cohesion in adolescents with brain injury. *Journal of Communication Disorders, 32*(5), 327–349.

Levenson, R.W., Ekman, P., & Friesen, W.V. (1990). Voluntary facial action generates emotion-specific autonomic nervous system activity. *Psychophysiology, 27*(4), 363–384.

Martin, I., & McDonald, S. (2005). Exploring the causes of pragmatic language deficits following traumatic brain injury. *Aphasiology, 19*, 712–730.

McDonald, S. (2007). Neuropsychological and social underpinnings of communication disorders after traumatic brain injury. In M.J. Ball & J. Damico (Eds.), *Clinical aphasiology – future directions*. Sussex, UK: Psychology Press.

McDonald, S., & Flanagan, S. (2004). Social perception deficits after traumatic brain injury: The interaction between emotion recognition, mentalising ability and social communication. *Neuropsychology, 18*, 572–579.

McDonald, S., Togher, L., & Code, C. (Eds.) (1999). *Communication disorders following traumatic brain injury*. Hove, UK: Psychology Press.

McDonald, S., Flanagan, S., Martin, I., & Saunders, C. (2004). The ecological validity of TASIT: A test of social perception. *Neuropsychological Rehabilitation, 14*, 285–302.

McDonald, S., Bornhofen, C., Shum, D., Long, E., Saunders, C., & Neulinger, K. (2006). Reliability and validity of "The Awareness of Social Inference Test" (TASIT): A clinical test of social perception. *Disability and Rehabilitation, 28*, 1529–1542.

Mesulam, M.-M. (2002). The human frontal lobes: Transcending the default mode through contingent encoding. In D.T. Stuss & R.T. Knight (Eds.), *Principles of frontal lobe function* (pp. 8–30). Oxford: Oxford University Press.

Milders, M., Fuchs, S., & Crawford, J.R. (2003). Neuropsychological impairments and changes in emotional and social behaviour following severe traumatic brain injury. *Journal of Clinical and Experimental Neuropsychology, 25*(2), 157–172.

Milne, E., & Grafman, J. (2001). Ventromedial prefrontal cortex lesions in humans eliminate implicit gender stereotyping. *Journal of Neuroscience, 21*(12), 1–6.

Phillips, M.L., Drevets, W.C., Rauch, S.L., & Lane, R. (2003). Neurobiology of emotion perception I: The neural basis of normal emotion perception. *Society of Biological Psychiatry, 54*, 504–514.

Saver, J.L., & Damasio, A.R. (1991). Preserved access and processing of social knowledge in a patient with acquired sociopathy due to ventromedial frontal damage. *Neuropsychologia, 29*, 1241–1249.

Shimokawa, A., Yatomi, N., Anamizu, S., Torii, S., Isono, H., & Sugai, Y. et al. (2001). Influence of deteriorating ability of emotional comprehension on interpersonal behavior in Alzheimer-type dementia. *Brain and Cognition, 47*(3), 423–433.

Togher, L., & Hand, L. (1998). Use of politeness markers with different communication partners: An investigation of five subjects with traumatic brain injury. *Aphasiology, 12*(7–8), 755–770.

Youse, K.M., & Coelho, C.A. (2005). Working memory and discourse production abilities following closed-head injury. *Brain Injury, 19*(12), 1001–1009.

Further Readings

Adolphs, R. (2002). Recognizing emotion from facial expressions: Psychological and neurological mechanisms. *Behaviour and Cognitive Neuroscience Reviews, 1*(1), 21–62.
Adolphs and colleagues have done an enormous amount of work in the area of social neurosciences. This article overviews the recognition of emotions and how this may be processed in the brain.

Bechara, A. (2002). The neurology of social cognition. *Brain, 125*(8), 1673–1675.
This review provides an interesting discussion of the extent to which social cognition can be functionally separated from non-social cognition

Grice, H.P. (1978). Further notes on logic and conversation. In P. Cole (Ed.), *Syntax and semantics: Pragmatics* **(Vol. 9). New York: Academic Press.**
The theoretician most famously cited for his views regarding conversational inference is Paul Grice. The interested reader may well find his original writings worth visiting

Halliday, M.A.K., & Hasan, R. (1985). *Language, context and text: Aspects of language in a social-semiotic perspective.* **Melbourne: Deakin University Press.**
The application of functional systemic linguistics to discourse production has been enormously valuable and the interested reader would do well to refer to Halliday and Hasan's book on the topic.

Shamay-Tsoory, S.G., Tomer, R., & Aharon-Peretz, J. (2005). The neuroanatomical basis of understanding sarcasm and its relationship to social cognition. *Neuropsychology, 19*, 288–300.
This article provides some interesting anatomical work looking at brain regions and sarcasm.

Sperber, D. & Wilson, D. (2002). Pragmatics, modularity and mind-reading. *Mind and Language, 17*(1–2), 3–23.
Sperber and Wilson have been influential with their discussion of the role of relevance in inferential communication via their book *Relevance* published in 1986 and in this paper they discuss the role of mind reading.

Stuss, D.T., & Knight, R.T. (Eds.) (2002). *Principles of frontal lobe function.* **New York: Oxford University Press.**
This text book presents a number of authorative chapters on different facets of frontal lobe function and constructs associated with executive function.

29

The Torque Defines the Four Quadrants of the Human Language Circuit and the Nuclear Symptoms of Schizophrenia Identify their Component Functions

TIMOTHY J. CROW

Warneford Hospital, Oxford, UK

ABSTRACT

Hearing "voices," the experience of thoughts as not one's own, and incoherent speech are symptoms of mental illness, but each can also be considered a deviation of language. Here I argue that (1) the "torque" is the feature that defines the human brain as four chambered by comparison with the two chambers of the generalized mammalian brain, (2) by separating "thought" from speech production in the frontal lobes, and "meaning" from speech perception in occipito–parieto–temporal association cortex the torque thereby confers on the species the capacity for language and (3) the phenomena of psychosis can be seen as "leakage" from one to another of the four quadrants of association cortex. Thus, schizophrenia is the "price that *Homo sapiens* pays for language," but also the key to its cerebral basis.

29.1. INTRODUCTION

In all populations a fraction of individuals in the third or fourth decades of life begins to hear voices when there is no-one to account for that experience, or becomes convinced that a conspiracy is afoot when none such exists. Others develop swings of mood into depression on the one hand and elation on the other. Such states are regarded as illnesses, schizophrenia in the first case, manic-depressive disorder in the second.

The core nuclear symptoms of schizophrenia according to Kurt Schneider are defined by the glossary of the present state examination (Wing *et al.*, 1974) as follows:

1. *Thought echo or commentary*: The subject experiences his own thought as repeated or echoed with very little interval between the original and the echo.

2. *Voices commenting*: A voice or voices heard by the subject speaking about him and therefore referring to him in the third person.

3. *Passivity* (*delusions of control*): The subject experiences his will as replaced by that of some other force or agency.

4. *Thought insertion*: The subject experiences thoughts *which are not his own* intruding into his mind. In the most typical case the alien thoughts are said to have been inserted into the mind from outside, by means of radar or telepathy or some other means.

5. *Thought withdrawal*: The subject says that his thoughts have been removed from his head so that he has no thoughts.

6. *Thought broadcast*: The subject experiences his thoughts actually being shared with others.

7. *Primary delusions*: Based upon sensory experiences (delusional perceptions) the patient suddenly becomes convinced that a particular set of events has a special meaning.

Schneider (1959) considered that when present these symptoms identify illnesses to which we agree to attach the label schizophrenia. In other words, he proposed an operational definition – this he suggested was the core, but he did not exclude that illnesses without these symptoms might also be a part of the same entity. Nor did he have a theory to account for these particular symptoms or their aggregation. But that they do aggregate together and can be used to define a core syndrome (nuclear schizophrenia) was well established in the World Health Organization Ten-Country study of the Incidence and Manifestations of Schizophrenia

(Jablensky *et al.*, 1992). Importantly this study also demonstrated that when Schneider's first rank symptoms are used in the definition the incidence is seen to be more uniform across populations than when wider definitions are used.

The argument of this chapter is that nuclear symptoms lead us to a concept of the structure of language in terms of areas of hetero-modal association cortex (see Chapter 5) – the four quadrant theory. But in evaluating this theory it must be borne in mind that schizophrenia is not a categorical disease entity but rather, alongside other psychoses and perhaps other non-psychotic conditions, it should be considered in dimensional terms. As such the nuclear symptoms are considered as a component of a dimension of positive symptoms, that is features that are pathological by their presence, to be contrasted with negative symptoms – features such as poverty of speech and affective flattening which are pathological by the absence of some normal function. Positive and negative symptoms have been contrasted in the two syndrome concept (Crow, 1980, 1985) according to which these syndromes have different pathophysiological bases. A further development is that thought disorder (considered by E. Bleuler to be the fundamental characteristic) is considered as a separate dimension. Moreover, it seems that schizophrenic psychoses merge into schizo-affective and then affective psychoses; therefore one has to consider the component of mood, yielding a possible five dimensions: positive, negative, thought disorder (disorganization), depression and elation.

The implication of a dimensional concept is that these factors somehow extend into the normal population. Kretschmer (1925, pp. 118–119) wrote:

"We can never do justice to the endogenous psychoses so long as we regard them as isolated unities of disease, having taken them out of their natural heredity environment, and forced them into the limits of a clinical system. Viewed in a large biological framework, however, the endogenous psychoses are nothing other than marked accentuations of normal types of temperament."

Such a concept emphasizes the relevance of psychotic phenomena to understanding variation in the general population including that relating to language. The challenge is to understand the character of dimensions and their relationship to variations in neural structure.

With what anatomical structures are we concerned? Paul Broca, best known for his localization of a component of language in the frontal lobes and on the left side, in 1877 had the further concept:

"Man is, of all the animals, the one whose brain in the normal state is the most asymmetrical. He is also the one who possesses most acquired faculties. Among these faculties – which experience and education developed in his ancestors and of which heredity hands him the instrument but which he does not succeed in exercising until after a long and difficult education – the faculty of articulate

language holds pride of place. It is this that distinguishes us the most clearly from the animals" (Broca, 1877, p. 528).

The concept that cerebral asymmetry is the characteristic that defines the human brain has been defended most tenaciously in recent decades by Marian Annett (2002) on the basis of her studies of the genetics of handedness. According to this view (the Broca–Annett axiom) the two hemispheres of the human brain are not equivalent in the way that they are in other mammals. Evidence from handedness of primates (McGrew & Marchant, 1997) is in agreement; directional asymmetry on a population basis is specific to man. (On cerebral asymmetry see Chapter 24).

That the asymmetry of the human brain has an anatomical basis was first suggested at the end of the nineteenth century but then largely forgotten until the rediscovery of the "cerebral torque," the bias from right frontal to left occipital across the anterior–posterior axis that apparently characterizes the human brain, by Yakovlev and Rakic in 1966 and of the asymmetry of the planum temporale by Geschwind and Levitsky (1968). According to a recently developed technique for analyzing the torque (Barrick *et al.*, 2005) the torque and asymmetry of the planum temporale are intercorrelated.

29.2. BI-HEMISPHERIC THEORY OF LANGUAGE

A number of authors have postulated that language is bi-hemispheric. Cook (2002) for example asks: If language is localized to the left hemisphere, what is the right hemisphere doing? Beeman and Chiarello (1998) have developed the theme that the right hemisphere plays a role in prosody, pragmatics and affect and that the remoter associations of phonological engrams are stored in the right hemisphere. But if we assume that the torque, a bias from right frontal to left occipital, is the only feature that distinguishes the brain of Homo sapiens from that of other primates, this gives us a route to understanding the function that distinguishes us from other mammals. The torque is the key to the neural basis of language.

A direct consequence of the fact that the torque constitutes a bias across the anteroposterior axis is that the human brain is divided into four quadrants: the dorsolateral prefrontal cortex (DLPFC) on the right and the left side, and occipito–parieto–temporal cortex (OPTC), again on both sides. The human brain is distinguished from that of all other mammals by having four quadrants of association cortex, right and left motor and right and left sensory. Thus functions are discernable in man that are not present in other mammals and this, it is argued, gives us the route to the neural basis of language. The first consequence is that transmission between areas of association cortex has directionality. If we assume that transmission is predominantly from the area of greater to lesser size then the direction is from left to right posteriorly, that is in sensory association cortex, and from right to left anteriorly in relation to motor function. Thus is identified a circuit from left to right OPTC to right DLPFC, and then to left DLPFC. This is the *sapiens*-specific speech circuit.

De Saussure (1916) argued that what is characteristic of human language is that the sign consists of two parts – the signifier, a phonological engram, and its associations, the signifieds or concepts and meanings. Later linguists, for example Paivio (1991), Wray (2002) have spoken of a duality of patterning or "a duality of representation in the brain." Paivio distinguished between logogens and imagens, that is between phonological and graphic representations; the key point being that for a given "sign" they are related and alternate forms of representation. An analogous concept is incorporated in popular models of working memory, that a "phonological loop" is coupled with a "visuospatial sketchpad" to form a whole controlled by an "executive." However, few theories have been related to hemispheric lateralization, and none it seems have taken account of the torque. Here it is argued that a simple formulation is possible and couples de Saussure's distinction between the signifier and the signified with specialization of hemispheric function (Figure 29.1).

The core feature of the torque as already noted is that it crosses the anteroposterior axis in a way that separates left and right motor and left and right sensory functions in the opposite sense. Thus, if we assume that the signifier is located in the left hemisphere and that it has two representations, one motor in Broca's and one sensory in Wernicke's areas, we must also assume that the signifieds have two

forms, one motor in right DLPFC and one sensory in right OPTC. Thus the "abstract concept" center of Lichtheim is located in the right hemisphere and divided in two. It is not difficult to see that the quadripartite schema thus arrived at, corresponds to Chomsky's (1995, p. 168) distinction in The Minimalist Program between articulatory–perceptual and conceptual–intentional, with the specification that the articulatory component is anterior as also is the intentional but on the opposite side, and the conceptual component of the signifieds is posterior and on the right. Thus, the anatomy of the torque dictates that there are four and only four compartments of language.

29.3. PRINCIPLES OF CONNECTIVITY OF HETERO-MODAL ASSOCIATION CORTEX

Three principles influence connectivity between the four quadrants and each is defined by the title of a canonical paper (Table 29.1): Ringo *et al.*'s (1994) principle that "time is of the essence," Witelson and Nowakowski's (1991) that "left out axons make men right" and Harasty *et al.*'s (2003) principle that "the left human speech cortex is thinner and longer than the right":

1. Ringo *et al.* propose that "time is of the essence," in other words that transmission between areas of association cortex has to be distinguished from transmission within such areas. This corresponds to the distinction made by Pulvermüller (2002) between "synfire chains" and interhemispheric conduction. The cortex must be assumed to function in terms of waves of patterned activity, and what occurs within a given area of association cortex is clearly distinct from what is transmitted between such areas. Interhemispheric messages must be more circumscribed, and *Ringo et al.*'s point is that they take time. Pulvermüller argues that phonology must depend upon intra-areal transmission, that is synfire chains, and that the requirement for rapid consecutive organization of motor activity entails that such engrams should be located on one side of the brain,

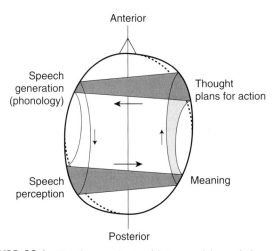

FIGURE 29.1 The four quadrants of hetero-modal association cortex as defined by the torque. Association cortex on one side is "ballooned" (thinned and broadened) relative to the homologous area on the other side. Each quadrant is postulated to have a primary function that relates to its efferent (motor) connexions in the anterior half or afferent (sensory) connexions in the posterior half of the brain, and is further defined by whether it is constituted of "ballooned" cortex (right frontal and left occipito–temporo–parietal) or not.

TABLE 29.1 Principles of Connectivity of Heteromodal Association Cortex

Authors	Title	Focus
Ringo *et al.* (1994)	"Time is of the essence"	Commissural transmission time
Witelson & Nowakowski (1991)	"Left out axons make men right"	Interhemispheric connectivity
Harasty *et al.* (2003)	"The left human speech cortex is thinner and longer than the right"	Planum temporale asymmetry

that is the left. It must also be the case that the motor engrams in Broca's area are qualitatively distinct from the sensory engrams in Wernicke's area, because the former are represented in cortex that is thickened relative to the latter.

2. Witelson and Nowakowski (1991) argued that "left out axons make men right" with the implication that greater degrees of asymmetry are associated with fewer interhemispheric connexions, and that such anatomical asymmetries are greater in males than females. There is a suggestion that there are sex differences in the corpus callosum consistent with greater asymmetry, that is lesser connectivity in the frontal lobes for females and in the occipito–parietal lobes for males (see Chapter 25). This frontal asymmetry is plausibly the correlate of greater verbal fluency in females and faster brain growth.

3. Most important of all, Harasty et al. (2003) have argued that "the left human speech cortex is thinner and longer than the right," that is that asymmetry of the planum temporale is a shape change – as they express it "ballooning" of the cortex – on the left compared to the right. This has functional implications given that the apical dendrites of the pyramidal cells are orientated perpendicular to the cortical surface – it suggests that the ratio of apical to basal dendritic trees is different on the two sides, being reduced in the ballooned cortex relative to the other side.

Together these principles constrain communication within and between the four quadrants: (i) within quadrants transmission is rapid and can sustain fine discriminations such as are involved in the motor and perceptual phonological engrams and perhaps "thought," (ii) transmission between quadrants is more circumscribed as it has to be according to Ringo et al., (iii) transmission between quadrants is also unidirectional and in the case of interhemispheric messages involves a transformation, a "distillation" as a consequence of the Harasty et al. principle and (iv) there is the probability that synfire chains within quadrants are associated with distinct and relatively segregated functions.

29.4. ANOMALIES OF ANATOMICAL ASYMMETRY AND SCHIZOPHRENIA

The consequences of this segregation are revealed by pathology. There are variations in anatomical asymmetry in the general population and the extremes are associated with the symptoms described as psychotic. Thus, individuals with a diagnosis of schizophrenia have been shown to have:

(a) Lesser or reversed asymmetries to the left of the parahippocampal, fusiform and superior temporal gyri, the latter two gyri representing hetero-modal association cortex and the parahippocampal gyrus the way station from association cortex into the limbic circuit.

(b) A reduction in leftward bias of the planum temporale.
(c) Reversal of the asymmetry to the left of density of pyramidal cells in DLPFC (Figure 29.2), together with losses of size and shape asymmetries of the cell bodies.

In a recent meta-analysis of voxel-based morphometry (VBM) studies Honea et al. (2005) found that the most frequent deficits recorded in schizophrenia were in the medial temporal lobe on the left side and the superior temporal gyrus bilaterally. However, in terms of lateralization the biggest deficits were in the left medial temporal lobe and in the right anterior cingulate gyrus, that is along the axis of the torque, although notably relating to limbic cortex rather than to the neocortex itself.

The anomalies of asymmetry are subtle, but arguably the hallmark. They are associated with sex differences as also are the changes in density of fibers in the corpus callosum that accompany them. Females have more fibers than males, and relative to same sex controls, female patients with schizophrenia show a decrease in density while males show an increase.

A sex difference in the manifestations of schizophrenia is well attested – onsets are earlier by 2–3 years and outcome is generally worse in males. Where do the sex differences come from? A plausible line of thought is that they are related to the sex difference in laterality – girls being more strongly right-handed than boys and acquiring words faster. But for both boys and girls there are disadvantages relating to ambidexterity (the point of "hemispheric indecision" – Crow

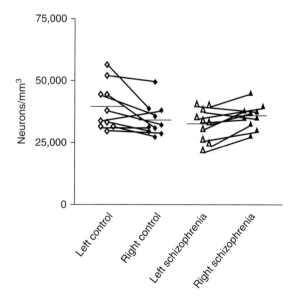

FIGURE 29.2 Neuronal density in area 9 of DLPFC in normal individuals and patients with schizophrenia. In normal individuals there is an asymmetry to the left accompanied by an asymmetry of size and shape of neurons. In patients the asymmetry of density is lost or reversed and the asymmetries of size and shape are lost. *Source*: Cullen et al., 2006.

et al., 1998) and those who later develop psychosis are closer to hemispheric indecision than the population as a whole, more precisely for a given degree of lateralization they have fewer words (Leask & Crow, 2005). They are also delayed in reading (Crow *et al.*, 1995). The relationships between lateralization and sex and verbal and spatial ability have recently been replicated in a sample of 0.25 million in the BBC Internet survey (Peters *et al.*, 2006).

29.5. THE CENTRAL PARADOX AND ITS RESOLUTION

This section outlines the hypothesis that the genetic bases of cerebral asymmetry, language and psychosis are related (see also Boxes 29.1 and 29.2). Two evolutionary theorists – Julian Huxley and Ernst Mayr – posed the paradox: Why if schizophrenia is genetic in origin and is

---

Box 29.1   The Xq21.3/Yp11.2 translocation

Between 6 and 4.5 million years ago a 3.5-Mb contiguous block of sequences on the X chromosome long arm (Xq21.3) was duplicated to the Y chromosome short arm (Yp11.2) creating a new region of homology (Sargent *et al.*, 1996). This event is a candidate genetic change for the split of the chimpanzee and hominid lineages.

The duplicated block was subsequently split by a paracentric inversion (reversal of a block of the chromosome) (Schwartz *et al.*, 1998) to give two blocks of homology in Yp11.2. This event has not been dated but is a conceivable correlate of the transition (speciation event) to modern Homo sapiens.

Genes within this region are therefore present on both the X and Y chromosomes in Homo sapiens but on the X alone in the great apes and other primates. Of the three genes expressed within the region, the ProtocadherinXY gene pair (PCDH11X and PCDH11Y) code for cell adhesion molecules of the cadherin superfamily expected to play a role in intercellular communication perhaps acting as axonal guidance factors and

influencing the connectivity of the cerebral cortex; these genes have been subject to accelerated evolution in the hominid lineage (Williams *et al.*, 2006).

Sargent, C.A., Briggs, H., Chalmers, I.J., Lambson, B., Walker, E., & Affara, N.A. (1996). The sequence organization of Yp/proximal Xq homologous regions of the human sex chromosomes is highly conserved. *Genomics*, *32*, 200–209.

Schwartz, A., Chan, D.C., Brown, L.G., Alagappan, R., Pettay, D., Disteche, C., McGillivray, B., De la Chapelle, A., & Page, D.C. (1998). Reconstructing hominid Y evolution: X-homologous block, created by X–Y transposition, was disrupted by Yp inversion through LINE–LINE recombination. *Human Molecular Genetics*, *7*, 1–11.

Williams, N.A., Close, J.P., Giouzeli, M., & Crow, T.J. (2006). Accelerated evolution of Protocadherin11X/Y: A candidate gene-pair for cerebral asymmetry and language. *American Journal of Medical Genetics. Part B, Neuropsychiatric Genetics*, *141*, 623–633.

Box 29.2   Implications for evolutionary theory

The human capacity for language raises problems for evolutionary theory. How can so complex and specific a function have arisen? Was the change gradual or was there discontinuity? The innovation of the torque and its dependence on the sex chromosomes casts light on the genetic mechanism. The original duplication (see Box 29.1) created a block of new genes on the Y (expressed only in a male). Such events may be common but generally deleterious and quickly selected out. Just occasionally the beneficiary of such a translocation (a "hopeful monster") succeeds because the distinctive feature is selected by females. Then that Y chromosome spreads within the population from fathers to sons. But an influence on daughters arises because X–Y homologous genes are protected from the "*epigenetic*" process in females by which most genes on one X are inactivated. But *sequence* change in PCDH X and Y has also occurred. We find 16 aminoacid changing substitutions in the Y sequence relative to the chimpanzee, while there have also been five aminoacid changing substitutions in the X sequence. These substitutions are explicable only on the basis that the X protein is operating in a new environment, that is the presence

of the Y protein (Williams *et al.*, 2006). Some changes are radical and these changes are proposed as critical for cerebral asymmetry and language, the features that distinguish Homo sapiens from Pan troglodytes and Pan paniscus.

Thus the evolution of language in man is a paradigm for the speciation process in other sexual organisms: a singular event on the heterogametic chromosome (the Y in mammals, the W in birds) establishes a feature that is selected by the homogametic sex (in mammals the females, in birds the males) and this is followed by selection on homologous genes on the X or Z chromosome to create a new "specific mate recognition system" (Paterson, 1993).

Paterson, H.E.H. (1993). *Evolution and the recognition concept of species*. Baltimore: Johns Hopkins Press.

Williams, N.A., Close, J.P., Giouzeli, M., & Crow, T.J. (2006). Accelerated evolution of Protocadherin11X/Y: A candidate gene-pair for cerebral asymmetry and language. *American Journal of Medical Genetics. Part B, Neuropsychiatric Genetics*, *141*, 623–633.

associated with a substantial fecundity disadvantage are the genes not selected out? The disadvantage is of the order of 30% in females and 70% in males and no doubt is a consequence of failure to establish a pair bond. There must be a balancing advantage Huxley *et al.* (1964) argued. But the answer they proposed – that the individual with psychosis is endowed with particular resistance to wound shock and stress – does not have empirical support, and as Kuttner argued is far removed from the essence of the psychopathology. Is it not more plausible that the genetic advantage is not confined to or even manifest in the affected individual but that the mechanism relates to variation concerned with the most recently evolved human capacities?

One can ask further: How old is the genetic variation? From the fact that psychotic illness is present with similar characteristics in all extant populations including those such as the Australian aborigines who have been separated from other populations for a significant fraction of the life of the species the predisposition must precede the dispersal following the origin, somewhere in Africa some 100–200,000 years ago. This immediately raises the question: Is not the disadvantage in some way related to the characteristic that defines the species, that is language? Could it not be that the balancing advantage is not to the individual with psychosis, nor as has been suggested to his first degree relatives but to Homo sapiens as a species, and that it is the capacity for language?

Progress depends upon our ability to give some account of the genetic origins of language and the species that is to identify, if such a thing exists, a "speciation event" (Crow, 2002). The account summarized in Boxes 29.3 and 29.1 relies upon:

1. The Broca–Annett axiom that the torque is specific to the human brain.

2. Evidence that individuals with sex chromosome aneuploidies (a deficit or excess of a sex chromosome) have delays in the development of hemispheric function and aspects of language.
3. The hypothesis that speciation events are at least sometimes related to chromosomal rearrangements and that the sex chromosomes play a particular role.

29.6. HOW DO THE NUCLEAR SYMPTOMS ARISE?

The nuclear symptoms of schizophrenia (see Section 29.1) tell us about the components of language. The autonomy of "thought" is attested by the evidence of individuals who suffer from thought insertion or withdrawal – autonomy is lost and this tells us something about the normal process. Thought is a function that is under the control of the individual and maybe identifies the self. It is separate on the one hand from speech production and on the other from meanings or concepts. These are sensory and located in the posterior half of the brain (Figure 29.3).

There is a degree of independence to the neural activity within each of the four quadrants. This follows from the distinction between neural transmission that occurs within and between areas of association cortex, the segregation and the differences depending on the cerebral torque and the Harasty *et al.* principle. But if one imagines that there are degrees of torque, and that some individuals have lesser development of aspects of the torque than the rest of the population, then these individuals are perhaps more at risk of psychotic symptoms. Nuclear symptoms can be considered as "leakages," as

---

Box 29.3   The evidence for an asymmetry determinant on both X and Y chromosomes

Individuals who lack an X chromosome (XO, Turner syndrome) have non-dominant hemisphere (spatial) deficits on cognitive testing, whereas individuals with an extra X (XXY, Klinefelter, and XXX syndromes) have verbal or dominant hemisphere deficits (see following table). But then the question arises of why males, who only have one X chromosome do not have spatial deficits such as seen in Turner syndrome. The answer must be that the copy of the gene on the X chromo-

some is complemented by a copy on the Y, that is, that the gene is in the X/Y homologous class (Crow, 1993). A hormonal explanation will not account for the similarity of the changes in XXY individuals, who are male, and XXX individuals, who are female. The case that the gene is present also on the Y chromosome is strongly reinforced by the verbal deficits/delays that are observed in XYY individuals.

	XX normal female	XY normal male	XO Turner syndrome	XXY Klinefelter syndrome	XXX	XYY
No of sex chromosomes	2	2	1	3	3	3
Verbal ability				Delayed	Delayed	Delayed
Spatial ability			Decreased			

the transmission of activity between areas of association cortex when, in normal individuals, such transmission does not occur. Thus alien thoughts can be conceived as an influence of right OPTC on right DLPFC. The individual experiences the thought as a thought but (as the activation has arisen from the sensory part of the brain) as not under his control. Conversely, thoughts spoken aloud represent leakage from right DLPFC to left OPTC, perhaps by back transmission on the right side and then across the posterior corpus callosum. "Running commentary" is a similar symptom with respect to plans for action. These are somehow converted into phonological engrams and presumably are experienced in auditory association cortex on the left side.

Furthermore, disorders of meaning – exaggerated significance attached to particular events or stimuli – plausibly arise from excess associations between phonological engrams on the left side of the brain and associations in the right posterior hemisphere – leading to exaggerated importance being attached to particular incoming messages. A summary of the principal symptoms of schizophrenia together with their relationship to the three syndromes and possible locations is included in Table 29.2.

Thus the nuclear symptoms of schizophrenia can be conceived in terms of the four quadrant theory as the route to understanding the neural organization of language – the symptoms are deviations that point to the normal functions of the differentiated compartments characteristic of the human brain – schizophrenia can be understood as the price that *Homo sapiens* pays for language.

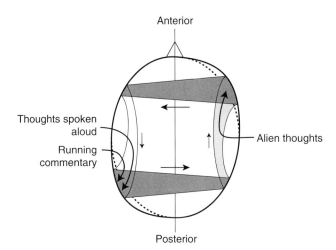

FIGURE 29.3 Schematic representation of the origin of nuclear symptoms. It is assumed that the four quadrants of hetero-modal association cortex have a degree of independence and that each has a distinct function (see also Figure 29.1). There are rules, presently obscure, for what is filtered by the long tracts between areas and hemispheres. The nuclear symptoms are postulated to reflect aberrant transmission (perhaps even backflow) between areas, that thus may be described as "leakage" – in the case of alien thoughts etc. leakage from posterior (sensory) to anterior (motor), and in the case of the specific auditory hallucinations (thoughts spoken aloud) from anterior to posterior.

29.7. CHALLENGES AND FUTURE DIRECTIONS

There is no clear dividing line between schizophrenic illnesses and major affective disorders, for example bipolar illness and unipolar depression. As for schizophrenic illnesses the symptoms of the latter disorders can be considered as language related – for example pressure of speech in mania, retardation in depression, appear to be simple rate changes in speech production. To conceive them as such raises the problem of the relationship between affect and other aspects of cognitive function including, particularly, language. This important boundary has to be addressed in neural terms. It would appear that some formulation that coordinates lateralized cortical function with the influence of ascending monoaminergic systems is required.

Further investigations in post-mortem and imaging studies are to be expected. A particularly challenging and recent finding (Mackay *et al.*, 2006) is that in bipolar disorder there are structural changes that are strictly sex dependent

TABLE 29.2   Symptoms of Schizophrenia as Disorders of Language

Symptom	Description	Dimension	Function	Location
Auditory hallucinations including thoughts spoken aloud and running commentary	"Voices" referred to the outside world	Positive	Perceived speech	?Superior temporal gyrus
Delusions	Convictions held "in the teeth of" contrary evidence	Positive	Meaning	OPTC association cortex – ? oon the right side
Affective flattening	Failure to express affect	Negative	Prosody	Right DLPFC
Poverty of speech	Too few words	Negative	Speech production deixis	?Right DLPFC
Thought insertion, withdrawal, broadcast	Thoughts experienced as "not one's own"	?Negative		Leakage into right DLPFC
Thought/speech disorder	Incoherent/disorganized speech	Disorganization	Discourse planning	Right hemisphere/arcuate, uncinate bundles

and related to counter-torque, that is the axis from left frontal to right occipital. Male probands have apparently greater volumes of cortex in these quadrants than male controls whereas female patients have lesser volumes than female controls (although apparent volume change may represent a shape change as argued in Section 29.3). Thus, bipolar disorder is clearly an anomaly of asymmetry and this could cast some light on the role of affective change but it is at present unclear precisely what. Further, post-mortem studies on the minicolumn structure of the planum temporale, the DLPFC and other areas of association cortex and related limbic areas (anterior cingulate and parahippocampal gyrus in particular) can be expected to illuminate this problem. In addition, it is clear that diffusion tensor imaging (DTI) is capable of revealing changes in white matter tracts including the major fiber bundles in each hemisphere (arcuate and uncinate bundles connecting the region including Broca's and Wernicke's areas and the temporal lobe) and the cingulum bundle as well as interhemispheric pathways of the corpus callosum and anterior commissure. Further information on these connections in normal individuals and those suffering from psychosis can be expected to cast considerable light on the role of asymmetry and its deviations in psychiatric disorder.

References

Annett, M. (2002). *Handedness and brain asymmetry: The right shift theory*. Hove, UK: Psychology Press.

Barrick, T.R., Mackay, C.E., Prima, S., Vandermeulen, D., Crow, T.J., & Roberts, N. (2005). Automatic analysis of cerebral asymmetry: an exploratory study of the relationship between brain torque and planum temporale asymmetry. *Neuroimage, 24*(3), 678–691.

Beeman, M., & Chiarello, C. (1998). Concluding remarks: Getting the whole story right. In M. Beeman & C. Chiarello (Eds.), *Right hemisphere language comprehension* (pp. 377–389). Mahwah, NJ: Erlbaum.

Broca, P. (1877). Rapport sur un memoire de M. Armand de Fleury intitulé: De l'inegalité dynamique des deux hemisphères cerébraux. *Bulletins de l'Academie de Medecine, 6*, 508–539.

Chomsky, N. (1995). *The Minimalist Program*. Cambridge, MA: The MIT Press.

Cook, N.D. (2002). Bihemispheric language: How the two hemispheres collaborate in the processing of language. In T.J. Crow (Ed.), *The speciation of modern Homo sapiens* (pp. 169–194). Oxford: Oxford University Press.

Crow, T.J. (1980). Molecular pathology of schizophrenia: More than one disease process? *British Medical Journal, 280*, 66–68.

Crow, T.J. (1985). The two-syndrome concept: Origins and current status. *Schizophrenia Bulletin, 11*, 471–486.

Crow, T.J. (1993). Sexual selection, Machiavellian intelligence, and the origins of psychosis. *Lancet, 342*, 594–598.

Crow, T.J. (Ed.) (2002). *The speciation of modern Homo sapiens*. Oxford: Oxford University Press.

Crow, T.J., Done, D.J., & Sacker, A. (1995). Childhood precursors of psychosis as clues to its evolutionary origins. *European Archives of Psychiatry and Clinical Neuroscience, 245*, 61–69.

Crow, T.J., Crow, L.R., Done, D.J., & Leask, S.J. (1998). Relative hand skill predicts academic ability: Global deficits at the point of hemispheric indecision. *Neuropsychologia, 136*, 1265–1275.

Cullen, T.J., Walker, M.A., Eastwood, S.L., Esiri, M.M., Harrison, P.J., & Crow, T.J. (2006). Anomalies of asymmetry of pyramidal cell density and structure in dorsolateral prefrontal cortex in schizophrenia. *British Journal of Psychiatry, 188*, 26–31.

Geschwind, N. & Levitsky, W. (1968). Human brain: Left–right asymmetry in temporal speech region. *Science, 161*, 186–187.

Harasty, J., Seldon, H.L., Chan, P., Halliday, G., & Harding, A. (2003). The left human speech-processing cortex is thinner but longer than the right. *Laterality, 8*(3), 247–260.

Honea, R., Crow, T.J., Passingham, D., & Mackay, C.E. (2005). Regional deficits in brain volume in schizophrenia: A meta-analysis of voxel-based morphometry studies. *American Journal of Psychiatry, 162*, 2233–2245.

Huxley, J., Mayr, E., Osmond, H., & Hoffer, A. (1964). Schizophrenia as a genetic morphism. *Nature, 204*, 220–221.

Jablensky, A., Sartorius, N., Ernberg, G., Anker, M., Korten, A., Cooper, J.E., Day, R., & Bertelsen, A. (1992). Schizophrenia: Manifestations, incidence and course in different cultures. A World Health Organization ten country study. *Psychological Medicine, Suppl 20*, 1–97.

Kretschmer, E. (1925). *Physique and character*. London: Kegan Paul, Trench, Trubner.

Leask, S.J., & Crow, T.J. (2005). Lateralization of verbal ability in pre-psychotic children. *Psychiatry Research, 136*, 35–42.

Mackay, C.E., Roddick, E., Barrick, T.R., Roberts, N., Crow, T.J., Lloyd, A.J., Young, A.H., & Ferrier, N. (2006). Sex differences in lobe volumes in bipolar disorder. *Schizophrenia Research, 81*, 6.

McGrew, W.C., & Marchant, L.F. (1997). On the other hand: Current issues in and meta-analysis of the behavioral laterality of hand function in nonhuman primates. *Yearbook of Physical Anthropology, 40*, 201–232.

Paivio, A. (1991). Dual coding theory: Retrospect and current status. *Canadian Journal of Psychology, 45*, 255–287.

Peters, M., Reimers, S., & Manning, J.T. (2006). Hand preference for writing and associations with selected demographic and behavioral variables in 255,100 subjects: The BBC internet study. *Brain and Cognition, 62*, 177–189.

Pulvermüller, F. (2002). *The neuroscience of language: On brain circuits of words and serial order*. Cambridge: Cambridge University Press.

Ringo, J.L., Doty, R.W., Demeter, S., & Simard, P.Y. (1994). Time is of the essence: A conjecture that hemispheric specialisation arises from interhemispheric conduction delay. *Cerebral Cortex, 4*, 331–343.

Saussure de, F. (1916). *Course in general linguistics* (translated by R. Harris, 1983 and published by Open Court, IL). Paris: Payot.

Schneider, K. (1959). *Clinical psychopathology* (translated by M.W. Hamilton & E.W. Anderson). New York: Grune and Stratton.

Wing, J.K., Cooper, J.E., & Sartorius, N. (1974). *The measurement and classification of psychiatric symptoms: An instruction manual for the PSE and Catego program*. Cambridge: Cambridge University Press.

Witelson, S.F., & Nowakowski, R.S. (1991). Left out axons make men right: A hypothesis for the origins of handedness and functional asymmetry. *Neuropsychologia, 29*, 327–333.

Wray, A. (2002). Dual processing and protolanguage performance without competence. In A. Wray (Ed.), *The transition to language* (pp. 113–137). Oxford: Oxford University Press.

Further Readings

Chaika, E. (1990). *Understanding psychotic speech. Beyond Freud and Chomsky*. Springfield, IL: CC Thomas.
This entertaining account reflects the attempts of a linguist to make sense of the abnormalities she perceived in the speech of psychotic individuals. It includes many examples and elucidates the nature of the abnormality in each case without straining to encompass them within a general theory.

Crow, T.J. (Ed.) (2002). *The speciation of modern Homo sapiens.* **Oxford: Oxford University Press.**
This volume presents the proceedings of the first conference to address the issue of the speciation of Homo sapiens and includes contributions from archaeology, palaeontology, linguistics, psychology, neuroanatomy, genetics and evolutionary theory.

Kasanin, J.S. (Ed.) (1946). *Language and thought in schizophrenia.* **Berkeley: University of California Press.**
This is a classic contribution to the literature that includes the theory of Norman Cameron of the origins of thought disorder, and a number of other early accounts.

McKenna, P.J., & Oh, T. (2005). *Schizophrenic speech: Making sense of bathroots and ponds that fall in doorways.* **Cambridge, UK: Cambridge University Press.**
This up to date volume gives a comprehensive account of approaches to thought disorder in the psychiatric literature. It represents collaboration between a psychiatrist and a linguist and is particularly good on the peculiarity of the clinical phenomena and the difficulty of achieving a unitary explanation.

Sims, A. (Ed.) (1995). *Speech and language disorders in psychiatry.* **London: Gaskell.**
This gem of a book includes 16 brief chapters by authors of widely differing expertise including linguistics, philosophy, psychiatry and neurology. The main focus is on schizophrenia. The viewpoints are diverse but representative of the field and there is an instructive historical introduction. Overall this is an excellent route into the relevant literature on language and psychosis.

Vygotsky, L.S. (1962). *Thought and language.* **Edited and translated by E. Haufmann & G. Vakar Cambridge, MA: The MIT Press.**
Edited and translated by E. Haufmann & G. Vak. This classic work represents an early attempt by a pioneer in psychiatric linguistics to describe the basic abnormalities he saw and to derive a developmental concept.

Stuttering and Dysfluency

DAVID B. ROSENFIELD

*Speech and Language Center, Neurological Institute, The Methodist Hospital/
Weill Cornell College of Medicine, Department of Communicative Disorders, University of Houston
Shepherd School of Music, Rice University Houston, TX, USA*

ABSTRACT

Stuttering, a global, pan-cultural disturbance of the speech-motor control system has afflicted human beings throughout time. An improved understanding of causes of stuttering would provide information about mechanisms of normal speech. This would translate into treatment for patients with speech afflictions due to other disturbances, such as tumor, stroke, vascular insult, and trauma.

As opposed to a disturbance in language, people who stutter have disturbed speech. Their mechanisms of oral-motor control are effectively intact when chewing, swallowing, breathing, or singing but when this control system integrates with processing of language, the system becomes dysfunctional, resulting in aberrant output.

An increasing body of neuroscience literature focuses upon the cause and treatment of stuttering. This research highlights genetics, neurophysiology, and auditory processing mechanisms, as well as clinical information. Understanding the role of these processes will assist investigators to provide newer and increasingly effective therapies for stuttering and other disturbances of fluency.

30.1. INTRODUCTION

Stuttering has been present throughout time. Ancient Mesopotamian clay tablets, Egyptian hieroglyphics (nit-nit), the Old Testament (Moses stuttered), and the Holy Koran refer to stuttering. Stuttering is a disturbance of speech motor production that occurs in all cultures, is referenced in all languages, and afflicts people throughout the world. As neuroscience investigations into the dynamics of speech motor control has expanded, there is increased focus upon those who speak abnormally. These individuals have normal language, meaning that their grammar, semantics and content of their communication efforts are normal, but are unable to say what they want to say. When they try, they stutter.

Understanding the enigma of stuttering would improve our knowledge of mechanisms of normal speech production which, in turn, would expand knowledge of normal processing of language and other aspects of human cognition. Understanding stuttering extends as a window into human language and cognitive processing.

People who have stuttered since childhood suffer considerable emotional pain and social stigma, with a prevalence slightly more than 1% of the adult population (Porfert & Rosenfield, 1978). When brain disease renders previously fluent individuals dysfluent, they are referred to as "acquired stutterers." These acquired stutterers have speech characteristics different from the previously described "developmental stutterers," those who have stuttered since childhood or adolescence. The prevalence of developmental stutterers among children is approximately 4% of the population. Eighty percent "outgrow" their stuttering but 20% do not, becoming adult stutterers.

There is increasing public focus on children unable to read, learn, or those suffering from abuse. Below, we discuss a sector of children as well as adults with an often overlooked problem – those with intact language yet unable to talk. These people, known as stutterers, know what they want to say but are unable to do so. They must uniquely and abnormally try to make conscious the normally unconscious "how" of the communicative process in order to achieve fluent sounds of speech.

30.2. DEFINITION AND CLINICAL CHARACTERISTICS OF DYSFLUENCY

There are multiple definitions of stuttering (Bloodstein, 1995; Conture, 2001). Given that the definitions are arbitrary or voted upon, they often shed more heat than light. Arguing definitions require minimal knowledge of science but can reflect beliefs gained from clinical experience. From a neuroscience perspective, the definition does have importance since investigators need to compare data and theories. Different definitions of how stuttering is ascertained (e.g., by the individual, confirmed by family members) can alter investigative paradigms, especially genetic studies in which family members state whether a deceased relative did or did not stutter, or whether an individual maintains he was or was not formerly a stutterer.

From a practical perspective, what is a "stutter"? If a child stumbles or bounces on a particular sound, or someone repeats a whole word, is that a stuttered dysfluency? Do bouncing, stumbling, or repetitions need to be performed with effort or should there be a struggle in achieving normal output of speech? Is the dysfluency with which we are familiar in adult stutterers isomorphic to the effortless dysfluency of what is commonly termed, "normal fluency of childhood"? Again, how we answer these questions very much affects our scientific inquiries into the etiology of dysfluent speech.

Among development stutterers, children who stutter have difficulty producing the intended sounds and words. They struggle with their output, often avoid speaking in school or in front of others, and become very upset with their abnormal speech, yet have normal elements of language. They know what they want to say but are unable to speak accordingly. Their language is normal but their speech is not.

The location of their stuttered dysfluencies is not random. Dysfluencies of children or adults occur at particular points in the surface structure of their sentence, usually at the beginning of phrases and the beginning of sentences. These dysfluencies are characterized by part-word repetitions and not usually associated with repetition of whole words or phrases. Sometimes, the individual will "back up" in the sentence in order to try again at fluent production of the difficult sounds. However, the primary disturbance of the dysfluent speech output remains: repetition of a single sound or a fragment of a word (Bloodstein, 1995).

In an effort to overcome these dysfluencies, the child or adult who stutters often engages in circumlocutions, substituting a word that can be fluently produced for a word that causes difficulty. For instance, a child who stutters might say, "Ni-ni-ni-good to see you" instead of continuing to stutter on the /ni/ of "nice." Acoustic analysis suggests that when the child attempts to say "ni-ni-nice," the initial /ni/ sound in the stuttered dysfluency differs from that in the final fluently produced target word. The individual is not saying the sound appropriately and may back up in his sentence output, often as a learned or non-voluntary strategy, to achieve fluent production. This backing up within the sentence and stumbling and repetition become the hallmark of that person's stuttering behaviors.

These disruptions in speech are often accompanied by "secondary characteristics" of stuttering, such as facial distortions or associated grimaces and even limb movements, all orchestrated in an effort to force out the appropriate sound. As a result, when someone who stutters wants to say something, knowing the intended sounds, words, and sentences yet is unable fluently to produce them, their final speech motor output represents the deficit in their speech production as well as their efforts to repair the deficit and achieve fluency.

Indeed, if the individual so chooses, she/he can stop stuttering simply by ceasing talking. When she/he tries to talk, the stutter is present and the attempts of achieving fluency, incorporating struggle and repetitions, become paramount. Depending upon the focus of that individual and how she/he has learned to wrestle with her or his dysfluent speech, she/he may elect to "wait out" the dysfluency until he can fluently produce the sound. She/he may have learned to avoid her or his struggling and bouncing of sounds until she/he feels she/he can fluently produce them, or may try different motor maneuvers (i.e., lips, jaw, eyes, hand, and so forth) to achieve her or his fluent target. Regardless, what she/he produces in speech reflects the deficit in speech motor production as well as the accompanying reparative process (Bloodstein, 1995; Rosenfield, 2001a).

It is difficult to identify the actual locus of a stuttered dysfluency. If a stutterer says "s-s-sound," where is the actual dysfluency? Many researchers previously contended that the dysfluency was on the /s/. Most now describe the deficit as on the *transition* from one sound to the next. The stutterer can say the /s/ but not the /ound/. The stutterer's strategy to achieve fluent output can result in repeating the /s/ until securing the transition into the following /ound/ or she/he may try some of the reparative maneuvers noted above (Rosenfield, 1997, 2001b).

30.2.1. Clinical Features of Stuttering

As noted above, the location of these stuttered dysfluencies is not random. The dysfluencies occur at the same location within the sentence where the occasional dysfluencies of fluent speakers also occur – at the beginning of sentences and phrases. Thus, one seldom hears a stutterer say, "Where is the book-k-k-k?" Rather, a stutterer says "Wh-Wh-Where is the book?" (Bloodstein, 1995) (Box 30.1).

Stuttering and Dysfluency   311

Box 30.1   Spectrogram of a stutterer saying, "domestic"

The /d/ sound is repeated several times before fluency is achieved. The horizontal axis reflects time and the vertical axis reflects frequency. Not that the dysfluency occurs at the beginning of the word.

Rosenfield, David B. (2005). Brain and language: disorders of fluency and voice. In Harry Whitaker (Ph.D. Ed.), *Encyclopedia of language and linguistics* (2nd edn). Oxford, England: Elsevier.

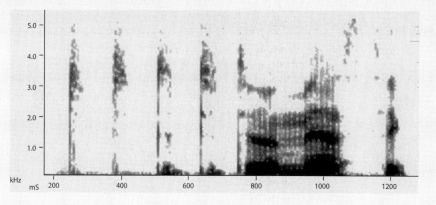

Just as all motor processing is affect-sensitive, signifying it worsens under stress, stuttering also worsens with emotional stress. Thus, the skill of correctly hitting a golf ball or swinging a baseball bat is compromised by distraction or heightened emotional vigilance, just as production of speech is affected by stress. Correct processing of motor control systems depends upon the gain and phase of the sensory-motor feedback loops, rendering their output sensitive to perturbations from multiple cerebral areas, especially the emotionally charged limbic system. The fact that a planned motor output worsens with psychological stress, such as speaking in front of a class (or hearing an automobile alarm go off while swinging a golf club) does not mean the disruption is "psychological," or the person could stop the interference if they so desired. Rather, it signifies the sensory-motor processing of the motor control is affect-sensitive. Motor systems are affect-sensitive systems and worsen with stress.

Stutterers seldom repeat whole words. Their dysfluent output usually consists of fragments of words, not the entire word. Acoustic analyses of these dysfluent productions, consisting of word fragments or parts of words, reveals that they differ from the target sounds the stutterer is trying fluently to produce (Bloodstein, 1995; Conture, 2001; Rosenfield, 2001b). This makes intuitive sense since were the target correctly produced, the speaker would not have difficulty and could proceed to the next sound. Instead, the target is incorrectly produced, necessitating that the speaker backs up in the sentence or struggles or circumlocutes in order to achieve normal production of the intended speech target.

It is within this context that the typical person who stutters presents himself struggling to produce correctly the intended sound, backing up within the sentence to achieve this output, grimacing, squeezing his eyes, and perhaps even banging arms against his side, all part of his struggle to achieve what most people simply take for granted – effortless, fluent speech.

Several maneuvers and paradigms increase fluency in stutterers, some more potent than others. The most potent fluency-producing maneuver in stuttering is singing. From a clinical perspective, this is an incredibly interesting finding. A young child who stutters might be terrified to speak in front of his class and deliver a book report, due to fear of stuttering. Yet, he may be as nervous or embarrassed to sing in front of the same class, yet is totally fluent. Stutterers do not stutter when they sing.

There are many theories of why stutterers are fluent when singing. On a clinical level, this must represent a different type of processing within the speech-motor control system. Singing certainly differs from speaking. People may be skilled in one of these domains yet not in the other. Different parts of the brain participate differently in singing and the rhythm of speech in singing versus speech. Further, the acoustic structure of sounds spoken versus sounds sung are very different (Peretz & Zatorre, 2003; Ross & Mannot, 2008).

Repeated reading of the same passage also lessens dysfluency. This process, termed "adaptation," refers to improved fluency with each verbalized repeated reading of the passage. This process is also accurate for the occasional dysfluencies normal fluent speakers encounter, explaining why people orally rehearse a speech or part of a play prior to their formal performance. From a motor control perspective, the effectiveness of this rehearsal is readily explained regardless of putative models one may entertain. Practice does make perfect.

Interestingly, the effectiveness of rehearsal in decreasing dysfluent output mandates that the output be vocalized, not just read silently or lip-read. Perkins (1973) nicely demonstrates that actual verbalization of the reading is more effective than whispering the reading and that lip-reading without sound or silent reading lack significant efficacy. When practicing speech tasks results in improvement of the speech-motor control system, that practice requires iteration of the actual task, not a portion of it.

Choral reading, consisting of reading a passage while others simultaneously read aloud the same passage, also improves but does not eliminate dysfluency. There is considerable documentation and experiments in this area, with multiple theories. This finding is fairly robust, suggesting that choral reading might affect auditory input and thus impact motor output, achieving improved performance (Conture, 2001).

Speaking in cadence with a metronome also improves fluent production. Again, there are multiple theories, including slowing down the motor task and altering the output such that the person is no longer actually producing normal speech but is adding a particular rhythm.

Another clinical aspect of stuttering is the effectiveness of loud broadband noise improving fluent production. Not only is this masking auditory feedback effective in ameliorating stuttering, but delayed auditory feedback (DAF) and frequency altered feedback (FAF) are also potent forms of alteration that improve the speech signal (Kalinowski *et al.*, 2000; Lincoln *et al.*, 2006).

As opposed to the developmental stutterers, some heretofore fluent individuals become dysfluent subsequent to acquired brain compromise. "Acquired stuttering" was initially reported by Rosenfield (1972), describing a patient with mild left brain compromise. Those findings remain substantiated by many other investigators. Acquired stutterers differ from developmental stutterers in that the former are dysfluent throughout the sentence, stutter when she/he sing, and are oftentimes not emotionally disturbed by his/her dysfluencies (see Table 30. 1). Acquired stuttering can result from injury to the brain in either hemisphere and in multiple locations. The injury is usually mild and prognosis for

recovery is good if the damage is unilateral but is not good if brain damage is bilateral. In some patients with acquired stuttering, there may be psychogenic factors (Helm *et al.*, 1978; Rosenfield & Barroso, 2000).

Whereas developmental stutterers have difficulty achieving the target sound (i.e., va-va-va-vision), the acquired stutterer achieves the target sound, repeating it (i.e., vi-vi-vi-vision). This difference is often clinically perceptible at the bedside if the examiner pays close attention to the actual output of sound.

Cluttering is a disturbance in speech production sometimes confused with stuttering (see Table 30.1). The speech of clutterers is characterized by excessive speed, repetition, interjections, drawling, disturbed prosody, and a monotonous sound. Some clutterers have articulatory disruption that is not consistent. Some investigators maintain clutterers' speech contains grammatical abnormalities and that these patients can have learning disturbances, be hyperactive, have poor concentration, and can have poorly integrated thought processes.

Many clutterers speak extremely rapidly, omitting sounds, syllables and whole words, and sometimes inverting the order of their sounds, repeating the initial sound, and prolonging several syllables of their intended words. Listeners often complain about the rapidity of the clutterer's speech. Unlike developmental Stutterers, people with cluttering frequently lack concern about their speech deficits (see Table 30.1) (Bloodstein, 1995; Rosenfield & Barroso, 2000).

The treatment for developmental stuttering is expansive, focusing primarily upon slowing down the speech process and teaching patients to achieve gentle onset of their vocalized targets. There are multiple commercial enterprises making vast claims treating stuttering, but to date there are no cures and individuals should exercise caution when engaging in new, non-validated therapies. Currently, there is certification among speech–language pathologists for skill in treating fluency disorders. Some also advocate pharmacologic therapies that focus upon different neurotransmitters (i.e., dopamine receptors, noradrenergic system) but these

TABLE 30.1   Characteristics of Stuttering and Cluttering

	Developmental stuttering	Acquired stuttering	Cluttering
Locus of lesion	Unknown but probably includes left perisylvian areas	Cortical or subcortical; left or right hemisphere	Unknown
Cause	Unknown	Physical trauma, vascular, metabolic, tumor	Unknown
Duration	80% of children outgrow: adults rarely outgrow stuttering	Good prognosis if unilateral; poor prognosis if bilateral	Varies
Locus of dysfluency	Beginning of sentence or phrase	Scattered throughout sentence	Varies
Singing	Induces fluency	May minimally improve fluency	Varies
Onset	Sub-acute	Sub-acute or acute	Gradual
Reaction to dysfluencies	Anxious	Not anxious	Not usually concerned

medications, if useful, are more helpful in association with formalized speech therapy. There is increased interest upon utilizing devices which alter auditory feedback, and these hold considerable promise in treatment (Kalinowski, *et al.*, 2000; Lincoln *et al.*, 2006; Rosenfield, 2001a).

30.3. THE PHYSIOLOGY OF STUTTERING

Understanding the physiology of stuttering involves neurological mechanisms as well as acoustics involved with processing normal and abnormal speech. At further issue is whether there is a genetic predisposition to whatever physiologic processes underlie output of stuttered speech.

There is debate whether genetics plays an important role in the cause or the maintenance of stuttering (Yairi *et al.*, 1996; Viswanath *et al.*, 2004; Suresh *et al.*, 2006). Regardless, identification of genetic factors underlying stuttering is very important. This information can lead to improved diagnosis and early identification of those at risk and permit possible therapeutic intervention at a young age. Identifying these genetic factors can lead to functional studies of proteins coded by these specific genes, providing bench-to-bedside clinical translation of underlying metabolic, signaling, transcription regulation, and cellular pathways involved in processing of speech signals. Identifying these genetic factors can also help establish a more informative nosology of disturbances in disturbances of speech, language, reading, and cognitive disorders, possibly even overlapping with disturbance of motor control systems affecting arms and legs.

All investigations of stuttering have identified a strong male prevalence among stutterers. As noted above, 80% of children "outgrow" their stutter but 20% do not. This variable underlines the importance of identifying current as well as former stutterers in any genetic investigation.

Many genetic investigations of stuttering have focused upon concordance rates among monozygotic versus dizygotic twin pairs, as well as analyses of pedigrees in different families. Many of these studies are compromised by small sample size, difficulty ascertaining twin status, and difficulty obtaining information from individuals in families pertaining to identifying speech disruption in those who are deceased or not obtainable to the investigator(s).

Howie (1981) directly examined twin pairs of stutterers and observed that 63% of 16 monozygotic twin pairs and 19% of 13 dizygotic twin pairs were concordant for stuttering. Andrews *et al.* (1991) investigated 3800 twins through self-employed reporting and discovered 20% concordance among 50 monozygotic twin pairs and 3% concordance among 85 dizygotic twin pairs. These investigations suggest an important factor in the cause of stuttering may be heredity but that heredity by itself may not be a sufficient condition to produce stuttering, since the concordance among homozygous (identical, same genetic code) twins was not 100%. Although twin studies suggest an important genetic role in the genesis of stuttering, genetic predisposition may be a necessary but not sufficient condition for stuttering to occur, suggesting the possible importance of environmental factors.

Some genetic investigations have focused upon aggregates of stuttering within individual families. Relatives of persons who stutter are at increased risk for stuttering, ranging from a three to a 10-fold increase. Segregation patterns of stuttering within these families do not suggest a single-gene, fully penetrable transmission model of inheritance, such as autosomal dominant, autosomal recessive, or x-linked. Rather, they suggest a diallelic model, the penetrance being affected by whether a parent stutters and also the sex of that parent who stutters. Stuttering is more common within the families of male stuttering parents (Cox *et al.*, 1984; MacFarlane *et al.*, 1991; Ambrose *et al.*, 1993; Suresh *et al.*, 2006).

If stutterers have altered physiology of speech processing, whether or not due to genetic predisposition, perhaps this can be further identified by analyzing altered structure of cerebral content. In other words, if one wants to investigate function, perhaps begin by investigating structure, recognizing the two can be related but not causally so.

Employing volumetric magnetic resonance imaging (MRI), Foundas *et al.* (2001) describe qualitative as well as quantitative differences between the brains of fluent speakers and stutterers. Qualitative findings were in frontal lobe gyral patterns involving Broca's area. Quantitative differences were in posterior temporal lobe regions involving Wernicke's area. Most right-handed fluent speakers have a larger left superior posterior temporal area (Brodmann's area 22, planum temporale), primarily consisting of Wernicke's area; these authors' most significant finding was that Wernicke's area, which is on the left side of the brain, and the area on the right corresponding to Wernicke's area were bilaterally larger but less asymmetric in stutterers than in fluent speakers.

Foundas *et al.* (2001) suggest these brain abnormalities permit normal development of language but can produce abnormalities in the speech-motor output of language. They query whether anatomic compromise of speech–language perisylvian cortex, including Broca's and Wernicke's areas, promotes instability in interaction between an outer "linguistic" loop and an inner "phonatory" loop, resulting in stuttering (discussed below in greater detail; Anderson *et al.*, 1999; Nudelman *et al.*, 1992). They also note that abnormalities in the anatomic substrates of language might differ among stutterers, some having difficulty processing speech motor control due to aberrancies in frontal opercular areas, whereas others have difficulty due to compromise near Wernicke's area.

Magnetoencephalography (MEG) has superb temporal resolution, with good accuracy for localizing active cortical areas in experimental paradigms (see Chapter 6). Salmelin *et al.* (2000) employed (MEG) to map cortical activation sequences during reading in fluent individuals and compared them to stutterers reading fluently. These authors observed altered brain function in developmental stutterers, finding essentially identical overt performance in the two groups but with different cortical activation patterns. Within the first 400 ms after seeing the word, processing in fluent speakers advanced from the left inferior frontal cortex (articulatory programing) to the left lateral central sulcus and dorsal premotor cortex (motor preparation). This sequence was reversed in developmental stutterers. Developmental Stutterers displayed early left motor cortex activation followed by a delayed left inferior frontal signal. Developmental stuttering (DS) thus appeared to initiate motor programs prior to preparation of the articulatory code. During speech production, the right motor/premotor cortex generated consistent evoked activation in fluent speakers but was silent in DS. These results, coupled with further experimental paradigms described in their report, suggest imprecise functional connectivity within the right frontal cortex and incomplete segregation between the adjacent hand and mouth motor representations in developmental stuttering during speech production. The authors maintain that a network including the left inferior frontal cortex and the right motor/premotor cortex, probably relevant in merging linguistic and affective prosody with articulation during fluent speech, may be partly dysfunctional in developmental stutterers.

Sommer *et al.* (2002) analyzed 15 DS and controls using diffusion tensor imaging (DTI) and voxel-based morphometry of the brain. DTI measures diffusion characteristics of water *in vivo* and establishes white matter fiber orientation, because water diffuses more rapidly when moving parallel rather than perpendicular to the longitudinal axis of axons. Fractional anisotropy (FA) of diffusion is a measure of coherence of the orientation of fibers within each voxel (the smallest distinguishable box-shaped part of a three-dimensional space). Their experiment, which was not focused on the entire brain, noted decreased FA immediately below the laryngeal and tongue representation in the left sensorimotor cortex, suggesting that DS results from disturbed timing of activation in speech-relevant brain areas.

In addition to the above investigations of abnormal brain structure in stuttering, which could cause or result from stuttered speech, there have been investigations of possible altered brain function in stuttering, primarily employing positron emission tomography (PET) or functional magnetic resonance imaging (fMRI). PET experiments, including reading aloud, reading silently, and choral reading suggest that people who stutter have different activation patterns than do fluent speakers (Fox *et al.*, 1996; Braun *et al.*, 1997; DeNil *et al.*, 2000). Despite differences in experimental design, these investigations similarly emphasize that right-handed stutterers lateralize speech-related activations to their right hemisphere in contrast to the fluent speakers who lateralize speech-related activations primarily to their left hemisphere.

Fox *et al.* (1996) noted that solo reading induced widespread activation of motor systems in both the cerebrum and cerebellum, with a right dominance, in people who stuttered. Similarly, Braun *et al.* (1997) observed stutterers that regional responses were absent, bilateral or lateralized to the right hemisphere when the subject was talking and stuttering. DeNil *et al.* (2000) support this conclusion by observing proportionately greater right hemisphere activation in stutterers while they read individual words overtly or covertly, as opposed to proportionately greater activation in the left hemisphere in non-stutterer controls. Fox *et al.* (1996) and Braun *et al.* (1997) reach similar conclusions when investigating fluency evoked by chorus reading and prolonged speech. These studies suggest brain activation patterns of stutterers differ from non-stutterers during stuttered speech and also when stutterers speak fluently.

30.4. THEORIES OF STUTTERING

Stutterers have difficulty controlling and coordinating their laryngeal sound source effectively with articulation. Laryngeal adductor and abductor muscles co-contract during dysfluencies, rendering normal speech output impossible. When stutterers try to "fake" these dysfluencies, or if fluent speakers try to produce these, they do not produce these co-contractions. The electromyographic relationship between laryngeal abductor and adductor muscles is abnormal during moments of the stutterer's dysfluency. Researchers have observed that stutterers who now have to use a sound source other than their larynx, such as stutterers who developed laryngeal cancer and subsequently used an electronic voice following their laryngectomy, became completely fluent with this electronic voicing apparatus (reviewed in Bloodstein, 1995).

There are numerous theories that abnormal cerebral laterality of language causes stuttering. Since right-handedness reflects language lateralized to the left hemisphere and since left-handedness reflects language lateralized to the left or to both hemispheres, many investigations have focused on the prevalence of right- and left-handedness among stutterers. This would imply stutterers lack normal language laterality for language.

Investigations exploring handedness, such as laterality testing incorporating dichotic listening and presentation of visual stimuli to competing hemispheres, have made this theory enticing but it has not been confirmed. Disruption of auditory processing and input has been described in investigations of dichotic presentation of meaningful linguistic stimuli; many stutterers lack the normal left brain (i.e., right ear) advantage. Being left- or right-handed is neither a necessary nor sufficient condition for stuttering (reviewed in Bloodstein, 1995 and in Rosenfield, 1997).

Some maintain that stuttering relates to abnormalities in auditory self-monitoring of speech. Stutterers become more fluent when loud, broadband noise prohibits hearing their own speech. This improvement also occurs with DAF, in which a stutterer hears what he says 250 ms after he speaks. FAF, in which the person who stutters hears his own speech but with an altered frequency, also promotes fluency (Kalinowski *et al.*, 2000; Lincoln *et al.*, 2006).

Early theories of motor disturbance also addressed stuttering. Some query whether stuttering reflected periodic irregularities in the timing of muscle movements within the speech motor control system. Background muscle tension appears to be elevated in developmental stuttering, perhaps making the high-precision adjustments needed during speech difficult to perform, resulting in movements that are neither smooth nor accurate. Stutterers lack good coordination of antagonist laryngeal muscles and appear slower in initiating phonation than are fluent speakers. Caruso (1991) suggests that the supplementary motor area is dysfunctional in stuttering, resulting in poor motor planning of the speech output.

The above theories describe differences between a group of stutterers and a group of non-stutterers. They describe mechanisms not common to all stutterers, since some of the stutterers do not have the abnormalities noted. One theory, which lacks anatomic or actual physiologic verification, does explain the necessary and sufficient conditions for a dysfluency to occur. It is complex, involves mathematical modeling, and incorporates control theory. We present a basic discussion of control theory and the necessary and sufficient conditions for stuttered dysfluencies to occur.

Employing control theory, researchers (Nudelman *et al.*, 1989; Rosenfield *et al.*, 1991) model the speech-motor control system as consisting of two nested loops, an inner phonatory loop producing sound and an outer linguistic loop selecting the sounds that are to be produced. Stuttering occurs when there is disruption of timing between these "functional loops." Evidence suggests that stutterers have slowing within the outer loop. Some authors (Anderson *et al.*, 1999; Foundas *et al.*, 2001) suggest that the perisylvian speech–language cortex (e.g., Broca's area, Wernicke's area) mediates the outer loop and that cortico-striatal-thalamo-cortical circuits mediate the inner loop. Anderson *et al.* (1999) further corroborate this model by noting that dopamine blockade and Parkinson disease slow the inner cortical–basal ganglia loop, reducing dysfluency in stuttering.

One of the problems in investigating stuttering and providing more effective therapies is lack of an animal model. We have developed such an animal model in Zebra finches, utilizing the phonatory iterations of their birdsong as a model to investigate the neurophysiology and neuropharmacology underlying fluent and dysfluent speech.

30.5. ANIMAL MODELING OF STUTTERING

Human beings are unique in that they communicate through a system of language. Non-human animals have a system of communication but lack language. Thus, a non-human animal might be able to signal food or danger or make sexual advances to another animal, but is not able to produce language akin to ours. However, many species share with us mechanisms for sound production.

Zebra finches (*Taeniopygia guttata*) share many features with human beings pertaining to communication and sound production. Both have critical periods during development during which they must hear the sounds of their respective communication/language they are to produce, and both have specific temporal periods during which they must hear as well as practice these sounds. For instance, a Zebra finch must hear songs from its adult tutors by day 65 and practice it by day 90. If this time course is prohibited by disruption in hearing or social isolation, the animal is unable to learn normal adult birdsong. Similarly, young children must hear and practice sounds of their environmental language if they are to acquire the language of their parents (Box 30.2).

The sound output of Zebra finches and humans is spectrally (i.e., multiple frequencies) and temporally (i.e., multiple changes in frequencies over time) diverse. Finches and humans both have an established hierarchy within their brains that controls these sound outputs. Our laboratory has demonstrated that ~5% of Zebra finches born to parents that sing normally and raised among normal singing parents, will have phonatory iterations similar to stuttering. When these "stuttering" birds raise other birds born to normal singing parents, approximately 60% of these birds also stutter. However, although these latter birds have multiple phonatory iterations, they do not repeat the spectral content of the birds from which they learned to stutter. In other words, they learn to make an error in the timing within their song production, but not in the frequency domain. There is considerable improvement in their abnormal repetitions when these birds born to normals but raised by stutterers (who were born to normal singing birds but developed stuttering despite their living among normal production of song) are placed in isolation or among normal singers (Helekar *et al.*, 2000; Botas *et al.*, 2001; Rosenfield *et al.*, 2001; Helekar *et al.*, 2003).

Thus, there may be a "rule to repeat," in birdsong. Some birds (~5%) are born repeaters, perhaps reflecting a genetic mandate. Others birds learn to repeat abnormally, and can improve toward normalcy with time, but still have the stuttering.

Zebra finches provide a good animal model of stuttering. Similar to human stutterers, the birds' stuttering is position-specific, although the iterations in birdsong reside at the end of the motif, whereas humans stutter at the beginning of sentences and phrases. The birds' and humans' errors are repetitive and non-voluntary, as suggested by the regularity of the iterations. The birds, similar to humans, seem to prefer not to stutter since the tutored repeaters improve significantly when they are placed among normal singing adults. It is not known whether the birds that stutter are distraught by their abnormal iterations or whether their colleagues perceive them as different (Helekar *et al.*, 2000; Botas *et al.*, 2001; Rosenfield *et al.*, 2001; Helekar *et al.*, 2003).

Box 30.2   Spectograms of normal Zebra finch birdsong

The horizontal axis reflects time and the vertical axis reflects frequency. The Zebra finch birdsong is spectrally complex, changes over time, and shares these properties with human speech. The spectograms show the phonatory iterations in a Zebra finch. Unlike stuttering, the repetition of the "syllable" occurs at the end of the motif.

Rosenfield, David B. (2005). Brain and language: disorders of fluency and voice. In Harry Whitaker (Ph.D. Ed.), *Encyclopedia of language and linguistics* (2nd edn). Oxford, England: Elsevier.

Normal song

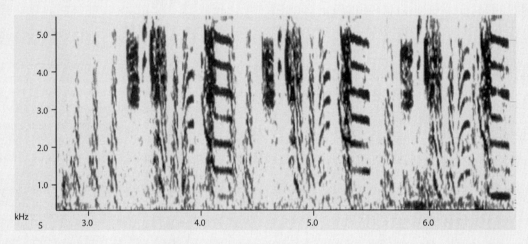

Repeater song

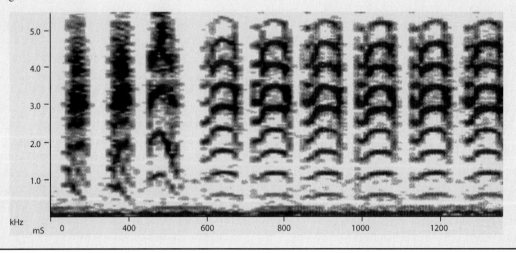

If one disregards the abstract linguistic and cognitive elements of speech, and focuses upon vocal-motor components, songbird vocalization provides substantial insight into human speech-motor control disorders. Similarities between birdsong and speech can help understand the pathophysiology of human dysfluencies, and possibly in testing neuromotor theories of stuttering.

processing of speech-related signals, whether input or output or both. Genetic investigations suggest a possible necessary genetic component to stuttering but this genetic predisposition may not be sufficient to cause stuttering. Current investigation, employing clinical, genetic, and brain imaging paradigms, coupled with animal investigations, may provide improved insight into this pan-cultural, global disturbance.

30.6. CHALLENGES AND FUTURE DIRECTIONS

Speech is a complex motor act. Stuttering is a complex disturbance of this process and can reflect abnormal cerebral

References

Ambrose, N.G., Yairi, E., & Cox, N. (1993). Genetic aspects of early childhood stuttering. *Journal of Speech and Hearing Research, 36,* 701–706.

Anderson, J.M., Hughes, J.D., Rothi, L.J., Crucian, G.P., & Heilman, K. (1999). Developmental stuttering and Parkinson's disease: The effects of levodopa treatment. *Journal of Neurology, Neurosurgery and Psychiatry*, 66, 776–778.

Andrews, G., Morris-Yates, A., Howie, P., & Martin, N. (1991). Genetic factors in stuttering confirmed. *Archives of General Psychiatry*, 48, 1034–1035.

Bloodstein, O. (1995). *A handbook on stuttering*. San Diego: Singular Publishing.

Botas, A., Espino, G., Rosenfield, D.B., & Helekar, S.A. (2001). Reduction of female-directed song motifs induced by repeated singing in laboratory-bred zebra finches. *Neuroscience Letters*, 297, 203–206.

Braun, A.R., Varge, M., Stager, S.G., Schulz, S., Selbie, J.M., Maisog, R.E., Carson, R.E., & Ludlow, C.L. (1997). Alerted patterns of cerebral activity during speech and language production in developmental stuttering: A PET study. *Brain*, 120, 762–784.

Caruso, A.J. (1991). Neuromotor processes underlying stuttering. In H.F.M. Peters, W. Hulstjin & C.W. Starkweather (Eds.), *Speech motor control and stuttering* (pp. 101–116). Amsterdam: Elsevier.

Conture, E. (2001). *Stuttering its nature, diagnosis, and treatment*. Needham Heights, MA: Allyn and Bacon.

Cox, N.J., Kramer, P.L., & Kidd, K.K. (1984). Segregation analyses of stuttering. *Genetic Epidemiology*, 1, 245–253.

DeNil, L.F., Kroll, R.M., Kapur, S., & Houle, S. (2000). A positron emission tomography study of silent and oral reading of single words in stuttering and nonstuttering adults. *Journal of Speech, Language, and Hearing Research*, 43, 1038–1053.

Foundas, A.L., Bollich, A.M., Corey, D.M., Hurley, M., & Heilman, K.M. (2001). Anomalous anatomy of speech–language areas in adults with persistent developmental stuttering. *Neurology*, 57, 207–215.

Fox, P.T., Ingham, R.J., Ingham, J.C., Roby, J., Martin, C., & Jerabek, P. (1996). A PET study of the neural systems of stuttering. *Nature*, 382, 158–161.

Helekar, S.A., Marsh, S., Viswanath, N.S., & Rosenfield, D.B. (2000). Acoustic pattern variations in the female-directed birdsongs of a colony of laboratory-bred zebra finches. *Behavioral Processes*, 49, 99–110.

Helekar, S.A., Espino, G.G., Botas, A., & Rosenfield, D.B. (2003). Development and adult phase plasticity of syllable repetitions in the birdsong of captive zebra finches (Taeniopygia guttata). *Behavioral Neuroscience*, 117, 939–951.

Helm, N.A., Butler, R.B., & Benson, D.F. (1978). Neurogenic acquired stuttering. *Neurology*, 5, 269–279.

Howie, P.M. (1981). Concordance for stuttering in monozygotic and dizygotic twin pairs. *Journal of Speech Hearing Research*, 24, 317–321.

Kalinowski, J., Dayalu, V.N., Stuart, A., Rastatter, M.P., & Rami, M.K. (2000). Stutter-free and stutter-filled speech signals and their role in stuttering amelioration for English speaking adults. *Neuroscience Letters*, 293, 115–118.

Lincoln, M., Packman, & Onslow, (2006). Altered auditory feedback and the treatment of stuttering: A review. *Journal of Fluency Disorders*, 31, 71–89.

MacFarlane, W.B., Hanson, M., & Mellon, C.D. (1991). Stuttering in five generations of a single family. *Journal of Fluency Disorders*, 16, 117–123.

Nudelman, H.B., Hoyt, B., Herbrich, K.E., & Rosenfield, D.B. (1989). A neuroscience model of stuttering. *Journal of Fluency Disorders*, 14, 399–427.

Nudelman, H.B., Herbrich, K.E., Hoyt, B.D., & Rosenfield, D.B. (1992). Phonatory response times of stutterers and fluent speakers to frequency-modulated tones. *Journal of the Acoustical Society of America*, 92, 1882–1888.

Peretz, I., & Zastorre, R. (2003). *The cognitive neuroscience of music*. New York, NY: Oxford University Press.

Perkins, W.H. (1973). Replacement of stuttering with normal speech. *Journal of Speech and Hearing Disorders*, 38, 283–294.

Porfert, A.R., & Rosenfield, D.B. (1978). Prevalence of stuttering. *Journal of Neurology, Neurosurgery, and Psychiatry*, 41, 954–956.

Rosenfield, D.B. (1972). Stuttering and cerebral ischemia. *New England Journal of Medicine*, 287, 991.

Rosenfield, D.B. (1997). In S. Schachter & O. Devinsky (Eds.), *Behavioral neurology and the legacy of Norman Geschwind* (pp. 101–111). Philadelphia, PA: Lippincott-Raven. (chapter 16).

Rosenfield, D.B. (2001a). Pharmacologic approaches to speech motor disorders. In D. Vogel & M.P. Cannito (Eds.), *Treating disordered speech motor control: For clinicians, by clinicians* (2nd edn, pp. 27–79). Austin, TX: Pro-Ed, Austin.

Rosenfield, D.B. (2001b). Do stutterers have different brains. *Neurology*, 57, 171–172.

Rosenfield, D.B., & Barroso, A.O. (2000). Difficulties with speech and swallowing. In W.G. Bradley, R.B. Darroff, G.M. Fenichel & C. David Marsden (Eds.), *Neurology in clinical practice: Principles of diagnosis and management* (3rd edn, pp. 171–186). Boston, MA: Butterworth-Heinemann.

Rosenfield, D.B., Viswanath, N.S., Callis-Landrum, L., DiDanato, R., & Nudelman, H.B. (1991). Patients with acquired dysfluencies: What they tell us about developmental stuttering. In H.F.M. Peters, W. Halstijn, & C.W. Starkweather (Eds.), *Speech motor control in stuttering* (pp. 277–284). Amsterdam: Elsevier.

Rosenfield, D.B., Viswanath, N.S., & Helekar, S.A. (2001). An animal model of stuttering. In H.G. Bosshardt, J.S. Yaruss & H.F.M. Peters (Eds.), *Fluency disorders: Theory, research, treatment and self-help. Proceedings of the Third World Congress on fluency disorders in Nyborg, Denmark* (pp. 119–122). Amsterdam: Nijmegen University Press.

Ross, E.D., & Monnot, M. (2008). Neurology of affective prosody and its functional–anatomic organization in right hemisphere. *Brain and Language*, 104, 51–74.

Salmelin, R., Schnitzler, A., Schmitz, F., & Freund, H.J. (2000). Single work reading in developmental stutterers and fluent speakers. *Brain*, 123, 1184–1202.

Sommer, M., Koch, M.A., Paulus, W., Weiller, C., & Büchel, C. (2002). Disconnection of speech-relevant brain areas in persistent developmental stuttering. *Lancet*, 360, 380–383.

Suresh, R., Ambrose, N., & Roe, C. (2006). New complexities in the genetics of stuttering: Significant sex-specific linkage signals. *American Journal of Human Genetics*, 78, 554–563.

Viswanath, N.S., Lee, H.S., & Chakraborty, R. (2004). Evidence for a major gene influence on persistence of developmental stuttering. *Human Biology*, 76, 401–412.

Yairi, E., Ambrose, N., & Cox, N. (1996). Genetics of stuttering: A critical review. *Journal of Speech and Hearing Research*, 48, 226–246.

Further Readings

Awad, M., Warren, J.E., Scott, S.K., Turkheimer, F.E., Wise, R.J.S. A common system for the comprehension and production of narrative speech. *Journal of Neuroscience*, 27(43), 11455–11464.
The authors excellently review our current understanding of input and output processing pertaining to cerebral processes underlying speech.

Bloodstein, O. (1995). *A handbook on stuttering*. San Diego, CA: Singular.
This handbook is an excellent review of stuttering, with a valuable chronology of research into this syndrome and a valuable list of references. The book is essential for anyone interested in stuttering.

Conture, E. (Ed.) (2001). Stuttering, its nature, diagnosis, and treatment. Needham Heights, MA: Allyn and Bacon.
This very good text highlights recent perspectives on treatment as well as cause.

Raphael, L.J., Borden, G.J., & Harris, K.S. (2006). *Speech science primer: Physiology, acoustics, and perception of speech.* **Philadelphia, PA: Lippincott Williams & Wilkins.**

The third edition of this hallmark textbook is essential for those interested in understanding mechanisms of speech production, especially acoustics, in addition to how the brain orchestrates sound production and what the oral-motor system does to produce sound waves. This textbook provides the necessary foundation to understand these mechanisms.

Mesial Temporal Lobe Epilepsy: A Model for Understanding the Relationship Between Language and Memory

JOSEPH I. TRACY and STEPHANIE B. BOSWELL

Department of Neurology, Comprehensive Epilepsy Center,
Thomas Jefferson University Hospital/Jefferson Medical College, Philadelphia, PA, USA

ABSTRACT

What is the role of medial temporal structures in language? While the hippocampus is traditionally considered a structure exclusively involved in memory, people with hippocampal damage do show language deficits. Mesial temporal lobe epilepsy (MTLE) provides a unique case study from which to examine the relationship between memory and language. While the pathology of MTLE is limited to the medial temporal lobe, these patients show language deficits in naming, fluency, and comprehension. Anatomical data show pathways exist that would allow medial temporal structures to influence lateral temporal and frontal language areas. Though MTLE can clearly affect language indirectly, there are cogent reasons to propose a direct role for the medial temporal lobe in language processing. Through the analysis of deficits seen in patients with MTLE and integrating this with current knowledge of hippocampal functioning, we attempt to unravel the complex relationship between language, memory, and the medial temporal lobe.

31.1. INTRODUCTION

Mesial temporal lobe epilepsy (MTLE) dramatically alters the cognitive, emotional, and practical lives of those who suffer it. When it occurs in childhood it affects development by compromising a key structure involved in memory (the hippocampus), and as a consequence hinders occupational and interpersonal growth. MTLE is the best characterized of the epilepsy syndromes in terms of seizure morphology, anatomic abnormalities, and neurocognitive deficits. Given the well-established role of the hippocampus and neighboring

structures in episodic memory, one would expect memory deficits to be the primary neuropsychological finding for these patients and, indeed, this is the case (see Box 31.1). However, naming, a language deficit, is the most frequently reported subjective complaint of patients with epilepsy and the vast majority of neuropsychological studies report that a naming deficit is prevalent in temporal lobe seizure patients when seizures emanate from the language-dominant hemisphere (Schefft *et al.*, 2003). Yet, naming, or the ability to specifically label objects, activities, or experiences, is not the only language skill found to be impaired in MTLE patients. Problems with verbal fluency (ease, speed, and flow of speech and self-generated ideas during discourse), and auditory comprehension (the ability to comprehend oral speech) are also widely reported. Even MTLE patients undergoing a more limited and selective surgical procedure such as a left amygdalohippocampectomy (selective removal of amygdala and hippocampus) show a decrease in expressive language skills post-surgery (Helmstaedter *et al.*, 1996), and the famous mesial temporal lobe amnesia patient, H.M., displayed language deficits. However, a recent model of the complex network involved in syntactic or lexical/semantic processing, in which numerous cortical and subcortical regions are cited as important, still omits the medial temporal lobe in its formulation (Ullman, 2001). Why would patients with strictly mesial temporal lobe damage display deficits in skills whose functional neuroanatomy appears to be outside this area of pathology?

One sure means may be that medial temporal lobe seizures are a unique form of pathology, and it is this unique feature that places much of the brain damage and dysfunction outside the temporal lobe. That is, seizures spread, and it is

Box 31.1   What is the primary deficit associated with MTLE?

The primary anatomic defect found in MTLE patients is hippocampal sclerosis (HS). It is characterized by atrophy of the hippocampal formation, specifically in CA1, CA3, and the granule cell layer of the dentate gyrus, with little effect on CA2. HS can be seen with magnetic resonance imaging (MRI)

with the primary findings involving reduced hippocampal volume, increased T2-weighted imaging signal intensity, and loss of normal internal architecture. The figure shows an MRI of a 42-year-old female with HS and complex partial seizures.

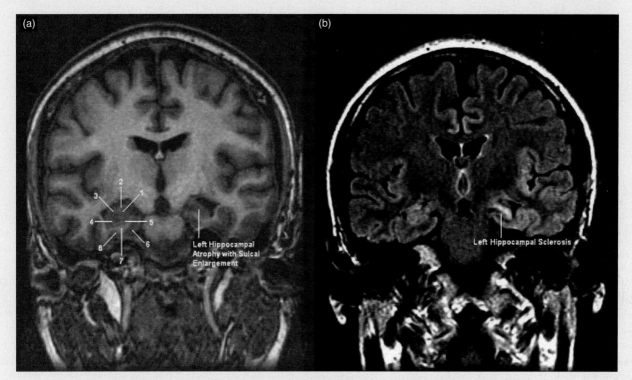

(a) The high resolution anatomical image shows important neuroanatomy in the healthy right hemisphere and hippocampal atrophy with sulcal enlargement in left hemisphere. (1) Dentate gyrus and CA4 region of hippocampal body; (2) CA1 region of hippocampal body; (3) CA2 region of hippocampal body; (4) CA3 region of hippocampal body; (5) subiculum; (6) entorhinal cortex; (7) parahippocampal gyrus; and (8) perirhinal cortex. (b) FLAIR Image of same individual showing HS.

It has been well documented that the hippocampus and parahippocampal gyri of the dominant hemisphere are necessary for the preservation of verbal information in episodic memory. Not surprisingly, episodic memory deficits are the best documented of the neuropsychological impairments in MTLE. The memory deficit encompasses both the impairment of long-term memory consolidation as well as the retrieval of newly acquired information. It is best demonstrated through word-list learning tasks or memory for abstract designs and prose passages. For example, neuron loss in the CA3 and hilar areas of the hippocampus, the characteristic anatomic finding in mesial temporal sclerosis, correlates with percent retention scores on a prose passage memory task, and the asymmetry between left- and right-hemisphere hippocampal volumes is a strong

predictor of both actual verbal memory performance and self-perceived ratings of memory skill in MTLE patients (Alessio *et al.*, 2006). These declarative memory problems in MTLE are caused by fairly specific hippocampal damage and do not require more global damage to the temporal lobe.

Margerison, J.H., & Corsellis, J.A. (1966). Epilepsy and the temporal lobes. A clinical, electroencephalographic and neuropathological study of the brain in epilepsy, with particular reference to the temporal lobes. *Brain*, 89(3), 499–530.

Jackson, G.D., Berkovic, S.F., Tress, B.M., Kalnins, R.M., Fabinyi, G.C., & Bladin, P.F. (1990). Hippocampal sclerosis can be reliably detected by magnetic resonance imaging. *Neurology*, 40(12), 1869–1875.

this generalization of seizures which causes burden and damage to cells outside the medial temporal lobe. Yet, there is strong evidence that the naming and other language deficits persist after surgical removal of the cells generating the seizure, the onset of good seizure control, and the elimination of the effects of seizure spread (Langfitt & Rausch, 1996). This suggests that aside from the medial temporal lobe's potential for disrupting known language areas there is the strong possibility that the medial temporal lobe plays a direct role in language. Does a dysfunctional medial temporal lobe have a direct effect on language abilities? Does language representation in the brain depend on an intact medial temporal lobe system? Does intact language depend on an intact memory system? As cognitive neuroscience moves away from strict modular approaches and tries to understand cognition by examining the interaction between brain structures, MTLE serves as an ideal case. It provides a unique opportunity to observe the knock out of specific brain structures, and to examine their impact on cognitive networks thought to lie mostly outside the medial temporal lobe.

31.2. DO MTLE PATIENTS SHOW LANGUAGE ABNORMALITIES?

Patients with left MTLE show a higher incidence of atypical language lateralization than both patients with right MTLE and healthy controls (Thivard et al., 2005). The percentage of atypical language lateralization, which is defined as both right-hemisphere dominant as well as bilateral representation, in normal controls is 4–6%, while the incidence is as high as 33% in patients with left MTLE (Adcock et al., 2003) but can vary depending on the language skill (Tracy et al., 2005). During functional magnetic resonance imaging (fMRI) exams involving verb generation (e.g., generating verbs when viewing a noun), temporal lobe epilepsy (TLE) patients with weaker left-hemisphere language lateralization tend to show either increased activation of the right hemisphere in areas homologous to the language regions of the left hemisphere or more anomalous patterns of activation throughout the brain compared to healthy controls (Thivard et al., 2005). MTLE is also associated with specific deficits in confrontation naming, verbal fluency, and comprehension with additional but less consistent reports of deficits in reading, repetition, spelling, writing, and speech quality.

31.2.1. Confrontation Naming

Visual confrontation naming is commonly tested by the Boston Naming Test (BNT). In the BNT, subjects are shown line drawings of common objects one at a time and asked to name them orally. It has been shown that patients with left (usually dominant lobe) TLE perform significantly worse on the BNT than patients with right TLE or with primary generalized epilepsy (Howell et al., 1994). These results are consistent with evidence that the BNT is a strong predictor of seizure focus and that left hippocampal volume predicts performance on the BNT (Alessio et al., 2006). While most of the literature focuses on visual confrontation naming, other research has focused on auditory naming (hearing a definition and naming the word). Through cortical stimulation by subdural electrodes, it was found that anterior temporal sites precluded auditory naming, while stimulation of posterior temporal lobe sites knocked out visual naming. Interestingly, these authors also found that sparing visual naming sites in the temporal lobe from resection did not prevent post-operative decline, but that leaving or removing auditory naming sites reliably predicted the status of both auditory and visual naming post-surgery (Hamberger et al., 2005). Thus, there may be both general and modality specific semantic systems that support naming, and the medial temporal lobe may be more important for the functioning of some types of naming than others.

31.2.2. Verbal Fluency

Patients with TLE also show deficits in verbal fluency. Verbal fluency encompasses both phonemic and semantic fluency, and is tested by requiring subjects to rapidly retrieve words that fit certain criteria. Phonemic fluency is measured by having the subject generate words that begin with a particular letter of the alphabet for a specified time period. Semantic fluency is typically measured by having the subject produce words that are members of a semantic category, such as fruits or animals. It is thought that semantic fluency is managed by the superior and middle temporal lobe, due to the need for accessing semantic knowledge, while phonemic fluency is controlled by inferior frontal regions, because there are high demands on search and retrieval. Both types of fluency are mediated by effective strategic search and retrieval through stores of verbal (semantic) knowledge. Generally, MTLE patients show greater deficits on semantic fluency as opposed to phonemic fluency tasks (Gleissner & Elger, 2001). When the TLE involves the dominant hemisphere, there are semantic fluency disruptions whether or not the patient has hippocampal sclerosis. In non-dominant (typically right) TLE, disruption of fluency only occurs when the hippocampus is damaged. Thus, in dominant TLE the lateral temporal lobe seems crucial, whereas in non-dominant TLE semantic fluency seems highly dependent on the hippocampus.

31.2.3. Comprehension

Comprehension deficits have been associated with hippocampal dysfunction in MTLE and in patients who undergo a selective amygdalohippocampectomy. Following anterior temporal lobectomy, 40% of patients show a significant decline in linguistic functions including comprehension and fluency (Bartha et al., 2004). Reading comprehension is generally intact. An examination of brain activation during reading following anterior temporal lobe resection found performance was intact but that in addition to the expected left superior temporal activation there was activation in the right hemisphere involving the hippocampus and the frontal lobe, suggesting possible recruitment of less normative areas to accomplish the task (Noppeney et al., 2005).

31.2.4. Other Language Skills

Repetition abilities (repeating words upon command) related to pronunciation, and writing skills generally appear intact in MTLE. Measures of macrolinguistic abilities during narrative discourse such as word count, speech rate, pause duration, number of non-communicative words, and descriptive details do show differences relative to controls. However, these may stem from underlying verbal fluency and semantic knowledge weaknesses, and perhaps even working memory difficulties (Waites et al., 2006).

31.3. WHAT SPECIFIC EVIDENCE IS THERE SHOWING MEDIAL TEMPORAL LOBE INVOLVEMENT IN LANGUAGE?

While traditional language processing areas include the inferior frontal gyrus (Broca's area), superior temporal and middle temporal gyri, supramarginal gyrus and angular gyrus (Wernicke's area), there is evidence that structures in the medial temporal lobe have a role in language processing.

Activation in the hippocampus has been found during meaningful language comprehension tasks but not if non-semantic language is involved (Friederici, 2002). Also, an fMRI study in MTLE showed hippocampal formation involvement during a language comprehension task (Bartha et al., 2005).

Some of the most compelling evidence of medial temporal lobe involvement in syntactic and semantic processing comes from studies measuring event-related potentials (ERPs) via intracranial electrodes in the rhinal cortex and the hippocampus while listening to sentences with syntactic and semantic errors. Semantic errors caused a response in the rhinal cortex, while syntactic errors caused a response in the hippocampus proper, as measured by large negative ERPs (McCarthy et al., 1995). This suggests the rhinal cortex plays a role in semantic integration, while the hippocampus proper is involved in syntactic processing. The exact role of the these structures is unclear, but it may involve switching to the non-preferred meaning of specific words or to the less-common syntactic structures found in ambiguous sentences. This switch requires connecting and binding different items,

and such binding is a special skill of the hippocampus. Meyer and colleagues proposed that the hippocampus may be sensitive to the syntactic aspects of speech involving re-analysis of mismatches in syntax (Meyer et al., 2005). In contrast, the rhinal cortex may be more sensitive to semantic mismatches. When sentences are regular in terms of grammar and semantic predictability the medial structures of the rhinal cortex and hippocampus appear coupled through gamma synchronization. When violations occur this coupling is disrupted and the material must be processed further by these structures before it can be represented and go further in the declarative memory system.

Language processing, both comprehension and expression, depend on the integrity of and access to semantic knowledge networks. Without properly functioning semantic networks, one loses this ability to appreciate the context of sentences and loses the ability to answer in a relevant fashion. While there is good evidence linking the lateral temporal cortex to the storage of semantic networks, the role of the medial temporal cortex is uncertain. Again, MTLE research contributes to clarify this issue. For instance, left MTLE patients are selectively impaired on memory tests invoking conceptual (semantic) processing (Blaxton, 1992). However, language systems not dependent on explicit, conscious access of semantic knowledge, such as implicit verbal memory, do not appear impaired in MTLE (Del Vecchio et al., 2004), suggesting awareness and mode of information processing (automatic versus controlled) may help determine medial temporal lobe involvement.

In the studies implicating medial temporal lobe involvement in semantic or syntactic processing, it cannot be ruled out that the role these structures play is simply related to the retrieval of semantic or syntactic information and not the actual storage or analysis of such information. In the case of MTLE, several studies have argued that the deficit in confrontation naming is related to a deficit in semantic knowledge networks rather than retrieval and word access (Bartolomei et al., 2001). For instance, the types of errors made during semantic fluency involve the ability to generate category exemplars; patients, however, retain their ability to generate category labels, a split that would not be observed if a general retrieval deficit was present. It is also not clear the degree to which medial involvement in language is simply the result of the need for more cognitive capacity. For instance, language discourse requires significant planning, use of detail, and a sense of social and communication pragmatics, all of which take cognitive resources. Some have reported that measures of macrolinguistic variables during discourse reveal that TLE patients have longer and more variable pause durations, difficulty producing a compact narrative, more repetition, more non-communicative words, reduced speech rate, inadequate detail, and excessive "repairs" (Waites et al., 2006). All of these difficulties may be reflective of over-stretched cognitive capacity or strained

Box 31.2   Potential causes of a breakdown in language processing

1. Impaired general cognitive capacities or resources.
2. Impaired non-language, cognitive skills often needed during language processing such as working memory or memory retrieval. These are necessary but do not represent core language processing components.
3. Deficiency in language-specific storage or processing skills involving syntactic and semantic knowledge, word search, word comprehension, fluency, repetition, or naming.
4. Impaired cognitive skills that are core components of successful language processing such as (1) binding, re-arranging, and re-analyzing of contextually based novel material for its potential match to existing knowledge, i.e., potential role of polysynaptic intrahippocampal pathways and (2) automatic access of semantic knowledge or anticipatory frameworks for incoming language, or identify violations based on more gestalt properties and concepts over a broad scale, i.e., potential role of direct intrahippocampal pathway.

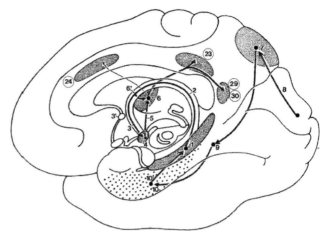

FIGURE 31.1   Cortical connections of the polysynaptic intrahippocampal pathway. Hippocampal outputs fibers to the cortex: arising from the hippocampus (1), fibers successively reach the body (2) and column (3) of fornix (3′, anterior commissure), the mamillary body (4), and then, via the mamillothalamic tract (5), the anterior thalamic nucleus (6); some fibers reach this nucleus directly (6′); from the anterior thalamic nucleus, the main cortical projections are the posterior cingulate (area 23) and retrosplenial (areas 29 and 30) cortices; some fibers may project to the anterior cingulate cortex (area 24). Input fibers from the cortex to hippocampus: the posterior parietal association cortex (7) in relation to the superior visual system (8) projects via the parahippocampal gyrus (9) to the entorhinal area (10); 10′, perforant fibers. Reproduced by permission of Springer Publishing. Duvernoy, H. (2005). *The human hippocampus: Functional anatomy, vascularization and serial sections with MRI* (3rd edn, p. 34). Berlin, Heidelberg, New York: Springer-Verlag.

working memory systems, factors that are not specific to language although still a necessary condition for effective language skills.

The upshot is that fluency, naming, and comprehension can breakdown for a variety of reasons, and distinguishing the nature of the breakdown is difficult (see Box 31.2). Temporal lobe pathology may have a direct impact on language, such as eliminating the input of the hippocampus or rhinal cortex to judgments about semantic or syntactic mismatch. Yet, other factors not specific to language processing can also be at work, such as low general cognitive capacity, weak working memory, or a breakdown in word search and retrieval.

31.4. ARE THERE ANATOMICAL CONNECTIONS THAT ALLOW THE MEDIAL TEMPORAL LOBE TO INFLUENCE LANGUAGE?

The way in which medial temporal lobe pathology accounts for declarative memory deficits is well known and is discussed in Box 31.1. How might perturbation of medial temporal lobe structures account for the language deficits? For this, we need to consider the broader brain connections that run to and from medial temporal lobe structures. If the hippocampus and related structures are involved in language, then it is likely that they would be involved through the influence of one or both of the following pathways.

The hippocampus and the entorhinal cortex form two functionally distinct pathways (Duvernoy, 2005) utilized

for declarative memory, but the connections they make with many of the structures involved in language provide a means for arguing that the hippocampus and other medial structures play a role in language processing. The polysynaptic pathway (see Figure 31.1) involves a long neuronal chain that courses through the entorhinal area, dentate, CA3 and CA1 fields, and the subiculum. The neurons are glutamatergic in nature. The outputs of this polysynaptic pathway to the cortex follow the fimbria and reach the anterior thalamic nucleus either directly or via the mamillary bodies. From the thalamus, fibers reach the posterior cingulate (BA 23) and retrosplenial cortices (BA 29 and 30), as well as the anterior cingulate. Inputs from the cortex to this polysynaptic pathway include the posterior parietal region (BA 7) and the temporal cortices (BA 40, 39, and 22), which make their way to the entorhinal area via the parahippocampus. In terms of memory, this pathway is considered crucial in perceiving and remembering objects in space and relating items in memory together.

A separate pathway, the direct intrahippocampal pathway (see Figure 31.2), contains fibers that reach the hippocampus and CA1 from the perirhinal and then entorhinal cortex without following the polysynaptic chain (e.g., skipping the subiculum). Note that the polysynaptic pathway enters the entorhinal cortex via the parahippocampus, as opposed to

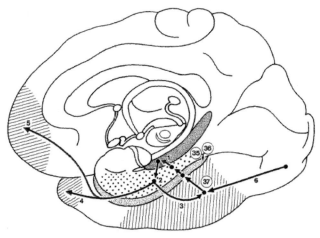

FIGURE 31.2 Cortical connections of the direct intrahippocampal pathway. (1) intrahippocampal circuitry. Hippocampal outputs fibers to the cortex: from the deep layers of the entorhinal cortex (2), fibers reach the inferior temporal association cortex (3), the temporal pole (4), and the prefrontal cortex (5). Inputs fibers from the cortex to hippocampus: the main origin of these fibers is the inferior temporal association cortex (area 37) in relation to the inferior visual system (6), reaching the entorhinal cortex through the perirhinal cortex (areas 35 and 36). Reproduced by permission of Springer Publishing. Duvernoy, H. (2005). *The human hippocampus: Functional anatomy, vascularization and serial sections with MRI* (3rd edn, p. 35). Berlin, Heidelberg, New York: Springer-Verlag.

the perirhinal area. The outputs of the direct pathway are the inferior temporal cortex, the temporal pole, and the prefrontal cortex. The inputs involve the inferior temporal cortex (BA 37 and 20) and the "what" pathway, involved in visual object recognition, object description, and the storage of object knowledge, which reach the entorhinal cortex via the perirhinal area (BA 35 and 36). With this anatomical groundwork laid out, we can consider the more specific role of the medial temporal lobe in language.

31.5. WHAT MIGHT THE MEDIAL TEMPORAL LOBE CONTRIBUTE TO LANGUAGE PROCESSING?

Models of hippocampal episodic memory provide clues as to how the medial temporal lobe might contribute to language processing. Several existing models suggest that the medial temporal lobe is necessary for the long-term storage of facts that will need to be consciously accessible at the time of later use, suggesting that conscious awareness during both learning and re-expression is necessary for the hippocampus to be recruited (Clark & Squire, 1998). A different model referred to as relational memory, argues that the hippocampus is crucial to associative learning and will be invoked whenever stimulus/stimulus or stimulus/response connections need to be encoded. The essence of the relational model is that the hippocampus is crucial, not

just to item storage *per se*, but to the binding of items, and then defining and storing the relation (association) between them. Other models of medial temporal lobe memory functioning exist and are still viable, including its role in responding to the familiarity (novelty, recency) of stimuli and its special role in navigational memory. A review of the literature in 1999 argued persuasively that extant evidence favors the relational model (Cohen *et al.*, 1999). This relational model points to a role for the hippocampus and related structures in cognition more broadly (i.e., in any type of associative learning), well beyond just their contribution to episodic memory.

As research has begun to parse out the distinct and separate functions of subregions in medial temporal cortex, there becomes even more fertile ground for implicating these regions in work of language. For instance, the rhinal cortex may operate in a fast, automatic manner, with its representational system limited to single items or over-learned combinations of items (O'Reilly & Norman, 2002). The rhinal cortex receives sensory input and by binding these sensory features it is well suited to object identification (Murray & Richmond, 2001). Less complex binding might also be utilized to form units of semantic information.

The hippocampus appears to work more slowly than the rhinal cortex. Its representational format is not limited to single items, and it can render in its space multiple items as well as inputs from different cortical regions and their relationship to each other (McClelland *et al.*, 1995). The hippocampus is well known to be sensitive to context and to bind these features (time, place, content) to an episode or input, such as a spoken utterance (Squire & Knowlton, 2000) toward the goal of integrating that information with existing knowledge structures to form more complex memories. This is consistent with the description of the hippocampus provided by Meyer *et al.* (2005) as analyzing "mismatched" sentence material ("controlled reanalysis and repair" p. 456) and it makes clear that this process is nonautomatic and dependent on declarative memory.

When one considers the inherent nature of grammar as a rule-based system for combining words, it makes sense that a structure such as the hippocampus, disposed to combining and binding, would be involved in grammar comprehension. If the items to be combined utilize well-worn, grammatically known relations, or part of the existing knowledge base, it is much less likely that the hippocampus will be invoked. In this case, controlled processing and declarative memory would not be needed to match informational elements, and the processing can occur more automatically. In fact, some studies have even shown activation of the hippocampus during the early stages of syntactic learning, but not during well-learned proficient syntax production or comprehension, where less effortful, more automatic processing can be used (Opitz & Friederici, 2003).

This quality of consciously binding, re-arranging, and re-analyzing contextually based novel material that does not already match existing knowledge suggests that medial temporal structures may not just be involved in memory and language comprehension, but also any brain activity requiring those skills. It also becomes clear that medial structures could handle this information in at least two ways; through a fast, automatic mode that is more limited to single or over-learned items, or a slower mode that in exchange for the less robust pace can handle multiple items.

31.6. WHAT ACCOUNTS FOR THE LANGUAGE DEFICITS IN MTLE?

The pathophysiologic processes that account for language deficits observed in mesial temporal lobe patients may be related to seizure spread to primary language areas and the connections between them, language reorganization, or disruption of the above computational skills needed for effective execution of language.

A simple mechanism to account for language deficits in MTLE is that the lateral temporal neocortex may be an area affected most immediately and reliably by seizure spread. For instance, Waites et al. (2006) found that functional connectivity between language areas was reduced in patients with MTLE. Thus, epileptic discharges (ictal or interictal) coursing through existing fiber tracts may disrupt their normal function, leading to compromised communication and a breakdown in networks. In chronic TLE, the years of seizure spread into these lateral language areas as well as other areas of the brain increase the likelihood of disruption.

In cases where the lateral neocortex is part of the epileptogenic zone, the integrity of the neural system there will likely have undergone detrimental changes associated with epileptogenesis. Studies supporting this involve those that show diffuse structural brain damage in chronic TLE (Moran et al., 2001). For instance, analyzed volumes of both mesial and lateral temporal lobe substructures in patients with MTLE have found prominent atrophy in both regions. The degree of atrophy in the temporal lobe was 8.3–18.4%, compared to controls, and the amount of atrophy showed a relationship to the degree of hippocampal atrophy. In addition, positron emission tomography (PET) studies have shown a functional decrease in not only in the hippocampus, but also in other neocortical regions including the lateral neocortex in patients with MTLE. For instance, patients with left TLE have shown hypometabolism in the lateral temporal cortex during PET scanning, actually exceeding that observed in the medial temporal cortex (Hammers et al., 2001). Hypometabolism in lateral neocortex both ipsilateral and contralateral to the hippocampus has also been demonstrated. Thus, while the struc-

tural abnormality of MTLE is in the medial temporal lobe, there is a detriment to structures outside this region from either epileptogenesis proper or seizure spread.

Another way of accounting for the language deficits in MTLE comes from our knowledge of the impact of early onset epilepsy (before age 6). Mesial temporal sclerosis (MTS) and chronic TLE may cause an anatomical reorganization of cognitive skills, and this reorganization is much more likely in early onset epilepsy (Springer et al., 1999). However, even adult TLE patients show more varied patterns of language lateralization and localization compared to normals (Pihlajamaki et al., 2000). The reorganization of language networks, either interhemispherically or intrahemispherically, may be causing the language deficits that are seen in MTLE, by changing the anatomical representation of these pathways from their normal sites. Thus, functional anatomical maps of normal language representation in the brain may not hold true in these patients, and the role of the hippocampus in language will, accordingly, be quite different in these patients. Seizures early in life may well be associated with the creation of noisy neural activity that interferes with the proper differentiation of cells and the development of categorical responses to phonemes and other language units. For instance, the work of Paula Tallal (Fitch & Tallal, 2003) has made clear that this noise often changes the sampling rate used to parse potential linguistic content. One could potentially see all thought as then influenced by these early disruptions of language. The strongest and most famous articulation of this impact is the Sapir–Whorf hypothesis, which argues that semantic categories and other language representations both structure and constrain thought.

One intriguing possibility is that the hippocampus determines whether reorganization of language is intrahemispheric or interhemispheric. Along those lines, MTS, common in early onset epilepsy, is correlated with a higher incidence of interhemispheric reorganization for receptive language than focal lesions in the primary language areas alone. Contrastingly, patients with lesions in language areas alone generally had an intrahemispheric shift, where the processing for those critical language skills was maintained in the same hemisphere in regions adjacent to the lesion (Pataraia et al., 2004). It may be that individuals with a more damaged hippocampus are more prone to re-organize language, and, if so, it is likely that these patients will evolve right-hemisphere representation of language. The nature of this effect is unclear but it may reflect the dependence of language processors in the brain on the parsing, binding, and re-analyzing capabilities of the medial structures in order to understand or produce complex speech. It also suggests there may be a dynamic force to re-organize, i.e., to seek out the computations typically provided by ipsilateral medial temporal structures, in order to make sure such skills are available.

Given that the polysynaptic and direct intrahippocampal pathways provide an anatomic basis for neural communication and interaction between the medial temporal lobe and neocortical language areas, and given that these medial structures can confer a needed skill in binding and re-analyzing language material, a direct role for these structures in language processing appears likely. This role may be as follows.

The entorhinal cortex plays a crucial role in both the direct and polysynaptic pathways as the main input to the hippocampus with the subiculum as the main output area. The polysynaptic pathway appears more strongly connected via the parahippocampus to the superior temporal gyrus and thus can receive inputs of graphemic (written, visual), phonemic, and morphologic material and utilize this pathway to bind these elements into wholes, either toward syntactic or semantic understanding. Because the polysynaptic pathway will carry inputs from structures sensitive to objects in space, this binding process may mark order and spatial arrangement when relating and associating the items in memory. In contrast, the direct intrahippocampal pathway receives inputs from the inferior temporal gyrus via the presubiculum and rhinal cortex (i.e., perirhinal area), suggesting that it may be used toward successful retrieval of pre-existing knowledge of facts and features of objects initiated by activity in the "what" pathway and object recognition system in order to make rich semantic knowledge available for language use and expression.

Taking these functional differences between the pathways a step further, one could speculate that the polysynaptic pathway is slower, non-automatic, and more sensitive to context, with its representational system limited to single items and their manipulation and arrangement. Its role in language may be best characterized as "controlled reanalysis and repair" (Meyer et al., 2005) and possibly more biased toward syntactic analysis, as the hippocampus has been proposed to be. The direct pathway may be faster, more automatic, and it may welcome multiple items as well as inputs from different cortical regions. Its role in language may be suited for less declarative and non-effortful tasks of language comprehension and expression. This skill is consistent with rapid object recognition and seamless, automatic access of semantic or syntactic knowledge. This would work for instance, to encode or retrieve objects in memory with a direct connection to their label in the lexicon. It may also guide anticipatory frameworks for incoming language or identify violations or problems in input based on its gestalt or holistic properties, taking into account context and coherence over a broader scale than would the polysynaptic pathway.

31.7. CHALLENGES AND FUTURE DIRECTIONS

We still do not know the functionality of specific regions of the medial temporal lobe and their exact contribution to language processing. The evidence certainly suggests that there is some dependency between memory and language, and there is also evidence that medial temporal structures, including the hippocampus, parahippocampal gyrus, and rhinal cortex have skills to contribute to language processing and do, in fact, participate to some degree. Because these structures are commonly damaged in MTLE, this pathology may be a direct cause of the language deficits that are seen in these patients, without drawing on explanations related to seizure spread, epileptogenicity, or reorganization of language representations. Also, unclear is the degree to which general aspects of cognitive capacity or deficits in processing components, not specific to language (retrieval, working memory), play a role in the language breakdowns seen in MTLE. MTLE patients provide excellent case studies for the study of complex networks involving both memory and language, and also put on display larger issues in cognitive neuroscience (see Box 31.3). More studies need to be undertaken to understand the clear role of medial structures in language processing networks so that care can be taken to spare these structures during temporal lobe resections whenever possible.

Box 31.3 Larger cognitive neuroscience issues implied by language findings in MTLE

1. Brain regions not specifically dedicated to language can nevertheless contribute computations or operations needed during language processing such as binding together units of information or re-analyzing potential mismatches/errors between language units (e.g., syntactic, semantic). These regions can be seen as necessary but not sufficient for successful language processing.

2. Hippocampal structures may be parsed into regions with distinct processing characteristics that vary according to the number of items that can be processed at once, or the speed, effort, resources, and monitoring demands inherent to the processing. May lead to clues about the principles that govern modes of information processing (automatic and controlled) more generally.

3. The potential reorganization of language representations following brain injury may depend on the integrity of areas not typically considered to be involved in language such as the hippocampus.

4. Role of medial temporal lobe structures in language may vary with the learning curve and the degree to which automaticity versus controlled processing, awareness, associative learning, or declarative memory are required.

References

Adcock, J.E., Wise, R.G., Oxbury, J.M., Oxbury, S.M., & Matthews, P.M. (2003). Quantitative fMRI assessment of the differences in lateralization of language-related brain activation in patients with temporal lobe epilepsy. *Neuroimage, 18*(2), 423–438.

Alessio, A., Bonilha, L., Rorden, C., Kobayashi, E., Min, L.L., Damasceno, B.P., & Cendes, F. (2006). Memory and language impairments and their relationships to hippocampal and perirhinal cortex damage in patients with medial temporal lobe epilepsy. *Epilepsy and Behavior, 8*(3), 593–600.

Bartha, L., Trinka, E., Ortler, M., Donnemiller, E., Felber, S., Bauer, G., & Benke, T. (2004). Linguistic deficits following left selective amygdalo-hippocampectomy: A prospective study. *Epilepsy & Behavior, 5*(3), 348–357.

Bartha, L., Marien, P., Brenneis, C., Trieb, T., Kremser, C., Ortler, M., Walser, G., Dobesberger, J., Embacher, N., Gotwald, T., Karner, E., Koylu, B., Bauer, G., Trinka, E., & Benke, T. (2005). Hippocampal formation involvement in a language-activation task in patients with mesial temporal lobe epilepsy. *Epilepsia, 46*(11), 1754–1763.

Bartolomei, F., Wendling, F., Bellanger, J.J., Regis, J., & Chauvel, P. (2001). Neural networks involving the medial temporal structures in temporal lobe epilepsy. *Clinical Neurophysiology, 112*(9), 1746–1760.

Blaxton, T. (1992). Dissociations among memory measures in memory-impaired subjects: Evidence for a processing account of memory. *Memory and Cognition, 20*, 549–562.

Clark, R.E., & Squire, L. (1998). Classical conditioning and brain systems: Tthe role of awareness. *Science, 280*, 77–81.

Cohen, N., Ryan, J., Hunt, C., Romine, L., Wszalek, T., & Nash, C. (1999). Hippocampal system and declarative (relational) memory: Summarizing the data from the functional neuroimaging studies. *Hippocampus, 9*, 83–98.

Del Vecchio, N., Liporace, J., Nei, M., Sperling, M., & Tracy, J. (2004). A dissociation between implicit and explicit verbal memory in left temporal lobe epilepsy. *Epilepsia, 45*(9), 1124–1133.

Duvernoy, H. (2005). *The human hippocampus: Functional anatomy, vascularization and serial sections with MRI* (3rd edn). New York: Springer.

Fitch, R.H., & Tallal, P. (2003). Neural mechanisms of language-based learning impairments: Insights from human populations and animal models. *Behavioral and Cognitive Neuroscience Reviews, 2*(3), 155–178.

Friederici, A.D. (2002). Towards a neural basis of auditory sentence processing. *Trends in Cognitive Sciences, 6*, 78–84.

Gleissner, U., & Elger, C.E. (2001). The hippocampal contribution to verbal fluency in patients with temporal lobe epilepsy. *Cortex, 37*(1), 55–63.

Hamberger, M.J., Seidel, W.T., McKhann, G.M., Perrine, K., & Goodman, R.R. (2005). Brain stimulation reveals critical auditory naming cortex. *Brain, 128*(11), 2742–2749.

Hammers, A., Koepp, M.J., Labbe, C., Brooks, D.J., Thom, M., Cunningham, V.J., & Duncan, J.S. (2001). Neocortical abnormalities of [11C]-flumazenil PET in mesial temporal lobe epilepsy. *Neurology, 56*(7), 897–906.

Helmstaedter, C., Elger, C.E., & Hufnagel, A. (1996). Different effects of left anterior temporal lobectomy, selective amygdalohippocampectomy, and temporal cortical lesionectomy on verbal learning, memory, and recognition. *Journal of Epilepsy, 9*, 39–45.

Howell, R.A., Saling, M.M., Bradley, D.C., & Berkovic, S.F. (1994). Interictal language fluency in temporal lobe epilepsy. *Cortex, 30*(3), 469–478.

Langfitt, J.T., & Rausch, R. (1996). Word-finding deficits persist after left anterotemporal lobectomy. *Archives of Neurology, 53*(1), 72–76.

McCarthy, G., Nobre, A.C., Bentin, S., & Spencer, D.D. (1995). Language-related field potentials in the anterior-medial temporal lobe: I. Intracranial distribution and neural generators. *Journal of Neuroscience, 15*(2), 1080–1089.

McClelland, J.L., McNaughton, B.L., & O'Reilly, R.C. (1995). Why there are complementary learning systems in the hippocampus and neocortex: Insights from the successes and failures of connectionist models of learning and memory. *Psychological Review, 102*(3), 419–457.

Meyer, P., Mecklinger, A., Grunwald, T., Fell, J., Elger, C.E., & Friederici, A.D. (2005). Language processing within the human medial temporal lobe. *Hippocampus, 15*(4), 451–459.

Moran, N.F., Lemieux, L., Kitchen, N.D., Fish, D.R., & Shorvon, S.D. (2001). Extrahippocampal temporal lobe atrophy in temporal lobe epilepsy and mesial temporal sclerosis. *Brain, 124*(1), 167–175.

Murray, E.A., & Richmond, B.J. (2001). Role of perirhinal cortex in object perception, memory, and associations. *Current Opinion in Neurobiology, 11*(2), 188–193.

Noppeney, U., Price, C.J., Duncan, J.S., & Koepp, M.J. (2005). Reading skills after left anterior temporal lobe resection: An fMRI study. *Brain, 128*(6), 1377–1385.

O'Reilly, R.C., & Norman, K.A. (2002). Hippocampal and neocortical contributions to memory: Advances in the complementary learning systems framework. *Trends in Cognitive Science, 6*, 505–510.

Opitz, B., & Friederici, A.D. (2003). Interactions of the hippocampal system and the prefrontal cortex in learning language-like rules. *Neuroimage, 19*(4), 1730–1737.

Pataraia, E., Simos, P.G., Castillo, E.M., Billingsley-Marshall, R.L., McGregor, A.L., Breier, J.I., Sarkari, S., & Papanicolaou, A.C. (2004). Reorganization of language-specific cortex in patients with lesions or mesial temporal epilepsy. *Neurology, 63*(10), 1825–1832.

Pihlajamaki, M., Tanila, H., Hanninen, T., Kononen, M., Laakso, M., Partanen, K., Soininen, H., & Aronen, H.J. (2000). Verbal fluency activates the left medial temporal lobe: A functional magnetic resonance imaging study. *Annals of Neurology, 47*(4), 470–476.

Schefft, B.K., Testa, S.M., Dulay, M.F., Privitera, M.D., & Yeh, H.S. (2003). Preoperative assessment of confrontation naming ability and interictal paraphasia production in unilateral temporal lobe epilepsy. *Epilepsy & Behavior, 4*(2), 161–168.

Springer, J., Binder, J., Hammeke, T., Swanson, S., Frost, J., Bellgowan, P., Brewer, C., Perry, H., Morris, G., & Mueller, W. (1999). Language dominance in neurologically normal and epilepsy subjects: A functional MRI study. *Brain, 122*, 2033–2045.

Squire, L.R., & Knowlton, B.J. (2000). The medial temporal lobe, the hippocampus, and the memory systems of the brain. In M.S. Gazzaniga (Ed.), *The new cognitive neurosciences* (pp. 765–780). Cambridge, MA: MIT Press.

Thivard, L., Hombrouck, J., du Montcel, S.T., Delmaire, C., Cohen, L., Samson, S., Dupont, S., Chiras, J., Baulac, M., & Lehericy, S. (2005). Productive and perceptive language reorganization in temporal lobe epilepsy. *Neuroimage, 24*(3), 841–851.

Tracy, J., Waldron, B., Vidanage, S., Sperling, M., Glosser, D., Sharan, A., & Siddqui (2005). *The organization of specific language skills in early and late onset temporal lobe epilepsy.* Paper presented at the American Epilepsy Society, Washington, DC.

Ullman, M.T. (2001). A neurocognitive perspective on language: The declarative/procedural model. *Nature Reviews Neuroscience, 2*(10), 717–726.

Waites, A.B., Briellmann, R.S., Saling, M.M., Abbott, D.F., & Jackson, G.D. (2006). Functional connectivity networks are disrupted in left temporal lobe epilepsy. *Annals of Neurology, 59*(2), 335–343.

Further Readings

Bartha, L., Marien, P., Brenneis, C., Trieb, T., Kremser, C., Ortler, M., Walser, G., Dobesberger, J., Embacher, N., Gotwald, T., Karner, E., Koylu, B., Bauer, G., Trinka, E., & Benke, T. (2005). Hippocampal

formation involvement in a language-activation task in patients with mesial temporal lobe epilepsy. *Epilepsia*, *46*(11), 1754–1763.

The fMRI study explores the effects of hippocampal dysfunction on the language comprehension network. Recommended for those who are interested in learning more about the role of the hippocampal formation in greater language networks.

Friederici, A.D. (2002). Towards a neural basis of auditory sentence processing. *Trends in Cognitive Sciences*, *6*(2), 78–84.

Argues that auditory sentence processing is accomplished through a temporofrontal network. The authors conclude that syntactic precedes semantic processing, and that they interact most in the later stages of language work.

Hoenig, K., & Scheef, L. (2005). Meditemporal contributions to semantic processing: fMRI evidence from ambiguity processing during semantic context verification. *Hippocampus*, *15*, 597–609.

The authors further delve into the role of the hippocampus in semantic processing. They review the role of the hippocampus in lexico-semantic ambiguity processing and its relation to a larger semantic network.

Kaan, E., & Swaab, T.Y. (2002). The brain circuitry of syntactic comprehension. *Trends in Cognitive Science*, *6*(8), 350–356.

Kaan and Swaab review the literature on syntactical processing in the brain and discuss the different types of tasks that have been used to tease out syntactic processes from other language processes, including complex versus simple sentences and jabberwocky and syntactic prose. They critically review each of these approaches for their benefits and drawbacks. Finally, they discuss the areas of the brain that are implicated in the syntactic network. Recommended for those who are interested in learning more about syntactic networks.

Meyer, P., Mecklinger, A., Grunwald, T., Fell, J., Elger, C.E., & Friederici, A.D. (2005). Language processing within the human medial temporal lobe. *Hippocampus*, *15*, 451–459.

The authors explore the connection of the rhinal cortex and the hippocampus to the networks involved in language processing using the event-related potential technique in patients with TLE. Recommended for those who are interested in the medial temporal lobe's function in normal language comprehension.

32

Subcortical Language Mechanisms

STEPHEN E. NADEAU

Geriatric Research, Education and Clinical Center, Brain Rehabilitation Research Center, Malcom Randall Veterans Administration Medical Center, and the Department of Neurology, University of Florida College of Medicine, Gainesville, FL, USA

ABSTRACT

In the 1970s the discovery of aphasia in patients with strokes apparently limited to the basal ganglia and the thalamus was greeted with considerable excitement. The discovery of subtle language impairment in patients with other disorders of the basal ganglia (Parkinson's and Huntington's diseases) provided additional converging evidence implicating the basal ganglia in language function. In the 1990s however, newer studies revealed that such cases of aphasia were, in fact mainly due to clandestine dysfunctions of the overlying cortex. In the meantime expanded knowledge in both the neuropathology of Parkinson's and Huntington's diseases and in the basic neurophysiology of the basal ganglia suggests that the only true subcortical aphasia is due to thalamic stroke. The neuroscience of the thalamus cannot yet provide us a cogent explanation for the linguistic manifestations of thalamic stroke, and clinical studies continue to contribute to our emerging understanding of thalamic function.

32.1. INTRODUCTION

Following the seminal nineteenth century discoveries of Paul Broca and Carl Wernicke, it became accepted that language was a function supported by the cerebral cortex and that aphasia was caused by cortical lesions. However, following the introduction into clinical practice of such advanced imaging techniques as computed tomography (CT) in the 1970s and magnetic resonance imaging (MRI) in the 1980s, a number of investigators reported aphasia associated with lesions, most often stroke, apparently limited to subcortical structures, including the basal ganglia, the

thalamus, and subcortical white matter pathways (Figure 32.1). This discovery was particularly startling because no one would have dreamt that such phylogenetically ancient structures as the basal ganglia and the thalamus could possibly have anything to do with such a recent evolutionary development as language (Box 32.1).

Since the first reports of "subcortical aphasia," neuroscience has advanced considerably, and with this has come a more nuanced understanding of the mechanisms underlying this disorder (Nadeau & Crosson, 1997). In particular, it has become evident that the contribution of basal ganglia lesions to aphasia is largely illusory, an artifact of vascular mechanisms underlying stroke, and that the contributions of thalamic lesions to aphasia is indirect, mediated by the roles the thalamus plays in cortical function. Nevertheless, the mechanisms underlying subcortical aphasia continue to be a source of controversy. There are two major reasons for this: (1) traditional conceptualizations of the extent of pathology accounting for language disorders in stroke and degenerative disorders such as Parkinson's disease, and (2) insufficient awareness of emerging neuroscience on the function of deep brain structures.

32.2. DISTRIBUTION OF PATHOLOGY IN SUBCORTICAL STROKE AND DEGENERATIVE DISORDERS ASSOCIATED WITH LANGUAGE IMPAIRMENT

In the disorders that have provided the major models for the study of subcortical aphasia (stroke, Parkinson's disease,

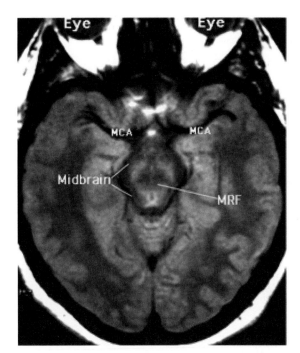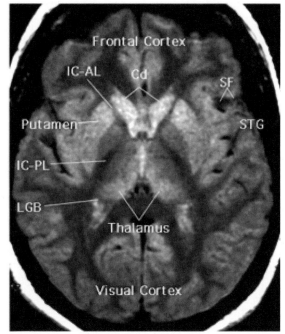

FIGURE 32.1   Major cerebral structures and landmarks relevant to the problem of subcortical aphasia (thalamus, caudate, putamen), anterior limb of internal capsule (ICAL), posterior limb of internal capsule (ICPL), midbrain reticular formation (MRF), visual cortex, lateral geniculate body (LGB), frontal cortex, superior temporal gyrus (STG), Sylvian fissure (SF), and middle cerebral artery (MCA).

Box 32.1   The basal ganglia and language: A functional imaging challenge

Notwithstanding the rising tide of evidence to the contrary, functional imaging (chiefly fMRI) seems to have given theories of basal ganglia involvement in language a new lease on life (see Crosson *et al.*, 2003; Gill Robles *et al.*, 2006). However, there are two potential non-linguistic sources of basal ganglia activity that must be ruled out by suitable control experiments before one can use functional imaging results to challenge the strong scientific framework established by other methodologies: (1) because there are direct projections from prefrontal cortex and premotor/ supplementary motor/motor areas to the caudate and putamen, respectively, one would expect cortical activity to frequently generate synaptic activity in these subcortical structures, demonstrated on functional images, that has to do with prefrontal and motor aspects of the task, not language *per se*; and (2) it is very difficult to be certain that the task employed

in a functional imaging study engages the brain in the way intended, and it is often difficult to rule out the possibility that subjects are doing additional, unintended, and perhaps inadvertent things in the scanner, as trivial as generating focal changes in motor tone, that could account for some or all of the imaging findings. Published functional imaging studies have not satisfactorily met these challenges.

Crosson, B., Benefield, H., Cato, M.A., Sadek, J.R., Moore, A.B., Wierenga, C.E., *et al.* (2003). Left and right basal ganglia and frontal activity during language generation: Contributions to lexical, semantic, and phonological processes. *Journal of the International Neuropsychology Society, 9,* 1061–1077.

Gill Robles, S., Gatignol, P., Capelle, L., Mitchell, M.-C., & Duffau, H. (2006). The role of dominant striatum in language: A study using intraoperative electrical stimulations. *Journal of Neurology, Neurosurgery and Psychiatry, 76,* 940–946.

Huntington's disease), the pathology is often considerably more extensive than is generally recognized.

32.2.1. The Case of Stroke

In the case of stroke, there has long been a widespread assumption that conventional imaging (CT and MRI) is highly sensitive in the detection of ischemic lesions. Thus,

in patients with aphasia associated with infarction apparently confined to subcortical structures, it has been assumed that the overlying cortex was completely normal. However, functional imaging studies that measure cortical blood flow (e.g., ^{99m}Tc-HMPAO single photon emission computed tomography, SPECT, $H_2^{15}O$ positron emission tomography, PET, and more recently perfusion-weighted MRI), or cerebral metabolic rate (e.g., ^{18}F-fluorodeoxyglucose PET) have

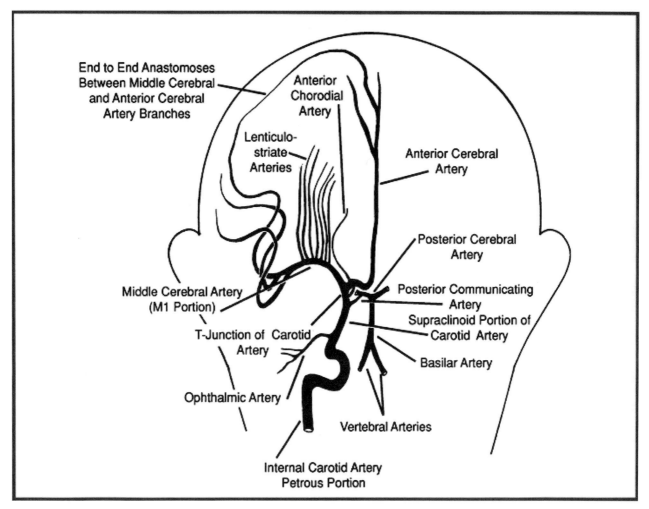

FIGURE 32.2 Anterior–posterior view of cerebral vascular anatomy relevant to the problem of non-thalamic subcortical aphasia. *Source*: From Nadeau, S.E., & Crosson, B. (1997). Subcortical aphasia. *Brain and Language, 58,* 355–402, 436–458. Imprint of Elsevier.

long shown that in patients with stroke apparently confined to deep structures, there are often severe abnormalities in cortical blood flow or metabolism, acutely and chronically, and that the distribution of these abnormalities correlates with the pattern of aphasia. Conversely, patients with acute stroke without evidence of cortical hypoperfusion do not exhibit impairment of language (Hillis *et al.*, 2002). This is not surprising. Small infarcts substantially restricted to the basal ganglia, as well as the larger subcortical lesions that implicate white matter pathways that have been linked to subcortical aphasia, are caused by extension of blood clot, or embolism of blood clot into either the tip of the internal carotid artery (the carotid T-junction) or the proximal portion of the middle cerebral artery (Figure 32.2). In either event, flow in the middle cerebral artery, the major source of blood to the hemisphere, is markedly reduced. Perfusion of the hemisphere becomes reliant upon blood flowing backward through end-to-end connections between distal middle cerebral artery branches and distal branches of other

major vessels supplying the cortex. In very few people is this residual blood flow adequate to the metabolic demands of neural tissue throughout the middle cerebral artery territory. The least well-collateralized regions (typically the subcortical structures) undergo infarction that is usually manifest on conventional imaging studies. However, there may be extensive damage, perhaps most affecting neural connections, throughout the middle cerebral artery territory, which may eventually become manifest only as diffuse hemispheric atrophy (Weiller *et al.*, 1993).

32.2.2. The Case of Parkinson's Disease

While it is often assumed that the clinical manifestations of Parkinson's disease are solely related to dopamine deficiency in the basal ganglia, it is clear that this disease is characterized by progressive widespread cortical degeneration (Braak *et al.*, 2003). A host of abnormalities in cognitive and linguistic function has been discovered in patients with Parkinson's disease. Dopamine deficiency could pro-duce linguistic impairment in Parkinson's disease via three mechanisms: (1) dysfunction of basal ganglia structures directly involved in language, (2) the production of noise in language cortices as a result of basal ganglia dysfunction, and (3) cortical dysfunction because of dopamine depletion in cortical projection targets, which include language cor-tex. In any event, such deficits should resolve with effective doses of dopaminergic medications. However, language defi-cits, like many other cognitive deficits in Parkinson's disease, are not dopamine sensitive (Malapani *et al.*, 1994; Nadeau & Crosson, 1997; Cools *et al.*, 2001, 2003; Lewis *et al.*, 2005).

A study of phonologic and semantic working memory and semantic and syntactic comprehension in patients with Parkinson's disease showed impaired sentence comprehen-sion relative to controls but normal working memory and no differences between performance on and performance off dopaminergic medications (Skeel *et al.*, 2001). These studies suggest that if there is a dopamine-related impact on either language or concept generation in Parkinson's dis-ease, it must be very subtle.

32.2.3. The Case of Huntington's Disease

Huntington's disease is widely assumed to be a disease predominantly affecting the basal ganglia, particularly the head of the caudate nucleus. However, many studies have demonstrated evidence of extensive cortical disease and the post-mortem brain of the patient with Huntington's disease is characterized by profound cortical atrophy. Consequently, language abnormalities in this disease cannot be attributed to disease of the head of the caudate nucleus.

32.3. EMERGING NEUROSCIENCE OF THE BASAL GANGLIA

The second major reason for the continuing controversy regarding the mechanisms underlying subcortical aphasia is that, in attempts to explain subcortical aphasia on the basis of damage to subcortical structures, there has been insuf-ficient awareness of data from other fields of neuroscience that bear upon the function of these structures.

This is most clearly the case for the putamen (Figure 32.1), which has been unequivocally established as a struc-ture involved in motor function. Two observations make the point. First, the major afferent connections of the putamen are with motor and premotor cortex and supplementary motor area. The major efferent projections, conducted via the globus pallidus and anterior portions of the thalamus, are to the supplementary motor area. Second, in its early stages, Parkinson's disease is a disorder caused almost exclusively by dopamine deficiency in the basal ganglia, which is most extreme in the putamen. Expectably, the initial manifesta-tions of Parkinson's disease are exclusively motoric and the

modest disorders of language that have been reported (pri-marily involving syntax) develop later in the disease, when cortical pathology has become significant.

Defining the precise role of the head of the caudate nucleus continues to pose a challenge. A number of inves-tigators have reported clinical findings in patients with infarcts of or hemorrhages into this region. These have included dysarthria, akinesia (at times referred to by the poorly defined term abulia), and possibly adynamic apha-sia (Kumral *et al.*, 1999), although most reported patients with infarcts of or hemorrhage into the head of the cau-date nucleus have not been aphasic (Nadeau & Crosson, 1997). However, lesions of the head of the caudate nucleus, whether infarcts or hemorrhages, invariably cause some damage to the anterior limb of the internal capsule. This is a major white matter conduit between the frontal cortex, the thalamus, and more caudal structures. Thus, in practice, one cannot easily dissociate deficits relatable to dysfunction of the head of the caudate nucleus from deficits relatable to dysfunction of the anterior limb of the internal capsule. It has been shown, for example, that small infarctions (lacunes) involving the ventral genu of the internal capsule, which carries fibers from the anterior limb to the thalamus, produce all the features of thalamic aphasia (Tatemichi *et al.*, 1992). The major afferent and efferent connections of the head of the caudate nucleus are with prefrontal cortex. Thus, if the function of the head of the caudate nucleus were impor-tant to language, one would expect infarction of this struc-ture to manifest as a disorder of endogenous engagement of concept representations (i.e., adynamic aphasia, Gold *et al.*, 1997) or a disorder in the modification and manipulation of concept representations (i.e., simplification of syntax and violation of rules of verb argument structure with preserved grammatic morphology (Nadeau, 1988; Nadeau & Rothi, 2004)). Studies of patients with Parkinson's disease on and off levodopa might still be able to make a useful contribu-tion to elucidating the impact of dysfunction of the head of the caudate nucleus on language and other cognitive proc-esses. However, because dopamine depletion in the head of the caudate nucleus is much less severe than in putamen, and becomes substantial only late in the disease course, such studies would have to focus on patients with advanced dis-ease and inevitably, substantial cortical pathology.

32.4. MECHANISMS OF NON-THALAMIC SUBCORTICAL APHASIA

Four mechanisms account for non-thalamic subcortical aphasia, that is hypoperfusion, ischemic damage, discon-nection, and pressure effect:

1. *Hypoperfusion in the setting of acute stroke*: A blood clot wedged in the proximal middle cerebral artery reduces blood flow to the cerebral cortex. Cortical blood flow is most severely reduced in cortical regions that receive the least

blood via other routes (e.g., collateral vessels from the anterior and posterior cerebral arteries). Ischemia in these poorly collateralized regions produces focal disorders of language function. This has been most elegantly demonstrated in perfusion-weighted MRI studies (Hillis *et al.*, 2002, 2004).

2. *Ischemic damage*: If the hypoperfusion cause by usually transient proximal middle cerebral artery occlusion is sufficiently prolonged, permanent damage to poorly collateralized cortical regions will occur. If sufficiently severe, this damage will appear as focal cortical abnormalities on CT or MRI studies. However, in cases of "subcortical aphasia," the cortical damage is not severe enough to appear on CT or MRI, even though it is sufficient to cause aphasia. This has been well-demonstrated on functional imaging studies performed long after stroke, in which reduced cerebral blood flow or metabolism reflects loss of synaptic activity because of cortical damage (Nadeau & Crosson, 1997). It is also evident in the hemispheric cortical atrophy that has been demonstrated long after infarcts seemingly limited to the deep structures (Weiller *et al.*, 1993).

3. *Disconnection*: Proximal middle cerebral occlusion or hemorrhage into the putamen can damage subcortical white matter pathways. This may disconnect the thalamus from language cortices, thereby adding features of thalamic aphasia to aphasia stemming from cortical damage.

4. *Pressure effects*: A large hemorrhage into the putamen can produce high pressure on overlying cerebral cortex. This pressure may be sufficient to impede blood flow to this cortex, thereby causing ischemic damage and resultant cortical aphasia.

32.5. SUMMARY: NON-THALAMIC SUBCORTICAL APHASIA

With a better understanding of the hemodynamics of acute stroke, the limited sensitivity of CT and MRI for ischemic damage to the cortex, and the functions of subcortical structures, most notably the basal ganglia, it has become apparent that non-thalamic subcortical aphasia is substantially an artifact. In actuality, it reflects inapparent cortical hypoperfusion, cortical damage, or thalamocortical disconnection. The implications of thalamocortical disconnection and the mechanisms underlying thalamic aphasia will be the topic of the next section.

32.6. EMERGING NEUROSCIENCE OF THE THALAMUS

The thalamus, through its complex anatomic and functional relationship with the cerebral cortex, is directly implicated in language function (Nadeau & Crosson, 1997). The precise neural mechanisms underlying thalamic aphasia pose the greatest and most important research challenge in the field of subcortical aphasia. For a better understanding of the role of the thalamus in language function it is necessary to have some familiarity with the anatomy and functions of the thalamus.

32.6.1. Functional Neuroanatomy of the Thalamus

The thalamus is located at the center of the cerebrum at its junction with the midbrain (Figures 32.1 and 32.3). Although its function is by no means simple, it is far simpler and far better understood than that of any other cerebral structure. In essence, it functions as a *relay device*. Specific groups of neurons within the thalamus (nuclei) relay sensory information from more peripheral neural way stations to the cerebral cortex. Other thalamic nuclei relay information from one part of the cortex to another.

Two thalamic nuclei are of particular interest with respect to language and language-related processes: the dorsomedial nucleus and the pulvinar-lateral posterior complex. The dorsomedial nucleus relays information from the prefrontal cortex, as well as several subcortical structures, back to the prefrontal cortex. The pulvinar-lateral posterior complex relays projections from the frontal, temporal, and parietal cortices (including those portions of the dominant hemisphere directly implicated in language function), as well as some subcortical structures, back to these same cortices. As we shall see below, there appears to be a logical relationship between dysfunction in these two nuclear groups and particular features of thalamic aphasias.

It is more accurate to characterize the thalamus as a regulated or *gated relay device*. This gating feature helps to address two questions. First, some might have asked what could conceivably be the purpose of long connections descending from, for example, temporo-parietal cortex to the pulvinar-lateral posterior complex, only to be sent right back to the very same cortex. One key byproduct of passing cortical information through the thalamus may be that this information (and by implication, the originating cortex) is subjected to the regulatory mechanism provided by the thalamic gate. Second, ischemic strokes involving either the dorsomedial nucleus or the pulvinar-lateral posterior complex (the thalamic regions most directly linked to language cortex) are extremely rare. How can we then explain thalamic aphasia in terms of dysfunction of these two regions? One possibility is to look for another cause of dysfunction than direct damage. It appears that thalamic strokes sparing these nuclear groups produce aphasia by damaging key components of the thalamic gating mechanism.

32.6.2. Processing by the Thalamic Gating Mechanism

Nearly the entire thalamus is enveloped by a paper-thin layer of cells comprising the nucleus reticularis of the thalamus (Figure 32.4). The cells of this nucleus, unlike

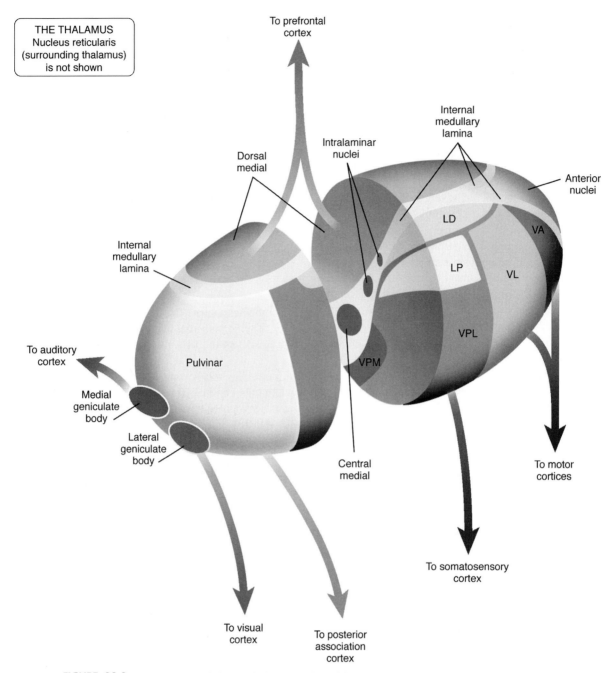

THE THALAMUS
Nucleus reticularis
(surrounding thalamus)
is not shown

To prefrontal cortex

Internal medullary lamina

Intralaminar nuclei

Anterior nuclei

Dorsal medial

LD

VA

Internal medullary lamina

LP

VL

To auditory cortex

VPL

Pulvinar

VPM

Medial geniculate body

Lateral geniculate body

Central medial

To motor cortices

To visual cortex

To posterior association cortex

To somatosensory cortex

FIGURE 32.3 Cartoon of the thalamus, depicting the loci of the major nuclei. Center median (Cm) is labeled "central medial" in this figure and parafascicularis (Pf), located anterior, ventral and medial to Cm, is not depicted. *Source*: From Nadeau, S.E., *et al.* (2004). *Medical neuroscience*. Philadelphia: Saunders. Imprint of Elsevier. See Plate 21.

those in the remainder of the thalamus, send their projections back into the thalamus, where they synapse both on thalamic relay neurons and on inhibitory thalamic interneurons. They employ an inhibitory neurotransmitter (gamma-aminobutyric acid, GABA). They are thus admirably suited to regulating thalamic transmission to the cortex, either by directly inhibiting thalamic relay neurons, or by indirectly potentiating their activity by inhibiting inhibitory interneurons within the thalamus.

The neurons of nucleus reticularis are regulated by a host of brain systems. Two systems appear to be particularly important. The *first* is the midbrain reticular formation, a complex network of neurons within the core of the midbrain immediately below the thalamus. The midbrain reticular formation defines level of wakefulness or arousal. With high levels of arousal, the midbrain reticular formation inhibits nucleus reticularis at the same time that it excites thalamic relay neurons, thereby allowing all neural

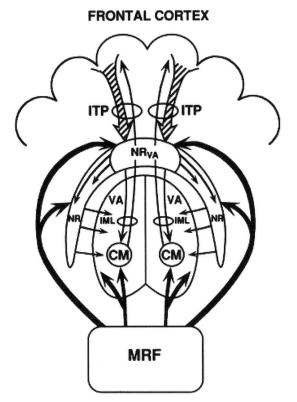

FRONTAL CORTEX

FIGURE 32.4 Schematic diagram of essential relationships involved in the regulation of the thalamic gating mechanism. See text for details. ITP, inferior thalamic peduncle; NR, nucleus reticularis; NR_{VA}, the portion of NR immediately anterior to ventral anterior nucleus (VA) that receives input both from the midbrain reticular formation and the prefrontal cortex via the ITP; Cm, center median nucleus; IML, internal medullary lamina, the white matter fascicle separating the medial from the lateral thalamus and containing the connections between prefrontal cortex and Cm. *Source*: From Nadeau, S.E., & Crosson, B. (1997). Subcortical aphasia. *Brain and Language*, *58*, 355–402, 436–458. Imprint of Elsevier.

transmission to pass readily through the thalamus. With low levels of arousal, as in deep sleep or coma, because of low input from the midbrain reticular formation, cortical relay of thalamic input is effectively blocked. In this way, the impact of the midbrain reticular formation on thalamic transmission is *global and non-selective*.

The *second* major regulatory system is provided by projections from the entire cerebral cortex to nucleus reticularis. Two subsystems can be defined within this set of projections: a *direct* system, involving modality-specific cortical projections to individual thalamic nuclei that largely reciprocate thalamocortical projections and synapse on nucleus reticularis neurons as they pass through this nucleus; and an *indirect* system, involving projections from the prefrontal cortex to nucleus reticularis via the inferior thalamic peduncle (ITP) (Figure 32.4), providing a potential basis for multimodal regulation of thalamic transmission. Both systems provide a mechanism by which thalamic transmission from specific regions of specific nuclei can be *selectively* gated to the cortex.

The mechanisms underlying selective nucleus reticularis regulation of thalamocortical transmission are currently the subject of intense study and much debate and the last 10 years have seen a surge of interest in the role of nucleus reticularis in attentional processes (Crabtree & Isaac, 2002; Pinault, 2004; Sherman, 2005; Bezdudnaya *et al.*, 2006; McAlonan *et al.*, 2006). This work has focused almost exclusively on the direct corticothalamic system.

The indirect corticothalamic regulatory system, involving a pathway from prefrontal cortex to nucleus reticularis via the ITP, provides a mechanism by which thalamic transmission to specific regions of the cortex might be volitionally or intentionally regulated. Because language is largely produced volitionally, it would seem logical that to the extent that the thalamus is involved in language, dysfunction of this system would lead to aphasia. Empirical studies bear this out (see below).

32.6.3. Impairment of Function of the Indirect Thalamic Gate

The projections from both the midbrain reticular formation and the prefrontal cortex to nucleus reticularis terminate in a dense neural complex near the anterior pole of the thalamus, immediately adjacent to the ventral anterior nucleus of the thalamus (NR_{VA}). This neural complex appears to have a role in regulating the entirety of nucleus reticularis. The mechanism by which the prefrontal cortex regulates the thalamic gate also appears to require the participation of center median-parafascicularis (CmPf), a nuclear complex buried deep within posterior portions of the thalamus (Figure 32.3). Fibers from the prefrontal cortex to and from CmPf pass within the thalamus in a neural fascicle, the internal medullary lamina (IML). Immediately in front of the thalamus, these fibers pass through the ITP, which is continuous with the anterior limb of the internal capsule and the deep frontal lobe white matter (Figures 32.1 and 32.3). Although the means we use to localize strokes within the thalamus are prone to error, it appears that all the thalamic strokes that have been reported to result in aphasia have involved either the ventral anterior nucleus at the nexus of the connections between the prefrontal cortex (the ITP), the midbrain reticular formation, and the regulatory neural complex of nucleus reticularis anterior, the ventral anterior nucleus (NR_{VA}); the IML carrying connections from the prefrontal cortex to CmPf; or CmPf itself. Thus, it appears that all ischemic strokes causing thalamic aphasia disrupt the regulatory system for the thalamic gate for the entire hemi-thalamus by damaging the prefrontal-ITP-NR_{VA}-IML-CmPf axis. This disruption would implicate transmission from all thalamic nuclei, including the pulvinar-lateral posterior complex, which is heavily connected to language cortices. Thalamic infarcts that are not associated with aphasia appear to spare these structures and pathways.

Hemorrhages, which do not respect vascular territories, may directly damage the pulvinar-lateral posterior complex or the dorsomedial nucleus.

32.6.4. Purpose of the Thalamic Gate

The brain contains approximately 100 billion neurons organized into hundreds of systems, each of which is nearly infinitely malleable. Given this situation, it appears logical that there should be some systems whose function is simultaneously to maintain order and to optimize the resources that are brought to bear on a particular problem, that is, systems that allocate processing demands among the multitude of different neural networks, or within parts of particular networks. A number of different lines of research have begun to delineate executive systems that serve this purpose.

Since William James, this process of allocating brain resources has often been referred to as attention. Attention appears to be an appropriate term when the particular neural networks or components of networks in cerebral association cortices that are selected act to focus sensory systems on some particular locus in the environment. However, there is now abundant reason to believe that precisely analogous processes occur in polymodal or supramodal cerebral cortices, such as those supporting language function, that correspond to thinking and the formulation of non-attentional behaviors. These processes have been the focus of considerable study over the past 20 years (Goldman-Rakic, 1990; Fuster, 1991). The term "working memory" is often used to refer to the information maintained by activation of a selected neural network, although we prefer the more neutral term *selective engagement*. In this conceptualization, attentional processes represent the subtype of selective engagement involved in allocating components of neural networks located in sensory association cortices in the service of attentional behavior.

The thalamus appears to play a major role in selective engagement processes. Non-human primate research suggests, however, that, at this stage in the evolution of the brain, very little of this regulation involves thalamic projections to primary sensory cortices. In humans, thalamic gating mechanisms appear to be most important in regulating selective engagement of association cortices (unimodal and polymodal), as well as supramodal association cortices supporting higher neural functions such as language.

32.7. MECHANISMS OF THALAMIC APHASIA

The neuroscience of thalamocortical interaction has not advanced to the point of providing a precise picture of how disorders of thalamic gating impact higher cortical function and we are primarily dependent upon human data in the inferences we draw on this subject. Thalamic aphasia is classically characterized by anomia in spontaneous language (at times severe), some impairment in naming to confrontation, apparently normal grammar, normal articulation, and usually flawless repetition. Naming deficits disproportionately affect low frequency words (Raymer *et al.*, 1997). In the worst cases, modest impairment in comprehension may be noted. Some patients make semantic paraphasic errors and rare patients, primarily those with thalamic hemorrhages, transiently produce neologisms. Can we relate this pattern of deficits to any fundamental functional attributes of the cerebral cortices involved? We have proposed that linguistic deficits wrought of thalamic dysfunction reflect selective functional impairment of cortices supporting declarative memories with sparing of cortices supporting procedural memories (Nadeau & Crosson, 1997) (Box 32.2).

Box 32.2   Declarative and non-declarative memory

Declarative or explicit memory consists of knowledge of facts and events and is ordinarily available to conscious recollection. It is represented in association cortices throughout the brain. New declarative memories are encoded by a system comprised of the hippocampus, cortices overlying the hippocampus, several structures in the limbic system (which presumably help to define which facts are worth remembering), the dorsomedial nucleus, and the anterior nuclear group of the thalamus (Squire, 1987, 1992).

Non-declarative memory consists of a heterogeneous collection of abilities, including skills and habits, implicit memory, and some forms of classical conditioning. Non-declarative memory is reflected only in behavioral change, as when one's tennis game improves with practice, or as when an experimental subject demonstrates an autonomic response or a correct decision when re-exposed to a stimulus previously presented too briefly to provide even a sense of familiarity, let alone a basis for conscious recognition. Skills, habits, and some forms of classical conditioning are represented in motor and premotor cortex, cerebellum, and possibly the basal ganglia. Unlike declarative memory, non-declarative memory does not appear to require a special auxiliary processing mechanism like the hippocampal system to be instantiated. Patients with impaired declarative memory generally have preserved non-declarative memory.

Squire, L.R. (1987). *Memory and the brain*. New York: Oxford University Press.
Squire, L.R. (1992). Declarative and non-declarative memory: Multiple brain systems supporting learning and memory. *Journal of Cognitive Neuroscience, 4,* 232–243.

The principal linguistic deficit exhibited by patients with thalamic aphasia is in lexical–semantic access (Nadeau & Crosson, 1997), a process that likely depends on declarative memories. In contrast, linguistic processes involving phonological processing and grammar are generally spared. Although we have declarative knowledge of the spelling of words, we rely on entirely automatic processes not available to conscious recollection in the actual production of spoken words. Although we have declarative knowledge of the concepts about which we plan to speak, the process of translating concepts into clause sequences and appropriate phrase structures is also largely an automatic one, largely unavailable to conscious recollection (except to the extent that we deliberately modify grammar to meet situational demands). Thus, the preponderance of data on phonological and grammatical processes in thalamic aphasia appears to indicate substantial sparing of procedural memory systems.

Unfortunately, this tidy picture cannot be completely reconciled with available data on phonologic and grammatic function in patients with thalamic lesions.

32.7.1. Phonological Dysfunction

Patients with thalamic hemorrhages have often been reported to produce phonemic paraphasic errors and neologisms. This phenomenon can also occasionally be seen with ischemic lesions (Raymer et al., 1997; Ebert et al., 1999). As it appears to represent dysfunction involving phonological sequence knowledge, which would seem to be procedural in nature (Nadeau, 2001; Nadeau & Rothi, 2004), this finding is not easily reconciled with a concept that thalamic aphasia primarily reflects impairment of declarative processes. The disproportionate occurrence of neologisms with acute hemorrhage raises a question of some kind of pathologic excitatory phenomenon. However, whether hemorrhage or ischemic lesion, we are still left without an explanation for the occasional occurrence of neologisms. We have offered two possibilities (Nadeau & Crosson, 1997): (1) the process of becoming literate serves to additionally define phonemes, inflectional grammatic morphemes, derivational morphemes, and syllables as lexical elements in their own right, consciously retrievable like any item in declarative memory, and therefore susceptible to errors of selection like whole words (Bertelson & De Gelder, 1989; Semenza et al., 1990). This "sublexicalization" may create susceptibility to the production of sublexical paraphasias manifesting as phonologic errors and neologisms in the presence of thalamic injury; and (2) thalamic lesions may disrupt mutually inhibitory relationships within cortices supporting semantic representations. As a result competing semantic entrees can vie simultaneously for translation into a phonologic sequence. The ultimate outcome is a word blend that includes phonologic components corresponding to both semantic entrees. This mixing of phonological components may yield a neologism.

32.7.2. Grammatic Dysfunction

An important recent paper has provided compelling evidence of syntactic simplification and possibly agrammatism in a patient with bilateral paramedian thalamic infarcts (De Witte et al., 2006). How might this be explained? A concept representation, corresponding to a noun phrase, can be represented in neural network terms as a specific pattern of activation of the neurons in association cortices representing the various features of that concept (Nadeau, 2001; Nadeau & Rothi, 2004). Production of a sentence invariably involves modification of that distributed concept representation and usually involves generation and reciprocal modification of two or more distributed concept representations and the binding of these representations into a super-distributed representation. For example, the sentence "the old man shot the burglar" involves simultaneously generating two distributed concept representations, one corresponding to an old man, another to a burglar. It also requires reciprocal modification of these representations to yield a new representation corresponding to "shooter old man" that is now linked to another new representation corresponding to "shot burglar." The use of verbs mandating three arguments would correspond to the simultaneous maintenance and reciprocal modification of three distributed concept representations. Addition of embedded clauses will demand still further active distributed concept representations. The production of normal syntax requires both the ability to generate and maintain multiple distributed concept representations (i.e., selective engagement/working memory) and skill in the modification of these various representations. Broca's aphasia occurs with lesions that damage the frontal apparatus that apparently is required to modify distributed concept representations, and certainly to reciprocally modify multiple distributed concept representations (hence, among other things, simplification of syntax). It may be that selective engagement mechanisms supporting needed working memory are integral to this apparatus, or that frontal networks involved in this selective engagement are to some degree anatomically separate. The report of De Witte et al. (2006) suggests that disorders of selective engagement caused by thalamic lesions can produce a pattern of language abnormality with some similarities to Broca's aphasia. Even some abnormalities of grammatic morphology might be explained in terms of working memory deficits, although we suspect that in Broca's aphasia, the loss of phonologic sequence knowledge involving words, their affixes, and short phrases (caused by the perisylvian component of the lesions) makes a major contribution to agrammatism (Nadeau & Rothi, 2004). For example, the use of articles is particularly dependent upon sustained working memory of what has just been said to determine whether the reference should be definite or indefinite. Absent working memory of what has been said, the absence of a concept of definiteness or indefiniteness may lead to failure to produce the article, as in the case described by De Witte and colleagues.

The notion that thalamic lesions can lead to disorders of endogenous generation of distributed concept representations invites comparison with adynamic aphasia (Gold *et al.*, 1997). In adynamic aphasia, subjects appear to be incapable of endogenous generation of any distributed concept representations, hence their extreme non-fluency. However, when distributed concept representations are generated from without, as in a picture description task, subjects with adynamic aphasia perform relatively normally. The report of De Witte *et al.* (2006) suggests that at least with certain uncommon thalamic lesions, distributed concept representations can be endogenously generated to the degree that the patient can translate them into words, but these representations cannot be sustained in the manner necessary for them to be related to each other and modified in the well-practiced ways that instantiate the rules of syntax (Nadeau & Rothi, 2004). Thus, we might employ the term "paradynamic" to describe their patient to indicate that the fundamental problem does involve endogenous selective engagement, but that rather than a complete failure of endogenous engagement, as in adynamic aphasia, it constitutes disjointed, excessively brief, and substantially dysfunctional engagement.

There are intimations of the problem observed by De Witte *et al.* in older literature. Nevertheless, De Witte *et al.* appear to be the first to provide such detailed evidence of syntax simplification and at least some evidence of agrammatism in thalamic aphasia. Does this merely reflect the thoroughness of their investigation, or might it reflect something unusual about their patient's lesion? The lesion in their case might have been somewhat unusual. In particular, the lesion was bilateral (fairly common with certain types of thalamic infarcts), it likely involved the prefrontal-ITP-NR$_{VA}$-IML-CmPf axis bilaterally, thus producing bilateral disorders in thalamic engagement involving the pulvinar-lateral posterior complex and dorsomedial nucleus, and it extended relatively anterior and deep, potentially destroying prefrontal to dorsomedial nucleus pathways that run beneath the thalamus before curving upward to extend into the dorsomedial nucleus (Steriade *et al.*, 1984; Nadeau & Crosson, 1997). In contrast, Raymer *et al.* (1997) noted normal syntax in a patient with a very large infarct in the same vascular territory that was, however, unilateral (left).

In summary, if these explanations for the findings of De Witte and colleagues find support in future investigations of syntactic disorders and agrammatism in subjects with thalamic aphasia, it may still be possible to account for thalamic aphasia as a disorder of declarative knowledge processing.

32.8. SUMMARY: THALAMIC APHASIA

Lesions of the thalamus producing aphasia appear to predominantly involve either the junction of the frontal projections to the thalamus and the anterior portion of nucleus reticularis, immediately anterior to the ventral anterior nucleus of the thalamus, NR$_{VA}$; or the pathway within the thalamus, embedded in the IML, that connects the frontal lobes to CmPf. Either lesion apparently disrupts function of the entire hemi-thalamus, including the pulvinar-lateral posterior complex, which by virtue of its cortical connectivity is logically implicated in lexical–semantic function. Lesions of pulvinar-lateral posterior complex (e.g., hemorrhages – infarcts are rare) could have the same effect on lexical–semantic function. In addition, midline infarcts extending anteroventrally could directly disconnect the dorsomedial nucleus and the prefrontal cortex. In such cases, dorsomedial nucleus function might be doubly compromised – by the direct disconnection and by thalamic hemidysfunction related to the prefrontal-ITP-NR$_{VA}$-IML-CmPf axis lesion – with the consequence of even worse impairment of prefrontal function. These various lesions might be expected to produce various disorders of selective engagement of cortex. The precise linguistic profile could depend on the particular combination of lesion loci and the specific topography of the damage within the various pathways involved. Acutely, the particular disorders of selective engagement produced by these lesions may be compounded by diaschisis: that is, projection targets of the thalamus become acutely dysfunctional because sudden loss of the excitatory input from thalamic projection neurons (which are glutamatergic) renders the cortex transiently hypoexcitable to all input, including input from other regions of the cortex. Diaschisis following stroke appears to resolve over days to weeks, and consistent with this, patients with thalamic infarcts and aphasia typically demonstrate substantial early improvement. Unfortunately, it is difficult experimentally to distinguish manifestations of diaschisis from manifestations of recovery of compromised but not fatally injured neurons and their processes, and from manifestations of neuroplastic processes.

The majority of reports on thalamic aphasia suggest that disorders of selective engagement caused by thalamic lesions predominantly involve lexical–semantic function. However, the occasional occurrence of phonemic paraphasic errors and a recent detailed report of impaired grammar demand a broader and more nuanced interpretation of thalamic aphasia and the selective engagement mechanisms disrupted by the lesions responsible.

32.9. CHALLENGES AND FUTURE DIRECTIONS

Our understanding of mechanisms of thalamic aphasia will be advanced by further studies of syntax, grammatical, and derivational morphology in subjects with thalamic aphasia; better correlation of the pattern of language abnormalities observed, the particular loci of thalamic lesions, and the topography of the damage; better mapping of the distribution of thalamic lesions on existing atlases (thick slice, uniplanar imaging cannot provide the requisite data); and better delineation of associated neuropsychological dysfunction. For example, the presence or absence of executive dysfunction may shed light on both the ramifications of particular thalamic lesions and the potential for prefrontal dysfunction as a partial explanation for the pattern of language impairment observed. The science of thalamic aphasia will also benefit from basic neuroscientific research on the mechanisms by which the frontal lobes regulate thalamic processing.

References

Bertelson, P., & De Gelder, B. (1989). Learning about reading from illiterates. In A.M. Galaburda (Ed.), *From reading to neurons* (pp. 1–25). Cambridge: MIT Press.

Bezdudnaya, T., Cano, M., Bereshpolova, Y., Stoelzel, C.R., Alonso, J.-M, & Swadlow, H.A. (2006). Thalamic burst mode and inattention in the awake LGNd. *Neuron, 49,* 421–432.

Braak, H., Del Tredici, K., Rüb, U., de Vos, R.A.I., Jansen Steur, E.N.H., & Braak, E. (2003). Staging of brain pathology related to sporadic Parkinson's disease. *Neurobiology of Aging, 24,* 197–211.

Cools, R., Barker, R.A., Sahakian, B.J., & Robbins, T.W. (2001). Enhanced or impaired cognitive function in Parkinson's disease as a function of dopaminergic medication and task demands. *Cerebral Cortex, 11,* 1136–1143.

Cools, R., Barker, R.A., Sahakian, B.J., & Robbins, T.W. (2003). L-dopa medication remediates cognitive inflexibility, but increases impulsivity in patients with Parkinson's disease. *Neuropsychologia, 41,* 1431–1441.

Crabtree, J.W. & Isaac, J.T.R. (2002). New intrathalamic pathways allowing modality-related and cross-modality switching in the dorsal thalamus. *Journal of Neuroscience, 22,* 8754–8761.

De Witte, L., Wilssens, I., Engelborghs, S., De Deyn, P.P., & Mariën, P. (2006). Impairment of syntax and lexical semantics in a patient with bilateral paramedian thalamic infarction. *Brain and Language, 96,* 69–77.

Ebert, A.D., Vinz, B., Görtler, M., Wallesch, C.-W., & Herrmann, M. (1999). Is there a syndrome of tuberothalamic artery infarction? A case report and critical review. *Journal of Clinical and Experimental Neuropsychology, 21,* 397–411.

Fuster, J.M. (1991). The prefrontal cortex and its relation to behavior. *Progress in Brain Research, 87,* 201–211.

Gold, M., Nadeau, S.E., Jacobs, D.H., Adair, J.C., Gonzalez-Rothi, L.J., & Heilman, K.M. (1997). Adynamic aphasia: A transcortical motor aphasia with defective semantic strategy formation. *Brain and Language, 57,* 374–393.

Goldman-Rakic, P.S. (1990). Cellular and circuit basis of working memory in prefrontal cortex of nonhuman primates. *Progress in Brain Research, 85,* 325–336.

Hillis, A.E., Wityk, R.J., Barker, P.B., Beauchamp, N.J., Gailloud, P., Murphy, K. *et al,* (2002). Subcortical aphasia and neglect in acute stroke: The role of cortical hypoperfusion. *Brain, 125,* 1094–1104.

Hillis, A.E., Barker, P.B., Wityk, R.J., Aldrich, E.M., Restrepo, L., Breese, E.L. *et al,* (2004). Variability in subcortical aphasia is due to variable sites of cortical hypoperfusion. *Brain and Language, 89,* 524–530.

Kumral, E., Evyapan, D., & Balkir, K. (1999). Acute caudate vascular lesions. *Stroke, 30,* 100–108.

Lewis, S.J.G., Slabosz, A., Robbins, T.W., Barker, R.A., & Owen, A.M. (2005). Dopaminergic basis for deficits in working memory but not attentional set-shifting in Parkinson's disease. *Neuropsychologia, 43,* 823–832.

Malapani, C., Pillon, B., Dubois, B., & Agid, Y. (1994). Impaired simultaneous cognitive task performance in Parkinson's disease: A dopamine-related dysfunction. *Neurology, 44,* 319–326.

McAlonan, K., Cavanaugh, J., & Wurtz, R.H. (2006). Attentional modulation of thalamic reticular neurons. *Journal of Neuroscience, 26,* 4444–4450.

Nadeau, S.E. (1988). Impaired grammar with normal fluency and phonology. Implications for Broca's aphasia. *Brain, 111,* 1111–1137.

Nadeau, S.E. (2001). Phonology: A review and proposals from a connectionist perspective. *Brain and Language, 79,* 511–579.

Nadeau, S.E., & Crosson, B. (1997). Subcortical aphasia. *Brain and Language, 58,* 355–402. 436–458.

Nadeau, S.E., & Rothi, L.J.G. (2004). Rehabilitation of language disorders. In J. Ponsford (Ed.), *Cognitive and behavioral rehabilitation: From neurobiology to clinical practice* (pp. 129–174). New York: Guilford.

Nadeau, S.E., Ferguson, T.S., Valenstein, E., Vierck, C.J., Petruska, J.C., Streit, W.J. *et al.* (2004). *Medical Neuroscience.* Philadelphia, PA: Saunders/Elsevier.

Pinault, D. (2004). The thalamic reticular nucleus: Structure, function and concept. *Brain Research Reviews, 46,* 1–31.

Raymer, A.M., Moberg, P., Crosson, B., Nadeau, S., & Gonzalez Rothi, L.J. (1997). Lexical–semantic deficits in two patients with dominant thalamic infarction. *Neuropsychologia, 35,* 211–219.

Semenza, C., Butterworth, B., Panzeri, M., & Ferreri, T. (1990). Word formation: New evidence from aphasia. *Neuropsychologia, 28,* 499–502.

Sherman, S.M. (2005). Thalamic relays and cortical functioning. *Progress in Brain Research, 149,* 107–126.

Skeel, R.L., Crosson, B., Nadeau, S.E., Algina, J., Bauer, R.M., & Fennell, E.B. (2001). Basal ganglia dysfunction, working memory, and sentence comprehension in patients with Parkinson's disease. *Neuropsychologia, 39,* 962–971.

Steriade, M., Parent, A., & Hada, J. (1984). Thalamic projections of nucleus reticularis thalami of cat: A study using retrograde transport of horseradish peroxidase and fluorescent tracers. *The Journal of Comparative Neurology, 229,* 531–547.

Tatemichi, T.K., Desmond, D.W., Prohovnik, I., Cross, D.T., Gropen, T. I., Mohr, J.P. *et al.* (1992). Confusion and memory loss from capsular genu infarction: A thalamocortical disconnection syndrome? *Neurology, 42,* 1966–1979.

Weiller, C., Willmes, K., Reichle, W., Thron, A., Insensee, C., Buell, U. *et al.* (1993). The case of aphasia or neglect after striatocapsular infarction. *Brain, 116,* 1509–1525.

Further Readings

Kandel, E.R. (2006). *In search of memory.* **New York: WW Norton and Company.**
This is the scientific autobiography of Eric Kandel, a Nobel Laureate. The eminently readable and exciting tour de force of a brilliant scientist who has led an extraordinarily rich life, should be required reading for every young scientist, as it at once inspires with the joy of pursuing science and provides invaluable guidance on how to pursue top notch science without getting lost along the way. It should also be required reading for university presidents who are puzzling over how to propel their institutions into the top ranks. Kandel's discoveries have provided the foundation for our modern understanding of the neural basis of memory.

Nadeau, S.E., Ferguson, T.S., Valenstein, E., Vierck, C.J., Petruska, J. C., Streit, W.J. *et al.* (2004). *Medical Neuroscience.* **Philadelphia, PA: Saunders/Elsevier.**
This textbook of neuroscience is written primarily for a medical student audience. It focuses heavily on function, invoking anatomy only as needed to explain function, and it makes extensive use of case vignettes to enable understanding of the scientific principles involved. The entire book, but

particularly the final chapter on higher neural function, can provide a useful introduction to basic neuroscience for those in fields of cognitive science that have not heretofore required grounding in neuroscience.

Sherman, S.M. (2005). Thalamic relays and cortical functioning. *Progress in Brain Research, 149,* **107–126.**
Sherman provides a relatively accessible review of recent neuroscientific progress in understanding thalamocortical interactions. This paper

focuses primarily on the direct corticothalamic regulatory mechanism (one we have suggested may not be as relevant to language function). However, it nicely illustrates the range of neuroscientific investigations that are underway and the ways in which they can elucidate the underlying mechanisms.

33

Language and Communication Disorders in Multilinguals

MICHEL PARADIS[1,2]

[1]*Department of Linguistics, McGill University, Montreal, Quebec, Canada*
[2]*Cognitive Neuroscience Center, Université du Québec à Montréal, Montreal, Quebec, Canada*

ABSTRACT

A number of multilingual language disorders can be understood by taking into account the differential roles of declarative and procedural memory in the acquisition and use of each of a patient's languages. Normal adults find it extremely difficult to acquire a second language. It should therefore not be surprising that patients with aphasia have difficulty re-acquiring the impaired aspects of their language, whether first or second. Fortunately, language therapy can take advantage of the natural propensity of late-onset multilingual individuals with aphasia to compensate for any lack of linguistic competence in their non-native languages by relying on both metalinguistic knowledge and pragmatics.

Until very recently, the vast majority of studies on bilingual aphasia were classic case studies. Breaking with tradition, most current studies explore theoretical and clinical questions experimentally, with the help of consenting patients as participants.

33.1. INTRODUCTION

A multilingual individual is a person capable of communicating verbally in more than one language. Multilingual individuals differ among themselves according to a number of parameters that may each play a role in shaping any given aphasic patient's recovery pattern (Box 33.1). These variables are of several different kinds and their various influences may interact in determining the patient's communicative abilities in each language before the damage took place, and, as we shall see, to some extent their relative manifestations in aphasia. Although these matters are also relevant to developmental disorders, we will focus only on acquired disorders.

The parameters that may differentially affect language impairment in multilingual aphasia, amnesia, Alzheimer's,

Huntington, and Parkinson's disease include the relative premorbid degree of proficiency and accuracy, the age and manner of acquisition, the contexts of use, the structural distance between the languages, and the kind of motivation and affect associated with each. Proficiency generally refers to fluency and accuracy:

- Fluency, or normal speech rate, may reflect either automatization, when performance is not under deliberate control, but relies on implicit competence as in one's native language(s), or speeded-up (but still conscious) controlled use, that is, controlling one's performance by the conscious application of learned knowledge, to the extent that a later-learned language has not been automatized. Speed alone is not an indication of automatic processing. To ensure that fluency in the performance of a language task reflects automatization, one must consider not only reaction times (RTs) but also variability of performance. This is done by calculating the coefficient of variation, that is, an individual's standard deviation of RTs divided by his or her mean RT. This coefficient decreases when faster RTs reflect an increase of automaticity and hence a decrease in controlled processing – a qualitative change; however, the coefficient of variation does not decrease in speeded-up controlled processing – a strictly quantitative change (Segalowitz, 2000). In other words, individuals whose speech is automatized have less variable performance.

- Accuracy reflects the degree of mastery of the native norms. It is no more an indication of internalization (and hence of automaticity) than fluency, as it too may reflect the conscious (but speeded-up) application of explicit rules. Here again, lack of variation (i.e., systematicity) is of the essence.

Box 33.1   Bilingual aphasia recovery patterns

The pattern is said to be *parallel* when recovery of each language is proportional to its premorbid proficiency; namely, if both languages were equally proficient before the aphasia, they are recovered at the same time and to the same extent; if one was more proficient than the other, it recovers better.

Recovery is called *differential* when it does not reflect premorbid proficiency: one language is recovered better than the other in unexpected ways; the difference between the two may be much greater than before or sometimes the previously least proficient is recovered best.

Successive recovery describes a pattern where one language remains unavailable until it is spontaneously recovered weeks or months after the other has reached a plateau.

Recovery is *selective* when one language is never recovered. Sometimes comprehension is retained but the language is unavailable for production.

In *antagonistic recovery*, one language is recovered first, but it regresses as a second language becomes available and progresses.

Sometimes this phenomenon recurs over a period of days, weeks, or months during which language availability alternates. This pattern is called *alternating antagonism*.

Some patients are unable to speak one language without continually switching back and forth. This has been termed *mixed* or *blending recovery* because they blend two languages in the same way that some unilingual aphasic patients blend words or syntactic constructions (e.g., *irregardless* for *irrespective* or *regardless*).

Selective aphasia is a special case of differential aphasia in which symptoms are observed in only one of the patient's languages. A new case was reported recently (Acqui *et al.*, 1998).

Recently published cases of bilingual aphasia (reviewed in Paradis, 2001) have confirmed the occurrences of the various types of recovery patterns reported in earlier literature. Additional cases of differential recovery have been reported since (Muñoz & Marquardt, 2003; Chengappa *et al.*, 2004).

Acqui, M., Caroli, E., Ruggeri, A.G., & Lunardi, P. (1998). Afasia in bilingue: Valutazione di un caso ed analisi dei meccanismi fisiopatologici. *Nuova Rivista di Neurologia, 8*, 31–34.

Chengappa, S., Bhat, S., & Padakannaya, P. (2004). Reading and writing skills in multilingual/multiliterate aphasics: Two case studies. *Reading and Writing, 17*, 121–135.

Muñoz, M.L., & Marquardt, T. (2003). Picture naming and identification in bilingual speakers of Spanish and English with and without aphasia. *Aphasiology, 17*, 115–132.

Paradis, M. (2001). Bilingual and polyglot aphasia. *Handbook of neuropsychology* (2nd edn, pp. 69–91). Oxford: Elsevier Science.

- Age of appropriation (acquisition or learning) will have an impact not only on pronunciation, which is generally the most noticeable and the most strongly affected feature, but also on morphology and syntax. These effects may increase with age of first contact with the subsequent languages (from about age 5 to adulthood). The most significant impact of age of appropriation on impairment patterns is probably due to the degree to which each language has been incidentally acquired and automatized rather than explicitly learned (Box 33.2). The later the second language was appropriated, especially if it was learned in a formal environment (i.e., through instruction), the more likely it is that individuals will gain greater amounts of metalinguistic knowledge that may be spared by the aphasia-causing brain damage and that may then be exploited in therapy.

33.2. LANGUAGE IMPAIRMENTS

Aphasia is the result of damage to the cognitive neurofunctional system that sustains implicit linguistic competence (see Box 33.2). Hence, a language will be vulnerable to aphasia to the extent that it was acquired incidentally and that it has been automatized. Lesions in the left basal ganglia, portions of the right cerebellum and the perisylvian cortical regions that constitute "the classical language areas" (broadly, Broca's and Wernicke's areas) will result in language-specific impairments (i.e., impairments in the grammar, namely phonology, morphology, syntax, and/or the lexicon). Amnesia is the result of damage to the declarative memory system that sustains all conscious memories, including metalinguistic knowledge (i.e., knowledge of pedagogical grammar rules and other facts about language). Hence, metalinguistic knowledge is vulnerable to amnesia. Lesions in the hippocampal system, including the parahippocampal gyri and the mesial temporal lobes, will result in the loss of previously known metalinguistic material (retrograde amnesia) or of the ability to gain new metalinguistic knowledge (anterograde amnesia). Implicit linguistic competence remains available in amnesia, while metalinguistic knowledge may remain available in aphasia. Words that cannot be accessed in the lexicon (word-finding difficulty) can sometimes be accessed through metalinguistic tasks such as translation.

Contexts of use will affect ease of access to the lexicon in various semantic domains and influence the relative proficiency in comprehension, production, reading, and writing – but also the degree of habitual language mixing. This should be taken into account when assessing the significance of language mixing in multilingual patients with aphasia and should not be confused with pathological manifestations. Two types of pathological language-switching problems may be

Box 33.2 Implicit linguistic competence and explicit metalinguistic knowledge

Implicit linguistic competence	Explicit metalinguistic knowledge
Is acquired incidentally	Is learned consciously
Is stored implicitly	Is stored explicitly
Is used automatically	Is consciously controlled when used
Is sustained by procedural memory	Is sustained by declarative memory
It involves parts of the right cerebellum, the left neostriatum, other basal ganglia, and circumscribed perisylvian cortex	It involves the hippocampal system: mesial temporal lobes, parahippocampal gyri, and anterior cingular cortex (conscious control)

Components of language that can be described in terms of rules (phonology, morphology, syntax, and the morphosyntactic properties of the lexicon) constitute implicit linguistic competence. Components of which the speaker is aware (the vocabulary and any consciously learned form) constitute metalinguistic knowledge.

Implicit memory is vulnerable to aphasia and Parkinson's disease. Bilingual Parkinsonian patients have been shown to exhibit greater syntactic impairments in their native language (Zanini *et al.*, 2004).

Explicit memory is vulnerable to amnesia and Alzheimer's disease. Bilingual individuals with Alzheimer's disease tend to use L1 either exclusively, or often in inappropriate contexts, or at least retain it better than their L2 (Hyltenstam & Stroud, 1993). Access to L2 can be lost in retrograde amnesia (Jamkarian, 2001).

There are thus two ways of speaking: the controlled way and the automatic way. Individuals with genetic dysphasia, autism, and Down syndrome rely greatly on the controlled use of metalinguistic knowledge, as do many incipient second language learners.

Fluency and accuracy are not necessarily indicators of implicit linguistic competence. The former may be the result of speedier processing and the latter the result of efficient monitoring of such speeded-up (*albeit* controlled) performance. Controlled processing is not only slower but more variable than automatic processing.

Hyltenstam, K., & Stroud, C. (1993). Second language regression in Alzheimer's dementia. In K. Hyltenstam & Å. Viberg (Eds.), *Progression and regression in language* (pp. 222–242). Cambridge: Cambridge University Press.

Jamkarian, T. (2001). Language amnesia heading, in the link "répond à vos questions," http://www.amnesique.com.

Zanini, S., Tavano, A., Vorano, L., Schiavo, F., Gigli, G.L., Aglioti, S.M., & Fabbro, F. (2004). Greater syntactic impairments in native language in bilingual Parkinsonian patients. *Journal of Neurology, Neurosurgery and Psychiatry*, 75, 1678–1681.

---

encountered: (1) On the one hand, one finds frequent violations of grammaticality constraints, that is, switching at junctures where non-brain-damaged individuals never do (e.g., between a clitic pronoun and a verb (*je* don't understand)), or borrowing function words in isolation (*je parle anglais* to *mes enfants*). Blending may occur at all levels of linguistic structure. At the phonological level, patients may speak one of their languages with phonological interference from (one of) the other(s), that is, they speak with a foreign accent that was not present premorbidly. At the word level, they may produce blended forms, for example, *twörpö* from German *Zwerg* and Hungarian *törpö* (dwarf). At the syntactic level, they may use structures from one language while speaking another (e.g., he lives in Toronto since 10 years), a type of error they never made before. (2) On the other hand, another potential problem involves mixing two languages when speaking to non-bilinguals. Patients may also switch more than they used to, as a conscious way of coping with word-finding difficulty, but also involuntarily and irrepressibly.

The typological distance between the patient's languages will influence the number and type of possible errors in each, as determined by its particular structure (Paradis, 2001). One language (say, inflectional-morphology-rich Farsi) may offer many more obligatory contexts than another (say, inflectional-morphology-poor English). Hence, a larger number of inflectional errors in one language does not necessarily mean that it is more affected than the other – there are merely more opportunities for errors to occur. For this reason, the assessment of some languages, because of features specific to their grammar, may be more revealing than it is for other languages. So, a simple error count will not do.

The kind of motivation (e.g., integrative (to fit in socially) versus instrumental (to get a promotion at work)) during the appropriation of a second language may influence the amount of actual meaningful communicative interaction, which can lead to acquisition through repeated use as opposed to sheer explicit learning, and thus affect the extent of implicit linguistic competence in that language. Motivation and affect are well known to influence language behavior of patients with aphasia: lack of motivation leads to dynamic aphasia; conversely, patients with global aphasia will blurt out a complex expletive when angered, even though they are unable to put two words together most of the time.

33.3. IMPLICIT AND EXPLICIT MEMORY

Over the past 10 years, the differential roles of declarative and procedural memories have become increasingly clear in multilingual language disorders. This difference is relevant to multilingual aphasia recovery patterns, and concerns about bilingual aphasia assessment and treatment. Within the framework of the implicit/explicit perspective (Box 33.2), all late-learned languages (L2, L3, Ln) are sustained to a large extent by declarative memory. As such, they are more likely to manifest dynamic interference from one another than from the native language(s). Goral *et al.* (2006) offer supportive evidence in their description of the case of a trilingual woman with aphasia who exhibited more interference between L2 and L3 than between either of those languages and her native language, which suggests that cerebral representation and processing are more similar among later-learned languages than between them and the native language. This may well be because non-native languages are sustained to a greater extent by declarative memory. Later-learned languages may nevertheless contain elements of static interference from L1 to varying degrees, partly depending on their respective grammatical structures. These properties, present in normal multilingual individuals, may color the languages of aphasic patients as well and should not necessarily be taken as evidence of pathology. This is why it is important to assess patients' performance against their premorbid accuracy, fluency, contexts of language use, and speech habits, with the help of an exhaustive questionnaire to this effect [(e.g., the History of Bilingualism questionnaire of the *Bilingual Aphasia Test*, described in Paradis & Libben, 1987), supplemented with additional questions such as those found in the *Language Use Questionnaire* (Muñoz *et al.*, 1999), which are easily adaptable to any pair of languages and any bilingual environment]. A relative or close friend can provide this information when the patient is unable to do so.

The most convincing clinical evidence so far of procedural acquisition of the native language and explicit learning of a later-learned second language, and of their respective cerebral substrates, comes from a case published by Moretti *et al.* (2001). The patient was a 46-year-old female Croatian–Italian bilingual who exhibited symptoms that, at first sight, looked like antagonistic recovery. The patient, tested 1 month post-onset, had developed a severe impairment of her native language, while her second language was relatively well preserved (subsequent to an infarct limited to the left caudate nucleus). Four months later, a dramatic improvement in the native language was accompanied by deterioration of the second language (following the extension of the ischemic lesion to the frontal and temporal cortices). The authors interpret these phenomena as providing evidence that the native language may be acquired and used through procedures sustained by implicit memory whereas a later-learned second language depends on explicit learning and use of metalinguistic knowledge, which requires the integrity of the declarative memory system. They also suggest that this case demonstrates the different roles of the subcortical and cortical structures that regulate implicit and explicit language functions.

Another piece of evidence comes from Meguro *et al.'s* (2003) report that their Japanese–Portuguese bilingual patients with Alzheimer's disease had great difficulty with kanji in Japanese and with irregularly spelled words in Portuguese, whereas they were able to read Japanese kana normally and regularly spelled Portuguese words nearly normally. Unlike kana script and regular Portuguese spelling, kanji and irregular Portuguese words have no one-to-one correspondence between grapheme and sound and must be memorized. In other words, for these patients, the items that rely on declarative memory were selectively or considerably more impaired than the items that are derivable by internalized rules.

33.4. PATHOLOGICAL LANGUAGE MIXING AND SWITCHING

There has been an attempt in the recent literature to distinguish pathological *mixing* (intermingling different languages within a single utterance) from pathological language *switching* (alternating between languages at utterance boundaries, even when the patient's interlocutors do not understand one of the languages). It has been suggested that these symptoms follow lesions in different brain areas: left temporoparietal cortex for mixing; left anterior cingulate and frontal lobe for switching (Fabbro *et al.*, 2000). Both pathological switching and mixing in a single patient have also been described (Mariën *et al.*, 2005). The authors infer from the parallelism between the evolution of the language symptoms and their single photon emission computed tomography (SPECT) findings that subcortical left frontal lobe circuitry may be crucially involved in both language switching and mixing.

Riccardi *et al.* (2004) describe a patient who still language switched frequently 5 years post-stroke but who nevertheless seemed to show sensitivity to the pragmatics of the switching situation as he was able to adapt to the proficiency of his interlocutors and switch in appropriate circumstances. Rossi *et al.* (2003) report the case of a patient who did not exhibit any phenomena resembling pathological mixing during standardized testing but who nevertheless constantly mixed her three languages during unilingual conversations, despite being persistently reminded that she had to use the language of the examiner. The authors conclude that a conversational setting may be more revealing of pathological mixing than a formal test. Abutalebi *et al.* (2000) describe a trilingual woman who would inadvertently switch from one language to another, even in conversation with unilingual speakers, despite being fully aware of her impairment. A circumscribed infarct was located in the periventricular white matter surrounding the left caudate nucleus. The authors therefore speculate that damage to the dominant basal ganglia may disrupt the inhibition/disinhibition process required to switch between languages, an area suspected of being involved in the access and control of lexical representations.

The neuroanatomical bases of cerebral language control are being intensively explored in an effort to understand not only mixing and switching but also the diversity of bilingual aphasia recovery patterns (Green, 2005).

33.5. EXPERIMENTAL STUDIES

A growing number of researchers have come to realize that bilingual speakers, and bilinguals with aphasia in particular, are ideal individuals for testing grammatical and psycholinguistic theories. These individuals offer an opportunity to study the impact of brain damage on different linguistic systems simultaneously. Depending on whether two languages are similar or not with respect to a particular grammatical parameter (i.e., adverb placement), one may predict that the deficits will be of the same type and/or degree in either languages or that they will be present in one language but not in the other. One may, for example, evaluate the validity of the predictions made by various production theories of agrammatism, such as the tree-pruning hypothesis versus the verb-movement deficit hypothesis: One would expect differences between Greek (a morphology-rich language) and English (a morphology-poor language) according to one theory, but similar impairments in both languages according to the other (Alexiadou & Stavrakaki, 2006). Two experimental studies investigating the performance of bilingual individuals with aphasia on cognate versus non-cognate words report better performance on naming and/or lexical decision tasks involving cognates in both languages (Detry et al., 2005; Lalor & Kirsner, 2001) but another found that cognate status increased the likelihood of correct picture naming by aphasic patients in their L2 but not in their L1. Cognates had no effect in the control group.

A fluent late Greek–English bilingual patient was tested on placement of different types of adverbs in each language (Alexiadou & Stavrakaki, 2006). Her performance differed in English and Greek but was highly dependent on the syntactic tree hierarchy in both languages. She exhibited a similar pattern in English and Greek on CP-related adverbs and had more problems in English than in Greek with modal, negative, and aspectual adverbs. The findings were explained as resulting from the specific linguistic properties of Greek versus English. Grammatical deficits in aphasia depend on how the system can break down. The patient's performance reflected a syntactic deficit in verb movement, which is a parameter of English but not Greek. Patients may perform significantly worse in one language because of factors intrinsic to the structure of the language (Paradis, 2001). A number of recent cases confirm that the type of error depends on the type of aphasia, but that potential errors in each case are constrained by the structural

characteristics of each language (Chengappa et al., 2004 (case 2); Lim & Douglas, 2000; Oren & Breznitz, 2005). Basically, in each circumstance, we have some universal principles whose parameters of realization depend on the language structure, the aphasia type, and the age and manner of acquisition – the latter determine whether declarative or procedural memory is involved (Paradis, 2004).

Diego Balaguer (de) et al. (2004) report on the performance of two Spanish–Catalan bilingual aphasic patients diagnosed with Broca's aphasia on a morphological transformation task in which the patients were asked to produce regular and irregular verb forms. Contrary to results obtained in English by Ullman et al. (1997), the authors found that both patients had more difficulty with irregular than regular verbs in both Spanish and Catalan (most errors were "no response"). They conclude that this pattern of performance is inconsistent with predictions derived from the procedural/declarative memory model. Because irregular forms are not generated by rule, it has been hypothesized that they must be learned and stored in declarative memory, from which they are retrieved as whole unanalyzed items. However, the predictions of the declarative/procedural model (Paradis, 1994; Ullman, 2001), while applicable to all languages, are nevertheless contingent upon the structural characteristics of each one, particularly the extent to which specific constructions are implicitly rule-governed. For example, Ullman et al. (1997) reported that patients with agrammatism performed better in the production of irregular than regular verb morphology in English (because irregular forms are learned declaratively, whereas regular morphology is derived by rule and generalizable to all regular verbs). However, the reverse is true in Catalan and Spanish (as well as in German (Penke et al., 1999), Greek (Tsapkini et al., 2001) and Italian (Laiacona & Caramazza, 2004)) because the irregular verb stem must be used together with regular tense inflections that vary in person and number. The irregular verb form thus becomes even more grammatically complex than regular forms, for at least two reasons: (1) Spanish and Catalan affixed irregular verbs are not retrieved from declarative memory (the way invariant forms are in English) and (2) regular rule-generated tense suffixes, marked with regular person and number inflection, must be added to the irregular stem (a stem that speakers generally do not know exists until linguists describe it as such). It is therefore not surprising that the much more complex system (including rule-generated suffixes) associated with irregular verbs in Spanish and Catalan should cause problems for individuals suffering from agrammatism, who exhibit a pattern opposite to that of English agrammatics.

Two studies have nevertheless reported that irregular forms *in English* also are less preserved (Faroqi-Shah & Thompson, 2003; Shapiro & Caramazza, 2003). However, the first concerns a patient whose brain damage was not restricted to the left frontal lobe, but also involved the parietal operculum and a portion of the temporal lobe. Too few details are given in the second study to substantiate the localization claim. More importantly, evidence suggests the existence of two parallel frontal/basal-ganglia circuits, one projecting from the basal ganglia to BA 44, which sustains procedural-memory-related functions, and the other projecting to BA 45/47, which sustains lexical/declarative memory retrieval/selection (Ullman, 2006). Anterior structures (most clearly around BA 45/47), as well as the basal ganglia, underlie the retrieval/selection of lexical knowledge. Thus, anterior aphasics have trouble not only with regular forms, but also with *retrieving* lexical items from declarative memory, including irregular verb forms (though they typically find it easier to recognize them, as was the case with Shapiro and Caramazza's patient, and as predicted by Ullman *et al.*, 1997).

33.6. ASSESSMENT OF LANGUAGE DISORDERS IN MULTILINGUAL SPEAKERS

In order to arrive at a proper diagnosis and plan appropriate treatment, it is important to be aware that manifestations of aphasia symptoms are different in various languages. These manifestations are constrained in any given language by the corollary to Murphy's Law that states that only that which can go wrong will go wrong. In other words, the structure of the language determines what types of errors may occur. The reason why a certain type of error is salient or conspicuous in a given language may be due to one or more of several factors: (1) the incidence of obligatory contexts, (2) the importance of the form for the derivation of meaning, (3) the frequency of use of the item in a language/culture, (4) the structural complexity of the item (e.g., number of deviations from the canonical form), (5) the presence or absence of redundancy (e.g., word order *and* agreement versus either word order *or* agreement), (6) the presence or absence of a zero morpheme and whether nouns and verbs exist as bare roots or must necessarily be inflected, (7) whether, when inflections are omitted, the remaining form is pronounceable or not, and (8) whether the form is memorized or derivable by rule, that is, whether it is regular or irregular. The form of the error will likely depend on the type of aphasia (e.g., omission versus substitution and/or type of substitution), but the pool of possible errors (i.e., what *can* be omitted or substituted) is restricted by the grammar of the language (Paradis, 2001).

Thus, the type of errors may depend on the type of aphasia, though potential errors are constrained by the structural characteristics of each language. An agrammatic patient may substitute a default morpheme (the zero morpheme when available, but when this is not an option because it would lead to an unpronounceable form or merely a non-word, some other, less marked, morpheme, such as the nominative case for nouns and the infinitive for verbs, again modulated by frequency of use). As long as agrammatism was defined

as the omission of grammatical morphemes in obligatory contexts and paragrammatism as their substitution, there was a risk of diagnosing agrammatism in English (or Spanish or Hungarian) but paragrammatism in Hebrew, in an English- (or Spanish- or Hungarian-) Hebrew bilingual patient who was actually agrammatic in both languages. Hebrew does not possess affixed inflections that could be dropped; instead, it has vowels that can be substituted within a template of consonants (dropping these vowels would result in unpronounceable sequences of consonants). The clinician and researcher should therefore be aware that what is generally true in their own language may not apply in the language of the patient under assessment.

Therefore, the evaluation instrument should not be a simple translation of a battery designed for, and standardized in, another language. Each task requires a different criterion of cross-language equivalence, depending on the rationale behind its design (i.e., what it is that we want to measure, and why) and on the structural distance between the languages and the cultural differences between the communities. Stimulus items may be culturally inappropriate. They may refer to objects that either are not part of the culture or may look or work differently, and hence would not be recognized or would be misunderstood. Some grammatical constructions may present very different levels of difficulty in a patient's language than in the language of the original battery. Most obviously, any task that is based on phonological minimal pairs or rhyming words will simply not work at all when translated into another language. In addition, words may have a frequency of use that varies widely from that of their translation equivalent. And some common syntactic constructions, such as the passive in English, are rarely if ever used in some other languages. Hence, translations will not usually yield interpretable results. Rather, corresponding items in another language must be selected so as to tap the same information as the original, in accordance with the rationale that led to the items' construction in the first place. Such a transposition may yield a test in which the items have no obvious resemblance to the original – and yet they are much more valid than a mere translation would have been (i.e., they test the same underlying characteristics as the original, *albeit* through different means). This is why *The Bilingual Aphasia Test* was devised with a different criterion of equivalence for each of its 32 tasks (see Paradis, 2004; Chapter 3, for details).

33.7. MULTILINGUAL APHASIA REHABILITATION

The number of publications dealing with bilingual aphasia rehabilitation has risen considerably since the beginning of the twenty-first century (six studies between 2001 and 2004 to be exact, more than over the past 50 years). A number of questions were formulated 15 years ago. One set of questions concerns whether therapy should be provided in both languages simultaneously (and if so, why? If not, why not?), or successively, and if in only one of them, which one (the native language, the one that was most fluent premorbidly, the best recovered, the language of the hospital or that of the home environment). Another set of questions relates to whether therapeutic benefits (if any) are transferred from the treated to the non-treated language, and if so, whether the transfer is a function of (1) structural distance between the languages, (2) order of acquisition, (3) dominance pre-onset, (4) dominance post-onset, (5) type of aphasia, (6) pattern of recovery, and/or (7) type of therapy. If there is an effect, is it directional, that is, from the native language to the second language – or the reverse; from the most to the least fluently recovered – or the reverse; from the most fluent to the least fluent pre-onset – or the reverse (Paradis, 1993)?

Recent studies have attempted to address some of these questions. The following data have been reported: Improvement in L1 after L1 treatment and in L2 after L2 treatment (though more extensive in the native language), without transfer of benefits in either direction, namely without improvement in the untreated language (Galvez & Hinckley, 2003); improvement in both languages subsequent to treatment in L1, and in both after treatment in L2, with much greater improvement in L1 irrespective of the language treated, (Hinckley, 2003; Gil & Goral, 2004), thus showing effects on both the treated and untreated languages; modest gains in both languages, but generalization gains only from L1 to L2 and for cognates only (Kohnert, 2004); generalization from treated words to semantically related ones in the native language without transfer to L2 words, without generalization to semantically related words in L2 but with transfer to L1 following treatment in L2 in a dominant bilingual, as well as within- and across-language generalization in a balanced bilingual (Kiran & Edmonds, 2004); transfer of gains into the untreated language only when common processes are targeted (Laganaro & Venet, 2001; Lalor & Kirsner, 2001); and finally, significant improvement in the treated L2, but also in L3 and L4 (Filiputti et al., 2002).

From this and earlier literature on bilingual aphasia rehabilitation we may conclude that it is quite possible for therapy to be effective in only one of a patient's languages. Hence, if therapy shows no effect in one language, that does not mean that it should not be attempted in the other. Moreover, translation may be available when word retrieval is not (Lalor & Kirsner, 2001; Detry et al., 2005). Lalor and Kirsner (2001) infer from the results of their experiment that treatment needs to be designed so that structural similarities between the patient's languages are exploited; they further suggest that treatment at points of structural similarity will provide a means by which one language can be treated via therapy in the other.

When linguistic competence is unavailable, compensatory strategies may be used: (1) metalinguistics, as discussed above, but also (2) pragmatics. Whereas there is no clinical or valid experimental evidence suggesting that L2 implicit competence is represented (in whole or in part) in the right hemisphere (in particular, there is no greater incidence of crossed aphasia in bilingual than unilingual individuals), bilinguals, subsequent to right-hemisphere lesions, may nevertheless, like unilinguals, be expected to exhibit impairment of affective prosody and the ability to handle humor, sarcasm, irony, inference, analogy, non-explicit speech acts, and in general, any non-literal meaning. But the linguistic aspects of prosody or any aspect of grammar will not be impaired any more than has been reported for unilinguals. Conversely, the intact right hemisphere may be of assistance to bilingual aphasic patients, as speakers of a second language generally become skilled at relying on pragmatic cues to compensate for the gaps in their L2 grammar.

33.8. CHALLENGES AND FUTURE DIRECTIONS: THE CEREBRAL ORGANIZATION OF LANGUAGES

From all of the above clinical evidence, we may infer that, to the extent that both languages have been internalized, each forms a subsystem of the larger language neurofunctional system, subserved by distinct neural circuits intertwined within the same gross anatomical areas, and uses the same processing mechanisms.

Because of the task-specificity of procedural memory, phonology constitutes a functional module independent of the syntax module, which has a different underlying structure and deals with different types of entities. The phonological systems of each language of a bilingual speaker, say, English and Spanish, are both instantiations of computational principles of *phonology* (as opposed to principles of syntax) and hence constitute subsystems (or modules) of the individual's phonology. As such, while quite distinct, English phonology has more in common with Spanish phonology than with either English or Spanish syntax. This explains how the phonology of *both* languages can be impaired in a patient, while syntax remains intact; or only the phonology of one of the languages may be impaired (the phonological subsystem of English *or* of Spanish).

On the other hand, the English language subsystem of a Spanish–English bilingual individual – made up, among other things, of phonology, morphology, and syntax, which cooperate in the comprehension and production of verbal messages – also forms a natural class: *a language* (as distinct from other cognitive functions such as mathematics or music). Thus, the English system as a whole is also susceptible to selective inhibition as evidenced by selective recovery and selective aphasia (see Trudeau *et al.*, 2003, for recent evidence bearing on languages as subsystems of the larger language system). But what does *not* occur is differential aphasia, namely the impairment of phonology in one language and syntax in the other. This would require the concurrent inhibition of two functional modules that do not form a natural class. To the extent that one of the languages has *not* been automatized, it will be sustained by declarative memory, involving the hippocampal system and anterior cingular cortex (Paradis, 2004).

The definition of multilingualism is therefore not a concern. Whether we consider a so-called perfect bilingual or a late-onset second language learner at any stage of development, the same principles apply: To the extent that a language has been internalized, its implicit grammatical competence is processed by procedural memory and forms a subsystem of the language function; to the extent that there are gaps in the implicit competence of one (or more) of the languages, the speaker will compensate by using explicit knowledge sustained by declarative memory.

Future research is likely to focus on the involvement of cerebral structures that support declarative and procedural memory in the acquisition, use, and loss of linguistic competence and the learning and efficient control of metalinguistic knowledge, as they relate to the array of communication disorders affecting multilingual individuals with aphasia, amnesia, developmental specific language impairment, and Alzheimer's, Huntington's, and Parkinson's diseases. We are also likely to witness an expansion of the search for explanations of the varied patterns of recovery of multilingual patients.

References

Abutalebi, J., Miozzo, A., & Cappa, S.F. (2000). Do subcortical structures control "language selection" in polyglots? Evidence from pathological language mixing. *Neurocase, 6*, 51–56.

Alexiadou, A., & Stavrakaki, S. (2006). Clause structure and verb movement in a Greek–English bilingual patient with Broca's aphasia: Evidence from adverb placement. *Brain and Language, 96*, 207–220.

Chengappa, S., Bhat, S., & Padakannaya, P. (2004). Reading and writing skills in multilingual/multiliterate aphasics: Two case studies. *Reading and Writing, 17*, 121–135.

Diego Balaguer (de), R., Costa, A., Sebastián-Galles, N., Juncadella, M., & Caramazza, A. (2004). Regular and irregular morphology and its relationship with agrammatism: Evidence from two Spanish–Catalan bilinguals. *Brain and Language, 91*, 212–222.

Detry, C., Pillon, A., & de Partz, M.-P. (2005). A direct processing route to translate words from the first to the second language: Evidence from a case of a bilingual aphasic. *Brain and Language, 95*, 40–41.

Fabbro, F., Skrap, M., & Aglioti, S. (2000). Pathological switching between languages after frontal lesions in a bilingual patient. *Journal of Neurology, Neurosurgery and Psychiatry, 68*, 650–652.

Faroqi-Shah, Y., & Thompson, C.K. (2003). Regular and irregular verb inflections in agrammatism: Dissociation or association? *Brain and Language, 87*, 9–10.

Filiputti, D., Tavano, A., Vorano, L., De Luca, G., & Fabbro, F. (2002). Nonparallel recovery of languages in a quadrilingual aphasic patient. *International Journal of Bilingualism, 6*, 395–410.

Galvez, A., & Hinckley, J.J. (2003). Transfer patterns of naming treatment in a case of bilingual aphasia. *Brain and Language, 87*, 173–174.

Gil, M., & Goral, M. (2004). Nonparallel recovery in bilingual aphasia: Effects of language choice, language proficiency, and treatment. *International Journal of Bilingualism, 8*, 191–219.

Goral, M., Levy, E.S., Obler, L.K., & Cohen, E. (2006). Cross-language lexical connections in the mental lexicon: Evidence from a case of trilingual aphasia. *Brain and Language, 98*, 235–247.

Green, D.W. (2005). The neurocognition of recovery patterns in bilingual aphasics. In J.F. Kroll & A.M.B. de Groot (Eds.), *Handbook of bilingualism: Psycholinguistic perspectives* (pp. 516–530). Oxford: Oxford University Press.

Hinckley, J.J. (2003). Picture naming treatment in aphasia yields greater improvement in L1. *Brain and Language, 87*, 171–172.

Kiran, S., & Edmonds, L.A. (2004). Effect of semantic naming treatment on crosslinguistic generalization in bilingual aphasia. *Brain and Language, 91*, 75–77.

Kohnert, K. (2004). Cognitive and cognate-based treatments for bilingual aphasia: A case study. *Brain and Language, 91*, 294–302.

Laganaro, M., & Venet, M.O. (2001). Acquired alexia in multilingual aphasia and computer-assisted treatment in both languages: Issues of generalization and transfer. *Folia Phoniatrica et Logopaedica, 53*, 135–144.

Laiacona, M., & Caramazza, A. (2004). The noun/verb dissociation in language production: Variety of causes. *Cognitive Neuropsychology, 21*, 103–123.

Lalor, E., & Kirsner, K. (2001). The role of cognates in bilingual aphasia: Implications for assessment and treatment. *Aphasiology, 15*, 1047–1056.

Lim, V., & Douglas, J. (2000). Impairment of lexical tone production in stroke patients with bilingual aphasia. *Brain and Language, 74*, 327–329.

Mariën, P., Abutalebi, J., Engelborghs, S., & De Deyn, P.P. (2005). Pathophysiology of language switching and mixing in a bilingual child with subcortical aphasia. *Neurocase, 11*, 385–398.

Meguro, K., Senaha, M., Caramelli, P., Ishizaki, J., Chubacci, R., Meguro, M., Ambo, H., Nitrini, R., & Yamadori, A. (2003). Language deterioration in four Japanese–Portuguese bilingual patients with Alzheimer's disease: A transcultural study of Japanese elderly immigrants in Brazil. *Psychogeriatrics, 3*, 63–68.

Moretti, R., Bava, A., Torre, P., Antonello, R., Zorzon, M., Zivadinov, R., & Gazzoto, G. (2001). Bilingual aphasia and subcortical–cortical lesions. *Perceptual and Motor Skills, 92*, 803–814.

Muñoz, M.L., Marquardt, T.P., & Copeland, G. (1999). A comparison of the codeswitching patterns of aphasic and neurologically normal bilingual speakers of English and Spanish. *Brain and Language, 66*, 249–274.

Oren, R., & Breznitz, Z. (2005). Reading processes in L1 and L2 among dyslexic as compared to regular bilingual readers: Behavioral and electrophysiological evidence. *Brain and Language, 18*, 127–151.

Paradis, M. (1993). Bilingual aphasia rehabilitation. In M. Paradis (Ed.), *Foundations of aphasia rehabilitation* (pp. 413–426). Oxford: Pergamon Press.

Paradis, M. (1994). Neurolinguistic aspects of implicit and explicit memory: Implications for bilingualism. In N. Ellis (Ed.), *Implicit and explicit learning of second languages* (pp. 393–419). London: Academic Press.

Paradis, M. (Ed.), (2001). *Manifestations of aphasia symptoms in different languages*. Oxford: Pergamon Press.

Paradis, M. (2004). *A neurolinguistic theory of bilingualism*. Amsterdam: John Benjamins.

Paradis, M., & Libben, G. (1987). *The assessment of bilingual aphasia*. Mahwah, NJ: Lawrence Erlbaum Associates.

Penke, M., Janssen, U., & Krause, M. (1999). The representation of inflectional morphology: Evidence from Broca's aphasia. *Brain and Language, 68*, 225–232.

Riccardi, A., Fabbro, F., & Obler, L.K. (2004). Pragmatically appropriate code-switching in a quadrilingual with Wernicke's aphasia. *Brain and Language, 91*, 54–55.

Rossi, E., Denes, G., & Bastiaanse, R. (2003). A single case study of pathological mixing in a polyglot aphasic. *Brain and Language*, *87*, 46–47.

Segalowitz, N. (2000). Automaticity and attentional skill in fluent performance. In H. Riggenbach (Ed.), *Perspectives on fluency* (pp. 200–219). Ann Arbor, MI: University of Michigan Press.

Shapiro, K., & Caramazza, A. (2003). Grammatical processing of nouns and verbs in left frontal cortex? *Neuropsychologia*, *41*, 1189–1198.

Trudeau, N., Colozzo, P., Sylvestre, V., & Ska, B. (2003). Language following functional left hemispherectomy in a bilingual teenager. *Brain and Cognition*, *53*, 384–388.

Tsapkini, K., Jarema, G., & Kehayia, E. (2001). Manifestations of morphological impairments in Greek aphasia: A case study. *Journal of Neurolinguistics*, *14*, 281–296.

Ullman, M.T. (2001). The neural basis of lexicon and grammar in first and second language: The declarative/procedural model. *Bilingualism: Language and Cognition*, *4*, 105–122.

Ullman, M.T. (2006). Is Broca's area part of a basal ganglia thalamocortical circuit? *Cortex*, *42*, 480–485.

Ullman, M.T., Corkin, S., Coppola, M., Hickok, G., Growdon, J.H., Koroshetz, W.J., & Pinker, S. (1997). A neural dissociation between language: Evidence that the mental dictionary is part of declarative memory, and that grammatical rules are processed by the procedural system. *Journal of Cognitive Neuroscience*, *9*, 266–276.

Further Readings

Fabbro, F. (1999). *The neurolinguistics of bilingualism: An introduction*. Hove: Psychology Press.
The first 10 chapters introduce the reader to the concepts related to the cerebral organization of language, and the remaining 18 deal with the neurolinguistics of bilingualism proper. All chapters are short (an average of a little over 8 pages each). The volume is a good basic introduction to the neuroanatomy and neurophysiology of language in general and of bilingualism in particular.

Fabbro, F. (Ed.) (2002). Advances in the neurolinguistics of bilingualism. Udine: Forum University Press.
Each contributor presents a chapter that reports current findings and discusses advances in the organization of concepts in bilingual memory, and the role of working memory and of implicit and explicit memory; clinical neurolinguistics of bilingualism, including assessment, mixing and switching; and rehabilitation issues.

Language and Communication in Aging

LORAINE K. OBLER[1] and SEIJA PEKKALA[2]

[1]Program in Speech, Language and Hearing Sciences, Graduate Center, City University of New York, New York, USA

[2]Department of Speech Sciences, University of Helsinki, Helsinki, Finland

ABSTRACT

A common observation in aging is problems with retrieving nouns. Although these lexical–retrieval problems frequently occur for proper nouns, they may also be seen for other nouns and even idioms. Comprehension problems can also be observed, especially for auditory material that consists of complex text or is produced in stressful (such as noisy) conditions. Discourse patterns may change as well, depending on the nature of the tasks eliciting them. For elders who are bilingual, these problems may be accompanied by attrition of a less-used language. More severe lexical–retrieval and comprehension problems are seen in elderly individuals who display mild cognitive impairment (MCI) or progress to Alzheimer's disease (AD), and all aspects of language and communication are impaired in the later stages of Alzheimer's dementia. Age-related language changes have been explained as language-specific or related to cognitive abilities such as memory and attention, and have been attributed to areas of the brain that undergo substantial age-related changes.

34.1. INTRODUCTION

Scientists agree that not all older individuals have language or communication problems. Thus, it is not aging *per se* that brings about the difficulties with communication that have been reported. However, at least in Western societies, people believe that older people "tend to run on," and older adults themselves report difficulties listening to speech in noisy situations and finding specific words they are searching for. Let us turn, first, to what we know about the language changes associated with healthy aging, then discuss the range of explanations for them that have been proposed in recent years. We conclude with a discussion of the language changes associated with the

dementing diseases that provide a complementary perspective on the factors that underlie language decline in adults.

34.2. LANGUAGE CHANGES IN HEALTHY AGING

Language changes associated with healthy aging have been reported for some aspects of lexical retrieval and comprehension, and for narrative and spelling abilities. Other areas of language show few or no declines with advancing age.

34.2.1. Lexical Retrieval

With respect to lexical–retrieval, age-related problems arise in finding a specific word (e.g., on tasks such as confrontation naming, or a task where one is given a definition and asked to tell what word it applies to). By contrast, being given a word and asked to define it (e.g., on a vocabulary test) shows few changes with advancing age, apart from the fact that, if not told explicitly to use single-word synonyms, older adults may use more two-word or longer responses. In an interesting study, however, Bowles *et al.* (2005) demonstrate that basic vocabulary shows a steeper decline with advancing age than does "advanced" vocabulary (items like "enconium" to be matched with one of the following: *repetition, friend, panegyric, abrasion,* and *expulsion*). Indeed, the "advanced" vocabulary from the Thorndike-Gallup Test shows little decline in performance in this cross-sectional study of individuals aged 35–70, whereas basic vocabulary scores peak at age 30, and then begin their decline.

Box 34.1   The transmission deficit hypothesis

Why do older adults have problems with remembering the names of things? Burke and her colleagues have conducted a series of studies to answer this question. The model they propose is called the transmission deficit hypothesis (TDH, Burke & Shafto, 2004). It argues that words in the lexicon are linked in a network via semantic, phonological, and orthographic "nodes" that together comprise our knowledge about a word. Cognitive aging reduces activation between these nodes, reducing one's ability to come up with the right word at the right time. This network operates in hierarchical fashion. First, a connection integrating different semantic representations takes place. Then irrelevant lexical representations get inhibited and the most-primed lexical representation gets excited. Subsequently, a similar phonological priming and activation process takes place until a target phonological form is selected. The rate of priming is determined by the strength of the connections which vary according to how recently and how frequently a node has been activated.

Aging affects the strength of the links between the semantic and phonological levels by weakening the connections between them. Phonological word-shapes tend to have sparser connections than semantic features. As a result, older adults tend to experience more word-finding problems than young people. Either they experience these as tip-of-the-tongue states, or they may omit elements in making slips-of-the-tongue. Tip-of-the-tongue instances occur when the phonological nodes of a target lexical item do not get sufficiently excited; the speaker knows the word, but is unable to produce it. Omission errors result in word-fragments, as some, but not all the phonological nodes are activated. In cases where other words sharing some or all of the same sounds get overtly "excited," a similar-sounding word may be selected for production, resulting in a phonologically related word-substitution error.

Burke, D.M., & Shafto, M.A. (2004). Aging and language production. *Current Directions in Psychological Science*, *13*, 21–24.

In the process of getting to a specific correct word during lexical access, it appears to be the phonological shape of the word that poses particular problems (see Box 34.1). Older adults do not benefit from semantic cues on a naming task (indeed, they may say *yeah, I know* if given one). However, they do benefit from phonological cues, for example the initial consonant-vowel combination or the first full syllable.

As older adults report, proper nouns are particularly problematic for them to recall, both in natural conditions and in experimental ones. Whether this is due to proper nouns' relatively low frequency or some other factor remains unclear.

In sum, then, while knowledge of vocabulary itself does not decline with advancing age, the ability to locate the specific word one wants to speak does decline. This holds true particularly for proper names, but also for general nouns and verbs, and is more evident for "basic" vocabulary than for very low-frequency items. Longitudinal studies reveal subtle naming problems at an earlier age than do cross-sectional ones.

34.2.2. Language Comprehension

Language comprehension problems associated with advancing age are not seen at the single-word level, but have been reported at the sentence and text levels. The factors that enter into age-related comprehension problems are such stressors as listening to language when it is speeded (Wingfield *et al.*, 2003) or even slowed (Bergman, 1980), when it is presented in an unfamiliar accent (e.g., Shah *et al.*, 2005), or when the syntax is complex. Text comprehension problems are seen for materials that require high-level inferencing such as when deducing the moral of a story or when the content is densely packed. Of course these problems are probably not exclusively language problems; most likely they arise from the interactions between language and other cognitive changes associated with aging, such as cognitive processing speed, attention, and memory.

34.2.3. Oral Language Narrative Production

In addition to language comprehension difficulties, age-related problems in language production have also been reported beyond the single-word lexical–retrieval level. These are reflected in narrative production tasks and depend on the type of task. If the task requires retelling a recently heard narrative, older adults are likely to give shorter responses than are younger adults. By contrast, when producing a free-narrative, compared to young adults (mean age 44 years), older adults (mean age 76 years) will usually give longer responses. They are also less "efficient," that is they produce less "dense" and more irrelevant information with fewer cohesive ties (Juncos-Rabadán *et al.*, 2005).

In oral speech, more frequent speech-errors are seen in old adults compared to young adults, although the elderly catch and correct virtually all of their speech-errors just as young individuals do (McNamara *et al.*, 1992). The types of speech-errors, however, may differ between older and younger adults. In a task designed to elicit phonological and morphological errors, older participants (mean age 72 years) made more errors of omission, but proportionately fewer "non-sequential substitution errors" than the younger adults (mean age 19 years). As in our discussion of comprehension above, we may assume that such errors arise as a consequence of general cognitive changes with age. In this instance, better self-monitoring may explain the decline in substitution errors, whereas poor immediate memory may explain the increased omissions.

34.2.4. Written Language

There is relatively little work on written language changes associated with advancing age. In fact, many writers continue to be quite productive into their later years. The most salient difficulties reported are those on spelling that mimic those on lexical retrieval mentioned above: producing the precisely correct item (e.g., spelling to dictation) becomes more difficult with advancing age, whereas recognizing the correctly spelled item does not (MacKay *et al.*, 1999).

Reading rates may slow down minimally, perhaps influencing comprehension, but such comprehension problems are more likely due to a decline in the ability to inhibit distractions and to keep materials in short-term and medium-term working memory than due to actual language-based changes.

34.3. AGING AND BILINGUALISM

People who master more than one language have advantages over those who master only one. Until recently, it appeared that these advantages were social and professional throughout life, but were cognitive only for a limited period in childhood. Today, we understand that cognitive advantages can be seen in adulthood as well. Bialystok *et al.* (2007) report that bilinguals are susceptible to Alzheimer's disease (AD) 4 years later than matched monolinguals, for example. Other studies indicate that those elderly bilinguals who have lifelong experience in using two different languages have developed better executive function and working memory which they can more effectively use for processing language than monolingual older adults (Bialystok *et al.*, 2006).

Attrition of language that is associated with advancing age can involve either the first language (L1) of the speakers if they predominantly use the second language (L2) or the L2 if they predominantly use the L1 (see Box 34.2). The language decline associated with advancing age in monolinguals shares certain patterns with the attrition that has been identified in bilinguals. Both groups tend to show impairment in retrieving target lexical items and changes in lexical organization. In particular, difficulties in naming target words fast and accurately can manifest as word substitutions, generalizations, simplifications, slowed speech production, hesitations, pauses, repairs, and false starts (Goral, 2004). In order to overcome such naming difficulties, both monolingual and bilingual adult speakers may use different strategies such as avoidance, rephrasing, and circumlocutions. Bilingual elders, as well, can borrow words from the other language or code-switch from one language to the other, if their interlocutor speaks both languages.

Box 34.2 First-language attrition in bilinguals

When bilinguals do not use one of their languages, or use it less frequently, they can show subtle changes that are called language attrition. Adult native speakers of English who had been living and communicating in a predominantly Hebrew-speaking environment for more than 8 years, showed subtle lexical-access problems when telling a story based on a picture-book (Olshtain & Barzilay, 1991). They substituted more general words for specific words (e.g., "water" for "pond") or used more imprecise words (e.g., "lake" for "pond").

Of course these adults were also getting 8 years older while they were immersed in a Hebrew-speaking environment. In order to tease out whether language changes seen in older bilinguals result primarily from age-linked phenomena or primarily from language attrition, three groups of Hebrew–English bilinguals were given a lexical-decision task in each of the two languages (Goral *et al.*, 2005). Two groups, an older one aged 55–64 and a younger one aged 19–38, were living in New York City, that is, in a primarily English-speaking environment. The third group, aged 18–40, was living in Israel, a primarily Hebrew-speaking environment. The older bilinguals living in the US showed longer response times for lexical-decisions on Hebrew words than did the younger bilinguals living in the US. However, they did not show longer response times on English words. It was thus concluded that first-language attrition dominates any age-related language lexical changes in bilingual participants living in a second-language environment.

Goral, M., Libben, G., Obler, L.K., & Jarema, G. (2005). Changes in L1 compound processing in L1-dominant bilinguals. Poster presented at the Second International Conference on First Language Attrition, Amsterdam, August 2005.

Olshtain, E., & Barzilay, M. (1991). Lexical retrieval difficulties in adult language attrition. In H.W. Seliger, & R.M. Vago (Eds.), *First language attrition* (pp. 139–150). Cambridge: Cambridge University Press.

34.4. DISTINGUISHING LANGUAGE IN HEALTHY AGING FROM THAT IN MILD COGNITIVE IMPAIRMENT AND AD

Neuropsychological research has recently focused on how to distinguish individuals who show the communication patterns of healthy aging from those individuals with mild cognitive impairment (MCI) and mild dementia, especially in the early stages of AD. Sometimes it is possible to distinguish these two conditions via language performance; other times it is not, either because the comparison has not been made, or the changes are subtle enough that they cannot be seen with the measures we use.

MCI is considered a transitional state between normal aging and dementia that presents with heterogeneous symptoms. Although MCI is predominantly associated with an impairment of episodic memory, other cognitive impairments have also been reported such as executive dysfunction and learning deficits. The language difficulties of normally aging individuals become more prominent if these individuals are afflicted by MCI or mild dementia. Individuals with MCI have shown significant impairment in language comprehension (e.g., as measured by the Token Test) and production (e.g., as tested with the Boston Naming Test) (e.g., Nordlund et al., 2005).

AD is a form of dementia that is associated with specific cellular-level deterioration that results in atrophy of the cortex, particularly in the frontal and temporal lobes. Language changes are often among the early signs of AD, along with memory changes, difficulty in applying knowledge, and behavioral changes. In fact, language changes are more salient in instances of earlier onset – in the 40s or 50s – and memory changes more salient in instances of later onset – in the 60s or later. Today, however, we do not distinguish presenile' from "senile" dementia as in previous centuries; the somewhat differing set of behaviors presented depending on the age-of-onset are considered to reflect the same underlying disease.

34.4.1. Lexical Retrieval in MCI and AD

Word retrieval is one of the most affected aspects of language in MCI and throughout AD. It is most often assessed by confrontation naming tasks and different types of verbal fluency tasks, that is, list-generation. Confrontation naming ability is significantly worse in MCI than in matched healthy control subjects (e.g., Nordlund et al., 2005). Naming of specific low-frequency words is particularly problematic for individuals with MCI (Adlam et al., 2006).

Patients with mild AD also perform significantly worse than normal controls on naming tasks. In addition to making the same types of errors as normals make, such as errors that are semantically related to the target (e.g., "racoon" for "beaver") they also make errors that suggest they have misidentified the picture they are naming (Nicholas et al., 1996) (e.g., "snake" for "pretzel"). Moreover, if given phonological cues, the patients may speak any word that begins with the sound, rather than one that might be appropriate for the picture, suggesting that self-monitoring is diminished. In mild AD, word-finding difficulties are more prominent than in MCI, and they tend to appear not only on confrontation naming tasks that require intentional access to the semantic knowledge, but also on tasks that require more automatic semantic processing, such as priming tasks (Duong et al., 2006). They can be seen on word-list generation tasks as well. Even patients with mild AD showed reduced success in generating semantically based lists of nouns and verbs

(e.g., clothes, animals, playing sports) relative to normal controls (Pekkala, 2004). Participants with mild AD made fewer switches between semantic subcategories, produced fewer words consecutively within semantic or phonological clusters, and activated fewer subcategories of semantic dimensions suggesting impaired executive and semantic processing.

One interesting way to distinguish MCI and early, mild AD is to note differences between processes involving nouns and verbs, suggesting that the lexicon is impaired differently in the two disease states. Östberg et al., (2005) indicated that semantic fluency (list-generation) performance in which action verbs were produced was significantly more impaired than noun or letter fluency in MCI, whereas noun fluency was more impaired than verb or letter fluency in mild to moderate AD. The authors speculated that in AD, the impaired noun fluency was due to atrophy in the entorhinal area as well as posterior (perceptual) association cortex which has been found to be responsible for visual associations related to nouns. Impaired verb fluency performance, by contrast, was attributed to the neocortical–hippocampal interaction and the perirhinal cortex (Brodmann area 35) which tend to be affected in MCI.

In sum, lexical access is impaired in MCI and AD relative to healthy controls, and the deficit involves both nouns and verbs. In MCI, word-finding problems concern words that are low in frequency and words that require intentional access to semantic memory. In mild AD, word-finding difficulty is more severe than in MCI, and problems arise with even automatic access to semantic representations. Lexical access deteriorates further during the course of AD resulting as increasing difficulty in word-finding in the moderate stage and inability to produce speech in the end-stage of the disease. People with AD make significantly more naming errors than healthy elderly controls, and phonological cuing tends not to help word retrieval. The reasons for these changes would appear to be linked to changes in language areas of the brain per se in conjunction with changes in semantic memory and executive function (for both lexical-search and self-monitoring) as well.

34.4.2. Comprehension in MCI and AD

While comprehension of everyday language tends to be well preserved in MCI and early stages of AD, comprehension of syntactically complex sentences has been found to be impaired relative to healthy participants (Nordlund et al., 2005). In MCI, comprehension problems have been linked to more difficult tasks in which, for example, additional conceptual or contextual knowledge must be generated by the listener in order to perform the task. Word-finding difficulties, as well as poor performance on assessing semantic knowledge of objects, suggest a semantic-memory deficit in MCI (e.g., Adlam et al., 2006; Duong et al., 2006).

In AD, language comprehension continues to decline across the progression of the disease. Understanding abstract language, such as humor, analogies, and sarcasm, can be difficult (Ripich & Ziol, 1998). In the early stages, conversational comprehension may appear relatively intact. More challenging discourse, with little redundancy or requiring high-level inferences, proves more troubling. By the mid-stages of AD, patients perform poorly on any standard comprehension task, and by the end-stages of AD, there is little evidence of either production or comprehension, though splinter pragmatic skills may remain, such as the ability to respond in common formulaic exchanges.

In sum, comprehension of everyday language is preserved in MCI and in mild AD, but difficulty in understanding syntactically complex sentences can appear. With the progression of AD, patients demonstrate increasing difficulties in comprehending language and following the topic of discourse, but maintain their ability to respond to highly stereotyped language until quite late in the disease. In both MCI and AD, deficits in executive function and, in particular, impaired access to or breakdown of semantic memory appear to underlie comprehension difficulties in AD (see Chapter 27).

34.4.3. Spontaneous Speech and Conversation in MCI and AD

Spoken and written picture-description tasks have been found to be a sensitive method to reveal differences in word retrieval and semantic processing between normal older adults and people with MCI, very early AD, and mild AD. While spontaneous, descriptive speech of healthy elderly adults may contain some errors, they seldom make semantic and phonemic paraphasias and they usually are able to produce the target word after a short delay. By contrast, the spontaneous speech of individuals with MCI, very early AD, tends to contain semantic paraphasias, word-finding delays, errors in monitoring their output, and decrease in information units (Forbes-McKay & Venneri, 2005), suggesting that self-monitoring of one's language performance is impaired.

As AD progresses into the mild stage, word-finding difficulty becomes more prominent: the number of errors increases significantly, indefinite and irrelevant terms predominate over more specific substantives, and individuals give up attempting to name more easily than they did in the earlier stage of AD. Conversational skills, however, remain largely intact in mild AD, but some patients may have difficulties with conversational initiation, bringing up and maintaining topics, and following the course of a conversation. Some patients may show reduction in their verbal abilities, especially in word retrieval. Semantically empty words (e.g., "thing," "stuff," "do," and deictic terms), circumlocutions, pronouns, gestures, and semantic paraphasias are used

to overcome the word-finding problems in order to maintain the flow of the conversation. Prosody, articulation, and rate of speech are relatively well preserved in mild AD (for review see Orange & Ryan, 2000).

In sum, the spontaneous, descriptive, speech of people with MCI or very early AD indicates more errors than that of healthy elderly adults. With further progression of AD, spontaneous speech production becomes more affected, and patients face increasing difficulties in finding correct words, making sense, and thus in participating in conversations.

34.5. EXPLANATIONS FOR LANGUAGE CHANGES IN HEALTHY AGING

Explanations for the language changes associated with advancing age have been suggested at many levels: neurological, language-related, cognitive, and social.

34.5.1. Neurological Functional–Anatomical Explanations

Some studies have suggested a "frontal-lobe" hypothesis for the cognitive declines associated with advancing age, and, by extension, for the language changes associated with them. Band et al., (2002) point out that a simple "frontal-lobe" hypothesis is misleading because it suggests a single brain basis for the cognitive and language changes associated with advancing age. Rather, it would appear that subareas of frontal lobe have different responsibilities for language as well as other cognitive-related tasks, and, of course, additional areas of brain are implicated in studies of the functional-anatomic levels associated with language processing in advancing age.

Other studies have reported that older adults recruit more areas of the brain even when they perform as well on cognitive tasks as younger adults do. This may mean that older adults employ more of left-hemisphere language areas, or more bilateral areas, within and beyond strictly language areas. Within the left-hemisphere, for example, Grossman et al. (2002) report somewhat different age-related activation of parietal and frontal lobes during a sentence-comprehension task for older (mean age 64 years) as compared to younger (mean age 23 years) adults. However, no difference was found in comprehension accuracy between the two groups. The older adults showed more activation in two areas of frontal lobe (premotor and dorsal inferior), but less in parietal areas. Differences were seen in right-hemisphere activation as well, for those sentences that required syntactic integration over many words: the older adults relied particularly on right parietal regions whereas the younger adults employed additional right posterolateral temporal regions.

34.5.2. Language-Level Explanations

A number of studies have found correlations between or among different language tasks, suggesting that the language problems associated with advancing age and/or dementia may be problems exclusive to language. Vocabulary ability, for example, is linked to sentence-length in a discourse-production task (Kemper & Sumner, 2001), and, in our own lab, performance on an idiom-production task and a picture-naming task likewise appear linked in older, as in younger, adults. At the more theoretical level, as discussed in Box 34.1, the transmission deficit hypothesis suggests that aging may cause reduction in activation between phonological, semantic, and orthographic systems, as a result of which age-related word-finding difficulties can emerge. Such intra-language links, of course, are consistent with problems being exclusively within language, but they are also consistent with cognitive-level explanations.

34.5.3. Cognitive-Level Explanations

Among cognitive measures, processing speed and working memory are the two that have been most implicated in studies of language in aging (e.g., Salthouse, 2000). The role of inhibition mechanisms or socio-economic factors has also been investigated.

Processing speed refers to the time it takes to conduct basic cognitive tasks, and it is known to slow down with advancing age. Consider how comprehension can decline over the course of a sentence or larger discourse unit if one misses intervals of content earlier on in the speech. Indeed, declining processing speed has been linked to reading comprehension. Perceptual speed has also been linked to performance on tasks of "verbal knowledge" such as vocabulary definition and identifying real words among a set of pronounceable non-words (Lövdén et al., 2004).

Working memory is another cognitive ability that is known to decline in healthy aging. It is thus not surprising that it has frequently been used to explain the language changes in advancing age. Clearly, working memory is required for comprehension of language and for self-monitoring production. Working memory measures are linked to syntactic complexity in a number of the studies of Kemper and her colleagues. For example, in Kemper et al. (2004), working-memory decline explains the absence of correlation between age and sentence-stem complexity in the older participants (mean age 73 years), although such a correlation was also obtained in the younger participants (mean age 20 years).

Another cognitive explanation for the language changes associated with aging argues that there is a decrease in the ability to suppress irrelevant information. Originally termed the inhibition deficit hypothesis (Hasher & Zacks, 1988), such distractibility may result in problems both with speech production and comprehension. In speech production, elderly individuals may show verbosity or inability to suppress incorrect words that come to mind (though normal elderly will self-correct these). In patients with early AD (and perhaps those with MCI), this may result in an increase in irrelevant, off-topic, ideas in conversation. During comprehension, distractibility is evidenced even at the basic perceptual level, with disproportionately greater difficulties for older adults than younger adults in noisy and other non-ideal listening conditions.

General intelligence may be linked to the amount of education one undertakes, of course, and has been linked to performance on cognitive tasks generally (Lövdén et al., 2004). However, Lövdén et al. pointed out that basic visual and hearing abilities are more predictive of intellectual functioning and verbal knowledge among older (31%) than younger people (11%). Among the social explanations for the language changes associated with healthy aging, most often discussed are years of education and socio-economic status. Of course, years of education is a coarse measure of participants' amount of education, as the quality of education differs so substantially across different schools and different times. Nevertheless, this coarse marker is often linked to performance on the full range of language tasks that have been administered: from lexical access (e.g., Connor et al., 2004), to syntactic structure and conceptual density in narrative discourse (see Box 34.3).

The other social factor that is less employed in language studies on elderly populations (although it is looked at considerably in studies of language development in children) is socio-economic status. Socio-economic status measures usually include education as one of their components. As well, they include the prestige-status of the participant's job, and sometimes that of the participant's spouse as well. The one study we know of linking language performance to socio-economic status is that of Connor et al. (2001) on aphasics who had suffered brain-damage as the result of a stroke resulting in language disturbance. Those of low socio-economic status (as measured by a combination of job-status and years of education, according to Hollingshead's measure) were more impaired initially than those of high socio-economic status. This remained true after therapy as well, although the rate of recovery was the same for both groups. Whether this finding results from health-risk differences or health-care ones between the two socio-economic groups that have implications for non-brain-damaged adults remains to be explored. However, in the Berlin Aging Study, socio-economic status did not predict age-linked changes in cognitive status, including verbal knowledge (Lövdén et al., 2004, p. 113).

34.6. CHALLENGES AND FUTURE DIRECTIONS

The challenges of studying language in healthy aging are many. First is the problem of defining what constitutes

Box 34.3   The nun study

An interesting question to pursue is whether there are any indicators in young adulthood that can predict cognitive decline in later life. Of course a big challenge is to find a population whose living conditions are as homogenous as possible. One such group is nuns from the same convent who share similar living conditions. Kemper and her students asked whether language measures in young adulthood could predict cognitive decline in later life, testing nuns from two convents who were aged 75–95 at the time of testing. Information about them included careful data about their health and education over their adult lifespan. In the original papers, the authors reported that the propositional density of the autobiographical essays nuns had written when they were to be admitted to the convent, aged 18–32, predicted which of the nuns would develop signs of AD (Snowden *et al.*, 1996), or otherwise die younger (Snowden *et al.*, 1999). In a follow-up study, Kemper *et al.* (2001) demonstrated that both propositional density and grammatical complexity in the nuns' young–adult essays predicted which members of the group would become demented.

At the later testing, the nuns were asked not only to write autobiographical essays about their childhood, but also to speak about "interesting or edifying" incidents from it. Mitzner

and Kemper (2003) analyzed those responses and demonstrate that the written essays distinguish poor- and good-performers better than the oral responses do. Measures of both types of discourse, moreover, are linked to the nuns' ability to perform "activities of daily living" in older adulthood as well as to the more obviously linked measure of the years of education each nun received.

Kemper, S., Greiner, L.H., Marquis, J.G., Prenovost, K., & Mitzner, T.L. (2001). Language decline across the life span: Findings from the nun study. *Psychology and Aging, 16*, 227–239.

Snowden, D.A., Greiner, L.H., Kemper, S.J., Nanayakkara, N., & Mortimer, J.A. (1999). Linguistic ability in early life and longevity: Findings from the nun study. In J.-M. Robine, B. Forette, C. Franceschi, & M. Allard (Eds.), *The paradoxes of longevity* (pp. 103–113). Berlin: Springer-Verlag.

Snowden, D.A., Kemper, S.J., Montimer, J.A., Greiner, L.H., Wekstein, D.R., & Markesbery, W.R. (1996). Linguistic ability in early life and cognitive function and Alzheimer's disease in late life. Findings from the nun study. *Journal of the American Medical Association, 275*, 528–532.

Mitzner, T.L., & Kemper, S. (2003). Oral written language in late adulthood: Findings from the nun study. *Experimental Aging Research, 29*, 457–474.

---

healthy aging: "normal" for who volunteers for studies can differ from "normal" for the population as a whole; "normal" for one culture may not be the same as "normal" for another. Should "normal" include individuals who may have had learning problems that were not diagnosed in childhood, for example? A second major challenge in studying language, particularly in aging, is finding ways to separate education effects from age-related effects. Years of education is such a coarse measure of education as it ignores quality of education altogether. A third challenge lies in determining how language-use effects interact in language performance in advancing age; those studying bilingualism are just starting to consider this issue, but there is no reason to believe that it does not apply to monolinguals as well. Indeed, Barresi *et al.* (1999) demonstrated a strong negative correlation between number of hours of television older monolingual participants reported watching per week and their confrontation naming scores.

With respect to the dementias, the biggest challenge is to differentiate between normal aging and mild forms of dementia. It could be helpful if language tasks could be used in making this diagnosis at an early stage of the dementing process. As well, creating therapies to enhance communication for patients with AD and other dementing diseases is a challenge worth pursuing further. Recent studies using group conversation therapy and cognitive–linguistic intervention programs have yielded promising outcomes in maintaining and even improving the communication and language skills

of people with dementia (e.g., Mahendra & Arkin, 2003). Additionally, counseling health-care staff and patients' family on the ways to facilitate communicate with an individual with dementia have proven successful.

References

Adlam, A.-L.R., Bozeat, S., Arnold, R., Watson, P., & Hodges, J.R. (2006). Semantic knowledge in mild cognitive impairment and mild Alzheimer's disease. *Cortex, 42*, 675–684.

Band, G.P.H., Ridderinkhof, K.R., & Segalowitz, S. (2002). Explaining neurocognitive aging: Is one factor enough? *Brain and Cognition, 49*, 259–267.

Barresi, B., Obler, L.K., Au, R., & Albert, M.L. (1999). Language-related factors influencing naming in adulthood. In H. Hamilton (Ed.), *Old age and language: Multidisciplinary perspectives* (pp. 77–90). New York: Garland.

Bergman, M. (1980). *Aging and the perception of speech*. Baltimore, MD: University of Maryland Press.

Bialystok, E., Craik, F.I.M., & Ruocco, A.C. (2006). Dual-modality monitoring in a classification task: The effects of bilingualism on aging. *The Quarterly Journal of Experimental Psychology, 59*, 1–16.

Bialystok, E., Craik, F.I.M., & Freedman, M. (2007). Bilingualism as a protection against the onset of symptoms of dementia. *Neuropsychologia, 45*, 459–464.

Bowles, R., Grimm, K.J., & McArdle, J.J. (2005). A structural factor analysis of vocabulary knowledge and relations to age. *Journal of Gerontology: Psychological Sciences, 60B*, 234–241.

Connor, L.T., Obler, L.K., Tocco, M., Fitzpatrick, P., & Albert, M.L. (2001). Effect of socio-economic status on aphasia severity and recovery. *Brain and Language, 78*, 254–257.

Connor, L.T., Spiro, A., Obler, L.K., & Albert, M.L. (2004). Change in object naming ability during adulthood. *Journal of Gerontology: Psychological Sciences, 59B*, 203–209.

Duong, A., Whitehead, V., Hanratty, K., & Chertkow, H. (2006). The nature of lexico-semantic processing in deficits in mild cognitive impairment. *Neuropsychologia, 44*, 1928–1935.

Forbes-McKay, K.E., & Venneri, A. (2005). Detecting subtle spontaneous language decline in early Alzheimer's disease with a picture description task. *Neurological Sciences, 36*, 243–254.

Goral, M. (2004). First-language decline in healthy aging: Implications for attrition in bilingualism. *Journal of Neurolinguistics, 17*, 31–52.

Grossman, M., Cooke, A., DeVita, C., Alsop, D., Detre, J., Chen, W., & Gee, J. (2002). Age-related changes in working memory during sentence comprehension: An fMRI study. *NeuroImage, 15*, 302–317.

Hasher, L., & Zacks, R.T. (1988). Working memory, comprehension, and aging: A review and a new view. In G.H. Bower (Ed.), *The psychology of learning and motivation* (Vol. 22, pp. 193–225). San Diego, CA: Academic Press.

Juncos-Rabadán, O., Pereiro, A.X., & Rodriguez, M. (2005). Narrative speech in aging: Quantity, information content and cohesion. *Brain and Language, 95*, 423–434.

Kemper, S., & Sumner, A. (2001). The structure of verbal abilities in young and older adults. *Psychology and Aging, 16*, 312–322.

Kemper, S., Herman, R.E., & Liu, C.-J. (2004). Sentence production by young and older adults in controlled contexts. *Journal of Gerontology: Psychological Sciences, 59B*, 220–224.

Lövdén, M., Ghisletta, P., & Lindenberger, U. (2004). Cognition in the Berlin aging study (BASE): The first 10 years. *Aging, Neuropsychology and Cognition, 11*, 104–133.

MacKay, D.K., Abrams, L., & Pedroza, M.J. (1999). Aging on the input versus output side: Age-linked asymmetrics between detecting versus retrieving briefly presented spelling patterns. *Psychology and Aging, 14*, 3–17.

Mahendra, N., & Arkin, S. (2003). Effects of four years of exercise, language, and social interventions on Alzheimer's discourse. *Journal of Communication Disorders, 36*, 395–422.

McNamara, P., Obler, L.K., Au, R., Durso, R., & Albert, M.L. (1992). Speech monitoring skills in Alzheimer's disease, Parkinson's disease, and normal aging. *Brain and Language, 42*, 38–51.

Nicholas, M., Obler, L.K., Au, R., & Albert, M. (1996). On the nature of naming errors in aging and dementia: A study of semantic relatedness. *Brain and Language, 54*, 184–195.

Nordlund, A., Rolstad, S., Hellström, P., Sjögren, M., Hansen, S., & Wallin, A. (2005). The Goteborg MCI study: Mild cognitive impairment is a heterogeneous condition. *Journal of Neurology, Neurosurgery, and Psychiatry, 76*, 1485–1490.

Orange, J.B., & Ryan, E.B. (2000). Alzheimer's disease and other dementias. Implications for physician communication. In R.D. Adelman & M.G. Greene (Eds). *Clinics in geriatric medicine. Communication between older patients and their physicians* (Vol. 16, pp. 153–167). Philadelphia, PA: W. B. Saunders Company.

Östberg, P., Fernaeus, S.-V., Hellström, Å., Bogdanovi, N., & Wahlund, L.-O. (2005). Impaired verb fluency: A sign of mild cognitive impairment. *Brain and Language, 95*, 273–279.

Pekkala, S. (2004). *Semantic fluency in mild and moderate Alzheimer's disease*. Publications of the Department of Phonetics, 47. Helsinki: University of Helsinki.

Ripich, D.N., & Ziol, E. (1998). Dementia: A review for the speech-language pathologist. In A.F. Johnson & B.H. Jacobson (Eds.), *Medical speech-language pathology: A practitioner's guide* (pp. 467–494). New York: Thieme Medical.

Salthouse, T.A. (2000). Aging and measures of processing speed. *Biological Psychology, 54*, 35–54.

Shah, A., Schmidt, B.T., Goral, M., & Obler, L.K. (2005). Age effects in processing bilinguals' accented speech. In J. Cohen, K. McAlister, K. Rolstad, & J. MacSwan (Eds.), *ISB4: Proceedings of the 4th International Symposium on Bilingualism* (pp. 2115–2121). Somerville, MA: Cascadilla Press.

Wingfield, A., Peelle, J.E., & Grossman M. (2003). Speech rate and syntactic complexity as multiplicative factors in speech comprehension by young and older adults. *Aging, Neuropsychology, and Cognition, 10*, 310–322.

Further Readings

Chapman, S.B., Zientz, J., Weiner, M., Rosenberg, R., Frawley, W., & Burns, M.H. (2002). Discourse changes in early Alzheimer disease, mild cognitive impairment, and normal aging. *Alzheimer Disease and Associated Disorders, 16*, 177–186.

In this study, the authors indicated that gist and detail levels of discourse processing are significantly impaired in individuals with AD and MCI compared with normal elderly adults.

Cullum, S., Huppert, F.A., McGee, M., Dening, T., Ahmed, A., Paykel, E.S., & Brayne, C. (2000). Decline across different domains of cognitive functions in normal ageing: Results of a longitudinal population-based study using CAMCOG. *International Journal of Geriatric Psychiatry, 15*, 853–862.

This article provides a description of the decline in different cognitive functions in a four-year follow-up study on people without dementia. It discusses the effects of sociodemographic variables, such as age, sex, education, social class, and baseline score on change in different domains of cognition.

Lindenboom, J., & Weinstein, H. (2004). Neuropsychology of cognitive ageing, minimal cognitive impairment, Alzheimer's disease, and vascular cognitive impairment. *European Journal of Pharmacology, 490*, 83–86.

In this brief article, neuropsychological changes in memory, language, visuo-spatial functions, and executive functions associated with each of the four different conditions are reviewed, and the relation between aging and cognition is summarized.

Wingfield, A., & Grossman, M. (2006). Language and the aging brain: Patterns of neural compensation revealed by functional brain imaging. *Journal of Neurophysiology, 96*, 2830–2839.

This article provides a two-part model for language comprehension based on the authors' review of fMRI studies. It suggests that language-region areas support comprehension, and are complemented by other regions (for example those responsible for working memory) to maintain effective comprehension with advancing age.

CLINICAL NEUROSCIENCE OF LANGUAGE

B. Language and Communication in Developmental Disorders

35

Acquired Epileptiform Aphasia or Landau–Kleffner Syndrome: Clinical and Linguistic Aspects

GIANFRANCO DENES

Department of Linguistics, Università Ca' Foscari, Venezia, Italy

ABSTRACT

The acquisition of language is usually a simple and effortless process. About 7% of the children show, however, a specific language impairment (SLI), with comparatively normal abilities in other areas. The etiology of SLI is complex, ranging from genetic to environmental factors, although in most cases a definite etiology is lacking. Studying developmental disorders can be viewed as true "experiments of nature," and besides the important clinical and rehabilitative aspects, may yield information about normal developmental processes and the organization of the language process. In this chapter we focus on a specific developmental disorder, the syndrome of acquired aphasia and epilepsy, also referred to as the Landau–Kleffner (LK) Syndrome. Children affected by LK syndrome, who already developed age-appropriate language, show a subacute regression of language, up to a total loss and flanked, in the majority of cases, by epilepsy. While the latter symptom subsides, in the majority of the affected children the aphasic disturbance persists throughout their lives. The clinical and neurolinguistic aspects of the syndrome are discussed and related to other, more pervasive developmental disorders, such as autistic regression with or without epilepsy.

35.1. INTRODUCTION

Patient C.S. was born after an unremarkable pregnancy and birth history (for a detailed discussion see Denes *et al.*, 1986). His developmental milestones were normal: he started to walk between the ages of 8 and 10 months, used single words before 1 year and was able to put words together at the age of two. At age three (at that time his language development was,

according, to his parents, normal) he became less responsive to oral speech and his expressive language deteriorated, up to an almost complete inability to produce and understand oral speech, despite intensive speech therapy. At the same time behavioral changes, consisting mainly of aggressive and oppositional traits, were noted. Apart from the language impairment, the neurological examination, the computerized tomography (CT) scan, screening for metabolic abnormalities, and pure tone audiometry were reported as normal. Auditory evoked potentials where normal at the brain stem level. At the cortical level the early primary responses to auditory synthetic stimuli were also normal while the late components showed abnormally long latency and they were distorted. Repeated electroencephalograms (EEG) showed the presence of bilateral paroxysmal discharges that were more evident in the temporal regions. Despite antiepileptic treatment, 6 months later the child developed a single epileptic seizure. In the following years he was seizure free and the EEG pattern improved to normal.

At the age of five his mother, in her effort to communicate, thought of asking him to match the picture of a common object to the corresponding written word. In the following months and years, C.S. developed a striking ability to handle written language, both in comprehension and production, and he was able to successfully attend primary school. In the following years his oral-auditory language impairment did not change, while his nonverbal abilities developed normally and appropriate to his age.

35.1.1. Neuropsychological Assessment of C.S.

At the age of 11 C.S. was submitted to a formal neuropsychological assessment that, apart from the language impairment, was normal: he was oriented in time and place and could recognize and use objects normally. His drawing ability was remarkably good and he was able to perform written calculations. He scored 34 out of 36 in the Raven Colored Matrices (a test of nonverbal intelligence) and his spatial memory span assessed with the Corsi Test was also normal (span of 4). While he showed a mild pattern of oral apraxia, that is an inability to imitate simple oral gestures (8 out of 20 correct), his performance on tests tapping the presence of limb apraxia was normal.

His behavior was characterized mainly by a certain degree of indifference to the external milieu, interrupted from time to time by mimic reactions or smiles when his gaze crossed that of familiar people. He showed no interest in words spoken to him, although he did occasionally show startle reactions to sudden noises. This was in sharp contrast to his mostly preserved ability to recognize familiar environmental nonspeech sounds (e.g., the ringing of the telephone): he scored 10/12 correct on a sound–picture matching test. However, the recognition of familiar tunes and the capacity to follow rhythms were practically absent.

35.1.2. Language Examination of C.S.

C.S. showed a practically zero performance in all tasks involving auditory modality, his performance being above chance only in a vowel discrimination task. Using written material, however, he scored almost at ceiling in all lexical tests (word identification, written naming of pictures, category naming). On the written version of the Token Test (De Renzi & Vignolo, 1962), a widely used test for assessing comprehension deficits in aphasic patients, his performance was only mildly impaired (25 out of 36 correct, mean normal score is 30), his errors involving only knowledge of function words. At the morphosyntactic level, C.S. showed a marked difficulty in handling free morphemes (articles, prepositions), while he did not show any particular difficulty in using bound morphemes. This was tested with an insertion task where he was presented with incomplete sentence that he had to fill in order to correctly describe a picture placed in front of him.

35.2. LANGUAGE REGRESSION AND EPILEPSY

Despite its remarkable complexity, the acquisition of language and speech is normally simple and effortless. By the age of three, regardless of the culture, the vast majority of children are able to master their native language up to the level of full sentences. About 7% of the children are, however, affected by developmental disorders of speech and language (Grizzle & Simms, 2005), in the absence of causal factors such as mental retardation, deafness, neurological deficits, or social deprivation. A genetic etiology has been

discussed in a considerable number of cases (for a review see Vargha-Khadem *et al.*, 2005).

A structural lesion of the language-related areas, following trauma, vascular lesions or other etiologies, can provoke epilepsy and language disturbances, frequently mimicking the adult pattern of acquired aphasia, but usually with a better prognosis (Basso & Scarpa, 1990). Aphasia can be transient as observed in the developmental period after a seizure that originates in the language areas, usually in the left temporal lobe of the dominant hemisphere. Transient aphasia has been considered an ictal or post-ictal phenomenon. In conjunction with prolonged ictal discharges, however, slow recovery has been described in some cases (van Hout, 1992).

Thornier to disentangle is the problem of the association of a persistent language regression and an abnormal EEG pattern, as first described in 1957 by Landau and Kleffner and subsequently referred to as Landau–Kleffner (LK) syndrome or, more recently, as acquired epileptiform aphasia (AEA) (Stefanatos *et al.*, 2002). More than 200 cases have been described (Paquier *et al.*, 1992; Stefanatos *et al.*, 2002) but the question of its etiology, and similarities and differences with other language regression disorders is far from being resolved.

35.2.1. Clinical Features of LK Syndrome or AEA

The age of onset varies from 18 months to late infancy with the peak incidence between 3 and 7 years (Bishop, 1985; Paquier *et al.*, 1992). At the time of onset, age-appropriate speech has already been developed, although, in exceptional cases, LK syndrome is seen in subjects with a history of developmental language disorder, Marien *et al.*, 1993). The onset may be subacute or "stuttering" and the first clinical manifestation of the language disturbance is an auditory inattention to both verbal and nonverbal stimuli. The symptom may progress up to a pattern of cortical or word deafness (verbal auditory agnosia) (Rapin *et al.*, 1977). At the same time, expressive language deteriorates with the appearance of phonological paraphasias and articulatory disorders. Eventually, the child may become completely mute, expressing himself or herself through unarticulated sounds or by the use of appropriate gestures. Rarely, fluent aphasia (Lerman *et al.*, 1991) and jargon have been observed (Landau & Kleffner, 1957; Rapin *et al.*, 1977) (Box 35.1).

Different etiologies, ranging from inflammatory demyelinising diseases to a low grade autoimmune encephalitis have been discussed, but in the majority of cases they have not been confirmed (for a review see Stefanatos *et al.*, 2002). Epilepsy is present in more than 80% of the cases (Stefanatos *et al.*, 2002). The seizures are heterogeneous, in both their manifestations and their frequency. Generalized seizures are of the motor type and consist of eye blinking or brief ocular deviations. In contrast to the language impairment, seizures have a benign course, are responsive cognitive abilities such as memory, orientation, and praxis are well within normal limits. There is only one patient described in the literature with an accompanying cognitive deficit, that is a mathematical impairment that persisted well after the aphasic impairment had dissolved (Papagno & Basso, 1993). In two thirds of the cases behavioral disorders ranging in severity from mild to pronounced and characterized by hyperactivity, sometimes coupled with aggressive traits, are reported (Stefanatos *et al.*, 2002). In some cases, particualry those with early onset and or severe and persisternt EEG abnormalities, an aberrant behavior reminiscent of the autistic spectrum disorder has been described. The presence of psychotic traits is an exception (Zivi *et al.*, 1990).

35.3. THE NATURE OF THE LANGUAGE DEFICIT

In the most severe cases, children, in the acute stage, are unresponsive to both verbal and nonverbal sounds, followed by a recovery of the ability to respond to nonverbal sounds and correctly identify the meaning of environmental sound (Denes *et al.*, 1986). Discrimination and identification of vowel and consonant sounds are severely impaired (Denes *et al.*, 1986). Equally impaired is performance on auditory comprehension tasks, such as the Peabody Picture Vocabulary Test or the Token Test. Oral production is also severely impaired to the point of almost total disappearance of speech and singing, and oral production being limited to meaningless sounds. In less severe cases, oral production, although defective, is present. A detailed analysis of a LK child, who showed a fluctuating course, reported that the total number of words and mean length of utterances matched the clinically observed fluctuations of aphasia (van de Sandt-Koenderman *et al.*, 1984). Analysis of paraphasias showed that most of them could be classified as phonemic paraphasias. Semantic paraphasias were absent, while neologisms, present in the stage of language breakdown, tended to disappear very fast during the recovery phase.

Compared to the severity of the language impairment in the oral-auditory modality, the ability to learn and use language through other modalities such as the visual modality (reading and writing) can be significantly superior as in the case just described. Teaching these children sign language can also be highly proficient as illustrated by Perez and Davidoff (2001). They reported a boy with acquired epileptiform aphasia who progressively lost speech comprehension and expression abilities from 3 years and 6 months of age to 7 years. From the age of 6 years onwards he was taught and educated in sign language. At the age of 13 years and 6 months his sign language was evaluated and compared to a child with congenital sensory-neural deafness. It was found that the boy had achieved the same proficiency in sign language as the control child with deafness.

to common antiepileptic standard treatment, and, by the age of adolescence, generally subside (Mantovani & Landau, 1980).

The spectrum of EEG abnormalities reported in LK syndrome may vary. The most common abnormality is represented by bilateral independent or synchronous temporal or temporo-parietal spikes and spike and wave complexes centered over temporal regions or generalized sharp waves. An EEG abnormality recorded during slow-wave stages of sleep and consisting of continuous spikes and waves (electrical status epilepticus) has been described in a number of LK patients (Tassinari *et al.*, 1992) and related lo the severity of their language impairment. Neurological examination, pure tone audiograms, brain stem auditory evoked potentials, cerebrospinal fluid and neuroimaging (CT scan and MRI) are normal. In the few metabolic studies performed single photon emission computed tomography (SPECT) and positron emission tomography (PET), a reduced brain glucose utilization, as well as hypoperfusion predominantly over perisylvian areas and sometimes also subcortical structures, were found (Maquet *et al.*, 1990).

The language impairment cannot be viewed as merely a consequence of a general cognitive deficit, since nonverbal

The preserved ability to learn written language and sign language suggests that higher-order language areas are mostly preserved and input from the visual route is preserved. One can thus safely assume that the core deficit is characterized by a high level auditory processing disorder with the effect of blocking the auditory path to the language area and the process of language acquisition in a child whose language is not yet fully developed.

According to *interference theory* the abnormal electrical brain activity disrupts the cortical (mostly temporal) language areas that mediate the processing of phonetic and phonological aspects of language. It is thus not surprising to find a selective impairment processing the morphosyntactic components of language irrespective of it being spoken, written, or sign language (see Bishop, 1982 and the previously described case): some form of phonological processing is in fact required in processing function words and, more generally, treating the morphosyntactic elements of language (Caramazza *et al.*, 1981).

Experimental support for the interference hypothesis seems to come from patients with LK syndrome who have shown a dramatic improvement or even total recovery of language following the surgical elimination of the epileptogenic discharge from the perisylvian region of the affected hemisphere (Morrell *et al.*, 1995; Grote *et al.*, 1999). We must, however, be reminded that in a consistent percentage of LK children, the language impairment, either in terms of regression of lack of development, persists well after the resolution of the EEG abnormalities. It has, therefore, been postulated that the neural development is impaired by the epileptiform discharges (Morrell *et al.*, 1995). More specifically, the superfluous neurons or synapses that are usually eliminated during the subtractive phase in neurodevelopment are sustained by the abnormal electrical activity resulting in atypical or inefficient pattern connectivity. While this hypothesis could nicely fit the lack of development of language in the early cases of LK syndrome, it is more problematic when applied to the case of children who manifest the syndrome at an age when language is fully developed.

35.4. THE SPECTRUM OF REGRESSIVE LANGUAGE DISORDERS AND EPILEPSY

LK syndrome is not the only developmental disorder combining EEG abnormalities and language regression. In a review of the syndrome, Stefanatos *et al.* (2002) list a number of developmental neurocognitive epileptiform disorders that are associated with speech and language deterioration (see Table 35.1). It is therefore not surprising to find discrepancies in the definitions of the clinical characteristics and nosological boundaries between LK and other related disorders. Among the latter, of particular relevance is autistic regression, that is, a disorder surfacing in the first or second year of life after normal development and characterized by a loss of social and emotional reciprocity, language and non verbal communicative disorders, and stereotyped behavior. The etiology is still unknown, neither metabolic nor structural abnormalities have been detected.

Autistic regression is part of the autistic spectrum disorder, a broad category which includes the various types of pervasive developmental disorders according to the Diagnostic and Statistic Manual of Mental Disorders (American Psychiatric Association, 1994). Abnormal EEG findings, with or without epileptic seizures, have been documented in a consistent minority of children affected by autistic regression. Given the number of shared characteristics of the two syndromes, a quest for defining the differential diagnostic criteria appears particularly difficult. In a few of the large scale studies on the autistic spectrum disorders, Kurita *et al.* (1992) found that in general language and speech were lost before the age of 30 months in autistic regression, while in LK children the onset is usually at 3–4 years of age, when language has already developed significantly. Another distinguishing criterion is found in the topographic abnormalities of the EEG: while in LK syndrome the epileptiform activity is mostly centered in the language-related temporal cortices, in autistic regression the abnormalities are widespread, affecting the entire cerebral cortex. Furthermore, nonverbal cognitive abilities are usually spared in LK syndrome while there is a general decline in autistic regression and autistic spectrum disorders. Finally, from a behavioral point of view, of paramount importance is that in most LK children the display of affection is preserved (Landau & Kleffner, 1957), although some of them can display persistent autistic traits (Rossi *et al.*, 1999).

35.5. CHALLENGES AND FUTURE DIRECTIONS

Like most neuropsychological disorders LK syndrome is "an experiment of nature": the pathological process, characterized by a central auditory processing disorder in infancy, disrupts the main input and output speech modality, thus allowing researchers to explore how alternative routes to the use of spoken language can develop and to gain important insights on the role of phonology (or lack of it) for the development of different cognitive and language components, such as morphosyntax and short term memory.

From a nosological point of view, 50 years after its recognition, many aspects of the syndrome are still unresolved. Is LK syndrome an autonomous entity as described originally? Or, as suggested by many convergent studies, is it part of a broader pathological spectrum (Stefanatos *et al.*, 2002; Canitano & Zappella, 2006)? For example, would processes sustained by abnormal brain electrical activity and confined to language cortices lead to LK syndrome? Or, if spreading to strictly connected adjacent brain structures such as limbic

TABLE 35.1   Summary of Epileptiform Neurodegenerative Disorders that are Linked with
Speech/Language Deterioration in Children

Epileptiform neurodegenerative disorders	Receptive language	Expressive language/speech	Cognitive	Social	Behavior	Reference
Autistic regression (with epileptiform EEG)	++(+)+	+++	++(+)	++(+)	++(+)	Nass and Devinsky (1999)
Childhood disintegrative disorder (with epileptiform EEG)	+++	+++	+++	++(+)	++(+)	Kurita et al. (1992)
Continuous spike and wave during slow wave sleep syndrome	+++	+++	+++	++(+)	++(+)	De Negri (1997)
Acquired epileptiform aphasia	+++	++(+)	(+)	(+)	+(+)	Landau and Kleffner (1957)
Congenital aphasia (with epileptiform EEG)	+(+)	++(+)	(+)	(+)	(+)	McKinney and McGreal (1974)
Acquired opercular syndrome	−	+++	(+)	?	?	Shafrir and Prensky (1995)
Atypical benign partial epilepsy	(+)	++	(+)	(+)	(+)	Hahn et al. (2001)

Note: Extent of impairment: +++severe, ++moderate, +mild, −normal to borderline, (+)variable, ?insufficient information. The severity ratings are estimates used here for illustrative purpose only to convey relative similarities and divergences between conditions. These conditions are extremely variable so there exception to the depicted patterns. References provide relevant case material. *Source*: Stefanatos, G.A., Kinsbourne, M., & Wasserstein J. (2002). Acquired epileptiform aphasia: A dimensional view of Landau–Kleffner syndrome and the relation to regressive autistic spectrum disorders. *Child Neuropsychology, 8*(3), 195–228.

cortices, temporal, frontal and subcortical areas, would it give rise to severe cognitive, behavioral and psychotic traits and include patterns such as autistic regression?

Equally unresolved is the problem of its etiology. In the absence of structural and metabolic abnormalities the hypothesis of an autoimmune disorder as the basis of LK syndrome, autistic regression and autistic spectrum disorder seems to find some support although no conclusive data are available (Stefanatos *et al.*, 2002).

Finally, from a linguistic point of view, further studies of the syndrome could shed light on some unresolved problems. For example, what is the difference in the language development of congenital peripheral deafness and central auditory processing impairment occurring after early language acquisition, where a "trace" of phonological mediation could be preserved?

Acknowledgment

Thanks to Dr. Maria Rosser, Coventry University, for her kind help in the improvement of the English text.

References

American Psychiatric Association (1994). Pervasive developmental disorders. In *Diagnostic and statistic manual of mental disorders* (4th edn., pp. 65–98). Washington DC: American Psychiatric Association.

Basso, A., & Scarpa, M.T. (1990). Traumatic aphasia in children and adults: A comparison of clinical features and evolution. *Cortex, 26*(4), 501–514.

Bishop, D.V. (1982). Comprehension of spoken, written and signed sentences in childhood language disorders. *Journal of Child Psychology and Psychiatry, 23*(1), 1–20.

Bishop D.V. (1985). Age of onset and outcome in acquired aphasia with convulsive disorders (Landau-Kleffner syndrome). *Developmental Medicine and Child Neurology, 27*, 705–712.

Canitano, R., & Zappella, M. (2006). Autistic epileptiform regression. *Functional Neurology, 21*(2), 97–101.

Caramazza, A., Berndt, R.S., Basili, A.G., & Koller, J.J. (1981). Syntactic processing deficits in aphasia. *Cortex, 17*(3), 333–348.

De Negri, M. (1997). Electrical status epilepticus during sleep (ESES). Different clinical syndromes: Towards a unifying view? *Brain and Development, 19*(7), 447–451.

De Renzi, E., & Vignolo, L.A. (1962). The Token Test: A sensitive test to detect receptive disturbances in aphasics. *Brain, 85*, 665–678.

Denes, G., Balliello, S., Volterra, V., & Pellegrini, A. (1986). Oral and written language in a case of childhood phonemic deafness. *Brain and Language, 29*(2), 252–267.

Grizzle, K.L., & Simms, M.D. (2005). Early language development and language learning disabilities. *Pediatric Review, 26*(8), 274–283.

Grote, C.L., van Slyke, P., & Hoeppner, J.A. (1999). Language outcome following multiple subpial transection for Landau–Kleffner syndrome. *Brain, 122*(3), 561–566.

Hahn, A., Pistohl, J., Neubauer, B.A., & Stephani, U. (2001). Atypical "benign" partial epilepsy or pseudo-Lennox syndrome Part I: Symptomatology and long-term prognosis. *Neuropediatrics, 32*(1), 8–15.

Kurita, H., Kita, M., & Miyake, Y. (1992). A comparative study of development and symptoms among disintegrative psychosis and infantile autism with and without speech loss. *Journal of Autism Developmental Disorders*, 22(2), 175–188.

Landau, W.M., & Kleffner, F. (1957). Syndrome of acquired aphasia with convulsive disorder in children. *Neurology*, 7(8), 523–530.

Lerman, P., Lerman-Sagie, T., & Kivity, S. (1991). Effect of early corticosteroid therapy for Landau–Kleffner syndrome. *Developmental Medicine and Child Neurology*, 33(3), 257–260.

Mantovani, J.F., & Landau, W.M. (1980). Acquired aphasia with convulsive disorder: Course and prognosis. *Neurology*, 30(5), 524–529.

Maquet, P., Hirsch, E., Dive, D., Salmon, E., Marescaux, C., & Franck, G. (1990). Cerebral glucose utilization during sleep in Landau–Kleffner syndrome: A PET study. *Epilepsia*, 31(6), 778–783.

Marien, O., Saerens, J., Versleger, W., Borrgreve, F., & De Deyn, P.P. (1993). Some controversies about type and nature of aphasic symptomatology in Landau–Kleffner's syndrome: A case study. *Acta Neurologica Belgica*, 93, 183–203.

McKinney, W., & McGreal, D.A. (1974). An aphasic syndrome in children. *Canadian Medical Association Journal*, 110(6), 637–639.

Morrell, F., Whisler, W.W., Smith, M.C., Hoeppner, T.J., de Toledo-Morrell, L., Pierre-Louis, S.J., Kanner, A.M., Buelow, J.M., Ristanovic, R., Bergen, D. *et al.* (1995). Landau–Kleffner syndrome. Treatment with subpial intracortical transection. *Brain*, 118(6), 1529–1546.

Nass, R., & Devinsky, O. (1999). Autistic regression with rolandic spikes. *Neuropsychiatry Neuropsychology and Behavioral Neurology*, 12(3), 193–197.

Paquier, P.F., van Dongen, H.R., & Loonen, C.B. (1992). The Landau–Kleffner syndrome or "acquired aphasia with convulsive disorder". Long-term follow-up of six children and a review of the recent literature. *Archives of Neurology*, 49(4), 354–359.

Papagno, C., & Basso, A. (1993). Impairment of written language and mathematical skills in a case of Landau–Kleffner syndrome. *Aphasiology*, 7, 451–462.

Perez, E.R., & Davidoff, V. (2001). Sign language in childhood epileptic aphasia (Landau–Kleffner syndrome). *Developmental Medicine and Child Neurology*, 43(11), 739–744.

Rapin, I., Mattis, S., Rowan, A.J., & Golden, G.G. (1977). Verbal auditory agnosia in children. *Developmental Medicine and Child Neurology*, 19(2), 197–207.

Rossi, P.G., Parmeggiani, A., Posar, A., Scaduto, M.C., Chiodo, S., & Vatti, G. (1999). Landau–Kleffner syndrome (LKS): Long-term follow-up and links with electrical status epilepticus during sleep (ESES). *Brain Development*, 21(2), 90–98.

Shafrir, Y., & Prensky, A.L. (1995). Acquired epileptiform opercular syndrome: A second case report, review of the literature, and comparison to the Landau–Kleffner syndrome. *Epilepsia*, 36(10), 1050–1057.

Stefanatos, G.A., Kinsbourne, M., & Wasserstein, J. (2002). Acquired epileptiform aphasia: A dimensional view of Landau–Kleffner syndrome and the relation to regressive autistic spectrum disorders. *Child Neuropsychology*, 8(3), 195–228.

Tassinari, C.A., Michelucci, R., Forti, A., Salvi, F., Plasmati, R., Rubboli, G., Bureau, M., Dalla Bernardina, B., & Roger, J. (1992). The electrical status epilepticus syndrome. *Epilepsy Research Supplement*, 6, 111–115.

Van Hout, A. (1992). Acquired aphasia in children. In S.J. Segalowitz & I. Rapin (Eds.), *Handbook of neuropsychology: Child neuropsychology* (Vol. 7, pp. 139–161). Amsterdam: Elsevier.

Van de Sandt-Koenderman, W.M., Smit, I.A., van Dongen, H.R., & van Hest, J.B. (1984). A case of acquired aphasia and convulsive disorder: Some linguistic aspects of recovery and breakdown. *Brain and Language*, 21(1), 174–183.

Vargha-Khadem, F., Gadian, D.G., Copp, A., & Mishkin, M. (2005). FOXP2 and the neuroanatomy of speech and language. *Nature Review Neuroscience*, 6(2), 131–138.

Zivi, A., Broussaud, G., Daymas, S., Hazard, J., & Sicard, C. (1990). Epilepsy-acquired aphasia syndrome with psychosis. Report of a case. *Annales Pediatrie*, 37(6), 391–394.

Further Readings

Marcus, G., & Rabagliati., H. (2006). What developmental disorders can tell us about the nature and origins of language. *Nature Neuroscience*, 9(10), 1226–1229.
The study of developmental disorders is well placed to provide important new clues on clarifying the problem of language evolution, but has been hampered by a lack of consensus on the aims and interpretation of the research project .In this provocative article it is suggested that the application of the Darwinian principle of "descent with modification" can help to reconcile much apparently inconsistent data.

Stefanatos, G.A., Kinsbourne, M., & Wasserstein, J. (2002). Acquired epileptiform aphasia: A dimensional view of Landau–Kleffner syndrome and the relation to regressive autistic spectrum disorders. *Child Neuropsychology*, 8(3), 195–228.
In this review article the authors argue that LK syndrome can be conceptualized on a spectrum with other epileptiform neurocognitive disorders that may share pathophysiological features. The implications of this viewpoint are discussed, with emphasis on parallels between LK and regressive autistic spectrum disorders.

Watkins, K.E., Dronkers, N.F., & Vargha-Khadem, F. (2002). Behavioural analysis of an inherited speech and language disorder: Comparison with acquired aphasia. *Brain*, 125(3), 452–464.
Genetic speech and language disorders provide the opportunity to investigate the biological bases of language and its development. In this article the authors report the investigations of the KE family, half the members of which are affected by a severe disorder of speech and language, which is transmitted as an autosomal-dominant monogenic trait. It is suggested that the verbal and nonverbal deficits arise from a common impairment in the ability to sequence movement or in procedural learning. Alternatively, the articulation deficit, which itself might give rise to a host of other language deficits, is separate from a more general verbal and nonverbal developmental delay.

36

Language and Communication in Williams Syndrome

MAYADA ELSABBAGH

*Centre for Brain and Cognitive Development, Birkbeck,
University of London, London, UK*

ABSTRACT

The neurodevelopmental disorder Williams Syndrome has sparked much debate in the scientific community. The cognitive profile in this genetic condition is usually described as strikingly uneven, where language is relatively proficient but visuospatial processing is impaired. Some argue that this profile demonstrates the independence of language from the rest of cognition, supporting a genetically prespecified language faculty. Others have challenged this view, pointing to the contribution of interactions between genes and the environment over development. Instead of describing the end products of development in terms of patterns of proficiency and impairment in higher-level domains, the study of Williams Syndrome can reveal how these complex interactions unfold over time giving rise to the phenotype characteristic of this population. Explaining this phenotype is achieved through tracing domain-relevant mechanisms of developmental change and understanding the alternative constraints which govern atypical development in condition.

36.1. INTRODUCTION

In the 1960s, clinical descriptions of the rare genetic condition Williams Syndrome (WS) captured the attention of geneticists, linguists, and cognitive scientists alike. Initial excitement about this syndrome was motivated by the inference that the behavioral profile in this syndrome demonstrates the independence of language from general cognition: despite having severe mental retardation, individuals with WS seemed to exhibit a surprising command of language. Since these initial descriptions, research on WS has mainly focused on describing the end products of development in terms of patterns of proficiency and impairment in higher-level domains. Frequently, the cognitive profile in this syndrome continues to be broadly described as strikingly uneven, where language, music, and face processing are areas of strength but visuospatial processing is an area of weakness (e.g., Bellugi *et al.*, 2000). The claim is that the cognitive pattern in WS provides evidence for a modular architecture of the human mind, where different domains can be fractionated. Some have even argued that this profile supports a genetically prespecified language faculty (Pinker, 1999). The underlying assumption of such theoretical models is that developmental disorders provide us with a unique opportunity to directly link genes, the brain, and behavioral outcomes.

Developmentalists have challenged the view that developmental disorders, including WS, support genetically prespecified modules. Instead, they argue that explaining the behavioral profile observed in this population, and more generally in typical and atypical development, must go beyond describing inter- or intra-domain dissociations. Rather, it must be done through examining behavior over developmental time and investigating more basic processes which might underlie performance in cognitive domains (Elman *et al.*, 1996; Karmiloff-Smith, 1998). This vision, notwithstanding considerable debate, paved the way for original and complex questions regarding the *developmental process* that links genes to behavioral outcomes. With the accumulation of evidence on the cognitive, behavioral, and neural profile in WS, and in view of parallel advances in research on typical development, we can begin to integrate key findings from different research areas into a coherent model of why and how language emerges as a relative strength in WS.

Box 36.1   Genetic diagnosis and individual variability

The issue of individual differences in neural and cognitive characteristics has been a serious concern in research on developmental disorders such as autism and specific language impairment. Interestingly, issues of population heterogeneity have not been addressed in great detail in research on WS. In fact, rather strong theoretical claims have been made with groups as small as three or four individuals. Why has the discussion of heterogeneity in WS been lacking? This is in part due to the fact that while other disorders are diagnosed on the basis of behavioral criteria, WS has in recent years been diagnosed using genetic criteria. In WS, some 28 genes are missing from one copy of chromosome 7 (Donnai & Karmiloff-Smith, 2000). When confirming a diagnosis of WS, the fluorescence *in situ* hybridization (FISH) test is used to probe for the deletion of the elastin gene which is in the WS critical region. But, does genetic diagnosis provide a golden standard guaranteeing population homogeneity? The answer to this is clearly no. While several physical and physiological characteristics are associated with WS, including specific facial features and heart abnormalities, there is substantial variability in the expression of these symptoms. For instance, cardiac abnormalities appear to be expressed on a continuum

of severity where, in some individuals, they are hardly apparent but in others they cause frequent hospitalization and in some cases are fatal. Moreover, neuroanatomical studies of the WS brain constituently report substantial variability among individuals, which surpasses that found in typical individuals. In view of this variability in physical and physiological characteristics, it is no surprise that what has been traditionally described as the WS characteristic cognitive phenotype is also exhibited with substantial variability (e.g., Porter & Coltheart, 2005). Although its origins are poorly understood, this variability most likely reflects probabilistic interactions between genes and the environment in producing behavioral outcomes (Elman *et al.*, 1996).

Donnai, D., & Karmiloff-Smith, A. (2000). Williams syndrome: From genotype through to the cognitive phenotype. *American Journal of Medical Genetics: Seminars in Medical Genetics, 97*, 164–171.
Elman, J.L., Bates, E., Johnson, M.H., Karmiloff-Smith, A., Parisi, D., & Plunkett, K. (1996). *Rethinking innateness: A connectionist perspective on development.* Cambridge, MA: MIT Press.
Porter, M.A., & Coltheart, M. (2005). Cognitive heterogeneity in Williams syndrome. *Developmental Neuropsychology, 27*, 275–306.

36.2. LANGUAGE IN WS: THE EVIDENCE

Before discussing theoretical models, it is essential to begin with a "theory-neutral" view of the cognitive profile in this developmental disorder. This is particularly important given that descriptions of the cognitive profile in WS can at times be confusing or even paradoxical. Readers of the literature are guaranteed to encounter euphemisms frequently used to describe the cognitive profile. These include descriptions of "intact versus impaired" abilities, striking patterns of "strengths and weakness," or "peaks and valleys" in performance. Others prefer to make more qualified claims of "relative proficiency" of some domains over others. And yet, several lines of evidence also refer to a more complex picture, occasionally reporting serious difficulties in areas that are generally described as proficient. So what do researchers actually mean when they draw these conclusions? Do these generalizations hold in the face of experimental evidence? While fine-grained research on language in WS is still lacking or un-replicated, the answer appears to be less clear-cut than these sweeping claims suggest.

When encountering a person with WS, few would dispute that there is in fact a single feature that could be described as "striking." Many of these individuals will engage in long and at times formulaic conversations which nevertheless successfully capture the interest and attention of their interlocutor. For instance, after a brief casual conversation with a young woman with WS, she enthusi-

astically expressed how much she and I had in common, *We're both wearing a black dress, you like chocolate ice-cream and I like chocolate ice-cream, you study Williams Syndrome and I have Williams Syndrome!*

Interestingly, this verbal fluency might mask their difficulties in several sub-domains of language. The crucial questions are whether the language produced or understood (a) is consistent with the individual's mental age, that is, it is delayed but not different from the rest of cognition, (b) surpasses what one might expect on the basis of overall mental age, that is, whether there is indeed a dissociation between language and general cognition, or (c) is altogether atypical; that is, both delayed and deviant. As a general rule, language in WS appears to be reasonably predicted on the basis of overall mental age (Brock, 2007) and performance within individuals across various domains is highly correlated (Mervis *et al.*, 1999). Furthermore, language profiles in WS are significantly variable across individuals (see Box 36.1). These considerations suggest that language is strictly speaking not "special." However, there are important exceptions to this general rule. For instance, on tests of vocabulary comprehension, performance of individuals with WS surpasses what is expected based on their mental age. Moreover, their pragmatic skills appear to deviate from what is expected for their mental age: they tend to make greater use of devices which engage their interlocutor (Reilly *et al.*, 2004). These patterns suggest that proficiency in the language domain in WS depends, at least in part, on the specific task and the developmental level of the individual. Therefore,

Box 36.2 Building developmental trajectories

The goal of most studies of developmental disorders is to assess how the experimental group differs from a norm that is whether differences in performance between groups can be attributed to the disorder itself. Different comparison or "control" groups are employed to examine this. Control groups can be normal individuals or individuals with other disorders that are matched on chronological age or mental age. Traditionally, these matching procedures have been used to assess questions regarding delay or deviance in the behavior of the experimental group relative to the control group. However, these procedures are associated with certain limitations. For example, if matched on chronological age, the WS group would be at a potential disadvantage given their mental retardation, which would mask specific proficiencies or deficits. Furthermore, given the substantial heterogeneity in mental age in the WS population, matching on the basis of mean verbal or spatial MA would not reflect the variability between the groups, since it frequently results in groups which are more restricted in age range relative to the WS group. Because of the frequent disparity between verbal and performance IQ, matching on performance IQ typically yields a control group of very young children with far less world experience. To answer to these challenges, some researchers have stressed the importance of building task-specific *developmental trajectories* (e.g., Karmiloff-Smith, 1998). The idea is that assessing change in performance level over developmental time is more useful in elucidating atypical trajectories than snapshots captured at different moments in time and averaged over the group. Performance of the target group can then be compared against a trajectory of normal controls, in which patterns of developmental change and developmental stability emerge more clearly. Building developmental trajectories is, however, challenging. Some of the problems frequently encountered are the disparity in age ranges if the sample size is small and the difficulty in devising tasks that are usable for a wide range of ages.

Karmiloff-Smith, A. (1998). Development itself is the key to understanding developmental disorders. *Trends in Cognitive Sciences*, 2, 389–398.

what is special about language in WS is not so much their general command of the system, but instead a pattern of communication which is both delayed and atypical, albeit very impressive!

36.3. THEORETICAL APPROACHES TO THE NEUROCOGNITIVE STUDY OF WS

In view of the complexity of the behavioral patterns in WS, What then is the basis for claims of strength in language?

Such claims are usually based on task- and population-specific comparisons. These selective comparisons are consistent with the aims and assumptions of different theoretical approaches which in turn shape the questions and conclusions regarding what cases of genetic developmental disorders such as WS can reveal about the architecture of the mind/brain. Extending approaches of adult neuropsychology to cases of developmental disorders, some researchers have focused on seeking selective patterns of impairment and sparing, mostly in late childhood and adulthood. These patterns are used like *double dissociations* that can then reveal whether two skills are separable, not only functionally, but also developmentally, neuroanatomically, and genetically. In these models, the original claim of dissociation between language and general cognition was replaced by more sophisticated models contrasting performance both across and within domains. Let us consider some concrete examples. Performance of individuals with WS on vocabulary measures is impressive when contrasted with their severe difficulties on visuospatial tasks (Bellugi *et al.*, 2000). Furthermore, individuals with WS outperform individuals with other developmental disorders such as Down Syndrome or Specific Language Impairment (SLI) on measures of phonology and syntax. In more strict applications of modular analysis, SLI and WS have been taken as a doubly dissociated pair of the rule-based (e.g., regular past tense in English such as *work-worked*) versus associative memory language systems (e.g., irregular past tense such as *eat-ate*), implying genetically specified modules for these two sub-domains (Pinker, 1999). Even further refinements of dissociations have been made within sub-domains of language in WS (see Brock, 2007 for a detailed review). For instance, non-spatial language has been found to be more proficient relative to spatial language (e.g., *above the box, below the table*).

The extension of the double dissociations' logic to cases of genetic developmental disorders has a number of strong implications (Elsabbagh & Karmiloff-Smith, 2006) including that a *static* model can explain patterns of behavioral outcomes, deeming the developmental process itself to be a minor contributor. When these behavioral dissociations are mapped onto the genetic deletions as in WS, a further implication is that the brain's modular architecture is universal across different individuals and genetically prespecified.

How tenable are the assumptions of static models and prespecified architecture when applied to genetic developmental conditions? In typical development, infants' brains become progressively specialized and localized in a relatively specific manner that might suggest predetermined modularity. On the other hand, cases of acquired lesions in infancy and childhood pose a major problem for claims of predetermined modularity. Although reorganization post injury is not viewed as an important factor in adult cases, plasticity and reorganization are the rule rather than the exception in childhood cases (Johnson, 2005). Experience-dependent plasticity appears to be a hallmark of our species' brains, whether they develop typically or atypically. Instead of demonstrating a prespecified brain architecture, it seems more likely that the case of WS actually reveals that mapping between genes, the brain and behavioral outcomes is indirect and highly experience dependant.

Considering some of these links clearly illustrates this point. Genetic abnormalities in WS have widespread effects on brain architecture at multiple levels including gross neural anatomy, morphology, and cytoarchitectonics. Compared to normal individuals, the overall brain volume in WS is reduced to about 80% (Reiss *et al.*, 2000). Cortical thickness and complexity, that is, cortical folding and sulcal patterning, differ in individuals with WS relative to normal controls (Thompson *et al.*, 2005). Similarly, instead of specifying domain-specific knowledge, genes contribute to the formation of domain-relevant learning mechanisms, which are gradually refined by the ontogenetic process (Karmiloff-Smith, 1998). These genetic and neuroanatomical abnormalities contribute to the complex pattern of behavioral and cognitive profile in this neurodevelopmental disorder. In WS, these atypicalities are evident in language development from outset, as will be discussed next. In adulthood, different neural correlates from those found in typical circumstances are involved in processing of syntax and semantics in WS (Mills *et al.*, 2000).

In sum, although traditionally WS has been presented a case for "intact" and "impaired" modules across various cognitive domains, several theoretical and empirical arguments have challenged this view. This emerging picture has gradually changed the view of WS from a case of independence of different domains, to one in which complex genetic and environmental interactions give rise over developmental time to uneven behavioral profiles in the phenotypic outcome, via alternative trajectories (Karmiloff-Smith, 1998). But what precisely are these alternative trajectories? What alterations in perception or in processing of input drive these trajectories and what constrains them? These are some of the questions addressed next.

36.4. CHARACTERIZING LANGUAGE AND COMMUNICATION IN WS

Models of typical and atypical development provide a more complex view of the emergence of language as the product of interaction of several domain-relevant mechanisms over developmental time (Elman *et al.*, 1996). Although the study of genetic developmental disorders is a promising direction for enriching our understanding of genotype–phenotype relations, this depends on the ability of researchers to capture and understand the developmental mechanisms relating genes to final outcomes (Karmiloff-Smith, 1998). At first blush, these propositions of the developmental

framework are certainly appealing. However, they do raise a multitude of difficult questions: What exactly is the nature of these domain-relevant developmental mechanisms? Do they rely on general principles common to all domains? How do they respond to the specific computational problems posed by each domain? Exactly how and when do these domain-relevant mechanisms give rise to the specialized and complex adult brain?

In the study of typical development, a variety of perceptual and social mechanisms have been proposed as important precursors to language. However, research on developmental disorders in general and on WS in particular, still lacks the questions which address developmental processes themselves rather than their final products in terms of patterns of proficiency or impairment. Hence, it is essential to integrate some of these advances from research on typical development in order to understand how language competence is achieved in WS.

In what follows, a number of domain-relevant processes will be discussed in relation to language in WS. The underlying assumption is that if domain-relevant processes provide the basis for behavioral outcomes over developmental time, understanding how individuals with developmental disorders perceive and organize input, and what alternative constraints govern this organization would clarify the *process* by which higher-level outcomes are achieved. Atypical strategies in organizing environmental input are likely to shape the emerging cognitive and behavioral profiles in WS. Hence, elucidating alterations in these domain-relevant mechanisms would make it possible to understand what gives rise to the uneven behavioral profile observed in this clinical group. The domain-relevant mechanisms described next include alterations in basic sensory and perceptual inputs (hearing and hyperacusis, auditory processing), atypical strategies in organizing input and forming generalizations (statistical learning), alterations in the environmental context of language acquisition (social communication), and finally, interactions within the language system.

36.4.1. Hearing and Hyperacusis

WS has presented a paradox to the scientist: whereas in adulthood language stands out as proficient relative to visuospatial skills, in infancy and toddlerhood WS language is seriously delayed. Yet young children with WS are fascinated by sound, while at the same time often suffering from hyperacusis. Hypersensitivity to sound is one of the most prevalent features of WS, affecting over 90% of individuals in this clinical population. Broadly termed *hyperacusis*, it encompasses a number of reported symptoms including loudness discomfort or fear of certain sounds, and to a lesser extent, fascination by sound or reduced hearing thresholds (Levitin *et al.*, 2005). Hyperacusis is very rarely seen in the general population and is usually linked to hearing loss. However, in WS the link between hyperacusis and hearing loss is unclear. By contrast, the effects of hyperacusis in WS appear to be modulated by factors like mood of the individual, tiredness, anxiety, and familiarity and expectancy. In view of the fact that hyperacusis seems to be context dependent and linked to higher-order emotion functions, it is unclear what effects of this symptom might have on auditory functioning, in general, and on language, in particular. It is possible that hyperacusis manifests itself in more complex ways than originally thought, alongside other non-auditory factors. It is possible that higher-order executive functions (e.g., attention) and emotions actually modulate the influence of perceptual sensitivity and hyperacusis, where the interaction of these factors exerts an influence on speech perception.

36.4.2. Auditory Processing

In the empirical literature, a number of proposals generally termed low-level deficit hypotheses, have investigated the role of perceptual mechanisms in the auditory domain as contributing factors in developmental disorders. For instance, impairments in rapid auditory processing have been linked to disorders of language such as in SLI and developmental dyslexia. Could the same auditory processing mechanisms deemed problematic in cases where language is an area of difficulty, such as in SLI and dyslexia, be relevant in explaining why language is a strength in WS?

In research on language impairment, some have rejected any causal links between problems in low-level auditory processing and language given that these impairments are not found in all or in most cases. Instead it is argued that auditory problems have an impact on the discriminability of input, leading to confusion among sounds without necessarily predicting specific problems in speech processing or in language (McArthur & Bishop, 2004). Consistent with previous claims (Elsabbagh & Karmiloff-Smith, 2006), the influence of auditory skills cannot be ruled out simply because differences in perceptual processing are not found in all or some participants mostly tested in late childhood and adulthood. This is because differences in perceptual processing need not exist throughout the entire life span. Crucially, their presence early in development, that is, in infancy and early childhood, may trigger cascading effects, the indirect results of which are found later in development. Moreover, even if not taken as singular explanatory factors, low-level perceptual skills may play an important role in altering the experience of the child in the environment, providing a possible mechanism by which developmental outcomes are reached.

Remarkably, there is very little work on auditory processing in WS. A couple of studies of brain neuroanatomy and neurophysiology examined the neural correlates of auditory skills. In the auditory cortex, abnormalities are found in neuronal size and cell packing density, coupled with less asymmetry than that observed in the typical brain (Holinger et al., 2005). These patterns are thought to reflect increased connectivity in WS auditory cortex, leading the authors to claim "relative sparing of language, music, and other auditory function" (Holinger et al., 2005). In another study using functional brain imaging when listening to music versus noise, patterns of activation were less distinct in individuals with WS than in controls, showing more variable and diffuse patterns of cortical activation, together with abnormally high levels of activation in the amygdala and cerebellum (Levitin et al., 2003). Neuroanatomical findings, however, remain limited by the very small sample sizes of 3–5 individuals in each study (but see Thompson et al., 2005). This is particularly problematic given that these studies usually report great variability within the WS groups (see Box 36.1). More importantly, attempting to relate behavioral profiles to brain morphometry is of limited use in the absence of rigorous behavioral data concerning central auditory function in WS.

Perhaps this important area of research has been overlooked because of claims that the domains for which auditory skills are relevant, that is, language and music, are preserved in WS. These low-level auditory abilities have been assumed to be normal (e.g., Holinger et al., 2005). However, claims of intactness of language in WS have been seriously challenged on theoretical and empirical grounds. Could assumptions of intactness of the auditory domain in WS turn out to be as flawed as assumptions of intactness of language and face processing in this population? Could fine-grained methods reveal subtle abnormalities in auditory processing as well? Testing the validity of the assumption of intact auditory skills is particularly important in view of the puzzling alterations in the auditory perceptual experience in WS, where some individuals with WS experience discomfort to sound and yet are fascinated by it. This would have important implications for language in WS.

36.4.3. Statistical Learning

While different domains vary substantially in the nature of their input and their specific computational goals, for example, recognize a face or parse a syntactic structure, in each case the brain has to form coherent percepts through discovering perceptual units and discerning patterns in the input. Some studies have investigated infants' abilities to exploit rich cues in natural language such as prosodic and rhythmic characteristics to learn about the units and the structure of their mother tongue (Jusczyk, 1999). Increasingly, the focus in the developmental literature has shifted toward understanding the remarkable statistical learning capacities both infants and adults employ in discovering patterns and regularities not only in linguistic input but also in the general auditory and visuospatial domains

(Newport & Aslin, 2000). We still have little understanding of how processes which structure or make use of structured input are used in the development and processing of higher-level domains such as language, and importantly of the constraints on this type of learning. However, these processes may well play an important role.

Although statistical learning has not been formally tested in WS, some findings have important implications in this regard. Several studies of WS have documented atypical strategies in processing various stimuli. When processing unfamiliar melodies, individuals with WS do not benefit from the same cues that are used by typically developing individuals. More specifically, they are able to use pitch cues, which rely on tracking the absolute value of individual notes but not contour cues, which rely on tracking the pattern of rise and fall of the notes (Deruelle et al., 2005). Hence, individuals with WS exhibit differences in their sensitivity to auditory pitch and contour cues.

How do these findings relate to language in WS? The role of such strategies used in organizing auditory input, for example, pitch versus contour cues, has been discussed in detail in several developmental models concerned with language acquisition and on-line processing. These models emphasize that while a modular organization of language may be plausible in adulthood, shared developmental mechanisms are very likely to come into play, including mechanisms of auditory pattern analysis (McMullen & Saffran, 2004). Prosodic contours provide infants with rich cues marking syntactic boundaries since these contours correlate probabilistically with syntactic structures in language. If infants with WS turn out to be less sensitive to contour cues, this is likely to contribute to the significant delay in language acquisition in this population. It is also consistent with findings specifically related to infants' segmentation of the speech stream revealing serious delay in toddlers with WS relative to typical controls (Nazzi et al., 2003). Perhaps in the clinical group a lack of sensitivity to these prosodic contour cues requires a more extended period of exposure to linguistic input for the discovery of structure to get off the ground.

More specific predictions can also be tested. Consistent with some hypotheses in research on language processing in typical adults (Patel et al., 1998), these differences in sensitivity to pitch versus contour cues suggest that syntactic relations, which rely on local dependencies, should be proficient in WS. Syntactic relations, however, that rely on long-distance dependencies where the integration of relational information over time is required, would be affected. Although no study to date has directly examined these structures in either domain, partial support can be found in studies examining the neural correlates of atypical language processing. A study using event-related potentials (ERP) examined the on-line processing of sentences ending either in a probable way or in an anomalous way (e.g., *my fingers are in the moon*). Individuals with WS did not

show the typical left hemisphere asymmetry for grammatical words which indexes the integration of the elements of the sentence nor did they show the typical processing difference between content words and structure words (Mills et al., 2000). These findings are also consistent with more general patterns of linguistic processing in this population where, although individuals with WS appear to master quite a few formal rules of phonology and syntax, their representations are nevertheless fragile, and impairments are often masked by their superficial fluency (Thomas et al., 2001; Grant et al., 2002).

Another intriguing pattern of results is that differences in sensitivity to pitch and contour information in the auditory domain are remarkably similar to findings *across* domains in WS. For instance, individuals with WS are better local than global processors as shown by paradigms of visuospatial processing of hierarchical stimuli (Deruelle et al., 1999). They rely more on featural properties than configural properties when processing faces (Deruelle et al., 1999; Karmiloff-Smith et al., 2004). This raises the possibility that similar underlying strategies have different implications for the emergence of higher-level abilities such as language and visuospatial processing. Further investigation of these cross-domain patterns can be incredibly revealing in relation to the role of shared underlying mechanisms across domains.

36.4.4. The Social Context

In addition to being a formal and hierarchically structured system, language is also a tool for communication. In infancy, sharing eye gaze, establishing joint attention with adults, using gestures, and coordinating all of these behaviors along with vocalizing, allow infants to make impressive progress in language learning in a relatively short period of time. For instance, infants acquire various object names through following the eye gaze or pointing of the caregiver who then names the object. This seemingly inborn desire to communicate, coupled with a rich set of cues available from communication-rich environments makes important contributions to language development.

Alterations in the communicative context appear to contribute to language acquisition and processing in this population. One of the more striking characteristics about people with WS is their profound interest in other people. They tend to be overly friendly and they are frequently able to accurately infer others' emotions and intentions (Tager-Flusberg & Sullivan, 2000), even though this friendliness is at times expressed inappropriately toward strangers as much as friends. Similar manifestations of this over sociability are apparent early on, where infants with WS appear to be fascinated by faces and voices (Jones et al., 2000).

This excessive sociability obviously provides the drive to communicate and interact with others. However, several lines of evidence have indicated that sociability in WS does not reflect typical patterns of social communication. For instance, performance of children and adults with WS on formal tests of theory of mind, including abilities such as understanding intentions and visual-perspective taking, is weak (Tager-Flusberg & Sullivan, 2000). Moreover, toddlers with WS also show deficits in shared attention, where they tend to follow the experimenter's pointing gestures less than typically developing toddlers (Laing et al., 2002). Interestingly, their deficits result from excessive interest in looking at the adult without shifting attention to the objects. The latter result is somewhat remarkable given that shared attention is one of the important contributors to vocabulary development in typical circumstances. Despite these early deficits in such mechanisms and delays in language acquisition, older children and adults with WS eventually catch up and their vocabulary surpasses the level predicted on the basis of their mental age. This suggests that they arrive at alternative solutions, albeit less efficient ones which take longer for the acquisition process to get off the ground. Hence, in the context of language acquisition, it seems that many communicative behaviors may reflect superficial understanding of the functional relevance of these communicative acts, giving rise to a qualitatively different process of acquisition.

36.4.5. Interaction among Language Systems

Thus far, the arguments presented rely on the assumption that language development is a dynamical system, which emerges as a result of interactions at multiple levels (Elman et al., 1996). Not only does this include interactions between language and other systems, but also interactions within the system itself. Several theoretical models of language acquisition have been concerned with the extents to which development of a given linguistic domain relies on and is facilitated by development in other subsystems. For instance, one of these "bootstrapping" hypotheses have suggested that one route to the development of syntax relies on the use of semantics (Gleitman, 1990).

We still have a very limited understanding of the interactions among different language systems in WS. Small alterations in language acquisition early on are likely to give rise to different interactions among subsystems. Several examples presented have already demonstrated that language in WS is not only delayed but in many ways deviant, suggesting alteration within the language system itself. One concrete hypothesis that has been advanced is that language development in this clinical population takes an alternative pathway for the development of syntax, involving greater reliance on phonological compared to grammatical or semantic information (Klein & Mervis, 1999; Thomas & Karmiloff-Smith, 2003). Support for such claims came from

non-word repetition tasks, in which participants with WS were significantly less constrained than controls by the phonological resemblance of the non-sense words with meaningful words, which seemed to point to greater reliance on phonology than on semantics (Grant et al., 1997). An alternative possibility is that problems in phonology are subtle, particularly because individuals with WS are able to use contextual, possibly syntactic or semantic cues to resolve ambiguities in the speech stream, suggesting that compensation for difficulties is bidirectional between different aspects of language in this clinical group. More elaborate theoretical models are required, including more specific hypotheses about exactly which aspects of phonology are important, at what stage of development these are used, and exactly what role they play in compensating for difficulties in other areas like syntax.

36.5. LANGUAGE IN WS: CONNECTING THE DOTS

Generally speaking, finding from WS across domains have pointed to patterns of atypical behavioral and brain organization. Research in this area have begun to elucidate atypical developmental pathways in this population through targeting domain-relevant mechanisms, including alterations in sensitivity to environmental input and differences in reliance on strategies for organizing input. Good discriminability of auditory input in WS is likely to contribute to the well-developed phonological system and the good articulatory capacities observed in this population. Overall, this could provide some basis for their relative proficiency in language. However, the impact of alteration in perceptual sensitivity such as hyperacusis on language acquisition and processing remains to be clarified. The discovery of structure in language for individuals with WS is probably affected not only by the alterations in low-level auditory skills, but also by inefficient strategies used in parsing input and forming generalizations, some of which appear to operate across domains. These alterations, combined by an atypical social communicative context give rise to a qualitatively different process of language acquisition in this clinical group (Table 36.1).

The notion that domain-relevant mechanisms help to explain the emerging behavioral phenotype in WS is not to conjecture a direct link between these two levels. Such mechanisms are important, particularly in early development, in altering the experience and the perception of input, providing a possible mechanism by which developmental outcomes are reached. Hence, multiple and converging levels of domain-relevant factors need to be invoked to understand how these exert an impact on higher-level behavioral domains, in an interactive rather than an additive fashion.

Language and Communication Disorders in Autism and Asperger Syndrome

LUCA SURIAN[1] and MICHAEL SIEGAL[2,3]

[1]*Department of Cognitive Sciences and Education, University of Trento, Rovereto (TN), Italy*
[2]*Department of Psychology, University of Sheffield, Sheffield, UK*
[3]*Department of Psychology, University of Trieste, Trieste, Italy*

ABSTRACT

Research on language and communication in autism and Asperger syndrome has often revealed delays and deficits in the acquisition of phonology, lexical semantics, and grammar. However, the most persistent linguistic deficit concerns pragmatic skills, the ability to produce utterances that are contextually adequate and to interpret speakers' intended meanings. Recent experimental investigations have pointed to the possible origins of such deficits. These include difficulties in executive functioning, in reasoning about mental states, in integrating local details of a situation with global, contextual information, and in attending specifically to the voices and speech of others. A more comprehensive understanding of the determinants of language and communication deficits will require advances in the classification of subgroups of the heterogeneous range of individuals diagnosed with autistic spectrum disorders (ASD).

37.1. INTRODUCTION

Billy is a 10-year-old boy attending fourth grade primary school. He is an unusual child who continually attracts curiosity from adults and his peers because of his peculiar style of communication. During recess, he stands out immediately. While everybody else is playing, he stays alone in a corner. From time to time, he paces up and down often on tip-toe and waves his hands like butterfly wings. When classes resume, and the teacher speaks to the pupils, Billy interrupts suddenly out of the blue quoting lines emphatically and repetitively

from a film as if he were acting. When asked to do arithmetic, he continues to speak out nonsensically in a loud voice. When asked to read a story, he has difficulties in comprehension and in the identification of the main characters.

Billy's development has been highly atypical. At 4 years of age, he used a rich vocabulary that included words rarely used by children of his age. However, he preferred to remain isolated and to play alone rather than interacting with other children, sometimes even choosing to spend hours reading the telephone books. Billy also rejected changes to his usual rigid routine that he sought to follow meticulously for days and weeks at a time. For example, on Sundays, he insisted that his father takes him to a place where he could spot cars. On Mondays, he insisted on doing crossword puzzles and Tuesday would be the day to visit his grandparents. Billy's parents are sensitive and loving and understood very early on in his development that he is different from other children. They were alarmed and sought professional help for their child who was diagnosed as having high-functioning autism.

As illustrated by the case of Billy, autism is a developmental disorder diagnosed by observing characteristic abnormalities in social interaction, communication, and repertoires of interests that emerge early in the development and persist throughout life. The condition is sometimes referred to as autistic spectrum disorders (ASD) – a label that is meant to emphasize the continuity between the different nosological categories and between clinical and preclinical conditions (Frith, 2003). Asperger syndrome is a condition similar to autism in the severity of social and communication deficits, but is not associated with mental retardation and language delays. (Klin *et al.*, 2000).

Language and communication development have always been a major focus of investigations on autism and related disorders. From the very start of autism research, communication behavior has been a core source of diagnostic symptoms and a window through which to look for a better understanding of the underlying causal mechanisms. Although the social and communication disorders that characterize children with ASD often appear very early in development (Osterling et al., 2002), and can be detected reliably in the second year of life, the behavioral stereotypies and abnormalities in interests can emerge somewhat later and are shared with other disorders such as Tourette syndrome and obsessive-compulsive disorder (Sigman & McGovern, 2005).

Children with autism show severe disorders of both verbal and non-verbal communication, including a deficit in understanding and producing non-requestive pointing. They exhibit a tendency to neglect the eye region of the face of participants in a conversation, thus loosing crucial cues to interpret social interactions (Klin et al., 2002). The failure to orient spontaneously toward voices is included in commonly used diagnostic tools such as the Autism Diagnostic Observation Schedule (ADOS; Lord et al., 2000). Language development is strongly associated with other aspects of the autistic phenotype – the behavioral manifestation of autism – such as difficulties in imitation (Carpenter et al., 2005) and in reasoning about others' beliefs, intentions, and emotions. A substantial language delay in toddlers involving, for example, the complete absence of words at 16 months or of two-word utterances at 24 months, should prompt screening for autism.

Whereas researchers once focused on language abnormalities and their determinants as shown in the general population of children diagnosed with ASD, there now appears to be a shift of attention to account for the high within-group variability in autism (Tager-Flusberg, 2004). This methodological choice should yield a description of the autistic language phenotype in terms of relatively homogeneous subgroups that may be more easily mapped to differences in the genotype – the genetic endowment – of individuals with autism. More emphasis is also given to developmental changes that are observed in different areas of language (Sigman & McGovern, 2005). In some areas, many children show few problems and apparently catch up with the rest of the population; yet in other aspects of linguistic competence, persistent deficits occur even in the articulate and professionally successful adults who have autism or Asperger syndrome.

37.2. LANGUAGE DEFICITS IN AUTISM

In the following section, we highlight the main differences that have been reported between people with ASD

and normal controls on both formal aspects of language competence (e.g., grammar) as well as the pragmatic aspects of language usage.

37.2.1. Phonology, Morphology, and Syntax

Several studies have addressed the issue of whether and how the development of syntax and morphology in children with ASD differs from typical development. One conclusion is that, in this respect, they are similar to the mental age or IQ matched controls. Similar conclusions have also been reached by looking at phonological development, the acquisition of morphemes, and the comprehension of active and passive sentences (Tager-Flusberg, 2004).

By comparing children with autism with control groups of mentally retarded children matched on mental age, many of these studies have aimed to disentangle aspects of children's deficits that are likely to result from the autistic disorder from those that should be seen, following a more parsimonious explanation, as resulting from mental retardation. According to Rapin and Dunn (2003), previous studies showing intact phonology and syntax in autism have usually been based on selected samples. They reported that 144 individuals (63%) of their cohort of 299 children with autism showed mixed expressive/receptive disorders in which both syntax and phonology are deficient. A similar high proportion (59%) of autistic children showing phonologic–syntactic disorder have also been reported in another cohort of 197 autistic children. When a large, unselected sample is considered, most children with autism do tend to show difficulties. Nevertheless, Frith (2003) and Tager-Flusberg (2004) maintain that similarities in the development of grammatical morphemes and structures found in autism, Down syndrome and typical development indicate that children with autism are neither necessarily nor selectively impaired in formal aspects of language acquisition. When delays in grammatical development are found they are not disproportionate with respect to the associated mental retardation.

A second challenge to the "intact grammar view" has come from proposals for a close link between syntactic knowledge and the ability to attribute mental states. Sentences that involve descriptions of mental states such as beliefs express the propositional attitude held by an agent. These typically take the following form: Sally believes that the marble is in the box ($<$S believes that $p>$, where p is a proposition represented by a subordinate sentence that occupies the object complement slot). The syntax of sentential complements is necessary to covey in English verbal representations of propositional attitudes. It has been claimed that this particular aspect of syntactic knowledge that may be delayed in atypically developing children is the key to understanding why they often do not pass theory of mind (ToM) tasks that involve predicting how a story character with a false belief will be misled about the location of an object or the contents of a box (Schick *et al.*, 2007). The positive correlations found between performance on ToM tasks and on tests designed to assess competence on sentential complementation suggest a close link between the acquisition of ToM competence and the acquisition of this specific aspect of syntactic knowledge. However, the proposal that syntax in this sense underpins the emergence of ToM reasoning in both healthy and atypically developing children is in need of clear support (Harris *et al.*, 2005; Siegal & Peterson, in press). For one thing, many of the typically developing 3-year olds who fail ToM tasks are nevertheless able to produce and comprehend sentence complements that take the structure [person]-[pretends]-[that *x*] (e.g., "He pretends that his puppy is outside").

37.2.2. Lexical Semantics

Some studies have reported that the development of knowledge about word meaning in children with autism follows what one would predict on the basis of children's mental age (see Box 37.1). Yet other features of autistic children's language, such as the idiosyncratic use of words, suggest a somewhat different conclusion. How could autistic children have normal lexical semantics when they often use words in a manner that is meaningless to others? Recent empirical studies have also revealed peculiarities in the acquisition of word meanings that differ from the pattern shown by children without autism. This work has been inspired by the metarepresentational theory of autism,

---

Box 37.1   Cognitive models of autism

Three main cognitive models of autism have often driven investigations. These concern *theory of mind* (ToM) – including the ability to represent and reason about mental states and actions, especially the ability to read emotions from the eyes and to understand that others may have false beliefs that differ from reality (Figure (a)), *central coherence* – the ability to process stimuli in context and attend and integrate contextual specified "local" information with global or configurational information (Figure (b)), and *executive functioning* (Figure (c)). The latter is an umbrella term that includes working memory, inhibitory skills, control of attentional shifts, and other cognitive skills that are fundamental when an individual has to confront and adapt to new situations and problems for which he or she has no previously stored solutions. Research on language and communication in autism became a means for testing and refining these models. Language turned out to be a crucial source of evidence for testing some hypotheses about the core cognitive deficits in children with autism (Tager-Flusberg *et al.*, 2005; Siegal & Surian, 2007).

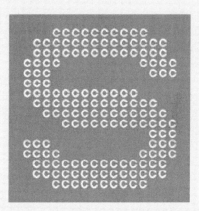

(b)   A test of "central coherence": in this "Navon figure" subjects have to report the information either at the global or at the local level. *Source*: From Dakin, S. & Frith, U. (2005). Vagaries of visual perception in autism. *Neuron*, 48, 497–507.

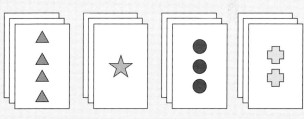

(c)   A test of "executive functioning": the Wisconsin Card Sorting Task that involves sorting cards by illustrated colors and shapes. (Note that in the figure the individual shapes are not illustrated in color.) (See Plate 22.)

Siegal, M., & Surian, L. (2007). Conversational understanding in young children. In E. Hoff & M. Shatz (Eds.), *Blackwell handbook of language development* (pp. 304–323). Oxford: Blackwell.

Tager-Flusberg, H., Paul, R., & Lord, C. (2005). Language and communication in autism. In F.R Volkmar, R. Paul, A. Klin & D. Cohen (Eds.), *Handbook of autism and pervasive developmental disorders*: *Diagnosis, development, neurobiology, and behavior* (Vol. 1, 3rd edn, pp. 335–364). Hoboken, NJ: John Wiley.

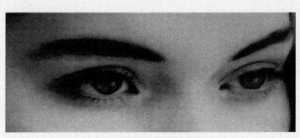

(a)   An advanced test of "theory of mind": subjects are asked to choose the correct adjective to describe pictures showing the ocular area of faces. *Source*: From Baron-Cohen, S., Wheelwright, S., Hill, J., Raste Y., & Plumb, I. (2001). The "reading in the eyes" test, revised version: A study with normal adults, and adults with Asperger syndrome or high-functioning autism. *Journal of Child Psychology and Psychiatry*, 42, 241–252.

namely the hypothesis that autism, at the cognitive level, is best explained by a ToM deficit, or, in other words, a selective difficulty in acquiring and using representations of mental states with propositional contents such as beliefs.

The straightforward prediction one could make from such theory regarding the acquisition of lexical semantics is that children with ASD should show difficulties in learning the meaning of psychological terms such as knowing and believing. Consistent with this prediction, children with ASD when asked to tell a story depicted in a series of drawings have often – but not always – been shown to produce mental states terms less frequently than do typically developing children or children with developmental delay. They also fail to differentiate properly between mental states verbs when they are asked to evaluate the logical implications of sentences that include them. For example, they often do not recognize that the sentence "he knows that p" implies the truth of p, but that the sentence "he believes that q" does implies neither that q is true, nor that it is false (Dennis *et al.*, 2002).

Children with autism also often fail to use speakers' gaze to interpret novel words, while typically developing infants apply this strategy at 18–24 months of age – a finding highlighted by recent research on the neural basis of gaze comparing children with autism with typically developing children (Box 37.2). Abnormalities in the activity of the occipital–temporal lobe of people with autism during eye direction processing have been recently reported in a study on event-related potentials (ERP) (Senju *et al.*, 2005). Occipito-temporal negativity was bilaterally distributed in children with autism whereas it was lateralized in the right hemisphere in the typically developing controls. Moreover the amplitude of this negativity was affected by eye direction

(averted versus direct) in typically developing children but not in children with autism.

37.2.3. Pragmatics

Pragmatics concerns knowledge of how language is used in communication. It involves the ability to extract meaning from utterances based on aspects of the communicative context that are relevant for correct interpretation. The context for utterance interpretation must include some information about the speakers involved in the conversation (e.g., part of their semantic and episodic memory, their attitudes, dispositions, and purposes), as well as a number of extra-personal features such as the time and the place of the conversation. Current models of utterance interpretation that focus on how speaker's meaning is inferred by assuming the relevance of the speaker's communication act (e.g., Wilson & Sperber, 2004) have inspired research on pragmatic development in children with autism.

In an investigation that sought to characterize a specific disorder of pragmatics in children with autism, Surian *et al.* (1996) reported a selective difficulty in recognizing utterances that violate pragmatic constraints concerning the quantity and the type of information that has to be conveyed in responses to simple questions. Children with autism found difficult to recognize when an utterance did not conform to conversational rules that mandate speakers to provide sufficient, relevant, and true information in a clear, orderly, and unambiguous way, and their ability to do so appeared related to their ability to pass ToM tasks. The positive association between performance on ToM tasks

---

Box 37.2   Word learning in children with ASD

Children with ASD do not appear impaired in exploiting some constraints that help them to learn novel words meanings. For example, if they are given a novel word in a context where there are two relevant objects, one familiar and one unfamiliar, they consistently know, as do typically developing children, that the novel word refers to the unfamiliar object (Preissler & Carey, 2004). This good performance is not surprising if one assumes that they based their reasoning on the mutual exclusivity constraint and that this constraint has little to do with taking into account speakers' mental states and communicative intentions. In this view, the mutual exclusivity constraint is specific to word learning and states, roughly, that there is only one label for an object. Even typically developing infants at 17 months have been shown to behave according to the mutual exclusivity constraint. However, this is a hotly debated topic, with many researchers proposing the opposite view, namely that word learning constraints, including the mutual exclusivity constraint are actually the result of general

inferential skills and pragmatic reasoning involving mentalistic attributions (Diesendruck *et al.*, 2003). The success of children with autism in using the mutual exclusivity constraint appear to be strong evidence that this constraint is not based on a complex set of premises that require the attribution of representational states to the speakers, since if that was the case the severe ToM deficit found in many children with autism would have been apparent also in their ability to apply the mutual exclusivity constraint (Marcus & Rabbagliati, 2006).

Diesendruck, G., Markson, L., & Bloom, P. (2003). Children's reliance on the creator's intent in extending names for artifacts. *Psychological Science, 14*, 164–168.

Marcus, G., & Rabbagliati, H. (2006). What developmental disorders can tell us about the nature and origins of language. *Nature Neuroscience, 9*, 1226–1229.

Preissler, M.A., & Carey, S. (2004). The role of inferences about the referential intent in word learning: Evidence from autism. *Cognition, 97*, B13–B23.

and pragmatic skills has been found also looking at autistic children's ability to understand ironic sarcastic and metaphorical speech that require a non-literal interpretation and assertions referring to mental states (Ziatas *et al.*, 2003).

Such pragmatic impairment cannot be explained mainly by a general delay in verbal skills. Rather, children with autism stand out in that they typically fail to use language to establish and maintain a focus of joint attention. Although high-functioning children with autism are reported to speak obsessively about a very narrow set of topics neglecting the lack of interest in their hearers, verbosity may be found only in some people with Asperger syndrome and such a feature may not be a reliable characteristic of these patients as a group.

Problems of children with autism in the use of linguistic reference emerge also in pronoun reversals. The use of personal pronouns in conversation depends on crucial contextual features such as who is the speaker and who is the listener at a given moment and for this reason it is central to pragmatic understanding. Children with autism often misuse the singular pronouns for the first and second person ("I" and "you"), but not the pronouns for referring to the third person. Although they use proper nouns correctly, they tend to use them also later in development when other children without autism would use pronouns instead. Children with autism may also show problems with deictic terms such as "this" and "that", or words like "come" and "go", "here" and "there"(Frith, 2003). Since the correct use of these words depends often on the personal perspective assumed by the speaker in relation to the hearer, such errors may be related to the ToM deficit.

Pragmatic difficulties in autism may ultimately prove to have a genetic basis. As reported by Folstein and Rosen-Sheidley (2001), parents of autistic children display language impairment specifically in the domain of pragmatics. Evidence comes from their participation in a short conversation designed to reveal their communication skills by talking about hobbies or professional activities. The parents' conversations were later analyzed to in terms of a tendency to be too frank in personal evaluations, or too formal, wordy, vague or confused. Some weaknesses in the conversational behavior of the parents of autistic children resembled those found in the clinical population, particularly showing lack of appropriate inhibition in communication, awkward expressions and odd verbal interaction.

37.2.4. Prosody

Related to pragmatics is the use of prosodic information in communication. Abnormal prosody in the language of autistic children has been first reported in clinical studies and, more recently, in experimental investigations. More distortion errors and less appropriate utterance phrasing, stress, and resonance have been noted in high-functioning persons with ASD than in controls (Shriberg *et al.*, 2001).

Clinical observations of children with autism often may include the presence of a "flat intonation" in speech that, in this respect, is similar to some vocal announcements generated by computer programs.

It is important to assess also how persons with ASD use prosody in language comprehension. Rutherford *et al.* (2002) found that adults with high-functioning autism and Asperger syndrome performed poorly in a task requiring inferences about the mental states of a speaker, for example, surprise or irritation, from his or her vocalizations.

Prosodic contours are also important in the interpretation of ironic remarks. Irony is often marked by a characteristic prosodic contour that helps the listener to interpret the ironic statement correctly. Wang *et al.* (2006) reported that children with ASD were less accurate than typically developing controls in deciding whether an utterance was meant to be ironic. In an fMRI investigation, they found greater activity in the right inferior temporal gyrus of children with ASD. This result contrasts with previous findings that reported hypo-activations of the same areas during ToM and pragmatic reasoning tasks. The hyper-activation found by Wang and colleagues suggest greater effort in performing an explicit task, while the hypo-activation found in previous studies could be due to the fact that subjects were not explicitly required to engage in mental states attributions.

37.2.5. Reading

There are a few studies on reading skills showing that performance is poor when individuals with autism have to read homographs that need to be disambiguated on the basis of contextual information, and that their ability to form bridging inferences appears less developed than mental age matched controls (Jolliffe & Baron-Cohen, 1999). The conversion of graphemes into phonemes shown by children with autism reveals the tacit knowledge of subtle rules, but standard tests of comprehension show that, although children often appear accurate in producing the sounds, they do not achieve a correct understanding of what they have read. These results have been interpreted to provide support for the weak central coherence hypothesis (Frith, 2003, see Box 37.1), which claims that children with autism have a deficit in interpreting incoming information by taking into account the relevant contextual elements in order to process content.

37.3. EXPLAINING LANGUAGE AND COMMUNICATION DEFICITS IN ASD

The theory inspired by psychoanalytic writings that communication deficits in autism are due to poor parenting – the "refrigerator mother" hypothesis – has never received empirical support. By contrast, in the last 30 years, evidence has accumulated for accounts that emphasize biological factors in the etiology of autism at the level of hormonal abnormalities, neural circuits and genetic bases (Folstein & Rosen-Sheidley, 2001). The similarities of linguistic phenotype in the close relatives of children with ASD provide strong evidence for the genetic origins of autism, setting its heritability at 90% or more (Pilowsky *et al.*, 2003). However, we are still a long way from pointing to the "autistic genotype" and from describing with some clarity the causal chain that leads from certain genetic features to the diagnostic peculiarities in language and communication.

Work on language and communication in autism may help in this very challenging enterprise in several ways. One possibility is that cognitive/linguistic phenotypes will help to form more homogeneous subgroups of autistic individuals and this will advance the search for the genetic markers of autism (Tager-Flusberg, 2004). But uncovering the genetic bases of ASD will hardly yield also a detailed account of the specific profile of selective deficits in language and communication skills. This account requires a cognitive model of how autistic children process information when they interpret and formulate verbal messages.

37.3.1. Do Communication Deficits in ASD Result from a Mindreading Impairment?

Current explanations of communication deficits in autism have aimed to show how these could have resulted from abnormalities in central coherence, attention, or ToM reasoning (Tager-Flusberg, 2004; Pellicano *et al.*, 2006). One straightforward cognitive explanation of communication disorders in autism derives naturally from the metarepresentational ToM theory of autism that has addressed the issue of whether a person who could not express ToM reasoning could nonetheless participate meaningfully in conversation with others. Accounts of human communication based on processing inferences of speakers' utterances predict that metarepresentations are necessarily required even in the simplest cases of utterance interpretation (Sperber & Wilson, 2002). If one says "nice day," you will form a representation of such utterance that is enriched with the propositional attitude held by the speaker toward what he said (He *believes* that today is a nice day). To do this you will employ, following Leslie *et al.* (2004), the ToM mechanism. More complex metarepresentations are needed to interpret sarcasm and irony and thus such cases should be particularly hard for persons with ASD, even for those that can pass simple "first order" ToM tasks.

The difficulties in forming metarepresentations and, more specifically, representing communication intentions appear to be a cognitive impairment that children with autism share, to some degree, with deaf children from hearing families who have late access to a sign language, blind children, non-vocal children with cerebral palsy, and patients that have sustained right hemisphere lesions (Siegal

& Peterson, in press). However, the communication deficit in autism emerges very early in development, well before the age of 3 years, when even typically developing children fail verbally based ToM tasks, and there are substantial numbers of children with ASD who do pass ToM tasks. So the challenge is now to clarify more precisely how various components of the mindreading system (e.g., including both domain specific processes required to represent core mentalistic concepts such as beliefs, and domain general processes required in communication tasks such as executive functions required in inhibition and set-shifting) contribute to the performance on both mental states reasoning and communication tasks.

The fact that autistic symptoms in communication emerge very early may imply that a ToM deficit is not a viable explanation for the behavioral manifestation of the disorder. Nevertheless, this conclusion rests on the assumption that only performance on standard ToM tasks provides an adequate assessment of the metarepresentational skills acquired by young children. Recent studies employing nonverbal tasks suggest exactly the opposite: standard false belief tasks are likely to underestimate young children's tacit knowledge of mental states in that there is evidence that even at 13 months of age infants can attend to what others know and believe (Surian *et al.*, 2007). Difficulties in mentalizing could then well be an adequate explanation at the cognitive level for the communication deficit shown by even very young children with autism. This would also account for variations in the performance of children with autism and normal 3-year olds on different versions of false belief tasks (Surian & Leslie, 1999).

The cognitive explanation we have discussed so far sees the ToM computational machinery as a necessary component of communicative competence, in that it provides the required formats to represent speakers' mental states, including their informative and communicative intentions (Surian *et al.*, 1996; Ziatas *et al.*, 2003 & Tager-Flusberg, 2004). The communication difficulties of children with autism may ultimately all derive from their deficit in mentalization skills. But another view seems to consider the relationship between communication skills and mentalizing competence in the reverse direction: early communication activities could provide a necessary input required in normal ToM development (Siegal & Peterson, in press). Despite their apparent differences, these two proposals are not mutually exclusive. The latter is mainly concerned with the *computational bases* of a communication difficulty at a certain time in development and seeks to answer the question: which one is the cognitive component whose malfunctioning causes these failures in on-line communication? The former view instead is more concerned with the *developmental origins* of the communication deficits. Both explanations may not prove satisfactory, however, unless we come up with viable fractionations of ToM and communication processes. Their relation, both at the computational and at the ontogenetic level, would ultimately be clarified only by clearly specifying their different components (Leslie *et al.*, 2004).

37.3.2. Enhanced Perceptual Functioning and Pragmatic Development

Studies on both visual and auditory processing have found a dissociation between tasks that normally involve implicit or spontaneous information processing and tasks that explicitly require subjects to process prespecified information (Wang *et al.*, 2006). On tasks requiring active processing and involving explicit instructions such as the identification of local details in a complex figure such as small Ss that make a large X, children with autism may perform on the normal range or at least at levels expected from their mental age. By contrast on tasks where the instructions are not explicit in directing attention preferentially to certain stimulus features such as the local or global characteristics of a stimulus, they perform poorly and show abnormal neurological correlates. This pattern of findings underscores the importance of specifying differences between children with autism and normal controls not only at the level of what they know or can represent, but also at the level of how knowledge in certain domains is attended to and processed. Many studies have shown that deaf children who are late learners of a sign language have

persistent difficulties in passing ToM tasks. As such children have been cut off from participation in conversations because of their deafness, deprivation in linguistic and conversational experience appears to affect proficiency in mental state attributions (Siegal & Peterson, in press). Children with autism are not, typically, deaf, but they have been reported to react to auditory stimuli in peculiar ways. For example, they cover their ears with their hands, they sometimes show very low tolerance for some noises, and they manifest peculiar brain activation patterns in response to some auditory stimuli. One hypothesis is that some abnormalities in audition may affect their propensity to attend to speech and engage in conversations. This then deprives the child from input that is pivotal in the acquisition of ToM concepts. This failure in ToM acquisition in turn would be responsible for the persistent pragmatic deficits even in the high-functioning individuals with ASD.

In this respect, there is a growing body of evidence on abnormal auditory processing in autism (see Box 37.3) and the very early failure to orient toward linguistic stimuli in these children (Osterling *et al.*, 2002; Bebko *et al.*, 2006). The performance of people with ASD on auditory tasks crucially depends on the type of task (Samson *et al.*, 2006); it tends to be poor, and abnormalities are more likely to emerge in ERP patterns, when the tasks involve spectrally and temporally dynamic materials. Temporally complex auditory stimuli, such as melodies, tend to manifest gestalt properties

---

Box 37.3   Neural substrates of attention to vocal and non-vocal sounds

In fMRI research, Gervais *et al.* (2004) have reported that a group of five adult males diagnosed with autism did not show activation in regions of the superior temporal sulcus (STS) in response to vocal sounds. As shown below, a healthy control group clearly showed activation (marked yellow) to vocal sounds in the STS region in the right and left hemispheres whereas the autistic group did not, although both groups displayed a normal activation pattern in response to non-vocal sounds .

These findings are consistent with those from a study on cortical event-related brain potentials (ERPs) in which school-aged children with autism displayed an attentional deficit in orienting to the "speechness" quality of sounds as represented by vowels (Ceponiene *et al.*, 2003). Such results indicate that this orienting deficit is speech–sound specific and cannot be the outcome of a deficit in more basic sensory processes.

Ceponiene, R., Lepisto, T., Shestakova, A., Vanhala, R., Alku, P., Naatanen, R., & Yaguchi, K. (2003). Speech–sound-selective auditory impairment in children with autism: They can perceive but do not attend. *Proceedings of the National Academy of Sciences, 100,* 5567–5572.

Gervais, H., Belin, P., Broddaert, N., Leboyer, M., Coez, A., Sfaello, I., Barthélémy C., Brunelle, F., Samson, Y., & Zilbovicius, M. (2004). Abnormal cortical voice processing in autism. *Nature Neuroscience, 7,* 801–802.

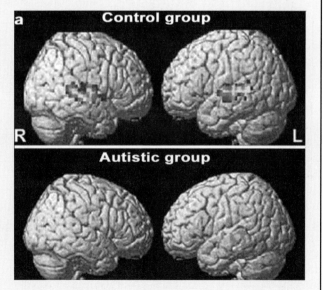

Individuals with autism, unlike individuals with typical development, failed to activate the STS in response to vocal sounds. *Source*: From Gervais *et al.*, 2004; Reprinted by permission from Macmillan Publishers Ltd.: *Nature Neuroscience*, copyright, 2004. (See Plate 23.)

that emerge from the whole sequence and are not reducible to the properties of the single constituents. These results are consistent with naturalistic observations showing avoidance reactions to spectrally complex stimuli of both social and non-social nature (e.g., cafeterias and vacuum cleaners). By contrast, the performance of individuals with autism on spectrally simple stimuli at a single frequency, such as pure tones, appears normal or even superior to normal.

Therefore, given that linguistic stimuli are typically spectrally and dynamically complex, at least some of the peculiarities in patterns of language and communication in children with autism could be mediated by an auditory processing disorder. More specifically, consistent with the absence of a preference for 'motherese' speech that is commonly used by mothers in their communication with children over non-speech analogs, ERP abnormalities in children with autism point to a deficit in the spontaneous orientation of attention toward linguistic stimuli (Kuhl *et al.*, 2005).

37.4. CHALLENGES AND FUTURE DIRECTIONS

Several priorities for future research emerge from this review. First, the characteristics of children termed as autistic are varied and heterogenous (Pellicano *et al.*, 2006). Despite extensive research in the last 20 years on difficulties in mental state reasoning, executive functioning, and central coherence, it is not known yet what components of cognitive functioning and linguistic competence are preserved or damaged at different ages. Second, the relation between language and cognition in autism, including ToM, executive functioning, and the processing of visual and auditory stimuli needs to be better understood. In this connection, a major priority must be to define groups of children with varying levels of verbal intelligence in relation to their non-verbal intelligence given indications that the children with autism who show the most severe symptoms are the ones who are disproportionately disadvantaged in language (Joseph *et al.*, 2002). Third, more research is required on the genetics of autism and language disorders in order to outline similarities and differences between the genetics of language disorders in autism and in other conditions such as Williams syndrome and specific language impairment.

Clearly, the roots of autism are established very early in development when attention to voices and sounds can be crucial for development. Therefore, a fourth research direction concerns the relation between early linguistic experience and the capacity for auditory attention in autism on the one hand and, on the other hand, the development of ToM and pragmatic competence. The hypothesis that abnormalities in auditory processing are a determinant of later ToM and pragmatic deficits assumes the primacy of auditory processing in autism, but is neutral with respect to the acquisition mechanisms underlying the expression of ToM reasoning and pragmatics. This perspective on autism is committed to the crucial role of conversational input and the effects of deprivations, but it is not committed to one particular type of acquisition mechanism, although a maturational one appears more plausible for core domains of cognition such as language and intuitive psychology.

It is unlikely that a single agent can be singled out as a cause of all ASD. However, if is established that auditory attention, for example, does make a causal contribution to autism in a substantial number of cases, an important step would be to train children with autism to attend to voices and speech. Such methods have been used to enable children at risk of dyslexia to read better through enhancing their ability to discriminate speech sounds and may also help children with autism to attend better to voices and speech, thus improving their ability to follow messages in conversation.

Overall, we emphasize the importance of pursuing an integrative approach that encompasses investigations targeted at genetic, neurobiological, and cognitive levels of explanations. This strategy is necessary to gain a comprehensive understanding of the wide range of language and communication difficulties shown in autism and Asperger syndrome, and their manifestation in infants, children, and adults.

References

Bebko, J.M., Weiss, J.A., Demark, J., & Gomez, P. (2006). Discrimination of temporal synchrony in intermodal events by children with autism and children with developmental disabilities without autism. *Journal of Child Psychology and Psychiatry, 47*, 88–98.

Carpenter, M., Tomasello, M., & Striano, T. (2005). Role reversal imitation and language in typically developing infants and children with autism. *Infancy, 8*, 253–278.

Dennis, M., Lazenby, A.L., & Lokyer, L. (2002). Inferential language in high function children with autism. *Journal of Autism and Developmental Disorders, 31*, 47–54.

Folstein, S.E., & Rosen-Sheidley, B. (2001). Genetics of autism: Complex aetiology for a heterogeneous disorder. *Nature Review Genetics, 2*, 943–955.

Frith, U. (2003). *Autism. Explaining the enigma* (2nd edn.). Oxford: Blackwell.

Harris, P.L., de Rosnay, M., & Pons, F. (2005). Language and children's understanding of mental states. *Current Directions in Psychological Science, 14*, 69–73.

Jolliffe, T., & Baron-Cohen, S. (1999). Linguistic processing in high-functioning adults with autism or Asperger syndrome: Can local coherence be achieved? A test of central coherence theory. *Cognition, 71*, 149–185.

Joseph, R.M., Tager-Flusberg, H., & Lord, C. (2002). Cognitive profiles and social-communicative functioning in children with autism spectrum disorder. *Journal of Child Psychology and Psychiatry, 43*, 807–822.

Klin, A., Volkmar, F.R., & Sparrow, S.S. (Eds.) (2000). *Asperger syndrome.* New York: Guilford.

Klin, A., Jones, W., Schultz, R., Volkmar, F., & Cohen, D. (2002). Visual fixation patterns during viewing of naturalistic social situations as predictors of social competence in individuals with autism. *Archives of General Psychiatry, 59*, 809–816.

Kuhl, P.K., Coffey-Corina, S., Padden, D., & Wilson, D. (2005). Links between social and linguistic processing of speech in preschool children with autism: Behavioral and electrophysiological measures. *Developmental Science, 8*, F1–F12.

Leslie, A.M., Friedman, O., & German, T.P. (2004). Core mechanisms in "theory of mind". *Trends in Cognitive Sciences, 8*, 528–533.

Lord, C., Risi, S., Lambrecht, L., Cook, E.H., Leventhal, B.D., DiLavore, P.C., Pickles, A., & Rutter, M. (2000). The autism diagnostic observation scale-generic: A standard measure of social and communication deficits associated with the spectrum of autism. *Journal of Autism and Developmental Disorders, 30*, 205–223.

Osterling, J.A., Dawson, G., & Munson, J.A. (2002). Early recognition of 1-year-old infants with autism spectrum disorder versus mental retardation. *Development and Psychopathology, 14*, 239–251.

Pellicano, E., Maybery, M., Durkin, K., & Maley, A. (2006). Multiple cognitive capabilities/deficits in children with an autism spectrum disorder: "Weak" central coherence and its relationship to theory of mind and executive control. *Development and Psychopathology, 18*, 77–98.

Pilowsky, T., Yirmiya, N., Shalev, R., & Gross-Tsur, V. (2003). Language abilities of siblings of children with autism. *Journal of Child Psychology and Psychiatry, 44*, 914–925.

Rapin, I., & Dunn, M. (2003). Update on the language disorders of individuals on the autistic spectrum. *Brain and Development, 25*, 166–172.

Rutherford, M.D., Baron-Cohen, S., & Wheelwright, S. (2002). Reading the mind in the voice: A study with normal adults and adults with Asperger syndrome and high functioning autism. *Journal of Autism and Developmental Disorders, 32*, 189–194.

Samson, F., Mottron, L., Jemel, B., Belin, P., & Ciocca, V. (2006). Can spectro-temporal complexity explain the autistic pattern of performance on auditory tasks. *Journal of Autism and Developmental Disorders, 36*, 65–76.

Schick, B., de Villiers, P., de Villiers, J., & Hoffmeister, R. (2007). Language and theory of mind: A study of deaf children. *Child Development, 78*, 376–396.

Senju, A., Tojo, Y., Yaguchi, K., & Hasegawa, T. (2005). Deviant gaze processing in children with autism: An ERP study. *Neuropsychologia, 43*, 1297–1306.

Shriberg, L., Paul, R., McSweeny, J., Klin, A., & Cohen, D. (2001). Speech and prosody characteristics of adolescents and adults with high-functioning autism and Asperger syndrome. *Journal of Speech, Language, and Hearing Research, 44*, 1097–1115.

Siegal, M., & Peterson, C.C. (in press). Language and theory of mind in atypically developing children: Evidence from studies of deafness, blindness, and autism. In C. Sharp, P. Fonagy & I. Goodyer (Eds.), *Social cognition and developmental psychopathology*. New York: Oxford University Press.

Sigman, M., & McGovern, C.W. (2005). Improvement in cognitive and language skills from preschool to adolescence in autism. *Journal of Autism and Developmental Disorders, 35*, 15–23.

Sperber, D., & Wilson, D. (2002). Pragmatics, modularity and mindreading. *Mind and Language, 17*, 3–23.

Surian, L., & Leslie, A. (1999). Competence and performance in false belief understanding: A comparison of autistic and three-year-old children. *British Journal of Developmental Psychology, 17*, 131–145.

Surian, L., Baron-Cohen, S., & van der Lely, H. (1996). Are children with autism deaf to Gricean maxims. *Cognitive Neuropsychiatry, 1*, 55–72.

Surian, L., Caldi, S., & Sperber, D. (2007). Attribution of beliefs by 13-month-old infants. *Psychological Science, 17*, 141–155.

Tager-Flusberg, H. (2004). Strategies for conducting research on language in autism. *Journal of Autism and Developmental Disorders, 34*, 75–80.

Wang, A.T., Lee, S.S., Sigman, M., & Dapretto, M. (2006). Neural basis of irony comprehension in children with autism: The role of prosody and context. *Brain, 129*, 932–943.

Wilson, D., & Sperber, D. (2004). Relevance theory. In L.R. Horn & G. Ward (Eds.), *Handbook of pragmatics* (pp. 607–632). Oxford: Blackwell.

Ziatas, K., Durkin, K., & Pratt, C. (2003). Differences in assertive speech acts produced by children with autism, Asperger syndrome, specific language impairment, and normal development. *Development and Psychopathology, 15*, 73–94.

Further readings

Charman, T. (2004). Matching preschool children with autism spectrum disorders and comparison children for language ability: Methodological challenges. *Journal of Autism and Developmental Disorders, 34*, 59–64.
Raises important issues that concern how to carry research on the language abilities of children with ASD.

Frith, U., & Hill, E. (Eds.) (2003). *Autism: mind and brain*. Oxford: Oxford University Press.
Along with Frith (2003) above, this provides an excellent collection of work on the neural basis of autism.

Happé, F., Ronald, A., & Plomin, R. (2006). Time to give up on a single explanation for autism. *Nature Neuroscience, 9*, 1218–1220.
Emphasizes the diversity of autistic symptomology, including the area of language and communication, and the implications for future research.

Mesibov, G.B., Shea, V., & Adams, L.W. (2001). *Understanding Asperger syndrome and high functioning autism*. New York: Kluwer.
A clear and concise introduction to the diagnosis and functional disorders in Asperger syndrome.

Tager-Flusberg, H. (2000). Language and understanding minds: Connections in autism. In S. Baron-Cohen, H. Tager-Flusberg & D.J. Cohen (Eds.), *Understanding other minds: Perspectives from autism and developmental cognitive neuroscience* (pp. 124–149, 2nd edn). Oxford: Oxford University Press.
A wide-ranging review of research on the relationships between language deficits and the core problems in mentalization and ToM research.

PART IV

CLINICAL NEUROSCIENCE OF LANGUAGE

C. Recovery from, Treatment and Rehabilitation of
Language and Communication Disorders

38

Spontaneous Recovery of Aphasia

STEFANO F. CAPPA

Vita-Salute University and San Raffaele Scientific Institute, Milan, Italy

ABSTRACT

The tissue damage caused by a stroke is typically stable after the acute stage. The fact that most patients affected by aphasia actually show a variable degree of recovery over time indicates that the brain has the potential to compensate the consequence of localized damage. Better recovery can be expected when the lesion is of limited size and the severity of the language impairment is mild. The impact of other factors, such as age, gender or handedness is less established. The mechanisms responsible for recovery are probably multiple, and play a different role according to the time after onset. In the early stage recovery may simply reflect the fact that the acute effect of a lesion reflects not only permanent damage, but also reversible dysfunction. In later stages, recovery may reflect the ability of the brain to reorganize to support a specific function (its "plasticity").

38.1. INTRODUCTION

People who become aphasic after a stroke, or a traumatic brain injury, typically recover their language function, in part or, sometimes, even completely. The functional consequences of acute, non-progressive pathological involvement of the nervous system (such as in dthe case of ischemic or hemorrhagic stroke, or trauma due to physical agents) generally undergo a variable degree of recuperation in the period following injury. The distinction between "spontaneous" and "treatment-induced" recovery is a matter of definition. Probably, there is no such thing as spontaneous recovery: the patient is always engaged in some form of "treatment," which can vary from totally unspecific to highly specialized. In this chapter, "spontaneous" is used in a broad sense, that is, by considering studies that have not attempted to define the contribution of professional rehabilitation to the recovery process.

It must be underlined that recovery can be defined in several ways. Most studies of recovery in aphasia are based on the assessment of changes in the level of impairment, typically in the scores on aphasia tests. However, what really matters for the individual patient is the recovery of disability, that is, an improvement of the ability to communicate, and the prevention of the handicap that may follow from aphasia. This general principle has led to the recommendation to include an assessment of functional disability (such as, in the case of aphasia, a functional communication scale) in any research investigation of recovery Box 38.1.

38.1.1. Can We Predict the Potential for Recovery?

One important aspect of research on aphasia recovery is to assess the contribution of multiple variables on the long-term prognosis of the individual patient. The findings of this class of studies are necessary in order to provide a baseline for the evaluation of treatment effects. There are several reviews of prognostic factors in aphasia (Cappa, 1998). The number of papers dedicated to this issue in the last few years is limited. This could be taken to indicate that the main questions are considered as solved. However, a perusal of the available evidence clearly indicates that the quality of many studies in this area is far from satisfactory. In particular, many recovery studies deal with small groups of selected patients, rather than with large consecutive series (Pedersen *et al.*, 2004). Given the large number variables that are entered into the outcome analysis, sample size plays a crucial role. There is also a large variability in the methods of assessment of aphasia, ranging

Box 38.1   Impairment, disability and handicap

An important distinction is made by the World Health Organization between impairment, disability and handicap:

Impairment: Any loss or abnormality of psychological, physiological, or anatomical structure or function.
Disability: Any restriction or lack (resulting from an impairment) of ability to perform an activity in the manner or within the range considered normal for a human being.
Handicap: A disadvantage for a given individual, resulting from an impairment or disability, that limits or prevents the fulfillment of a role that is normal, depending on age, sex, social and cultural factors, for that individual.

These concepts have been recently revised in the International Classification of Functioning, Disability and Health, which replaces the term "disability" with "activity limitation," and "handicap" with "participation restriction."

World Health Organization (1980). *The international classification of impairments, disabilities and handicaps – a manual of classification relating to the consequences of disease.* Geneva: World Health Organization.
World Health Organization (2001). *International classification of functioning, disability and health.* Geneva: World Health Organization.

from clinical scales to standardized aphasia tests. This factor has a considerable impact on the results, because a low-sensitivity assessment results in an underestimation of the impairment, and an overestimation of spontaneous recovery. In addition, the available evidence about the role of many factors remains inconclusive, and thus in need of further investigation.

38.2. SOUNDLY ESTABLISHED FACTORS AFFECTING RECOVERY

There is convincing empirical evidence that lesion size, lesion site, severity of the disease and the time that has passed since the incidence (time post-onset) play a crucial role in recovery.

38.2.1. Lesion Size

It is hardly surprising that the size of the cerebral lesion, which, with some remarkable exceptions (see below), is closely related to aphasia severity, is the strongest negative predictor of recovery in all studies (for a summary, see Cappa, 1998). In the chronic stage after brain damage,

lesion size has usually been estimated on the basis of volumetric measurements of the damaged brain regions with computerized tomography (CT) or magnetic resonance imaging (MRI). More recently, it has become feasible to assess *in vivo* the area of hypoperfusion in the acute stage after a stroke using perfusion magnetic resonance. It has thus been possible to show that the extent of hypoperfusion is a strong predictor of recovery in the chronic stage after a brain lesion (Croquelois *et al.*, 2003) (Box 38.2).

Box 38.2   The lesion overlap method

A traditional method used to investigate the relationship between lesion site and a specific brain function is based on the identification of shared areas of brain damage in patients who show a particular deficit, that is, for example, a disorder of single word comprehension. There are several variants of this method, ranging from the simple tracing of the lesions on a standard brain template on the basis of the results of each subject's CT of MR scan, to relatively objective semi-automatic methods. However, the logic behind the approach remains the same: the area(s) that are affected by all the patients showing the deficit (overlap) support the normal operation of the affected function. This assumption has been criticized on several grounds. In the first place, the shared area may actually be the most vulnerable to the specific pathology (e.g., the core of the territory of the middle cerebral artery in the case of ischemic strokes). Second, in the case of chronic patients, the shared area of damage found in patients who do not recover a specific function may be the brain region that is responsible for mediating language improvement.

Ogar, J., Willock, S., Baldo, J., Wilkins, D., Ludy, C., & Dronkers, N. (2006). Clinical and anatomical correlates of apraxia of speech. *Brain Language, 97*(3), 343–350.
Hillis, A.E., Work, M., Barker, P.B., Jacobs, M.A., Breese, E.L., & Maurer, K. (2004). Re-examining the brain regions crucial for orchestrating speech articulation. *Brain, 127*, 1479–1487.

38.2.2. Lesion Site

While lesion size clearly matters, there is also ample evidence that the location of a lesion has an important impact on the possibility to recover specific aspects of linguistic function. In general, the negative influence of a lesion localized to a specific cerebral area on the recuperation of a definite aspect of language function has been taken to indicate a crucial role for that function (see Cappa, 1998).

Investigating the acute stage after a stroke and using dynamic contrast magnetic resonance perfusion weighted imaging (PWI), Hillis and her coworkers showed that the left midfusiform gyrus region (BA 37) is necessary for object naming. The area was consistently hypoperfused in patients with defective oral and written naming in the acute stage, and its reperfusion resulted in improved performance (Hillis *et al.*, 2006). This is an excellent example of an anatomical correlation supported by imaging studies.

38.2.3. Clinical Picture

While all studies concur in indicating that initial severity, defined with global impairment measures, is the strongest predictor of recovery (Pedersen *et al.*, 2004), the type of aphasic syndrome, according to the traditional taxonomy (Broca's aphasia, global aphasia, and so forth), is not a good independent predictor of recovery. The assignment of a patient to a specific syndrome (such as Broca's aphasia) is based on a cluster of symptoms, which are defined in terms of performance in different language modalities (typically, oral expression, auditory comprehension, and repetition). Different rates of recovery for different linguistic functions result in some patients evolving, so to speak, from one clinical syndrome to another.

38.2.4. Time Post-Onset

As in the case of the recovery of other functional consequences of stroke, the rate of spontaneous recovery is maximal in the first 6 months, with a very steep curve in the first 6 weeks (Pedersen *et al.*, 1995). It must however be underlined that there is evidence that significant improvement can be observed in severely aphasic patients up to 2 years post-onset (Nicholas *et al.*, 1993).

38.3. FACTORS THAT MAY AFFECT SPONTANEOUS RECOVERY

While there is ample evidence that lesion size, lesion site, severity of disease and time after lesion onset influence recovery, it is less clear to what degree etiology of the disease, personal and external factors contribute to recovery.

38.3.1. Etiology

It is obvious that aphasia associated with a progressive disease of the nervous system, such as a malignant tumor or a degenerative condition, has a negative prognosis. The evidence about other etiologies is, however, limited. In the case of vascular aphasia, there is some evidence that hemorrhagic strokes are associated with a more severe clinical picture in the acute stage but they result in better long-term

outcome compared to ischemic strokes (Nicholas *et al.*, 1993). Several explanations have been advanced to account for this difference. Hemorrhagic strokes have a less destructive impact on brain tissue than ischemic strokes. Moreover, patients with hemorrhagic stokes are less likely to have diffuse white matter damage than patients with ischemia. There is also some evidence that aphasia following traumatic brain injury recovers better than aphasia due to cerebrovascular lesion (Cappa, 1998). The difference may, however, be due to the on average younger age of head injury patients (Box 38.3).

Box 38.3   The physiopathology of ischemic stroke

The functional consequences of an ischemic stroke do not reflect only the site and extent of irreversible tissue damage (necrosis), but also other lesion-related factors. Among the most studied are ischemic penumbra and diaschisis. The ischemic penumbra is the area of brain tissue surrounding the "core" ischemic region, in which blood flow is decreased to a level that results in irreversible damage to nerve cells (i.e., below 10% of the normal level). In contrast, the level of ischemia is less severe in the penumbra and the cells may remain viable for several hours. If the blood flow to the penumbra is restored, the nervous cells can return to normal function, that is, the tissue can be salvaged.

Diaschisis refers to a loss of function in a portion of the brain that is at a distance from the site of tissue damage, but is connected to it by means of fibre pathways. The phenomenon was originally conceived by von Monakow as a "functional shock" due to the sudden interruption of the connections, but it has been found to be potentially long-lasting.

Fisher, M., & Ginsberg, M. (2004). Current concepts of the ischemic penumbra. *Stroke*, *35*, 657.

Reggia, J.A. (2004). Neurocomputational models of the remote effects of focal brain damage. *Medical Engineering and Physics*, *26*, 711–722.

38.3.2. Handedness

Left-handed patients as well as patients with "atypical" hemispheric language dominance, such as "crossed" aphasics, have been reported to recover better or faster (Cappa, 1998). However, the available evidence is inconclusive and in need of further investigation (Marien *et al.*, 2004).

38.3.3. Hemispheric Asymmetries

Functional imaging of language in normal subjects can reveal asymmetric activation of the two hemispheres, reflecting language dominance (Harrington *et al.*, 2006).

Unfortunately, however, information about pre-stroke dominance is not available in most patients. Some surrogate measures have thus been used, such as anatomical asymmetries between the hemispheres using *in vivo* structural MR, and they have been related to both handedness and language lateralization (see Josse & Tzourio-Mazoyer, 2004, for a review). However, a recent study using the Wada test with epileptic patients failed to find a correlation between asymmetry of the planum temporale and language lateralization. Only the asymmetry in gray matter density in Broca's area, measured *in vivo* with voxel-based morphometry, had a direct relationship with language lateralization (Dorsaint-Pierre *et al.*, 2006). There is some evidence of gender-related differences in hemispheric asymmetry, possibly mediated by the impact of sexual hormones on neural differentiation (Josse & Tzourio-Mazoyer, 2004). The hypothesis of a less lateralized pattern of hemispheric specialization in females has received some support by functional imaging studies (Shaywitz *et al.*, 1995), but has not always been replicated (Frost *et al.*, 1999). Decreased lateralization has been suggested to result in improved recovery, on the basis of the limited evidence of a better prognosis in left-handers (see above). However, the global outcome of stoke appears to be worse in females (Niewada *et al.*, 2005). Gender did not affect aphasia recovery in the study by Laska *et al.* (2004).

38.3.4. Age

Increasing age has been generally associated with poorer recovery (Laska *et al.*, 2004) because of age-associated factors that typically play a negative role in recovery, such as hypertension, diabetes, Alzheimer's disease, and so on. It is thus possible that co-morbidity, rather than age *per se*, is a major negative prognostic factor for aphasia recovery. Given the economical constraints in the provision of rehabilitation services in most countries, this point has important practical implications: an otherwise healthy elderly person has probably good chances to benefit from aphasia therapy.

The effect of age on recovery is clear-cut at the other end of the developmental spectrum. While a fast and relatively complete recovery can in general be expected in children, subtle linguistic deficits can be persistent at follow-up (Nass & Trauner, 2004).

38.3.5. Education

Education is usually not included as a variable in outcome analyses. Clinical experience, however, suggests that education does indeed play a role, although it is unclear if the positive effects of higher education are independent from those of other potentially related variables, such as socio-economic status and general intelligence. There is in

fact evidence that a high socio-economic status has a preventive effect on stroke mortality (Arrich & Mullner, 2005).

38.3.6. Multilingualism

There is a considerable literature about different recovery patterns in patients who know more than one language. However, no systematic pattern emerge from the available case studies (see a review in Green, 2005). It is worthwhile stressing here that the traditional doctrine, that is, the first acquired language recovers first, is not supported by clinical nor empirical observations.

38.3.7. Mood, Motivation, and Social Support

Despite the obvious clinical relevance of these factors, no systematic evidence is available on their impact on aphasia recovery. A negative effect of social isolation on the outcome of stroke patient has been reported (Boden-Albala *et al.*, 2005). A possible speculation is that social isolation may be one of the factors that are responsible for the increased mortality observed in aphasic patients at 1 year after a stroke (Laska *et al.*, 2004). Depression, which is frequently observed in aphasic patients, has been found to be an independent predictor of long-term handicap after a stroke, and thus should be treated adequately (Sturm *et al.*, 2004).

38.3.8. Summary

Clinically established severity of the disease, lesion size and site and time post-onset are all established negative prognostic factors for spontaneous aphasia recovery. Lesion site has predictive value for specific aspects of linguistic recovery. The role of other variables, such as gender, age, handedness, and linguistic background is not firmly established.

38.4. WHY SPONTANEOUS RECOVERY?

Given the fact that aphasia recovers, the interesting question is what are the responsible neural mechanisms. An improvement of communicative abilities can sometimes be due to the consequence of the spontaneous or training-induced adaptation of the patient to his or her impairment. However, the fact that specific aspects of linguistic impairment, such as the ability to retrieve lexical items, undergo spontaneous recovery with time suggests that changes are taking place not only at the behavioral, but also at the neurophysiological level. These modifications are considered to reflect plastic changes in the damaged brain. The concept of

38.4.1. Neurogenesis

The traditional teaching that neural cells represent a fixed endowment at birth, which progressively decreases with age, has been substantially revised in recent years. The formation of new nerve cells has been observed in the mature brain, in particular in two areas, the hippocampus and the subventricular regions. Moreover, multipotent stem cells have been isolated in several areas of the nervous system (Emsley et al., 2005). Although the actual contribution of neurogenesis to recovery after brain damage is not known, if "neuroreparative" approaches become feasible, they should have an important impact on neurorehabilitation. A review of this topic can be found in Savitz et al. (2004).

38.4.2. Regression of Hypoperfusion and Diaschisis

The recovery period can be divided into at least three stages, based on different physiological mechanisms responsible for clinical improvement: an early stage (about 2 weeks post-onset), a lesion stage (up to 6 months post-onset), and a chronic stage (after 6 months). During the early period, the likely mechanisms for clinical improvement are the disappearance of cerebral edema and intracranial hypertension, the decrease of local inflammation, the normalization of hemodynamics in ischemic areas surrounding the region of irreversible tissue damage (penumbra), and the regression of diaschisis effects.

The contribution of techniques allowing the measurement of ischemic penumbra, such as perfusion computed tomography (p-CT), and dynamic contrast magnetic resonance PWI technique, has already been mentioned. The observation that restoration of blood flow may result in recovery of specific aspects of linguistic function suggests that at least part of the spontaneous improvement of language function in the early stage after a stroke can be attributed to the rescue of the ischemic areas which are not irreversibly damaged (Croquelois et al., 2003).

An additional mechanism of recovery is the regression of functional impairment in structurally unaffected brain regions connected to the damaged area, both in the ipsilateral and in the contralateral hemisphere. A correlation between regression of hypoperfusion and hypometabolism in the post-acute stage and clinical improvement of aphasia has been reported in several studies of subcortical strokes (Hillis et al., 2002). A similar mechanism seems to apply to patients with aphasia due to cortical lesions. Several studies with positron emission tomography (PET) have shown that in the early months after a stroke there is recovery of perfusion and metabolism in cerebral areas, which are not affected by permanent structural damage. The correlation of these changes with language recovery has been investigated in aphasic patients. For example, patients with mild aphasia due to a small stroke were studied using PET 2 weeks after onset and again 3 months later, when extensive recovery can be expected (Cappa et al., 1997). Metabolism was widely depressed in structurally unaffected areas in the acute stage and the reduction was not limited to the ipsilateral hemisphere, but extended contralaterally. The functional deactivation in structurally intact regions that are connected with the area of anatomical damage correlated to spontaneous recovery.

38.4.3. Remapping

The mechanisms underlying recovery at later stages are a matter of debate. Recovery may be achieved by adopting novel cognitive strategies for performance, which, at the neural level, implies the recruitment of uninjured cerebral areas.

In general, the recovery of linguistic functions has been attributed to a reorganization of the cerebral areas underlying language processing. The concept that functional localization in cortical areas is not completely fixed but can undergo plastic changes (remapping) has received much support from experimental literature about brain changes following peripheral deafferentation (for a review, see Kaas & Collins, 2003). In experimental animal models it has been shown that after resection of a peripheral nerve the corresponding cortical representation starts to respond very soon to input that typically activates neighboring cortical regions. It is, however, evident that these plastic changes in the organization of sensorimotor brain areas occur in experimental situations that are very different from the pathophysiological changes occurring with a stroke. Here, the damage affects directly the specialized cortical regions, rather than the input or output connections.

In the nineteenth century, it was already proposed that the homotopic (i.e., homologous) areas of the hemisphere contralateral to the language-dominant one have compensatory functions. The most influential hypothesis about the cortical changes associated to recovery in the nineteenth century. This hypothesis was based on clinical observations of recovered aphasic patients whose linguistic abilities worsened after a second stroke affecting the right hemisphere. Another hypothesis suggests that recovery may be due to the recruitment of additional areas that are unaffected by the lesion within the language-dominant hemisphere. Both contralateral and ipsilateral areas may be part of redundant language networks that are inhibited by the "primary" language areas in the intact brain (Heiss et al., 2003).

The possibility to measure brain activity in vivo with neuroimaging techniques allows the direct testing of this set of hypotheses. The basic design is to study brain activation in recovered aphasic subjects, while they are engaged in a language task and compare the activation pattern to the one of normal controls engaged in the same task. The first study adopting this approach used the PET technique to investigate the cerebral activations in response to two tasks (nonword repetition and verbal fluency) in patients who had recovered from Wernicke's aphasia (Weiller *et al.*, 1995). Increased right hemisphere activation was observed and interpreted as reflecting a compensatory function of the right hemisphere, taking over the function of the damaged dominant hemisphere. While this finding has been replicated in many subsequent studies, the functional significance of the contralateral activation remains a matter of debate. There are now a number of studies that ascribe the better long-term recovery to the preserved areas of the ipsilateral (left) hemisphere. For example, it has been shown that in recovered aphasic patients even limited preservation of perilesional tissue seems to have an important impact on recovery.

Follow-up studies of aphasic patients at different stages of recovery can provide important insights about the functional significance of right-hemisphere activation. Some investigations have revealed the existence of a temporal dynamics in the cerebral reorganization mechanisms after stroke. The initial engagement of non-damaged homologous areas of the right hemisphere is followed over time by a gradual decrease and a concomitant significant increase of activity in left hemispheric perilesional areas. The transfer of functional competence from the right to the left hemisphere is associated with an improvement of linguistic performances (Heiss & Thiel, 2006). Some authors have argued that "right hemispheric recruitment" in functional recovery may simply reflect the reliance on additional cognitive and linguistic resources, which are not required by healthy individuals during linguistic processing. A recent study has indicated in the subacute stage (on average 12 days post-stroke) there is a correlation between the bilateral recruitment of the language areas and language improvement. However, the same patients, re-examined in the chronic stage, showed a right-to-left shift of activations, associated with further improvement (Saur *et al.*, 2006). These findings are compatible with the idea that right hemispheric engagement reflects a sort of "emergency" measure to support language performance, leaving room, so to speak, for the specialized left hemisphere to make long-term recovery.

Additional insights about hemispheric dynamics come from the technique of repetitive transcranial magnetic stimulation (rTMS). Low-frequency rTMS has been applied to the areas shown to be active during language tasks in order to assess their contribution to task performance (Heiss & Thiel, 2006). Interference with language tasks was observed when the right inferior frontal gyrus was stimulated indicating a functional role of the right homolog of Broca's area in language performance. Another interesting observation has been reported in patients with chronic aphasia (Naeser *et al.*, 2005). A significant improvement in a naming task was found in two out of three patients after a series of ten, 20-min, 1-Hz rTMS

stimulations to the pars triangularis of the right homolog of Broca's area. The authors suggest that this effect is due to the suppression of an inhibitory effect exerted by this area on the controlateral hemisphere, via the corpus callosum.

38.4.4. Summary

There is considerable evidence that both intrahemispheric and interhemispheric reorganization play a role in the process of spontaneous recovery of aphasia. It may be hypothesized that in the first months after a stroke, when recovery proceeds at a fast rate, the regression of functional depression in ipsilateral and controlateral areas is the main mechanism at work. At the clinical level, this period is usually associated with a prevalent improvement of auditory comprehension. The subsequent phase, characterized by a much less steep recovery function, is related to the process of functional reorganization, with a complex interplay of the undamaged regions of the left hemisphere and of the healthy right hemisphere.

38.5. CHALLENGES AND FUTURE DIRECTIONS

There is no doubt that a deeper understanding of the factors that affect spontaneous recovery and of the underlying biological mechanisms is relevant for the development of rational therapeutic intervention for aphasia. Large-scale clinical studies of recovery after stroke are difficult to organize and require substantial financial support that may not be easily obtained. In recent years, there has been a prevalence of smaller scale studies, focusing on selected patients and using sophisticated technological tools. Functional imaging can address the crucial issues of cerebral plasticity and reorganization of function *in vivo* in ways that could not have been predicted a few years ago. However, we should never forget that the quality of the information that can be derived from technology depends crucially on the quality of the questions that are formulated by the researcher. While we have certainly learnt a lot in the last decade, many of the basic questions remain unresolved. Stroke units that have access to all the relevant technologies and expertises in the acute stage are probably an ideal setting to assess in a systematic way the impact of patient- and lesion-related variables on the recovery process. A multimodal imaging approach, combined with an up-to-date assessment of language abilities going beyond the traditional clinical evaluation and include an approach so far reserved to case studies (Doesborgh *et al.*, 2003) could be extended to large clinical populations. These kind of studies are required in order to progress to the next step, that is, to the investigation of the possible impact of behavioral and/or pharmacological interventions on the spontaneous recovery process.

References

Arrich, J.L.W., & Mullner, M. (2005). Influence of socioeconomic status on mortality after stroke: Retrospective cohort study. *Stroke, 36*, 310–314.

Boden-Albala, B., Litwak, E., Elkind, M.S., Rundek, T., & Sacco, R.L. (2005). Social isolation and outcomes post stroke. *Neurology, 64*, 1888–1892.

Cappa, S.F. (1998). Spontaneous recovery from aphasia. In B. Stemmer & H.A. Whitaker (Eds.), *Handbook of neurolinguistics* (pp. 535–545). San Diego, CA: Academic Press.

Cappa, S.F., Perani, D., Grassi, F., Bressi, S., Alberoni, M., Franceschi, M., Bettinardi, V., Todde, S., & Fazio, F. (1997). A PET follow-up study of recovery after stroke in acute aphasics. *Brain and Language, 56*, 55–67.

Croquelois, A., Wintermark, M., Reichhart, M., Meuli, R., & Bogousslavsky, J. (2003). Aphasia in hyperacute stroke: Language follows brain penumbra dynamics. *Annals of Neurology, 54*, 321–329.

Doesborgh, S.J.C., van de Sandt-Koenderman, W.M.E., Dippel, D.W.J., van Harskamp, F., Koudstaal, P.J., & Visch-Brink, E.G. (2003). Linguistic deficits in the acute phase of stroke. *Journal of Neurology, 250*, 977–982.

Dorsaint-Pierre, R., Penhune, V.B., Watkins, K.E., Neelin, P., Lerch, J.P., & Bouffard, M. et al (2006). Asymmetries of the planum temporale and Heschl's gyrus: Relationship to language lateralization. *Brain, 129*, 1164–1176.

Emsley, J.G., Mitchell, B.D., Kempermann, G., & Macklis, J.D. (2005). Adult neurogenesis and repair of the adult CNS with neural progenitors, precursors, and stem cells. *Progress in Neurobiology, 75*, 321–341.

Frost, J.A., Binder, J.R., Springer, J.A., Hammeke, T.A., Bellgowan, P.S., Rao, S.M., & Cox, R.W. (1999). Language processing is strongly left lateralized in both sexes. Evidence from functional MRI. *Brain, 122*, 199–208.

Green, D.W. (2005). The neurocognition of recovery patterns in bilingual aphasics. In J.F. Kroll & A.M.B. de Groot (Eds.), *Handbook of bilingualism* (pp. 516–530). Oxford: Oxford University Press.

Harrington, G.S., Buonocore, M.H., & Farias, S.T. (2006). Intrasubject reproducibility of functional MR imaging activation in language tasks. *American Journal of Neuroradiology, 27*, 938–944.

Heiss, W.-D., & Thiel, A. (2006). A proposed regional hierarchy in recovery of post-stroke aphasia. *Brain and Language, 98*, 118–123.

Heiss, W.-D., Thiel, A., Kessler, J., & Herholz, K. (2003). Disturbance and recovery of language function: Correlates in PET activation studies. *Neuroimage, 20*(Suppl 1), S42–S49.

Hillis, A.E., Wityk, R.J., Barker, P.B., Beauchamp, N.J., Gailloud, P., Murphy, K., Cooper, O., & Metter, E.J. (2002). Subcortical aphasia and neglect in acute stroke: The role of cortical hypoperfusion. *Brain, 125*, 1094–1104.

Hillis, A.E., Kleinman, J.T., Newhart, M., Heidler-Gary, J., Gottesman, R., Barker, P.B., Aldrich, L., Llinas, R., Wityk, R., & Chaudhry, P. (2006). Restoring cerebral blood flow reveals neural regions critical for naming. *Journal of Neuroscience, 26*, 8069–8073.

Josse, G., & Tzourio-Mazoyer, N. (2004). Hemispheric specialization for language. *Brain Research Reviews, 44*, 1–12.

Kaas, J.H., & Collins, C.E. (2003). Anatomic and functional reorganization of somatosensory cortex in mature primates after peripheral nerve and spinal cord injury. *Advances of Neurology, 93*, 87–95.

Laska, A.C., Hellblom, A., Murray, V., Kahan, T., & Von Arbin, M. (2004). Aphasia in acute stroke and relation to outcome. *Journal of Internal Medicine, 249*, 413–422.

Marien, P., Paghera, B., De Deyn, P.P., & Vignolo, L.A. (2004). Adult crossed aphasia in dextrals revisited. *Cortex, 40*, 41–74.

Naeser, M.A., Martin, P.I., Nicholas, M., Baker, E.H., Seekins, H., Helm-Estabrooks, N., Cayer-Meade, C., Kobayashi, M., Theoret, H., Fregni, F., Tormos, J.M., Kurland, J., Doron, K.W., & Pascual-Leone, A. (2005). Improved picture naming in Broca's aphasia after TMS to part of right Broca's area: An open protocol study. *Brain and Language, 93*, 95–105.

Nass, R.D., & Trauner, D. (2004). Social and affective impairments are important recovery after acquired stroke in childhood. *CNS Spectrum, 9*, 420–434.

Nicholas, M.L., Helm-Estabrooks, N., Ward-Lonergan, J., & Morgan, A.R. (1993). Evolution of severe aphasia in the first two years post onset. *Archives of Physical Medicine and Rehabilitation, 74*, 830–836.

Niewada, M., Kobayashi, A., Sandercock, P.A., Kaminski, B., & Czlonkowska, A. (2005). Influence of gender on baseline features and clinical outcomes among 17,370 patients with confirmed ischaemic stroke in the international stroke trial. *Neuroepidemiology, 24*, 123–128.

Pedersen, P.M., Joergensen, H.S., Nakayama, H., Raaschou, H.O., & Skyhoj Olsen, T. (1995). Aphasia in acute stroke: Incidence, determinants and recovery. *Annals of Neurology, 38*, 659–666.

Pedersen, P.M., Vinter, K., & Olsen, T.S. (2004). Aphasia after stroke: Type, severity and prognosis. The Copenhagen aphasia study. *Cerebrovascular Disorders, 17*(1), 35–43.

Saur, D., Lange, R., Baumgaertner, A., Schraknepper, V., Willmes, K., Rijntjes, M., & Weiller, C. (2006). Dynamics of language reorganization after stroke. *Brain, 129*, 1371–1384.

Savitz, S.I., Dinsmore, J.H., Wechsler, L.R., Rosenbaum, D.M., & Caplan, L.R. (2004). Cell therapy for stroke. *NeuroRx, 1*, 406–414.

Shaywitz, B.A., Shaywitz, S.E., Pugh, K.R., Constable, R.T., Skudlarski, P., Fulbright, R.K., Bronen, R.A., Fletcher, J.M., Shankweiler, D.P., Katz, L., & Gore, J.C. (1995). Sex differences in the functional organization of the brain for language. *Nature, 373*, 607–609.

Sturm, J.W., Donnan, G.A., Dewey, H.M., Macdonell, R.A., Gilligan, A.K., & Thrift, A.G. (2004). Determinants of handicap after stroke: The North East Melbourne stroke incidence study (Nemesis). *Stroke, 35*, 715–720.

Weiller, C., Isensee, C., Rijntjes, M., Huber, W., Mueller, S., Bier, D., Dutschka, K., Woods, R.P., Noth, J., & Diener, H.C. (1995). Recovery from Wernicke's aphasia – a PET study. *Annals of Neurology, 37*, 723–732.

Further Readings

Catani, M. (2006). Diffusion tensor magnetic resonance imaging tractography in cognitive disorders. *Current Opinion in Neurology, 19*(6), 599–606.
This paper shows the potential of DTI techniques, which will probably play an important role in studies of the neurological correlates of recovery in the next few years.

Hillis, A.E. (2007). Magnetic resonance perfusion imaging in the study of language. *Brain and Language, 102*, 165–175.
An excellent review of the application of the PWI technique to the study of language including the neural correlates of recovery.

Price, C.J., & Crinion, J. (2005). The latest on functional imaging studies of aphasic stroke. *Current Opinion in Neurology, 18*, 429–434.
A recent update on functional MR studies of the neural mechanisms of aphasia recovery.

Price, C.J., Mummery, C.J., Moore, C.J., Frakowiak, R.S., & Friston, K.J. (1999). Delineating necessary and sufficient neural systems with functional imaging studies of neuropsychological patients. *Journal of Cognitive Neuroscience, 11*, 371–382.
An influential theoretical discussion of the methodological aspects of recovery studies.

Therapeutic Approaches in Aphasia Rehabilitation

LUISE SPRINGER[1,2]

[1]*Collaborative Research Centre "Media and Cultural Communication",
University of Cologne, Köln, Germany*
[2]*School of Speech and Language Therapy, University Hospital, Aachen, Germany*

ABSTRACT

The ultimate aim of aphasia treatment is to improve patients' oral and written language abilities and to facilitate their participation in everyday communication. Over the last decades, a range of approaches in aphasia therapy have been introduced, based on cognitive neuropsychological models, psycholinguistic theories, and socio-pragmatic approaches. Planning therapy for aphasic individuals involves selecting the therapeutic approach most appropriate for dealing with a particular type of disorder, degree of severity, stage of recovery, the extent of the patients' participation in social life, while, at the same time, considering each individual patient's cognitive and pragmatic resources as well as the individual needs and goals. This chapter discusses the main therapeutic approaches as they relate to the stages of recovery: activation therapy at the acute stage, impairment-specific treatment at the post-acute stage, social participation and consolidation at the chronic stage of recovery.

39.1. INTRODUCTION

Aphasia is a relatively common language disorder, occurring in about 25% of all stroke patients. Aphasic disorders hinder communication, place restrictions on family and social life, and constitute a considerable barrier to professional rehabilitation. In our modern-day world, even manual workers need to have high-level oral and written language skills.

At a linguistic level, aphasia is regarded as an impairment of the components of the language system (semantics, syntax, phonology, morphology). The language limitations include both expressive and receptive modalities (comprehension, speaking, reading and writing) to varying degrees. Traditional linguistic learning approaches are aimed primarily at correcting specific language deficits and concentrate on particular linguistic units, regularities, and language modalities. Therapy materials based on these approaches thus focus on specific linguistic deficits without considering the communicative aspects of language. For the aphasic patient, however, communication and social participation in everyday life is more important than linguistic correctness thus making the communicative resources and limitations of the aphasic individual another focus of rehabilitation.

Recent years have seen the emergence of two distinct sets of approaches for rehabilitation of language-impaired individuals. These approaches differ according to the theoretical assumptions they make about the nature of aphasia, about recovery and learning mechanisms and therapeutic goals (for an overview, Stark *et al.*, 2005). The first type of approach is based on cognitive neuropsychological models or linguistic theories and addresses the language impairment directly while the second approach is based on socio-pragmatic models and focuses on the communicative resources and social consequences of the impairment. I will argue that a combination of both approaches is required to profit from the re-learning capacity of the aphasic individual and to facilitate the transfer of the improved language abilities to everyday life (e.g., Lesser & Algar, 1995; Byng & Duchan, 2005).

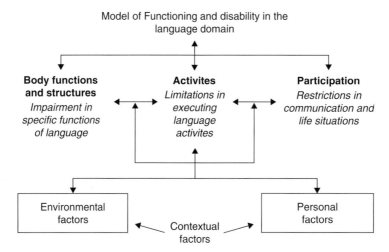

FIGURE 39.1   Model of functioning and disability in the language domain. Functional integrity or impairment of body functions include cognitive functions such as language, memory, and attention, emotional functions and sensory functions such as seeing and hearing. Disabilities at the activities level are difficulties an individual may have in executing language activities. Participation refers to the way an individual can communicate and be involved in a life situation. Environmental factors make up the physical, social and attitudinal environment in which people live and conduct their lives. Personal factors are the particular background of an individual's life, and comprise features of the individual that may play a role in disability at any level such as health states, lifestyle, habits, upbringing, coping styles, social background, education, and profession. *Source*: Based on the International Classification of Functioning, Disability and Health (WHO, 2001). Modified and applied to language functions by L. Springer.

39.2. HOW TO APPROACH APHASIA THERAPY

In view of the range of different approaches and findings of therapy studies, the challenge for therapists is to decide which approach is the most appropriate one for an aphasic individual at a specific stage of recovery. A comprehensive resource for defining goals and targeting approaches to aphasia therapy is the conceptual framework of the World Health Organisation's "International Classification of Functioning, Disability and Health (ICF)" (see Figure 39.1).

Central to the classification are functions and disabilities such as integrity or impairment of language function, capacities and limitations to language activity and restrictions in participation in a range of different areas of life. Additionally, there are context factors covering both material and social environmental aspects plus aspects relating to personal biography, attitude, and experience.

An impairment is defined as a divergence from or loss of normal bodily or (neuro-) psychological functions. Examples are impoverished verb representations in the mental lexicon or a breakdown of automatic retrieval processes. Activity limitations are described as difficulties an individual may have in executing language tasks such as in reading and writing words and sentences. Participation restrictions refer to problems an individual may experience when getting involved in conversation such as starting, sustaining, and ending an interchange of thoughts. Such thoughts can be communicated by way of spoken or written signs, to one or more familiar persons or strangers, and in formal or in casual settings. A lack of devices that ensure outdoor mobility or no availability of specifically designed communicative devices are examples for environmental barriers.

Although optimal rehabilitation should consider both function and disability aspects, most therapeutic approaches focus entirely on the impairment level while others start from the standpoint of reducing the barriers in social context and using the verbal and non-verbal resources available to achieve participation in everyday life (Byng & Duchan, 2005; Howe *et al.*, 2004).

When planning and setting the therapeutic goals for aphasia therapy, it is important to use diagnostic tools that provide an overview of an aphasic person's abilities and disabilities in both the language and associated cognitive domains such as attention, working memory, auditory discrimination, or visual recognition. It is equally important to also include the socio-pragmatic level, that is the aphasic person's participation in everyday life. Essential to setting therapeutic goals is to make certain that their achievement can be assessed quantitatively or by qualitative analysis.

During the course of therapy, the goals defined at the outset are subject to change and this requires adaptation to the changes in the nature of the dysfunction, learning skills and motivation, and also to social needs. The goal-setting should involve both the patients and their family circles to ensure successful transfer of what is (re-)learned in therapy to the everyday setting.

In the following, a framework of aphasia treatment is outlined which considers the rehabilitation process at the various stages of recovery. Different approaches are combined into a comprehensive treatment regimen which differentiates three stages of recovery: the acute, post-acute, and chronic stage.

39.3. A STAGE-ORIENTED FRAMEWORK: FITTING THE THERAPEUTIC APPROACH TO THE STAGE OF RECOVERY

In terms of neurophysiology, the basic mechanisms of functional recovery in the brain are restitution, substitution, and compensation. Each mechanism appears to correlate in different ways with changes in language (re-)learning and communicative behavior at different post-onset times. Consequently, the therapy needs to be adapted to the mechanisms of functional recovery at the different stages in time. Three therapeutic phases can be distinguished: (1) activation therapy at the acute stage, (2) impairment-specific therapy starting at the early post-acute stage, and (3) social participation and consolidation at the postacute and chronic stage (see Box 39.1).

Support for the stage-oriented approach comes from a functional magnetic resonance imaging (fMRI) study (Saur *et al.*, 2006) which described three distinct reorganizational phases: (1) an early phase (0–4 days post-stroke) characterized by acutely diminished activation of the non-damaged left hemisphere speech region, (2) an early post-acute phase (approximately 2 weeks post-infarct) in which an upregulation of neuronal activation that correlates with improved performance occurs in homologous (primarily anterior) speech regions of the right hemisphere, and (3) a consolidation phase (4–12 months post-infarct) in which further improvements correlates with a reduction in right hemisphere activation and increasing activation of the intact left hemisphere speech regions (Saur *et al.*, 2006).

In the acute stage, stimulation techniques are usually applied to support the restitution of temporarily blocked processes. In the postacute and chronic stage, intensive and specific treatment is viewed as supporting a gradual functional reorganization of the impaired language system. This may be achieved by relearning language knowledge (such as morphological agreement or word order in complex sentences) and procedures (such as phoneme–grapheme-conversion), or by compensating for impaired brain functions via verbal and non-verbal (communicative) strategies.

Another crucially important factor for optimal reorganization is the intensity of therapy, that is short intensive treatment intervals (10–20 h/week over 2–4 weeks) that are interspersed with no-treatment phases and often within group settings (see the principle of massed practice, Pulvermüller *et al.*, 2001; Bhogal *et al.*, 2003).

39.3.1. Activation Therapy

Restitution of impaired language functions typically occurs during the first 4–6 weeks and leads to complete recovery in about one-third of patients. At this stage, aphasia therapy aims at enhancing temporarily impaired language functions and at hindering the development of maladaptations, such as automatisms or inadequate verbal and non-verbal compensatory strategies. On the one hand, treatment of individuals in the acute phase should start as soon as patients can be sufficiently stimulated. In principle, attempts to implement early language use in patients can influence the restitution of language functions in a positive way. On the other hand, stressful or incorrect attempts at activation may force pathological language development, often in the form of automatically recurring utterances.

The main approaches and techniques in the acute and post-acute phase are:

- Basic stimulation of neuropsychological and sensory–motor functions
- Deblocking and cueing techniques
- Multimodal stimulation
- Inhibition of maladaptations
- Non-verbal compensation

Box 39.1   Stages of recovery and treatment

Principles and methods of aphasia therapy must accommodate the patient's stage of recovery. Stages of recovery and corresponding phases of therapy are listed below. Early activation therapy aims at the restitution of language functions via increasing or strengthening of connections between the remaining neurons in and around a damaged neural network. During the phase of impairment-specific therapy a gradual functional reorganization of the impaired language system is intended. For the late post-acute and chronic stages of recovery therapy aims at consolidation of achieved language abilities and further social participation.

Stage of recovery	Acute	Early postacute/late postacute	Chronic
Time post-onset	0–4 weeks	1–4/6 month/ 6 month–1 year	>1 year
Phase of treatment	Activation Therapy	Impairment-specific therapy / Social participation and consolidation	
	Family counseling > self-help groups		

In cases of severe, acute aphasia, before actual speech and language therapy is initiated, the seriousness of accompanying neuropsychological symptoms may necessitate basic stimulation aimed at enhancing attention capacity and increasing motivation. Without establishing these basic functions, it is difficult to enlist the cooperation of the patient for speech, language, and motor therapy. For example, if severe speech and swallowing disorders are combined, the first step will be to restore the vital functions of chewing and swallowing. Or, in cases of hemiplegia, it may be necessary at the outset of therapy to aim at improving symmetrical body and head posture. Targeted exercises are required if specific modes of primary sensory abilities, such as central hearing and vision, are severely affected.

Deblocking techniques are used to facilitate temporarily disturbed language functions and modalities by employing more intact capacities and modalities. The deblocking effect occurs at an unconscious level. Semantic deblocking makes use of priming effects based on similarity relationships between the target word and a preceding item. Not only a semantically related lexical item can serve as a prime, however. An intact oral/written modality or action may also deblock a related language activity. For example, speaking, reading, acting out or singing in unison with the therapist may facilitate independent (autonomous) language output. So-called cueing techniques stimulate the target name by presenting the initial phoneme or syllable, a semantically related word, an incomplete sentence, or a description of functions and attributes of the object to be named.

In direct stimulation the patient is asked to imitate non-verbal or verbal acts by mimicking gestures, speaking or singing in unison, or by oral repetition. These imitation tasks require the patient to only provide direct and immediate responses without the necessity for internal activations of broader linguistic knowledge and conscious selection among alternatives. Typically, phonemic, semantic or syntactic errors are neither corrected nor explicitly pointed out. On the contrary, when errors occur, other, simpler, stimuli are given.

Inhibition of automatic utterances can be supported by enhancing language comprehension and by concentrating on non-verbal and/or written modes of expression. Perseverations (unintentional repetition of words and phrases) are prevented by immediate interruption, introduction of a pause, intentional distraction or simplification of the language requirements.

Non-verbal compensation in the acute and post-acute phase means encouraging basic body signs for communication. This can be a simple non-verbal yes/no code such as squeezing hands, blinking the eyes, lifting a finger, and so forth, to enable a dialog with clinical staff and relatives.

39.3.2. Impairment-Specific Therapy

Once the patient is medically stable and the aphasia can be thoroughly examined, impairment-specific treatment can

start. In aphasia arising from vascular etiology, following the acute stage, four main standard syndromes are described: global, Broca's, Wernicke's, and anomic aphasia. Alongside the standard syndromes, a distinction is also made between conduction aphasia and transcortical aphasia (see Chapter 1). Aphasic syndromes, however, reflect the functional architecture of language only in a limited sense. Also, it can be assumed that more complex language functions are distributed over a number of different cortical areas, or mapped by neural nets. Should these have suffered partial damage, functions may be substituted or compensated. Ascribing aphasic syndromes (and more especially individual symptoms) to strictly defined areas of the perisylvian region therefore remains a matter of controversy.

In clinical practice, syndrome description is often used to communicate a first rough picture of the linguistic deficit. For targeted therapy of aphasia, however, a description of the aphasic syndrome is too imprecise. Diagnostic procedures are needed which provide a more detailed picture of the nature of the linguistic and neuropsychological dysfunctions, of unimpaired skills and strategies, and also of the degree to which social participation is restricted (such as the Comprehensive Aphasia Test (CAT), Swinburn et al., 2004).

When language processing models are available to describe and explain symptoms in terms of underlying mechanisms (as is the case for reading, writing, naming and sentence-processing) these models may guide the planning of therapy (e.g. example, Whitworth et al., 2005).

Impairment-specific training aims primarily at relearning degraded linguistic knowledge, reactivating impaired linguistic modalities such as oral and written comprehension, and learning explicit compensatory linguistic strategies. Stimulus material focuses on particular linguistic units and regularities of the language and on formal linguistic properties of phrases, sentences and text-structures (see Box 39.2). The expectation here is that a good performance in specific linguistic tasks should generalize to everyday communication.

39.3.3. Social Participation and Consolidation

When the late postacute or early chronic stage of recovery is reached, the impairment-specific treatment should be complemented by augmenting the patient's degree of social participation and consolidation of the gains made so far. This phase focuses on three therapeutic goals: (1) maintenance of (re-)learned language abilities, (2) further verbal and non-verbal compensation, and (3) optimal participation in communication and social life. To achieve these goals, socio-pragmatic approaches should be applied. These focus directly on increasing the degree of participation in social life to improve communication in natural settings and to reduce communicative barriers in social contexts (for a

Box 39.2   Standard techniques of impairment-specific therapy

The following table illustrates selected standard techniques of impairment-specific therapy.

Language domains	Tasks based on psycholinguistic and/or neuropsychological models and theories
Lexical systems – Semantics	Multimodal semantic matching tasks with semantically related distractors (hyponyms, whole-part relations, object-action) Metalinguistic tasks such as – Semantic judgements with wide or narrow semantic distractors – Descriptions of word- and sentence meanings – Identification of semantic categories and relations Circumscribing words via verbal and non-verbal means – Description of the semantic features of the target item, for example, location, action, properties and other associations of the target word
– Phonological word form	Oral and/or written word-to-picture-matching with phonological distractors Picture naming using a cueing hierarchy 1. Sentence completion 2. Onset-syllable 3. Onset-phoneme 4. Onset-grapheme of the target word 5. Oral/written rhyme of the target word
Sublexical phonology	Identification, segmentation and combination of – Syllables and onset-rhyme-structures – Phoneme strings in words and pseudowords
Prosody	Identification of the prosodic features of utterances – Word accent – Intonation
Syntax and morphology	Mapping thematic roles onto syntactic structures (agent/patient > subject/object) – Mapping active versus passive sentences to pictures – Metalinguistic judgments of verbs and thematic roles in sentences via wh-questions Producing phrase structures and sentences on different levels in the "syntactic tree." – Constituent arrangement tasks – Sentence completion – Picture description with simple and later more complex sentences Discrimination and production of morphological marking – Completion tasks of agreement inflection – Completion tasks of tense inflection
Text/discourse	Comprehension and production of oral and written text/discourse-types (e.g., monologues versus dialogs) – Analyzing the main topics of a story – Analyzing the coherence of texts – Story (re-)telling or story completion using oral, written or pictorial contexts – Working on the stylistic combination of sentences (cohesion) – Production of different text/discourse types and related styles (e.g., informal versus formal, monologues versus interactive text production) – Communication with single partners or in a group
Written language – Grapheme system – Orthohraphic lexicon	Written-word-to-pictures and oral-word-to-picture matching with deficit-specific distractors (homophone pairs, orthographical similar pairs) Delayed copying words/sentences
– Phoneme–grapheme-conversion	Identification of graphemes Phoneme–grapheme-matching and vice versa Anagram sorting Writing words/sentences to dictation via a letter-by-letter strategy

summary see Carlomagno *et al.*, 2000). There are several socio-pragmatic methods supporting the patient's participation in everyday communication. Examples are:

- Promoting Aphasics Communicative Effectiveness (PACE) (Glindemann & Springer, 1995; Carlomago *et al.*, 2000);
- Group activities with aphasics in therapy settings (e.g., "Constraint-induced therapy", Pulvermüller *et al.*, 2001) and in self-support groups;
- Counseling for people with aphasia and relatives (e.g., van der Gaag *et al.*, 2005);
- Conversational coaching (Hopper, Holland, & Rewega, 2002);
- Training volunteers (e.g., Kagan *et al.*, 2001);
- Reducing environmental barriers (Howe *et al.*, 2004).

Most of these socio-pragmatic methods concentrate on what the aphasic individual is actually able to achieve with her or his limited formal language skills – and just as importantly, on what the aphasic speaker can achieve in cooperation with the dialog partner. The situational context and the communicative strategy of both the aphasic speaker and her or his communication partners are, therefore, the primary focus of treatment. In severe cases of aphasia, the acquisition of non-verbal compensation strategies has to be supported, for example, via descriptive gestures, drawings or device-supported aids to communication (see Chapter 42). In contrast to most of the pragmatic approaches, in the "Constraint-induced therapy" (during which language games are practiced at frequent intervals within group settings) the use of non-verbal compensatory strategies needs to be suppressed in favor of verbal communication (Pulvermüller *et al.*, 2001).

Achieving a better quality of life requires not only that the patients themselves change their social attitude and expectations, but also that their families and their careers adapt to the new life plan. Therapy thus helps to remove social barriers and contributes to successful communication and reinforcement of social engagement and inclusion.

Of the therapeutic approaches previously described, impairment-specific therapy – based on neuropsychological models and/or psycholinguistic theories – is the best equipped with methods and techniques and has provoked numerous studies as well as controversies. It is typically used in the post-acute and early chronic stage. Because of the large field of such model-driven approaches only two examples concerning the lexical and syntactic disorders are discussed in more detail in what follows.

39.4. SPECIFIC NEUROPSYCHOLOGICAL AND PSYCHOLINGUISTIC APPROACHES TO THE TREATMENT OF LEXICAL AND SYNTACTIC DISORDERS

The cognitive neuropsychological approach seeks to locate the language impairment within a hypothesized model of normal language processing at the word and sentence level (Bock & Levelt, 1994; Whitworth *et al.*, 2005). The derivation of therapeutic methods from these models, however, is not straightforward. It is more the case that the models help to identify the nature of the disorders and to make the decision concerning which processing strategies might best be used by the aphasic individual. The therapist then has to choose among the various treatment methods available, or, sometimes, the therapist must even design specific materials and new procedures specifically for the individual patient. As Basso and Marangolo (2000) have pointed out, the advantage of model-based approaches to aphasia therapy lies in an optimization of methods and techniques.

39.4.1. Therapy of Lexical Disorders in Aphasia

Lexical disorders are very common with aphasic patients and consequently the therapy of lexical disorders is an important research field (see table in Box 39.2; for a summary see Nickels, 2002). Symptoms include word-finding difficulties and the production of lexical errors (paraphasias) such as *bag* instead of *suitcase* or *water* instead of *fish*. These can occur in spontaneous speech or may be provoked by specific tasks such as naming objects.

Whereas in classic therapy approaches, patients with lexical disorders were treated irrespective of differences in the underlying psycholinguistic mechanisms of their impairment, nowadays approaches have been developed that attempt to explain the impairment within a psycholinguistic framework (see Figure 39.2) and target the impaired process in the therapeutic approach. Within such a framework the underlying impairment is inferred by analyzing the nature of errors, the linguistic characteristics affecting performance (e.g., word frequency, level of abstractness, word length, morphological complexity, living/non-living, graphemic regularity), and convergent phenomena from different tasks with common underlying processing mechanisms.

Model-based therapy aims to improve impaired lexical processing at the semantic and/or phonological word form level (Nickels, 2002). Most of the semantic approaches try to activate the meaning of words in aphasic patients (Doesborgh *et al.*, 2003). Typical tasks are word-picture-matching, sorting words in categories (e.g., vehicles and buildings, clothes and furniture), and producing synonyms. Phonological treatment is aimed at improving the retrieval of the word form and the selection and sequencing of speech sounds. Examples are oral and written naming of pictures via repetition, reading aloud or using a phonological cueing hierarchy such as prompting with the onset of the word or the first syllable.

To assist an aphasic patient's immediate retrieval of a word corresponding to the picture of an object (prompting), semantic and or phonological cueing methods are helpful. Cueing can also improve lexical access in the long term

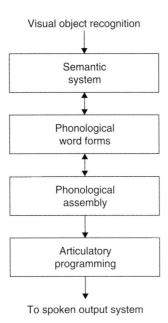

Visual object recognition

Semantic
system

Phonological
word forms

Phonological
assembly

Articulatory
programming

To spoken output system

FIGURE 39.2 A simplified model of spoken word production. There are four main processing levels involved in spoken word production such as in picture naming: (1) Activation of word meanings from the semantic system in response to an idea or concept. Characteristic disorders at this level are semantic errors in all written and oral modalities. (2) Retrieval of the phonological word form. Typical errors here are delays and failures in word retrieval, verbal and non-verbal circumlocutions, and the "tip of the tongue" phenomenon. (3) Phonological assembly of phoneme strings. A disorders at this level is reflected in phonological paraphasias; (4) Articulatory programming to convert phonemes into neuromuscular commands. Disorders result in speech apraxia. *Source*: The simplified model is based on the logogen model of Patterson & Shewell (1987).

(Howard, 2000). The oral or written cue delivered by the therapist contains semantic, phonological and/or syntactic information about the target word and hence acts as a prime. The effectiveness of various priming techniques in treating naming disorders has been shown in group studies and (model-oriented) single-case studies. Semantic priming refers to providing information about the meaning of a target. It is used to facilitate the retrieval of concept or object names and, if repeated several times, can be a very effective treatment method. Likewise, phonological priming of target word forms also has long lasting effects (e.g., Miceli *et al.*, 1996; Hickin *et al.*, 2002; Wambaugh, 2003; Abel *et al.*, 2005).

Some authors stress the importance of treating semantic impairments with semantic tasks (e.g., Nettleton & Lesser, 1991; Doesborgh *et al.*, 2003) and using phonological tasks to improve the retrieval of phonological word forms (Miceli *et al.*, 1996). The difference between semantic and phonological tasks may, however, be overstated. Indeed, semantic as well as phonological therapies are often beneficial for all patients, possibly due to the fact that (1) patients often exhibit mixed rather than pure disorders, (2) lexical processing is interactive in nature, and (3) both therapies entail semantic

and phonological aspects, which can hardly be avoided given the nature of the naming task (Wambaugh, 2003).

39.4.2. Therapy of Syntactic Disorders

Syntactic disorders are typical for people with agrammatism. For the major European languages, agrammatism is characterized by incomplete or incorrectly constructed phrases and sentences, impoverished syntactic structures, a reduced repertoire of verbs, an omission of function words, incorrect inflectional morphology, and non-fluent speech. Discrepancies in degree and kind of agrammatic disturbance within one individual are common across receptive and expressive tasks. For example, when describing a picture, the sentence patterns produced may show more constituents and more morphosyntactic marking than in spontaneous language.

The varying ability in comprehension, judgement and production tasks, and the differences between talk-in-interaction versus monologues are the reason for current controversies about the underlying causes of agrammatism and appropriate treatment approaches (for summaries see Kean, 1995; Kolk, 1998). Agrammatic symptoms may reflect directly an impairment of (morpho-)syntactic knowledge and/or a specific restriction of grammatical processing capacities. Alternatively to a syntactic deficit, some authors have postulated impairment at the canonical word form level, or, in other words, of lemmas. A lemma contains the lexical item's meaning and syntactic information, such as word class (e.g., verbs or nouns). In case of a verb, the lemma specifies syntactic information, such as whether a verb is intransitive (e.g., *sleep*) or transitive (e.g., *drink*) and if transitive, what thematic roles (such as agent, theme) it takes.

It has also been claimed that agrammatic patients rely on simple and automatized utterances reflecting adaptation to their reduced capacities (Kolk, 1998). In this view agrammatic surface-symptoms do not simply reflect an impairment but the general adaptation of an impaired system to varying cognitive and communicative demands. As in normal speakers, in agrammatism the computation and particularly the style of language used (e.g., an informal simple or more formal complex style) depend on the interaction with the dialog partner and on the properties of oral versus written tasks. For example, interactive factors include the presence or distance of the partner, and oral or written tasks refer to online or off-line conditions of a given task, or modality-specific features such as visibility, audibility, simultaneity, sequentially, or revocability.

When planning therapy for agrammatic disorders, the observed deficits can be mapped onto a syntactic processing model to discern the level of processing at which the impairment occurs (see Figure 39.3).

Two types of therapeutic approaches are distinguished in the syntactic processing model: one focusing on the

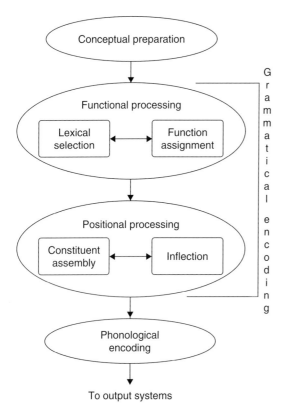

FIGURE 39.3 Syntactic processing model. Bock & Levelt (1994) distinguished four different levels of processing: (1) The level of conceptual preparation generating what is intended to be said. (2) The functional level involving lexical selection (a lexical item's lemma information) and assigning functional (thematic) roles to the selected verb (e.g., functional roles include the agent of an event, the entity on whom the action is performed such as the theme, recipient, or goal). (3) The positional level involved in assembling words into a sentence frame and building the inflectional endings, and (4) the phonological encoding level which serves to build a phonetic plan for the utterance. *Source*: Simplified and adapted version from Bock & Levelt (1994).

positional level and another on the functional level. Based on the functional-level approach several variants of treatment have been developed to retrieve the lemma of verbs (i.e., the meaning and syntax of verbs) with one focusing on the level of lexical selection of verbs with increasing number of arguments. An example for a typical practice sequence is the following: (1) activate the meaning of the verb, (2) build a sentence frame around the verb, and (3) map thematic roles onto the grammatical roles (e.g., for a simple transitive verb (*drink*) the agent (*woman*) should be mapped onto the participant role and the theme (*tea*) onto the object role as in the sentence *The woman is drinking tea*).

Another functional-level method is based on the hypothesis of a mapping deficit. This hypothesis assume that in agrammatism the projection of functional onto positional level information is impaired whereas information on either level remains relatively accessible (Mitchum *et al.*, 2000). These impaired mapping processes are treated primarily by metalinguistic and comprehension tasks. Patients learn

to identify functional (thematic) roles (e.g., agent, theme, patient) irrespective of word order. In active sentences, the agent role and the grammatical subject are the same and therefore often easy to process. In passive and reversible sentences, however, mapping of thematic roles to syntactic functions is often impaired in agrammatism. (e.g., passive reversible sentence: *Paul is kissed by Mary?* Task: *Which one is doing the kissing?* – incorrect answer: *Paul*). With this treatment approach the expectation is that sentence production is improved, even without explicit training. A positive transfer from metalinguistic and comprehension tasks to production does, however, not always occur.

The positional level approach assumes that (morpho-) syntactic structures (such as verb inflection, morphological agreement) and constituent assembly are specifically affected in agrammatism. Consequently, the therapeutic goal is to work on linguistic parameters such as syntactic functions (function words), morphosyntactic marking and also complex phrases and sentences. Within the framework of the positional level approach a specific linguistic approach in the spirit of the generative Chomsky tradition aims at the syntactic processing deficit. This deficit is described as an inability to access the high nodes of the syntactic tree (Friedmann, 2005; see Chapter 13, this volume). In syntactic trees, content and function words are represented in different nodes. The lowest node is the verb phrase, the nodes above are the agreement phrase and the tense phrase, and the complementizer phrase is placed at the highest point of the syntactic tree. The hierarchical order is used to analyze the severity of impairment and to develop treatment tasks. An example of these types of approaches is the Treatment of Underlying Forms (TUF), developed by Thompson & Shapiro (2005). In TUF, the focus is on the practice of complex, non-canonical sentence structures (e.g., object clefts: *It was the artist who the thief chased*, object relatives: *The man saw the artist that the thief chased*, wh-question: *Who did the thief chase?*). This treatment approach showed positive effects insofar as generalization to similarly structured but untrained sentences was observed. In addition, there was also an improvement in spontaneous discourse in most of the mild to moderately agrammatic individuals.

Although in the functional as well as the positional approach, (re-)learning of trained sentences and transfer to similar untrained stimuli seems to be achieved in moderate agrammatism, frequently no generalization to everyday communication is reported. This may arise from training isolated words and/or sentences in highly structured exercises without considering the discourse level. Additional training of these linguistic units (word and/or sentences) in situational contexts and in conversational dialogs is thus required to attain generalization to everyday language use.

For the treatment of severe chronic agrammatism, a compensatory method is the Reduced Syntax Therapy (REST) (Springer *et al.*, 2000). This approach comprises three major features: (1) activation of functional-level information, (2) mapping of thematic roles onto simplified syntactic structures, and (3) omitting morphosyntactic markings. In REST, only basic syntactic processes are considered that can also be found in normal language such as contextual ellipses (e.g., *drinking coffee, running fast, gone to Vienna*). The aphasic individual is instructed to expand one-word utterances into two- and three-constituent utterances in several steps, while ignoring function words and inflectional endings. At the most basic level, REST consists of two-word utterances such as a non-finite verb and a noun functioning as direct object: *shake hands, wash dishes, eat(ing) cake*. Thus, REST encourages patients to employ a simpler style of language in conversation. This enables patients to communicate more fluently and with less effort. Even after extensive left hemisphere damage and more than 3 years of aphasia, a short but intensive REST-treatment improved spontaneous speech in 9 of 11 patients with severe agrammatism. It has been hypothesized that REST enhances elementary language functions of labeling and sequencing in the unimpaired right hemisphere. If the learning abilities of the aphasic patient are adequate, this enhancement can become the starting point for a gradual (re-)learning of syntactic structures.

In sum, impairment-focused treatment based on psycholinguistic and/or neuropsychological models should be applied conjointly with approaches involving social participation of the patient in activities of daily living. This degree of participation will, of course, depend on the severity of the patient's disorder, his or her learning capability and motivation, and a complex mix of social and personal factors. For example, the degree of participation will be compromised if the patient is neither motivated nor able to undergo intensive repetitive language training despite his or her demonstrated capability for relearning. In such cases, a more adequate approach would be to work on participation in real life such as making use of remaining language abilities with the assistance of communication partners. This is in contrast to patients who are highly motivated and ready to practice oral and written language in exercises as well as in everyday communication. These patients seem to profit most from specific impairment-based treatment with intensive computer-assisted hometraining in combination with pragmatic approaches.

39.5. EFFICACY STUDIES OF APHASIA THERAPY

Many follow-up studies have demonstrated that specific aphasia therapy leads to more improvement in linguistic performance than spontaneous recovery alone (Robey, 1998). Although there has been some discussion in the past surrounding the efficacy of speech and language therapy, all studies showing negative results have been demonstrated to be seriously flawed with respect to the selection of the patient population and with respect to the quality and frequency of treatment. There is compelling evidence that intensive therapy performed as frequently as possible and specifically targeted at the linguistic dysfunction is a key pre-condition for success (Pulvermüller *et al.*, 2001; Bhogal *et al.*, 2003). Brain imaging studies have shown that the linguistic reorganization of a damaged brain is aided by intensive speech therapy even in the chronic phase. It has been demonstrated that the reorganization process primarily consists of activation of homologous right hemisphere regions and of intact areas of the dominant left hemisphere (e.g., Musso *et al.*, 1999; Thompson & Shapiro, 2005; Saur *et al.*, 2006).

39.6. CHALLENGES AND FUTURE DIRECTIONS

There is currently insufficient research that investigates why one kind of therapy works and another fails. We also know very little about the neural mechanisms that underlie the various therapeutic approaches, whether therapy influences these mechanisms and if so in which way. Rather than studying whether aphasia treatment is generally effective, future research needs to aim more specifically at the following questions: What is the best timing for treatment? What is the most appropriate approach for a particular case and in the acute, post-acute and chronic stage of recovery? To what degree is the approach influenced by the type and severity of the impairment and the individual goals of the patient, considering her or his specific learning strategies and socio-pragmatic resources? What is the impact of non-linguistic cognitive abilities and disabilities on the outcome of therapy? How can environmental barriers be reduced to support the rehabilitation process?

There is currently a lack of studies that adequately monitors both the nature of aphasia and the impact of therapy. Future therapy studies may provide the key to identify the factors governing functional reorganization after cerebral damage. This knowledge would be of immense value in developing sound neuroscientifically based approaches to the rehabilitation of patients suffering from aphasia.

Acknowledgments

Supported by funding from the German Research Foundation (SFB/FK 427; A1). I am grateful to Dr. F.A. Rodden for his helpful comments and suggestions throughout the creation of this work.

References

Abel, S., Schultz, A., Radermacher, I., Willmes, K., & Huber, W. (2005). Decreasing and increasing cues in naming therapy. *Aphasiology, 19*(9), 831–848.

Basso, A., & Marangolo, P. (2000). Cognitive neuropsychological rehabilitation: The emperor's new clothes. *Neuropsychological Rehabilitation*, *10*(3), 219–229.

Bhogal, S.K., Teasell, R., & Speechley, M. (2003). Intensity of aphasia therapy, impact on recovery. *Stroke*, *34*, 987–992.

Bock, K., & Levelt, W. (1994). Language production: Grammatical encoding. In M.A. Gernsbacher (Ed.), *Handbook of psycholinguistics* (pp. 945–984). San Diego, CA: Academic Press.

Byng, S., & Duchan, J.F. (2005). Social model philosophies and principles: Their application to therapies for aphasia. *Aphasiology*, *19*(10/11), 906–922.

Carlomagno, S., Blasi, V., Labruna, L., & Santoro, A. (2000). The role of communication models in assessment and therapy of language disorders in aphasic adults. *Neuropsychological Rehabilitation*, *10*(3), 337–363.

Doesborgh, S.J.C., van de Sandt-Koenderman, M.W.E., Dippel, D.W.J., van Harskamp, F., Koudstaal, P., & Visch-Brink, E.G. (2003). Effects of semantic treatment on verbal communication and linguistic processing in aphasia after stroke. *Stroke*, *35*, 141–146.

Friedmann, N. (2005). Degrees of severity and recovery in agrammatism: Climbing up the syntactic tree. *Aphasiology*, *19*(10/11), 1037–1051.

Glindemann, R., & Springer, L. (1995). An assessment of PACE therapy. In C. Code & D. Müller (Eds.), *The treatment of aphasia: From theory to practice* (pp. 90–107). London: Whurr.

Hickin, J., Best, W., Hebert, R., Howard, D., & Osborne, F. (2002). Phonological therapy for word-finding difficulties. A re-evaluation. *Aphasiology*, *16*, 981–999.

Hopper, T., Holland, A., & Rewega, M. (2002). Conversational coaching: Treatment outcomes and future directions. *Aphasiology*, *16*(7), 745–761.

Howard, D. (2000). Cognitive neuropsychology and aphasia therapy: The case of word retrieval. In I. Papathanasiou (Ed.), *Acquired neurogenic communication disorders. A clinical perspective* (pp. 76–99). London: Whurr.

Howe, T., Worrall, L., & Hickson, L. (2004). What is an aphasia-friendly environment? A review. *Aphasiology*, *18*(11), 1015–1037.

Kagan, A., Black, S.E., Duchan, J., Simmons-Mackie, N., & Square, P. (2001). Training volunteers as conversation partners using "Supported Conversation for Adults with Aphasia" (SCA): A controlled trial. *Journal of Speech, Language and Hearing Research*, *44*, 624–638.

Kean, M.L. (1995). The elusive character of agrammatism. *Brain and Language*, *50*, 369–384.

Kolk, H. (1998). Disorders of syntax in aphasia: Linguistic–descriptive and processing approaches. In B. Stemmer & H.A. Whitaker (Eds.), *Handbook of neurolinguistics* (pp. 250–260). San Diego Academic Press.

Lesser, R. & Algar, L. (1995). Towards combining the cognitive neuropsychological and pragmatic in aphasia therapy. *Neuropsychological Rehabilitation*, *5*(1/2), 67–92.

Miceli, G., Amitrano, A., Capasso, R., & Caramazza, A. (1996). The treatment of anomia resulting from output lexical damage: Analysis of two cases. *Brain and Language*, *52*, 150–174.

Mitchum, C.C., Greenwald, M.L., & Sloan Berndt, R. (2000). Cognitive treatments of sentence processing disorders: What have we learned? *Neuropsychological Rehabilitation*, *10*, 311–336.

Musso, M., Weiller, C., Kiebel, S., Müller, S.P., Bülau, P., & Rijntjes, M. 1999). Training induced brain plasticity in aphasia. *Brain*, *122*, 1781–1790.

Nettleton, J., & Lesser, R. (1991). Therapy for naming difficulties in aphasia: Application of a cognitive neuropsychological model. *Journal of Neurolinguistics*, *6*, 139–159.

Nickels, L. (2002). Therapy for naming disorders: Revisiting, revising and reviewing. *Aphasiology*, *16*(10/11), 935–979.

Patterson, K.E., & Shewell, C. (1987). Speak and spell: Dissociations and word-class effects. In M. Coltheart, R. Job, & G. Sartori (Eds.), *The cognitive neuropsychology of Language* (pp. 273–295). Hillsdale, NJ: Lawrence Erlbaum.

Pulvermüller, F., Neininger, B., Elbert, T., Mohr, B., Rockstroh, B., Koebbel, P., & Taub, E. (2001). Constraint-induced therapy of chronic aphasia after stroke. *Stroke*, *32*, 1621–1626.

Robey, R.R. (1998). A meta-analysis of clinical outcomes in the treatment of aphasia. *Journal of Speech, Language and Hearing Research*, *41*, 172–187.

Saur, D., Lange, R., Baumgärtner, A., Schraknepper, V., Willmes, K., Rijntjes, M., & Weiller, C. (2006). Dynamics of language reorganization after stroke. *Brain*, *129*, 1371–1384.

Springer, L., Huber, W., Schlenck, K.-J., & Schlenck, C. (2000). Agrammatism: Deficit or compensation? Consequences for aphasia therapy. *Neuropsychological Rehabilitation*, *10*(3), 279–309.

Stark, J., Martin, N., & Fink, R.B. (2005). Current approaches to aphasia therapy: Principles and applications. *Aphasiology*, *19*(10/11), 903–905.

Swinburn, K., Porter, G., & Howard, D. (2004). *Comprehensive Aphasia Test*. Hove: Psychology Press.

Thompson, C.K., & Shapiro, L.P. (2005). Treating agrammatic aphasia within a linguistic framework: Treatment of underlying forms. *Aphasiology*, *19*(10/11), 1021–1036.

van der Gaag, A., Smith, L., Davis, S., Moss, B., Cornelius, C., Laing, S., & Mowles, C. (2005). Therapy and support services for people with long-term stroke and aphasia and their relatives: A six-month follow-up study. *Clinical Rehabilitation*, *19*, 372–380.

Wambaugh, J.L. (2003). A comparison of the relative effects of phonologic and semantic cueing treatment. *Aphasiology*, *17*, 433–441.

Whitworth, A., Webster, J., & Howard, D. (2005). *A cognitive neuropsychological approach to assessment and intervention in aphasia (a clinicians guide)*. New York: Psychology Press.

World Health Organisation (WHO), (2001). *International Classification of Functioning, Disability and Health (ICF)*. Albany, NY/Geneva: WHO Press, from http://www.who.int/classification/icf/en/

Further Readings

Caplan, D., Dede, G., & Michaud, J. (2006). Task-independent and task-specific syntactic deficits in aphasic comprehension. *Aphasiology*, 20, 893–920.

The article presents 42 cases studies of syntactic comprehension in aphasics in the two tasks: sentence-picture matching and object manipulation. The results show that in both tasks no deficits affected performance on all sentence types from a particular syntactic structure. The implications of the pattern of performance for the nature of aphasic deficits are discussed.

Elman, R.J. (2006). Evidence-based practice: What evidence is missing? *Aphasiology*, 20, 103–109.

The article discusses the possible sources of bias within the "Evidence-Based Practice" process, resulting in "missing" evidence. Specific examples are provided from various healthcare fields.

Horton, S. & Byng, S. (2002). "Semantic therapy" in day-to-day clinical practice: Perspectives on diagnosis and therapy related to semantic impairments in aphasia. In A.E. Hillis (Ed.), *The Handbook of adult language disorders. Integrating cognitive neuropsychology, neurology, and rehabilitation* (pp. 229–249). New York: Psychology Press.

The article offers a review of semantic therapy methods and tasks used in research and in clinical practice. The need for more systematic and specific therapies is discussed.

CHAPTER

40

The Pharmacological Treatment of Aphasia

ANDREW W. LEE and ARGYE E. HILLIS

Department of Cerebrovascular Neurology, Johns Hopkins University School of Medicine, Baltimore, MD, USA

ABSTRACT

Aphasia is a disorder of language resulting from injury to the frontal, temporal, or parietal cortex and/or thalamus, or connections between primary sensory cortex and these association areas of the "language dominant" hemisphere. Stroke is the most common cause of aphasia. There are exciting new developments in the treatment of stroke, whereby language dysfunction may be reversed in the acute phase using elevation of blood pressure or clot dissolving agents. In addition, a number of therapies are being investigated in the chronic recovery phase to potentiate neural plasticity. Investigations of the effect of pharmacological treatment to reduce language decline in cases of progressive aphasia due to neurodegenerative disease are still in its infancy. This chapter aims to review the recent developments in the pharmacological treatment of aphasia and how this may augment language therapy.

40.1. INTRODUCTION

Aphasia is defined as a disorder of language due to damage to eloquent cortical and subcortical brain regions. The most common etiology is stroke, which provides a template for studying the progression of acute to subacute to chronic changes in language dysfunction and recovery as a result of neural injury. In the acute phase of stroke, exciting new research in the last decade has demonstrated that it is possible to limit neural cell death from ischemia by either clot busting agents or by opening collateral blood flow by elevations in blood pressure.

In the subacute to chronic phase, the undamaged brain tissue shows a remarkable ability to subsume the functions of damaged brain tissue by increasing synaptic function in undamaged parts of the brain. This "rewiring" process is termed neural plasticity. There are a number of pharmacological agents that may augment neural plasticity.

The aim of this chapter is to review the state of play of pharmacotherapy in aphasia as it applies to stroke specifically including both acute reperfusion therapies as well as agents that augment neural plasticity.

40.2. STROKE: HOW DOES IT CAUSE LANGUAGE DEFICITS?

The common public perception of stroke is the sudden onset of a devastating loss of function, characterized by loss of speech and complete paralysis. The term "stroke" was initially coined as it was perceived that the symptoms were caused by some higher divinity visiting on the unfortunate victim a form of punishment for their inequity. For many years the outcome of stroke was poor, and there was a degree of pessimism from both the public as well as neurologists that anything could be done. However, in the last 30 years significant developments in brain research have shown that therapy can limit the damage to the brain from stroke and even reverse tissue dysfunction.

Stroke is usually caused by a blood clot blocking a blood vessel leading to a specific part of the brain. This interrupts the flow of blood and nutrients such as oxygen to brain tissue and as a result neurons die (Figure 40.1). The symptoms of stroke reflect the types and locations of neurons that are affected. For example, in left posterior frontal cortical strokes, there is often

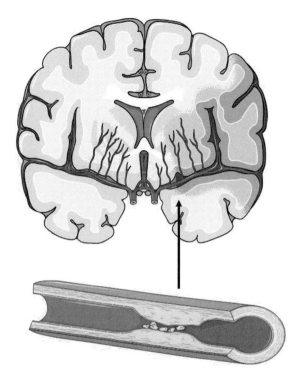

FIGURE 40.1 Stroke is caused by a blockage of a blood vessel supplying a specific part of the brain (black arrow). As a result brain tissue undergoes ischemic necrosis (shaded blue region of the brain). Figure reproduced by consent of Servier (http://www.servier.com/SMART/home_smart.asp). (See Plate 24.)

involvement of Broca's area and motor areas leading to nonfluent aphasia in addition to paralysis.

40.2.1. Do Neurons Die Immediately?

Much like a swimmer floundering in a heavy ocean, neurons have mechanisms to try and keep afloat when their blood supply is reduced. The first thing that occurs is a halt in neuron protein synthesis. This is followed in turn by cessation of electrical activity and finally by a loss of cell membrane integrity and neuron death. In primate models of stroke, Astrup et al. (1977) showed that when blood flow was reduced to 30% of normal, electrical function of neurons stopped. At levels of blood flow of 10% of normal, failure of the Na^+/K^+ ATPase pump occurred, which is responsible for keeping cell membrane integrity, and heralded impending cell death. But what happens between the thresholds of membrane and electrical failure? It was shown by Astrup and colleagues that restoring blood flow by raising the blood pressure, and therefore opening collateral vessels would result in restored nerve function, analogous to jump starting a car with an almost dead battery. The areas of brain tissue which could be seen as being in a hibernating state where function can potentially be restored is termed the *ischemic penumbra* (Box 40.1).

40.2.2. The Ischemic Penumbra Defined by the DWI/PWI Mismatch

Aphasia is the result of involvement of an area of eloquent cortex from a stroke. However, the volume of brain tissue is not only composed of densely infarcted tissue that will never recover, but also of tissue that has the potential to recover. Due to a failure of cellular membrane integrity, brain tissue that has undergone irreversible infarction tends to accumulate water as the Na^+/K^+ ATPase pump responsible for pumping water out of the cell can no longer function. In much the same way as a boat with a hole in its bottom will accumulate water, so too does the dying cell. The movement of water from the extracellular to the intracellular space results in a *restriction* of water *diffusion* as it is now confined within the cell, rather than in the larger extracellular space. *Diffusion* is the free movement of water molecules in a given volume by Brownian motion. Using our analogy of the ocean, if we were to put a piece of wood into the ocean, this piece of wood could drift anywhere, given the ocean is sdo large. On the other hand if you place the same piece of wood in a bath tab, it will remain in one place. If we liken water molecules to our piece of wood, and the "drifting" of the wood to Brownian motion, then one is able to demonstrate how *diffusion* of water molecules is restricted once there is a net movement of water into cells. This *restricted diffusion*, which is a surrogate marker of impending cell death, can be visualized using a *diffusion weighted MRI* (DWI).

Brain tissue within the penumbra has not had a sufficient reduction of blood flow to cause dysfunction of the Na^+/K^+ ATPase pump, and hence DWI is typically normal. However, measurement of cerebral blood flow using a *perfusion weighted MRI* (PWI) will show reduced blood flow not only in areas where irreversible infarction has occurred but also in the penumbra. Therefore, the subtraction of the volume of the DWI from the PWI will yield the volume of the tissue that approximately corresponds to the ischemic penumbra (Figure 40.2; Schlaug et al., 1999). The ischemic penumbra can also be visualized with positron emission tomography (PET), computerized tomography (CT) perfusion scans, or other imaging modalities that can measure regional cerebral blood flow and/or metabolism (Croquelois et al., 2003).

40.2.3. The Diffusion/Perfusion Mismatch in Aphasia

In subjects with dominant/left hemispheric stroke, error rate on language tests correlates well with the volume of abnormality on PWI (Hillis et al., 2000; Reineck et al., 2005). Furthermore, Reineck et al. (2005) showed that using language tests to estimate the volume of perfusion

Box 40.1 How does MRI image brain tissue in stroke

Magnetic resonance imaging (MRI) is a modality that can image body tissue in more detail. It is done by generating a strong magnetic field of 1.5 Tesla or more. At this strength, protons (H^+) can be excited by a magnetic pulse to a higher energy level and are oriented either perpendicular or parallel to a magnetic field, depending on the amount of energy that they absorb from the magnetic pulse. In much the same way that it is easier to keep a canoe parallel rather than perpendicular to a river current, protons that are oriented perpendicular to the magnetic field are at a higher energy state than those that are parallel. When the magnetic pulse terminates these protons move back to their resting state by emitting energy and the time it takes to go back to the resting state can be measured. As water is the most abundant source of protons in the human body, the relative distribution of water within the body can be measured to develop an image of tissues. MRI can also detect an overall Brownian motion, or *diffusion* of water in tissues. With imminent cell death there is a net movement of water from the intracellular to the extracellular space resulting in restricted water diffusion. This is shown as a "bright" area on diffusion weighted image (DWI).

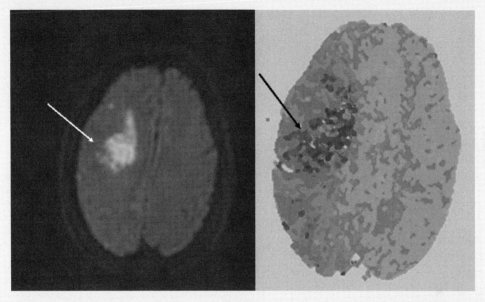

This figure shows a diffusion weighted image (DWI) on the left and a perfusion weighted image (PWI) on the right of a subject present with a middle cerebral artery stroke. Note that while the area of presumed irreversible infarction is relatively smaller on the DWI (white arrow) the perfusion deficit represented by the blue color on the right is significantly larger (see Plate 25).

abnormality and DWI to calculate the volume of diffusion abnormality allows one to calculate a "diffusion-clinical mismatch" that predicts the potential for recovery of language with reperfusion when PWI is not available. Other studies have demonstrated that either language tests or PWI (or other blood flow imaging system) can be used to demonstrate successful restoration of blood flow in acute stroke (Hillis, 2002a; Hillis, 2002b; Croquelois *et al.*, 2003).

40.2.4. Restoring Language Function by Reperfusion

It is widely agreed that blood flow must be restored to the ischemic penumbra for the tissue to become functional again, although the window of time to intervene successfully depends on the degree of blood flow reduction and other factors. If no intervention occurs, eventually the ischemic penumbra proceeds to complete tissue infarction and no degree of reperfusion therapy can restore lost neuronal function. Disruption of the clot in ischemic stroke through intravenous or intraarterial thrombolysis (breaking up the clot in the artery), opening the artery with stents or surgery, or improving collateral blood flow to the salvageable tissue, have all been shown to improve language in acute ischemic stroke (Hillis & Heidler-Gary, 2002; Wityk *et al.*, 2002; Hillis *et al.*, 2002a; Hillis *et al.*, 2003a; Hillis *et al.*, 2006). A large, multicenter, randomized, placebo-controlled trial demonstrated that a thrombolytic agent, tissue plasminogen activator (rTPA) given within 3 h of ischemic stroke onset results in an improved outcome, presumably by restoring perfusion and cerebral function via lysis of clot (NINDS, 1995). Recently, the use of diffusion–perfusion mismatch has been shown to successfully select patients who can undergo thrombolysis up to 6 h after the onset of stroke symptoms (Thomalla *et al.*, 2006).

Studies have shown that restoration of brain perfusion can also be achieved by elevating blood pressure increasing blood flow through collateral vessels to ischemic brain tissue, resulting in a reversal of neurological deficits. Improved stroke outcome with this type of intervention was first demonstrated in animal studies (Drummond *et al.*, 1989). Several human investigations have since demonstrated a significant improvement in language and other cortical functions when phenylephrine, a peripheral vasoconstrictor was used to raise systemic blood pressure in ischemic human stroke within 12 h of symptom onset (Rordorf *et al.*, 2001). A pilot randomized trial conducted by our group demonstrated that the use of induced hypertension with phenylephrine combined with either volume expansion or oral medications (fludrocortisone, midodrine, and/or salt tablets) resulted in an elevation of mean arterial pressure. This elevation in mean arterial pressure was associated with an improvement in measures of cortical language function for dominant hemisphere stroke (as well as spatial attention in right hemisphere stroke), as well as a resolution in perfusion deficits as measured by PWI post-treatment (Hillis *et al.*, 2003b). Conversely, other researchers have demonstrated a detrimental effect on cerebral function when blood pressure was reduced in the acute phase (Oliveira-Filho *et al.*, 2003). Both functional imaging outcomes were significantly better in the treated group compared to the control group that received conventional management. See Figures 40.2(a) and 40.2(b) for an example of this therapy in a subject with thalamic aphasia.

While there is experimental evidence as well as biological plausibility for the use of blood pressure augmentation to restore language function in stroke, no large prospective, randomized, double-blind placebo-controlled trials have been done to evaluate this therapy in everyday clinical practice. What is evident is that there is a certain proportion of ischemic stroke patients that are pressure dependent as evidenced by a resolution of the perfusion deficit with elevated blood pressure. However, therapy to elevate blood pressure may not be appropriate for patients with pre-existing heart disease or peripheral vascular end organ compromise.

40.2.5. The Potential for Neuroprotection

While many neuroprotective agents have been shown to be effective in reducing stroke volume in animal studies, most have shown no benefit on outcome in human studies. However, there is a renewed interest in using neuroprotective agents to extend the duration of the ischemic penumbra,

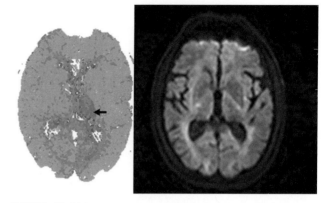

FIGURE 40.2(a)   PWI (left) and DWI (right) of a subject with a left posterior cerebral artery stenosis, resulting in significant hypoperfusion to the left thalamus (black arrow) and a thalamic aphasia syndrome. This subject was placed on blood pressure elevating therapy, increasing his baseline mean arterial pressure by 10% (see Plate 26(a)).

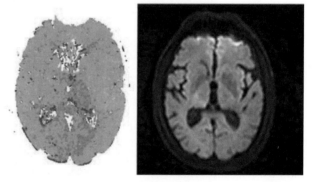

FIGURE 40.2(b)   One day later, his aphasia has resolved, and the perfusion deficit in his left thalamus is less pronounced compared to previously, but is still present. As there is no corresponding left thalamic DWI lesion, the entire area of perfusion deficit (blue) in Figure 40.2(a), can be considered the ischemic penumbra, and shown to be resolving in Figure 40.2(b) (see Plate 26(b)).

pending definitive reperfusion therapy. A novel agent, NXY-059 was initially shown to have efficacy in improving outcome in a randomized, placebo-controlled trial of ischemic stroke (Lees *et al.*, 2006). NXY-059 is thought to be a free radical scavenger. Free radicals are generated in ischemic brain tissue and are responsible for neurovascular tissue injury following reperfusion and may lead to intracerebral reperfusion hemorrhages. In addition, the presence of free radicals in an already ischemic environment can further exacerbate the rate of neuron death. NXY-059 was shown to improve the modified Rankin Score at 90 days in treated patients compared to placebo. The effect was small, but statistically significant in this very large study. However, a follow-up study failed to show significantly more improvement in outcome of the treated patients. At this stage no specific studies looking at restoring language function and neuroprotectants have been performed. However given the potential for restoring general cerebral function, larger scale trials in acute aphasia should be done to explore this potential intervention.

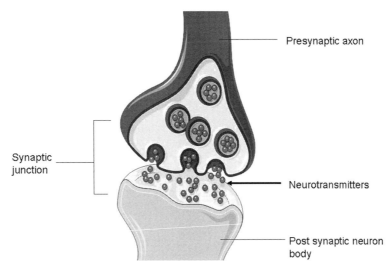

Presynaptic axon

Synaptic
junction

Neurotransmitters

Post synaptic neuron
body

FIGURE 40.3   Neural plasticity is thought to occur at the level of the synaptic junction whereby either more neurotransmitter is released or alternatively more synapses are formed. Figure reproduced with consent of Servier (http://
www.servier.com/SMART/home_smart.asp). (See Plate 27.)

40.3. SUBACUTE (AND CHRONIC) STROKE: ENHANCING NEURAL TRANSMISSION AND NEUROPLASTICITY

So far we have looked at restoring language function by restoring blood flow in the acute phase and thus preventing neuron death. However, what happens once neurons have died? It is known that neuronal networks have an ability to compensate for areas that are lost. In much the same way as an electricity company can get other power stations to take over the load during a blackout, in the subacute and chronic phases of stroke, further recovery depends on other parts of the brain assuming the function of the damaged area. This reorganization of structure/function relationships likely takes place through formation of new synapses or strengthening of synaptic transmission (see Figure 40.3). The primary mechanisms by which synaptic transmission changes are long-term potentiation (LTP) and long-term depression (LTD). LTP and LTD depend on the availability of neurotransmitters such as norepinephrine and acetylcholine. Maintaining changes in synaptic transmission depends on new protein synthesis through a process that requires dopamine and serotonin. Pharmacological interventions that manipulate these neurotransmitters thus may facilitate synaptic plasticity triggered by stimulation such as language therapy (Box 40.2).

40.3.1. Cholinesterase Inhibitors

Cholinesterase inhibitors increase the concentration and duration of action of the neurotransmitter acetylcholine by inhibiting the enzyme involved in its breakdown. These agents were first tested in Alzheimer's Disease, which results in a progressive loss of neurons whose transmission

---

Box 40.2   Synapses explained

The human nervous system can be likened to a network of electrical wires. Unlike the hardwired network of our electricity grid, each neuron is not in direct continuity to the next. In order for an impulse or message to be transmitted from one neuron to the next, this message must be able to "jump" across a given space, the "synaptic junction" (see Figure 40.3). In the same way that inter-company couriers are dispatched to carry documents from one building to the next, the neuron also has couriers that can cross the synaptic junction. These are chemicals termed neurotransmitters. They have a relatively short lifespan as they are either taken back up by their parent neuron or broken down. Antidepressants and antidementia agents work by preventing either the reuptake or breakdown of these neurotransmitters and thus synapses are made to work harder to compensate for damaged areas within the brain.

---

depends on acetylcholine. Alzheimer's Disease is characterized by progressive dysfunction of bilateral temporal and parietal lobes resulting in predominant deficits in memory and in at least one other cognitive domain (e.g., visuospatial skills, language, executive functions). A number of agents have been tested in Alzheimer's Disease with variable effects on cognition. The most consistently effective medications are the central cholinesterase inhibitors. In trials of central cholinesterase inhibitors, very small improvements or reduced rate of decline in memory, language, and/or visuospatial functions have been consistently demonstrated in treatment groups relative to placebo groups of patients with Alzheimer's Disease (Cummings, 2004).

Several studies have investigated the effects of cholinesterase inhibitors on aphasia. The application of the reversible central cholinesterase inhibitor, donepezil, has been found to result in improvement of aphasic deficits in single case studies of subjects with vascular lesions without Alzheimer's Disease (Hughes *et al.*, 2000). Similarly, in trials for the treatment of vascular dementia, donepezil was associated with a small, but statistically significant, improvement in cognition, overall function, and activities of daily living (Salloway *et al.*, 2002, 2003).

A small pilot, open label, prospective study of incremental doses of donepezil in chronic left hemisphere stroke was performed by Berthier *et al.* (2003) over a 20-week period. A series of 10 participants with left hemisphere stroke with aphasia were given increasing doses of donepezil at 5 and 10 mg at weeks 4 and 16, respectively. This was followed by a washout period of 4 weeks. At each increment in donepezil dose and at the end of the washout period, the Aphasia Quotient of the Western Aphasia Battery was measured and compared with the subject's Aphasia Quotient at baseline. There was a mean increase in the Aphasia Quotient associated with the use of donepezil and also with increasing dose. At the same time the mean Aphasia Quotient declined during the washout phase, suggesting that in this cohort, the use of donepezil was associated with some improvement in aphasia. The mechanism of action is thought to be due to the facilitation of neural plasticity via LTP and enhancement of neural activity (Tsz-Ming & Kaufer, 2001).

40.3.2. Piracetam

Piracetam belongs to the pyrrolidone family of chemicals and first came to prominence in the 1970s in France as a nootropic agent, a term meaning to "enhance learning and memory." Interest in this group of drugs has been renewed in their use in mild cognitive impairment and epilepsy. The pyrrolidone group of compounds are unique in that subtle changes in their chemical structure can result in very different effects on the central nervous system. An example of this group is levetiracetam, which is related to piracetam but has a profound antiepileptic effect. However, all have a common mechanism of action of potentiating acetylcholine by either increasing its production or decreasing its breakdown. In addition piracetam may potentiate neuron function with added neuroprotective functions such as enhancing cellular metabolism and oxidative glycolysis, as well as optimizing cerebral blood flow and reducing the risk of thrombosis by improving erythrocyte function and decreasing platelet aggregation. For a comprehensive dissertation on the pyrrolidone family of agents, the reader is directed to the review of Shorvon (2001).

An additional pilot study by Kessler *et al.* (2000) showed that in 12 patients assigned to piracetam for 6 weeks, that there was an increase in blood flow in eloquent cortex in the treatment versus the placebo group. Specifically, there was an increase in perfusion to Heschl's gyrus and both Wernicke's and Broca's areas. In contrast, the placebo group showed an increased perfusion, but only to the inferior part of the left precentral gyrus. Improvement in cortical perfusion was also associated with an improvement of neuropsychological functions. However, it is unclear whether the improvement of neuropsychological functions was the result of improved blood flow, or whether piracetam resulted in the potentiation of neural plasticity, followed by improved function, reflected in increased blood flow. Either way, the changes in blood flow to language cortex shown by Kessler *et al.* (2000) provides a biologically plausible mechanism for the effect of piracetam on the language improvement found by Huber *et al.* (1997).

40.3.3. Bromocriptine

The ascending dopamine pathways arise from the substantia nigra and ventral tegmental area and project to the supplemental motor area and prefrontal agranular regions. These pathways are part of the network involving the anterior cingulate gyrus, dorsolateral prefrontal and inferior frontal cortices. Vascular lesions in these areas can produce nonfluent aphasia and deficits in focusing attention and concentration in language production. It is from these anatomical correlates that the use of bromocriptine, a dopamine receptor agonist, was postulated to have an effect on aphasia rehabilitation. More specifically, given the multiple projections from the midbrain to the cortex, it was hypothesized that bromocriptine could be used for the treatment of transcortical aphasia (Albert *et al.*, 1988).

Several case reports and small open label trials of bromocriptine in stroke have demonstrated a partial recovery of speech or language after treatment with bromocriptine. However, these results have not been confirmed in subsequent, larger, randomized placebo-controlled trials. The most recent of these trials by Ashtary *et al.* (2006) showed that aphasia improved following intervention with either bromocriptine and language therapy or language therapy alone. However, there was no added benefit of bromocriptine over language therapy. (For a comprehensive review of bromocriptine and aphasia, see Berthier, 2005.)

40.3.4. Dexamphetamine

Dexamphetamine facilitates the presynaptic release of dopamine, serotonin, and norepinephrine, while inhibiting reuptake from the synaptic cleft. In studies of hemiplegic stroke, the concurrent administration of dexamphetamine and physical therapy has been shown to facilitate motor recovery. In an initial unblinded pilot study, dexamphetamine was shown to have a positive effect on recovery of language in patients with aphasia due to subacute stroke (Walker-Batson *et al.*, 1992). A subsequent randomized, double-blind, placebo-controlled trial of dexamphetamine and language therapy versus language therapy alone enrolled 21 subacute stroke patients (Walker-Batson *et al.*, 2001). A group of 12 subjects were assigned to the treatment arm of 10 mg dexamphetamine prior to language therapy; nine patients were assigned to the placebo arm of language therapy alone. Subjects had to be enrolled between 16 and 45 days following an ischemic left cerebral hemisphere stroke. After a 5-week period, the dexamphetamine treatment group showed more improvement in language function as tested by the Porch Index of Communicative Abilities. This difference between the treatment and placebo arms was maintained and even further improvement occurred at 6 months after dexamphetamine withdrawal. However, no effect has been yet documented in chronic aphasia (beyond 6 months), which may suggest that neural plasticity has a ceiling effect (McNeil *et al.*, 1997).

40.3.5. Antidepressants

As mentioned in the previous discussion on synapses and neural plasticity, serotonin and norepinephrine are important synaptic neurotransmitters. Specific serotonin reuptake inhibitors can increase the amount of serotonin available for synaptic transmission and are commonly used as antidepressants. Similarly, monoamine oxidase inhibitors prevent the breakdown of serotonin increasing its duration of action at the synaptic junction. The question is whether these agents can potentiate neural plasticity in language post-stroke.

While antidepressant agents show potential in improving neural plasticity in animal models of brain injury, there is a paucity of human trials in stroke rehabilitation and language recovery. Both, Pariente *et al.* (2001) and Loubinoux *et al.* (2005) have shown independently that the specific serotonin reuptake inhibitor, fluoxetine, results in improved motor function. In the former study, there was improvement in a cohort of hemiparetic subjects, while in the latter there was an improvement in motor function tests in a group of normal controls.

However, when moclobemide, a monoamine oxidase inhibitor, was used in addition to language therapy in individuals with aphasia, no improvement was found with moclobemide administration (Laska *et al.*, 2005). One of the confounding factors in this study was that out of the 45 subjects randomized to the moclobemide and control arms, approximately half the subjects in each group did not receive ongoing speech and language therapy throughout the study period. The importance of this finding is that while pharmacological modulation may augment language therapy, recovery may depend on ongoing language therapy. Although there is controversy regarding which pharmacological agents potentiate language recovery, there is wide agreement that the intervention that most consistently results in language improvement in aphasia is language therapy.

40.4. SUMMARY: THE STATE OF PLAY OF PHARMACOLOGICAL INTERVENTION IN APHASIA

Most of the research on the treatment of aphasia has been conducted in aphasia due to stroke – the most common etiology of aphasia. The initial, acute phase of stroke results in a degree of neuronal dysfunction that depends on the degree of ischemia (reduction in blood flow). This initial acute phase can be corrected by removal of the clot from the obstructed cerebral artery, via thrombolysis, stent insertion, or surgery. If these interventions are not possible or hazardous, then the improving collateral blood flow around the obstructed blood vessel, via volume expansion or blood pressure elevating agents shows promise.

In subacute and chronic phases of stroke, recovery depends on neural plasticity or reorganization of structure–function relationships. It is during these subacute and chronic phases that language therapy, in conjunction with some pharmacological agents, has been shown to result in improvement in function, presumably through reorganization brought about by synaptic plasticity. What is not known is which agents will consistently result in augmentation of language therapy and/or improvement in language function on their own, nor for what duration these agents should be given. More importantly, it is not known whether a combination of agents that potentiate different synaptic neurotransmitters and pathways might result in benefit when single agents do not.

Aphasia can also result from neurodegenerative disease, such as primary progressive aphasia (PPA). PPA is a clinical condition, caused by several different pathological conditions, including frontotemporal lobar degeneration (tau-positive or ubiquitin-positive), Pick's Disease, corticobasal degeneration, and unusual focal forms of Alzheimer's Disease (see Davies *et al.* 2005 for review). PPA is characterized by decline in language, without deficits in other cognitive domains, for at least 2 years. Treatment trials for PPA that include pharmacological interventions are just being initiated (Reed *et al.*, 2004). It is plausible that language therapy with or without medications (e.g., donepezil, piracetam) may result in temporary improvement or reduction in the rate of decline in language in a subset of these patients; effects may depend on the pathological etiology.

40.5. CHALLENGES AND FUTURE DIRECTIONS

In the acute and subacute phase of stroke therapy, the biggest question that remains to be answered is whether the diffusion/perfusion mismatch can be used to extend reperfusion therapies beyond the 3 h window for thrombolysis. While observational studies suggest that this is possible, a randomized placebo control trial has yet to be published. Similarly, it remains to be seen whether induced elevation in blood pressure will also pass the test of a large, randomized, placebo-controlled clinical trial.

In the subacute and chronic phases of stroke aphasia recovery, further research will probably concentrate on whether a combination of agents in addition to language therapy will have an additional impact on aphasia recovery. Newer MRI modalities, such as diffusion tensor imaging (DTI), may also help in the prediction of aphasia recovery. Current research in stem cells in animal recovery models of stroke are promising, but much remains to be done before these possible therapies are deployed for human use over the next quarter of a century.

Pharmacological intervention alone probably has the most impact during the acute treatment of stroke with the use thrombolytics or surgical intervention and perhaps blood pressure elevation to restore brain perfusion. In the subacute to chronic phase, pharmacological agents may provide useful adjuvant therapy, but probably only in the presence of continued speech and language therapy. Identifying the optimal duration of such pharmacological interventions and whether or not combinations of agents will work where single agents have previously failed require further investigation.

Acknowledgment

The authors' work is supported by NIH (NIDCD), through RO1 DC05375.

References

Albert, M.L., Bachman, D.L., Morgan, A., & Helm-Estabrooks, N. (1988). Pharmacotherapy of aphasia. *Neurology*, 8, 877–879.

Ashtary, M., Janghorbani, M., Chitsaz, A., Reisi, M., & Bahrami, A. (2006). A randomized double blind controlled trial of bromocriptine efficacy in nonfluent aphasia after stroke. *Neurology*, 66, 914–916.

Astrup, J., Symon, L., Branston, N.M., & Lassen, N.A. (1977). Cortical evoked potential and extracellular K^+ and H^+ levels at critical levels of brain ischemia. *Stroke*, 8, 51–57.

Berthier, M.L. (2005). Poststroke aphasia: Epidemiology, pathophysiology and treatment. *Drugs Ageing*, 22(2), 163–182.

Berthier, M.L., Hinjosa, J., Del Carmen, M., & Fernandez, I. (2003). Open-label study of donepezil in chronic post stroke aphasia. *Neurology*, 60, 1218–1219.

Croquelois, A., Wintermark, M., Reichhart, M., Meuli, R., & Bogousslavsky, J. (2003). Aphasia in hyperacute stroke: Language follows brain penumbra dynamics. *Annals of Neurology*, 54, 21–29.

Cummings, J.L. (2004). Alzheimer's disease. *New England Journal of Medicine*, 351, 56–67.

Davies, R.R., Hodges, J.R., Kril, J.J., Patterson, K., Halliday, G.M., & Xuereb, J.H. (2005). The pathological basis of semantic dementia. *Brain*, 128, 1984–1995.

Drummond, J.C., Oh, Y.S., Cole, D.J., & Shapiro, H.M. (1989). Phenylephrine-induced hypertension reduces ischemia following middle cerebral artery occlusion in rats. *Stroke*, 20, 1534–1538.

Hillis, A.E., & Heidler-Gary, J. (2002). Mechanisms of early aphasia recovery: Evidence from MR perfusion imaging. *Aphasiology*, 16, 885–896.

Hillis, A.E., Barker, P.B., Beauchamp, N.J., Gordon, B., & Wityk, R.J. (2000). MR perfusion imaging reveals regions of hypoperfusion associated with aphasia and neglect. *Neurology*, 55, 782–788.

Hillis, A.E., Kane, A., Tuffiash, E., Ulatowski, J.A., Barker, P.B., Beauchamp, N.J., & Wityk, R.J. (2002a). Reperfusion of specific brain regions by raising blood pressure restores selective language functions in subacute stroke. *Brain and Language*, 79, 495–510.

Hillis, A.E., Wityk, R.J., Barker, P.B., Beauchamp, N.J., Gailloud, P., Murphy, K., Cooper, O., & Metter, E.J. (2002b). Subcortical aphasia and neglect in acute stroke: The role of cortical hypoperfusion. *Brain*, 125, 1094–1104.

Hillis, A.E., Ulatowski, J.A., Baker, P.B., Torbey, M., Ziai, W., Beauchamp, N.J., Oh, S., & Wityk, R. (2003a). A pilot randomized trial of induced blood pressure elevation: Effects on function and focal perfusion in acute and subacute stroke. *Cerebrovascular Diseases*, 16, 236–246.

Hillis, A.E., Wityk, R., Barker, P.B., & Carmazza, A. (2003b). Neural regions essential for writing verbs. *Nature Neuroscience*, 6, 19–20.

Hillis, A.E., Kleinman, J.T., Newhart, M., Heidler-Gary, J., Gottesman, R., Barker, P.B., Aldrich, E., Llinas, R., Wityk, R., & Chaudhry, P. (2006). Restoring cerebral blood flow reveals neural regions critical for naming. *Journal of Neuroscience*, 26(31), 8069–8073.

Huber, W., Willmes, K., Poeck, K., Van Vleyman, B., & Deberdt, W. (1997). Piracetam as an adjuvant to language therapy for aphasia: A randomized, double blind, placebo controlled pilot study. *Archives of Physical Medicine and Rehabilitation*, 78, 245–250.

Hughes, J.D., Jacobs, D.H., & Heilman, K.M. (2000). Neuropharmacology and linguistic neuroplasticity. *Brain and Language*, 71, 96–101.

Kessler, J., Thiel, A., Krabe, J., & Heiss, H. (2000). Piracetam improves activated blood flow and facilitates rehabilitation of post stroke aphasic patients. *Stroke*, 31, 2112–2116.

Laska, A.C., von Arbin, M., Kahan, T., Hellblom, A., & Murray, V. (2005). Long term antidepressant treatment with moclobemide for aphasia in acute stroke patients: A randomized, double-blind, placebo-controlled study. *Cerebrovascular Diseases*, 19, 125–132.

Lees, K.R., Zivin, J.A., Ashwood, T., Davalos, A., Davis, S.M., Diener, H.-C., Grotta, J.C., Lyden, P., Shuaib, A., Hardemark, H.-G., & Wasiewski, W. (2006). NXY-059 for acute ischemic stroke. *New England Journal of Medicine*, 354, 588–600.

Loubinoux, I., Tombari, D., Pariente, J., Gerdelat-Mas, A., Franceries, X., Cassol, E., Rascol, O., Pastor, J., & Chollet, F. (2005). Modulation of behaviour and cortical motor activity in healthy subjects by a chronic administration of a serotonin enhancer. *Neuroimage*, 27, 299–313.

McNeil, M.R., Doyle, P.J., & Spencer, K.A. (1997). A double-blind, placebo-controlled study of pharmacological and behavioural treatment of lexical–semantic deficits in aphasia. *Aphasiology*, 11(4/5), 385–400.

NINDS (National Institute of Neurological Disorders and Stroke) (1995). Tissue plasminogen activator for acute ischemic stroke. *New England Journal of Medicine*, 333(24), 1581–1588.

Oliveira-Filho, J., Silva, S.C.S., Trabuc, C.C., Pedreira, B.B., Sousa, E.U., & Bacellar, A. (2003). Deterimental effect of blood pressure reduction in the first 24 hours of acute stroke onset. *Neurology*, 61, 1047–1051.

Pariente, J., Loubinoux, I., Carel, C., Albucher, J.-F., Leger, A., Manelfe, C., Rascol, O., & Chollet, F. (2001). Fluoxetine modulates motor performance and cerebral activation of patients recovering from stroke. *Annals of Neurology*, 50, 718–729.

Reed, D.A., Johnson, D.A., Thompson, C., Weintraub, S., & Mesulam, M.M. (2004). A clinical trial of bromocriptine for treatment of primary progressive aphasia. *Annals of Neurology*, 56(5), 750.

Reineck, L., Aggarawal, S., & Hillis, A.E. (2005). "Diffusion clinical mismatch" is associated with potential for early recovery of aphasia. *Neurology, 64*, 828–833.

Rordorf, G., Korosh, W.J., Ezzedine, M.A., Segal, A.J., & Buonanno, F.S. (2001). A pilot study of drug induced hypertension for treatment of stroke. *Neurology, 56*, 1210–1213.

Salloway, S., Pratt, R.D., & Perdomo, C.A. (2002). Donepezil treated patients with vascular dementia demonstrate cognitive and global benefit. Results from study 308, a 24 week, randomized, double blind, placebo controlled trial. *Neurobiology of Aging, 23*(Suppl. 1), s57.

Salloway, S., Pratt, R.D., & Perdomo, C.A. (2003). A comparison of the cognitive benefits of donepezil in patients with cortical versus subcortical vascular dementia: A subanalysis of two 24 week randomized double blind placebo controlled trials. *Neurobiology, 60*(5 Supp. 1), A141–A142.

Schlaug, G., Benfield, A., Baird, A.E., Siewert, B., Lovblad, K.O., Parker, R.A., Edelman, R.R., & Warach, S. (1999). The ischemic penumbra operationally defined by diffusion and perfusion MRI. *Neurology, 53*, 1528–1537.

Shorvon, S. (2001). Pyrrolidone derivatives. *Lancet, 358*, 1885–1892.

Thomalla, G., Schwark, C., Sobesky, J., Bluhmki, E., Fiebach, J.B., Fiehler, J., Zaro Weber, O., Kucinski, T., Juettler, E., Ringleb, P.A., Zeumer, H., Weiller, C., Hacke, W., Schellinger, P.D., & Rother, J. (2006). Outcome and symptomatic bleeding complications of intravenous thrombolysis within 6 hours in MRI-selected stroke patients: Comparison of a German multicenter study with the pooled data of ATLANTIS, ECASS, and NINDS tPA trials. *Stroke, 37*(3), 852–858.

Tsz-Ming, C., & Kaufer, D.J. (2001). Effects of donepezil on aphasia, agnosia and apraxia (abstract). *The Journal of Neuropsychiatry and Clinical Neurosciences, 13*, 140.

Walker-Batson, D., Unwin, H., & Curtis, S. (1992). Use of amphetamine in the treatment of aphasia. *Restorative Neurology and Neuroscience, 4*, 47–50.

Walker-Batson, D., Curtis, S., Natarajan, R., Ford, J., Dronkers, N., Salmeron, E., Lai, J., Unwin, D.H., & Feeney, D.M. (2001). A double-blind, placebo-controlled study of the use of amphetamine in the treatment of aphasia editorial comment. *Stroke, 32*(9), 2093–2098.

Wityk, R., Hillis, A.E., Beauchamp, N.J., Barker, P.B., & Rigamonti, D.J. (2002). Perfusion weighted MRI in adult moyamoya syndrome: Characteristic patterns and change after surgical intervention – case report. *Neurosurgery, 51*, 1499–1506.

Further Readings

Astrup, J., Symon, L., Branston, N.M., & Lassen, N.A. (1977). Cortical evoked potential and extracellular K⁺ and H⁺ levels at critical levels of brain ischemia. *Stroke, 8*, 51–57.

This seminal paper describes the relationship between cerebral blood flow and markers of electrical function and cell death as it relates to the ischemic penumbra. It also demonstrates the relationship between blood pressure and cerebral blood flow that forms the basis of induced hypertension as a reperfusion therapy.

Berthier, M.L. (2005). Poststroke aphasia: Epidemiology, pathophysiology and treatment. *Drugs Ageing, 22*(2), 163–182.

Provides a comprehensive review of the state of play of clinical studies in the pharmacological treatment of aphasia due to stroke.

NINDS (National Institute of Neurological Disorders and Stroke) (1995). Tissue plasminogen activator for acute ischemic stroke. *New England Journal of Medicine, 333*(24), 1581–1588.

This landmark article describes the first randomized, double-blind, placebo-controlled trial for the use of a thrombolytic in the treatment of stroke.

Schlaug, G., Benfield, A., Baird, A.E., Siewert, B., Lovblad, K.O., Parker, R.A., Edelman, R.R., & Warach, S. (1999). The ischemic penumbra operationally defined by diffusion and perfusion MRI. *Neurology, 53*, 1528–1537.

Provides a description of the ischemic penumbra according to MRI criteria.

Shorvon, S. (2001). Pyrrolidone derivatives. *Lancet, 358*, 1885–1892.

Provides a comprehensive review of these agents in the treatment of both memory and language disorders as well as epilepsy with a description of their early history of use.

41

Recovery and Treatment of Acquired Reading and Spelling Disorders

ANNA BASSO

Department of Neurological Sciences, Milan University, Milan, Italy

ABSTRACT

In this chapter, acquired reading and spelling disorders are described with reference to a dual-route model of lexical processing. We report selected cases of treatment of cognitive reading and writing syndromes – deep, surface and phonological dyslexia and deep, surface and phonological dysgraphia. This is followed by a critical evaluation of the methodology. It is argued that these syndromes are not a rational starting point for therapeutic intervention since a given syndrome can arise following damage to different functional components. Surface dyslexia, for instance, can follow damage to any of the lexical components necessary for reading known words. To initiate a well-motivated treatment, one must know what components are damaged. Some therapeutic suggestions for treating all of the components involved in reading and writing are offered.

41.1. INTRODUCTION

The past 30 years have seen important progress in our understanding of the normal processing and acquired disorders of reading (dyslexias) and spelling (dysgraphias); diagrammatic models of lexical processing have become commonplace in the cognitive neuropsychological literature. The model of lexical processing referred to in this chapter is the dual-route model (Figure 41.1).

In the dual-route model here proposed, reading or spelling can be achieved through a lexical route, which allows the correct pronunciation and spelling of stored words, and a non-lexical route, which allows conversion of sub-lexical units of phonemes or graphemes into sequences of graphemes or phonemes, respectively. Only the conversion mechanisms are dedicated to reading and spelling; the lexical routes utilize parts of the lexical–semantic system which are also used in other tasks, such as auditory comprehension and naming (for a detailed description of the structure and processing of the lexical–semantic system see Basso, 2003).

Reading and spelling disorders are generally discussed in the framework of seven syndromes: deep dyslexia (DD) and deep dysgraphia, surface dyslexia and surface dysgraphia, phonological dyslexia and phonological dysgraphia and letter-by-letter dyslexia. Letter-by-letter dyslexia is characterized by disruption early in the reading process (component F in Figure 41.1); reading is slow and laborious and frequently performed by first naming the letters that make up the string. DD is due to the lesion of two components of the reading system: the non-lexical route (component M) and the lexical route (components G, C, H). The locus of damage to the lexical route can vary although the semantic system itself is frequently damaged. The most important defining symptoms of DD are impairment of nonword reading and semantic paraphasias in word reading. In surface dyslexia the nonword reading route is (relatively) preserved and the lexical route impaired. The preserved non-lexical route allows correct reading of nonwords and regularly spelled words that honor the grapheme-to-phoneme correspondence rules, but the application of sub-lexical conversion rules in reading irregular words spawns regularization errors such as reading PINT as MINT, PRINT and FLINT. In transparent orthographies like Spanish and Italian, in

Auditory input   Written input

FIGURE 41.1   Schematic representation of the structure of the dual-route model of the lexical–semantic system.

which almost all words can be read correctly by applying conversion rules, surface dyslexia is difficult to demonstrate and regularization errors will occur only in reading unknown foreign words which have come to be regularly used in the oral language, such as, for instance, the word *jeans*. Phonological dyslexia follows damage to the non-lexical route (component M) and is the only dyslexia syndrome that does not entail other language deficits. Reading of nonwords is impaired and reading of known words preserved. The dissociation between preserved and impaired processes is never complete, in this as well as in any of the other forms of aphasia.

The dysgraphia syndromes correspond fairly well to the dyslexia syndromes. In deep dysgraphia, as in DD, damage to the non-lexical route (component L) causes difficulty in nonword writing and damage to the lexical–semantic route (components B, C, D) causes errors in word writing, the most characteristic being semantic paragraphias. Surface dysgraphia results from damage to the lexical route (components B, C, D). Writing of nonwords is preserved (or significantly better than writing of words); writing of regular words that honor the conversion rules is similar to writing of nonwords; the most frequent errors in the spelling of irregular words that cannot be rendered through the phoneme-to-grapheme conversion rules are phonologically plausible errors, such as writing YACHT as YOT. Finally,

phonological dysgraphia follows damage to the non-lexical writing route (component L). Nonword writing is impaired or impossible whereas writing of known words, whether regular or irregular, is still possible (for a detailed description of the mechanisms of reading and writing see Chapter 20).

Below, illustrative rehabilitation cases for each dyslexia and dysgraphia syndrome are reported. I shall then argue that the analysis of errors subjects make on reading and spelling regular and irregular words and nonwords is not a sound basis for treatment. The diagnostic process should end only when the underlying cause of the reading or writing deficits is disclosed, that is when the damaged component/s – input buffers, input lexicons, semantic system, output lexicons, output buffers or conversion mechanisms – is/are identified. A brief description of suggested treatment procedures for each component involved in reading and spelling follows.

41.2. ILLUSTRATIVE CASES

41.2.1. Letter-by-Letter Dyslexia (or Pure Alexia)

Two treatment approaches have been shown to be effective for people with letter-by-letter dyslexia: cross-modality cueing and brief exposure of words. In cross-modality cueing subjects are taught to trace the contour of the letters they are unable to name when visually perceived and to utilize kinesthetic information to name the letter. The second approach, brief exposure of words, is meant to discourage letter-by-letter reading. The motivation for such a treatment comes from the observation that some letter-by-letter dyslexics can derive meaning from words they are unable to name.

Maher *et al.* (1998) reported successful treatment for subject VT, a 43-year-old woman with chronic and stable pure alexia. She had difficulty naming letters when seeing them but could name them when allowed to pretend to copy them by tracing them with her finger. Using this motor cross-cueing strategy she reached a faster reading rate in 4 weeks.

Rothi and Moss (1992) presented single words on a computer screen and asked their letter-by-letter dyslexic subject to make a semantic decision about the word (e.g., is this an animal?); correct responses were above chance, notwithstanding the subject reporting that he could not read the word. After 20 therapy session reading rate had improved.

41.2.2. Deep Dyslexia

Therapy for DD may address either the conversion rules (component M) or the semantic system (component C) or both. Bachy-Langedock and de Partz (1989) described SP who suffered a left cerebral hemorrhage, surgically evacuated. Three months post-onset his speech was fluent with anomia, phonological and semantic errors; comprehension of single words was relatively well preserved. Reading of nonwords was impaired; reading of real words was superior but also impaired and characterized by a grammatical effect (nouns were read better than verbs) and an effect of imageability; semantic errors were present in written word-to-picture matching. His reading disorders were classified as DD.

The aim of the reading therapy was to re-teach grapheme-to-phoneme correspondences. In the first stage SP was taught to generate code words for individual letters; he first had to say the word when seeing the letter, then the first phoneme only. He was then trained to read simple nonwords in which each letter corresponded to a phoneme. Stage 2 therapy targeted groups of letters that correspond to one phoneme; finally, in stage 3, three specific conversion rules which had caused most of his errors were specifically trained. The whole therapy lasted 2 years, after which time SP could read aloud, albeit slowly, his errors in reading aloud were significantly reduced.

Re-learning of grapheme-to-phoneme conversion rules and of phoneme blending was the focus of the successful treatment of another deep dyslexic subject (Yampolsky & Waters, 2002).

Patient JJ (Hillis & Caramazza, 1994), with semantic damage and impaired letter-to-sound conversion, underwent treatment directed at the semantic impairment (SI). A written word was presented along with 40 pictures from two semantic categories and JJ was asked to point to the corresponding picture. If the response was incorrect, JJ was shown the correct picture and the same item was re-presented after a short delay until the response was correct. Comprehension and oral reading of the treated words improved rapidly.

Box 41.1   The case of DD

DD was a 55-year-old right-handed woman with 13 years of education who experienced a left hemisphere stroke. A CT scan performed 8 months post-onset showed an infarct in the territory of the middle cerebral artery. A first neuropsychological evaluation carried out 6 months post-onset disclosed severe Wernicke aphasia. For 6 months DD was individually followed by a speech therapist five times a week. At 1-year post-stroke she was seen at a different hospital. DD still showed Wernicke aphasia with reading disorders. She was an avid reader and was particularly motivated to improve her reading abilities. Careful assessment of DD's reading was carried out.

She rarely produced errors in letter reading; in a visual lexical decision task she correctly identified 90% of real words and 85% of nonwords. Nonword reading was severely impaired; she produced frequent phonological errors and some lexicalizations such as reading *albero* (tree) for *alfero* (nonword). Single written words were presented for reading; DD correctly read 25/92 words. Apparently, short words were read better than long words and concrete words better than abstract words; errors were visual and semantic; she read, for instance, *glass* for *bottle* and *violin* for *orchestra*.

Results of oral reading tasks indicated that DD's single-word oral reading was severely impaired with semantic errors; grapheme-to-phoneme conversion was also impaired as indicated by her impaired nonword reading. The reading pattern and evidence of disruption of both lexical and non-lexical routes are consistent with a diagnosis of DD and a rehabilitation program for the reading disorder was instantiated.

41.2.3. Surface Dyslexia

Before starting therapy of a surface dyslexic subject it is necessary precisely to locate the damage; it can be at the level of the orthographic input lexicon (component G), the semantic system (component C) or the phonological output lexicon (component H). The most frequently cited case of rehabilitation of a surface dyslexic subject is EE (Coltheart & Byng, 1989) with damage to the orthographic input lexicon.

Testing 4 months post-onset revealed better reading of regular than irregular words, regularization errors and relatively spared reading of nonwords; this pattern is consistent with surface dyslexia and EE's impairment was located at the level of the orthographic input lexicon.

Therapy was aimed to improve the reading of irregular words using a whole-word training approach. Irregular written words were presented alongside a picture representing meaning; 24 words were presented during Phase 1, 54 during Phase 2 and 101 during Phase 3. There was significant improvement of the treated words.

Weekes and Coltheart (1996) reported a similar successful treatment for subject NW with surface dyslexia and surface dysgraphia.

41.2.4. Phonological Dyslexia

In the few published cases of therapy for phonological dyslexia (e.g., Kendall *et al.*, 1998) treatment has generally been directed toward recovery of the conversion rules (component M); grapheme-to-phoneme conversion training was the focus of therapy for MM with severely impaired sublexical reading (Greenwald, 2004).

Phonological dyslexic subjects, however, often have difficulty reading function words and abstract words in addition to a major difficulty in reading nonwords. Friedman *et al.* (1998) report treatment of two subjects with phonological dyslexia and difficulty reading function words. Function words were paired with homophone (or nearly so) content words, such as *bee* for *be*; subjects were then instructed to inhibit the content word when it was not a homophone (e.g., *meat* for *me*) and produce the function word.

41.2.5. Deep Dysgraphia

As in DD, treatment can be directed to the conversion rules (component L) or the lexical damage, frequently located in the semantic system (component C), or both.

HG (Hillis, 1991), a 22-year-old woman 7 years post-onset still presented with pervasive semantic errors and difficulty in writing nonwords; treatment aimed to improve semantics. HG was shown a written word and pictures of objects in the same semantic category. She was asked to indicate the picture corresponding to the written word; when an error occurred, the differences between the selected and the correct item were pointed out. Improvement generalized to other tasks and to untrained items in the same semantic category.

Sound-to-letter conversion teaching was used by Hillis Trupe (1986) with another subject, JS, who showed severe aphasia and a pattern of writing impairment consistent with deep dysgraphia. A treatment hierarchy for teaching phoneme-to-grapheme conversion was implemented. Improved writing of single letters served as a self-cue in spelling and to block the semantic errors.

41.2.6. Surface (or Lexical) Dysgraphia

The most frequently cited case of a treatment for a surface dysgraphic subject is de Partz *et al.*'s (1992) case, even though LP was not a pure case of surface dysgraphia because writing of nonwords was impaired; the first therapy phase was in fact directed toward recovery of the conversion rules.

LP contracted encephalitis when 24 years old. One-year post-onset nonword spelling was relatively well preserved; word spelling was more impaired and showed a frequency effect; he was classified as a surface dysgraphic due to impairment of the lexical route and better preserved non-lexical route.

Initially therapy aimed to improve the non-lexical writing route; this treatment lasted 6 months and only after this period was treatment directed to improve spelling of irregular words and homophones. This was done using a visual imagery strategy; 240 words misspelled by LP were selected and 120 were treated. An image was generated for each word and embedded in the written word; the word "flamme" (flame), for instance, was written with the 2 letter M's represented by a flame. LP was asked to copy the word with the embedded image and then to produce the word and the image in response to the spoken word. After 9 months, writing of the treated and some untreated words improved significantly.

41.2.7. Phonological Dysgraphia

This is a pure spelling disorder and therapy has generally been directed toward recovery of the conversion rules (component L). Luzzatti *et al.* (2000) described two subjects with Broca aphasia, RO and DR; they both had agrammatic speech, phonological dyslexia and severely impaired writing. RO was unable to write any word whereas DR could write around 25% of regular words. In both cases nonword spelling was severely impaired and treatment aimed to improve phonological-to-grapheme conversion rules. The subjects were first trained to segment words into syllables and syllables into phones and then to write to dictation single phonemes and short words with one-to-one phoneme-to-grapheme conversion. Finally, more complex phoneme-to-grapheme rules were introduced. At the end of therapy RO was able to write 90% of regular words and DR showed near normal spelling performance.

41.3. LIMITS OF THE SYNDROME-BASED APPROACH

The (relatively) new syndromes of reading and spelling disorders are based on an analysis of errors in reading out loud and writing to dictation known words and nonwords; however, normal subjects read for comprehension, be it the newspaper, a book, a note, or writing letters, filling out a

Box 41.2   The case of SI

Subject SI suffered an ischemic stroke at the age of 55. An acute CT scan and a subsequent one performed at 8 months post-onset revealed a large infarct in the left temporal and parietal lobes. When evaluated at our aphasia unit she was 1-year post-onset and presented with Wernicke aphasia; her speech was fluent with frequent semantic errors that were also present in writing, repetition and reading aloud. Phonological errors were also present but were less frequent. Her husband complained of her comprehension disorders but she was 1-year post-onset and had already received a 6-month treatment. Her husband wondered whether it was still possible for her to ameliorate.

A language evaluation was carried out in order to locate SI's functional impairment. Processing of input phonemes and graphemes was grossly spared; processing of the input

lexicons was slightly impaired as shown by results of the auditory and visual lexical decision tasks (87% and 90% correct, respectively). In auditory and written word-picture matching tasks she made semantic errors that were also frequent in oral and written naming; she said, for instance, *orange* for *cherry*, *key* for *lock* and wrote *nose* for *mouth*; phonological and orthographic errors were present but less frequent. Semantic errors were also present in repetition, reading aloud and writing to dictation. Nonword reading and spelling were severely impaired; she could read or write almost none.

The ubiquity of the semantic errors allows a diagnosis of severe SI; the sub-lexical conversion routines (grapheme-to-phoneme, phoneme-to-grapheme) are also impaired. It was decided to treat the SI.

questionnaire and do not frequently read out loud or write to dictation. In the cognitive architecture of the reading and spelling processes here illustrated, reading for comprehension and spontaneous writing make use of lexical components (input lexicons, semantic system and output lexicons) that are not dedicated solely to reading and writing. Damage to any of these components will give rise to impairments that are not circumscribed to reading or writing and that may be more distressing for the aphasic subject. Damage to the output phonological lexicon as in surface dyslexia, for instance, will also cause problems in naming and in spontaneous speech production.

A diagnosis based on the location of the impaired component is more appropriate for a rational and motivated treatment than classification of the subject in one or the other syndrome. If a subject is diagnosed as having "surface dyslexia," we do not know whether his or her reading errors arise from damage to the input lexicon, the semantic system or the output lexicon. On the other hand, if we know which component is damaged we know both the locus of damage and its consequences on all language tasks, including reading and spelling. In front of a subject with DD who has difficulty reading, comprehending, writing and speaking, the clinician must answer the question "what is the best treatment for a semantic damage?" and not "how can reading aloud be improved?" the "best" treatment may or may not include reading.

Careful reading of Boxes 41.1 and 41.2 should clarify the differences between a syndrome-based and a functional approach. DD and SI are in fact the same subject. In Box 41.1, the aim of the clinician was to analyze the reading disorder in order to be able to treat it, as requested by the subject herself; in Box 41.2, aim of the clinician was to locate the subject's functional damage in order to be able to instantiate a treatment rationally linked to her deficits. It

is well possible that the conclusion reached in both cases would be to treat the semantic damage but this would probably be pursued in different ways. The first clinician would resort to reading tasks whereas the second clinician could indifferently use naming or reading tasks as well as images instead of words (see below).

There are, however, hints that things are changing. In some of the more recent studies no mention is made of the reading and writing syndromes and therapy is directed to the damaged component (e.g., Cardell & Chenery, 1999; Beeson *et al.*, 2002; Ska *et al.*, 2003; Schmalzl & Nickels, 2006).

41.4. SUGGESTED TREATMENT OF THE LEXICAL COMPONENTS

In this section some suggestions for the rehabilitation of the several lexical components (e.g., input lexicon, output buffer, semantic system) involved in reading and spelling are presented. Even partial recovery of the processing of the damaged component will show up in all the tasks that involve that component. None of the proposed interventions is new; they all have been previously described and shown to be efficacious (for a review, see Basso, 2003). With respect to the cases illustrated above, the main difference lies in the fact that reading and spelling tasks may sometimes be omitted since recovery of the damaged component may be achieved through other tasks.

41.4.1. Orthographic Input Buffer (Abstract Letter Identification)

Damage to the orthographic input buffer (component F) results in slow and inaccurate grapheme identification.

Computerized programs tackling the reading difficulties of letter-by-letter readers can easily be implemented. The use of computers allows an individual to work independently at home and the program can be manipulated so as to adapt to each individual's needs. Single letters, words or nonwords appear on the screen and the subject is required to write them on the keyboard. If the response is correct the computer presents another stimulus; if the response is not correct, this is signaled to the subject and the stimulus represented. The duration of the exposure as well as the size, character and font of the stimulus can be manipulated.

41.4.2. Conversion Mechanisms

The phoneme-to-grapheme (component L) and grapheme-to-phoneme (component M) conversion routes are dedicated to reading and spelling. The two conversion routes are functionally independent and can be impaired separately but in most subjects they are both impaired. A third conversion route, the input-to-output phoneme conversion, which allows for the repetition of nonwords, is more resistant to functional damage, probably because the relationship between input and output phonemes is always one-to-one, without exception.

The rehabilitation program here illustrated involves all three routines at the same time. It may seem to be a loss of time to include a process that is intact, as is frequently the case for the input-to-output phoneme conversion, but the time spent by the subject to perform a task they can easily perform is trivial compared to the advantage of a varied and stimulating therapy in which subjects are encouraged by their successes.

In order to involve only the conversion mechanisms, it is suggested one work with nonwords that should be short in case of severe damage to any of the three routes and become longer and more difficult (from an orthographic and phonological point of view) as the impairment becomes less severe. If the subject has difficulty repeating, reading or spelling even single phonemes or letters, then simple CV syllables, where only the consonant varies whereas the vowel [A] is kept constant, should be used.

The subject is first asked to repeat a syllable; if he fails the stimulus is presented again. The subject is then required to write the syllable and to check whether what he has written corresponds to the syllable he has just repeated. If this is not the case and the subject does not spot the error, the therapist should attract the subject's attention to the error and ask him to read what he has written. If the subject fails, the therapist provides the correct answer and helps him to write the syllable. The subject is then invited to pay attention to the correct spelling and to copy the syllable after a short delay. A new stimulus is then given and the whole procedure is started again. After having repeated and written 3–4 syllables, the subject is asked to read them out loud in random order. Correct repetition of the syllable ensures that the subject has correctly identified the heard phonemes and

can translate them from input-to-output phonemes. Writing the syllable requires the conversion of phonemes into graphemes, and reading it requires the conversion of graphemes into phonemes. After the subject is able to spell most of the single syllables, two-syllable nonwords are introduced.

Phonological awareness is held to be important for reading acquisition and is frequently impaired in phonological dyslexics. For this reason when two-syllable nonwords are introduced, exercises for phonological awareness are also proposed. After hearing the nonword, the subject has to repeat either the first or the second syllable, say what the last letter is, and so on. After this, the nonword should be written and then read aloud as illustrated before.

An interesting aspect of such a program is that it can be carried out at home with the help of any naïve person who has been adequately instructed and this allows for a more intensive treatment.

The program is easily applicable in Italian and Spanish whose orthographies are transparent; very few phonemes must be rendered by two letters and very few letters correspond to more than one phoneme. Other languages, such as English and French, have more opaque orthographies with less transparent conversion rules that must be explained and trained one by one. Carrying out the program in an opaque orthography will probably require more time for the learning of specific conversion rules.

41.4.3. Semantic System

Damage to the semantic system (component C) is frequent in DD and deep dysgraphia. Because of its centrality, damage to the semantic system necessarily involves impairment in comprehension and production of spoken and written words. The internal organization of the semantic system has been a subject of much debate; a widely accepted opinion is that meaning is represented as a set of semantic features and that the semantic information associated with an object can be accessed from both words and pictures. This has an important consequence for rehabilitation since to address the semantic system language is not required. Pictures, compared to words, have a privileged access to the semantic system because they represent directly some of the semantic features of the corresponding concept whereas the relationship between a word and its meaning is arbitrary. The picture of a lion, for example, represents an animal but the word *lion* could be used to refer to a different concept. Severe semantic damage can therefore be tackled more easily with pictures.

Some examples of exercises involving the semantic system are the following: category sorting using increasingly specific categories (animals, → mammals, → ferocious), the odd-one-out (within increasingly semantically associated objects), semantic associations and relationship judgments. Otherwise the therapist may choose to work on a single concept at a time until the subject has demonstrated to have a clear idea of the concept. At this point a second concept pertaining to the same category may be introduced and the differences and similarities between the two illustrated. Knowing what a hammer is can facilitate knowledge of what saws are.

The same exercises can then be carried on using words instead of pictures.

41.4.4. Input Lexicons (Components B, G)

Research papers on rehabilitation of word comprehension disorders are rare. One possible reason for this is the observation that comprehension disorders are the first to recover spontaneously in many subjects; in moderate and mild aphasic subjects rehabilitation of single word comprehension may not be necessary; only severe aphasic subjects have such impaired comprehension of single words as to need therapy. In these cases, however, a less impairment-based and a more pragmatic approach is better justified (Basso, 2003).

In general, tasks devised to evaluate processing of a given component can also be used to treat that same component. Oral and written lexical decision tasks are the task of choice for evaluation of the input lexicons and can be used therapeutically. A subject with damage to the orthographic input lexicon may be asked to recognize whether a written string of letters corresponds to a known word or not; if not, they may be asked to spot the error. Otherwise a written word may be presented with 3 or 4 very similar strings of letters and the subject has to identify the correct one. The same holds for the phonological input lexicon; given a string of phonemes the subject has to say whether it corresponds to a real word or not.

These exercises present an advantage compared to classic word-picture matching tasks. Most importantly, the choice of words is not limited to picturable objects or actions; all words can be used. Furthermore, in word-picture matching the correct picture may be selected without the subject having really understood the word but simply because they know that one of the pictures corresponds to the word and the selected picture seems the most probable choice.

Finally, treatment for the orthographic input lexicon may be carried out by the subject alone with the aid of a small dictionary containing only frequent words. The subject must look up in the dictionary words that sound familiar, pay attention to the spelling, read the definition (comprehension of one or two key words is sufficient), write the word down and look up in the dictionary to check if spelling is correct.

41.4.5. Output Lexicons (Components D, H)

The literature on therapy for naming disorders is quite rich. This is not unexpected since almost all aphasic subjects present with more or less severe naming disorders. Besides naming difficulties, damage to the output phonological lexicon will prevent correct reading aloud of words in

case of concomitant damage of the conversion rules; if these are undamaged, only reading of irregular words will be difficult. Similarly, damage to the orthographic output lexicon will prevent correct writing of words in case of concomitant damage of the conversion rules; if these are undamaged only spelling of irregular words will be difficult.

In most cases anomic subjects have been required to produce the target word in a naming-to-confrontation task but the strategies used have differed; the most frequently used strategies have been phonemic or semantic cueing, repetition and reading. When the different cues have been compared, the phonological cue (saying the first phoneme or syllable of the to-be-named word) has been found to be the most efficacious, but its facilitation effect is short-lived.

The limited amount of picturable words, however, suggests resorting as soon as possible to a task that would allow learning of a large corpus of words. A small dictionary can be the answer.

Briefly, subjects look up in the dictionary a word that they believe was familiar to them, read the definition, write the word down in an exercise book and say the word aloud; each day they add 3–4 new words. The day after, the subject should first try to conjure up the words already written, check whether they have retrieved all words and then go on to learn a few new words and so on, as long as learning of new words can be demonstrated.

41.4.6. Output Buffers (Components E, I)

Buffers are working memory components assigned to the temporary storage of lexical and non-lexical representations for successive elaboration. Damage to a buffer will therefore contribute a length effect in the processing of words and nonwords.

Therapy for the output buffers is straightforward: repetition (for the phonological) and dictation (for the orthographic) of words and nonwords of such length that the subject makes many errors but does sometimes produce a correct answer. However, any task requiring the production of a string of phonemes (for the phonological) or graphemes (for the orthographic) necessarily involves activation of the corresponding output buffer.

41.5. CHALLENGES AND FUTURE DIRECTIONS

Continued advances in our understanding of acquired reading and spelling disorders provide new insights into the cognitive structure of normal processing.

A general but very important question that has been extensively studied concerns the efficacy of treatment for aphasia. There is now sufficient experimental evidence based on group studies to argue that language disorders evolve towards amelioration in many aphasic subjects and that rehabilitation has a positive effect (Basso, 2003).

However, subjects were rarely assigned randomly to treated or untreated groups and heterogeneous methods were used to treat heterogeneous subjects. Robey and Schultz (1998) remind us that outcome research should be programmatic and evolve through the traditional five-phases model illustrated in Box 41.3.

Box 41.3   Five-phase outcome research model (Robey & Schultz, 1998)

Phase 1. Goal: Develop hypotheses to be tested in later stages, study whether subjects improve, define target population. Experiments are brief with small groups, no control subjects are required. Research continues with Phase 2 if the results of the studies in Phase 1 are positive.

Phase 2. Goal: Refine hypotheses, standardize protocols, determine treatment dosage and subject selection criteria, find an explanation for the effect of the treatment. Utilizes small-group and single-subject experiments; no controls are required. Research continues if results are positive.

Phase 3. Goal: Test the efficacy of the treatment developed in Phases 1 and 2 in a randomized controlled trial with random subject assignment to treatment and no-treatment conditions.
   Large sample sizes are necessary. If efficacy is demonstrated research continues with Phase 4.

Phase 4. Goal: Test efficacy of treatment under typical conditions with typical subjects. Self-selected controls or subjects receiving a different treatment are adequate. Large samples are required.

Phase 5. Goal: Continue research on efficacy comparing, for instance, different intensity and duration of treatment, collecting subject's satisfaction data and so forth.

Robey, R.R., & Schultz, M.C. (1998). A model for conducting clinical-outcome research. An adaptation of the standard protocol for use in aphasiology. *Aphasiology, 12*, 787–810.

From single-subject studies we know that almost all functional impairments, including reading and spelling disorders, are amenable to partial recovery. The problem with the single-case study is generalization, since no other aphasic subject will present exactly the same impairments as the subject being treated. Single-case studies pertain to Phases 1 and 2 and we must now proceed to Phases 3, 4 and 5.

A group of rationally similar subjects should be collected, minimal requirements on implementation should be established, a treatment successfully tested in single cases should be implemented and recovery of treated subjects should be compared to recovery of a group of randomly assigned untreated subjects.

In this chapter only knowledge gained from behavioral research has been considered. Further progress in the area of reading and spelling is likely to be best realized through extensions of current multidisciplinary research and clinical application. An important field of research that can in the future offer a lot to therapy is computer modeling.

Computer simulation models represent possible ways in which the brain may support reading and spelling processes and attempts have been made to simulate the effects of brain damage. The most important contribution of connectionist models to aphasia therapy is the importance given to learning and re-learning. Different treatment approaches have already been compared (Plaut, 1996) and it is possible that in the near future treatment approaches can be "tested" to find out what treatment are most effective for a given subject.

References

Bachy-Langedock, N., & de Partz, M.-P. (1989). Co-ordination of two reorganization therapies in a deep dyslexic patient with oral naming disorders. In X. Seron & G. Deloche (Eds.), *Cognitive approaches in neuropsychological rehabilitation* (pp. 211–247). Hillsdale, NJ: Lawrence Erlbaum.

Basso, A. (2003). *Aphasia and its therapy.* New York: Oxford University Press.

Beeson, P.M., Hirsch, F.M., & Rewega, M.A. (2002). Successful single-word writing treatment: Experimental analyses of four cases. *Aphasiology, 16*, 473–491.

Cardell, E.A., & Chenery, H.J. (1999). A cognitive neuropsychological approach to the assessment and remediation of acquired dysgraphia. *Language Testing, 16*, 353–388.

Coltheart, M., & Byng, S. (1989). A treatment for surface dyslexia. In X. Seron & G. Deloche (Eds.), *Cognitive approaches to neuropsychological rehabilitation* (pp. 159–174). Hillsdale, NJ: Lawrence Erlbaum.

De Partz, M.-P., Seron, X., & van der Linden, M. (1992). Re-education of a surface dysgraphia with a visual imagery strategy. *Cognitive Neuropsychology, 9*, 369–401.

Friedman, R.B., Lott, S.M., & Sample, D. (1998). A reorganization approach to treating phonological alexia. *Brain and Language, 65*, 196–198.

Greenwald, M. (2004). "Blocking" lexical competitors in severe global agraphia: A treatment of reading and spelling. *Neurocase, 10*, 156–174.

Hillis, A.E. (1991). Effects of a separate treatment for distinct impairments within the naming process. In T. Prescott (Ed.), *Clinical aphasiology* (Vol. 19, pp. 255–265). Austin, TX: Pro-Ed.

Hillis, A.E., & Caramazza, A. (1994). Theories of lexical processing and rehabilitation of lexical deficits. In M.J. Riddoch & G. Humphreys (Eds.), *Cognitive neuropsychology and cognitive rehabilitation* (pp. 450–484). Hillsdale, NJ: Lawrence Erlbaum.

Hillis Trupe, A.E. (1986). Effectiveness of retraining phoneme to grapheme conversion. In R.H. Brookshire (Ed.), *Clinical aphasiology* (pp. 163–171). Minneapolis, MN: BRK.

Kendall, D.L., McNeil, M.R., & Small, S.L. (1998). Rule-based treatment for acquired phonological dyslexia. *Aphasiology, 12*, 587–600.

Luzzatti, C., Colombo, C., Frustaci, M., & Vitolo, F. (2000). Rehabilitation of spelling along the sub-word routine. *Neuropsychological Rehabilitation, 10*, 249–278.

Maher, L.M., Clayton, M.C., Barrett, A.M., Shober-Peterson, D., & Gonzales-Rothi, L.J. (1998). Rehabilitation of a case of pure alexia: Exploiting residual abilities. *Journal of the International Neuropsychological Society, 4*, 636–647.

Plaut, D.C. (1996). Relearning after damage in connectionist networks: Toward a theory of rehabilitation. *Brain and Language, 52*, 25–82.

Robey, R.R., & Schultz, M.C. (1998). A model for conducting clinical-outcome research. An adaptation of the standard protocol for use in aphasiology. *Aphasiology, 12*, 787–810.

Rothi, L.J.G., & Moss, S. (1992). Alexia without agraphia: Potential for model assisted therapy. *Clinical Communication Disorders, 2*, 11–18.

Schmalzl, L., & Nickels, L. (2006). Treatment of irregular word spelling in acquired dysgraphia: Selective benefit from visual mnemonics. *Neuropsychological Rehabilitation, 16*, 1–17.

Ska, B., Garneau-Beaumont, D., Chesneau, S., & Damien, B. (2003). Diagnosis and rehabilitation attempt of a patient with acquired deep dyslexia. *Brain and Cognition, 53*, 359–363.

Weekes, B., & Coltheart, M. (1996). Surface dyslexia and surface dysgraphia: Treatment studies and their theoretical implications. *Cognitive Neuropsychology, 13*, 277–315.

Yampolsky, S., & Waters, G. (2002). Treatment of single oral reading in an individual with deep dyslexia. *Aphasiology, 16*, 455–471.

Further Readings

Beeson, P.M., & Hillis, A.E. (2001). Comprehension and production of written words. In R. Chapey (Ed.), *Language intervention strategies in aphasia and related neurogenic communication disorders* **(pp. 572–595). Baltimore, MD: Lippincott Williams & Wilkins.**

The chapter outlines the processes necessary for the comprehension and production of familiar and unfamiliar written words. Acquired impairments of reading and spelling are extensively described with reference to a dual-route model. The most frequently applied intervention strategies are described and selected cases are reported. The chapter ends with a glossary and an extensive reference list.

Hillis, A.E. (Ed.) (2002) *The handbook of adult language disorders.* **New York: Psychology Press.**

The handbook is organized in such a way as to present for each cognitive field a first chapter describing the cognitive architecture of the normal cognitive function under consideration, a second chapter describing the neuroanatomical correlates, and a third chapter about treatment. Reading is covered by Hillis, Hillis and Tuffiash, and Friedman; spelling by Rapp, Rapcsak and Beeson, and Beeson and Rapcsak. Taken together, the six chapters make a complete survey of the current state of reading and writing disorders.

Whitworth, A., Webster, J., & Howard, D. (2005). *Assessment and intervention in aphasia.* **Hove: Psychology Press.**

The book is divided into three sections. In Part 1 the cognitive neuropsychological approach is described. Part 2 is devoted to assessment. Part 3 provides a review of the therapy literature on naming, comprehension disorders, reading and writing. The most important studies on therapy of reading and spelling disorders are described and detailed summaries of the treatment interventions are reported. Summary tables offer a clear overview of the studies reviewed and a comprehensive reference list closes the book.

42

The Role of Electronic Devices in the Rehabilitation of Language Disorders

BRIAN PETHERAM[1,2] and PAM ENDERBY[3]

[1]*Faculty of Computing, Engineering, & Mathematical Sciences, University of the West of England, Bristol, UK*
[2]*Speech & Language Therapy Research Unit, Frenchay Hospital, Bristol, UK*
[3]*ScHARR HSR Department, University of Sheffield, Sheffield, UK*

ABSTRACT

In an era when advances in our knowledge of language and the brain open up the possibility of achieving greater results from rehabilitation, it is ironic that most health services worldwide are experiencing ever tighter resource constraints which make it difficult to deliver the amount of treatment that the research says is necessary to be efficacious. One way out of this impasse is to use technology as an assistant to the therapist, enabling the delivery of controlled and targeted treatment, often without the therapist's actual presence. This engages the client as an active partner in the process which can also have beneficial effects on morale and well-being. In addition, new technology enables people with communication disorders to live more fulfilling lives by giving them means by which the negative effects of their impairment are minimized.

This chapter gives an overview of these exciting possibilities.

42.1. INTRODUCTION

Technology is changing every aspect of our lives. It is therefore not surprising that it should and inevitably will affect all aspects of rehabilitation. In practice this means the use of *computer systems* for rehabilitation since it is the system as a whole rather than the individual devices that has the effect. The only real exception to this is some specialized devices which are used as communication aids for people with impaired speech or language. Boxes 42.1 and 42.2 give an introduction to some of the issues most relevant to the technology *per se* in language rehabilitation.

On a professional level, computer technology can already assist with the administrative and financial management of clinics; the assessment of patients; recording of histories; diagnostic profiling and treatment selection. Thus, computers can make services more clinically responsive as well as provide therapy to supplement or even replace the direct intervention of the therapist. They can sensitively monitor change over time and provide more cost effective and accessible therapy. However, the use of technology is not

---

Box 42.1   Why use computers in language rehabilitation?

There are several general characteristics of computer systems that make them particularly useful for language rehabilitation:

- *Vast storage*: A huge volume of stimuli including sounds, pictures, and videos can be stored, searched, and retrieved with no effort – no more shoe boxes full of flash cards and photos!
- *Tireless patience*: Once a computer system has been programmed it can administer the same tasks endlessly.
- *Remote communication*: Clinicians can view results and update tasks on a client's home machine from their desk in the clinic – less time spent in traffic jams!
- *Adherence to protocols*: You can control the way the task is undertaken when the client is working independently – handle wrong answers, provide cues, etc.
- *Configurable to the individual*: Systems can easily be set up to suit the individual client so they have the most appropriate mix of tasks, use a suitable input device, etc.

---

always well integrated into the portfolio of competencies of therapists at undergraduate and graduate level.

On another level, the process of rehabilitation is increasingly seen as a partnership between client and professional and there is a developing focus on the quality of life of the client as well as specific ameliorations of the impairment *per se*. This latter may involve a process of adaptation or substitution rather than restitution in respect of lost capabilities that it may not be possible to restore by treatment. Thus, computer technology can offer aids to daily living to people with language impairments in a similar manner to the ways in which it is becoming a part of everyone's daily life.

42.2. PROVISION OF THERAPY

This section addresses computers as a mechanism for delivering rehabilitation, paying particular attention to settings and client groups. It is helpful to distinguish between computer only therapy (COT) and computer assisted therapy (CAT). CAT can enable the therapist to deliver improved quality of treatment by offering the possibility of using richer multimedia stimuli and also by automatically recording details of performance within the session as well as the client's responses and any relevant results or analyses. The boxes list some of the most relevant features. However, it can be argued that COT is the mode of use where computers may potentially make the greatest contribution. With an emphasis on improved efficiency of provision of therapy, they provide an increased opportunity for practice without a therapist present, allowing for greater autonomy of the person, who can choose the time, frequency, setting, and type of treatment (Petheram, 1996; Pederson *et al.*, 2001). It also allows a motivated client to practice more frequently. This has major implications. In an era when resource constraints limit the amount of "knee to knee" treatment this offers the opportunity to deliver the volumes of specified treatment and stimulation that have the potential to actually take advantage of mechanisms such as brain plasticity and exploit learning theory-based approaches. It also means that the client can take some responsibility for his or her own recovery and may gain a morale boost from this increasing independence. This does not mean that therapists are superfluous, but that rather than spending time administering repetitive drill and practice type exercises, they can focus on developing and directing the strategy for the treatment and configuring the systems accordingly.

Whilst the principle of COT has been long established, technical developments have meant that rather than standalone systems which required visits from the therapists for updating the system and capturing the results, the exploitation of communication technologies and the Internet have led to a new generation of systems. Research has tested the potential of Internet usage to allow the home-based therapy remotely supervised by the therapist (Mortley *et al.*, 2001; Mortley *et al.*, 2004). Results indicated that this mode of therapy had potential benefits and was acceptable to patients, it increased practice time by patients and exercises could be adjusted to respond to patients' progress. Follow-up interviews of the people who used the systems supported the conclusion that this mode of therapy for a broad range of language deficits was efficacious, acceptable, and gave subjects a high degree of independence. This independence can lead to positive gains in morale as the subjects stated that they feel less dependent on the therapist and more involved in treatment. Many of the clients were keen to continue using the systems beyond the trial period.

Most studies have focused on clients with chronic acquired language disorders but Laganaro *et al.* (2003) evaluated the effects and feasibility of an unsupervised computer-based therapy for anomia in a study that included stroke patients with recently acquired aphasia and who were still in-patients. They found that individually adapted CAT can be effective as an adjunct to clinical therapy for anomia, not only with chronic aphasic out-patients but also in acute in-patients.

42.3. COMPUTER-BASED TREATMENT AND LANGUAGE

This section addresses ways in which computers may be efficacious from a linguistic point of view and for specific language impairments. There is an increasing body of evidence to indicate that the use of computers can be effective in language rehabilitation. One of the few large-scale studies was carried out by Katz and Wertz (1997). They examined the effects of computer-provided reading activities on language performance in persons with longstanding chronic aphasia. Fifty-five aphasic adults were assigned randomly to one of three conditions: computer reading treatment, computer stimulation, or no treatment. Subjects in the computer groups used computers 3 h each week for 26 weeks. Computer reading treatment software consisted of visual matching and reading comprehension tasks. Computer stimulation software consisted of nonverbal games and cognitive rehabilitation tasks. Significant improvement over the 26 weeks occurred on five language measures for the computer reading treatment group, on one language measure for the computer stimulation group, and on none of the language measures for the no treatment group. The results suggest that computerized reading treatment can be administered with minimal assistance from a clinician, improvement on the computerized reading treatment tasks generalized to noncomputer language performance, and improvement resulted from the language content of the software and not stimulation provided by a computer. Thus, it can be reasonably claimed that the computerized reading treatment they provided to chronic aphasic patients was efficacious.

In another study on the subject Fink *et al.* (2005) compared face to face clinician guided aphasia therapy with computer supported treatment. They used a single case series design to evaluate two different methods of cued naming therapy. The participants were six people with aphasia who were not in spontaneous recovery. One method involved a clinician providing cued naming therapy to three of the subjects three times weekly. The other method involved three different subjects receiving cued naming therapy once weekly by a clinician which was supplemented with two sessions of self-guided therapy using a computer with specially designed cueing tasks. The results showed that all subjects improved on the set of words that

had been cued during therapy. When comparing the baseline scores with the scores taken from assessments directly following treatment, there was a large effect size for the group receiving all clinician guided therapy. However, the effect size was medium and large for the group who had received partially self-guided therapy. Results taken after the follow-up period showed that improvement had been maintained and that the effect size was large for all subjects. For some subjects there was also generalization to untreated items. The software used in this study was based on the type of therapy that is used in theoretically driven clinician guided therapy for cued naming. This highlights how crucial it is that software being used for therapy must be supported by theoretical models and principles. Any evaluation of computer-delivered therapy is directly related to the quality of the software and how appropriately it is applied and integrated into the rehabilitation process. The authors suggest that the results show that computer-delivered therapy can be a useful adjunct to clinician guided therapy. They also stress how important this might be in relation to cost effectiveness of services and clients' access to these. Many studies now take advantage of the well-established capabilities of this mode of treatment in a way that is almost taken for granted and the computer mediated aspect may only be mentioned in passing.

Most of the work (including the Mortley studies cited above) has focused on anomia or word finding. This is clearly valuable work as word finding is the most common difficulty suffered by people with aphasia. There are some studies which have worked at the sentence level by presenting disordered sentences for evaluation by the user, thus addressing syntactic deficits. However, it is worth pointing out that the focus is usually at the single word level rather than whole utterances or sentences. It is still largely the case that the tasks and exercises presented by the systems are decontextualized, that is they focus on language which is not embedded in a day to day communication context. Recent developments in the multimedia capabilities of computers are likely to lead to computer-based treatment programs that address these aspects.

42.4. DIAGNOSIS AND ASSESSMENT

We now consider the ways in which computers may help to overcome some of the limitations of traditional paper-based assessment methods. Whilst there are many sophisticated and psychometrically robust assessments for persons with disorders of communication, there are also many barriers to their effective use in the clinic. Paper-based assessments are frequently adjusted and modified to accommodate time restraints, personal preferences, or for other logistical reasons. This compromises the standardization in that they may be administered in a way that makes them less reliable and accurate than they were designed to be. Computerized assessments can provide discipline and structure, which restricts modification, and has the additional benefit of being able to analyze information more speedily and effectively, in order to inform diagnosis and treatment planning. A further constraint of paper-based assessment is that they give a snap shot of a client's performance and do not necessarily reflect the range of abilities which may be displayed on different days, at different times, or in different circumstances whereas computers can capture data continuously, and thus enable a more comprehensive and reliable view of the client's capability. One of the strongest contributions to the assessment of communication disorders by technology will be the ability to have detailed objective evidence of types and degrees of errors over considerable number of interactions over long periods of time of patients using computerized therapy materials.

It is often difficult to judge whether a person with a chronic – longstanding – communication disorder would benefit from speech and language therapy and whether one particular therapeutic approach would be preferable to another. Computer technology could well assist this difficult clinical decision. In one study a speaker-dependent voice recognition system (that is the voice recognition software was trained to recognize the individual's particular voice) was used to compare participants' practice attempts with a model of a word that they had produced and was judged to be their best attempt at that word (Palmer *et al.*, 2004). Variations of these attempts were identified indicating potential range of ability. Furthermore, three conditions of stimulating the target word were compared: the target word was presented on the screen for reading but no feedback was given, or the client was given feedback as to how close they were to their best attempt, and finally the written target word was given along with an auditory model followed by visual feedback of their closeness to their best attempt. This study reports that the eight subjects with longstanding dysarthria gained the ability to offer more consistent speech production at their best level and improved their level of performance. There was a differential effect of the three conditions with four normal speaking participants showing different preference compared to participants with dysarthria who benefited particularly from an auditory reinforcement to the target. This illustrates the strength of speech and language samples elicited and analyzed by computers in demonstrating the retained skills and variation in performance of users which can inform the clinician of the potential of the subject to improve and the most effective routes for intervention.

42.5. OUTCOME MEASUREMENT AND AUDIT

Therapists worldwide are increasingly called upon to justify their work in terms of benefit to clients and to subject their practice to audit. Computer technology can also play a role in this sphere. Despite the evidence that speech and language therapy is effective in improving the function of persons with certain forms of dysphasia and dysarthria, all studies indicate that there are a proportion of people who do not progress so well, fueling debate with funding agencies who frequently demand specific evidence related to services for particular clients. Collecting longitudinal data on computers from computer driven exercises provides not only the therapist and patient with information regarding progress but also can be used to add objective evidence, and inform service delivery or lead to modifications.

It is likely that some therapists and/or services are more effective than others. It would be valuable for us as a profession as well as to our clients if we were able to identify which therapists or services were most effective with different client groups so that overall quality of services could be driven up. This can only be achieved by undertaking benchmarking exercises. This requires the collection of data on consecutive patients as they enter and leave a service in order to make comparisons between therapist's caseloads (internal benchmarking) or services (external benchmarking). Of course the data has to be collected in systematic and reliable fashion and appropriate comparisons made in order to provide valid and robust information.

The collection of such patient data was undertaken in a benchmarking exercise of 11 services for persons with dysphasia in the UK using the Therapy Outcome Measure (Enderby *et al.*, 2006) and associated software. The results indicate that some services attract persons with more severe dysphasia than others which begs the question of why this may be and should stimulate discussion with referrers. Furthermore some services were associated with improving the language disorder whilst others were associated with improving the language function of similarly impaired clients. Data such as this will become increasingly available as more patient records become computerized. However, collecting more data will not necessarily provide us with meaningful information if we are not assertive in influencing data sets and process required for appropriate clinical audit aimed at quality improvement.

42.6. ALTERNATIVE COMMUNICATION

It has been demonstrated that electronic devices have the potential to actually replace lost communication functionality. One example of this is the use of voice recognition software to assist people who have preserved oral communication but impaired writing skills. A study by Wade *et al.* (2001) has given encouraging indications that by using adapted approaches to training, "off the shelf" voice recognition software can be used by persons with both dysphasia and dysarthria to improve their use of the computer for communication and corresponding. This allows the person with dysphasia to compose and create e-mails and letters without assistance and because of the nature of the technology they can take as long as they like and have as many tries as they like until they are satisfied with the results. A related point is that many people with language problems after stroke also suffer from hemiplegia of their preferred upper limb. This can lead people to be reluctant to produce written output even if it is linguistically correct. Technology thus can avoid potential embarrassments arising from both excessive delays and corrections and inappropriate appearance of the output. Communication aids have mainly been associated with helping people who have severe speech impairment but relatively intact language skills. However, some recent work has led to the production of a communication aid specifically designed for people with aphasia. Van de Sandt-Koenderman *et al.* (2005) undertook a study which examined the communicative needs of people with aphasia and led to the development of a communication aid based on a hand-held computer which supports the finding of the desired utterance as well as the delivery of it. This is now commercially available and being used in everyday life by people with aphasia.

42.7. SUPPORT FOR EVERYDAY LIFE

Technology is a two edged sword for those with disabilities. Whilst some will benefit from systems which facilitate independence and environmental control previously unattainable, others will be further marginalized by being unable to access technology which is becoming available to the majority of the population. It is clear that over the past few years people who are not online are disadvantaged in many aspects of their daily life compared to those who are. More and more basic aspects of daily life are migrating on to the web and it is certain that in the near future critical services such as social security benefits and voting will migrate online to a greater or lesser extent.

Persons with dysphasia may be excluded further by being unable to use e-mail or the Internet if this requirement is not considered by therapists and computer professionals. Teaching persons unfamiliar with computer technology to surf the web takes particular skills. Teaching persons unfamiliar with technology who are also aphasic to surf requires very specialist skills, probably best achieved by computer specialists and therapists working together and respecting each others contributions.

An emerging aspect of the information society is the increasing use of the Internet as a social medium; people are increasing communication via chat rooms, forums, blogs, and home pages. A study at the Connect Centre in London (Moss *et al.*, 2004) facilitated a group of people with aphasia in learning to use the Internet and, at their request, setting up their own web site (www.aphasiahelp.org). A striking finding was that as well as wanting to use the web for shopping and booking holidays, etc. the participants were very keen to learn how other people with aphasia were coping. This resulted in a personal home page which gave their perspective on their condition. Given that there is evidence that people with communication disorders can feel isolated and may never even encounter someone else with the same problem, this potential for community building could have a significant positive impact.

42.8. CHALLENGES AND FUTURE DIRECTIONS

The pace of change in technology is not likely to slow up any time soon and the chances are it will increase. For professionals and others involved in language rehabilitation it means there is going to be a dilemma about how soon to adopt new technologies and how close to the "bleeding edge" of technical progress is appropriate. New technologies bring new possibilities and clients and the profession should not miss out on these, but the nature of the industry is such that products are typically released before they are fully debugged and this can lead to harmful consequences, especially for vulnerable people with communication disorders. This issue is likely to be continually recurring for the foreseeable future. To some extent language rehabilitation is sheltered from this in that the devices used are often adapted from devices produced for other purposes. This means that by the time we get to use the devices they are relatively well proven in other contexts. However, it can be frustrating in some ways: for example relatively minor modifications to a popular voice recognition package would have made it much more useful in language treatment but the developers were totally focused on the commercial market and were not at all interested in what they saw as a commercially insignificant application. Fortunately, many manufacturers are more enlightened and see "design for all" as both something they should be doing as responsible corporate citizens, and also as a way of improving their products for all types of users.

Arguably the most significant challenge is not so much the technology itself but the things that need to be done in order for it to be effectively integrated into our practice. This requires a focus on the needs of the situation as much as on the technology itself. A telling example of this is the ongoing failure to computerize medical records in the UK National Health Service in spite of many years of effort and billions of pounds of expenditure. The problem is not that the computers cannot store and process medical records but that they need to be told what data to store, who can use it, and in what circumstances. These are organizational issues that have so far proved to be insoluble and are something that requires a new academic study of organizational and human complexity which acknowledges that variables are not always predictable and interact in ways that compound each other.

As more of everybody's life moves online so we must ensure that people with communication disorders do not miss out and become literally disenfranchised. Technical developments may actually help in this as computing power physically moves out of the box and becomes embedded in myriad everyday devices (a recent example – though of dubious utility – is the intelligent refrigerator), or even in our clothes as the growing interest in wearable computing indicates. We will also be untethered from the wires as wireless networks become ubiquitous. This will open up opportunities for new forms of treatment including an expansion of independent self-treatment in the client's home setting and possibly some blurring of the boundaries between treatment for restitution of language capability and treatment for compensation of missing capabilities as more "intelligence" is built into devices.

However barriers still remain. One of the most significant is the information and communication technologies skill level of practitioners who need to be adept at identifying appropriate software and in structuring the integration of technology within their therapy to improve people's functioning. They will need to appreciate that new technology can be an opportunity rather than a threat. Indeed, it may be useful to view technology as a prosthesis for the therapist as much as the client. There is nothing that technology does that human therapists cannot do, given infinite time and resources; unfortunately that is not the situation in which most therapy takes place! By taking away many of the time consuming and often tedious tasks such as analyzing complex assessment scores, managing appointment bookings, and administering repetitive drill and practice exercises, the therapist can focus on higher level tasks such as developing treatment strategies and tailoring treatment to the needs of individual clients. The therapist is also more likely to see greater gains in treatment effects as technology enables clients to undertake greater volumes of exercises, etc. than resource constraints normally allow.

References

Enderby, P., John, A., & Petheram, B. (2006). *Therapy outcome measures for rehabilitation professionals* (2nd edn). Chichester: John Wiley.

Fink, R., Brecher, A., Sobel, P., & Schwatz, M. (2005). Computer assisted treatment of word retrieval deficits in aphasia. *Aphasiology, 19*, 943–954.

Katz, R., & Wertz, R.T. (1997). The efficacy of computer-provided reading treatment for chronic aphasic adults. *Journal of Speech, Language, and Hearing Research, 40*, 493–507.

Laganaro, M., Di Pietro, M., & Schnider, A. (2003). Computerised treatment of anomia in chronic and acute aphasia: An exploratory study. *Aphasiology, 17*, 707–721.

Mortley, J., Enderby, P., & Petheram, B. (2001). Using a computer to improve functional writing in a patient with severe dysgraphia. *Aphasiology, 15*, 443–461.

Mortley, J., Wade, J., & Enderby, P. (2004). Superhighway to promoting a client-therapist partnership: Using the Internet to deliver word-retrieval computer therapy monitored remotely with minimal speech and language therapy input. *Aphasiology, 18*, 193–211.

Moss, B., Parr, S., Byng, S., & Petheram, B. (2004). "Pick me up and not a down down, up up": How are the identities of people with aphasia represented in aphasia, stroke and disability websites. *Disability and Society, 19*, 753–768.

Palmer, R., Enderby, P., & Cunningham, S.P. (2004). The effect of three practice conditions on the consistency of chronic dysarthric speech. *Journal of Medical Speech and Language Pathology, 12*, 183–189.

Pederson, P.M., Vintner, K., & Olson, T.S. (2001). Improvement of oral naming by unsupervised computerized rehabilitation. *Aphasiology, 15*, 151–169.

Petheram, B. (1996). The behaviour of stroke patients in unsupervised computer administered aphasia therapy. *Disability and Rehabilitation, 16*, 61–66.

Van de Sandt-Koenderman, M., Wiegers, J., & Hardy, P. (2005). A computerised communication aid for people with aphasia. *Disability and Rehabilitation, 27*(9), 529–533.

Wade, J., Petheram, B., & Cain, R. (2001). Voice recognition and aphasia: Can computers understand aphasic speech. *Disability and Rehabilitation, 23*(14), 604–613.

Further Readings

Beukelman, D., & Mirenda, P. (2005). *Augmentative and alternative communication: Management of severe communication disorders in children and adults.* **Baltimore, MD: Paul H. Brookes Publishing.**
This revised and updated third edition incorporates the most up-to-date research and developments in the augmentative and alternative communication field.

LoPresti, E.F., Mihailidis, A., & Kirsch, N. (Eds.) (2004). Assistive technology for cognitive rehabilitation: State of the art. *Special Issue of Neuropsychological Rehabilitation, 14*(1/2)**.**
A comprehensive review of literature in assistive technology for cognition (ATC). ATC interventions address a range of functional activities requiring cognitive skills as diverse as complex attention, executive reasoning, prospective memory, self-monitoring for either the enhancement or inhibition of specific behaviors, and sequential processing. These technologies address the needs of individuals with information processing impairments that may affect visual, auditory, and language ability, or the understanding of social cues. Many of the techniques may be adapted for more linguistically based interventions.

Müller, D. (Ed.) (2002). *Disability and Rehabilitation – Special Issue on Assistive Technology, 24*(1–3)**.**
A collection of eighteen articles on technology and rehabilitation in a broad sense but most of them are relevant to language rehabilitation and some specifically focus on that topic.

Petheram, B. (Ed.) (2004). Computers and aphasia: Their role in the treatment of aphasia and the lives of people with aphasia. *Special issue of Aphasiology, 18*(3)**.**
A collection of seven papers that specifically address the use of computers in aphasia rehabilitation from a range of perspectives.

PART V

RESOURCES

43

Resources in the Neuroscience of Language: A Listing

BRIGITTE STEMMER

Faculty of Arts and Science and Faculty of Medicine, Université de Montréal, Montréal, Quebec, Canada

The objective of this listing is to provide reference material for those new to the field of the neuroscience of language and to experts who want a brief overview of current resources. The chapter lists journals, books, and sourcebooks with a brief description taken from the journal or book; no evaluation of the quality of the material is given. To make the vast amount of material manageable, we have only considered books published since 1999 and that are directly related to the field or, in a somewhat broader perspective, point the reader to other fields or disciplines that have an impact on the neuroscience of language. Similarly, only international journals written in the English language that regularly publish articles in the field are included. The book and journal listing is based on searches in the PsychInfo database. Publishers' catalogs were also consulted. The final listing concerns associations and societies, with their editor in chief, web address and a brief description of their missions. The reader is encouraged to browse these web sites as the associations and societies frequently provide a host of information on such topics as conferences, news in the field, funding opportunities, student stipends, job offers and so on.

JOURNALS

Aphasiology (Chris Code (Ed.); Psychology Press)
http://www.tandf.co.uk/journals/titles/02687038.asp
Aphasiology is concerned with all aspects of language impairment and disability and related disorders resulting from brain damage. Aphasiology includes papers on clinical, psychological, linguistic, social, and neurological perspectives of aphasia.

Brain. **A journal of neurology** (Alastair Compston (Ed.); Oxford University Press)
http://brain.oxfordjournals.org/
The journal publishes contributions in neurology and related clinical disciplines.

Brain and Behavioural Sciences (BBS) (Paul Bloom & Barbara L. Finlay (Eds.); Cambridge University Press)
http://www.bbsonline.org/
The journal publishes significant and controversial pieces in any area of psychology, neuroscience, behavioral biology or cognitive science. Each article is accompanied by open peer commentary from specialists within and across these disciplines.

Brain and Cognition (Sidney J. Segalowitz (Ed.); Elsevier)
http://www.elsevier.com/wps/find/journaldescription.cws_home/622798/description#description
Brain and Cognition publishes contributions relevant to all aspects of human neuropsychology other than language or communication. Coverage includes, but is not limited to: memory, cognition, emotion, perception, movement, or praxis, in relationship to brain structure or function.

Brain and Language (Steven Small (Ed.); Elsevier)
http://www.elsevier.com/wps/find/journaldescription.cws_home/622799/description#description
The journal publishes papers relevant to human language or communication in relation to any aspect of the brain or brain function.

Cerebral Cortex (Pasko Rakic (Ed.); Oxford University Press)
http://cercor.oxfordjournals.org/
Cerebral Cortex publishes papers on the development, organization, plasticity, and function of the cerebral cortex, including the hippocampus, thalamocortical relationship, or cortico–subcortical interactions.

Cognition (Gerry T.M. Altmann (Ed.); Elsevier)
http://www.elsevier.com/wps/find/journaldescription.cws_home/505626/description#description
Cognition covers all the different aspects of cognition, ranging from biological and experimental studies to formal analysis. Contributions are from a wide range of disciplines that have some bearing on the functioning of the mind.

Cognitive Neuropsychiatry (Anthony S. Davis & Peter W. Halligan (Eds.); Psychology Press)
http://www.tandf.co.uk/journals/titles/13546805.asp
The journal publishes papers that address issues in clinical and cognitive neuropsychiatry, and related fields of clinical psychiatry, behavioral neurology, and cognitive neuropsychology.

Cognitive Neuropsychology (Alfonso Caramazza (Ed.); Psychology Press)
http://www.tandf.co.uk/journals/pp/02643294.html
The journal covers neuropsychological work bearing on the understanding of normal and pathological cognitive processes at any stage of lifespan. It also covers neuroimaging and computational modeling research that is informed by consideration of neuropsychological phenomena.

Cortex (Sergio Della Sala (Ed.); Masson)
http://www.cortex-online.org/
Cortex focuses on the study of the inter-relations of the nervous system and behavior, particularly as these are reflected in the effects of brain lesions on cognitive functions.

Developmental Neuropsychology (Dennise L. Molfese (Ed.); Lawrence Erlbaum Ass.)
http://www.tandf.co.uk/journals/titles/8756-5641
The journal publishes scholarly papers on the appearance and development of behavioral functions, such as language, perception, and cognitive processes as they relate to brain functions and structures.

Human Brain Mapping (Peter T. Fox & Jack L. Lancaster (Eds.); Wiley)
http://www3.interscience.wiley.com/cgi-bin/jhome/38751?CRETRY=1&SRETRY=0
The journal publishes peer-reviewed basic, clinical, technical, and theoretical research in the field of human brain mapping including research derived from non-invasive brain imaging modalities used to explore the spatial and temporal organization of the neural systems supporting human behavior.

Journal of Clinical and Experimental Neuropsychology (Louis Costa &Byron P. Rourke (Eds.); Psychology Press)
http://www.tandf.co.uk/journals/titles/13803395.asp
The journal publishes papers that address theoretical and methodological papers, critical reviews of content areas, and theoretically relevant case studies. Emphases of interest include the impact of injury or disease on neuropsychological functioning; validity studies of psychometric and other procedures; empirical evaluation of behavioral, cognitive and pharmacological approaches to treatment/intervention; psychosocial correlates of neuropsychological dysfunction; theoretical formulation and model development; methodological issues.

Journal of Cognitive Neuroscience (Mark D'Esposito (Ed.); MIT Press)
http://jocn.mitpress.org/
The journal provides a forum for research in the biological bases of mental events and publishes papers that bridge the gap between descriptions of information processing and specifications of brain activity and drawing on developments in neuroscience, neuropsychology, cognitive psychology, linguistics, computer science, and philosophy.

Journal of Communication Disorders (Wilfred G. van Gorp & Daniel Tranel (Eds.); Psychology Press)

http://www.elsevier.com/wps/find/journaldescription.cws_home/505768/description#description
The journal publishes original articles, reports of experimental or descriptive investigations, theoretical or tutorial papers, case reports, or brief communications on topics related to disorders of speech, language, and hearing.

Journal of Fluency Disorders (A. Craig (Ed.); Elsevier)
http://www.elsevier.com/wps/find/journaldescription.cws_home/505771/description#description
The journal publishes research and clinical reports; methodological, theoretical, and philosophical articles; reviews; short communications and is devoted specifically to fluency. It covers clinical, experimental, and theoretical aspects of stuttering, including the latest remediation techniques.

Journal of Memory and Language (K. Bock (Ed.); Elsevier)
http://www.elsevier.com/wps/find/journaldescription.cws_home/622888/description#description
The journal publishes papers that contribute to the formulation of scientific issues and theories in the areas of memory, language comprehension and production, and cognitive processes.

Journal of Neurolinguistics (Henri Cohen (Ed.); Elsevier)
http://www.elsevier.com/wps/find/journaldescription.cws_home/866/description#description
The journal publishes novel, peer-reviewed research into the interaction between language, communication and brain processes. Contributions cover neurology, communication disorders, linguistics, neuropsychology, and cognitive science in general as well as interdisciplinary work on any aspect of the biological foundations of language and its disorders resulting from brain damage and studies of normal subjects, with clear reference to brain functions.

Journal of Phonetics (G. Docherty (Ed.); Elsevier)
http://www.elsevier.com/wps/find/journaldescription.cws_home/622896/description#description
The journal publishes experimental or theoretical papers with phonetic aspects of language and linguistic communication processes.

Journal of Speech, Language, and Hearing Research (JSLHR) (Craig Champlin, Karla McGregor, Katherine Verdolini (Eds.); The American Speech-Language-Hearing Association)
http://jslhr.asha.org/
The journal publishes papers pertaining to the processes and disorders of hearing, language, and speech, and to the diagnosis and treatment of such disorders.

Laterality. Asymmetries of Body, Brain and Cognition (Chris McManus, Mike Nicholls, Giorgio Vallortigara (Eds.); Psychology Press)
http://www.tandf.co.uk/journals/pp/1357650X.html
The journal publishes research on all aspects of lateralization in humans and non-human species with a special interest in the psychological, behavioral and neurological correlates of lateralization.

Nature (Philip Campbell (Ed.) Nature Publishing Group)
http://www.nature.com/nature/index.html
Nature is a weekly interdisciplinary journal of science and one of the most cited weekly science journal. It publishes research in all fields of science and technology and provides news and interpretation of topical and coming trends affecting science, scientists, and the wider public.

Nature Neuroscience (Sandra Aamodt & Kalyani Narasimhan (Eds.); Nature Publishing Group)

http://www.nature.com/neuro/index.html

This journal publishes news and views, reviews, editorials, commentaries, perspectives, book reviews, and correspondence in all areas of neuroscience.

Nature Reviews Neuroscience (Claudia Wiedemann (Ed.) Nature Publishing Group)

http://www.nature.com/nrn/index.html

The journal publishes reviews on areas of neuroscience including cellular and molecular neuroscience, development of the nervous system, sensory, motor systems and behavior, regulatory systems, higher cognition and language, computational neuroscience, and disorders of the brain.

Neurocase (Bruce L. Miller, Hans J. Markowitsch, & Argye E. Hillis (Eds.); Psychology Press)

http://www.tandf.co.uk/journals/titles/13554794.asp

The journal publishes both adult and child case studies in neuropsychology, neuropsychiatry and behavioral neurology, group studies of subjects with brain dysfunction that address issues relevant to the understanding of human cognition, reviews of important topics in the domains of neuropsychology, neuropsychiatry and behavioral neurology; and brief reports.

Neuroimage (K.J. Friston (Ed.); Elsevier)

http://www.elsevier.com/wps/find/journaldescription.cws_home/622925/description#description

The journal focuses on the understanding of the mechanisms of brain function and how this function depends on its structure and architecture using imaging and modeling techniques.

Neuropsychologia (R. Mayes & S. Bentin (Eds.); Elsevier)

http://www.elsevier.com/wps/find/journaldescription.cws_home/247/description#description

The journal focuses on contributions that address functional aspects of the brain and use data to link in theory neural processes in the brain with perception, attention and awareness, action and motor control, executive functions and cognitive control, memory, language, and emotion and social cognition.

Neuropsychology (Stephen M. Rao (Ed.); American Psychological Association, APA)

http://www.apa.org/journals/neu/

The journal focuses on basic research, the integration of basic and applied research, and improved practice in the field of neuropsychology including empirical research on the relation between brain and human cognitive, emotional, and behavioral function.

Neuroscience Letters (S.G. Waxman (Ed.); Elsevier)

http://www.elsevier.com/wps/find/journaldescription.cws_home/506081/description#description

The journal publishes short papers in all areas of neuroscience including molecular, cellular, developmental, systems, behavioral and cognitive, and computational mechanisms.

Neuroreport (G. Gabella (Ed.); Lippincott Williams & Wilkins)

http://www.neuroreport.com/pt/re/neuroreport/home.htm;jsessionid=GBpbYTNkTHTGBbSGFr378TcrFz12TF26B3dR7h8TWXPXhQvfhjJ5!-260396143!181195628!8091!-1

The journal focuses on the rapid communication of new findings in neuroscience. Unlike other journals that publish contributions free of charge, the journal charges $75 per page for the rapid publication of articles.

Science (Donald Kennedy (Ed.); American Association for the Advancement of Science, AAAS)

http://www.sciencemag.org/

Science is a weekly, peer-reviewed journal that publishes original scientific research, plus reviews and analyses of current research and science policy.

The Mental Lexicon (Gonia Jarema & Gary Libben (Eds.); John Benjamins)

http://www.benjamins.com/cgi-bin/t_seriesview.cgi?series=ML

The Mental Lexicon is an interdisciplinary journal that publishes original research and reviews on research that bears on the issues of the representation and processing of words in the mind and brain.

Trends in Cognitive Science (Shbana Rahman (Ed.); Elsevier)

http://www.trends.com/tics/

The journal provides concise reviews, summaries, opinions and discussion of research in all aspects of cognition, the mind and the brain for experts and newcomers.

Trends in Neurosciences (Sian Lewis (Ed.); Elsevier)

http://www.trends.com/tins/default.htm

The journal publishes review articles, opinions on controversy, debate and hypothesis and updates on current research findings in all areas of neuroscience. The targeted audience includes students, researchers, and teachers.

BOOKS

Ahlsén, E. (2006). *Introduction to neurolinguistics.* **John Benjamins.**

The book is a basic introduction to neurolinguistics and addressed to students of linguistics and communication disorders. It covers theories, models, and frameworks underlying neurolinguistics, aspects of different components of language, reading and writing, bilingualism, the evolution of language, and multimodality.

Arbib, M., & Grethe, J. (2001). *Computing the brain.* **Academic Press. 380 pages.**

The book provides readers with an integrated view of current informatics research related to the field of neuroscience and targets a multidisciplinary audience with introductory chapters for the non-expert reader. The book includes an introduction to informatics technologies and the use of these technologies in neuroscience.

Avanzini, G., Lopez, L., Koelsch, S., & Manjno, M. (Eds.) (2006). *The neurosciences and music II: From perception to performance.* **New York Academy of Sciences.**

The work investigates the relationships between music and human neurological makeup, and the ways in which music can influence neurological development and addresses topics such as evaluation of neurological disorders and music, the relationship of music to development and language, and musical perception, and the use and impact of music therapy.

Bailey, D.B. Jr., Bruer, J.T., Symons, F.J., & Lichtman, J.W. (Eds.) (2001). *Critical thinking about critical periods.* **Paul H. Brookes Publishing.**

The book discusses what is known, what is thought to be known, and what is not yet known about critical periods and how they relate to the young child's visual system, social and emotional development, language acquisition, and early childhood education.

Ball, M.J., & Damico, J.S. (Eds.) (2007). *Clinical aphasiology: Future directions, a Festschrift for Chris Code.* **Psychology Press.**

This tribute to Chris Code presents a collection of work on topics within clinical aphasiology.

Baltes, P.B., Reuter-Lorenz, P.A., & Rosler, F. (Eds.) (2006). *Lifespan development and the brain: The perspective of bio-cultural co-constructivism.* **Cambridge University Press.**

Various theoretical models in psychology have emphasized the social foundation of the mind and the role that social interactions play in human development. Topics from a variety of fields are addressed such as biological aspects of co-operation, the role of social interaction in learning, the conceptualization of linguistic knowledge, and peer problem solving.

Banich, M.T., & Mack, M. (Eds.) (2003). *Mind, brain, and language: Multidisciplinary perspectives.* **Lawrence Erlbaum.**

The book provides an overview of how the structure of language influences the way we think and how the organization of the brain influences language.

Bialystok, E., & Craik, F.I.M. (Eds.) (2006). *Lifespan cognition: Mechanisms of change.* **Oxford University Press.**

The book discusses specific cognitive functions, such as attention, executive functioning, memory, working memory, representations, language, problem solving, intelligence, and individual differences from a developmental and aging perspective.

Broeder, P., & Murre, J. (2002). *Models of language acquisition: Inductive and deductive approaches.* **Oxford University Press. 302 pages.**

This book presents advances in computational modeling of language acquisition. The book considers the extent to which linguistic structure is readily available in the environment, the degree to which language learning is inductive or deductive, and the power of different modeling formalisms for different problems and approaches.

Brown, C.M., & Hagoort, P. (2001). *The neurocognition of language.* **Oxford University Press. 424 pages.**

The book provides a critical overview of the cognitive neuroscience of language, Discussed are the representations and structures of language, the cognitive architectures that underlie speaking, listening, and reading. The book reviews brain imaging literature on word and sentence processing and contributions from brain lesion data. It also explains the prospects and problems of brain imaging techniques for the study of language and contains a review of the neuroanatomical structure of Broca's language area.

Cacioppo, J.T., Visser, P.S., & Pickett, C.L. (Eds.) (2006). *Social neuroscience: People thinking about thinking people.* **The MIT Press.**

The book focuses on the neurobiological underpinnings of social information processing, particularly the mechanisms underlying "people thinking about thinking people" using methods such as functional brain imaging, studies of brain lesion patients, comparative analyses, and developmental data contribute to are brought to bear on social thinking and feeling systems.

Carlson, L., & Van der Zee, E. (2005). *Functional features in language and space. Insights from perception, categorization, and development.* **Oxford University Press.**

The "language and space" area is a relatively new research area in cognitive science and focuses on how language and spatial representation are linked in the human brain.

Coch, D., Dawson, G., & Fischer, K.W. (Eds.) (2007). *Human behavior, learning, and the developing brain: Atypical development.* **Guilford Press.**

Brain-behavior relationships are discussed in atypical developmental pathways such as in children with autism, dyslexia, specific language impairment, attention-deficit disorder, dyscalculia, Williams syndrome, or in children growing up in stressful environments.

Emmorey, K. (2001). *Language, cognition and the brain: Insights from sign language research.* **Lawrence Erlbaum.**

The book investigates sign language as a tool for investigating the nature of human language and language processing, the relations between cognition and language, and the neural organization for language.

Fabbro, F. (2004). *Neurogenic language disorders in children.* **Elsevier.**

The volume presents neurogenic language disorders including acquired epileptiform aphasias and autism as well as language disorders due to malformation or tumor lesions. Crossed aphasia in children, the modality and types of aphasia recovery in children and persistent acquired childhood aphasia are also discussed.

Farah, M.J., & Feinberg, T.E. (Eds.) (2006). *Patient-based approaches to cognitive neuroscience* **(2nd edn). The MIT Press.**

The new and updated edition covers the history and methods of cognitive neuroscience, perception and attention, language, memory and prefrontal function, dementias and developmental disorders, mental retardation, attention deficit hyperactivity disorder (ADHD), autism, and the molecular genetics of cognitive disorders.

Frackowiak, R., Friston, K., Frith, C. Dolan, R., Price, C., Zeki, S., Ashburner, J., & Penny, W. (2003). *Human brain function.* **Academic Press.**

The second updated edition provides a state of the art perspective of the theory, practice, and application of modern imaging methods employed in exploring the structural and functional architecture of the normal and diseased human brain.

Galaburda, A.M., Kosslyn, S.M., & Yves, C. (Eds.) (2002). *The languages of the brain.* **Harvard University Press.**

The authors argue that many possible "languages of thought" play different roles in the life of the mind. The contributors discuss topics such as learning of second languages, recovering language after brain damage, sign language, mental imagery, representations of motor activity, and the perception and representation of space.

Grodzinsky Y., & Amunts K. (Eds.) (2006). *Broca's region.* **Oxford University Press.**

The book is based on a workshop in 2004 that discussed Broca's region from all sorts of angles including neuroanatomy, physiology, evolutionary biology, cognitive psychology, clinical neurology, functional imaging, speech and language research, computational biology, and psycho-, neuro-, and theoretical linguistics.

Grodzinsky Y., Shapiro L., & Swinney D. (Eds.) (2000). *Language and the brain. Representation and processing.* **Academic Press.**

This book is a comprehensive look at sentence processing as it pertains to the brain and covers everything from language acquisition to lexical and syntactic processing, speech pathology, memory, neuropsychology, and brain imaging.

Held, C., Knauff, M., & Vosgerau, G. (Eds.) (2006). *Mental models and the mind: Current developments in cognitive psychology, neuroscience, and philosophy of mind.* **Elsevier.**

Building a bridge between the philosophy of mind and the empirical sciences of the mind/brain, the book develops a new perspective on the concept of mental models.

Howard, H. (2004). *Neuromimetic semantics* **Elsevier.**

This book attempts to relate truth-conditional semantics, cognitive linguistics and computational neuroscience. The book contains mini-summaries of biological visual processing, computational models of neural signaling, and the reduction of the Hodgkin–Huxley equations to the connectionist and integrate-and-fire neurons; Hebbian learning rules and the elaboration of learning vector quantization; the linguistic pathway in the left hemisphere; memory and the hippocampus; truth-conditional versus image-schematic semantics; objectivist versus experiential metaphysics; and mereotopology.

Ingram J.C. (2007). *Neurolinguistics: An introduction to spoken language processing and its disorders.* **Cambridge University Press.**

The book is divided into five parts covering foundational concepts and issues, speech perception and auditory processing, lexical semantics; sentence comprehension and discourse in healthy individuals and those with language disorders.

Jenkins L. (Ed.) (2004). *Variation and universals in biolinguistics.* **Elsevier.**

This book provides an interdisciplinary overview of work on the biology of language. Areas investigated and reviewed include: the micro-parametric theory of syntax, models of language acquisition and historical change, dynamical systems in language, genetics of populations, pragmatics of discourse, language neurology, genetic disorders of language, sign language, and evolution of language.

Levitin, D.J. (2006). *This is your brain on music: The science of a human obsession.* **Dutton/Penguin Books.**

The author investigates the role of music in human evolution and everyone's daily lives by synthesizing psychology, neuroscience, and musical examples from Mozart to Eminem.

Lieberman, P. (2000). *Human language and our reptilian brain: The subcortical bases of speech, syntax, and thought.* **Harvard University Press.**

The author discusses the neurological bases of human language and its evolution. The author argues that language is not an instinct coded in a discrete cortical "language organ," but a learned skill, based on a functional language system (FLS) distributed over many parts of the human brain.

Loritz, D. (1999). *How the brain evolved language.* **Oxford University Press.**

In the first half of the book the author retraces the steps by which Darwinian evolution selected first one-celled animals which could communicate among themselves and then multi-celled organisms which could communicate within themselves. The second half of the book explores the particular ways in which universal evolutionary designs – universal minimal neural networks – have been adapted for human language.

Mithen, S. (2006). *The singing Neanderthals: The origins of music, language, mind, and body.* **Harvard University Press.**

Drawing together strands from archeology, anthropology, psychology, neuroscience, and musicology, the author explains why we are so compelled to make and hear music.

Moldin, S.O., & Rubenstein, J.L.R. (Eds.) (2006). *Understanding autism: From basic neuroscience to treatment.* **CRC Press.**

The book covers diagnosis, epidemiology, genetics and genomics, neuroanatomy, neurophysiology, neuropsychology, and the neural systems underlying the autistic disorders spectrum.

Papathanasiou, I. (Ed.) (2003). *The sciences of aphasia.* **Pergamon.**

The book provides review chapters on controversial research and clinical issues in aphasia and aphasia therapy. Contributions cover the range of disciplines involved in aphasia, including neurology of aphasia, cognitive and linguistic approaches to aphasic therapy, psychosocial approaches, aphasia research methodology, and efficacy of aphasia therapy.

Paradis, M. (1998). *Pragmatics in neurogenic communication disorders.* **Elsevier Science.**

The author investigates the various types of deficits of pragmatic competence in different neurogenic communication disorders, and the use of pragmatic features as a compensatory strategy to supplement impaired linguistic competence. Information is provided on pragmatic deficits subsequent to right hemisphere damage, pragmatic disorders in dementias, deficits in discourse organization in right and left hemisphere damaged patients, as well as on the use of pragmatic features by fluent and non-fluent aphasic patients.

Paradis, M. (2001). *Manifestations of aphasia symptoms in different languages.* **Pergamon.**

The book stresses the importance to become aware of the manifestations of aphasia in languages other than one's own as the same underlying deficit may cause different surface manifestations in different languages. The characteristic symptoms of aphasia in 14 languages are described.

Paradis, M. (2004). *A neurolinguistic theory of bilingualism.* **John Benjamins.**

The author reviews the literature of the neurolinguistic aspects of bilingualism, provides a critical assessment of bilingual neuroimaging studies and proposes hypotheses about the representation, organization, and processing of two or more languages in one brain.

Posner, M.I., & Rothbart, M.K. (2007). *Educating the human brain.* **American Psychological Association.**

The authors discuss brain functions necessary for learning such as attending to information; controlling attention through effort; regulating the interplay of emotion with cognition; and coding, organizing, and retrieving information. They relate these aspects to school readiness, literacy, numeracy, and expertise.

Rogers, S.J., & Willams, J.H.G. (Eds.) (2006). *Imitation and the social mind: Autism and typical development.* **Guilford Press.**

The book discusses the role of imitation in both autism and typical development. Topics cover the neural and evolutionary bases of imitation, its connections to language development and relationships, and how early imitative deficits in autism might help to explain the more overt social and communication problems of older children and adults.

Rosen, G.D. (Ed.) (2006). *The dyslexic brain: New pathways in neuroscience discovery.* **Lawrence Erlbaum.**

The book is based on presentations of a 2004 symposium and discusses the neural components and functions involved in reading, the possible sources of breakdown, and makes suggestions for intervention.

Schmalhofer, F., & Perfetti, C.A. (Eds.) (2007). *Higher level language processes in the brain: Inference and comprehension processes.* **Lawrence Erlbaum.**

The book explains how behavior research, computational models, and brain imaging results can be unified in the study of human comprehension. The book is intended for advanced undergraduate and graduate cognitive science students, as well as researchers and practitioners.

Schumann, J., Gullberg, M., & Indefrey, P. (Eds.) (2006). *The cognitive neuroscience of second language acquisition.* **Blackwell Publishing.**

The contributions explore the cognitive neuroscience of second language acquisition from the perspectives of critical/sensitive periods, maturational effects, individual differences, neural regions involved, and processing characteristics.

Smolensky, P., & Legendre, G. (2006). *The harmonic mind: From neural computation to optimality-theoretic grammar. Volume II: Linguistic and philosophical Implications.* **The MIT Press.**

The book provides an integrated connectionist/symbolic architecture of the mind/brain, applied to neural/genomic realization of grammar; acquisition, processing, and typology in phonology and syntax; and foundations of cognitive explanation.

Stamenov, M.I., & Gallese, V. (Eds.) (2002). *Mirror neurons and the evolution of brain and language.* **John Benjamins.**

The volume discusses the nature of mirror neurons and the implications to our understanding of the evolution of brain, mind, and communicative interaction in non-human primates and man.

Verhoeven, L., & van Balkom, H. (Eds.) (2004). *Classification of developmental language disorders: Theoretical issues and clinical implications.* **Lawrence Erlbaum.**

The contributions address the question: "How can the child's linguistic environment be restructured so that children at risk can develop important adaptive skills in the domains of self-care, social interaction, and problem solving?" Target audience are students, researchers, and practitioners in the field of developmental language disorders.

Washburn, D.A. (Ed.) (2007). *Primate perspectives on behavior and cognition.* **American Psychological Association.**

The volume covers research methodology and subsequently a wide range of topics including language ability among higher primates and understanding human intelligence, cognitive processes, and motivation through date from primate research.

Witruk, E., Friederici, A.D., & Lachmann, T. (Eds.) (2002). *Basic functions of language, reading and reading disability.* **Kluwer Academic Publishers.**

The contributions of the book are based on a conference on basic mechanisms of language and language disorders in 1999 and cover the psychology and neurophysiology of language, reading and dyslexia.

Zatorre, R.J., & Peretz, I. (Ed.) (2001). *The biological foundations of music.* **New York Academy of Sciences.**

The volume explores the neurobiological basis for the creation and appreciation of music and includes contributions on both the structure and the function of brain pathways involved in the processing of complex sounds that lead to the perception of music. Evolutionary aspects of music recognition are examined, and theories of sound appreciation across cultures are presented.

SOURCE BOOKS

Birren, J.E., & Schaire, K.W. (Eds.) (2006). *Handbook of the psychology of aging* **(6th edn). Elsevier.**

The handbook is part of the *Handbooks on aging series* and is organized into four parts: concepts, theories, and methods in the psychology of aging; biological and social influences on aging; behavioral processes and aging; and complex behavioral concepts and processes in aging.

Boller, F., & Grafman, J. (Series editors) (1993–2003). *Handbook of neuropsychology.* **Volumes 1–11. Elsevier.**

Essential reference source to all topics related to neuropsychology. Several volumes contain sections on the neuroscience of language such as Berndt, R.S. (Ed.) (2001). *Handbook of neuropsychology* (Vol. 3, 2nd edn), *Language and aphasia* and Segalowitz, S.J., & Rapin, I. (Eds.) (2003). *Handbook of child neuropsychology* (Vol 8, 2nd edn).

Brown, K. (2006). *Encyclopaedia of language and linguistics* **(Volume set 1–14, 2nd edn). Elsevier.**

This totally revised 14 volumes set of the encyclopedia targets students, researchers and professionals who are seeking an authoritative source of information about any particular aspect of linguistics or its applications. Of particular interest is the section on "Language pathology and neurolinguistics" (section editor is H.A. Whitaker) and the extensive glossary.

Cohen, H., & Lefebvre, C. (2005). *Handbook of categorization in cognitive science.* **Elsevier.**

Categorization, the basic cognitive process of arranging objects into categories, is a fundamental process in human and machine intelligence and is central to investigations and research in cognitive science. Categorization is a key concept across the range of cognitive sciences, including linguistics and philosophy. The handbook provides an interdisciplinary approach to synthesize knowledge from the different fields.

Gazzaniga, M.S. (Ed.) (2004). *The cognitive neurosciences III* **(3rd edn). The MIT Press.**

This is the third revised and updated edition of this comprehensive classic reference covering all aspects of cognitive neuroscience, now also including the study of emotion and the social brain.

Hoff, E., & Shatz, M. (Eds.) (2007). *Blackwell handbook of language development.* **Blackwell Publishing.**

The handbook covers a wide range of areas such as brain development, computational skills, bilingualism, education, and cross-linguistic comparisons. Part I discusses the basic foundations and theoretical approaches to language development, part II language development in infancy, part III language development in early childhood, part IV language development after early childhood, and part V atypical language development.

Kroll, J.F., & De Groot, A.M.B. (Eds.) (2005). *Handbook of bilingualism. Psycholinguistic approaches*. Oxford University Press.

The handbook provides treatments of central issues related to bilingualism and second language acquisition including the neural mechanisms underlying bilingualism.

Loring, D.W. (1998). *INS Dictionary of Neuropsychology*, Oxford University Press.

This dictionary is a practical resource for neuropsychologists, neurologists, speech pathologists, psychiatrists, clinical psychologists, and occupational therapists whose work or research involves patients with nervous system disorders. The book provides concise information on terminology for neurobehavioral abnormalities, diseases affecting the nervous system, clinical syndromes, neuropsychological tests, rehabilitation methods, medical procedures, and basic neuroscience.

MacArthur, C.A., Graham, S., & Fitzgerald, J. (Eds.). (2006). *Handbook of writing research*. Guilford Press.

The volume covers knowledge on writing development in children and adolescents and the processes underlying successful learning and teaching. The first part focuses on theories and models of writing while the second part covers writing development.

Ritchie W., & Bhatia T. (Eds.) (2003). *Handbook of second language acquisition*. Academic Press.

The comprehensive reference source discusses the research, theory, and applications specific to second language acquisition and addresses issues such as maturation and modularity, language transfer between first and second languages, neuropsychology of second language acquisition, research and methodological issues in the study of second language acquisition, language contact, and its consequences.

Snyder, P. (Ed.) (2005). *Clinical neuropsychology: A pocket handbook for assessment* (2nd edn). American Psychological Association, APA.

Although mainly a practical reference source for neuropsychologists, interns, and trainees working in hospitals, this book is also interesting for those in clinical and psychology research. Besides quick-reference tables, lists, diagrams, photos, and decision trees, the book covers a broad array of illnesses including their neurochemical bases, where appropriate. It also provides a summary of neuroimaging technologies and accepted pharmacologic treatment approaches.

Spreen, O., & Risser, A.H. (2003). *Assessment of aphasia*. Oxford University Press.

The authors present tests for traditional use in the diagnosis of aphasia and in functional communication, childhood language development, bilingual testing, pragmatic aspects of language in everyday life, and communication problems in individuals with head injury or with lesions of the right hemisphere. The book is a comprehensive and practical resource for speech and language pathologists, neuropsychologists, and their students and trainees.

Traxler M., & Gernsbacher M. (Eds.) (2006). *Handbook of psycholinguistics*. Academic Press.

The second revised and updated edition of the *Handbook of psycholinguistics* represents a comprehensive survey of psycholinguistic theory, research and methodology, with special emphasis on the best empirical research conducted in the past decade. Target audience are professional researchers, graduate students, advanced undergraduates, university and college teachers, and other professionals in the fields of psycholinguistics, language comprehension, reading, neuropsychology of language, linguistics, language development, and computational modeling of language.

PROFESSIONAL ASSOCIATIONS AND SOCIETIES

American Psychological Association (APA)

http://www.apa.org/

APA is the largest association of psychologists worldwide.

American Speech-Language-Hearing Association (ASHA)

http://www.asha.org/default.htm

ASHA is the professional association for speech–language pathologists, audiologists, and speech, language, and hearing scientists in the United States and internationally.

Association for Computational Linguistics (ACL)

http://www.aclweb.org/

The Association is the professional society for people working on problems involving natural language and computation.

Cognitive Neuroscience Society (CNS)

http://www.cogneurosociety.org/

The Society (CNS) is committed to the development of mind and brain research aimed at investigating the psychological, computational, and neuroscientific bases of cognition.

International Association of Logopedics and Phoniatrics (IALP)

http://www.ialp.info/joomla/

An international organization representing persons involved with scientific, educational, and professional issues related to speech–language, hearing, and voice and swallowing disorders.

International Brain Research Organisation (IBRO)

http://www.ibro.org/

Obectives of IBRO are to develop, support, co-ordinate and promote scientific research in all fields concerning the brain.

International Society for the History of the Neurosciences (ISHN)

http://www.bri.ucla.edu/nha/ishn/

The mission of the society is to improve communication between individuals and groups interested in the history of neuroscience and promote research and education in the history of neuroscience.

Linguistic Society of America (LSA)

http://www.lsadc.org/

LSA is the largest linguistic society in the world and the only umbrella professional linguistics organization in the United States.

The Academy of Aphasia

http://www.academyofaphasia.org/

Members of the Academy are researchers who study the language problems of people who have neurological diseases.

The International Neuropsychological Society (INS)

http://www.the-ins.org/

The International Neuropsychological Society is a multi-disciplinary organization dedicated to enhancing communication among the scientific disciplines which contribute to the understanding of brain–behavior relationships.

Glossary

As this handbook reflects, the neuroscience of language is such a highly interdisciplinary field that even the expert often has to go online to decipher terminologies, concepts, and abbreviations in neighboring fields. The impossibility of constantly and sometimes redundantly explaining chapter-specific terminology in this book without exceeding space limits and interrupting the flow of the text, led some authors to make a suggestion that we have followed: we have included a glossary. This glossary is merely intended as a fast, albeit imperfect, reference guide. It was compiled with the help of all handbook authors, who contributed explanations and definitions of terms and concepts relevant to their chapters, and supplemented by the editors. It is not always possible to provide a concise explanation or precise definition of a term or concept, particularly one whose nature is being debated or currently investigated. The reader is thus advised to refer to the individual chapters or to specialized literature for more detailed information or deeper discussion. Terms written in italics indicate that the item is listed at another location in the glossary.

A

Abulia: A state in which an individual seems to have lost will or motivation.

Acetylcholine: A chemical that acts as a neurotransmitter and that plays important roles in the hippocampus, in declarative memory, and in learning.

Acquired stuttering: Stuttering behavior occurring in a heretofore fluent individual, usually an adult, from acquired brain disease or, sometimes, from functional disturbances.

Address space or addressing: A method used by any computational device to locate memories that are stored at particular locations. In the computer, addressing uses a structured, binary logic. It is not yet known in detail how the brain implements addressing of memories.

Affordances: An affordance of a visual scene is some property of the scene that provides parameters for action without necessary involvement of recognition of objects in the scene. Contrast, when walking down a crowded street avoiding bumping into people (using affordances) from recognizing a friend in the crowd. To a first approximation, affordances are computed in the *dorsal stream* of the primate cortex (passing forward from visual cortex through parietal cortex) while object recognition is computed in the *ventral stream* (passing forward from visual cortex through inferotemporal cortex). It has been posited that a similar dichotomy is relevant to language, with the dorsal stream mapping sound onto articulatory-based representations and the ventral stream mapping sound onto meanings.

Agrammatism: Impairments of grammar. Agrammatic speech is syntactically and/or morphologically simplified or incorrect. Patients with agrammatic speech often also have receptive agrammatism, that is, difficulties using syntactic structure to understand the meaning of sentences.

AIP (anterior intra-parietal sulcus): A region in the parietal lobe of macaque involved in computing *affordances* for the control of manual actions.

Alzheimer's disease: The most common form of dementia, associated with memory loss and a gradual decline of other cognitive functions and changes in personality or behavior.

Amygdala: Nucleus in the brain with significant contribution to emotional responses.

Anomia: See *dysnomia.*

Anterograde amnesia: The inability to form new long-term memories, resulting in a lack of awareness of what has occurred since the onset of the condition.

Aphasia: The loss of expressive and/or receptive language abilities following brain damage (see also *agrammatism, Broca's aphasia, Wernicke's aphasia, logopenic or mixed progressive aphasia, primary progressive aphasia, progressive non-fluent aphasia*).

Apraxia of speech: An impairment in motor speech planning functions occurring after phonemic processing (sound selection and sequencing) but before phonetic processing (neuromuscular specification and movement execution). Symptoms include disturbances in speech timing that cannot be attributed to a primary weakness or dyscoordination of the speech mechanism.

Arithmetical conceptual knowledge: The understanding of arithmetical operations, laws, and principles. Also labeled "adaptive expertise," it may or may not be explicitly known as such to the calculator, but allows us to make inferences and can be flexibly adapted to new tasks. Examples of this sort of knowledge are potentially infinite as demonstrated by calculators that answer the question "given X, what is the result of Y?"

Arithmetical facts: Single-digit, over-learned addition (e.g., 4 + 2) or multiplication problems (e.g., 5 × 6) listed in the so-called arithmetical tables. Arithmetical facts are thought to constitute, once learned, an independently stored system of ready-to-use arithmetical notions, retrieved by rote.

Articulatory programming: Generation of a plan and program for articulation; phoneme strings are converted into neuromuscular commands.

ASL: American sign language.

Asperger syndrome: A mental disorder, discovered by Hans Asperger, that is similar to autism in social and communicative impairments, but is not associated with delay in language acquisition or generalized intellectual impairment.

Association cortices: These are the regions of the brain that are involved in the processing and integration of sensory and motor information. For example, sensory information such as acoustic information is received by the primary (auditory) cortex and then send to association areas for further processing. Unimodal association areas receive information involving only one sensory modality while polymodal association areas receive information from more than one sensory modality. Unimodal and polymodal association cortices project to supramodal association cortices which are not well defined in terms of any given sensory modality.

Associative learning: A centuries-old model of learning whereby knowledge is acquired through repeated exposure to internal or external stimuli. Modern connectionist learning systems operate by the same principle, acquiring new behaviors by registering frequencies of stimuli co-occurrence and degrees of similarity among the stimuli they process. Associative learning systems can induce the patterns that characterize the stimuli to which they are exposed.

Asymmetric tonic neck reflex (ATNR): Characteristic posture of neonates when placed on their back with their head turned to one side. The arm and foot are extended on the side to which the infant is facing, and the contralateral limbs are flexed.

B

BA: See *Brodmann's area.*

Basal ganglia: A group of highly interconnected nuclei deep in the brain. These structures include the putamen, the caudate nucleus, the globus pallidus, and the substantia nigra. The basal ganglia are associated with some aspects of motor control, e.g. posture, and also contribute to cognitive and emotional processing. Although clearly involved in motor speech, there is a longstanding debate about the role of the basal ganglia in language.

BDNF: See *brain derived neurotrophic factor.*

Bihemispheric redundancy gain: Tendency for performance to be better when exactly the same information is presented to both brain hemispheres than when information is presented to only one brain hemisphere.

Bilateral: Involving both sides of the brain or body.

Biochemical: Relating to chemicals involved in biological functions (e.g., neurotransmitters).

Bolus-tracking perfusion-weighted imaging (PWI): A form of MRI used to show changes in regional cerebral blood flow reflected in the relative time to peak arrival and/or clearance of a bolus of contrast to each voxel in the image. (Also see *perfusion-weighted imaging.*)

Brain derived neurotrophic factor (BDNF): A protein found in the brain and elsewhere. It plays several functional roles, including in learning and memory.

Broca's aphasia: Brain damage following a stroke or some other event can compromise function. The function that is compromised depends on the location and extent of the lesion. Quite often, damage to a region known as Broca's area (after the nineteenth century physician, Paul Broca) in the left inferior frontal cortex can result in a partial loss of language function. Though traditionally associated with effortful speech, the loss of some syntactic elements, and relatively preserved comprehension, the precise nature of the pattern of sparing and loss is a topic of current research.

Broca's area: A classical brain language area consisting of the opercular part and the triangular part of the inferior frontal gyrus – which in turn correspond largely to Brodmann's areas (BA) 44 and 45, respectively. The area is named for the French physician Paul Broca, who first suggested its involvement in language. Today we know it is also involved in other functions such as sign language, manual sequencing, and music.

Broca's region: A brain region that plays important roles in language. It is generally taken to comprise Broca's area and certain nearby regions, including the orbital part of the inferior frontal gyrus (Brodmann's area 47) and the frontal operculum.

Brodmann's area (BA): An anatomically (i.e., cytoarchitectonically) defined area of the brain. The cytoarchitectonic map was developed by Korbinian Brodmann in the early 1900s.

C

Canonical neuron: See mirror neuron system.

Central coherence (CC): According to Frith (2002), CC involves the ability to extract and integrate meaning from contextual details together with a global appreciation of scenes or events. One proposal is that autism is characterized by weak CC compared to mental age matched control subjects.

Cerebellum: The part of the brain beneath the back of the cerebrum, below the membrane known as the tentorium. Although the cerebellum has long been implicated in movement, it is now also associated with memory, language, and other functions.

Cerebrum: The large, main part of the brain, made up of two hemispheres.

Childhood amnesia: Inability of most people to remember events from the earliest two or three years of life.

Cognitive neuropsychological approach: Model-based approach locating the language impairment within a hypothesized model of normal cognitive processing.

Coherence: In linguistics it refers to those elements that establish semantic continuity of a text or discourse (such as logically connected arguments or presuppositions and implications related to general world knowledge). The purely linguistic elements that contribute to text coherence are subsumed under the term *cohesion.*

Cohesion: In linguistics cohesion involves the selection of a lexical item that is in some way related to one occurring previously in a sentence or text, for example the simple repetition of a word, or its superordinate (*banana – fruit*), the use of pronouns to establish reference (*John – he*), and so on.

Cohort effect: Potential confounding variable in cross-sectional studies of aging. Older and younger individuals may differ in characteristics other than age *per se*, such as education, socialization, life experiences, healthcare during childhood, and so forth.

Compensation strategies: Therapy strategies attempting to optimize the use of retained verbal and non-verbal abilities, without focusing on the impaired functions.

Compounding: The process of word formation by which a word is created from two or more existing words (*lampshade, rock-climb, class action suit*).

Computational machinery: The set of computational devices involved in certain cognitive process.

Conceptual semantics: Having to do with the meanings of concepts, including concepts associated with words (e.g., living thing, furniture, hammer).

Confrontation naming: A neuropsychological procedure where subjects are visually presented with a stimulus or auditorily provided a heard definition or description and are then asked to specifically label the objects, activity, or experience.

Construction grammar: A grammatical system which, rather than combining words according to a few highly general syntactic rules, employs a large number of *constructions* of varied applicability, but each of which links a form of syntactic combination with the derivation of new meanings from the meanings of the constituents. The specialized constructions needed to express idioms blur the division between grammar and lexicon.

Continuity hypothesis: This hypothesis proposes that there is a continuum linking behaviors exhibited by healthy individuals with those exhibited by individuals with disease. For example, slips of the tongue produced by normal speakers are not unlike the semantic or phonemic paraphasias produced by speakers with aphasia.

Contralateral: Involving the opposite side of the brain or body.

Core domains of knowledge: These are domains such as syntax, numbers, physical causality, intentional actions, and mental states, for which it has been hypothesized that there are specific innate constraints to process relevant input and trigger cognitive processes. These activate, early in children's development, abstract notions or set parameters that are critical in acquiring an adult competence.

Corpus callosum: The largest of several fiber tracts connecting the left and right brain hemispheres, consisting of between 200 and 800 million fibers.

Cortex: The outermost layer of the cerebrum, consisting largely of neuronal cell bodies.

Cortical column: A population of neurons that are arranged vertically (i.e., perpendicular to the cortical laminae) in the cerebral cortex and thought to function as a single unit.

CPS: The "closure positive shift" is a large positive-going ERP component at central electrode sites that is elicited by prosodic speech boundaries. The CPS reflects the listener's segmentation of the acoustic signal into prosodic phrases and, therefore, seems to play an important role in language learning and syntactic analyses.

Cytoarchitecture: Structural analysis of the cerebral cortex, based on cellular features (size and shape of cells, cell packing density in different cortical layers, width of layers) as identified in cell body-stained specimens.

D

Deblocking techniques: Therapy that aims to reactivate access to impaired language functions via intact oral and written abilities.

Declarative memory: The memory system for knowledge of facts (knowledge "that") and personal experiences. Evidence suggests that this knowledge is largely, though not completely, available to conscious awareness. See also *procedural memory* (knowledge "how").

Deep dysgraphia: A writing disorder characterized by semantic errors and poor spelling of non-words.

Deep dyslexia: A reading disorder characterized by semantic errors and poor reading of non-words.

Default network: Certain brain regions, including inferior parietal, posterior cingulate, medial temporal/parahippocampal, and medial prefrontal cortices, that are active while subjects rest quietly and are less active during attention-demanding tasks.

Derivational morphology: The process of word formation by which a word is changed into a different word with a different syntactic function, either by adding an affix [nation→national → nationality] or by changing the stress pattern and/or vowel (sing→ song; permit (noun)→permit (verb)].

Developmental stuttering: A non-voluntary disturbance in speech-motor output, usually characterized by repetitions and prolongations of sound, often associated with facial grimacing and evidence of struggle. The person who stutters knows what he wants to say but has difficulty so doing. Developmental stuttering usually begins in early childhood.

Dichotic listening: An experimental technique of simultaneously presenting one sound or word to the right ear and a different, competing sound or word to the left ear and then asking the subject to identify the sounds or words. Over repeated trials the set of sounds or words presented to the right ear are typically processed more accurately than those presented to the left ear, in groups of subjects. This is typically interpreted as reflecting left hemisphere dominance for speech-language processing.

Diffusion tensor imaging (DTI): A magnetic resonance technique allowing the measurement of the restricted diffusion of water in cerebral tissue. It is used to image the location, orientation, and direction of the main fiber tracts in white matter. Also referred to as diffusion tensor magnetic resonance tractography.

Diffusion-weighted imaging (DWI): A form of MRI used to show accumulation of water molecules within the intracellular compartment (with restricted diffusion), which is a marker of cellular metabolic compromise that leads to cell death.

Direct stimulation: Tasks to imitate verbal and non-verbal acts by speaking or singing in unison or by oral repetition.

Dishabituation paradigm: Method of studying the ability of infants to discriminate between stimuli. A stimulus is presented repeatedly until a selected behavioral or physiological index (e.g., cardiac activity) decreases to a steady-state level (habituation). Then, the stimulus is replaced by a different stimulus. A corresponding increase in activity (dishabituation) indicates that the difference between stimuli was detected. Also known as habituation–dishabituation paradigm.

Dispersion: The distribution of items along a distance continuum. In the sonority continuum (Obstruent, Liquid, Nasal, Glide, Vowel), equal spacing among sequenced phonemes is preferred (e.g., OLV), and maximal spacing is most preferred (e.g., OV). Dispersion as a metric of optimal sound sequences can be calculated by a relatively simple formula.

Domain-relevant mechanisms: Mechanisms important for the development of a given domain but not exclusive to that domain, which increase in specialization over time giving rise to specialized domain-specific mechanisms.

Domain specificity: The degree to which a brain area or other biological substrate subserves only one cognitive function (i.e., the degree to which it is domain specific as opposed to domain general).

Dopamine: See *neurotransmitters.*

Dorsal stream: See *affordances.*

Double dissociation: A principle that suggests that two concepts or functions are dissociable and can occur independently of each other. For example, one brain region may show more activation during grammatical than lexical processing, while another region shows greater activation for lexical than grammatical processing.

Down syndrome: A genetic neurodevelopmental disorder affecting approximately 1 in 600 births associated with trisomy of chromosome 21.

DTI: See *diffusion tensor imaging.*

Dual-route model: A model of the structure and processing of the lexical–semantic system with two routes to reading and spelling, a lexical and a sub-lexical one.

Dysarthria: Speech difficulties resulting from a disruption of motor pathways.

Dysgraphia: An acquired writing disorder caused by cerebral damage. Following contemporary cognitive models, writing disorders are further distinguished in *phonological, surface, deep, direct, letter-by-letter* and *peripheral* dysgraphia. Acquired dysgraphia must be differentiated from *developmental* writing disorders.

Dyslexia: A developmental or an acquired reading disorder. Acquired reading disorders occur, for example, after a stroke or traumatic brain injury. Following contemporary cognitive models, reading disorders are further distinguished in *phonological, surface, deep, allographic* and *neglect* dyslexia.

Dysnomia: Also referred to as "anomia"; dysnomia is a difficulty in naming.

E

Ear advantage: An asymmetry often obtained in auditory laterality studies, especially studies using dichotic listening. A right-ear advantage (REA) is interpreted as evidence of left-hemisphere processing of the input, and a left-ear advantage (LEA) is interpreted as evidence of right-hemisphere processing. See *dichotic listening.*

EEG (electroencephalogram): Scalp-based recording of the brain's ongoing electrical activity.

Effective connectivity: The functional strength of a specific anatomical connection during a particular cognitive task; the influence that one region has on a second. This quantity can be estimated using a technique such as structural equation modeling.

ELAN: See *LAN.*

Epileptogenic zone: Area of the brain that serves as the origin and primary generator of hypersynchronous neural activity (i.e., seizures). Symptoms and behaviors can emerge from this zone, but as other brain regions are recruited into the

hypersynchronous activity a wide range of clinical phenomena can be observed. See also *mesial temporal lobe epilepsy* and *mesial temporal lobe seizure.*

Episodic memory: Memory for personally experienced events occurring at specific times and places.

ERPs: Event-related brain potentials (ERPs) are negative and positive deflections (ERP components) in the scalp-recorded electroencephalogram (EEG) that occur time-locked to an observable event (such as the presentation of a target word or syllable). It is usually assumed that ERP components reflect specific cognitive processes in the brain.

Executive functions: "Supervisory" cognitive functions that control and regulate other cognitive abilities. Executive functions, often attributed to the frontal lobes (but not exclusively), enable thought and behavior to proceed in a goal directed fashion and the individual to adapt to changing circumstances.

Extraoperative cortical stimulation: A direct cortical electrical stimulation technique in which stimulation is performed through implanted electrode arrays that allow testing after the patient has recovered from the initial surgery. (See also *intraoperative cortical stimulation.*)

F

F5: A region of macaque frontal cortex including neurons for the preparation of manual and orofacial actions, some of which are mirror neurons (see *mirror neuron system*). F5 is considered homologous to Brodmann's area 44 in humans.

Faithfulness: Now a technical term in psycholinguistics, this word refers to how close a response is to an intended target. In optimality theory, it refers to how closely a produced form resembles its predicted unmarked or optimal form. In connectionism, it refers to how closely inputs and outputs match. Faithfulness is an important concept in constraint-based theories of language processing.

Flynn effect: A large intergenerational rise in scores on intelligence tests that has been observed in various areas of the world. The magnitude of the effect is greater for non-verbal, "culture-reduced" tests than for tests that require learned information and skills.

Frontal lobes: The frontal lobes are responsible for the development, selection, and execution of plans for action. Their function is most crucial in novel situations that require the development of tailor-made plans. They are organized, from anterior to posterior, into prefrontal cortex, premotor cortex, and motor cortex (at the most posterior margin of the frontal lobes). This anatomy correlates with a hierarchy of function. Prefrontal cortex is involved in executive functions, working memory engagement, and certain aspects of language. Motor cortex, through its projections to the brainstem and spinal cord, actually produces movement. Premotor cortex can be viewed as a "motor association" cortex that is involved primarily in the planning and organization of movements. A specialized portion of premotor cortex located on the medial surface of the hemispheres – supplementary motor area (SMA) – in conjunction with anterior portions of the cingulate gyrus, influences movements according to internal demand.

Functional connectivity: The correlation between functional activity in two brain regions during a specific cognitive task; this can be evaluated either across individuals or by correlating time series in the two regions.

Functional imaging: Imaging techniques that focus on information about brain function as opposed to brain anatomy. There is often a trade-off between functional and structural information obtained from brain imaging procedures, e.g. *ERP*s provide better functional than structural information.

G

Gamma amino butyric acid (GABA): See *neurotransmitters.*

Geschwind model: A classic model of brain-language relationships that views Broca's area as the seat of expressive language (later interpreted as 'syntax'), Wernicke's area as the seat of receptive language (later interpreted as 'lexical processing') and the arcuate fasciculus as the connecting pathway between the two. In this model, lesions to Broca's area led to Broca's aphasia, lesions to Wernicke's area, Wernicke's aphasia and lesions to the arcuate fasciculus led to conduction aphasia.

Gestalt properties: In vision or audition, these are structural properties by which a stimulus is perceived as whole rather than in terms of component parts.

Glutamate: See *neurotransmitters.*

Grammatical encoding of a message: The grammatical encoding process consists of procedures for accessing lemmas (meaning and syntax of words), and of syntactic building procedures.

Grapheme: In the writing system of a given language, a written element that cannot be analyzed into smaller units; in languages with alphabetic script, the letter or letter groups that represent a phoneme: for instance, the phoneme /f/ may be represented by several letters or letter groups: for example, f, gh, or ph, in *scarf*, *rough*, and *graph*.

H

Habituation–dishabituation paradigm: See *dishabituation paradigm.*

Handedness: Manual asymmetry defined either as hand preference or as a difference between the hands in the performance of a motor task.

Hemiparesis: Incomplete or partial paralysis of one side of the body.

Hemodynamic: Concerned with the characteristics of blood flow.

Heschl's gyrus: Also referred to as transverse temporal gyrus. Considered the location of the primary auditory cortex in the human.

H.M.: A famous patient suffering from anterograde amnesia as a result of a 1953 brain surgery to address his epilepsy. In this surgery most of his medial temporal lobe structures were removed in both hemispheres.

Hippocampal sclerosis (Ammon's horn sclerosis): Neuron cell loss primarily in the hippocampus. Histopathological findings include segmental loss of pyramidal neurons, granule cell dispersion, and reactive gliosis.

Hippocampus: A structure in the medial temporal lobe that underlies learning in declarative memory.

Holophrase: A single "protoword" or formula without meaningful parts that is semantically akin to the phrases or sentences of a modern language.

Homophones: Words with same phonology but a different meaning ("two" versus "too").

Huntington's disease: A rare dominant genetic neurodegenerative disease characterized by progressive movement disorder (chorea) and dementia that may first appear in childhood but most often appears in early to mid-adulthood. The disease is marked by degeneration in the basal ganglia and the cerebral cortex, most particularly the orbitofrontal cortex, leading to change in personality and deterioration of interpersonal behavior.

I

Impairment-specific therapy: In the context of aphasia, a therapy that aims primarily at relearning degraded linguistic processing and/or knowledge.

Individual differences: Normal variability across individuals in a characteristic such as brain volume or localization of language or handedness. True individual differences are distinct from group differences (e.g., differences associated with sex, age, or linguistic group).

Inferences (inferencing): A mental process by which we reach a conclusion based on specific evidence such as logical reasoning from premises; drawing conclusions about the unsaid based on what is actually said.

Inflectional morphology: The process by which the form of a word is altered by the addition of an affix or by a stem-internal change that signals grammatical features such as person, number, tense, and mood (e.g., *flower-s, greet-ed, teach-taught*).

Insula: The insular cortex is also known by the name Island of Reil and is subjacent to the Ssylvian fissure and to its cortical rim composed of temporal, parietal, and frontal cerebral cortex. Its anterior extremity approaches Broca's area from below and its posterior extremity approaches Wernicke's from below. It is implicated in many language and speech functions.

Interactive activation: A neurocomputational model that emphasizes the spread of activation between neurons and the summing of activation from multiple sources to induce the firing of a receiver neuron.

Interhemispheric interaction: Coordination of processing between the left and right brain hemispheres.

Interrater reliability: Extent to which two or more raters provide a similar rating.

Intraoperative cortical stimulation: A form of the direct cortical electrical stimulation technique in which stimulation is applied at the time of the craniotomy (in the operating room). (See also *extraoperative cortical stimulation.*)

Invariant lateralization: Hypothesis that brain asymmetries, or the precursors of brain asymmetries, are present early in development and do not change substantially across the life span.

Ipsilateral: Involving the same side of the brain or body.

Ischemic penumbra: An area of brain tissue around an infarct that is getting just sufficient blood to survive for a limited time, but not enough to function.

Island of Reil: See *insula.*

J

Jabberwocky (from Lewis Carroll): Meaningless speech or writing that nevertheless meets criteria for phonemic and syntactic legitimacy in English.

Joint attention behaviors: A behavioral pattern characterized by alternating the gaze toward an object or event of interest and a social partner. Its function appears to be that of checking the attention of the social partner and/or directing it toward a shared focus.

L

LAN/ELAN: Left anterior negativities (LANs) are ERP components between 100 and 500 ms that have been found both for syntax violations and in sentences imposing high working memory demands. Early LANs (ELANs) are typical for morphologically marked word category violations in highly constrained sentence structures.

Language lateralization: The organization and neural representation of language systems to different sides of the brain generally established in early childhood (and most commonly left hemisphere in nature), but can be affected by acquired brain insult and pathology. Language skills that change their neural representation and come to be implemented by a different region or set of regions can be said to have reorganized. (See also *laterality, lateralization of function.*)

Language-ready brain: This refers to whatever brain properties unique to humans make possible the modern human child's ability to acquire the language of the community in which it is reared. The term is agnostic as to whether these properties are describable by a Universal Grammar.

Laterality: Preference for one side of the body (e.g. hand preference) or for one side of the brain (e.g. hemispheric asymmetry) over the other. Asymmetries may be spontaneous (e.g., hand preference) or asymmetric performance on specially devised visual, auditory, tactual, or motor tasks (e.g., a right-ear advantage in dichotic listening).

Lateralization of function: An asymmetry in the brain representation of a particular function; may refer to a categorical (all-or-none) difference between left and right sides or to a difference in the degree to which a function depends on one side. Also, with respect to progressive lateralization theories, the term may refer to the process through which the brain becomes lateralized.

LEA: Left-ear advantage; see *ear advantage.*

Lexical decision task: A psycholinguistic technique for measuring lexical access. Participants see words and non-words presented briefly on a computer screen. They are asked to decide whether it is a word or not as quickly and accurately as possible by pressing either a "yes" or "no" response button.

Lexis: The vocabulary or total stock of words in a language. Also known as the lexicon.

Logopenic or mixed progressive aphasia: Involves a disorder of naming and slowed speech with poor repetition, but relatively preserved semantic representations of concepts and modest grammatical difficulty.

Long-term memory (LTM): Memory which may persist indefinitely.

LTM: See *long-term memory.*

M

Maturational gradient: Differential rate of development with respect to a specified dimension, for example, right-to-left side or primary-to-secondary-to-tertiary cortex.

MEG/ERMF: Magnetoencephalography (MEG) is similar to the EEG/ERP technique but measures magnetic fields associated with brain activity. Event-related magnetic field (ERMF) components in the MEG are usually labeled with the letter "m" (e.g., N1m, N400 m).

Mental lexicon: The term used to describe the store of words in the mind. The term can also be used to refer to the cognitive system that constitutes the capacity for conscious and unconscious lexical activity.

Mentalizing: A terms that refers to cognitive processes that result in attributing mental states to agents. See also *theory of mind.*

Mesial temporal lobe epilepsy (MTLE): Recurrent epileptic seizures arising from mesial temporal lobe structures, including the hippocampus, parahippocampal gyrus, rhinal cortex, and amygdala of either one or both temporal lobes of the brain. It is among the most common of epilepsy syndromes and is well characterized in terms of seizure appearance, impact on functioning, and underlying pathology.

Mesial temporal lobe seizures: Partial seizures, which are seizures that arise from abnormal electric activity only in a localized part of the brain, can be simple or complex, distinguished by the degree to which alertness is compromised and consciousness is lost. Simple partial seizures cause sensory distortions or other sensations, but do not interrupt consciousness or memory. In contrast, complex seizures disrupt both.

Mild cognitive impairment (MCI): A condition in which a person has problems with memory, language, or another mental function severe enough to be noticeable and to show up on tests, but not serious enough to interfere with daily activities or to meet the criteria for the diagnosis of dementia.

Mirror neuron system (MNS): A brain region involved in both the generation and recognition of a class of actions. In monkeys, the "mirror property" has been observed even at the single neuron level. Mirror neurons are complemented by *canonical neurons* which fire when the monkey performs a specific action but not when it observes a similar action.

Mirror System Hypothesis: The claim that brain mechanisms for language evolved atop a mirror system for grasping through the successive emergence of systems for imitation, pantomime, protosign, and protospeech.

MMN: The mismatch negativity (MMN) is an early, fronto-centrally distributed negative ERP component reflecting the discrimination of sounds, including phonemes.

Model-based approaches: Assessment and therapy approaches based on cognitive neuropsychological or psycholinguistic models.

Modularity: A neurocomputational theory that emphasizes the linkage between particular psychological functions and specific neural systems, perhaps located in specific regions of the brain. Modularity requires that the information involved be processed inside a single module with minimal interaction with other modules.

Morpheme: Minimal meaningful unit of language. It can consist of a word (e.g., *book*, *heat*), or of part of a word (e.g., *-s* in *books* or *re-* and *-ed* in *reheated*).

Morphology: In linguistics, the internal structure of words and its study. Morphological processes include inflection, derivation, and compounding. (See also *inflectional morphology, derivational morphology,* and *compounding.*)

Motor cortex: See frontal lobes.

Motor schema: See *schema theory.*

MTLE: See *mesial temporal lobe epilepsy.*

Multimodal association cortices: The multimodal association cortices of the parietal, temporal, and frontal lobes integrate somatosensory, auditory, and visual information for higher order cognitive processing.

Multimodal microstructural approach: An approach to characterize fine-scale anatomy using multiple histological and neurochemical techniques, revealing different aspects of cellular organization or molecular composition.

Myelin: Fatty sheath around neurons that greatly increases conduction speed of nerve impulses.

Myeloarchitecture: Subdivisions (named or numbered) of the cerebral cortex that are based on features (e.g., stria of Gennari in the visual cortex) of myelinization (differential density of myelinated fibers and fiber bundles in different cortical layers), and identified in myelin-stained histological specimens are called myeloarchitectonical maps.

N

N100: An early negative *ERP* component around 100 ms linked to feature analyses of a stimulus and attention mechanisms. It has also been suggested that the N100 is implicated in processes crucial in word segmentation.

N400: A negative ERP component peaking around 400 ms at centro-parietal electrodes that reflects lexical retrieval and/or semantic integration.

Negative cortical stimulation effects: Instances in which direct cortical electrical stimulation results in the cessation or disruption of an ongoing behavior. (See also *positive cortical stimulation effects*.)

Neocortex: The top layer of the cerebral hemispheres is made up of six layers (I–VI). Other names for the neocortex include neopallium and isocortex (equal rind).

Neologisms: Words, terms, or phrases that have been newly created. The term *e-mail*, as used today, is an example of a neologism. In thought disorders or aphasia the term is usually used to describe the creation of words which have no apparent meaning.

Neurocognitive: Relating to the neural, cognitive, and computational correlates of mental functions.

Neurocomputational models: Accounts of the neural bases of psychological processes that are formulated in terms specific enough to be implemented as computer programs.

Neurodegenerative disease: A disease involving the progressive loss of brain tissue.

Neuroimaging: The construction of images that display information from actual brains. Using different methods (such as MRI, fMRI, PET, EEG, MEG), different types of images can be created, containing different types of information. These include "structural images" (which show anatomical structures) and "functional images" (which show brain activity during cognitive processing) and images which show both structural and functional components.

Neurotransmitters: Chemical substances in the brain that transmits information between neurons. They are released at the ends of axons, rapidly diffuse across a small space, the synapse, and bind to specialized receptors on the receiving end of the synapse. Important neurotransmitters are glutamate (a major excitatory neurotransmitter), and gamma amino butyric acid (GABA) (a major inhibitory neurotransmitter). Glutamatergic neurons often have very long axons that enable them to transmit information to distant regions while GABAergic neurons tend to have short axons that project to immediately adjacent excitatory neurons; they are therefore often referred to as interneurons. Other neurotransmitters that are largely modulatory are opioid neurotransmitters such as endorphins and enkephalins, acetylcholine, dopamine, norepinephrine, epinephrine, serotonin, and histamine.

Noise-vocoded speech: A synthesis technique where the amplitude variation across a number of different frequency channels modulates noise filtered to the same bandwidth as the channels; speech resynthesized this way sounds like a harsh whisper; temporal information is preserved, but spectral information is reduced; the more channels there are, the more detail is available in the spectral domain, and the more intelligible the speech.

Nonce words: Nonsense words that could be "possible words" in a language. They are constructed of a language's permissible phonemes and phoneme sequences and are frequently unwittingly produced by speakers with fluent aphasia (e.g., chipicters).

Number transcoding: Transforming a number from a given representational format into another format. Thus, reading aloud presupposes transcoding Arabic or alphabetically written numbers into spoken number words; in writing to dictation the reverse transcoding process is at work.

O

Obsessive-compulsive disorder (OCD): People who suffer from OCD have persistent thoughts, images, and desires that are experienced as intrusive and inappropriate (*obsessions*). They exhibit repetitive behaviors, such as overly frequent hand-washing and attention to tidiness, or mental acts, such as counting and repeating words silently (*compulsions*). These behaviors and mental acts may be performed in order to reduce the distress that typically accompanies the obsessions.

Optimality theory: A recent theory of phonological competence developed to account for regularities within the sound patterns of a language. It employs faithfulness and markedness constraints to ensure that produced word forms match stored word forms and to ensure that the phonological structures of produced word forms are permissible in a language.

Ordinality: The ability to judge which of two different quantities is larger.

P

P600/SPS: A positive-going centro-parietal ERP component around 600 ms that has been linked to the costs of structural processing difficulties and may consist of multiple subcomponents (SPS, syntactic positive shift).

Paragraphia: An error in writing consisting in the production of another (semantically or phonologically related) word.

Paraphasia: A substitution error, for example producing another semantically or phonologically related word (*magazine* instead of *book*, or, *pill* instead of *bill*).

Parkinson's disease: A slowly progressing neurodegenerative disease due to the loss of dopamine-producing brain cells in the basal ganglia leading to motor and cognitive impairments.

Pars opercularis: See *Broca's area*.

Pars triangularis: See *Broca's area*.

Perceptual schema: See *schema theory*.

Perfusion-weighted imaging (PWI): A magnetic resonance technique providing information about blood flow to a specific region of the brain.

Perseveration: This term refers to the inappropriate persistence of a behavior beyond the point at which it should have stopped. Major types of perseveration, including continuous, stuck-in-set, and recurrent forms, have been used to describe linguistic errors in aphasia.

PET: See *positron emission tomography*.

Phoneme: A sound unit in a language that cannot be further analyzed into smaller linear units and that distinguishes one word from another (e.g., /p/ and /b/ in the English minimal word pair *pull* and *bull*).

Phonological assembly: The generation of a metrically specified phoneme string for production.

Phonological cueing: Providing a phonological cue to facilitate the production of a word, usually the initial sound or syllable.

Phonological dysgraphia: A writing disorder characterized by difficulty in non-word writing.

Phonological dyslexia: A reading disorder characterized by difficulty in non-word reading.

Phonological encoding: Retrieval or building of a phonetic plan for words and for the utterance as a whole.

Phonological paraphasia: Phonological errors related to the target word (*boat* instead of *goat*).

Plasticity: Capacity of the nervous system to change in response to normal or pathological influences. Examples include the learning of new information, stimulation-dependent development of sensory systems during critical periods, and recovery of functions following brain damage. Also known as neuroplasticity and neural plasticity.

PMN: The fronto-central "phonological mapping negativity" (previously, "phonological mismatch negativity") is an ERP component that precedes the N400 and peaks around 300 ms. It seems to reflect a comparison (mapping) between phonological expectations and the actual input.

Polymodal association cortex: See *association cortices*.

Polysynaptic intrahippocampal pathway: A neuronal pathway with multiple interneurons and long neuronal chains that courses through the entorhinal area, dentate, CA3 and CA1 fields, and the subiculum. The polysynaptic pathway enters the entorhinal cortex via the parahippocampus as opposed to the perirhinal area.

Population-level asymmetries: Characteristics such as right-handedness and left-lateralized expressive language representation that are found in the majority of humans. Other asymmetries may occur in individuals without reflecting a similar asymmetry in the general population.

Positive cortical stimulation effects: Instances in which direct cortical electrical stimulation results in the elicitation of motor movements, sensations, or behaviors. (See also *negative cortical stimulation effects*.)

Positron emission tomography (PET): A functional imaging technique; see *functional imaging*.

Pragmatic inference: Information that is not stated explicitly, but can be inferred from the context or situation in which the message is conveyed. For example, upon hearing "It's hot in here" we may infer that the speaker wants us to open the window, if the speaker and listener are both in a closed room. (See also *inference*.)

Praxicon: Just as language builds sentences from items in the lexicon, so praxis builds skilled behavior from items in the praxicon, which may be considered as a set of well-practiced motor schemas.

Praxis: Comprises the class of actions involving instrumental action upon objects of the physical world. To be contrasted with communication.

Prefrontal cortex: See *frontal lobes*.

Premotor cortex: See *frontal lobes*.

Primary cortex: See *association cortices*.

Primary progressive aphasia (PPA): Refers to a progressive decline in language functioning that is not due to an obvious structural lesion (such as a stroke or trauma) and is not accompanied by changes in other cognitive domains (e.g., memory or visual perceptual-spatial difficulty) for about 2 years. Recent formulations suggest that there are three forms of primary progressive aphasia: *progressive non-fluent aphasia*, *semantic dementia*, and *logopenic or mixed progressive aphasia*.

Probability maps: *Cytoarchitectonic maps*, three-dimensional (stereotaxic) maps of cortical areas, subcortical nuclei, and fiber tracts showing the frequency with which a certain structure can be found at a certain position in the stereotaxic space of a reference brain. These maps are based on (cyto-)architectonic mapping in a sample of postmortem brains.

Procedural memory: The memory system that underlies the learning and control of motor and cognitive skills (knowledge "how") such as riding a bicycle. This knowledge seems to be entirely implicit. The system is rooted in brain circuits passing through the basal ganglia and frontal cortex.

Progressive lateralization: Hypothesis that the brain initially develops as a symmetrical organ, and that structural and functional asymmetries emerge with further maturation.

Progressive non-fluent aphasia: Form of aphasia in which speech is sparse and effortful, often accompanied by a disorder of the motor speech apparatus such as apraxia of speech, together with difficulty understanding long-distance grammatically mediated relations in a sentence.

Propositional attitude: Is the attitude taken by an agent toward the meaning of a sentence (a proposition). For example, an agent may have the attitude of *believing* that a proposition is true or may *pretend* that is true.

Prosody: Melody of speech, made up of fundamental frequency, amplitude, timing, and voice quality, and communicating emotions, attitudes, and personal information in speech.

Protolanguage: The word is used in at least three senses: (i) the ancestral language for a family of related languages (as in Proto-Indo-European); (ii) the precursor to language exhibited in the speech or signing of a young child (up to age 2 years or so); and (iii) a presumed precursor for human language exhibited by protohumans who have achieved an open system of communication which did not yet employ the syntax and compositional semantics which distinguishes language. For (ii) and (iii), but not (i), a protolanguage is a reduced form of communication that has an open-ended collection of "protowords" to symbolize objects, actions, or events, but no syntax to form subtle combinations of these protowords.

Protosign: A system of conventionalized manual gestures constituting a *protolanguage* in the sense of a presumed precursor for human language. Some claim that protosign provided the necessary scaffolding for *protospeech*, i.e., protolanguage that is spoken.

Protospeech: See *protosign*.

Psycholinguistic approach: Locating the language impairment within a model of normal word and/or sentence processing (model of sentence comprehension and production).

Pseudowords: Words that are made up but are spelled and can be pronounced as if they were real words. Examples are *blub*, *fungmok*, and *soped*.

PWI: See *perfusion-weighted imaging*.

R

Random generator: A metaphorical mechanism proposed to describe how speakers can use their knowledge of phonemes and phonotactics to construct (randomly generated) nonsense words. Used to describe how some, but not all, neologisms may be produced in aphasia.

rCBF: Regional cerebral blood flow, the measure of regional functional activity obtained using PET.

REA: Right-ear advantage, see *ear advantage*.

Receptor architecture: Subdivisions (named or numbered) of the cerebral cortex that are based on differences in the molecular composition of cortical and subcortical brain regions are called receptorarchitectonic maps. Regional and laminar differences can be assessed using receptor autoradiography by revealing subdivisions of the brain with distinctive distribution patterns of concentrations of receptor binding sites of transmitter receptors.

Repetition suppression: It refers to the reduced neural response associated with repeated processing of a stimulus.

S

Schema theory: Schema theory views concepts as abstracted from the perceptual and motor schemas of an agent interacting with its world. Knowledge is mediated through activity in a network of interacting schema instances. Perceptual schemas mediate the recognition of objects or actions in the world; motor schemas provide the units of action. Schema instances, which will include those encoding relevant knowledge and goals as well as the current situation, compete and cooperate to develop schema assemblages and coordinated control programs which support the animal or human's autonomous pursuit of goals whether praxic or communicative.

Self-organizing maps: Neurocomputational models that encode new information in terms of a multidimensional address space. On the emergent map, units that are neighbors occupy similar places in the multidimensional space. These models are often intended to simulate properties of cell assemblies in the cortex.

Semantic cueing: Providing semantic information to facilitate the production of a word.

Semantic dementia: A fluent form of progressive language decline that initially involves a disorder of naming and lexical comprehension, and progresses over time to involve difficulty understanding object concepts regardless of the modality or material of representation.

Semantic paraphasia: Semantically related errors to the target word (*sheep* instead of *goat*).

Sentence picture matching task: A comprehension task often given to aphasic subjects. Subjects view two or more pictures and at the same time hear a sentence. The task is to point to the picture that best matches the sentence they have heard. Often, the pictures minimally contrast some simple action that is easy to draw, such as *kicking*. Thus, in one picture a woman might be kicking a giraffe, while in a contrasting picture the roles are reversed and the giraffe is kicking the woman. In certain types of aphasias, it can be quite difficult to determine who is doing what to whom.

Short-term memory (STM): Memory which endures for a short time. However, there are two views (at least) of STM: memory with a decay rate that may yield its loss over a few seconds; and memory (*working memory*, WM) which lasts as long as it is needed, which may be for minutes or hours, as some set of tasks is completed.

Single photon emission computed tomography (SPECT): A functional imaging technique, see *functional imaging*.

SLI: See *specific language impairment*.

Social cognition: In social psychology it refers to the study of how the self interacts with the social environment and how people use social knowledge to manage daily life and influence behavior. The neuroscience of social cognition seeks to understand the relationship between the brain and the social mind.

Socio-pragmatic approaches: In language rehabilitation they refer to therapeutic approaches focusing on the communicative resources and social consequences of the impairment.

Sodium amobarbital: An anesthetic substance used in the intra-carotid sodium amobarbital procedure (the Wada test), which is injected into the vascular bed of one hemisphere, to determine language lateralization.

Somatic responses: Responses from the autonomic nervous system such as increased heart rate, skin conductance changes, and so forth.

Sonority sequencing principle: A universal phonetic principle that ranks syllable markedness in terms of the relative sonority values of phoneme classes (Obstruent, Nasal, Liquid, Glide, Vowel). Least-marked syllables display a rise in the sonority values of sequenced phonemes from the outermost onset to the peak (e.g., OLV) and a fall in the sonority values of phoneme sequences from the peak to the outermost coda (e.g., VLO).

Specific language impairment (SLI): A developmental language disorder associated with syntactic and other language impairments, as well as a range of motor and other non-language deficits.

SPECT: Single photon emission computed tomography; a functional imaging technique.

Stage-oriented approach: Aphasia therapy adapted to the mechanisms of functional recovery at the different stages in time (acute, post-acute and chronic stage of recovery).

Stem cells: Primal cells that are present in all multicellular organisms. They can renew themselves through cell division and have the potential to differentiate to specialized cell types, for example, neurons.

Stereotaxic space: A technique used in neurosurgery and brain imaging studies to localize a specific anatomical locus using standardized three-dimensional coordinates, for example for directing the tip of a surgical instrument (such as a needle or electrode) to a known location.

Stimulation therapy: In language rehabilitation it refers to reactivation of language processing through intensive auditory or multimodal (oral and written) stimulation.

STM: See *short-term memory*.

Stroke: Damage to the brain caused either by occlusion of a blood vessel by a blood clot (ischemic stroke or infarct), or hemorrhage into the brain from a site of damage to a blood vessel. In ischemic strokes, a portion of the brain dies because of lack of oxygen and sources of energy (e.g., glucose). In hemorrhages, a portion of the brain dies because the compression of brain by the hemorrhage causes ischemia of surrounding tissue.

Stuttering: See *acquired stuttering*, *developmental stuttering*, and *person who stutters*.

Subitizing: The ability to correctly appreciate small quantities (less than five elements, for example, dots in a given array) without serial processing. Thus reaction times in appreciating quantities appear to be about the same for each quantity inferior to five, which is the range of subitizing, and increase exponentially thereafter, when counting becomes necessary.

Supramodal cortices: See *association cortices*.

Surface dysgraphia: A writing disorder characterized by difficulty in writing irregular words.

Surface dyslexia: A reading disorder characterized by difficulty in reading irregular words.

Susceptibility artifact: A problem that occurs when using fMRI to image regions near tissue interfaces (e.g., brain areas near air-filled sinuses), where rapid changes in magnetic susceptibility induce large internal static local field gradients that result in both a distortion of the image and a loss of the fMRI signal. The result of the susceptibility artifact is a reduced signal-to-noise ratio in certain regions, including the temporal pole and the lateral–posterior surface of the middle and inferior temporal gyri.

Syntactic "trace": In linguistic theory, *trace* is the name given to an unpronounced element in the syntax. On one standard view, a trace is left behind in the position a noun phrase, for example, has moved from. In the sentence, *Bill*$_i$, *John likes t*$_i$ (*but Fred, he hates*), the trace (*t*) coindexed with *Bill* occupies the object position that *Bill* is "understood" in, that is, *Bill* is understood to be the object of *likes*. The noun phrase *Bill* has moved from the object position to the front of the sentence but its relation to the object position it moved from is marked by the trace.

T

Tachistoscope: Apparatus used to present visual stimuli for very brief durations. Tachistoscopes are used to present stimuli into the lateral half-fields in laterality studies because the stimuli can be made to disappear before a fixating eye movement can occur.

Theory of mind (ToM): The capacity to attribute mental states, such as thoughts, beliefs, desires, and intentions to others. First order ToM: what another person believes; second order ToM: what one person thinks about another person's thoughts.

Thrombolysis: A process by which a platelet-fibrin clot is broken down by a thrombolytic agent such as tissue plasminogen activator (tPA), often resulting in the restoration of blood flow.

"Tip-of-the-tongue" state: The inability to recall a word or phrase accompanied by the subjective impression that access to the word is being blocked or is just below the threshold of recollection.

ToM: See *theory of mind*.

Torque: A combination of anatomical asymmetries that gives the brain the appearance of being twisted, for example, a larger right hemisphere anteriorly combined with a larger left hemisphere posteriorly gives the impression of a counterclockwise twist.

TMS: See *transcranial magnetic stimulation*.

Tourette syndrome: A developmental brain disorder involving recurring motor movements and vocal sounds (tics).

Transcranial magnetic stimulation (TMS): A technique that causes local interfere with cortical neural activity using weak electric currents induced by rapidly changing magnetic fields.

U

Unilateral: Involving one side of the brain or body.

Universal Grammar: Those who hold that general principles of grammar are innately specified in the human genome refer to this innate system as Universal Grammar.

V

Validity: In statistics and psychometrics, it implies that a test is measuring what it is intended to measure.

VBM: See *voxel-based morphometry*.

Ventral stream: See *affordances*.

Verbal fluency: The ease, speed, and flow of oral speech. The neuropsychological procedure testing this requires subjects to rapidly retrieve words that fit certain criteria. Phonemic fluency is measured by having the subject generate words that begin with a particular letter of the alphabet for a specified time period. Semantic fluency is typically measured by having the subject produce words that are members of a semantic category, such as fruits or animals.

Voxel-based morphometry (VBM): A quantitative method of analyzing structural imaging data (typically acquired with magnetic resonance). It is generally used to measure local gray matter density.

W

Wernicke's aphasia: Brain damage following a stroke or some other event can compromise function. The function that is compromised depends on the location and extent of the lesion. Quite often, damage to a region known as Wernicke's area (after the nineteenth century neurologist) in the left superior temporal cortex can result in a partial loss of language function. Damage to this area is traditionally associated with fluent nonsensical speech, preserved syntax, and poor comprehension. The precise nature of the language disorder continues to be a topic of inquiry.

White matter: Nerve tissue that is paler in color than gray matter because it contains nerve fibers with large amounts of insulating material (myelin). In contrast to gray matter, the white matter does not contain nerve cells.

Working memory (WM): The notion of a "mental scratchpad" where information that is needed for a particular mental operation is kept active until that operation is completed. See also *short-term memory*.

Index

A

Abstract letter identification, 421–422
Abstract words, 149
Abulia, 204
Accuracy, in speech, 341
Acoustic evoked potentials, 59
Acoustic–phonetic processing, 109, 110
Acquired epileptiform aphasia, 361–365
Acquired reading disorders, 151
Acquired stuttering, 312
Action representation, in motor system and frontal cortex, 121–122
Action sentences, 86
Action–perception cycle, 240
Acute aphasias
  chronic aphasias vs., 270
  classification of, 271–274
  clinical presentation of, 269
  functional magnetic resonance imaging study of, 270
  neuropsychological variables, 270–271
  pathophysiologic variables of, 269–270
  stroke as cause of, 269
  studies of, 271–274
Acute conduction aphasia, 274–275
Acute paraphasia, 275–276
Acute transcortical motor aphasia, 275
Adaptation, 311
Affect lexicon, 205
Affective expression, 202
Aging
  aphasia and, 251–252, 392
  bilingualism and, 353
  cognitive deficits secondary to, 251, 355
  comprehension difficulties and, 352
  corpus callosum changes, 263
  language changes associated with, 351–353, 355–357
  language lateralization changes associated with, 252–253
  lexical retrieval changes, 351–352
  normal, 251
  noun retrieval problems associated with, 351–352
  oral language narrative production affected by, 352–353

socio-economic status, 356
  transmission deficit hypothesis, 352
  working memory declines, 356
Agnosia, 4
Agrammatic aphasia
  description of, 140, 141
  reduced syntax therapy for, 404–405
  treatment of, 403
  verb deficit in, 149
Agrammatic speech, 195
Agranular cortex, 47
Agraphia, 210–211
Alexia, 210
Alexithymia, 206
Allographic dysgraphia, 215
Alternating antagonism, 342
Alzheimer's dementia
  category-specific deficits in, 282
  cellular-level deterioration associated with, 354
  description of, 8, 139
  language comprehension in, 354–355
  lexical retrieval in, 354
  macro-level processing in, 284
  mild cognitive impairment vs., 353–354
  semantic categorization in, 283
  semantic knowledge in, 283
  spontaneous speech in, 355
  word-finding difficulties associated with, 355
Alzheimer's disease
  aging changes, 251
  characteristics of, 411
  cortical atrophy in, 281
  declarative memory in, 196
  description of, 204–205
  executive functioning limitations in, 10
  language lateralization and, 252–253
  lexical disorders in, 148, 196
  theory of mind in, 177–178
American Sign Language, 29
Amnesia
  interhemispheric interaction and, 264
  pathophysiology of, 342
Amsterdam–Nijmegen everyday language test, 8

Anomia, 147
  theory of, 133
Anomic aphasia, 4t
Anosognosia, 214
Anterior fusiform gyrus, 75
Anterograde amnesia, 196–197
Antidepressants, 413
Aphasia
  acquired epileptiform, 361–365
  acute. *See* Acute aphasias
  aging and, 251–252
  agrammatic. *See* Agrammatic aphasia
  assessment of
    batteries, 6, 7, 14
    bedside tests, 6, 7
    clinical-neuroanatomical approach to, 5
    comprehensive examinations, 7
    functional communication assessments, 8
    screening tests, 6, 7
  bilingual, 346–347
  Broca's. *See* Broca's aphasia
  classification of, 4
  definition of, 6, 195, 407
  diffusion/perfusion mismatch in, 408–409
  discovery of, 329
  explicit memory sensitivity to, 343
  fluent, 4t
  grammatical gender in, 142
  implicit memory sensitivity to, 343
  incidence of, 397
  inflectional morphology impairments in, 140–143
  jargon, 135
  language effects of
    description of, 397
    reperfusion used to restore, 409–410
  metalinguistic knowledge sensitivity to, 342
  multilingual
    recovery patterns for, 342
    rehabilitation, 346–347
  naming abilities in, assessment of, 6
  neologisms in, 133–135
  non-fluent. *See* Non-fluent aphasia
  pathophysiology of, 342
  phonemic paraphasias in, 133
  primary progressive, 279, 413

Aphasia (*continued*)
  pure motor, 276
  recovery patterns for, 342
  selective, 342
  serial models vs. connectionist models,
      132–133
  spontaneous recovery of, 389–394
  stroke-related, 74, 394, 407–410
  subcortical
    description of, 329–330
    mechanisms of, 332
    non-thalamic, 332–333
  symptoms of, 127
  thalamic, 336–339
  transcortical motor, 4t, 275
  transcortical sensory, 4t, 108
  transcranial magnetic stimulation studies in,
      122–123
  treatment of. *See* Aphasia treatment
  vascular etiology of, 400
Aphasia rehabilitation
  cognitive neuropsychological model, 397
  description of, 346–347, 397
  socio-pragmatic model, 397
Aphasia Screening Test, 14
Aphasia treatment
  activation therapy, 399–400
  antidepressants, 413
  bromocriptine, 412
  cholinesterase inhibitors, 411–412
  cognitive neuropsychological approach, 402
  deblocking techniques, 400
  dexamphetamine, 412–413
  direct stimulation, 400
  efficacy studies of, 405
  framework for, 399–402
  future research, 405
  goal setting, 398
  impairment-specific approaches, 400
  intensity of, 399
  lexical disorders, 402–403
  methods of, 398–399
  non-verbal compensation, 400
  piracetam, 412
  recovery stage considerations, 399–402
  rehabilitation. *See* Aphasia rehabilitation
  social participation and consolidation, 400–402
  spontaneous recovery, 389–394
  syntactic disorders, 403–405
  targeted, 400
Aphasic agraphia, 210–211
Aphasiology, 230
Apraxia of speech
  computed tomography findings, 277
  definition of, 276
  description of, 4–5
  motor nature of features of, 129
  "pure motor aphasia, " 276
  symptoms of, 127
Apraxic agraphia, 211
Apraxic dysgraphia, 215
Areal borders, 38
Arithmetical facts, 225
Arithmetical tables, 223–224
Asperger syndrome, 291, 377–384

Associative learning, 324
Asymmetric tonic neck reflex, 249
Attention
  definition of, 336
  neural substrates of, 383
Attitudinal meaning, 201–202
Audiovisual speech comprehension, 82
Auditory cortex, 82
Autism
  auditory processing abnormalities, 383–384
  case study of, 377–378
  cognitive models of, 379
  communication deficits in, 377, 382–383
  description of, 193–194, 377
  future research of, 384
  language deficits in
    causes of, 381–384
    lexical semantics, 379–380
    mindreading impairment as cause of,
        382–383
    morphology, 378–379
    phonology, 378
    pragmatics, 380–381, 383–384
    prosody, 381
    syntax, 378–379
  metarepresentational theory of, 379–380
  perceptual functioning in, 383–384
  reading deficits in, 381
  theory of mind and, 378–379, 382
  word learning in, 380
Autistic regression, 364

B

Back propagation, 233
Backward inference, 167
Basal ganglia
  Huntington's disease effects on, 195
  language and, 330
  in language processing, 40
  motor portions of, 190
  neuroscience of, 332
  Parkinson's disease effects on dopamine-
      producing neurons in, 195
Bilingual aphasia, 346–347
Bilingualism
  aging and, 353
  first-language attrition, 353
  language disorders. *See* Multilingual
      language disorders
Blending, 343
Blending recovery, from aphasia, 342
Blood-oxygen-level-dependent (BOLD) effect,
    62, 66
Boston Diagnostic Aphasia Examination, 5,
    14, 206
Boston Naming Test, 10, 321
Brain
  anatomy of, 248–249
  computer and, 229–230
  cytoarchitectonically distinct regions of, 46
  development of, 248–251
  hierarchy of, 46
  immature, asymmetries in, 248–249

language-ready, 238
  levels of, 46, 47f
  medial temporal lobe of, 48
  number representation by, 220–221
  paralimbic areas of, 46
  quadrants of, 300
  "torque", 299
Brain damage. *See also* Traumatic brain injury
  communication difficulties secondary to, 13
  pragmatic deficits following, 205–206
Brain hemispheres. *See also* Right hemisphere
  asymmetry, 258–260
  description of, 257
  functional differences between, 247
  interhemispheric interaction
    bihemispheric redundancy gain, 262–263
    childhood amnesia and, 264
    cognition effects, 263–264
    cognitive deficits effect on, 264
    costs and benefits of, 261
    description of, 258
    gender-based differences, 264–265
    individual variations, 263–265
    life span variations, 263
    mechanisms of, 260–263
    memory effects, 263–264
    mixing stimuli, 261–262
    in schizophrenia, 264
  in language comprehension, 73–74
  mood and meaning, 53
  transcranial magnetic stimulation studies of,
      for language specialization, 116–117
Broca–Annett axiom, 304
Broca's aphasia
  assessment approaches for, 5t
  comprehension failure in
    slow syntax hypothesis of, 159
    trace deletion hypothesis of, 157–160
  definition of, 4t
  selective pattern of loss in, 155
  syntactic comprehension in, 157
Broca's area/region
  anatomy of, 34, 35–36, 108, 129
  Brodmann's area 44 and 45
    activation of, during language production,
        77f, 78
    anatomy of, 35
    cytoarchitecture of, 36, 37f, 40–42
    description of, 129
    dorsal border of, 37
    in language dominant hemispheres, 36
    layers of, 36, 37f
    localization of, 36, 37
    mapping of, 78
    receptorarchitecture, 38–39
    variability of landmarks in, 37
  computational complexity of, 129
  damage to, 148
  description of, 33
  in Geschwind model, 231
  motor cortex and, 78
  speech, 78
  Wernicke's region and, 39–40
"Broca's complex", 36, 129
Brodmann, 33, 34

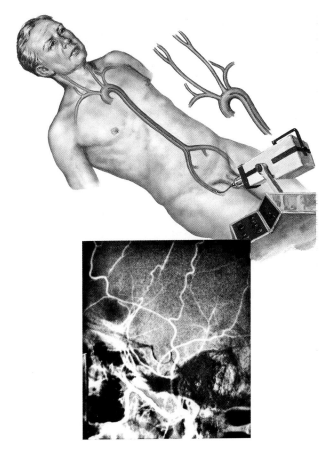

PLATE 1 Femorocerebral angiography is the means by which amobarbital is delivered to the patient's ICA, allowing anesthetization of one cerebral hemisphere. *Source*: Reprinted from Frank H. Netter. The Netter Collection of Medical Illustrations – Nervous, © 1984, Elsevier Inc. All Rights Reserved. (see Figure 3.1, page 24).

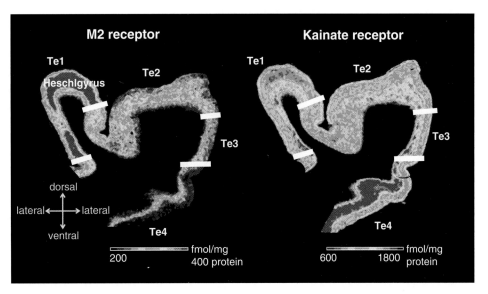

PLATE 2 Receptorarchitecture of the superior temporal gyrus. Distribution of the M2 receptor for acetylcholine and the kainite receptor for glutamate in a coronal section through the superior temporal gyrus and the Heschl gyrus (cryostat section, $20\,\mu m$). The color coding indicates receptor densities in fmol/mg protein (Zilles *et al.*, 2002a, b). The white lines indicate the borders between cytoarchitectonic areas Te1–Te4 (Morosan *et al.*, 2005) (see Figure 4.2, page 39).

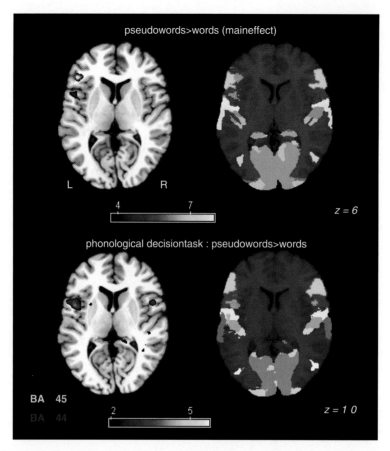

PLATE 3 Superimposition of functional activations obtained during an fMRI study analyzing the activations in Broca's region (left BAs 44 and 45) during the processing of visually presented words and pseudowords and cytoarchitectonic probability maps (Heim *et al.*, 2005). Subjects had to perform either a lexical or a phonological decision task. The upper row shows the activation for the lexical decision task (the contrast pseudowords versus words), the lower row shows the effect for the phonological decision task. The left images display SPM{t} maps rendered on the MNI reference brain; the right shows the same projected on the cytoarchitectonic maximum probability maps (Eickhoff *et al.*, 2005) of BA 44 and 45 (Amunts *et al.*, 2004) (see Figure 4.3, page 41).

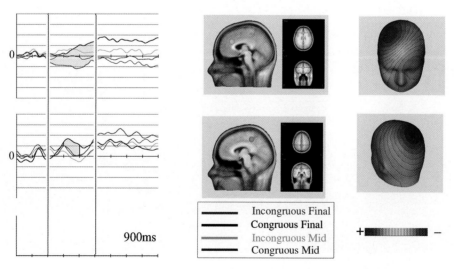

PLATE 4 Top row, time course of electrical activity in ACC. Bottom row, time course of electrical activity in PCC. Left panel, regional source waveforms for ACC (top) and PCC (bottom). Gray areas mark intervals from 200 to 500 ms when semantic effect is statistically significant. Middle panel, source location. Right panel, scalp projection of ACC (top) and PCC (bottom) sources. *Source*: Data from Frishkoff *et al.*, 2004 (see Figure 5.4, page 52).

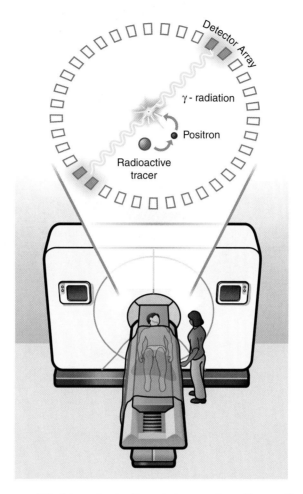

PLATE 5 Image courtesy of The Science Creative Quarterly, http: scq.ubc.ca, illustrator Jiang Long (see figure in Box 6.2, page 63).

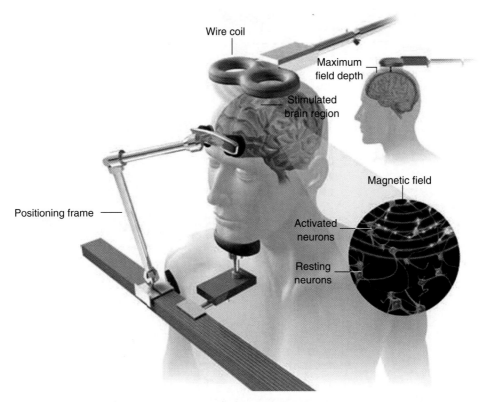

Wire coil

Maximum field depth

Stimulated brain region

Positioning frame

Magnetic field

Activated neurons

Resting neurons

PLATE 6 *Source*: With permission by Bryan Christie Design (http://www.bryanchristiedesign.com).
(see figure in Box 6.3, page 65)

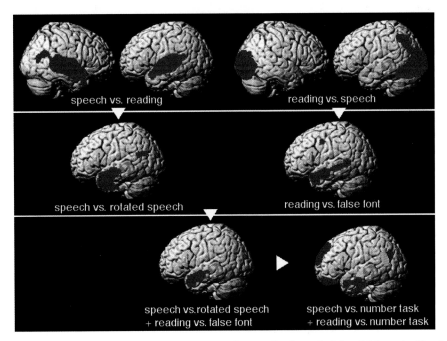

PLATE 7 Activity evoked by narratives, both spoken and written, rendered onto the left and right cerebral hemispheres (group averaged data, n = 11 normal subjects). Solid red regions are located over the lateral and inferior surfaces of the hemispheres, hatched red regions are located over the medial surfaces. The contrasts of speech with reading and reading with speech demonstrated bilateral, symmetrical activity in the superior temporal gyri and the occipital lobes, respectively. The asymmetry in posterior parietal cortex (left > right) during reading is the consequence of visual attention and reading saccades being directed to the right in left-to-right readers. Contrasting speech with its modality-specific baseline condition of spectrally inverted (rotated) speech, and reading with its modality-specific baseline condition of text-like arrays of false font, demonstrated activity centered around the superior temporal sulcus, predominantly lateralized to the left. The conjunction of activity for these two contrasts was centered over left anterolateral temporal cortex – a region that responded to intelligible language independent of modality. By using an alternative baseline condition ("number task"), an explicit task on simple number semantics (an odd/even decision on randomly presented numbers, 1–10), activity was also demonstrated in the anterior fusiform gyrus (the "basal" language area) and just ventral to the angular gyrus. There was also prominent activity in the left superior frontal gyrus, orbito-frontal cortex and in retrosplenial cortex (hatched region). The rationale for using the number task as an alternative baseline condition is described in the text. Data from Spitsyna *et al.* (2006) (see Figure 7.1, page 75).

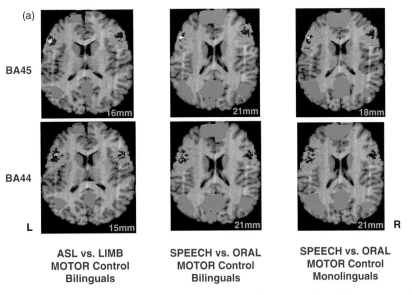

PLATE 8(a) Activations of BA45 (top row) and BA44 (bottom row) during production of language narratives compared to a motor control task (see Figure 7.2(a), page 77).

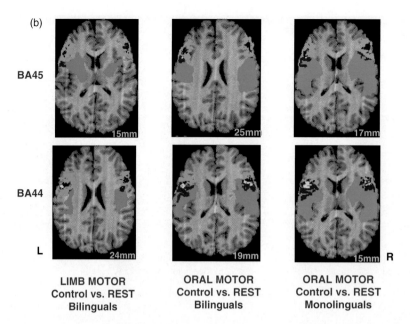

PLATE 8(b) Activation of BA45 (top) and BA44 (bottom) comparing each motor control task to a resting condition shown are representative horizontal slices (left side of each image corresponds to the left side of the brain; the level in mm superior to the AC-PC plane (z-coordinate of Talairach & Tournoux atlas, 1988, is indicated on each slice). Images displayed in the two columns on the left are from the bilingual (English and ASL) subjects, and those in the column on the right are from the monolingual English speakers. Voxels in dark blue correspond to core parts of the specific Brodmann area, those in light blue to peripheral voxels. Voxels significantly more active in one condition compared to a second (Z > 2.33) are shown in green. Voxels in the peripheral part of a Brodmann area that had a significant PET activation are displayed in red, and core voxels that were significantly activated are shown in yellow. From Horwitz *et al.* (2003) [Talairach, J., & Tournoux, P. (1988). *Co-planar stereotaxic atlas of the human brain* (M. Rayport, Trans.). New York: Thieme.] (see Figure 7.2(b), page 77).

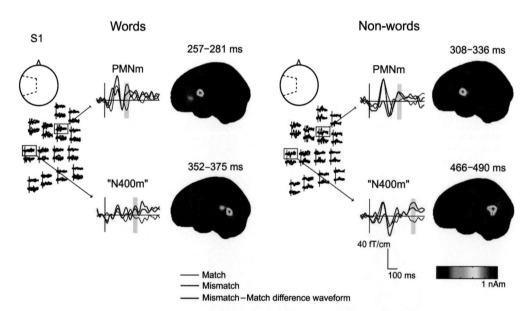

PLATE 9 Phonological mapping negativity (PMN) and semantic N400. MEG responses to words (left) and non-words (right) for one participant for those left-hemisphere channels showing maximum amplitude for the magnetic PMN (PMNm) and the N400-like response. The corresponding estimates of the PMN- and N400m-like response sources (over a 25 ms time window centered at the peak of the response) are depicted in the brain images. The gray vertical bars indicate the 50 ms time periods within which significant PMNm- and N400m-like responses occurred. *Source:* Modified after Kujala *et al.*, 2004 (see Figure 9.1, page 93).

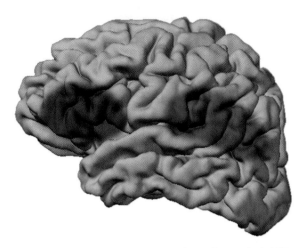

PLATE 10   Approximate extent of cortical damage to agrammatic non-fluent aphasic FCL and anomic fluent aphasic JLU (reproduced from Ullman *et al.*, 2005). (© Elsevier.) (See figure in Box 13.1, page 140.)

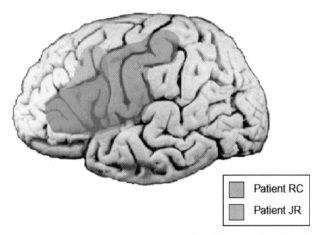

Patient RC
Patient JR

PLATE 11   Two patients reported by Shapiro and Caramazza showing different brain regions implicated in naming of nouns and verbs. Patient RC (lesion shown in red) is relatively impaired in naming verbs. Patient JR (lesion shown in blue) is more impaired in noun production. *Source*: From Shapiro and Caramazza, 2003, p. 204. (© Elsevier.) (See Figure 14.2, page 150.)

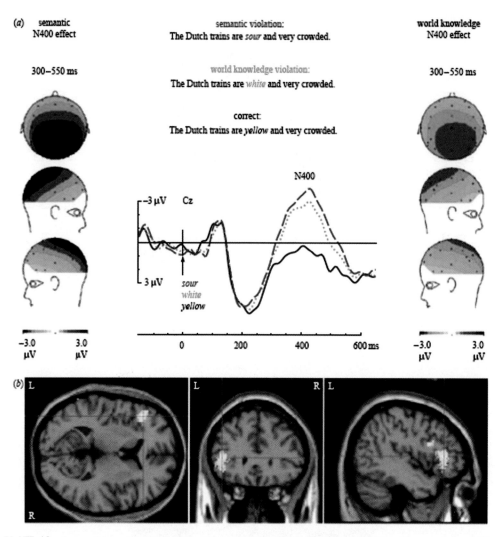

(a) semantic
N400 effect

300–550 ms

semantic violation:
The Dutch trains are *sour* and very crowded.

world knowledge violation:
The Dutch trains are *white* and very crowded.

correct:
The Dutch trains are *yellow* and very crowded.

world knowledge
N400 effect

300–550 ms

−3.0 μV   3.0 μV

N400

Cz

−3 μV

3 μV

sour
white
yellow

0   200   400   600 ms

−3.0 μV   3.0 μV

(b)

PLATE 12 *Source*: Figure taken from Hagoort & van Berkum (2007) (Fig. 4, p. 805). Printed by permission of The Royal Society (see figure in Box 17.1, page 180).

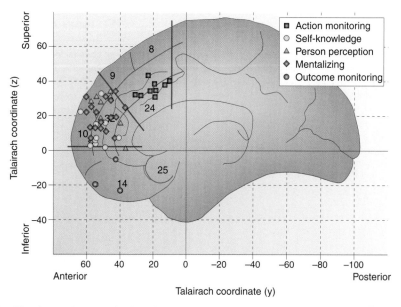

PLATE 13 The figure shows activations in the medial frontal cortex during action monitoring, social cognition and outcome monitoring. The meta-analysis suggests that social cognition tasks (including self-knowledge, person perception, mentalising) activate areas in the anterior rostral medial frontal cortex (arMFC). Monitoring of actions activate the posterior rostral region of the MFC (prMFC) while monitoring of outcomes involves the orbital MFC (oMFC). Figure reprinted by permission from Macmillan Publishers Ltd: *Nature Review Neuroscience*, D.M. Amodio & C.D. Frith (2006) (see Figure 17.1, page 185).

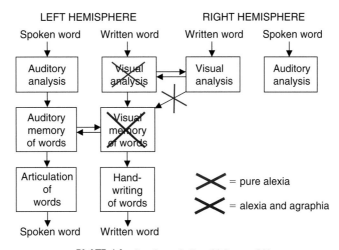

LEFT HEMISPHERE   RIGHT HEMISPHERE

Spoken word   Written word   Written word   Spoken word

Auditory analysis | Visual analysis | Visual analysis | Auditory analysis

Auditory memory of words | Visual memory of words

Articulation of words | Hand-writing of words

Spoken word   Written word

✕ = pure alexia

✕ = alexia and agraphia

PLATE 14   See figure in Box 20.2, page 211.

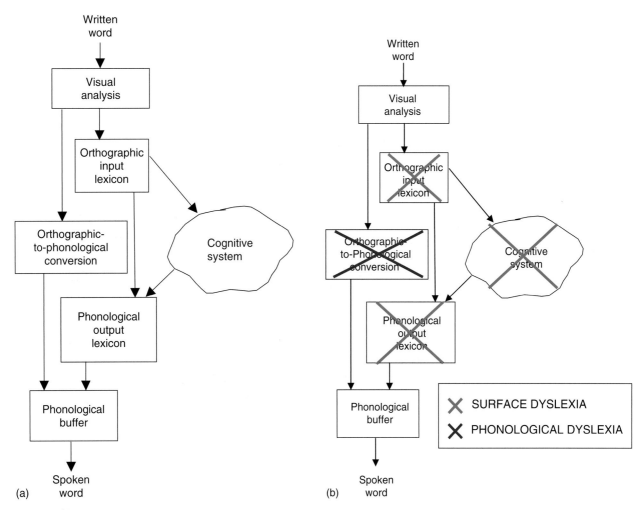

(a) lexical route:
Written word → Visual analysis → Orthographic input lexicon → Orthographic-to-phonological conversion → Cognitive system → Phonological output lexicon → Phonological buffer → Spoken word

(b) functional lesions:
Written word → Visual analysis → Orthographic input lexicon → Orthographic-to-Phonological conversion → Cognitive system → Phonological output lexicon → Phonological buffer → Spoken word

✕ SURFACE DYSLEXIA
✕ PHONOLOGICAL DYSLEXIA

PLATE 15   Dual-route model of reading: (a) lexical and SWL reading routes; (b) functional lesions causing phonological and surface dyslexia. (see Figure 20.1(a),(b), page 213).

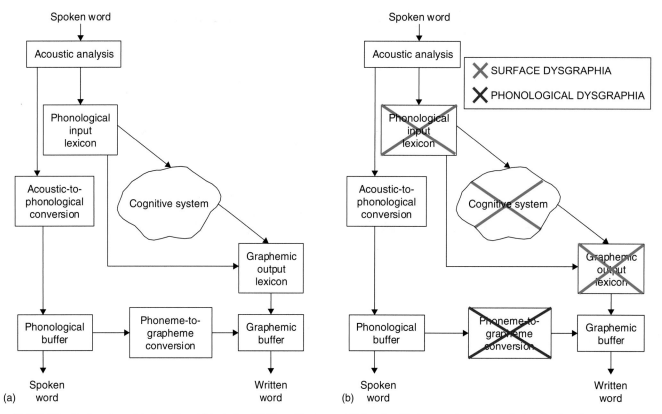

PLATE 16 Dual-route model of spelling: (a) lexical and SWL spelling routes; (b) functional lesions causing phonological and surface dysgraphia (see Figure 20.2(b), page 215).

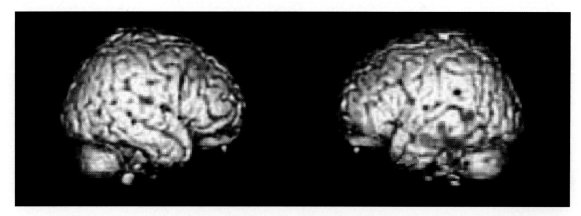

PLATE 17 Significant cortical atrophy in PNFA (*n* = 7, green), semantic dementia (*n* = 8, red), and FTD patients with a disorder of social comportment and executive functioning (*n* = 14, blue) using voxel-based morphometry relative to healthy seniors (*n* = 11) (see figure in Box 27.1, page 281).

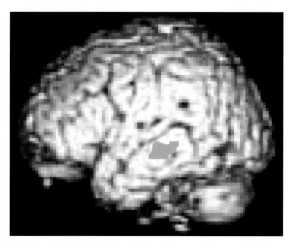

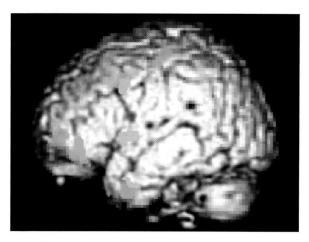

PLATE 18   VBM correlation of semantic categorization accuracy across pictures and words of natural and manufactured categories (green area) and correlation for pictures and words of natural kinds (red area) (see figure in Box 27.2, page 283).

PLATE 19   VBM correlation of grammatical comprehension in sentences with cortical atrophy in PNFA (see figure in Box 27.3, page 285).

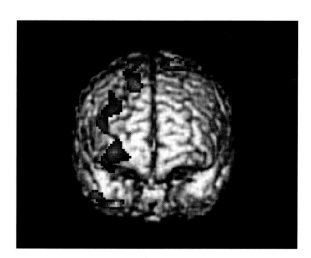

PLATE 20   See figure in Box 27.4, page 286.

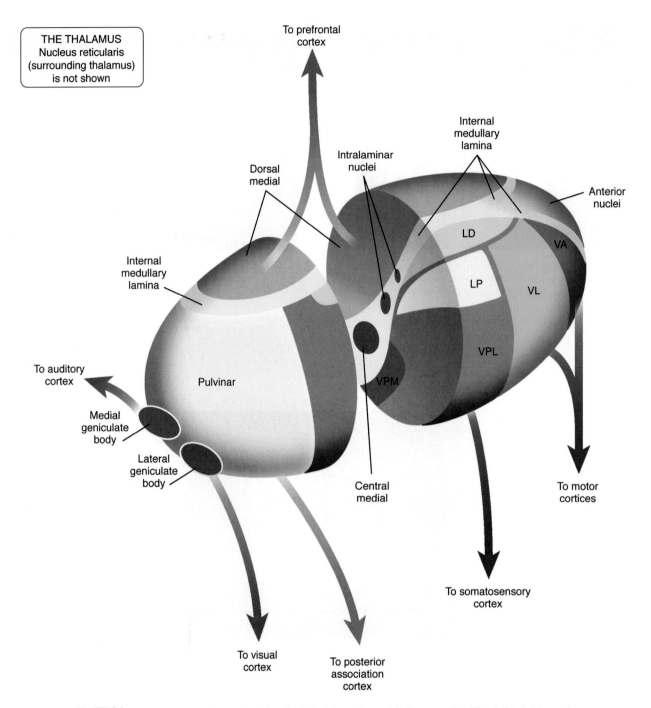

THE THALAMUS
Nucleus reticularis
(surrounding thalamus)
is not shown

To prefrontal
cortex

Dorsal
medial

Intralaminar
nuclei

Internal
medullary
lamina

Anterior
nuclei

LD

VA

Internal
medullary
lamina

LP

VL

VPL

To auditory
cortex

Pulvinar

VPM

Medial
geniculate
body

Lateral
geniculate
body

Central
medial

To motor
cortices

To somatosensory
cortex

To visual
cortex

To posterior
association
cortex

PLATE 21   Cartoon of the thalamus, depicting the loci of the major nuclei. Center median (Cm) is labeled "central medial" in this figure and parafascicularis (Pf), located anterior, ventral and medial to Cm, is not depicted. *Source*: From Nadeau, S.E., *et al.* (2004). *Medical neuroscience*. Philadelphia: Saunders. Imprint of Elsevier (see Figure 32.3, page 334).

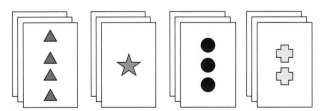

PLATE 22   A test of "executive functioning": the Wisconsin Card Sorting Task that involves sorting cards by illustrated colors and shapes (see figure (c) in Box 37.1, page 379).

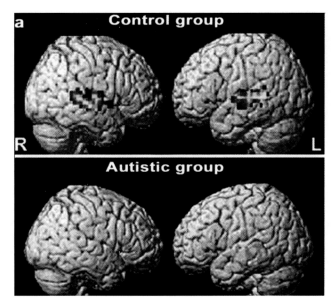

PLATE 23 Individuals with autism, unlike individuals with typical development, failed to activate the STS in response to vocal sounds. *Source*: From Gervais *et al.*, 2004; Reprinted by permission from Macmillan Publishers Ltd.: *Nature Neuroscience*, copyright, 2004) (see figure in Box 37.3, page 383).

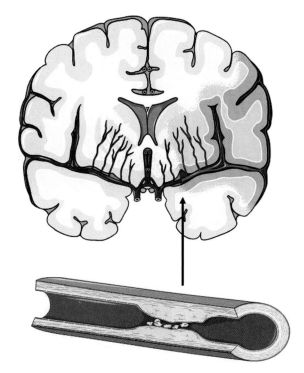

PLATE 24 Stroke is caused by a blockage of a blood vessel supplying a specific part of the brain (black arrow). As a result brain tissue undergoes ischemic necrosis (shaded blue region of the brain). Figure reproduced by consent of Servier (http://www.servier.com/SMART/home_smart.asp) (see Figure 40.1, page 408).

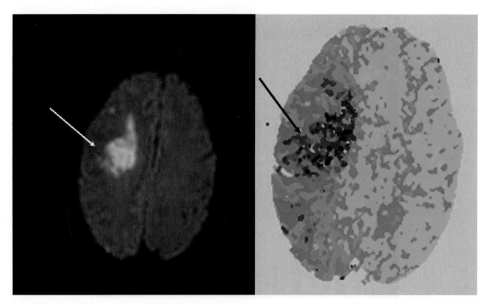

PLATE 25   This figure shows a diffusion weighted image (DWI) on the left and a perfusion weighted image (PWI) on the right of a subject present with a middle cerebral artery stroke. Note that while the area of presumed irreversible infarction is relatively smaller on the DWI (white arrow) the perfusion deficit represented by the blue color on the right is significantly larger (see figure in Box 40.1, page 409).

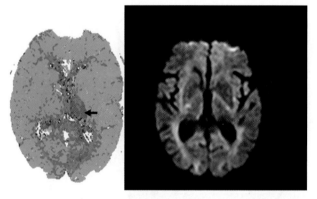

PLATE 26(a)   PWI (left) and DWI (right) of a subject with a left posterior cerebral artery stenosis, resulting in significant hypoperfusion to the left thalamus (black arrow) and a thalamic aphasia syndrome. This subject was placed on blood pressure elevating therapy, increasing his baseline mean arterial pressure by 10% (see Figure 40.2(a), page 410).

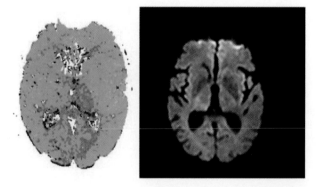

PLATE 26(b)   One day later, his aphasia has resolved, and the perfusion deficit in his left thalamus is less pronounced compared to previously, but is still present. As there is no corresponding left thalamic DWI lesion, the entire area of perfusion deficit (blue) in Plate 26(a), can be considered the ischemic penumbra, and shown to be resolving in Plate 26(b) (see Figure 40.2(b), page 410).

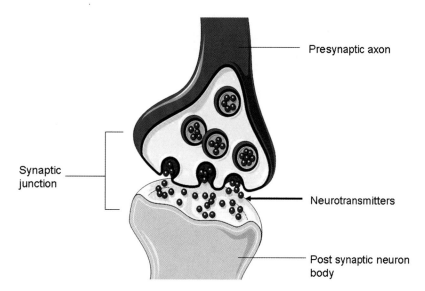

Presynaptic axon

Synaptic junction

Neurotransmitters

Post synaptic neuron body

PLATE 27   Neural plasticity is thought to occur at the level of the synaptic junction whereby either more neurotransmitter is released or alternatively more synapses are formed. Figure reproduced with consent of Servier (http://www.servier.com/SMART/home_smart.asp) (see Figure 40.3, page 411).